MATHEMATICS:

APPLICATIONS AND INTERPRETATION

STANDARD LEVEL
COURSE COMPANION

Paul Belcher
Jennifer Chang Wathall
Suzanne Doering
Phil Duxbury
Panayiotis Economopoulos
Jane Forrest
Peter Gray
Tony Halsey
David Harris
Lorraine Heinrichs
Ed Kemp
Paul La Rondie
Palmira Mariz Seiler
Michael Ortman
Nuriye Sirinoglu Singh
Nadia Stoyanova Kennedy
Paula Waldman

OXFORD
UNIVERSITY PRESS

Great Clarendon Street, Oxford, OX2 6DP, United Kingdom

Oxford University Press is a department of the University of Oxford. It furthers the University's objective of excellence in research, scholarship, and education by publishing worldwide. Oxford is a registered trade mark of Oxford University Press in the UK and in certain other countries

First published in 2019

British Library Cataloguing in Publication Data
Data available

978-0-19-842699-8

9 10

Paper used in the production of this book is a natural, recyclable product made from wood grown in sustainable forests. The manufacturing process conforms to the environmental regulations of the country of origin.

Printed in India by Multivista Global Pvt. Ltd

Acknowledgements

The author would like to thank the following authors for contributions to digital resources:

Alexander Bradley
Ingrid Delange
Ben Donaldson
Tom Edinburgh
Jim Fensom
Jane Forrest
David Harris
Georgios Ioannadis
Vilda Markeviciute
Martin Noon
Nadia Stoyanova Kennedy
Paula Waldman

The publisher and authors would like to thank the following for permission to use photographs and other copyright material:

All images from Shutterstock.com

Course Companion definition

The IB Diploma Programme Course Companions are designed to support students throughout their two-year Diploma Programme. They will help students gain an understanding of what is expected from their subject studies while presenting content in a way that illustrates the purpose and aims of the IB. They reflect the philosophy and approach of the IB and encourage a deep understanding of each subject by making connections to wider issues and providing opportunities for critical thinking.

The books mirror the IB philosophy of viewing the curriculum in terms of a whole-course approach and include support for international mindedness, the IB learner profile and the IB Diploma Programme core requirements, theory of knowledge, the extended essay and creativity, activity, service (CAS).

IB mission statement

The International Baccalaureate aims to develop inquiring, knowledgable and caring young people who help to create a better and more peaceful world through intercultural understanding and respect.

To this end the IB works with schools, governments and international organisations to develop challenging programmes of international education and rigorous assessment.

These programmes encourage students across the world to become active, compassionate, and lifelong learners who understand that other people, with their differences, can also be right.

The IB learner profile

The aim of all IB programmes is to develop internationally minded people who, recognising their common humanity and shared guardianship of the planet, help to create a better and more peaceful world. IB learners strive to be:

Inquirers They develop their natural curiosity. They acquire the skills necessary to conduct inquiry and research and show independence in learning. They actively enjoy learning and this love of learning will be sustained throughout their lives.

Knowledgeable They explore concepts, ideas, and issues that have local and global significance. In so doing, they acquire in-depth knowledge and develop understanding across a broad and balanced range of disciplines.

Thinkers They exercise initiative in applying thinking skills critically and creatively to recognise and approach complex problems, and make reasoned, ethical decisions.

Communicators They understand and express ideas and information confidently and creatively in more than one language and in a variety of modes of communication. They work effectively and willingly in collaboration with others.

Principled They act with integrity and honesty, with a strong sense of fairness, justice, and respect for the dignity of the individual, groups, and communities. They take responsibility for their own actions and the consequences that accompany them.

Open-minded They understand and appreciate their own cultures and personal histories, and are open to the perspectives, values, and traditions of other individuals and communities. They are accustomed to seeking and evaluating a range of points of view, and are willing to grow from the experience.

Caring They show empathy, compassion, and respect towards the needs and feelings of others. They have a personal commitment to service, and act to make a positive difference to the lives of others and to the environment.

Risk-takers They approach unfamiliar situations and uncertainty with courage and forethought, and have the independence of spirit to explore new roles, ideas, and strategies. They are brave and articulate in defending their beliefs.

Balanced They understand the importance of intellectual, physical, and emotional balance to achieve personal well-being for themselves and others.

Reflective They give thoughtful consideration to their own learning and experience. They are able to assess and understand their strengths and limitations in order to support their learning and professional development.

Contents

Digital contents

Digital content overview

Click on this icon here to see a list of all the digital resources in your enhanced online course book. To learn more about the different digital resource types included in each of the chapters and how to get the most out of your enhanced online course book, go to page ix.

Syllabus coverage

This book covers all the content of the Mathematics: applications and interpretation SL course. Click on this icon here for a document showing you the syllabus statements covered in each chapter.

Practice exam papers

Click on this icon here for an additional set of practice exam papers.

Worked solutions

Click on this icon here for worked solutions for all the questions in the book.

Introduction

The new IB diploma mathematics courses have been designed to support the evolution in mathematics pedagogy and encourage teachers to develop students' conceptual understanding using the content and skills of mathematics, in order to promote deep learning. The new syllabus provides suggestions of conceptual understandings for teachers to use when designing unit plans and overall, the goal is to foster more depth, as opposed to breadth, of understanding of mathematics.

What is teaching for conceptual understanding in mathematics?

Traditional mathematics learning has often focused on rote memorization of facts and algorithms, with little attention paid to understanding the underlying concepts in mathematics. As a consequence, many learners have not been exposed to the beauty and creativity of mathematics which, inherently, is a network of interconnected conceptual relationships.

Teaching for conceptual understanding is a framework for learning mathematics that frames the factual content and skills; lower order thinking, with disciplinary and non-disciplinary concepts and statements of conceptual understanding promoting higher order thinking. Concepts represent powerful, organizing ideas that are not locked in a particular place, time or situation. In this model, the development of intellect is achieved by creating a synergy between the factual, lower levels of thinking and the conceptual higher levels of thinking. Facts and skills are used as a foundation to build deep conceptual understanding through inquiry.

The IB Approaches to Teaching and Learning (ATLs) include teaching focused on conceptual understanding and using inquiry-based approaches. These books provide a structured inquiry-based approach in which learners can develop an understanding of the purpose of what they are learning by asking the questions: why or how? Due to this sense of purpose, which is always situated within a context, research shows that learners are more motivated and supported to construct their own conceptual understandings and develop higher levels of thinking as they relate facts, skills and topics.

The DP mathematics courses identify twelve possible fundamental concepts which relate to the five mathematical topic areas, and that teachers can use to develop connections across the mathematics and wider curriculum:

Approximation	Modelling	Representation
Change	Patterns	Space
Equivalence	Quantity	Systems
Generalization	Relationships	Validity

Each chapter explores two of these concepts, which are reflected in the chapter titles and also listed at the start of the chapter.

The DP syllabus states the essential understandings for each topic, and suggests some concept specific conceptual understandings relevant to the topic content. For this series of books, we have identified important topical understandings that link to these and underpin the syllabus, and created investigations that enable students to develop this understanding. These investigations, which are a key element of every chapter, include factual and conceptual questions to prompt students to develop and articulate these topical conceptual understandings for themselves.

A tenet of teaching for conceptual understanding in mathematics is that the teacher does not **tell** the student what the topical understandings are at any stage of the learning process, but provides investigations that guide students to discover these for themselves. The teacher notes on the ebook provide additional support for teachers new to this approach.

A concept-based mathematics framework gives students opportunities to think more deeply and critically, and develop skills necessary for the 21st century and future success.

Jennifer Chang Wathall

Investigation 1

A Measuring a potato

1. Make a list of all the physical properties of a potato. Which of these properties can you measure? How could you measure them? Are there any properties that you cannot measure? How do we determine what we can measure?
2. **Factual** What does it mean to measure a property of an object? How do we measure?
3. **Factual** Which properties of an object can we measure?
4. **Conceptual** Why do we use measurements and how do we use measuring to define properties of an object?

In every chapter, investigations provide inquiry activities and factual and conceptual questions that enable students to construct and communicate their own conceptual understanding in their own words. The key to concept-based teaching and learning, the investigations allow students to develop a deep conceptual understanding. Each investigation has full supporting teacher notes on the enhanced online course book.

Gives students the opportunity to reflect on what they have learned and deepen their understanding.

Reflect Why might extrapolation be risky?

Developing inquiry skills

Apply what you have learned in this section to represent the first opening problem with a tree diagram.

Hence find the probability that a cab is **identified** as yellow.

Apply the formula for conditional probability to find the probability that the cab was yellow **given that** it was identified as yellow.

How does your answer compare to your original subjective judgment?

Every chapter starts with a question that students can begin to think about from the start, and answer more fully as the chapter progresses. The developing inquiry skills boxes prompt them to think of their own inquiry topics and use the mathematics they are learning to investigate them further.

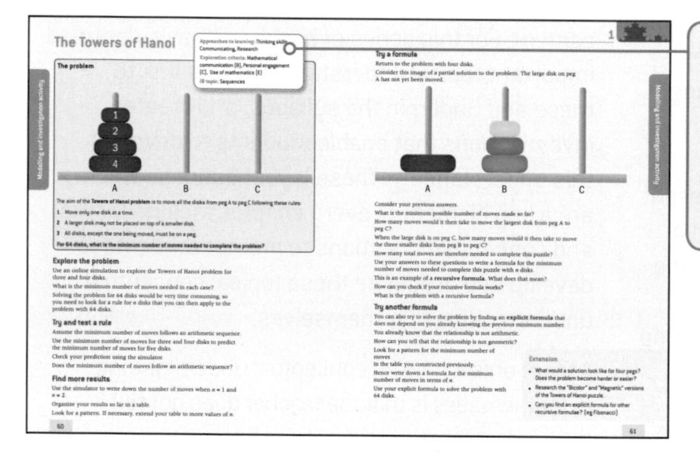
The Towers of Hanoi

Approaches to learning: Thinking skills, Communicating, Research
Exploration criteria: Mathematical communication (B), Personal engagement (C), Use of mathematics (E)
IB topic: Sequences

Modelling and investigation activity

The problem

The aim of the **Towers of Hanoi problem** is to move all the disks from peg A to peg C following these rules:

1. Move only one disk at a time.
2. A larger disk may not be placed on top of a smaller disk.
3. All disks, except the one being moved, must be on a peg.

For 64 disks, what is the minimum number of moves needed to complete the problem?

Explore the problem

Use an online simulation to explore the Towers of Hanoi problem for three and four disks.

What is the minimum number of moves needed in each case?

Solving the problem for 64 disks would be very time consuming, so you need to look for a rule for n disks that you can then apply to the problem with 64 disks.

Try and test a rule

Assume the minimum number of moves follows an arithmetic sequence.

Use the minimum number of moves for three and four disks to predict the minimum number of moves for five disks.

Check your prediction using the simulator.

Does the minimum number of moves follow an arithmetic sequence?

Find more results

Use the simulator to write down the number of moves when $n = 1$ and $n = 2$.

Organize your results so far in a table.

Look for a pattern. If necessary, extend your table to more values of n.

60

1

Try a formula

Return to the problem with four disks.

Consider this image of a partial solution to the problem. The large disk on peg A has not yet been moved.

Consider your previous answers.

What is the minimum possible number of moves made so far?

How many moves would it then take to move the largest disk from peg A to peg C?

When the large disk is on peg C, how many moves would it then take to move the three smaller disks from peg B to peg C?

How many total moves are therefore needed to complete this puzzle?

Use your answers to these questions to write a formula for the minimum number of moves needed to complete this puzzle with n disks.

This is an example of a **recursive formula**. What does that mean?

How can you check if your recursive formula works?

What is the problem with a recursive formula?

Try another formula

You can also try to solve the problem by finding an **explicit formula** that does not depend on you already knowing the previous minimum number.

You already know that the relationship is not arithmetic.

How can you tell that the relationship is not geometric?

Look for a pattern for the minimum number of moves in the table you constructed previously.

Hence write down a formula for the minimum number of moves in terms of n.

Use your explicit formula to solve the problem with 64 disks.

Extension

- What would a solution look like for four pegs? Does the problem become harder or easier?
- Research the "Bicolor" and "Magnetic" versions of the Towers of Hanoi puzzle.
- Can you find an explicit formula for other recursive formulae? (eg Fibonacci)

Modelling and investigation activity

61

The modelling and investigation activities are open-ended activities that use mathematics in a range of engaging contexts and to develop students' mathematical toolkit and build the skills they need for the IA. They appear at the end of each chapter.

The chapters in this book have been written to provide logical progression through the content, but you may prefer to use them in a different order, to match your own scheme of work. The Mathematics: applications and interpretation Standard and Higher Level books follow a similar chapter order, to make teaching easier when you have SL and HL students in the same class. Moreover, where possible, SL and HL chapters start with the same inquiry questions, contain similar investigations and share some questions in the chapter reviews and mixed reviews – just as the HL exams will include some of the same questions as the SL paper.

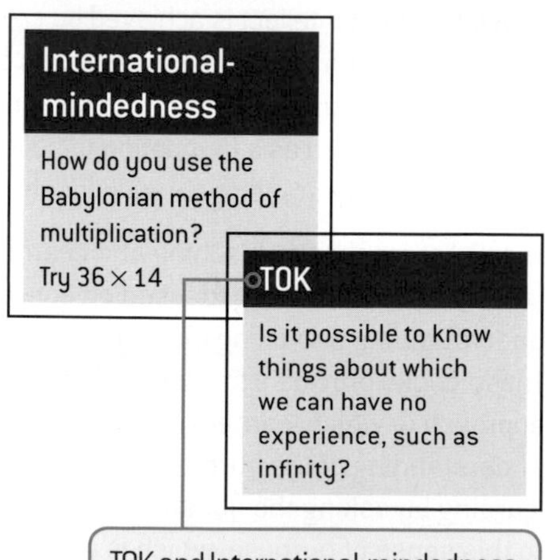

TOK and International-mindedness are integrated into all the chapters.

How to use your enhanced online course book

Throughout the book you will find the following icons. By clicking on these in your enhanced online course book you can access the associated activity or document.

Prior learning

Clicking on the icon next to the "Before you start" section in each chapter takes you to one or more worksheets containing short explanations, examples and practice exercises on topics that you should know before starting, or links to other chapters in the book to revise the prior learning you need.

Additional exercises

The icon by the last exercise at the end of each section of a chapter takes you to additional exercises for more practice, with questions at the same difficulty levels as those in the book.

Animated worked examples

This icon leads you to an animated worked example, explaining how the solution is derived step-by-step, while also pointing out common errors and how to avoid them.

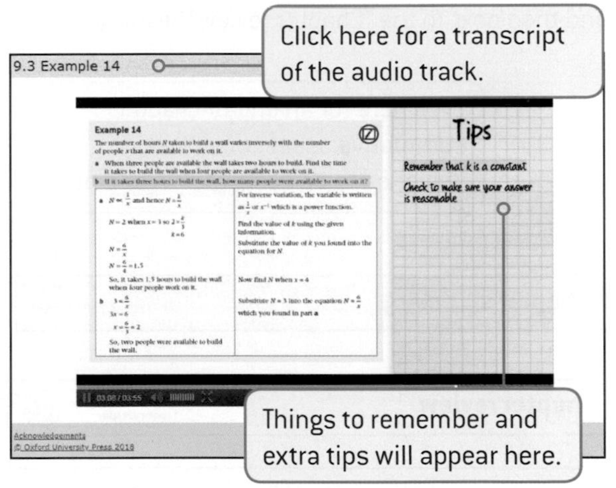

Click on the icon on the page to launch the animation. The animated worked example will appear in a second screen.

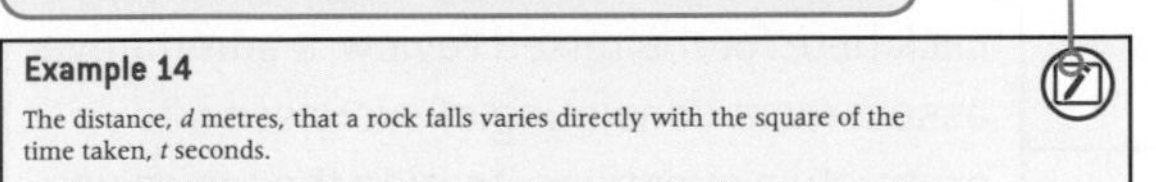

Example 14

The distance, d metres, that a rock falls varies directly with the square of the time taken, t seconds.

a If the rock falls 6 metres in 2 seconds, write an equation for d in terms of t.

b Find the distance the rock has fallen after 5 seconds.

Graphical display calculator support

Supporting you to make the most of your TI-Nspire CX, TI-84+ C Silver Edition or Casio fx-CG50 graphical display calculator (GDC), this icon takes you to step-by-step instructions for using tecÜology to solve specific examples in the book.

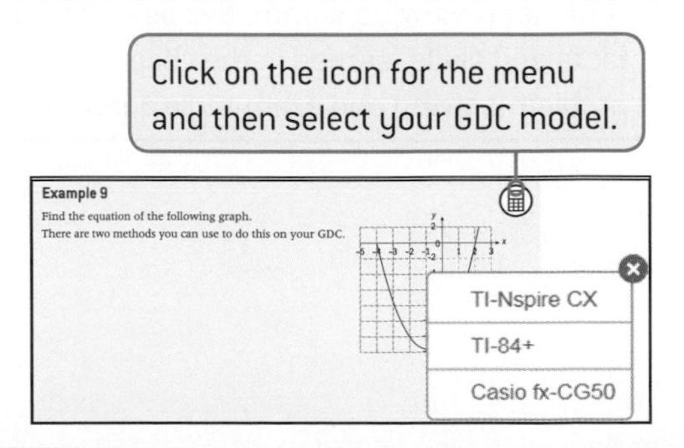

Data sets

Access a spreadsheet containing a data set relevant to the text associated with this icon.

Teacher notes

This icon appears at the beginning of each chapter and opens a set of comprehensive teaching notes for the investigations, reflection questions, TOK items, and the modelling and investigation activities in the chapter.

Assessment opportunities

This Mathematics: applications and interpretation enhanced online course book is designed to prepare you for your assessments by giving you a wide range of practice. In addition to the activities you will find in this book, further practice and support are available on the enhanced online course book.

End of chapter tests and mixed review exercises

This icon appears twice in each chapter: first, next to the "Chapter summary" section and then next to the "Chapter review" heading.

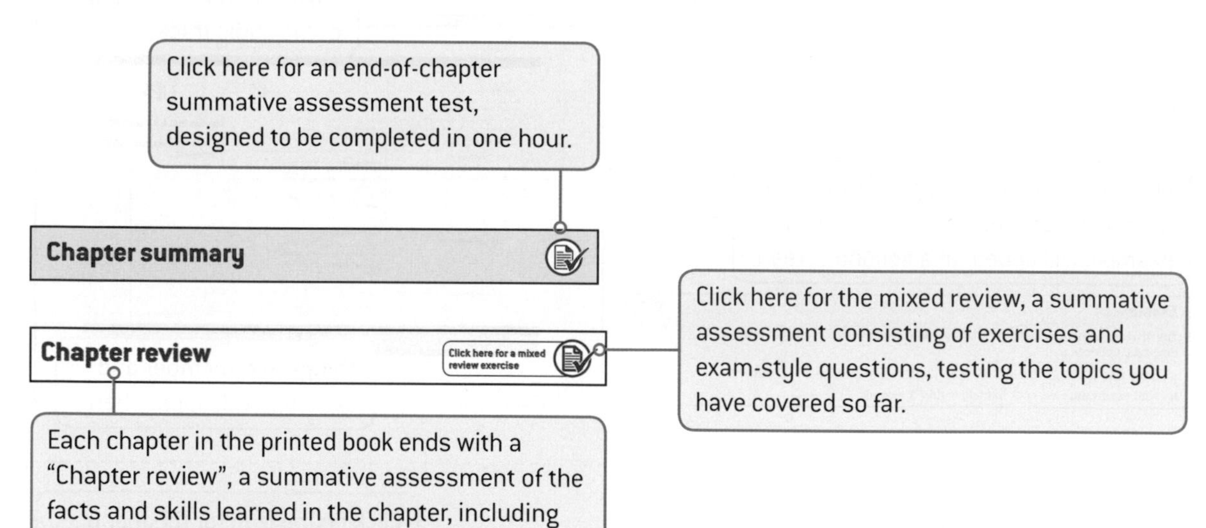

Exam-style questions

Exam-style questions

21 P1: a Find the binomial expansion of $\left(1-\frac{x}{4}\right)^5$ in ascending powers of x. (3 marks)

Plenty of exam practice questions, in Paper 1 (P1) or Paper 2 (P2) style. Each question in this section has a mark scheme in the worked solutions document found on the enhanced online course book, which will help you see how marks are awarded.

The number of darker bars shows the difficulty of the question (one dark bar = easy; three dark bars = difficult).

Exam practice exercises provide exam style questions for Papers 1 and 2 on topics from all the preceding chapters. Click on the icon for the exam practice found at the end of chapters 4, 9 and 13 in this book.

Answers and worked solutions

Answers to the book questions

Concise answer to all the questions in this book can be found on page 608.

Worked solutions

Worked solutions for **all** questions in the book can be accessed by clicking the icon found on the Contents page or the first page of the Answers section.

Answers and worked solutions for the digital resources

Answers, worked solutions and mark schemes (where applicable) for the additional exercises, end-of-chapter tests and mixed reviews are included with the questions themselves.

1 Measuring space: accuracy and 2D geometry

Almost everything we do requires some understanding of our surroundings and the distance between objects. But how do we go about measuring the space around us?

Concepts

- Quantity
- Space

Microconcepts

- Numbers
- Algebraic expressions
- Measurement
- Units of measure
- Approximation
- Estimation
- Upper/lower bound
- Error/percentage error
- Trigonometric ratios
- Angles of elevation and depression
- Length of arc

How does a sailor calculate the distance to the coast?

How can you measure the distance between two landmarks?

How can you calculate the distance travelled by a satellite in orbit?

How do scientists measure the depths of lunar craters by measuring the length of the shadow cast by the edge of the crater?

How far can you see? If you stood somewhere higher or lower, how would that affect how much of the Earth you can see?

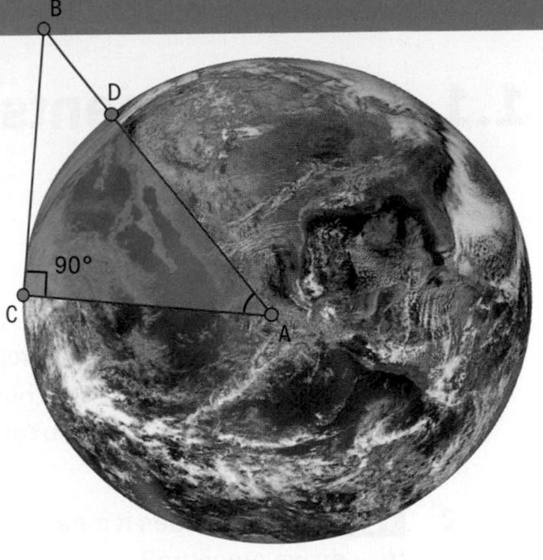

Think about the following:

- If you stood 10 metres above the ground, what would be the distance between you and the farthest object that you can see?
- One World Trade Center is the tallest building in New York City. If you stand on the Observatory floor, 382 m above the ground, how far can you see?
- How can you make a diagram to represent the distance to the farthest object that you can see? How do you think the distance you can see will change if you move the observation point higher?

Developing inquiry skills

Write down any similar inquiry questions that might be useful if you were asked to find how far you could see from a local landmark, or the top of the tallest building, in your city or country.

What type of questions would you need to ask to decide on the height of a control tower from which you could see the whole of an airfield? Write down any similar situations in which you could investigate how far you can see from a given point, and what to change so you could see more (or less).

Think about the questions in this opening problem and answer any you can. As you work through the chapter, you will gain mathematical knowledge and skills that will help you to answer them all.

Before you start

You should know how to:	Skills check
1 Find the circumference of a circle with radius 2 cm. eg $2\pi(2) = 12.6\text{ cm}$ (12.5663...)	**1** Find the circumference of a circle with radius $r = 5.3\text{ cm}$.
2 Find the area of a circle with radius 2 cm. $\pi(2)^2 = 12.6\text{ cm}^2$ (12.5663...)	**2** Find the area of a circle with radius 6.5 cm.
3 Find the area of: **a** a triangle with side 5 cm and height towards this side 8.2 cm $A = \dfrac{5\times 8.2}{2} = 20.5\text{ cm}^2$ **b** a square with side 3 cm $3^2 = 9\text{ cm}^2$ **c** a trapezoid with bases 10 m and 7 m, and height 4.5 m. $\dfrac{10+7}{2}\times 4.5 = 8.5\times 4.5 = 38.25\text{ m}^2$	**3** Find the areas of the following shapes. **a** 15.4 cm, 5.5 cm **b** 6.4 cm **c** 12 cm, 6.5 cm, 20 cm

Click here for help with this skills check

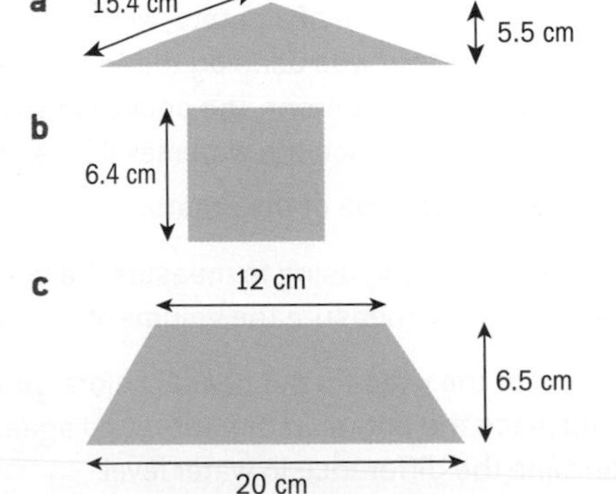

1.1 Measurements and estimates

Investigation 1

A Measuring a potato

1 Make a list of all the physical properties of a potato. Which of these properties can you measure? How could you measure them? Are there any properties that you cannot measure? How do we determine what we can measure?

2 **Factual** What does it mean to measure a property of an object? How do we measure?

3 **Factual** Which properties of an object can we measure?

4 **Conceptual** Why do we use measurements and how do we use measuring to define properties of an object?

B Measuring length

5 Estimate the length of the potato.

6 Measure the length of the potato. How accurate do you think your measurement is?

C Measuring surface area

Recall that the area that encloses a 3D object is called the surface area.

7 Estimate the surface area of the potato and write down your estimate.

Use a piece of aluminium foil to wrap the potato and keep any overlapped areas to a minimum. Once the potato is entirely wrapped without any overlaps, unwrap the foil and place it over grid paper with 1 cm^2 units, trace around it and count the number of units that it covers.

8 Record your measurement. How accurate do you think it is?

9 Measure your potato again, this time using sheets of grid paper with units of 0.5 cm^2 and 0.25 cm^2. Again, superimpose the aluminium foil representing the surface area of your potato on each of the grids, trace around it on each sheet of grid paper, and estimate the surface area.

10 Compare your three measurements. What can you conclude?

11 **Factual** How accurate are your measurements? Could the use of different units affect your measurement?

D Measuring volume

You will measure the volume of a potato, which has an irregular shape, by using a technique that was used by the Ancient Greek mathematician Archimedes, called displacement. The potato is to be submerged in water and you will measure the distance the water level is raised.

12 Estimate the volume of the potato.

13 What units are you using to measure the volume of water? Can you use this unit to measure the volume of a solid?

Note the height of the water in the beaker before you insert the potato. Slowly and carefully place the potato in the water, and again note the height of the water. Determine the difference in water level.

International-mindedness

Where did numbers come from?

14 Record your measurement for the volume of the potato.

15 **Conceptual** If you used a beaker with smaller units, do you think that you would have a different measurement for the volume?

E Measuring weight

16 Estimate the weight of the potato and write down your estimate.

17 Use a balance scale to measure the weight of the potato. Which units will you use?

18 **Conceptual** Could the use of different units affect your measurement?

F Compare results

19 Compare your potato measurements with the measurements of another group. How would you decide which potato is larger? What measures can you use to decide?

20 **Conceptual** How do we describe the properties of an object?

Measurements help us compare objects and understand how they relate to each other.

Measuring requires approximating. If a smaller measuring unit is chosen then a more accurate measurement can be obtained.

When you measure, you first select a property of the object that you will measure. Then you choose an appropriate unit of measurement for that property. And finally, you determine the number of units.

Investigation 2

Margaret Hamilton worked for NASA as the lead developer for Apollo flight software. The photo here shows her in 1969, standing next to the books of navigation software code that she and her team produced for the Apollo mission that first sent humans to the Moon.

1 Estimate the height of the books of code stacked together, as shown in the image. What assumptions are you making?

2 Estimate the number of pages of code for the Apollo mission. How would you go about making this estimate? What assumptions are you making?

3 **Factual** What is an estimate? What is estimation? How would you go about estimating? How can comparing measures help you estimate?

4 **Conceptual** Why are estimations useful?

Estimation (or estimating) is finding an approximation as close as possible to the value of a measurement by sensible guessing. Often the estimate is used to check whether a calculation makes sense, or to avoid complicated calculations.

Estimation is often done by comparing the attribute that is measured to another one, or by sampling.

Did you know?

The idea of comparing and estimating goes way back. Some of the early methods of measurement are still in use today, and they require very little equipment!

The logger method

Loggers often estimate tree heights by using simple objects, such as a pencil. An assistant stands at the base of the tree, and the logger moves a distance away from the tree and holds the pencil at arm's length, so that it matches the height of the assistant. The logger can then estimate the height of the tree in "pencil lengths" and multiply the estimate by the assistant's height.

The Native American method

Native Americans had a very unusual way of estimating the height of a tree. They would bend over and look through their legs!

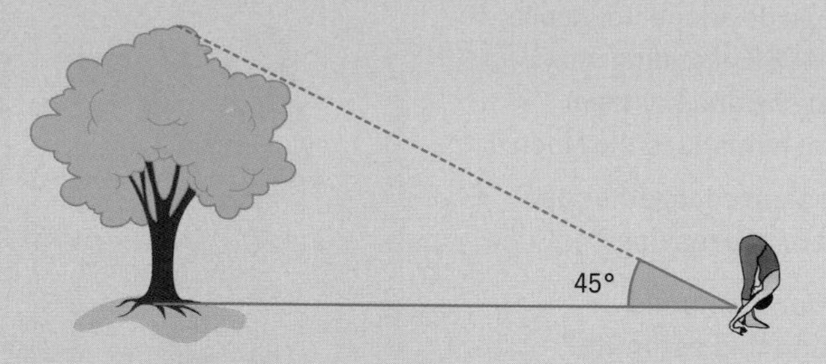

This method is based on a simple reason: for a fit adult, the angle that is formed as they look through their legs is approximately 45 degrees. Can you explain how this method works?

1.2 Recording measurements, significant digits and rounding

Investigation 3

When using measuring instruments, we are able to determine only a certain number of digits accurately. In science, when measuring, the **significant figures** in a number are considered only those figures (digits) that are definitely known, plus one estimated figure (digit). This is summarized as "all of the digits that are known for certain, plus one that is a best estimate".

1 Read the temperature in degrees Fahrenheit from this scale.

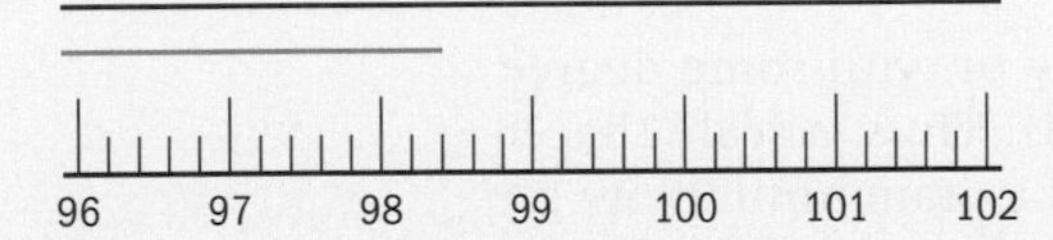

What is the best reading of the temperature that you can do? How many digits are significant in your reading of this temperature?

2 A pack of coffee is placed on a triple-beam balance scale and weighed. The image below shows its weight, in grams.

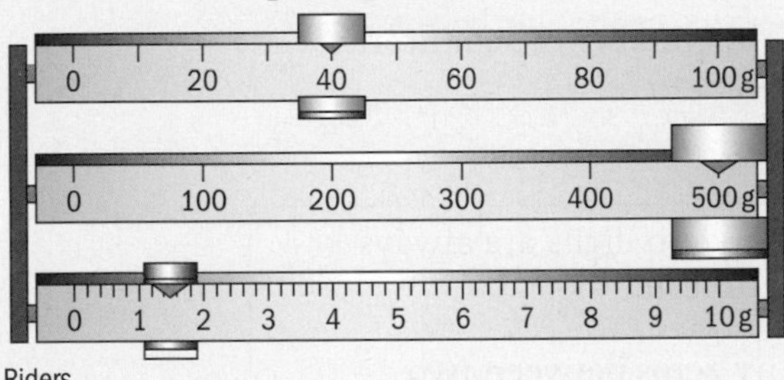

Find the weight of the pack of coffee by carefully determining the reading of each of the three beam scales and adding these readings. How many digits are significant in your reading of this weight?

3 **Factual** What is the smallest unit to which the weight of the pack of coffee can be read on this scale?

4 A laboratory technician compares two samples that were measured as 95.270 grams and 23.63 grams. What is the number of significant figures for each measurement? Is 95.270 grams the same as 95.27 grams? If not, how are the two measurements different?

5 **Conceptual** What do the significant figures tell you about the values read from the instrument? What do the significant figures in a measurement tell you about the accuracy of the measuring instrument?

6 **Conceptual** How do the reading of the measuring instrument and the measuring units limit the accuracy of the measurement?

TOK

What might be the ethical implications of rounding numbers?

Decimal places

You may recall that in order to avoid long strings of digits it is often useful to give an answer to a number of decimal places (dp). For example, when giving a number to 2 decimal places, your answer would have exactly two digits after the decimal point. You round the final digit up if the digit after it is 5 or above, and you round the final digit down if the digit after it is below 5.

Significant figures

Measuring instruments have limitations. No instrument is advanced enough to determine an unlimited number of digits. For example, a scale can measure the mass of an object only up until a certain decimal place. Measuring instruments are able to determine only a certain number of digits precisely.

The digits that can be determined accurately or with some degree of reliability are called significant figures (sf). Thus, a scale that could register mass only up to hundredths of a gram until 99.99 g could only measure up to 4 digits with accuracy (4 significant figures).

Example 1

For each of the following, determine the number of significant figures.

21.35, 1.25, 305, 1009, 0.00300, 0.002

21.35 has 4 sf and 1.25 has 3 sf.	Non-zero digits are always significant.
305 has 3 sf and 1009 has 4 sf.	Any zeros between two significant digits are significant.
In 0.003**00** only the last two zeros are significant and the other zeros are not. It has 3 sf.	A final zero or trailing zeros in the decimal part **only** are significant.
0.002 has only 1 sf, and all zeros to the left of 2 are not sf.	

Rounding rules for significant figures

The rules for rounding to a given number of significant figures are similar to the ones for rounding to the nearest 10, 100, 1000, etc. or to a given number of decimal places.

EXAM HINT

In exams, give your answers exact or accurate to 3 sf, unless otherwise specified in the problem.

Example 2

Round the following numbers to the required number of significant figures:

a 0.1235 to 2 sf **b** 0.2965 to 2 sf

c 415.25 to 3 sf **d** 3050 to 2 sf

a 0.<u>12</u>35 = 0.12 (2 sf)	Underline the 2 significant figures. The next digit is less than 5, so delete it and the digits to the right.
b 0.<u>29</u>65 = 0.30 (2 sf)	The next digit is greater than 5 so round up. Write the 0 after the 3, to give 2 sf.
c <u>415</u>.25 = 415 (3 sf)	Do not write 415.0, as you only need to give 3 sf.
d <u>30</u>50 = 3100 (2 sf)	Write the zeros to keep the place value.

Rounding rule for n significant figures

If the $(n + 1)$th figure is less than 5, keep the nth figure as it is and remove all figures following it.

If the $(n + 1)$th figure is 5 or higher, add 1 to the nth figure and remove all figures following it.

In either case, all figures after the nth one should be removed if they are to the right of the decimal point and replaced by zeros if they are to the left of the decimal point.

Exercise 1A

1 Round the following measurements to 3 significant figures.

a 9.478 m **b** 5.322 g

c 1.8055 cm **d** 6.999 in

e 4578 km **f** 13 178 kg

2 Round the numbers in question 1 parts **a** to **d** to 2 dp.

3 Determine the number of significant figures in the following measurements:

a 0.102 m **b** 1.002 dm

c 105 kg **d** 0.001020 km

e 1 000 000 μg

4 Find the value of the expression $\frac{12.35 + 21.14 + 1.075}{\sqrt{3.5} - 1}$ and give your answer to 3 significant figures.

Example 3

A circle has radius 12.4 cm. Calculate:

a the circumference of the circle

b the area of the circle.

Write down your answers correct to 1 dp.

Continued on next page

a $C = 2 \times \pi \times 12.4$ $= 77.9\text{ cm (1 dp)}$	Use the formula for circumference of a circle, $C = 2\pi r$. The answer should be given correct to 1 dp, so you have to round to the nearest tenth.
b $A = \pi(12.4)^2$ $= 483.1\text{ cm}^2\text{ (1 dp)}$	Use the formula for area of a circle, $A = \pi r^2$. The answer should be given correct to 1 dp, so you have to round to the nearest tenth.

Investigation 4

The numbers of visitors to the 10 most popular national parks in the United States in 2016 are shown in the table.

10 Most Visited National Parks (2016)	
Park	***Recreational Visits***
1. Great Smoky Mountains NP	11312786
2. Grand Canyon NP	5969811
3. Yosemite NP	5028868
4. Rocky Mountain NP	4517585
5. Zion NP	4295127
6. Yellowstone NP	4257177
7. Olympic NP	3390221
8. Acadia NP	3303393
9. Grand Teton NP	3270076
10. Glacier NP	2946681

1 Which park had the most visitors? How accurate are these figures likely to be?

2 Round the numbers of visitors given in the table to the nearest 10 000.

3 Are there parks with an equal number of visitors, when given correct to 10 000? If so, which are they?

4 Round the number of visitors, given in the table, to the nearest 100 000.

5 Are there parks with an equal number of visitors, when given correct to 100 000? If so, which are they?

6 Round the numbers of visitors given in the table to the nearest 1 000 000.

7 Are there parks with an equal number of visitors, when given correct to 1 000 000? If so, which are they?

8 Determine approximately how many times the number of visitors of the most visited park is bigger than the number of visitors of the least visited park. Which parks are they?

9 Determine approximately how many times the number of visitors of the most visited park is bigger than the number of visitors of the second most visited park.

10 **Factual** Determine approximately how many times the number of visitors of the most visited park is bigger than the number of visitors of the third most visited park.

11 **Conceptual** How did rounding help you compare the numbers of park visitors?

12 **Conceptual** What are the limitations of a measurement reading in terms of accuracy?

13 **Conceptual** How is rounding useful?

Exercise 1B

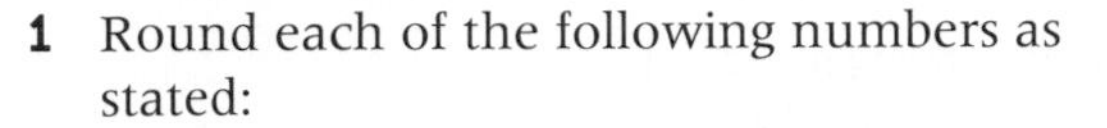

1 Round each of the following numbers as stated:

- **a** 8888 to 3 sf
- **b** 3.749 to 3 sf
- **c** 27 318 to 1 sf
- **d** 0.00637 to 2 sf
- **e** $\sqrt{62}$ to 1 dp

2 Round the numbers in the table to the given accuracy.

	Number	Round to the nearest ten	Round to the nearest hundred	Round to the nearest thousand
a	2815			
b	75 391			
c	316 479			
d	932			
e	8253			

3 Round the following amounts to the given accuracy:

- **a** 502.13 EUR to the nearest EUR
- **b** 1002.50 USD to the nearest thousand USD
- **c** 12 BGN to the nearest 10 BGN
- **d** 1351.368 JPY to the nearest 100 JPY

4 A circle has radius 33 cm. Calculate the circumference of the circle. Write down your answer correct to 3 sf.

5 The area of a circle is 20 cm^2 correct to 2 sf. Calculate the radius of the circle correct to 2 sf.

6 Calculate the volume of a cube with side 4.82 cm. Write down your answer correct to 2 sf.

1.3 Measurements: exact or approximate?

Accuracy

The accuracy of a measurement often depends on the measuring units used. The smaller the measuring unit used, the greater the accuracy. If I use a balance scale that measures only to the nearest gram to weigh my silver earrings, I will get 11 g. But if I use an electronic scale that measures to the nearest hundredth of a gram, then I get 11.23 g.

TOK

To what extent do instinct and reason create knowledge?

Do different geometries (Euclidean and non-Euclidean) refer to or describe different worlds? Is a triangle always made up of straight lines? Is the angle sum of a triangle always 180°?

The accuracy would also depend on the precision of the measuring instrument. If I measured the weight of my earrings three times, the electronic scale might produce three different results: 11.23 g, 11.30 g and 11.28 g. Usually, the average of the available measures is considered to be the best measurement, but it is certainly not exact.

Each measuring device (metric ruler, thermometer, theodolite, protractor, etc.) has a different **degree of accuracy**, which can be determined by finding the smallest division on the instrument. Measuring the dimensions of a rug with half a centimetre accuracy could be acceptable, but a surgical incision with such precision most likely will not be good enough!

A value is **accurate** if it is close to the **exact value** of the quantity being measured. However, in most cases it is not possible to obtain the exact value of a measurement. For example, when measuring your weight, you can get a more accurate measurement if you use a balance scale that measures to a greater number of decimal places.

Investigation 5

The **yard** and the **foot** are units of length in both the British Imperial and US customary systems of measurement. Metal yard and foot sticks were the original physical standards from which other units of length were derived.

In the 19th and 20th centuries, differences in the prototype yards and feet were detected through improved technology, and as a result, in 1959, the lengths of a yard and a foot were defined in terms of the metre.

In an experiment conducted in class, several groups of students worked on measuring a standard yardstick and a foot-long string.

1 Group 1 used a ruler with centimetre and millimetre units and took two measurements: one of a yardstick and one of a foot-long string. Albena, the group note taker, rounded off the two measurements to the nearest centimetre and recorded the results for the yard length as 92 cm and for the foot length as 29 cm. Write down the possible values for the unrounded results that the group obtained. Give all possible unrounded values for each measure as intervals in the form $a \le x < b$.

2 Group 2 used a Vernier caliper, which is able to measure lengths to tenths of a millimetre. They also took measurements of a yardstick and of a piece of string a foot long. Velina, the group note taker, rounded off the two measurements to the nearest millimetre and recorded the results as 91.5 cm and 31.5 cm. Write down the possible values for the unrounded results that the group obtained. Give all possible unrounded values for each measure as intervals in the form $a \le x < b$.

International-mindedness

SI units

In 1960, the International System of Units, abbreviated SI from the French, “systeme International”, was adopted as a practical system of units for international use and includes metres for distance, kilograms for mass and seconds for time.

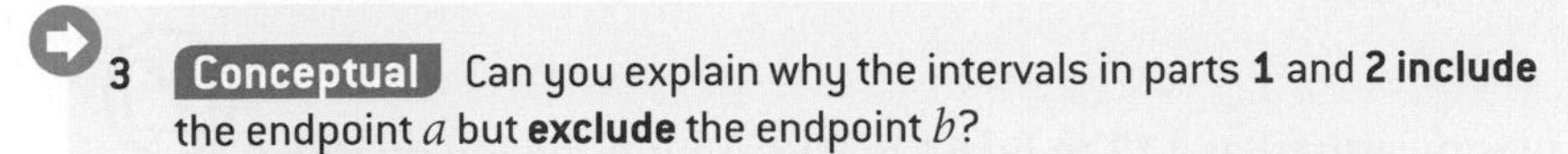

3 **Conceptual** Can you explain why the intervals in parts **1** and **2 include** the endpoint a but **exclude** the endpoint b?

4 **Conceptual** Based on your conclusions in parts **1** and **2**, make a conjecture about the interval in which the exact value should lie. How big is this interval in relation to the unit used? How would you determine the left and the right ends of the interval?

The left and the right ends of an interval in which an exact value of a measurement lies are respectively called the **lower bound** and the **upper bound**.

The lower bound and the upper bound are half a unit below and above a rounded value of a measurement. Thus the upper bound is calculated as the rounded measurement + 0.5 unit, and the lower bound is found as the rounded measurement − 0.5 unit.

Did you know?

The exact values of one yard and one foot are defined by an international agreement in 1959. A yard was defined as 0.9144 metres exactly, and a foot was defined as 0.3048 metres exactly.

Example 4

a Jane's weight is 68 kg to the nearest kg. Determine the upper and lower bounds of her weight.

b Rushdha's height is measured as 155 cm to the nearest cm. Write the interval within which her exact height lies.

a The upper and lower bounds are 68.5 and 67.5, respectively.	The range of possible values for Jane's weight is 68 ± 0.5 kg.
b $154.5 \leq h < 155.5$	The range of possible values for Rushdha's height is 155 ± 0.5 cm.

Example 5

Majid ran 100 metres in 11.3 seconds. His time is measured to the nearest tenth of a second. Determine the upper and lower bounds of Majid's running time.

Lower bound = 11.3 − 0.05 = 11.25	Lower bound is 11.3 − 0.5 unit, and upper bound is 11.3 + 0.5 unit.
Upper bound = 11.3 + 0.05 = 11.35	Majid's time is given to the nearest tenth of a second. A unit is a tenth of a second or 0.1 sec, and 0.5 unit is 0.05 sec.

Example 6

A rectangular garden plot was measured as 172 m by 64 m. Determine the lower and upper bounds of its perimeter.

$171.5 \le L < 172.5$ $63.5 \le W < 64.5$ Then the lower bound of the perimeter is $2 \times (171.5 + 63.5) = 470$, and the upper bound of the perimeter is $2 \times (172.5 + 64.5) = 474$. Thus $470 \le P < 474$.	For the lower bound of the perimeter, use the shortest possible lengths of the sides: the measured values – 0.5 m; for the upper bound, use the longest possible lengths: the measured values + 0.5 m.

Exercise 1C

1 Find the range of possible values for the following measurements, which were rounded to the nearest mm, tenth of m, and hundredth of kg respectively:

a 24 mm **b** 3.2 m **c** 1.75 kg

2 A triangle has a base length of 3.1 cm and corresponding height 4.2 cm, correct to 1 decimal place. Calculate the upper and lower bounds for the area of the triangle as accurately as possible.

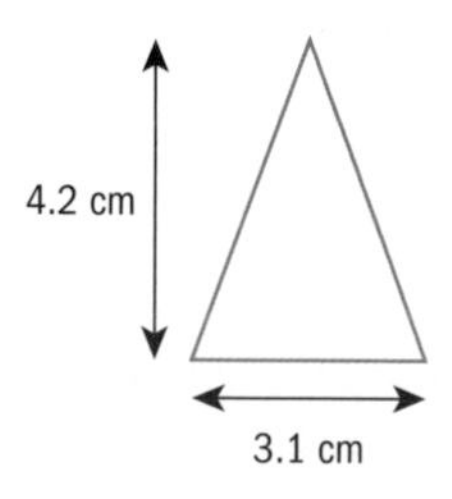

3 With 72 million bicycles, correct to the nearest million, Japan is at the top of the list of countries with most bicycles per person. On average, Japanese people travel about 2 km by bicycle, correct to the nearest km, each day. Calculate the upper bound for the total distance travelled by all the bicycles in Japan per day.

4 To determine whether a business is making enough profit, the following formula is used:

$$P = \frac{S - C}{S}$$

where P is relative profit, S is sales and C is costs. If a company has \$340 000 worth of sales and \$230 000 of costs, correct to 2 significant figures, calculate the maximum and minimum relative profit to the appropriate accuracy.

Since measurements are approximate there is always error in the measurement results. A measurement error is the difference between the exact value (V_E) and the approximate value (V_A), ie

$$\text{Measurement error} = V_A - V_E$$

Investigation 6

Tomi and Massimo measured the length of a yardstick and the length of a foot-long string and obtained 92.44 cm for the length of a yard and 31.48 cm for the length of a foot.

TOK

Do the names that we give things impact how we understand them?

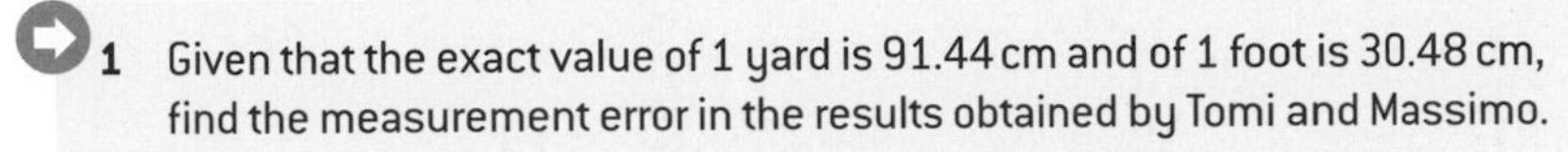

1 Given that the exact value of 1 yard is 91.44 cm and of 1 foot is 30.48 cm, find the measurement error in the results obtained by Tomi and Massimo.

Tomi thinks that the two measurements were equally inaccurate. Massimo thinks that one of the two measurements is more accurate than the other.

2 Who do you agree with: Tomi or Massimo? Why? Explain.

Massimo decides to find the magnitude of the measurement error as a percentage of the exact length.

3 Write down the error in measuring the length of 1 yard as a fraction of the exact length of 1 yard. Give your answer as a percentage.

4 Write down the error in measuring the length of 1 foot as a fraction of the exact length of 1 foot. Give your answer as a percentage.

5 **Conceptual** In what ways could expressing measurement errors as a percentage of the exact value be helpful?

6 **Conceptual** How could measurement errors be compared?

The percentage error formula calculates the error as a percentage of the measured quantity. For example, a weight measured as 102 kg when the exact value is 100 kg gives a measurement error of 2 kg and percentage error of $\frac{2}{100} = 2\%$. A weight measured as 27 kg when the exact value is 25 kg gives the same measurement error of 2 kg but a percentage error of $\frac{2}{25} = 8\%$.

Percentage error $= \left|\frac{V_A - V_E}{V_E}\right| \times 100\%$, where V_A is the approximate value and V_E is the exact value.

Example 7

The fraction $\frac{22}{7}$ is often used as an approximation of π.

a How close (to how many decimal places) does $\frac{22}{7}$ approximate π?

b Find the percentage error of this approximation, giving your answer to 2 dp.

a $\frac{22}{7} - \pi = 0.001264...$ Thus $\frac{22}{7}$ approximates π to 2 dp.	Measurement error $= V_A - V_E$, where V_A is the approximate value and V_E is the exact value.

Continued on next page

b Percentage error $= \left\| \dfrac{\frac{22}{7} - \pi}{\pi} \right\| \times 100\%$ $= 0.04\%$	Percentage error $= \left\| \dfrac{V_A - V_E}{V_E} \right\| \times 100\%$ Be careful to take the absolute value of the fraction as the percentage error is always a positive number!

In multistep calculations, you must be careful not to round figures until you have your final answer. Premature rounding of initial or intermediate results may lead to an incorrect answer.

TOK

How does the perception of the language being used distort our understanding?

Example 8

Calculate the density of a cube of sugar weighing 2.45 grams, where the side of the cube is 1.2 cm, correct to 1 dp.

Volume $= (1.2)^3 = 1.728\,\text{cm}^3$ Density $= \dfrac{2.45}{1.728} = 1.41782\ldots\ \text{g/cm}^3$ $= 1.4\ \text{g/cm}^3$ (correct to 1 dp)	If you first rounded the mass to 1 dp, then calculated the volume to 1 dp and then divided, you would obtain: $\dfrac{2.5}{1.7} = 1.47058\ \ldots\ \text{g/cm}^3,$ $= 1.5\ \text{g/ cm}^3$ (correct to 1 dp) This gives a percentage error of $\left\| \dfrac{1.5 - 1.41782\ldots}{1.41782\ldots} \right\| \times 100\% = 5.80\%$ Make sure to avoid premature rounding!

Rounding of intermediate results during multistep calculations reduces the accuracy of the final answers. Thus, make sure to round your final answer only.

Exercise 1D

1 Find the percentage error in using 3.14 instead of π.

2 In 1856, Andrew Waugh announced Mount Everest to be 8840 m high, after several years of calculations based on observations made by the Great Trigonometric Survey. More recent surveys confirm the height as 8848 m. Calculate the percentage error made in the earlier survey.

3 Eratosthenes estimated the circumference of the Earth as 250 000 stadia (the length of an athletic stadium). If we assume he used the most common length of stadia of his time, his estimate of the circumference of the Earth would be 46 620 km. Currently, the accepted average circumference of the Earth is 40 030.2 km. Find the percentage error of Eratosthenes' estimate of the circumference of the Earth.

4 The temperature today in Chicago is 50°F. Instead of using the standard conversion formula $°C = \frac{5}{9} \times (°F - 32)$, Tommaso uses his grandmother's rule, which is easier but gives an approximate value: "*Subtract* 32 *from the value in* °F *and multiply the result by* 0.5."

a Calculate the actual and an approximate temperature in °C, using the standard formula and Tommaso's grandmother's rule.

b Calculate the percentage error of the approximate temperature value in °C.

1.4 Speaking scientifically

Investigation 7

1 **a** Complete the table, following the pattern:

Number							
Written as powers of 10	10^3	10^2	10^1	10^0		10^{-2}	10^{-3}
Written as decimals	1000						
Written as fractions	$\frac{1000}{1}$				$\frac{1}{10}$		

b When you move from left to right, from one column to the next, which operation would you use?

c How would you write 10^{-4} as a decimal and as a fraction?

d Write 10^{-n} as a fraction.

2 Complete the table, following the pattern:

Number							
Written as powers of 2	2^3	2^2	2^1	2^0		2^{-2}	2^{-3}
Written as decimals	8						
Written as fractions	$\frac{8}{1}$		$\frac{2}{1}$		$\frac{1}{2}$		

a Find 2^0.

b Find 10^0.

c Find x^0.

d How are algebraic expressions similar to and different from a numerical expression?

Numerical expressions consist only of numbers and symbols of operations (addition, subtraction, multiplication, division and exponentiation), whereas **algebraic expressions** contain numbers, variables and symbols of operations.

Exponentiation is a mathematical operation, written as a^n, where a is called the base and n the exponent or power.

If the exponent, n, is a **positive integer**, it determines **how many times** the base, a, is multiplied by itself. For example, 8^2 means 8×8.

If the exponent, n, is a **negative integer** it determines how many times **to divide** 1 by the base, a. For example, 8^{-2} means $\frac{1}{8^2}$ or $\frac{1}{8 \times 8}$.

The base, a, is the factor in the expression $a \times a \times a \times \ldots \times a$ if the exponent, n, is positive and is the factor in the expression $\frac{1}{a \times a \times a \times \ldots \times a}$ if the exponent, n, is negative.

International-mindedness

Indian mathematician Brahmagupta is credited with the first writings that included zero and negative numbers in the 7th century.

Investigation 8

1 a Complete the first and the third columns of the table. The middle column you can choose to finish or not.

Expression	Expanded expression	Written as power of 10
$10^2 \times 10^3$	$(10 \times 10) \times (10 \times 10 \times 10)$	10^5
$10^4 \times 10^5$		
	$(10 \times 10 \times 10 \times 10) \times (10 \times 10 \times 10 \times 10 \times 10)$	
		10^8
$10^5 \times 10^6$		
$10^1 \times 10^{10}$		
$(10^2)^3$	$(10 \times 10) \times (10 \times 10) \times (10 \times 10)$	

b Follow the pattern from the table and rewrite $10^m \times 10^n$ as a single power of 10.

c Rewrite $10^2 \times 10^0$ as a single power of 10.

d What can you conclude for the value of 10^0? Why?

e Follow the pattern and rewrite $x^m \times x^n$ as a single power.

f Use powers and multiplication to write three expressions whose value is 10^{11}.

g Rewrite $(10^2)^3$ as a single power.

h Follow the pattern and rewrite $(10^m)^n$ as a single power.

i Follow the pattern from part **h**, and rewrite $(x^m)^n$ as a single power.

j Write the expanded expression for $(2 \times 3)^5$. Then rewrite each term of the expanded expression as a product of two single powers.

k Follow the pattern in part **j**, and rewrite $(x \times y)^n$ as a product of two single powers.

2 **a** Complete the first and the third columns of the table. The middle column you can choose to finish or not.

Expression	Expanded expression	Written as power of 10
$\frac{10^3}{10^2}$	$\frac{10\times10\times10}{10\times10}$	10^1
$\frac{10^5}{10^3}$		
	$\frac{10\times10\times10\times10\times10\times10}{10\times10\times10}$	
		10^4
$\frac{10^6}{10^6}$		
$\frac{10^3}{10^0}$		

b Follow the pattern from the table and rewrite $\frac{10^m}{10^n}$ as a single power of 10.

c Rewrite $\frac{10^2}{10^0}$ as a single power of 10. What would that be?

d Follow the pattern and rewrite $\frac{x^m}{x^n}$ as a single power.

e Use powers and division to write three expressions whose value is 10^5.

f Write the expanded expression of $\left(\frac{2}{3}\right)^5$. Rewrite the expanded expression as a quotient of two powers.

g Write the expanded expression of $\left(\frac{x}{y}\right)^n$. Rewrite the expanded expression as a quotient of two powers.

Laws of exponents

Law	Example
$x^1 = x$	$6^1 = 6$
$x^0 = 1$	$7^0 = 1$
$x^{-1} = \frac{1}{x}$	$4^{-1} = \frac{1}{4}$
$x^m x^n = x^{m+n}$	$x^2x^3 = x^{2+3} = x^5$

Continued on next page

International-mindedness

Archimedes discovered and proved the law of exponents, $10^x \times 10^y = 10^{x+y}$.

Law	Example
$\frac{x^m}{x^n} = x^{m-n}$	$\frac{x^6}{x^2} = x^{6-2} = x^4$
$(x^m)^n = x^{mn}$	$(x^2)^3 = x^{2\times3} = x^6$
$(xy)^n = x^n y^n$	$(xy)^3 = x^3 y^3$
$\left(\frac{x}{y}\right)^n = \frac{x^n}{y^n}$	$\left(\frac{x}{y}\right)^2 = \frac{x^2}{y^2}$
$x^{-n} = \frac{1}{x^n}$	$x^{-3} = \frac{1}{x^3}$

International-mindedness

Abu Kamil Shuja was called al-Hasib al-Misri, meaning "the calculator from Egypt", and was one of the first mathematicians to introduce symbols for indices in the 9th century.

Example 9

Use the laws of exponents to express each of the following in terms of powers of a single number:

a $\frac{2}{2^3} + (2^2)^3$ **b** $\frac{4^3 \times 4^{-7}}{4^2}$

a $2^{-2} + 2^6 = 64.25$	Use your GDC, or use $\frac{x^m}{x^n} = x^{m-n}$ for $\frac{2}{2^3}$ and $(x^m)^n = x^{mn}$ for $(2^2)^3$.
b $\frac{4^{-4}}{4^{-2}} = 4^{-2} = \frac{1}{4^2} = \frac{1}{16} = 0.0625$	Use your GDC, or use $x^m x^n = x^{m+n}$ to simplify the numerator and then $\frac{x^m}{x^n} = x^{m-n}$ to simplify the quotient.

Example 10

Write the following expressions using a single power of x:

a $(x^2)^3$ **b** $\left(\frac{1}{x^2}\right)^4$ **c** $(x^3)^{-1}$ **d** $\frac{x^3 \times x^{-1}}{x^5}$

a x^6	Use $(x^m)^n = x^{mn}$.
b $\left(\frac{1}{x^2}\right)^4 = \left(x^{-2}\right)^4 = x^{-8}$	Use $x^{-n} = \frac{1}{x^n}$ to write $\frac{1}{x^2}$ as x^{-2}, and then apply $(x^m)^n = x^{mn}$.
c $(x^3)^{-1} = \frac{1}{x^3} = x^{-3}$	Use $(x^m)^n = x^{mn}$, or use $x^{-n} = \frac{1}{x^n}$ twice.
d $\frac{x^2}{x^5} = x^{-3}$	Use $\frac{x^m}{x^n} = x^{m-n}$.

Exercise 1E

1 Calculate the following numerical expressions, giving your answer as a single power of an integer.

a $2^3 \times 2^3$

b $5^2 \times 5^1$

c $\frac{6^7}{6^5}$

d $\frac{4^2}{5^{-2}}$

e $8^6 \times 8^{-3}$

f $(3^2)^4$

g $\frac{3^{-4}}{3^2 \times 3^9}$

h $\frac{2^7 \times 2^{-4}}{2^3}$

i $5^4 \times 3^4$

j $\frac{20^3}{4^3}$

2 Simplify the following algebraic expressions:

a $x^{-2} \times (x^2)^3 \times (x^2 \times x^6)$

b $\frac{x^0 \times (x^2)^3}{x^2 \times x^{-3}}$

Standard form

In standard form (also known as scientific notation), numbers are written in the form $a \times 10^n$, where a is called the **coefficient** or **mantissa**, with $1 \leq a < 10$, and n is an integer.

Scientific notation is certainly economical; a number such as *googol*, which in decimal notation is written as 1 followed by 100 zeros, is written simply as 10^{100} in scientific notation.

With scientific notation, **Avogadro's constant**, which is the number of particles (atoms or molecules) contained in 1 mole of a substance, is written as $6.022140857 \times 10^{23}$. If not for scientific notation it would take 24 digits to write!

EXAM HINT

Note that in an exam you cannot write 3E5, as this is calculator notation and not the correct scientific notation. The correct notation is 3×10^5.

You may have noticed that your graphic display calculator gives any results with lots of digits in scientific notation. However, instead of writing the results in the form $a \times 10^n$, it gives them in the form $a\text{E}n$, where n is an integer. For example, 3×10^5 will be given as 3E5, and 3.1×10^{-3} as 3.1E−3.

To convert from decimal to scientific notation on your GDC, change the Mode from Normal to Scientific. Thus, all of your number entries will be immediately converted to scientific calculator notation. To convert to correct scientific notation, you will need to replace E with 10, eg 5E2 should be written as 5×10^2.

Example 11

Find the volume of a computer chip (a cuboid) that is 2.44×10^{-6} m wide, 1.5×10^{-7} m long and 2.15×10^{-4} m high.

The volume of a cuboid = length × width × height	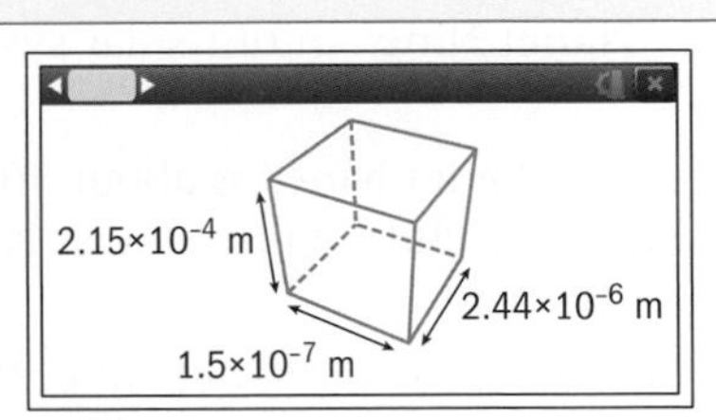 Make sure to give your answer in standard form.

Continued on next page

The volume is 7.869×10^{-17} m³.	To check your answer is sensible, multiply the rounded values $2 \times 2 \times 2 = 8$, and check the powers of 10 come to 10^{-17}.

Exercise 1F

1 Find the measurements below in the form $a \times 10^n$, where $1 \le a < 10$ and n is an integer:

- **a** the density of air at 27°C and 1 atm pressure, 0.00161 g/cm³
- **b** the radius of a calcium atom, 0.000 000 000 197 m
- **c** one light-year, 9 460 000 000 000 km
- **d** the mass of a neutron, 0.000 000 000 000 000 000 000 001 675 g.

2 Write down the following numbers found on a calculator display in scientific notation:

- **a** 1.2E–1
- **b** 5.04E7
- **c** 4.005E–5
- **d** 1E–3

3 The image of a speck of dust viewed through an electron microscope is 1.2×10^2 millimetres wide. The image is 5×10^2 times as large as the actual size. Determine the width, in millimetres, of the actual speck of dust.

4 **a** Convert the following from decimal to scientific notation:

age of the Earth = 4 600 000 000 years.

b Convert the following from scientific to decimal notation: 5×10^3.

5 One millilitre has about 15 drops. One drop of water has 1.67×10^{21} H_2O (water) molecules. Estimate the number of molecules in 1 litre of water. Write down your answer in standard form.

6 Scientists announced the discovery of a potential "Planet Nine" in our solar system in 2016.

The so-called "Planet Nine" is about 5000 times the mass of Pluto. Pluto's mass is 0.01303×10^{24} kg.

- **a** Calculate the mass of Planet Nine. Write down your answer in standard form.
- **b** The Earth's mass is 5.97×10^{24} kg. Find how many times Planet Nine is bigger or smaller compared to the Earth. Write down your answer correct to the nearest digit.

TOK

What do we mean when we say that one number is larger than another number?

7 The table below shows the population of different regions in 1985 and in 2005.

Place	Population in 1985	Population in 2005
Entire world	4.9×10^9	6.4×10^9
China	1.1×10^9	1.3×10^9
India	7.6×10^8	1.1×10^9
United States	2.4×10^8	3.0×10^8
Bulgaria	8.9×10^6	7.2×10^6

- **a** Find, in the form n:1, the ratio of the population of the entire world to that of China in 2005, giving n to the nearest integer.
- **b** Find the increase in the world's population between 1985 and 2005.
- **c** Calculate the percentage change in the population of each of the four countries, giving your answers accurate to 3 sf. List the countries according to their percentage change from highest to lowest.
- **d** State whether or not you agree that it is always the case that the country with a bigger percentage change also has a bigger population increase between 1985 and 2005. Justify your answer.

1.5 Trigonometry of right-angled triangles and indirect measurements

Investigation 9

1 Puzzle One:

What do you notice about the three squares built on the sides of the right-angled triangle ΔABC, where angle $\hat{C} = 90°$?

Can you state the relationship between the sides [AB], [BC] and [AC] of the triangle? If you cannot copy the shapes exactly, then use scissors to cut out copies of the red, blue and green squares, and see whether you can fit the green and blue squares exactly over the red one. You may cut up the blue and green squares along the internal lines to create smaller squares or rectangles.

A

C

B

2 Puzzle Two:

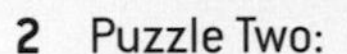

There are 11 puzzle pieces: 8 right-angled triangles and 3 squares of different sizes.
Cut out copies of the puzzle pieces and arrange them in the two frames.

Frames:

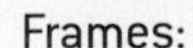

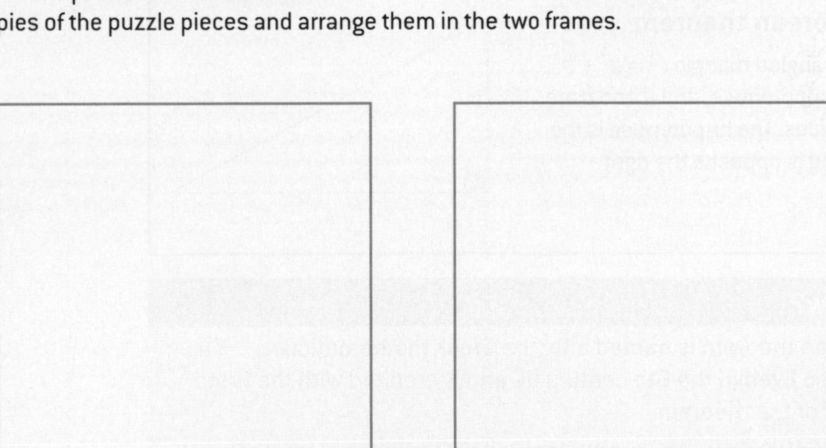

Continued on next page

Pieces:

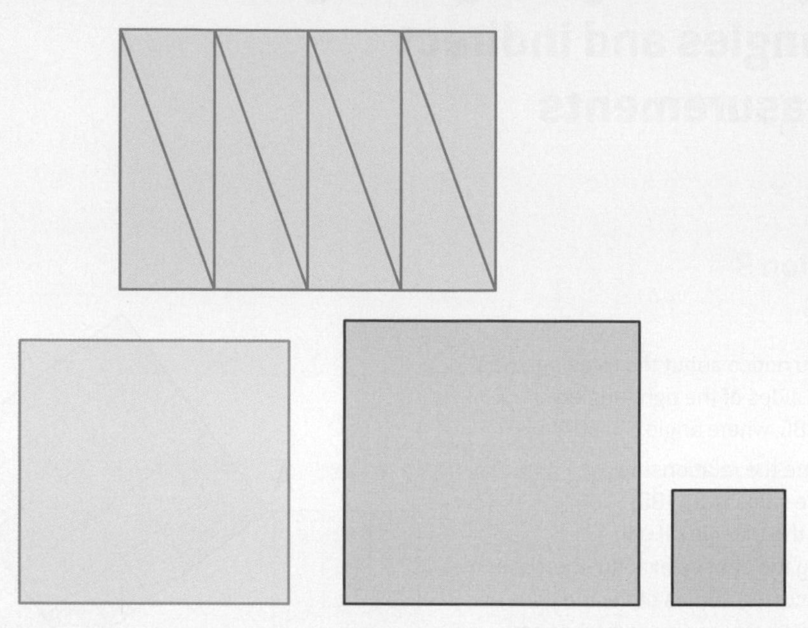

Is there more than one way to arrange the puzzle pieces into the two frames? What is the relationship between the areas of the frames?

What is the relationship between the areas of the three puzzle squares? What relationship between the sides of the right-angled triangles is revealed by the two puzzles?

3 **Conceptual** What is the relationship between the sides of any right-angled triangle?

The Pythagorean theorem

For every right-angled triangle $c^2 = a^2 + b^2$, where c is the hypotenuse, and a and b are the other two sides. The hypotenuse is the longest side and is opposite the right angle.

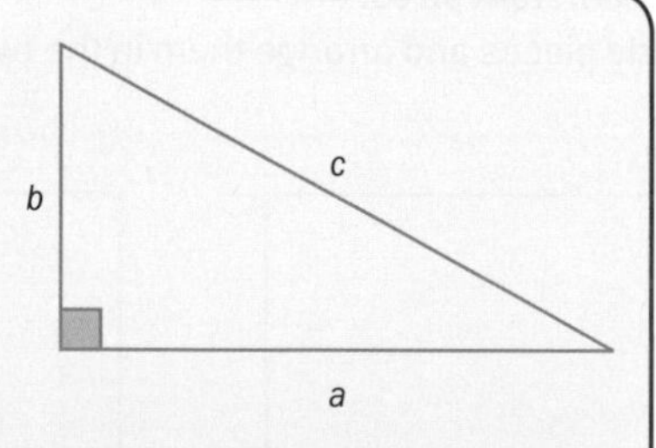

Did you know?

The Pythagorean theorem is named after the Greek mathematician Pythagoras, who lived in the 6th century BC and is credited with the first recorded proof of the theorem.

The theorem was known to some cultures well before Pythagoras; the ancient Egyptians knew about the 3-4-5 right-angled triangle (a triangle with sides 3, 4 and 5 units), and they knew how to measure a right angle by stretching a rope with equally spaced knots. The Egyptian surveyors (called "rope stretchers") used knotted cords to make measurements for building and restoring the boundaries of fields after the flooding of the River Nile.

If you know the lengths of two of the sides of a right-angled triangle you can always determine the length of the third side.

Example 12

Calculate the length of AB in the right-angled triangle ΔABC, where $CB = 15$ cm and $AC = 7$ cm.

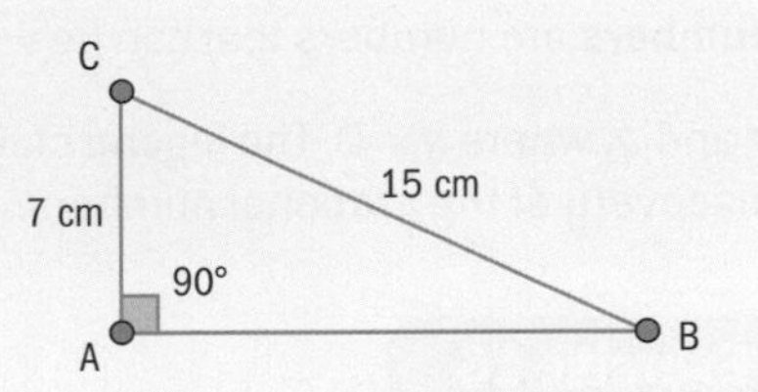

Using the Pythagorean theorem: $AB^2 + 7^2 = 15^2$ $AB = \sqrt{15^2 - 7^2}$ $= 13.3$ cm (13.2666...)	Rearrange to find AB. Remember to give the final answer to 3 sf unless a different accuracy is specified in the question.

The converse of the Pythagorean theorem states: If for a triangle $c^2 = a^2 + b^2$, where a, b and c are its sides, then it must be a right-angled triangle.

The converse of the Pythagorean theorem is also useful, as if you know the three sides of a triangle you can determine whether the triangle has a right angle or not.

Example 13

Determine whether or not a triangle with sides 6 cm, 8 cm and 9 cm is a right-angled triangle.

Is $9^2 = 6^2 + 8^2$ true? Since $81 \neq 100$ we conclude that this triangle does not have a right angle.	To determine whether a triangle with sides 6 cm, 8 cm and 9 cm has a right angle, use the converse of the Pythagorean theorem. Check whether $9^2 = 6^2 + 8^2$.

Example 14

Sam's vegetable garden is a triangle. His plan shows that the plot is a right-angled triangle. Sam measures the sides of his vegetable garden as 17 m, 144 m and 145 m. Show that the plot is indeed a right-angled triangle.

Is $145^2 = 144^2 + 17^2$ true? Since $21\,025 = 20\,736 + 289$, Sam can conclude that his plot is a right-angled triangle.	To determine whether his plot is a right-angled triangle with sides 17 m, 144 m and 145 m, use the converse of the Pythagorean theorem. Check whether $145^2 = 144^2 + 17^2$.

Did you know?

The existence of irrational numbers is attributed to a Pythagorean, Hippasus (5th century BC). He proved that $\sqrt{2}$, the length of the hypotenuse of a right-angled triangle with sides 1 and 1, is an irrational number.

The **irrational numbers** are numbers that are not rational. And **rational numbers** are numbers that can be written as a quotient $\frac{p}{q}$ of two integers p and q, where $q \neq 0$. The legend claims that Hippasus was exiled for his discovery of the irrational numbers.

TOK

Is it ethical that Pythagoras gave his name to a theorem that may not have been his own creation?

Exercise 1G

1 Find the length of the unknown side in each of the following right-angled triangles. Round your answers to 1 dp.

a

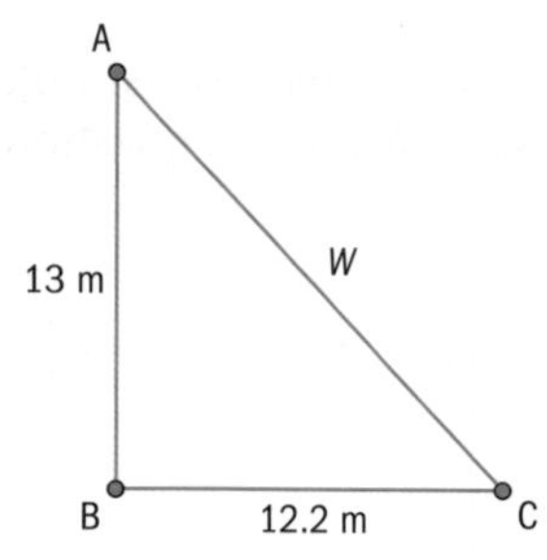

b

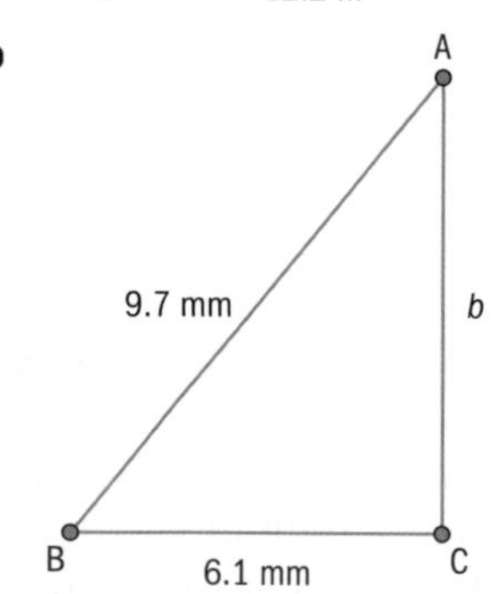

c

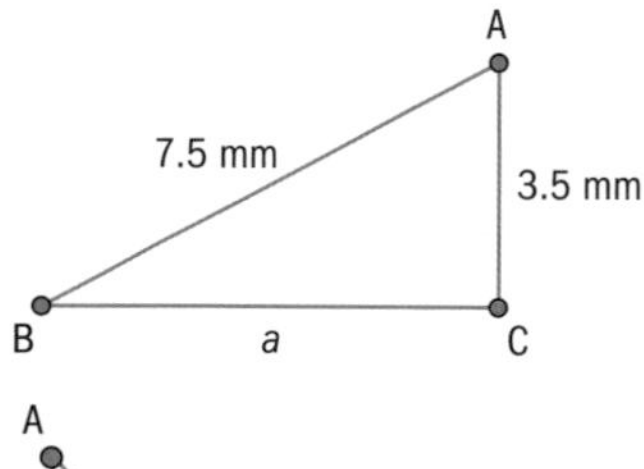

d

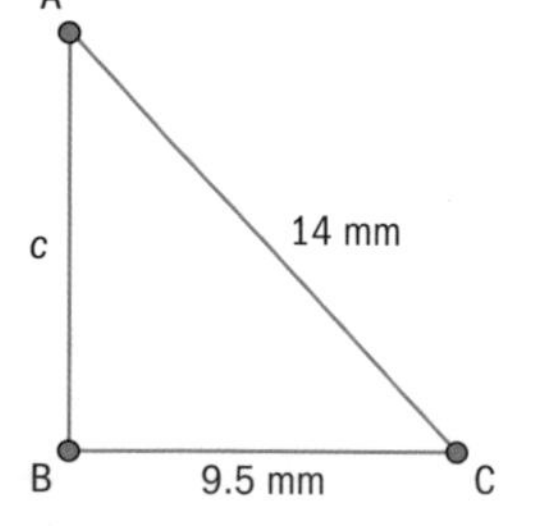

2 Determine whether or not the triangles with the following sides are right-angled triangles:

a 9 cm, 40 cm, 41 cm **b** 10 m, 24 m, 26 m

c 10, 10, $\sqrt{200}$ **d** 11.2, 7.5, 8.3

3 The spiral in the figure is made by starting with a right-angled triangle with both legs of length 1 unit.

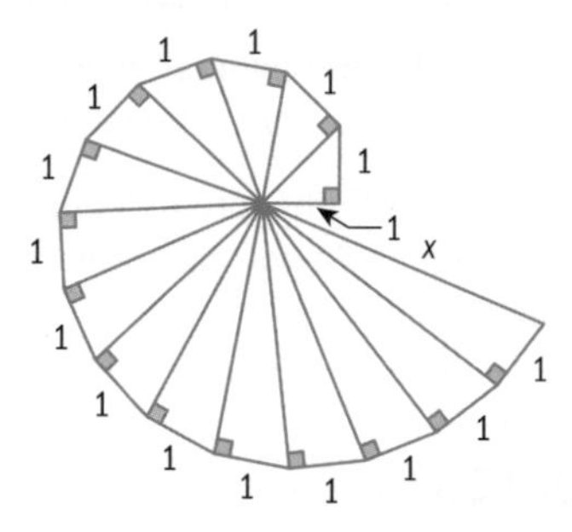

a Find the hypotenuse of the first triangle.

Then, the second right-angled triangle is built with one leg measuring 1 unit and the other leg being the hypotenuse of the first triangle.

b Find the hypotenuse of this second triangle.

A third right-angled triangle is built on the second triangle's hypotenuse, again with the other leg measuring 1 unit.

c Find the hypotenuse of the third triangle.

The process is continued in the same fashion and the hypotenuse of the final triangle is denoted by x.

d Find the length of x.

4 A right-angled triangle has a hypotenuse of 8.2 cm and another side length of 4.3 cm. Draw a diagram. Calculate the area of the triangle. Give your answer correct to 1 dp.

Investigation 10

The right-angled triangles ΔABC and ΔADE shown in the diagram below are such that $CB \perp AB$, $ED \perp AD$ and $B\hat{A}C = 30°$. The lengths of the sides are $AC = 15.0$ cm, $CB = 7.5$ cm, $AB = 13.0$ cm, $AE = 10.0$ cm, $ED = 5.0$ cm and $AD = 8.7$ cm, given to 1 dp.

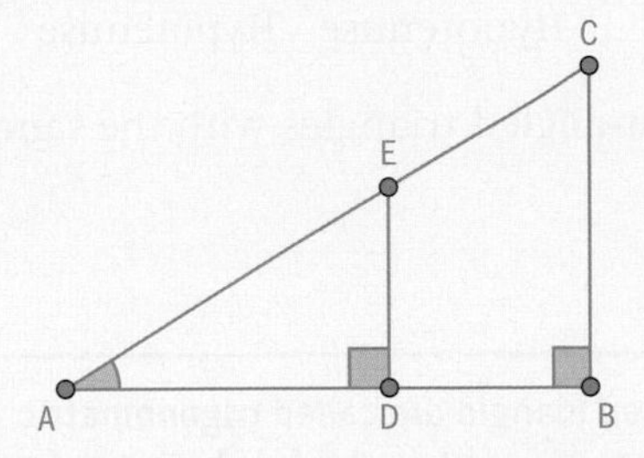

Diagram not to scale.

1 **Factual** For each side of ΔABC and ΔADE determine whether it is a hypotenuse, opposite to $B\hat{A}C$ or adjacent to $B\hat{A}C$. On your diagram, label each side of the two triangles with Opp, Adj or Hyp.

2 Place each measure of the sides of ΔABC and ΔADE in the appropriate cell in the table.

	Opp	Adj	Hyp	Opp/Hyp	Adj/Hyp	Opp/Adj
ΔABC						
ΔADE						

3 Calculate the ratios Opp/Hyp, Adj/Hyp and Opp/Adj for each triangle. Give your answers correct to 1 dp. Place each ratio in the appropriate cell in the table.

4 Compare the corresponding ratios for the two triangles. Write down your conclusions.

5 The right-angled triangles ΔFGH and ΔFIJ shown in the diagram on the right are such that $HG \perp FG$, $JI \perp FI$ and $G\hat{F}H = 60°$. The lengths of the sides are $FH = 15.0$ cm, $FG = 7.5$ cm, $FJ = 10.0$ cm and $FI = 5.0$ cm, given to 1 dp.

Diagram not to scale.

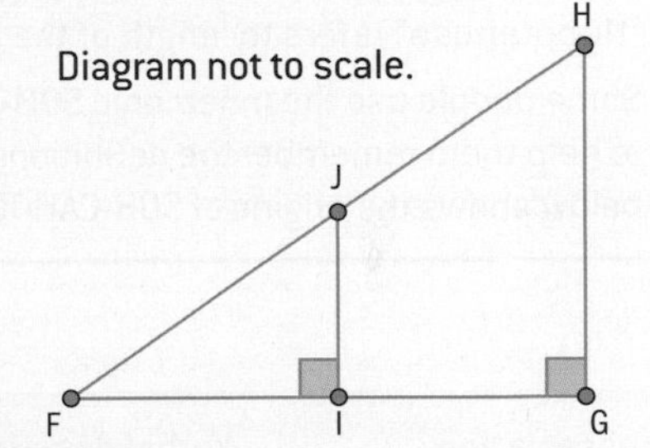

6 Find the lengths of HG and JI.

7 **Factual** For each side of ΔFGH and ΔFIJ, determine whether it is a hypotenuse, opposite to $G\hat{F}H$ or adjacent to $G\hat{F}H$. On your diagram, label each side of the two triangles with Opp, Adj or Hyp.

8 Place each measure of the sides of ΔFGH and ΔFIJ in the appropriate cell of the table below.

9 Calculate the ratios Opp/Hyp, Adj/Hyp and Opp/Adj for each triangle. Give your answers correct to 1 dp. Place each ratio in the appropriate cell in the table.

	Opp	Adj	Hyp	Opp/Hyp	Adj/Hyp	Opp/Adj
ΔFGH						
ΔFIJ						

10 Compare the corresponding ratios for ΔFGH and ΔFIJ. Compare these ratios with the ratios that you calculated for ΔABC and ΔADE. Is the conclusion different from the one you arrived at for triangles ΔFGH and ΔFIJ? If so, how?

11 **Conceptual** What can you say about the ratios of corresponding pairs of sides in similar triangles? How does changing the acute angles in the triangle affect the ratios of the sides?

Trigonometric ratios

Trigonometry (from Greek *trigōnon*, "triangle", and *metron*, "to measure") studies relationships between sides and angles of triangles. Astronomers in the 3rd century noted for the first time what you saw in your investigation, that the ratios $\frac{\text{Adjacent}}{\text{Hypotenuse}}$, $\frac{\text{Opposite}}{\text{Hypotenuse}}$ and $\frac{\text{Opposite}}{\text{Adjacent}}$ are constant for all right-angled triangles with the same acute angles.

TOK

How certain is the shared knowledge of mathematics?

The ratios of the sides of a right-angled triangle are called **trigonometric ratios**. Three common trigonometric ratios are the **sine (sin), cosine (cos)** and **tangent (tan)**. These are defined for acute angle $\hat{A}$ in the right-angled triangle below:

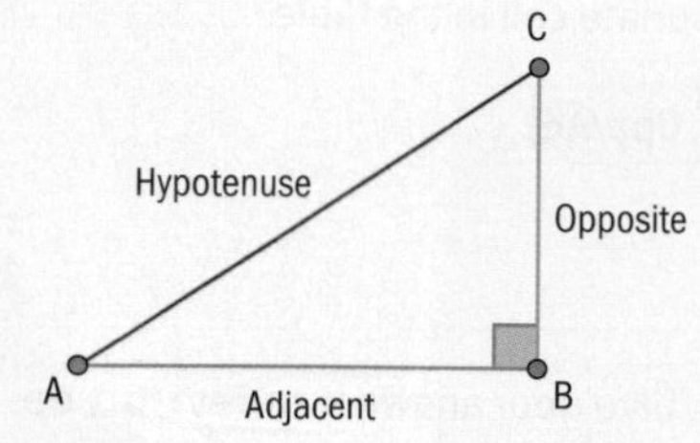

$$\sin \hat{A} = \frac{\text{Opposite}}{\text{Hypotenuse}}$$

$$\cos \hat{A} = \frac{\text{Adjacent}}{\text{Hypotenuse}}$$

$$\tan \hat{A} = \frac{\text{Opposite}}{\text{Adjacent}}$$

In the definitions above, "Opposite" refers to the length of the side opposite angle $\hat{A}$, "Adjacent" refers to the length of the side adjacent to angle $\hat{A}$, and "Hypotenuse" refers to length of the side opposite the right angle.

Some people use the mnemonic **SOH-CAH-TOA**, pronounced "soh-kuh-toh-uh", to help them remember the definitions of sine, cosine and tangent. The table below shows the origins of SOH-CAH-TOA.

Abbreviation	Verbal description	Definitions
SOH	**S**ine is **O**pposite over **H**ypotenuse	$\sin \hat{A} = \frac{\text{Opposite}}{\text{Hypotenuse}}$
CAH	**C**osine is **A**djacent over **H**ypotenuse	$\cos \hat{A} = \frac{\text{Adjacent}}{\text{Hypotenuse}}$
TOA	**T**angent is **O**pposite over **A**djacent	$\tan \hat{A} = \frac{\text{Opposite}}{\text{Adjacent}}$

Reflect What are trigonometric ratios?

Example 15

Find the size of $\hat{A}$ in triangle ΔABC, where angle $A\hat{B}C = 90°$.

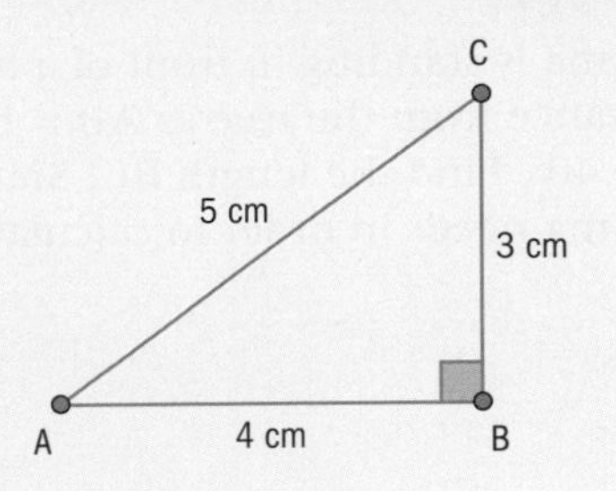

$\sin\hat{A} = \frac{\text{Opposite}}{\text{Hypotenuse}}$ $\sin\hat{A} = \frac{3}{5}$ $\hat{A} = \sin^{-1}\left(\frac{3}{5}\right)$ $\hat{A} = 36.9°$, correct to 3 sf.	Use $\sin\hat{A} = \frac{\text{Opposite}}{\text{Hypotenuse}}$. Once we know the trigonometric ratio, we can calculate angle $\hat{A}$ by using the $\sin^{-1}$ function on the GDC and finding $\sin^{-1}\left(\frac{3}{5}\right)$. Remember to give your answer to 3 sf unless otherwise specified.

Example 16

Shown here is the right-angled triangle ΔFGH, where FH = 14 cm, $F\hat{H}G = 52°$ and angle $F\hat{G}H = 90°$. Find the length of side GH.

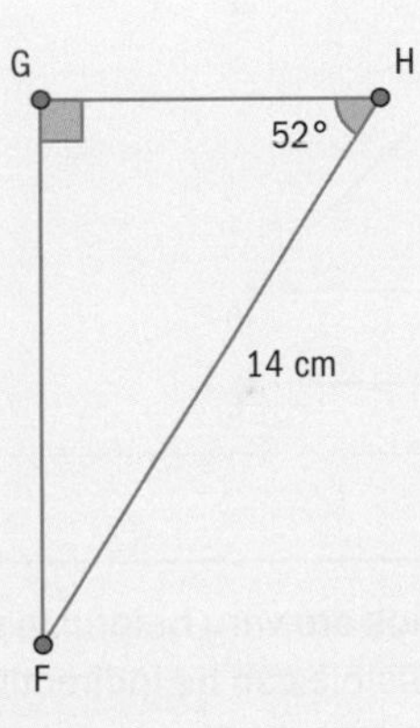

$\cos F\hat{H}G = \frac{\text{Adjacent}}{\text{Hypotenuse}}$ $\cos 52° = \frac{GH}{14}$ $GH = 14\cos 52°$ $= 8.62$ cm (3 sf)	Since we want to find the length of side GH we use the cosine. We use $\cos F\hat{H}G = \frac{\text{Adjacent}}{\text{Hypotenuse}}$. We substitute the values for $F\hat{H}G$ and the hypotenuse in the cosine ratio and do the calculations.

If at least one side and one non-right angle are known in a right-angled triangle, then all other angles and side lengths can be determined.

Example 17

Emma is standing in front of a big tree. She measures her distance from the tree as AB = 15 m. She also measures $\hat{A} = 40^\circ$. Find the length BC. State what other information Emma needs in order to calculate the height of the tree.

AB = 15 m and $\hat{A} = 40^\circ$.

$$\tan 40^\circ = \frac{CB}{15}$$

$$CB = 15 \tan 40^\circ$$

$$= 12.6\,\text{m (3 sf)}$$

Note that to obtain the height of the tree Emma will have to add to CB her height, which is represented by EA and DB.

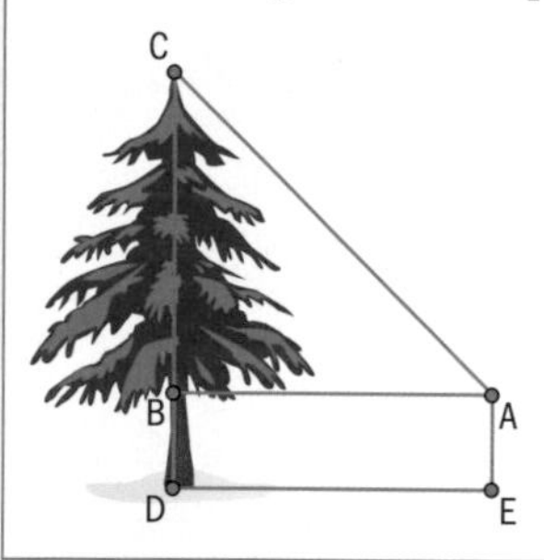

$\hat{A}$ can be measured with an instrument called a clinometer, as shown in the image below.

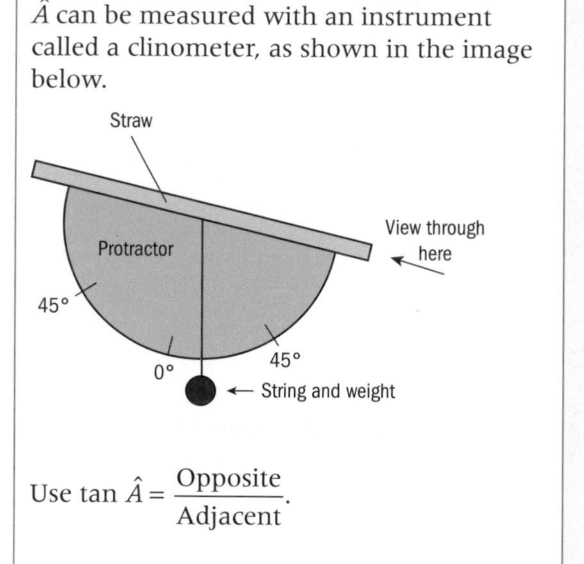

Use $\tan \hat{A} = \dfrac{\text{Opposite}}{\text{Adjacent}}$.

Trigonometric ratios are very helpful in solving practical problems. Distances which are inaccessible can be indirectly measured through use of trigonometry.

Investigation 11

The distance between two landmarks A and B (trees or other) cannot be measured directly as there is a building standing between them. How could you find the distance AB?

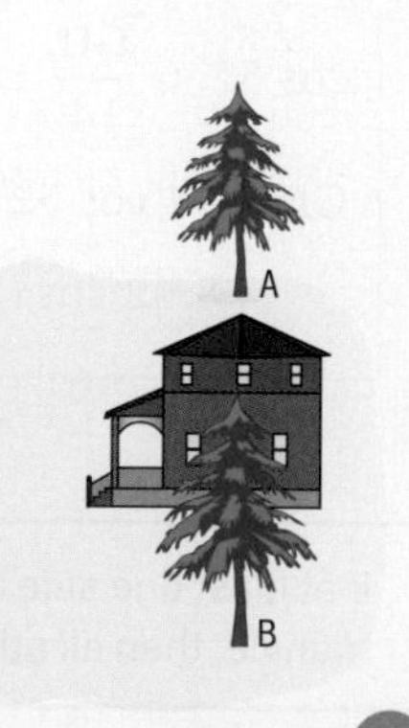

Group work:

1 Find two landmarks A and B (trees or other) such that you cannot measure the distance between A and B directly. For example, there may be a building or a tall wall between them and only the top of A is visible from B.

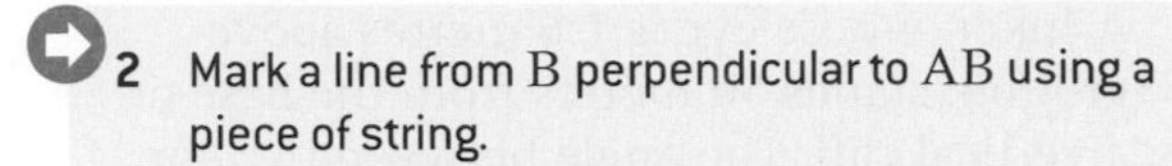

2 Mark a line from B perpendicular to AB using a piece of string.

3 Mark two points C and D on this line. Stand at point D, with your partner standing at point C. Your partner now walks parallel to AB (or perpendicular to BD) until you decide that your partner is directly between yourself and landmark A. Mark this point F.

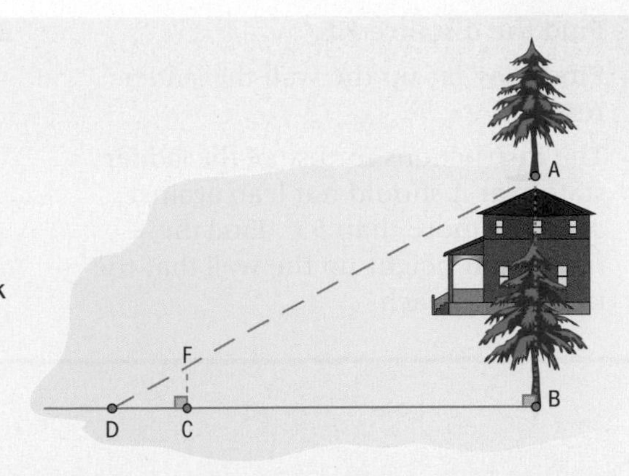

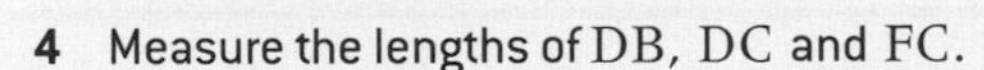

4 Measure the lengths of DB, DC and FC.

5 Calculate the length of AB from $\frac{DB}{DC} = \frac{AB}{FC}$.

Choose two different points for C and D, and repeat steps **3** to **5**, this time swapping roles with your partner.

Did you get the same result for the distance between A and B when you chose different points for C and D? Think of some sources of error in this experiment that may explain any differences you obtained.

How could you check whether the person at C indeed walked perpendicularly to DB in step **3**?

6 Now, measure $A\hat{D}B$ and use trigonometric ratios to find the distance AB. Can you find AB by using different trigonometric ratios? If so, calculate AB twice. Did you get the same result for AB?

What might be some sources of error in finding AB using trigonometric ratios?

7 Compare the values for AB found in parts **5** and **6**. Which of the two methods do you think produces a more accurate result for the distance AB? Why?

Could you think of other methods to find the distance AB indirectly?

8 What kind of real-life questions can be answered by using trigonometric ratios?

Exercise 1H

1 Determine the length of the unknown sides for each of the right-angled triangles below:

a

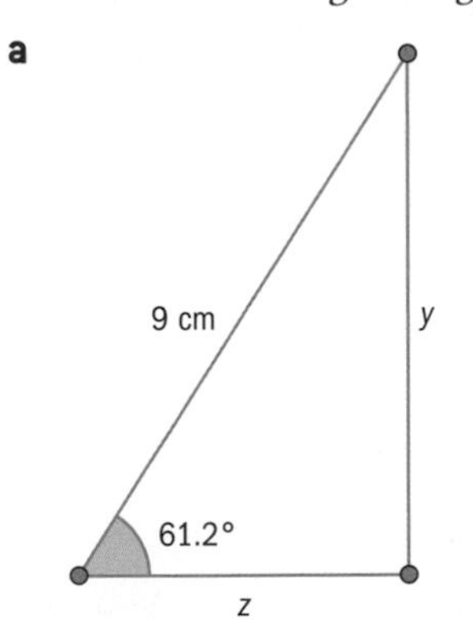

b

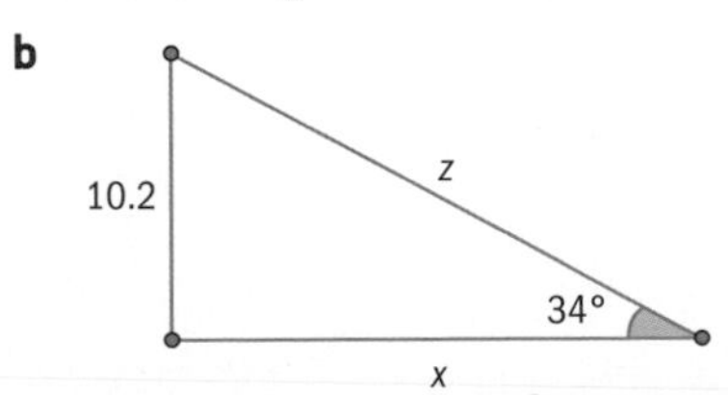

c

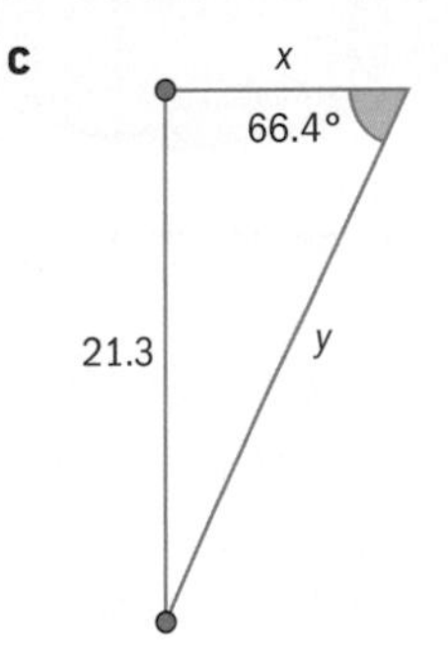

2 A ladder [KM] is 8.5 m long. It leans against a wall so that $L\hat{K}M = 30°$ and ΔKLM is a right-angled triangle.

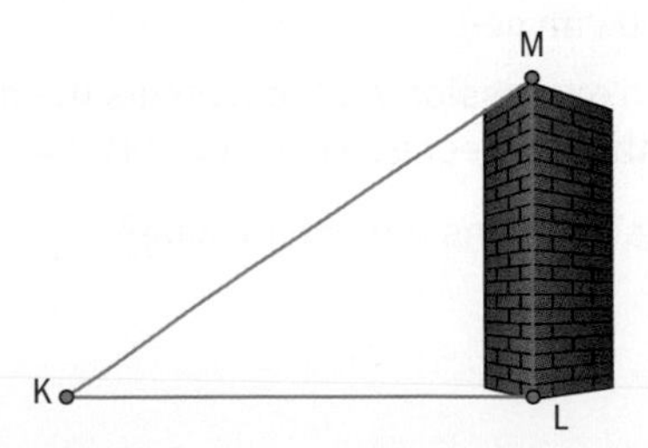

a Find the distance KL.

b Find how far up the wall the ladder reaches.

c The instructions for use of the ladder state that it should not lean against a wall at more than 55°. Find the maximum height up the wall that the ladder can reach.

3 A hiker, whose eye is 1.6 metres above ground, stands 50 metres from the base of a vertical cliff. The angle between the line connecting her eye and the top of the cliff and a horizontal line is 58°.

a Draw a diagram representing the situation.

b Find the height of the cliff.

Developing inquiry skills

Imagine that you are standing at point B, which is the length of BD above the ground.

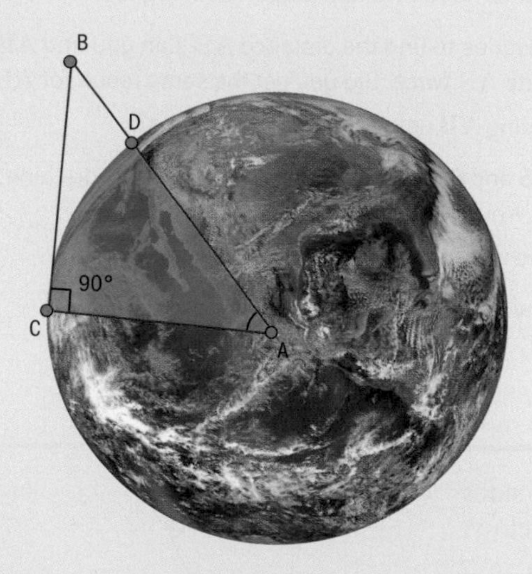

Which line segment in ΔABC do you think represents your line of sight to the horizon?

Why do you think ΔABC is a right-angled triangle?

Assume that the radius of the Earth is 6370 km.

How can you use what you have learned in this section to calculate the distance to the farthest object that you can see?

a If $BD = 10$ m, what would be the distance to the farthest object that you can see?

b If $BD = 382$ m, what would be the distance to the farthest object that you can see?

c Write an expression that represents the distance to the farthest object you can see, if $BD = h$ km.

What other assumptions are you making?

International-mindedness

The word "sine" started out as a totally different word and passed through Indian, Arabic and Latin before becoming the word that we use today.

1.6 Angles of elevation and depression

Since ancient times, sailors and surveyors have measured angles to determine measurements that are inaccessible directly, such as the heights of trees, clouds or mountains, and distances to coasts or buildings.

In the problem about measuring the height of a tree from the previous section, angle BÂC is the angle between the line of sight of the observer to the top of the tree and a horizontal line AB from their eye level. This angle is called the **angle of elevation**.

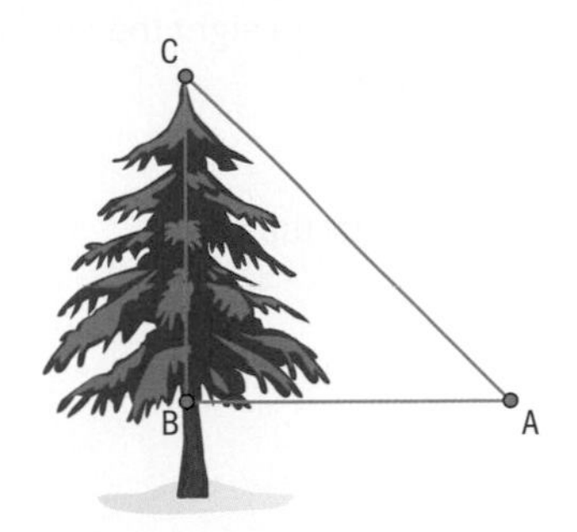

Similarly, when the object is below the horizontal line at eye level, an **angle of depression** is formed, as shown on the diagram below.

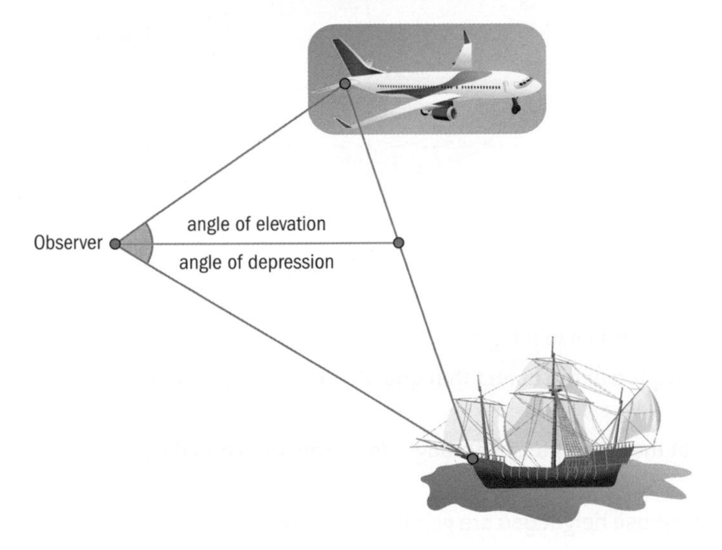

> The **angle of elevation** is the angle between the horizontal and the observer's line of sight to the object when the observer is looking upwards.
>
> The **angle of depression** is the angle between the horizontal and the observer's line of sight to the object when the observer is looking downwards.

Reflect Why are angles of elevation and depression useful?

Investigation 12

In this investigation you will take measurements of accessible distances and angles, and then use them to calculate inaccessible heights. You will measure angles with a device called a clinometer and then perform the calculations. But first you have to make the clinometer yourself.

1 Making a clinometer

A clinometer is an instrument for measuring angles above or below a horizontal line (angles of elevation or depression). A metal tube allows the user to sight the top (or the base) of an object, and a protractor-like device on the side measures the angle.

To construct a clinometer you will need:

- drinking straw (instead of a metal tube)
- protractor
- string
- washer
- sticky tape
- tape measure.

2 Clinometer assembly

Attach the drinking straw to the top straight edge of the protractor with tape.

Attach the washer to one end of the string, and attach the string to the 0 marking on the top edge of the protractor and let it hang at least 5 cm below the curved edge of the protractor so that it can swing freely.

You are done! Now try it out.

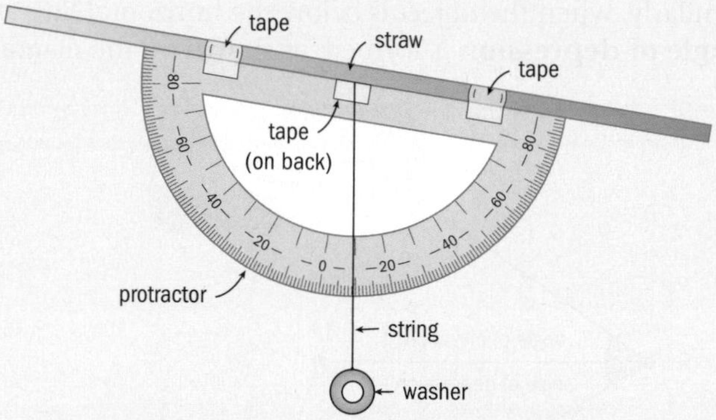

3 Using your clinometer

Start by holding the clinometer so that the straw is parallel to the ground.

Slowly tilt the straw up until you can see the very top of the object that you are measuring through the straw.

Have your partner record the angle measure that the string passes through. Your measure should be between 0° and 90°.

a Choose a tree or other tall object nearby whose height you are going to measure.

b Move away from the object until you can see the top of the object through the straw of your clinometer.

c One person should look through the clinometer at the top of the object you are measuring. The other person reads the angle that the string defines. Record this measurement. Remember that the number will be between 0° and 90°.

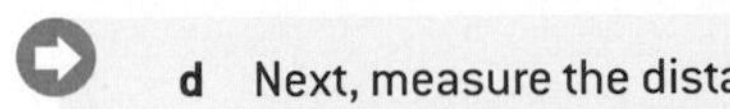

d Next, measure the distance to the object from the point on the ground at which the angle was measured. Record this measurement.

Measurement	Measure of angle of elevation	Measure of distance to the object	Calculated height of the object	Eye-level height	Total height of the object
1					
2					
3					

e Measure the height from the ground to the eye level of the person measuring the angle.

f Draw a clear and labelled diagram and write on it the information you have collected.

g Use the measurements you obtained to do the relevant calculations to find the height.

h After you have found out how high the object is from eye level, add the person's eye-level height to that number. This should give you the total height of the object.

i You can do a few sets of measurements and compare your results. What factors do you think could have affected the accuracy of your measurements and final result?

4 Suppose you are standing on the third floor of a building and want to find the distance to a building across the street. What information would you need and what measurements would you need to take in order to calculate this distance? Suppose you do have the necessary information and have obtained the measurements. Draw a diagram and explain how you would find the distance. How would you calculate the angle of depression if you know the side measurements?

5 **Factual** How are angles of elevation/depression calculated?

6 **Conceptual** Why are angles of elevation/depression useful?

Example 18

A ramp, BC, is to be constructed to give wheelchair access to the door of a house 1.35 m above the ground. Safety regulations require that the angle of the ramp be less than 8°.

a Draw a diagram and label the vertices and the sides.

b Find the length of the ramp BC, where $A\hat{B}C = 8°$.

c Find the distance AB, giving your answer to 1 dp.

a

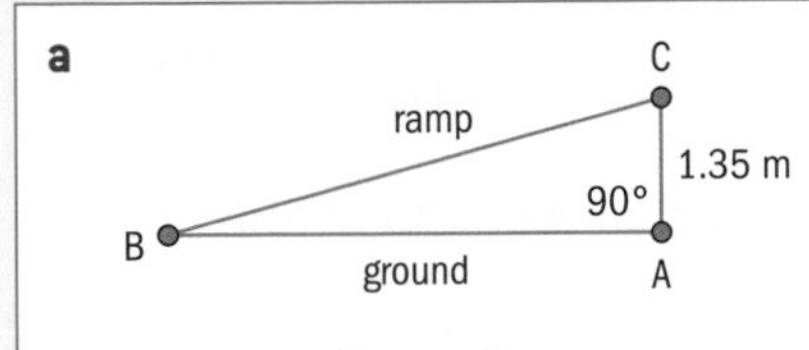

b $\sin \hat{B} = \dfrac{\text{Opposite}}{\text{Hypotenuse}}$

Use $\sin \hat{B} = \dfrac{\text{Opposite}}{\text{Hypotenuse}}$ and solve for BC.

$\sin 8° = \dfrac{1.35}{\text{BC}}$

Continued on next page

$BC = \dfrac{1.35}{\sin 8°}$ $BC = 9.70$ m (3 sf)	
c $\tan \hat{B} = \dfrac{\text{Opposite}}{\text{Adjacent}}$ $\tan 8° = \dfrac{1.35}{AB}$ $AB = \dfrac{1.35}{\tan 8°}$ so AB = 9.61 m (3 sf)	Use $\tan \hat{B} = \dfrac{\text{Opposite}}{\text{Adjacent}}$ and solve for AB.

Example 19

From his studio window at point A, which is 6.50 m above the ground, Tomchik views the top and the base of a building on the opposite side of the street. The angles of elevation and depression of the top and the base of the building are 55° and 18° respectively. Find the height of the building he is viewing, giving your answer correct to 2 dp.

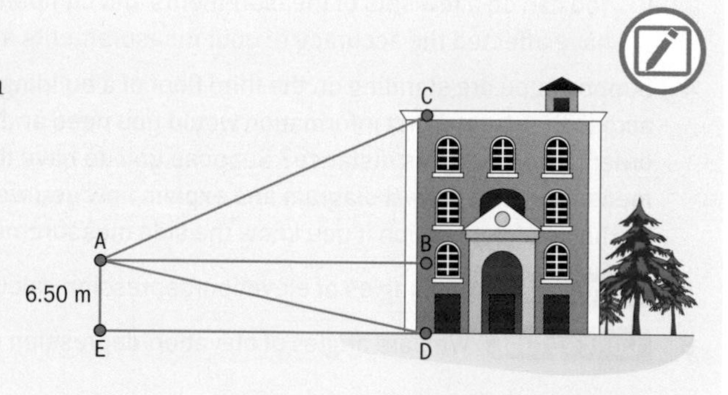

From ΔABD, $\tan B\hat{A}D = \dfrac{\text{Opposite}}{\text{Adjacent}}$ $\tan 18° = \dfrac{6.50}{AB}$ AB = 20.00 m (2 dp)	For ΔABD, we use $\tan B\hat{A}D = \dfrac{\text{Opposite}}{\text{Adjacent}}$, as we know that BD = AE = 6.50 m and $B\hat{A}D = 18°$.
From ΔABC, $\tan B\hat{A}C = \dfrac{\text{Opposite}}{\text{Adjacent}}$, $\tan 55° = \dfrac{CB}{20}$ CB = 20 tan 55° CB = 28.57 m (2 dp)	For ΔABC, we use also $\tan B\hat{A}D = \dfrac{\text{Opposite}}{\text{Adjacent}}$, as we now know that AB = 20 m and $B\hat{A}C = 55°$.
Then the height of the building is 28.57 + 6.50 = 35.07 m	To find the height of the building you need to add the lengths of sides [CB] and [DB].

Exercise 1I

1 The angle of depression from the top of a cliff to a boat at sea is 17°. The boat is 450 m from the shore.

  a Draw a diagram representing the situation.

  b Find the height of the cliff, giving your answer rounded to the nearest metre.

2 A family wants to buy an awning for a French window that will be long enough to keep out the sun when it is at its highest point in the sky. The height of the French window is 2.80 m. The angle of elevation of the sun from this point to the bottom of the window is 70°. The awning will be 90° to the window. Find how long the awning should be. Write down your answer correct to 2 dp.

3 Scientists measure the depths of lunar craters by measuring the length of the shadow cast by the edge of the crater using photos. In a photograph, the length of the shadow cast by the edge of the Moltke crater is about 606 metres, given to the nearest metre. The sun's angle of elevation is 65°. Find the depth of the crater, giving your answer rounded to the nearest metre.

4 The height of a building is 72 m. Find the angle of elevation from a point on ground level that is 55 m away from the base of the building.

5 From a boat 160 m out at sea, the angle of elevation to the coast is 18°. The angle of elevation to the top of a lighthouse on the coast is 22°.

  a Draw a diagram representing the situation.

  b Find the height of the lighthouse.

Investigation 13

1 Complete the following table.

Angle	Section of a circle with this angle	Length of the corresponding arc
360°		$2\pi r$
270°		
180°		
90°		
45°		
1°		
$\alpha°$		

Continued on next page

2 A window, which is a semicircle of radius 12 cm, is divided into six congruent sections, as shown in the diagram.
Find the length of the arc of the entire semicircular window.

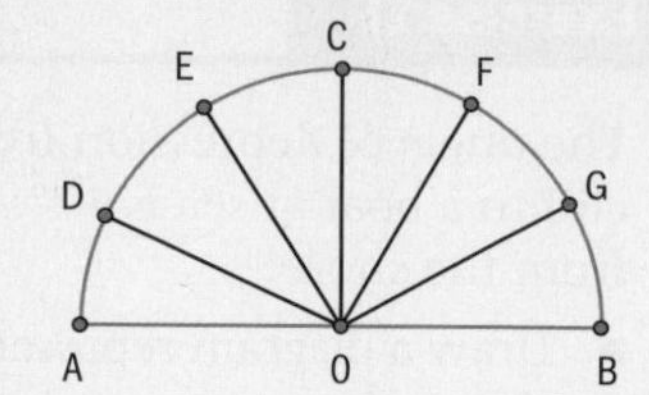

3 Find the length of the arc of one of the six sections of the window.

4 If the window is divided into sections with a central angle of 1°, find the length of the arc of such a window section.

5 Find the length of the arc of a window section if its central angle were $\alpha°$.

6 Generate a formula for finding the length of the arc of a circle with central angle $\alpha°$ and radius r, based on your conclusions above.

7 **Conceptual** How are the arc length and circumference of a circle related?

Length of arc formula

The length of the arc of a sector of a circle with radius r and central angle $\alpha°$ is $\alpha \times \frac{2\pi r}{360}$ or $\alpha \times \frac{\pi r}{180}$.

TOK

What does it mean to say that mathematics is an axiomatic system?

Example 20

Find the length of the arc of a circle with radius 2 cm and central angle 120°.

The length of the arc can be found in two ways.	
a $\frac{1}{3} \times 2\pi(2)$ $= 4.19$ cm (3 sf)	One way would be to use the fact that 120° is $\frac{1}{3}$ of 360°, which means that the length of the arc will be $\frac{1}{3}$ of the circumference of a circle with radius 2 cm, thus $\frac{1}{3} \times 2\pi(2)$ or 4.19 cm (3 sf).
b $120 \times \frac{\pi r}{180} = 120 \times \frac{2\pi}{180}$ $= 4.19$ cm (3 sf)	Another way would be to directly use the formula: $120 \times \frac{\pi r}{180}$ or $120 \times \frac{2\pi}{180}$, which again results in 4.19 cm (3 sf).

Developing your toolkit

Now do the Modelling and investigation activity on page 46.

Exercise 1J

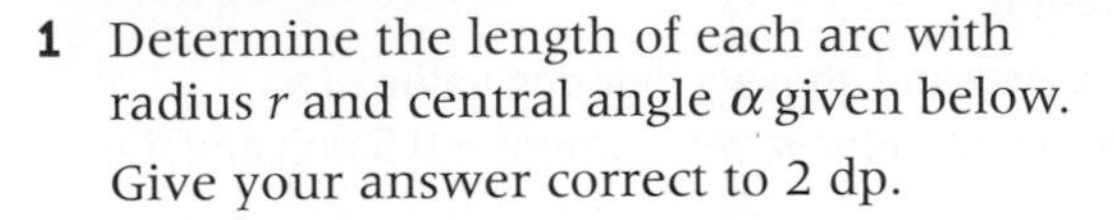

1 Determine the length of each arc with radius r and central angle α given below. Give your answer correct to 2 dp.

a $r = 5\text{ cm}, \alpha = 70°$

b $r = 4\text{ cm}, \alpha = 45°$

c $r = 10.5\text{ cm}, \alpha = 130°$

2 A clock is circular in shape with diameter 25 cm. Find the length of the arc between the markings 12 and 5, rounded to the nearest tenth of a centimetre.

3 The London Eye is a giant Ferris wheel in London. It is the tallest Ferris wheel in Europe, with a diameter of 120 m. The passenger capsules are attached to the circumference of the wheel, and the wheel rotates at 26 cm per second.

Find:

a the length that a passenger capsule would travel if the wheel makes a rotation of 200°

b the time, in minutes, that it would take for a passenger capsule to make a rotation of 200°

c the time, to the nearest whole minute, that it would take for a passenger capsule to make a full revolution.

4 A door with width 4.20 m has an arc as shown in the diagram. Find:

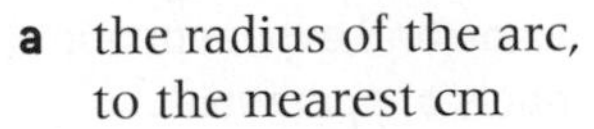

a the radius of the arc, to the nearest cm

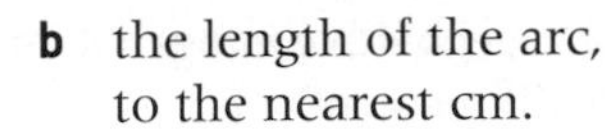

b the length of the arc, to the nearest cm.

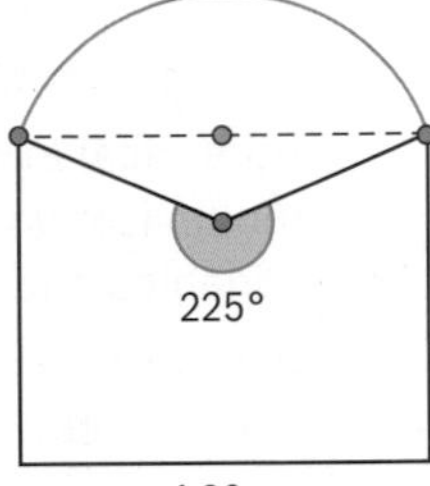

Chapter summary

Measurement

- Measurements help us to compare objects and understand how they relate to each other.
- Measuring requires approximating. If a smaller measuring unit is chosen then a more accurate measurement can be obtained.
- When you measure, you first select a property of the object that you will measure. Then you choose an appropriate unit of measurement for that property. And finally, you determine the number of units.

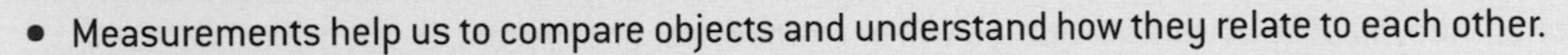

Estimation

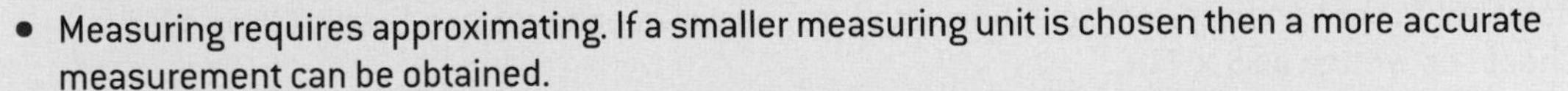

- Estimation (or estimating) is finding an approximation as close as possible to the value of a measurement by sensible guessing. Often the estimate is used to check whether a calculation makes sense or to avoid complicated calculations.
- Estimation is often done by comparing the attribute that we measure to another one or by sampling.

Accuracy

- A value is **accurate** if it is close to the **exact value** of the quantity being measured. However, in most cases it is not possible to obtain the exact value of a measurement. For example, when measuring your weight, you can get a more accurate measurement if you have a balance scale that measures to a greater number of decimal places.

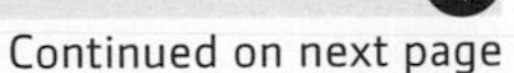

Continued on next page

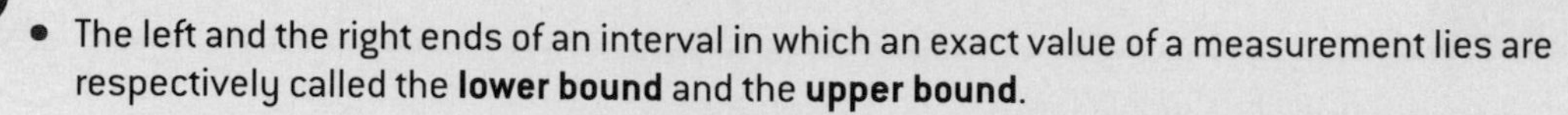

- The left and the right ends of an interval in which an exact value of a measurement lies are respectively called the **lower bound** and the **upper bound**.
- The lower bound and the upper bound are half a unit below and above a rounded value of a measurement. Thus the upper bound is calculated as the rounded measurement + 0.5 unit and the lower bound is the rounded measurement − 0.5 unit.
- Since measurements are approximate there is always error in the measurement results. A measurement error is the difference between the exact value and the approximate value, ie

$$\text{Measurement error} = V_A - V_E$$

- **Percentage error** $= \left|\frac{V_A - V_E}{V_E}\right| \times 100\%$, where V_A is the approximate value and V_E is the exact value.
- Rounding of intermediate results during multistep calculations reduces the accuracy of the final answers. Thus, make sure to round your final answer only.

Speaking scientifically

- **Numerical expressions** consist only of numbers and symbols of operations (addition, subtraction, multiplication, division and exponentiation), whereas **algebraic expressions** contain numbers, variables and symbols of operations.
- If the exponent, n, is a **positive integer**, it determines **how many times** the base, a, is multiplied by itself. For example, 8^2 means 8×8.
- If the exponent, n, is a **negative integer** it determines how many times **to divide** 1 by the base, a. For example, 8^{-2} means $\frac{1}{8^2}$ or $\frac{1}{8 \times 8}$.

Standard form

- In standard form (also known as scientific notation), numbers are written in the form $a \times 10^n$, where a is called the **coefficient** or **mantissa**, with $1 \le a < 10$, and n is an integer.
- You may have noticed that your graphic display calculator gives any results with lots of digits in scientific notation. However, instead of writing the results in the form $a \times 10^n$, it gives them in the form $a\text{E}n$, where n is an integer. For example, 3×10^5 will be written as 3E5, and 3.1×10^{-3} as 3.1E−3.
- To convert from decimal to scientific notation on your GDC, change the Mode from Normal to Scientific. Thus, all of your number entries will be immediately converted to scientific calculator notation. To convert to correct scientific notation, you will need to replace E with 10, eg 5E2 should be written as 5×10^2.

Trigonometry of right-angled triangles

- **Pythagorean theorem**: For every right-angled triangle $c^2 = a^2 + b^2$, where c is the hypotenuse, and a and b are the other two sides. The hypotenuse is the longest side and is opposite the right angle.

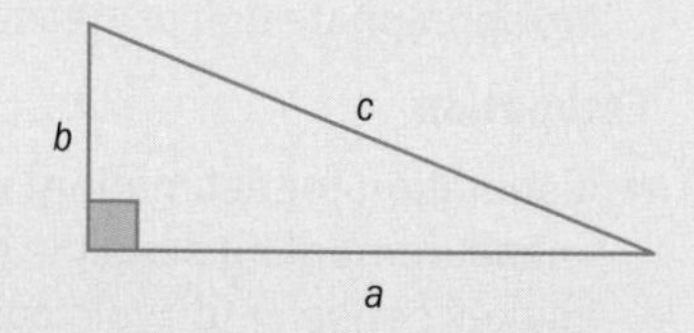

- If you know two of the sides of a right-angled triangle you can always determine the third side.
- **The converse of the Pythagorean theorem**: If for a triangle $c^2 = a^2 + b^2$, where a, b and c are its sides, then it must be a right-angled triangle.
- The converse of the Pythagorean theorem is useful, as if you know the three sides of a triangle you can determine whether the triangle has a right angle or not.

Trigonometric ratios

- The ratios of the sides of right-angled triangles are called trigonometric ratios. Three common trigonometric ratios are the sine (sin), cosine (cos) and tangent (tan). These are defined for acute angle $\hat{A}$ in the right-angled triangle below:

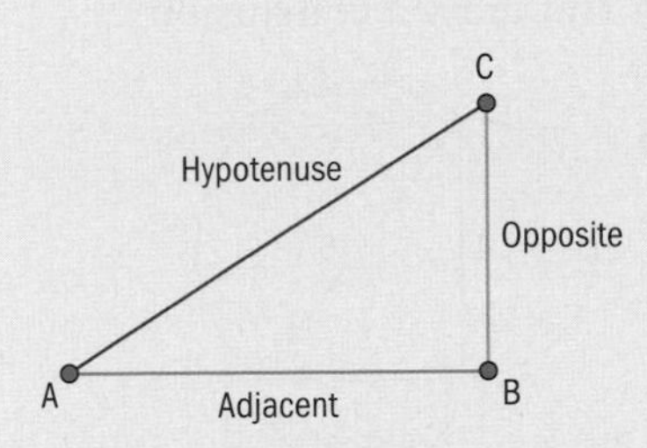

$$\sin \hat{A} = \frac{\text{Opposite}}{\text{Hypotenuse}}$$

$$\cos \hat{A} = \frac{\text{Adjacent}}{\text{Hypotenuse}}$$

$$\tan \hat{A} = \frac{\text{Opposite}}{\text{Adjacent}}$$

- If at least one side and one non-right angle are known in a right-angled triangle, then all other angles and side lengths can be determined.
- Trigonometric ratios are very helpful in solving practical problems. Distances that are inaccessible can be indirectly measured through use of trigonometry.

Angles of elevation and depression

- The **angle of elevation** is the angle between the horizontal and the observer's line of sight to the object when the observer is looking upwards.
- The **angle of depression** is the angle between the horizontal and the observer's line of sight to the object when the observer is looking downwards.
- The **length of the arc** of a sector of a circle with radius r and with central angle $\alpha°$ is $\alpha \times \frac{2\pi r}{360}$.

Developing inquiry skills

Does the measure of $\text{B}\hat{\text{A}}\text{C}$ depend on the length of BD or how high the observer stands above the ground?

How can you use $\text{A}\hat{\text{B}}\text{C}$ to find the measure of $\text{B}\hat{\text{A}}\text{C}$?

Which is the arc of visibility for an observer at point B?

Does the length of the arc of visibility depend on $\text{B}\hat{\text{A}}\text{C}$?

Assume that the radius of the Earth is 6370 km.

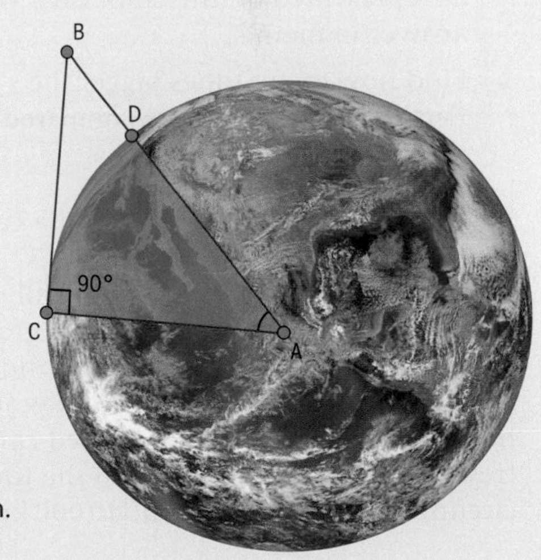

a Calculate the measure of $\text{B}\hat{\text{A}}\text{C}$ if $\text{BD} = 10$ m.

b Calculate the measure of $\text{B}\hat{\text{A}}\text{C}$ if $\text{BD} = 382$ m.

c Express the measure of $\text{B}\hat{\text{A}}\text{C}$ if $\text{BD} = h$ km.

d Calculate the length of the arc of visibility for $\text{BD} = 10$ m.

e Calculate the length of the arc of visibility for $\text{BD} = 382$ m.

f Calculate the length of the arc of visibility for $\text{BD} = h$ km.

What other assumptions are you making?

Chapter review

1 (Group work) Make a list of 10 human activities that are impossible to perform without measuring. Share the list with your peers. Discuss in your small group why measurement matters and how measuring and measurements are used in these activities.

2 Find the value of the following algebraic expressions when $a = 1.25$ and $b = 3.365$. Give your answers correct to 3 sf.

a $2a + 2b$

b $\dfrac{\sqrt{a+b}}{ab}$

c $\sqrt{a^2 + b^2 - 2ab}$

3 Researchers at Ghent University in Belgium have etched a tiny world map (on a scale of 1 to one trillion) onto an optical silicon chip. They reduced the Earth's 40 000 km circumference at the equator down to 40 micrometres, or about half the width of a human hair, to fit it on the chip.

a Using scientific notation, write down the length of the Earth's circumference as represented on the chip. Give your answer in metres.

b Find how many times bigger the Earth's circumference is compared to its length on the chip.

4 In 2008, the European Union set a carbon emissions target of 130 grams of CO_2 per km for passenger cars. Assuming all cars met this regulation to the nearest gram, calculate the upper and lower bounds of the amount of CO_2 in grams that would have been released if people used cars to travel a distance equivalent to the total circumference of the Earth, 40 000 km.

5 We used the relationship between the squares built on the sides of a right-angled triangle to discover the Pythagorean theorem. Does the same relationship apply if we draw semicircles on the sides of the right-angled triangle? Find the area of each semicircle shown in the following diagram and draw a conclusion.

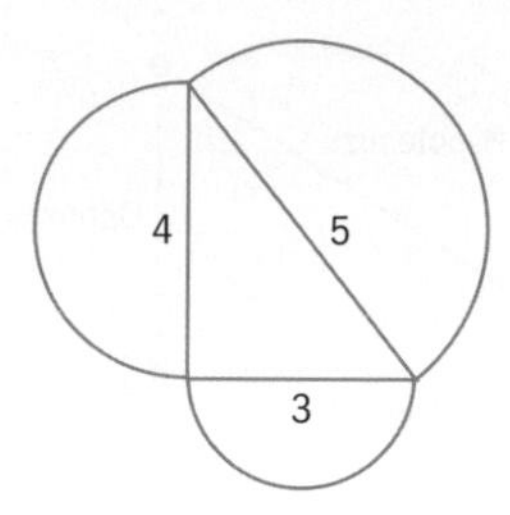

6 A rectangular pool measures 7.5 m by 15 m.

a Find the length of the diagonal, correct to the nearest cm.

Vasil measures the diagonal to be 16 m.

b Find the percentage error in his measurement. Give your answer to 3 sf.

7 Raja is laying tiles on a path that forms the diagonal of a square garden. If Raja is told that the length of the diagonal path is $100\sqrt{2}$ ft, determine the perimeter of the garden, in feet.

8 The roof of a house seen from the front is an isosceles triangle ΔABC with base 14.5 m and sides of length 8.5 m. Find how high above line [BC] the vertex A is.

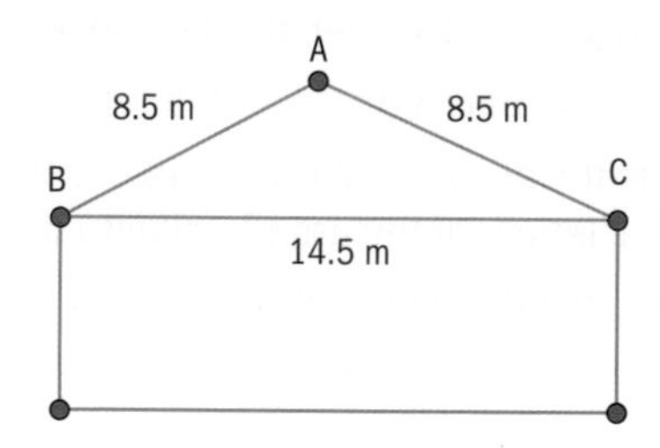

9 A ramp is placed from a ditch to a main road, which is 0.60 m above the ditch. If the length of the ramp is 3.65 m, determine the distance between the bottom of the ramp and the road.

10 A satellite is orbiting the Earth at a height of 800 km. Assume the radius of the Earth to be 6370 km.

a Draw a diagram representing the situation.

b Find the distance from the satellite to the Earth's horizon, giving your answer to the nearest km.

11 When fire occurs in high-rise buildings, the fire-fighters cannot use the regular stairs or lifts. They can reach higher floors using ladders. If the angle of a ladder to the ground can be at most 50°, determine:

a the length of the ladder that can be extended from the ground and reach a height of 20 m

b the height that a 15 m ladder can reach if extended from a fire truck. The height of the fire truck is 4.15 m.

Calculate your answers to the nearest cm.

12 From the top of a 150 m high-rise building, two cars on the same road below the building are seen at an angle of depression of 45° and 60° on each side of the tower. Find the distance between the two cars, correct to the nearest metre.

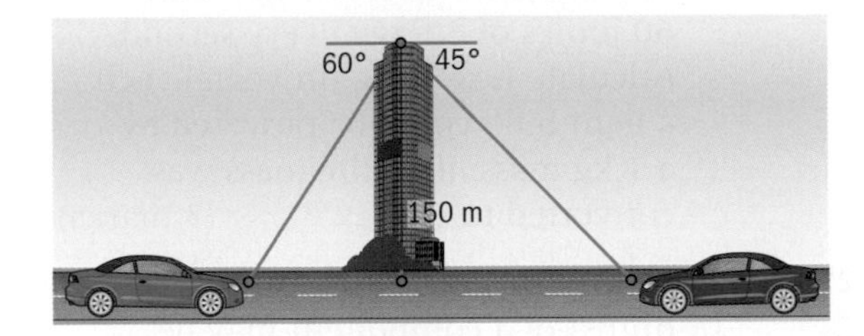

13 A surveyor wanted to measure the height of a mountain peak. She took a straight road leading to the mountain and measured the angle of elevation from point A on the ground to the top of the mountain D as 33°. She drove 0.5 km further towards the mountain and measured the angle of elevation again as 40°.

a Make a labelled diagram that represents the situation.

b Find the height of the mountain to the nearest metre.

14 A customer brings a special window frame, shown in the following diagram, to a glass shop. The wood frame has inside dimensions FG = 80 cm and HF = 136 cm. Calculate the length of the arc $\overset{\frown}{HI}$, if the centre of the circle of the arc is the midpoint of FG.

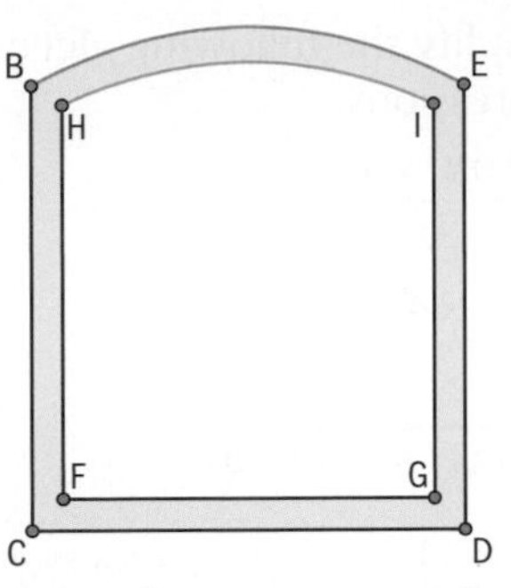

Exam-style questions

15 **P1:** The diagram below shows a cuboid of dimensions 5 cm, 6 cm and 12 cm.

X is the midpoint of side HG, and Y is the midpoint of side DC.

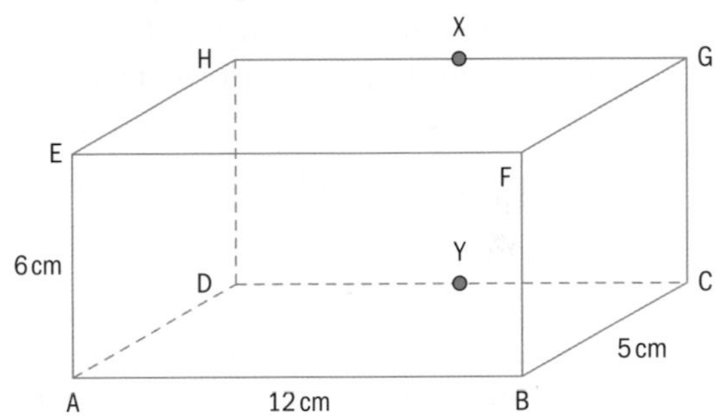

a Find the length AY. (2 marks)

b Find the size of angle $X\hat{A}Y$. (2 marks)

16 **P1:** A cylinder has radius 1.8 cm and height 14.5 cm.

a Calculate the volume of the cylinder, giving your answer to 3 sf. (2 marks)

b Calculate the total surface area of the cylinder, giving your answer to 3 sf. (2 marks)

17 **P1:** A pendant is made in the shape of a circular sector OAB, with angle $A\hat{O}B = 40°$ and radius 7 cm.

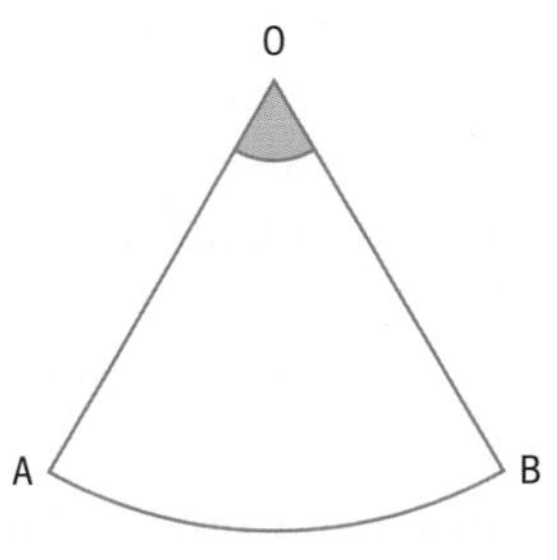

Find the perimeter of the pendant. (4 marks)

18 **P2:** Simplify the following algebraic expressions.

a $\frac{10x^3 \times 3x^4}{2x^{-6}}$ (3 marks)

b $\frac{x^2 \times 4x^{-3}}{x^{-1}}$ (3 marks)

c $\sqrt{3x^3 \times 12x^0 \times 4x^5}$ (3 marks)

d $\frac{\left(x^{-2}\right)^5}{\left(x^3\right)^{-4}}$ (3 marks)

19 **P1:** Elea is 1.6 m tall. She stands on the same horizontal level as the base of a tree. The height of the tree is 23.5 m, and it is 100 m away from Elea.

Find the angle of elevation of Elea's line of sight to the top of the tree. (4 marks)

20 **P1:** A mathematics student decides to use an approximation to π of $\frac{355}{113}$.
Calculate his percentage error in using this value, giving your answer in standard form. (3 marks)

21 **P1:** Sharon stands on the top of a cliff 90 m high.

The angle of elevation from Sharon to a flying kittiwake is 15°.
The angle of depression from Sharon to a yacht on the sea is 19°.
Given that the kittiwake is flying directly above the yacht, find the distance between the yacht and the kittiwake. (6 marks)

22 **P1:** A straight ladder of length 7.1 m rests against a vertical wall.

A person climbing the ladder should be "safe" as long as the foot of the ladder makes an angle of between 70° and 80° with the horizontal ground.

Determine the minimum and maximum heights that the ladder can safely lie against the wall. (4 marks)

23 **P1:** The speed of light is approximately 3×10^8 m s^{-1}.

The distance from the Earth to the Sun is 1.496×10^8 km.

The distance from the Earth to Proxima Centauri, the nearest star, is 4.014×10^{13} km.

a Find the time it takes (in minutes) for light from the Sun to reach the Earth. (3 marks)

b Find the time (in years) it takes for light from Proxima Centauri to reach the Earth. Assume that one year is equivalent to 365 days. (3 marks)

c Given that light from the Andromeda galaxy takes approximately 2.5 million years to reach us, find how far (in km) Andromeda is from the Earth. (3 marks)

24 **P1:** Einstein's equation relating mass and energy is given by the formula $E = mc^2$, where m is the mass of an object and $c = 3 \times 10^8$ m s^{-1}, the speed of light.

a Find how much energy is present in a 1 kg mass. (1 kg m s^{-1} = 1 joule) (2 marks)

b Given that a standard light bulb uses 60 joules of energy every second, calculate how long (in years) such a light bulb could be powered by a 1 kg mass, if all the mass was converted to energy. (3 marks)

25 **P1:** In an electric circuit, the resistance (R ohms) of a component may be calculated using the formula $R = \frac{V}{I}$, where V is the potential difference across the component and I is the current through the component.

The potential difference is measured as 6 V to the nearest volt, and the current is measured as 0.2 A to the nearest tenth of an ampere.

a Calculate lower and upper bounds for the resistance. (3 marks)

b The actual value of the resistor is 30 Ω. Calculate the maximum possible percentage error that could be obtained. (2 marks)

26 **P1:** From the top of a cliff 120 m high, Josie sees two small boats in the distance.

One boat lies at an angle of depression of 39° and the other at an angle of depression of 31°. Given that Josie and both ships lie in the same vertical plane, how far apart are the ships? (6 marks)

27 **P2:** The formula for the volume of a sphere is given as $V = \frac{4\pi r^3}{3}$, where r is the radius of the sphere.

a If the radius of a particular sphere is 3.5 cm, measured to the nearest tenth of a centimetre, find lower and upper bounds for its volume. (4 marks)

b Another sphere is designed to have a volume of 500 cm³, to the nearest 10 cm³.

Find the boundaries between which its radius must lie. (6 marks)

28 **P1:** The shape of an *E. coli* bacterium may be approximated by a cylinder of radius 1.3×10^{-6} m and length 4.5×10^{-6} m.

a Find the volume of the bacterium. (2 marks)

b For practical purposes, the volume is assumed to be 2×10^{-17} m.

Find the percentage error when assuming this value. (2 marks)

29 **P1:** The following table gives the masses (in kg) of the various particles which make up an atom.

Particle	Mass (kg)
Neutron	1.675×10^{-27}
Proton	1.673×10^{-27}
Electron	9.109×10^{-31}

a Find the average (mean) mass of all three particles, giving your answer in standard form to 3 sf. (2 marks)

b A helium atom consists of two protons, two neutrons, and two electrons.

Find the mass of a helium atom, giving your answer in standard form to 3 sf. (2 marks)

c Determine the ratio of electron mass to neutron mass, giving your answer in the form $1:x$ where x is accurate to 3 sf. (2 marks)

d Calculate the percentage error in taking the mass of an electron to be 1×10^{-30} kg. (2 marks)

30 **P1:** In the diagram shown above, *ABCD* is a plane figure, where both *BC* and *AD* are arc lengths of a circle centred at *O*. Length *OA* = 5 cm and *AB* = 8 cm.

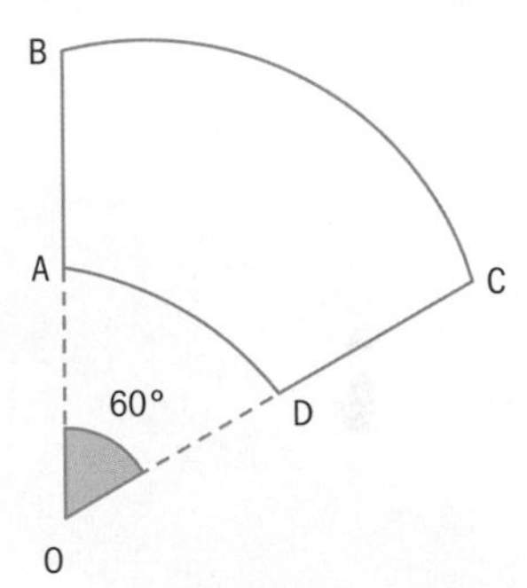

Find the perimeter of the figure ABCD. (6 marks)

31 **P1:** For parallel resistors in an electrical circuit, the total resistance R_{TOT} is given by the formula $\frac{1}{R_{TOT}} = \frac{1}{R_1} + \frac{1}{R_2}$.

$R_1 = 7.2\,\Omega$, measured to the nearest tenth of an ohm, and $R_2 = 3.6\,\Omega$, also measured to the nearest tenth of an ohm.

a Find the lower and upper bounds for the total resistance. (6 marks)

b If the actual total resistance is $2.40\,\Omega$, find the range of percentage errors that could be obtained for R_{TOT}. (4 marks)

Back of an envelope

Approaches to learning: Research, Thinking skills: Evaluating, Critiquing, Applying
Exploration criteria: Personal engagement (C), Reflection (D), Use of mathematics (E)
IB topic: Approximation, Estimation and Errors

How much water do you use?

How much water could you save?

If you combine simple approximations, such as the ones in this chapter, and elementary arithmetic, you can find reasonably good estimates to answer all sorts of interesting questions.

Did you know?

This type of calculation is often called a "back of the envelope" calculation because it is so simple that you could write it down on the back of an envelope!

How much water do you use?

Discuss:

- Which of your daily activities use water?
- How much water does each use?
- In an average week what does your household use water for? Write down all the uses you can think of.
- How many times does your family do each activity in an average week?

Research how many **litres** of water are used for each activity that you and your family do.

Use the figures collected so far to estimate the average number of litres of water used by your household in an average week.

Comment on your findings so far.

How reliable are these estimates?

What may have caused some errors? Are these errors significant? How could you adjust your figures to account for this?

Now calculate the estimated average amount of water used by households in your class in a **week**.

How does your household's water usage compare to others in your class?

You could also now estimate the average amount of water used by a household in one **year** in **your school**.

What information would you need for this?

Comment on your findings.

Estimate how many households there are in your country.

How might you do this?

Estimate the amount of water used for domestic tasks in your country in **one day, one week** and **one year**.

Check your estimates against official figures.

How accurate were you?

What is your percentage error?

How could you make your results more accurate?

What assumptions have you made in calculating your figures?

How do you think the official figures are calculated?

How much water can you save?

Discuss:

- Which activity uses the most water in your household?
- Is this what you expected?
- What could you do to reduce the amount of water used in your household?

Select one of the things you could do.

Estimate how much water your household could save in **one year** if it did this.

Estimate how much water could be saved in **your country** in **one year** if every household in your country did this.

Extension

Devise your own "back of the envelope" style question.

What other questions could you answer in a similar way?

2 Representing space: non-right angled trigonometry and volumes

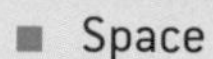

Concepts

- Space
- Representation

Microconcepts

- Area of a triangle and other 2D figures
- Area of a sector
- 3D solids: right cylinder; pyramid; right cone; sphere; hemisphere
- An angle between two lines, and between a line and a plane
- Volume
- Surface area of 3D figures

This chapter deals with finding measures of angles and distances in right and non-right triangles, and calculating the area of 2D and the volume of 3D geometric shapes. These techniques are used in astronomy to measure the distance to nearby stars, in cartography to measure distances between landmarks and in navigation to determine the courses of ships and aircraft.

How are the exact locations of landmarks established in mapmaking?

What part of your lawn can a sprinkler water and how does it depend on the angle of the sprinkler's rotation?

How can the distance to a nearby star be measured?

In 1856 The Great Trigonometric Survey of India measured the height of Mount Everest (known in Napali as Sagarmatha and in Tibetan as Chomolungma). The surveyors measured the distance between two points at sea level and then measured the angle from each of these two points to the top of the mountain.

The summit of Mount Everest is labeled E, the two points A and B are roughly at sea level and are 33 km apart.

If $\hat{E} = 90°$, what would be the maximum possible height of Mount Everest?

- Why do you think it took so long to determine the elevation of Mount Everest?
- Why do you think surveyors prefer to measure angles, but not lengths of sides?
- What assumptions are made?

Developing inquiry skills

Write down any similar inquiry questions you might ask if you were asked to find the height of a tree, the distance between two towns or the distance between two stars.

What questions might you need to ask in these scenarios that differ from the scenario where you are estimating the height of Mount Everest?

Think about the questions in this opening problem and answer any you can. As you work through the chapter, you will gain mathematical knowledge and skills that will help you to answer them all.

Before you start

You should know how to:

1 Use the properties of triangles, including the Pythagorean theorem.

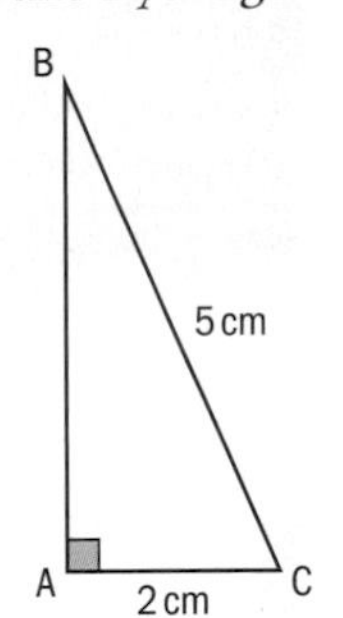

eg Find the length AB in this triangle.

$AB^2 = BC^2 - AC^2$

$AB^2 = 5^2 - 2^2$

$AB^2 = 21$

$AB = \sqrt{21}$

$= 4.58\,\text{cm}$

2 Trigonometric ratios in right triangles

eg $\sin\hat{B} = \frac{2}{5}$, $\cos\hat{B} = \frac{\sqrt{21}}{5}$, $\tan\hat{B} = \frac{2}{\sqrt{21}}$

3 Find the area of a triangle if a side and the height towards it are given.

4 Determine the third angle in a triangle if two angles are known.

Skills check

Click here for help with this skills check

1 In each of these right triangles, find the length of side x.

a

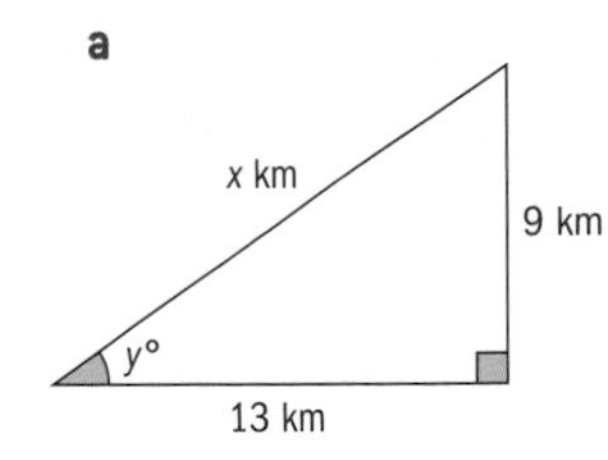

b

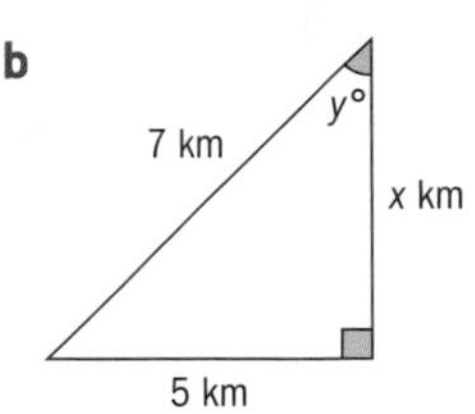

2 For the triangles in question 1, find the size of angle y.

3 Find the area of a triangle with base 6 cm and height 12.5 cm.

4 In ΔABC, $\hat{B} = 135°$ and $\hat{C} = 25°$. Find $\hat{A}$.

2.1 Trigonometry of non-right triangles

Right triangle trigonometry can be used to solve problems involving right triangles. But many geometry problems involve non-right triangles.

For example, ΔABC where $AB = 10\,cm$, $\hat{A} = 45°$ and $\hat{B} = 30°$ is one such triangle.

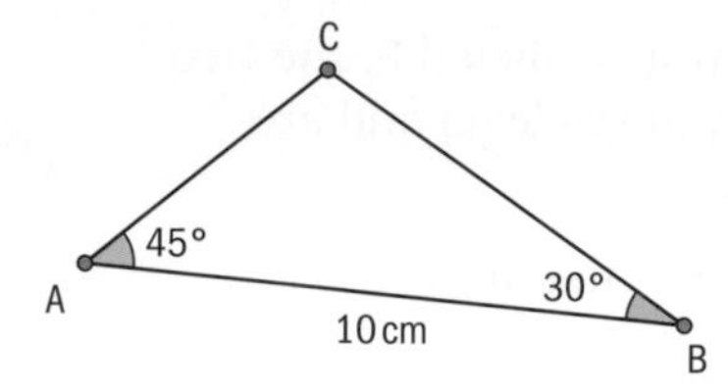

Reflect How can you determine the measures of the unknown angle and sides?

Investigation 1

1 A surveyor measures a plot of land in the form of a triangle ΔABC as shown in the diagram. She measures $a = 5\,m$, $b = 7\,m$ and $c = 8\,m$, and $\hat{A} = 38.2°$, $\hat{B} = 60.0°$ and $\hat{C} = 81.8°$, given to 1 dp.

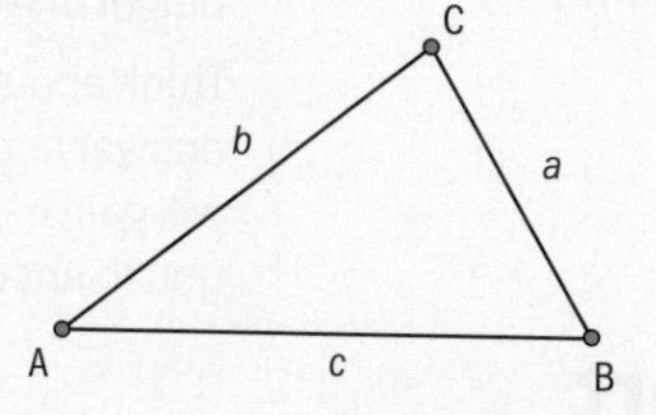

Place all relevant information at the appropriate places in the diagram above.

2 Calculate the ratios $\frac{a}{\sin \hat{A}}$, $\frac{b}{\sin \hat{B}}$ and $\frac{c}{\sin \hat{C}}$. Give your answers to 1 dp and place them in the table below.

ΔABC	$\frac{a}{\sin \hat{A}}$	$\frac{b}{\sin \hat{B}}$	$\frac{c}{\sin \hat{C}}$

Compare the three ratios. What do you notice?

3 ΔABC is shown in the diagram below.

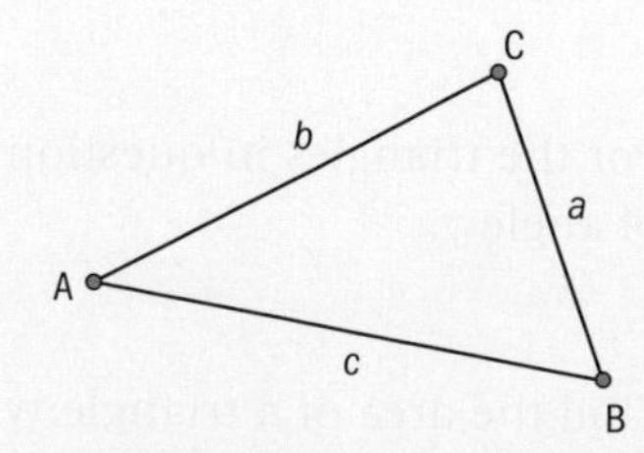

Draw the height of CH from vertex C towards side AB. Label the height h.

4 The height [CH] divides ΔABC into two right triangles: ΔACH and ΔBCH. Note that $\hat{A}$ belongs to ΔACH, and $\hat{B}$ belongs to ΔBCH.
Write an equation in terms of $\sin \hat{A}$ and h using right triangle trigonometry:
$\sin \hat{A} =$

5 Write an equation in terms of $\sin \hat{B}$ and h using right triangle trigonometry:
$\sin \hat{B} =$

6 Write each of the equations from parts **4** and **5** in the form.
$h = \ldots$

7 Equate the right-hand side expressions from the two equations in part **6**.

8 Write an equation equivalent to your equation in part **7**, regrouping first with $\sin \hat{A}$ and then with $\sin \hat{B}$.

9 **Conceptual** Do you think that if you had drawn the height from B to AC instead, you would have come to the same conclusion about the ratios $\frac{c}{\sin \hat{C}}$ and $\frac{a}{\sin \hat{A}}$? Why or why not?

10 **Conceptual** What is the relationship between the sine rule and the sine ratio in a right triangle?

The sine rule

For any ΔABC, where a is the length of the side opposite $\hat{A}$, b is the length opposite $\hat{B}$, and c is the length of the side opposite $\hat{C}$:

$$\frac{a}{\sin \hat{A}} = \frac{b}{\sin \hat{B}} = \frac{c}{\sin \hat{C}} \text{ or } \frac{\sin \hat{A}}{a} = \frac{\sin \hat{B}}{b} = \frac{\sin \hat{C}}{c}$$

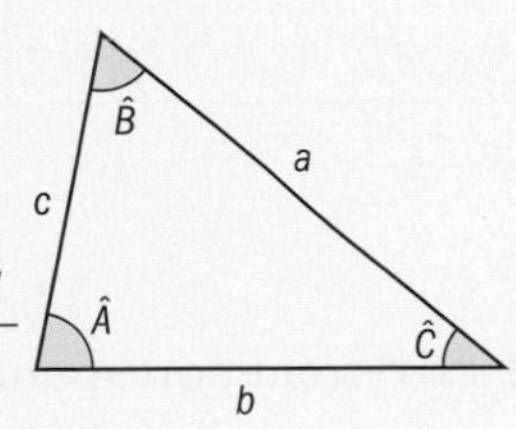

TOK

If sine, cosine and tangent are ratios in a right triangle, how can we use them on angles greater than 90 degrees?

The $\frac{a}{\sin \hat{A}} = \frac{b}{\sin \hat{B}} = \frac{c}{\sin \hat{C}}$ form of the sine rule is used to calculate the length of a side given two angles and an opposite side.

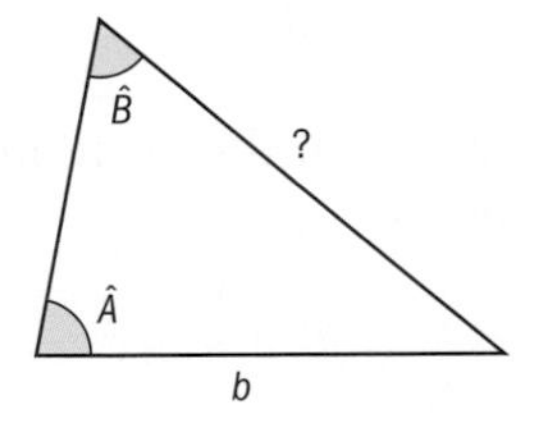

The $\frac{\sin \hat{A}}{a} = \frac{\sin \hat{B}}{b} = \frac{\sin \hat{C}}{c}$ form of the sine rule is used to calculate an angle given two sides and an opposite angle.

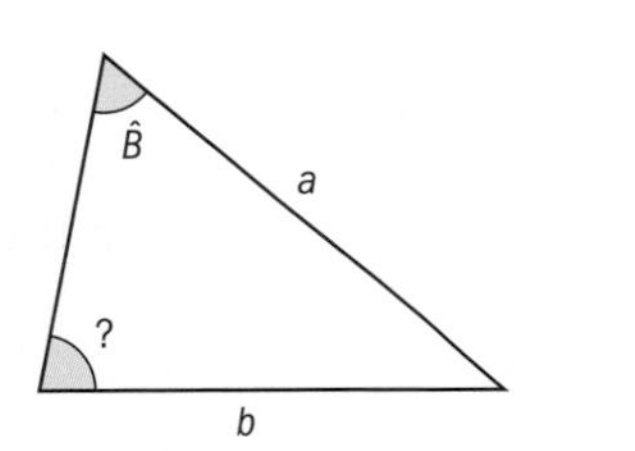

The sine rule is true for any triangle, including right-angled triangles.

Example 1

In ΔQRS, QR = 15 m, SR = 6.5 m and $\hat{S} = 125°$. Find $\hat{Q}$. Give your answer correct to 3 s.f.

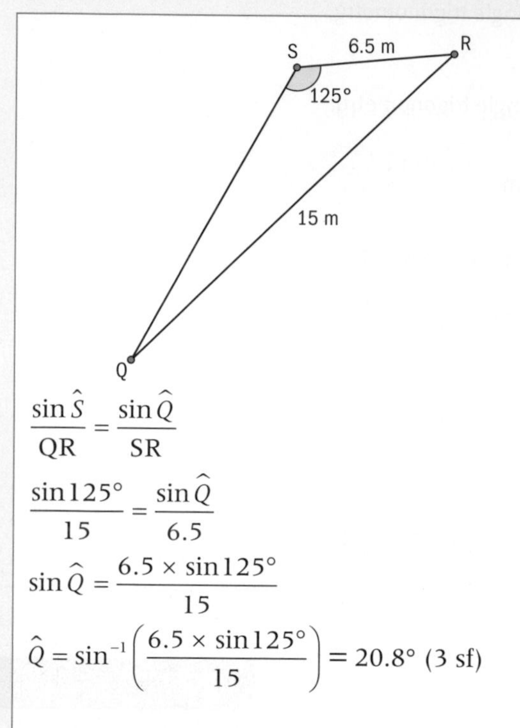

If you are not given a diagram, draw one, label the points and place the given information on your diagram.

$$\frac{\sin \hat{S}}{QR} = \frac{\sin \hat{Q}}{SR}$$

$$\frac{\sin 125°}{15} = \frac{\sin \hat{Q}}{6.5}$$

$$\sin \hat{Q} = \frac{6.5 \times \sin 125°}{15}$$

Identify the unknown and known sides and angles before you apply the sine rule.

$$\hat{Q} = \sin^{-1}\left(\frac{6.5 \times \sin 125°}{15}\right) = 20.8° \text{ (3 sf)}$$

Round your answer to the accuracy specified in the question. If not specified, round your answer to 3 sf.

Example 2

A ship S is on a bearing of 120° from port A and 042° from port B.

The distance between ports A and B is 15.2 miles.

Find the distance between the ship and each port.

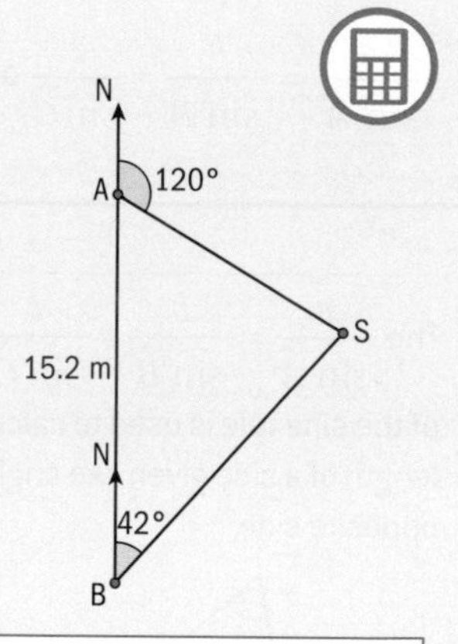

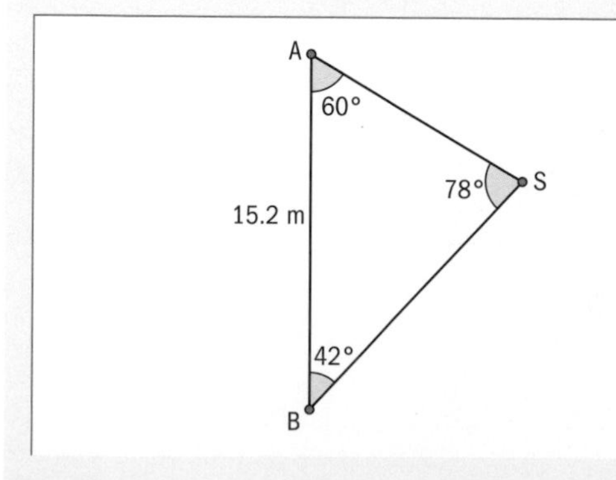

Draw a diagram to represent the positions of the ship and the ports.

$\hat{A} = 180° - 120° = 60°$ $\hat{S} = 180° - (42° + 60°) = 78°$	Use angle rules to find $\hat{A}$ and $\hat{S}$.
$\frac{AS}{\sin 42°} = \frac{BS}{\sin 60°} = \frac{15.2}{\sin 78°}$ $AS = \sin 42° \times \frac{15.2}{\sin 78°} = 10.4$ miles (3 sf) $BS = \sin 60° \times \frac{15.2}{\sin 78°} = 13.5$ miles (3 sf)	Use the sine rule to find the lengths.

Exercise 2A

1 Find the unknown angle for each ΔABC. Give your answers correct to 3 significant figures.

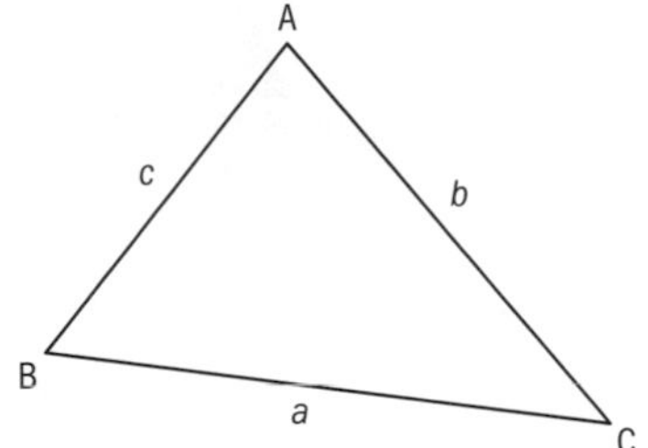

a Find $\hat{A}$ where $a = 12.5$ cm, $b = 15.4$ cm and $\hat{B} = 68°$.

b Find $\hat{B}$ where $b = 42.8$ cm, $c = 30.6$ cm and $\hat{C} = 41°$.

c Find $\hat{C}$ where $a = 7.2$ m, $c = 5.1$ m and $\hat{A} = 70°$.

2 In ΔABC, AC = 8 cm, $\hat{C} = 101°$ and $\hat{B} = 32°$. Find the lengths of [AB] and [BC].

3 A ship is sailing due north. The captain sees a lighthouse 12 km away on a bearing of 036°. Later, the captain observes that the bearing of the lighthouse is 125°. How far did the ship travel between these two observations?

The sine rule can be used to solve practical problems involving distances which are impossible to measure.

Example 3

A park service plans to build a bridge [AC] over a reservoir. A and B are two points on the shore of the reservoir, such that the distance AB = 14 m, $\hat{A} = 102°$ and $\hat{B} = 58°$.

Calculate the length of the bridge [AC] to the nearest metre.

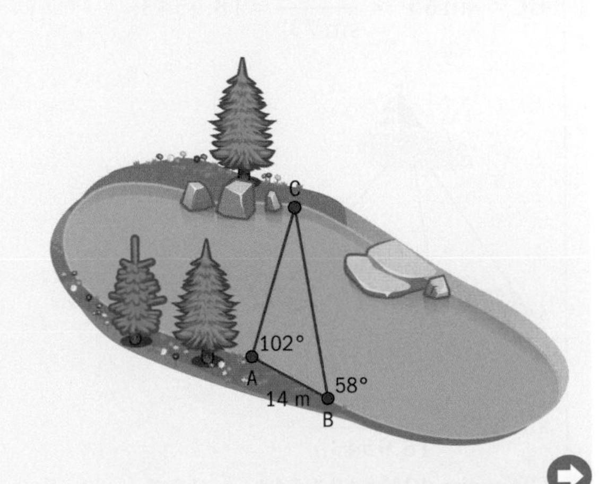

Continued on next page

$\hat{C} = 180° - (58° + 102°) = 20°$	Before you use the sine rule, you need to find $\hat{C}$.
$\frac{AC}{\sin 58°} = \frac{14}{\sin 20°}$	Now apply the sine rule.
$AC = \sin 58° \times \frac{14}{\sin 20°} = 35$ m (nearest metre)	The answer is rounded to the nearest metre at the accuracy level set by the question.

Did you know?

In surveying, this method of determining the location of a point and distance to it by measuring the angles from two known points at either end of a fixed segment is called **triangulation.**

The surveyor's instrument for measuring horizontal and vertical angles is called a **theodolite**.

International-mindedness

Triangulation was used in an argument between England and France over the curvature of the Earth. Research the curvature of the Earth calculated in the 17th century.

Example 4

Two people at points A and B observe a sailboat at position C. The distance between points A and B on the shoreline is 20 m.

The first observer measures $\hat{A} = 65°$ and the second observer measures $\hat{B} = 42°$.

Find the distance from C to the shoreline (AB).

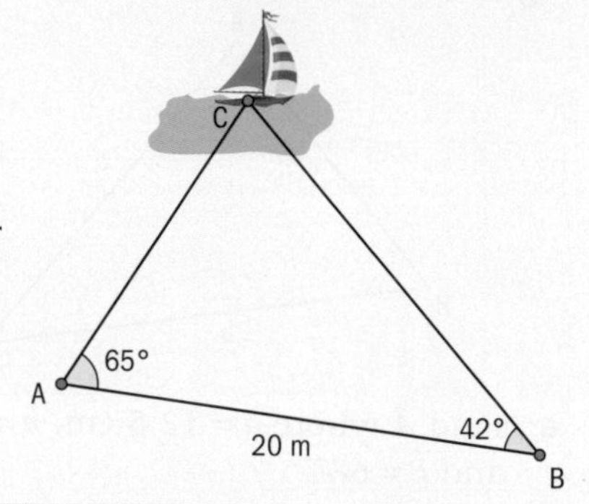

$\hat{C} = 180° - (65° + 42°) = 73°$	Before you apply the sine rule you need to find $\hat{C}$.
$\frac{BC}{\sin 65°} = \frac{20}{\sin 73°}$	Apply the sine rule to obtain the distance from C to B or from C to A.
$BC = \sin 65° \times \frac{20}{\sin 73°} = 18.9543...$	
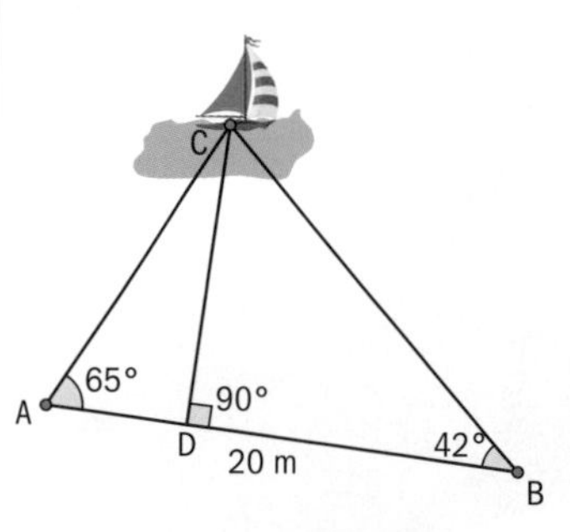	To find the distance from C to [AB], use the right triangle ΔBCD to find CD, where [CD] ⊥ [AB].
$\sin 42° = \frac{CD}{18.9543...}$ $CD = \sin 42° \times 18.9543... = 12.7$ m (3 sf)	Use the sine ratio in the right triangle ΔBCD to calculate CD.

Investigation 2

Triangulation is often used in surveying. It involves measuring two angles and the length of one side in a triangle and then using these measurements to calculate the lengths of the other two sides.

Each of the calculated distances is then used as sides in another triangle to calculate the distances to another point, which in turn can start another triangle. This is done to form a chain of triangles connecting the origin point to another point as needed. The chain includes as many triangles as necessary to reach the destination point or segment whose measure has to be obtained.

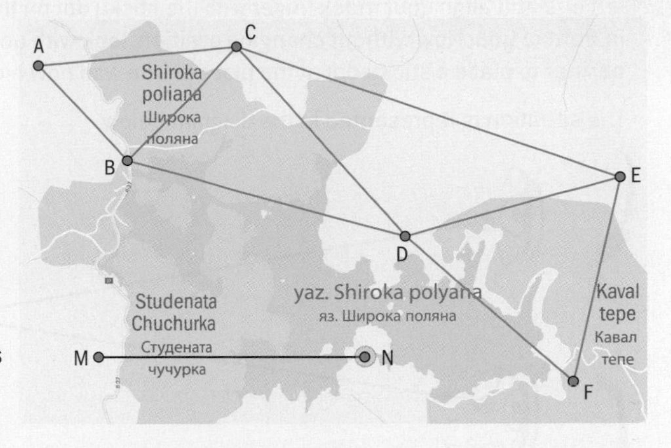

1 A surveyor measures the distance between two points A and B to be 350 m. She identifies visible landmarks at points C, D and E, shown on the map. She wants to find the distance DF.

Use a protractor to find $A\hat{B}C$ and $B\hat{A}C$ on the map. Calculate the length of [BC].

2 Use a protractor to measure $B\hat{C}D$ and $C\hat{B}D$. Use these measurements and the length of [BC] from part **1** to find the lengths of [CD] and [BD].

3 In ΔCDE, measure two angles. Find the length of [DE].

4 Measure two angles and calculate the length of [DF].

5 Are all of the calculated side lengths in parts **1–3** needed to find DF? If not, which one is not needed?

6 MN is a distance between two other landmarks, which cannot be directly measured. Use your protractor and information from above to find MN.

7 Although triangulation is a smart method to determine distances and is much used by surveyors, it is not very useful when the objects are far away, as the stars are.

Why do you think that triangulation might not work for places that are too far away, like the stars?

Investigation 3

Hold your hand out in front of you and hold up your finger and view it with your left eye first and then with your right eye, while your other eye is closed. Your finger will appear to move against the background. This is a phenomenon called **parallax**.

Next you will apply the method of parallax to measure the distance from your eyes to a finger that you hold up with arm stretched out in front of you. (Instead of measuring the distance from your eyes to your finger, you could measure the distance to a tree or another landmark outside.)

Continued on next page

1 Place a sticky dot on the wall in front of you roughly at eye level. Close your right eye and look with your left eye and align your index finger with the sticky dot on the wall as you hold your arm outstretched in front of you. Now without changing position, look with your right eye at your index finger. Ask your partner to place a sticky dot at the place on the wall now hidden by your finger.

The situation is represented in the diagram below.

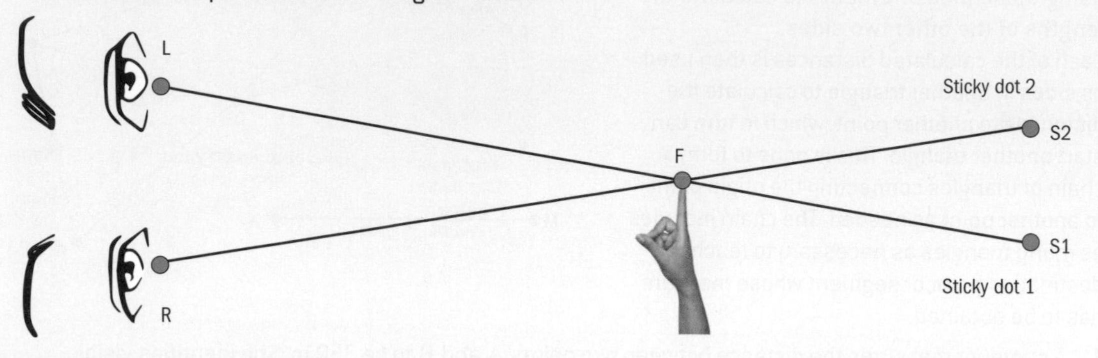

2 Measure $R\hat{L}F$ and $L\hat{R}F$ with your angle-measuring tool. What do you notice about them?

3 Measure the distance LR between your left and right eyes.

4 Calculate the distance between F and [LR].

Investigation 4

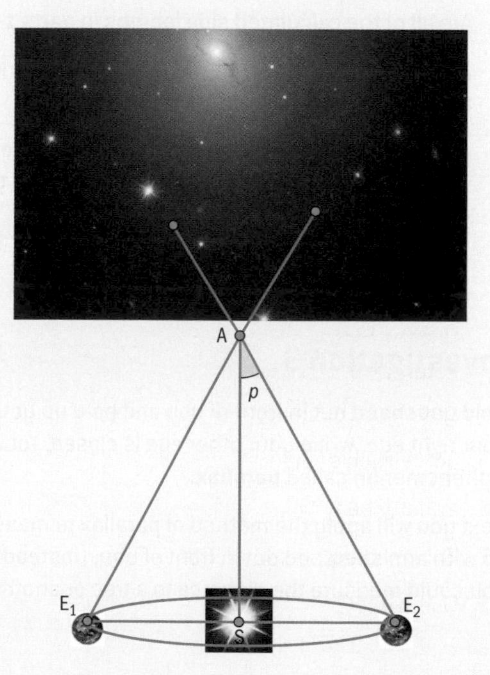

1 Astronomers use the parallax effect to measure distances to nearby stars. To measure the distance from the Earth to a star A, they would use a diagram such as the one shown.
S represents the position of the Sun, and E_1 and E_2 represent the positions of the Earth, as it orbits around the Sun, approximately six months apart (say the Earth's position in June and December). $E_1\hat{A}S$ (or $E_2\hat{A}S$) is called the **parallax angle** (p). It is half $E_1\hat{A}E_2$, which can be measured from the shift of the star A against the background of more distant stars.
Why do you think that astronomers measure the parallax angle $E_1\hat{A}S$ rather than $A\hat{E}_1S$ or $A\hat{E}_2S$?

2 Astronomers measure distances to stars by using the distance between the Sun and the Earth as a basic unit. They call it an astronomical unit (1 AU).
Show that when the parallax angle $p = 45°$, the distance from the Sun to the star A is 1 AU.

3 Use ΔAE_1S to write down the distance AS in terms of the parallax angle p and the distance between the Sun and the Earth, SE_1 (1 AU).

4 Show that when a star A has parallax angle $p = 1°$, the distance from the Sun to A is about 57 AU, rounded to the nearest integer.

5 A star A has a parallax angle of $1 \times 10^{-4°}$. Find the distance of the star to the Earth. Give your answer to the nearest thousand AU.

6 Why is it that the parallax effect can be used to measure the distances to nearby stars, but is not suitable for measuring distances to stars that are very far away?

Three satellites are located at points A, B and C where, $a = 7$ km, $b = 12$ km and $c = 14$ km, as shown on the diagram of ΔABC.

How can you determine the angles of the triangle? How would you go about it?

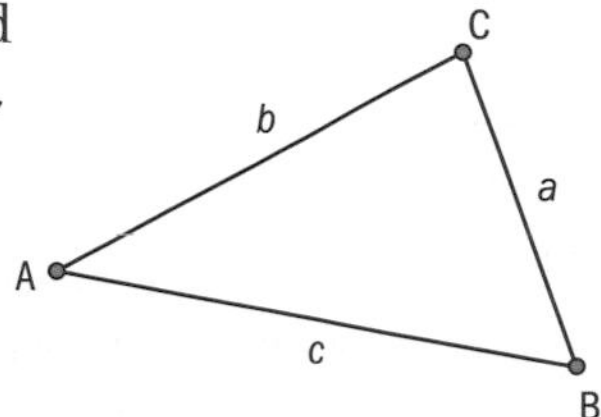

Investigation 5

1 A surveyor measures a plot of land in the form of a triangle ΔABC as shown in the diagram. She measures $a = 5$ km, $b = 7$ km and $c = 8$ km, and $\hat{A} = 38.2°$, $\hat{B} = 60.0°$ and $\hat{C} = 81.8°$, given to 1 dp.

Use the given information to complete the table below:

a^2	b^2	c^2	$2ac\cos\hat{B}$	$2ab\cos\hat{C}$	$2bc\cos\hat{A}$	$a^2 + b^2 - 2ab\cos\hat{C}$	$b^2 + c^2 - 2bc\cos\hat{A}$	$a^2 + c^2 - 2ac\cos\hat{B}$

2 Compare the entries in the different columns. Do you notice a pattern? What can you conclude?

3 Look at ΔABC shown in the diagram on the right.

The height of ΔABC, $[CH]$, is drawn from vertex C towards side $[AB]$ and labelled with h.

Point H divides side c into two segments, x and $c - x$.

Use the Pythagorean theorem to write an equation in terms of h, b and x.

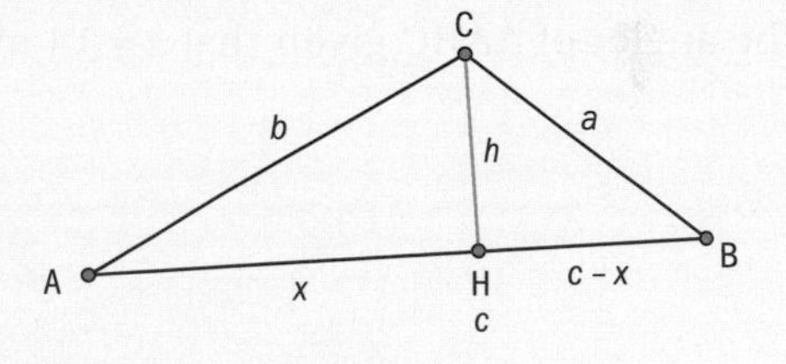

4 Use the Pythagorean theorem to write another equation in terms of h, a and $c - x$.

5 Solve each equation from parts **3** and **4** for h^2.

6 Equate the two right-hand side expressions from part **5**.

7 Solve the equation from part **6** in terms of a^2

8 Use ΔAHC to write an expression for x in terms of b and $\hat{A}$.

9 Substitute your expression for x from part **8** in the equation from part **7**. The result is called the cosine rule.

10 **Conceptual** What is the relationship between the cosine rule and the Pythagorean theorem in a right triangle?

Cosine rule

For ΔABC, where a is the length of the side opposite $\hat{A}$, b is the length of the side opposite $\hat{B}$, and c is the length of the side opposite $\hat{C}$:

$a^2 = b^2 + c^2 - 2bc\cos\hat{A}$

$b^2 = a^2 + c^2 - 2ac\cos\hat{B}$

$c^2 = a^2 + b^2 - 2ab\cos\hat{C}$

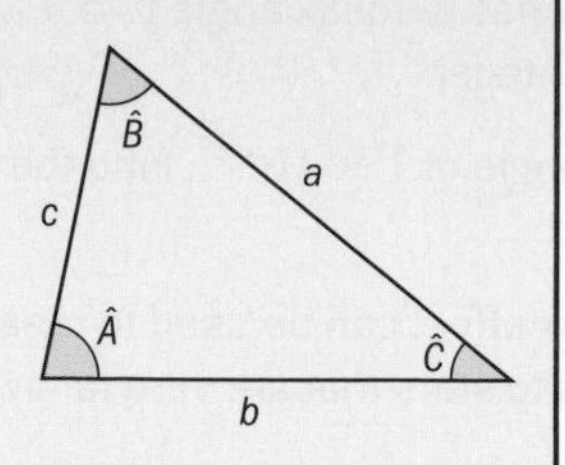

TOK

It has often been said that the Euler relation, $e^{i\pi} + 1 = 0$, is the most beautiful equation in all of mathematics. What is meant by beauty and elegance in mathematics?

The $a^2 = b^2 + c^2 - 2bc\cos\hat{A}$ form of the cosine rule is used to calculate the length of a side given two sides and the angle between them.	The $\cos\hat{A} = \frac{b^2 + c^2 - a^2}{2bc}$ form of the cosine rule is used to calculate an angle given all three sides.
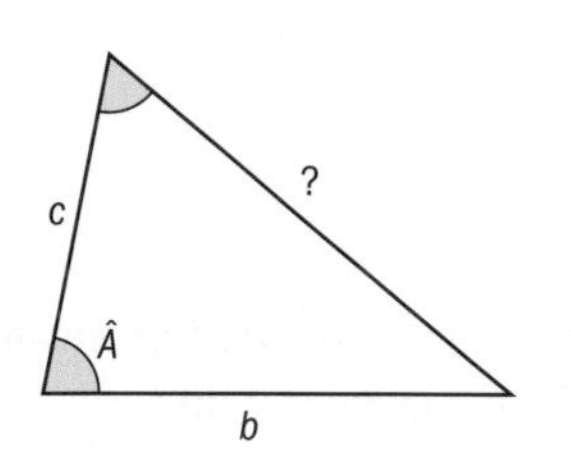	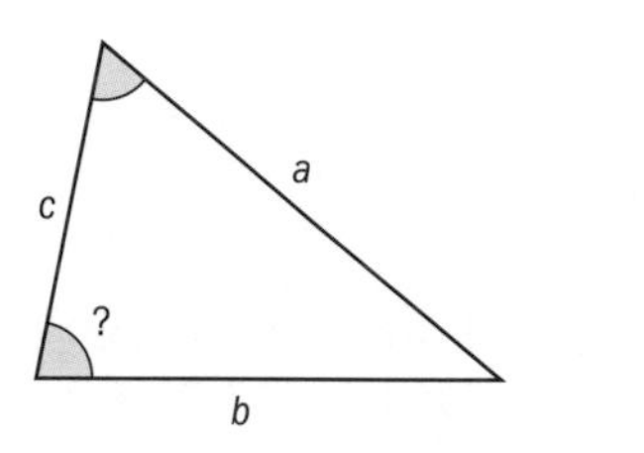

The cosine rule is true for any triangle, including right-angled triangles.

Example 5

Find the angles of ΔABC given that $a = 14$ m, $b = 12$ m and $c = 7$ m.

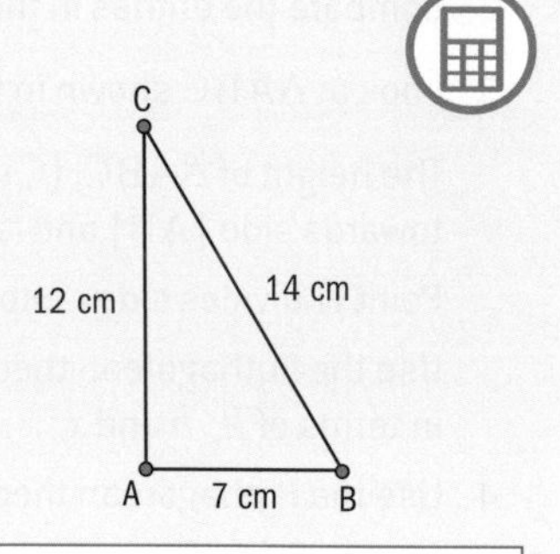

$\cos\hat{A} = \frac{12^2 + 7^2 - 14^2}{2\times 12\times 7}$ $\hat{A} = \cos^{-1}\left(\frac{12^2 + 7^2 - 14^2}{2\times 12\times 7}\right) = 91.0°$ (3 sf) $\cos\hat{B} = \frac{14^2 + 7^2 - 12^2}{2\times 14\times 7}$	To find the angles of ΔABC use the $\cos\hat{A} = \frac{b^2 + c^2 - a^2}{2bc}$, $\cos\hat{B} = \frac{a^2 + c^2 - b^2}{2ac}$ and $\cos\hat{C} = \frac{a^2 + b^2 - c^2}{2ab}$ forms of the cosine rule. Since the accuracy level has not been set in the question, all angle measures are given to 3 sf.

$\hat{A} = \cos^{-1}\left(\frac{14^2+7^2-12^2}{2\times14\times7}\right) = 59.0°$ (3 sf) $\cos\hat{C} = \frac{14^2+12^2-7^2}{2\times14\times12}$ $\hat{C} = \cos^{-1}\left(\frac{14^2+12^2-7^2}{2\times14\times12}\right) = 30.0°$ (3 sf)	

Example 6

Find the length of a lake, CB, given the distances AB = 290 m and AC = 225 m and the angle $\hat{A} = 72°$.

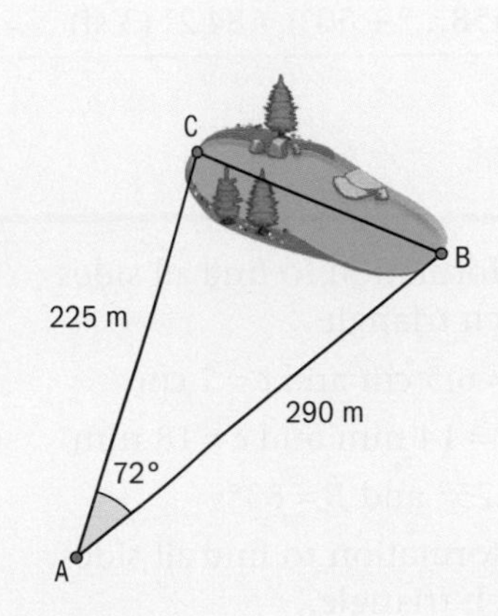

$CB^2 = 290^2 + 225^2 - 2 \times 225 \times 290 \cos 72°$ $CB^2 = 94\,398.2...$ $CB = \sqrt{94\,398.2...} = 307$ m	Use the $a^2 = b^2 + c^2 - 2bc\cos\hat{A}$ form of the cosine rule as you are given two sides and the included angle. Give the final answer to 3 sf.

In some cases, you may need to apply both the cosine and sine rules to find the information you need to solve problems.

Example 7

Towns A, B, and C form ΔABC with sides BC = 8 km and AC = 13 km and angle $\hat{C} = 50°$.

Find $\hat{A}$ and $\hat{B}$.

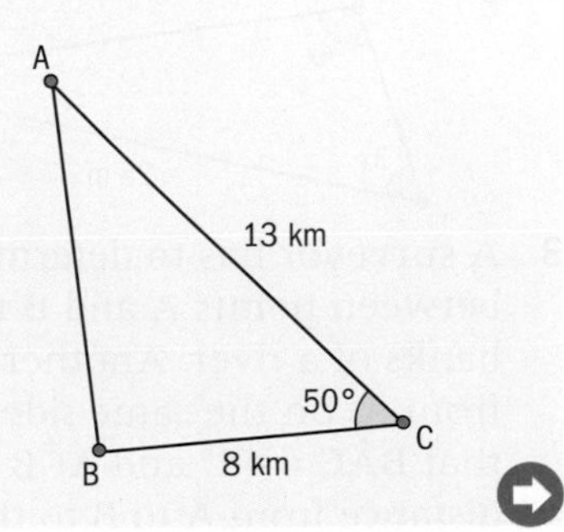

Continued on next page

$AB^2 = 8^2 + 13^2 - 2 \times 13 \times 8 \cos 50° = 99.3001...$ $AB = \sqrt{99.3001...}$ $AB = 9.96494... = 9.96$ km (3 sf)	First use the $c^2 = a^2 + b^2 - 2ab \cos \hat{C}$ form of the cosine rule to calculate the length of [AB].
$\frac{\sin \hat{A}}{8} = \frac{\sin 50°}{9.96494...}$ $\sin \hat{A} = 8 \times \frac{\sin 50°}{9.96494...}$ $\hat{A} = \sin^{-1}\left(8 \times \frac{\sin 50°}{9.96494...}\right) = 45.8°$ (3 sf) $\hat{B} = 180° - (45.7558...° + 50°) = 84.2°$ (3 sf)	Once you find AB, we can use the sine rule to determine $\hat{A}$ and $\hat{B}$. Remember not to use rounded values in intermediate calculations as that may result in an inaccurate final answer. Finally use the fact that angles in a triangle add up to 180°.

Exercise 2B

1 Use the given information to find all sides and angles in each triangle.

a $a = 12$ cm, $b = 6.5$ cm and $c = 7$ cm

b $a = 11$ mm, $b = 14$ mm and $c = 18$ mm

c $c = 22$ m, $\hat{A} = 25°$ and $\hat{B} = 83°$

2 Use the given information to find all sides and angles in each triangle.

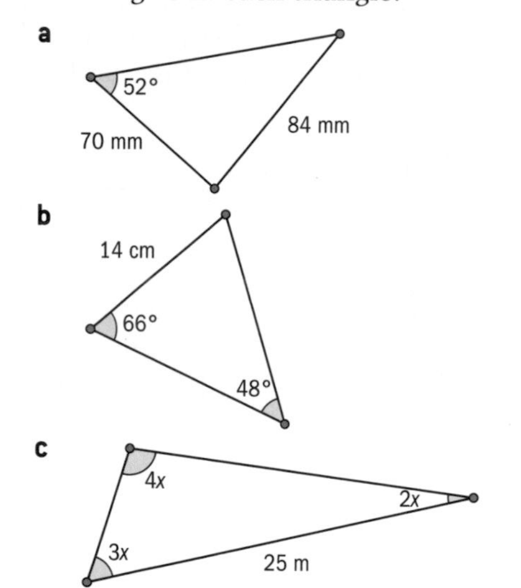

3 A surveyor has to determine the distance between points A and B that lie on opposite banks of a river. Another point C is 450 m from A, on the same side of the river, such that $B\hat{A}C = 45°$ and $A\hat{C}B = 55°$. Find the distance from A to B to the nearest metre.

4 A hiker leaves camp and walks 5 km on a bearing of 058°. He stops for a break, then continues walking for another 8 km on a bearing of 103°. He stops again before heading straight back. Find the distance back to camp from the second stop.

5 The straight-line distance between two cities A and B is 223 km. The straight-line distance between cities A and C is 152 km. The straight-line distance between cities B and C is 285 km.

Find the angles between each of the three cities.

6 To find the third side of ΔABC with AB = 40 cm, BC = 25 cm and $B\hat{A}C = 35°$, Velina and Kristian offered the following suggestions:

Kristian suggested: "Use the cosine rule as you are given two sides and one angle."

Velina suggested: "Use the sine rule as you are given two sides and an angle opposite to one of them."

State whose method is correct and justify your statement. Hence, solve the triangle.

Developing inquiry skills

Look again at the opening problem. The Great Trigonometric Survey used an instrument called a theodolite to measure the angles from points A and B to the summit E.

$B\hat{A}E = 30.5°$ and $A\hat{B}E = 26.2°$.

Points A and B are 33 km apart.

Find the height of Mount Everest to the nearest metre.

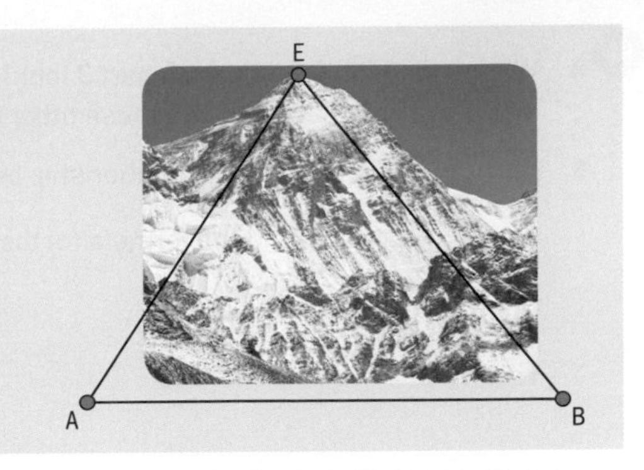

2.2 Area of a triangle formula: applications of right- and non-right angled trigonometry

Four triangles are given below. Determine the area of each triangle if possible. If the area cannot be determined, state what additional information is needed so that it can be done.

1

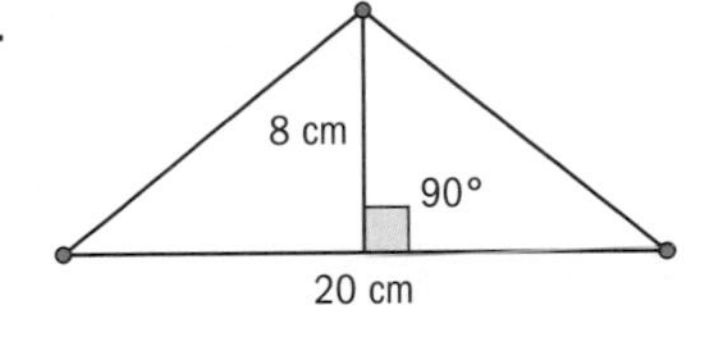

2

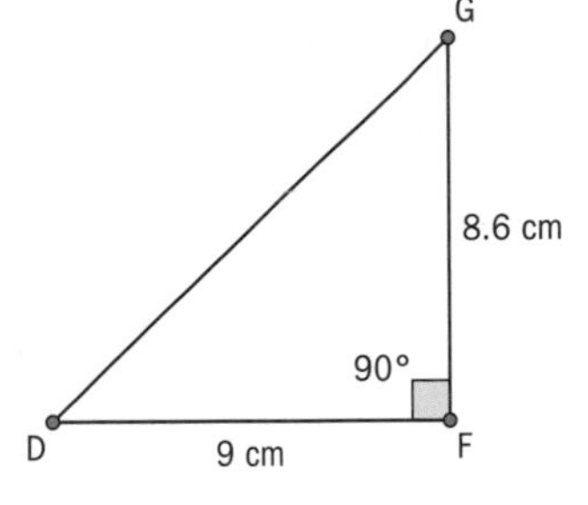

3

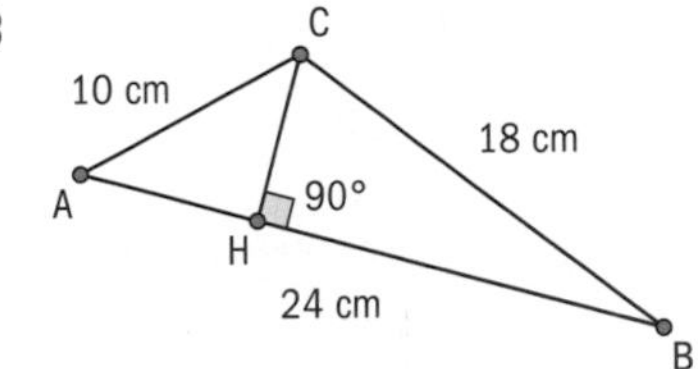

4

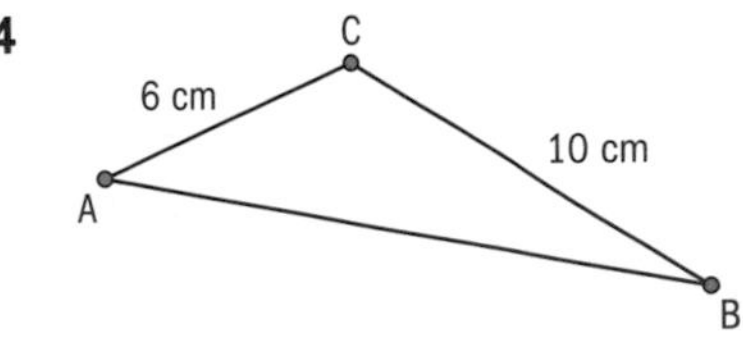

Investigation 6

Consider a plot of land in the shape of ΔABC with given sides a and c and angle $\hat{B}$.

The height h is drawn from C towards [AB], so that $h \perp [AB]$.

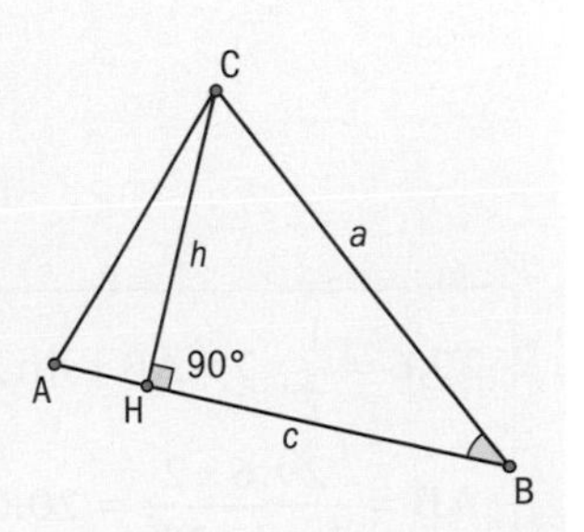

1 Write an expression for the area of ΔABC in terms of side c and the height h.

2 Consider ΔBHC. Write an expression for h in terms of a and angle B.

3 Substitute your expression for h from part **2**, in your expression for the area from part **1**, and hence show that.

area of $\Delta ABC = \frac{1}{2}ac\sin\hat{B}$

Continued on next page

4 What would the formula from part **3** look like if b, c and $\sin\hat{A}$ or a, b and $\sin\hat{C}$ are given? Write down these other forms of the area formula.

5 **Conceptual** What is the relationship between the formula

$\text{area} = \frac{1}{2}ac\sin\hat{B}$ and the formula for the area of a right triangle:

$\text{area} = \frac{1}{2}\times\text{base}\times\text{height}$?

International-mindedness

Area can be measured in square units, but also in "packets" like hectares or acres.

Which countries use bigha, mou, feddan, rai, and tsubo?

The area of any triangle ΔABC is given by the formulae

$\text{area} = \frac{1}{2}bc\sin\hat{A}$ or $\text{area} = \frac{1}{2}ac\sin\hat{B}$ or $\text{area} = \frac{1}{2}ab\sin\hat{C}$

Example 8

A farmer owns a triangular plot ΔABC, where the lengths are AB = 120 m and BC = 240 m, and $\hat{B} = 130°$.

Find the area of the plot to the nearest integer.

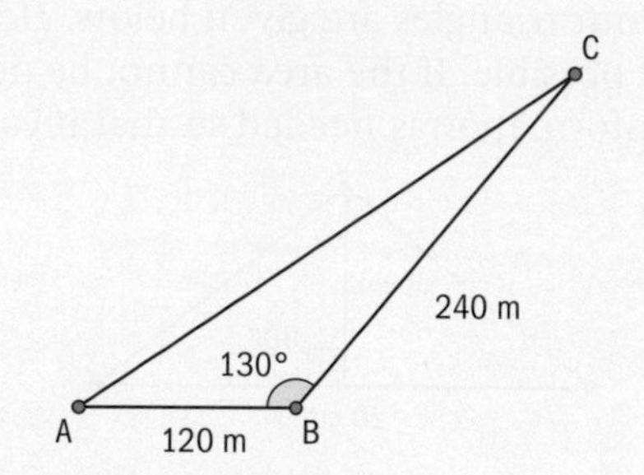

$\text{area} = \frac{1}{2}\times 120\times 240\times\sin 130°$ $\text{area} = 11\,031\text{ m}^2$	Use the formula $\text{area} = \frac{1}{2}ac\sin\hat{B}$. Round the area to the nearest integer.

Example 9

The area of ΔABC is 29.6 cm^2, $\hat{B} = 25°$ and BC = 7 cm. Find the length of side [AB].

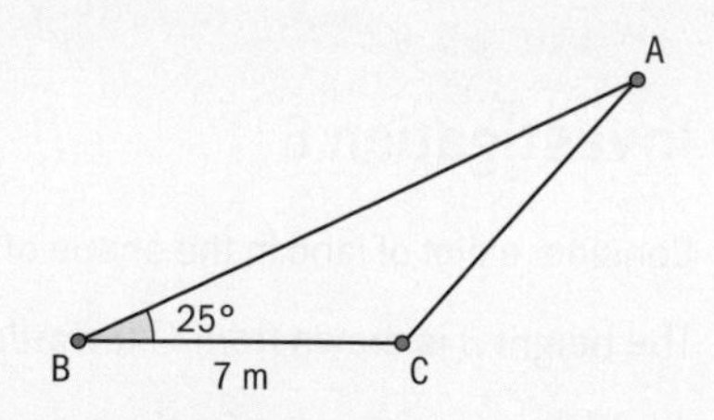

$29.6 = \frac{1}{2}\times 7\times\text{AB}\times\sin 25°$ $\text{AB} = \frac{29.6\times 2}{7\times\sin 25°} = 20.0\text{ cm (3 sf)}$	Use the formula $\text{area} = \frac{1}{2}ac\sin\hat{B}$. Round the answer to 3 sf.

Example 10

Find the area of a parallelogram ABCD with BC = 7.4 cm, AB = 12.3 cm and $\hat{B} = 39°$.

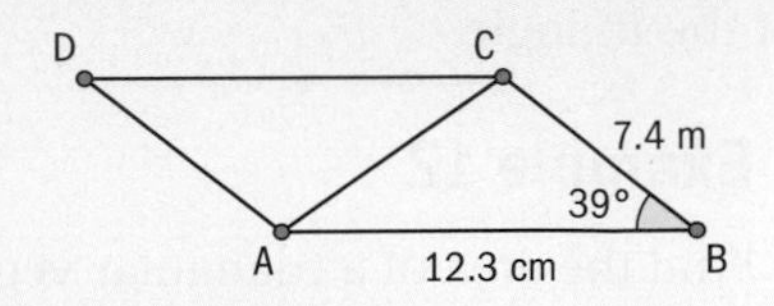

Area of $\Delta ABC = \frac{1}{2} \times 12.3 \times 7.4 \times \sin 39°$ Area of $\Delta ABC = 28.6403...\ \text{cm}^2$	You can find the area of ΔABC by using the area of a triangle formula. Since ΔABC = ΔADC (because of three equal sides) they will have equal areas.
Area of ABCD = $2 \times 28.6403... = 57.3\ \text{cm}^2$ (3 sf)	The area of the parallelogram will be twice the area of ΔABC.

The area of any parallelogram ABCD is given by the formula

$$\text{area} = ab \sin \hat{C}$$

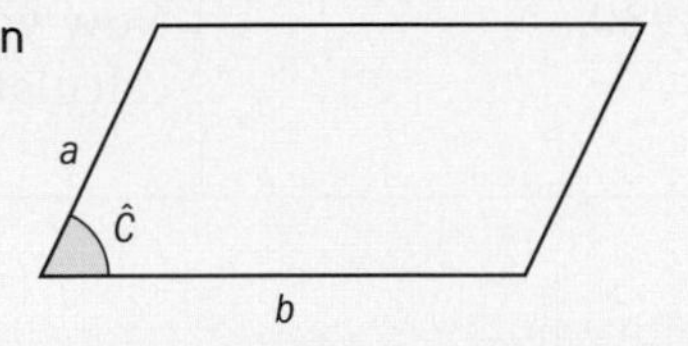

Example 11

A roof is in the shape of an isosceles triangle ΔPRQ where PQ = RQ, $\hat{P} = 61°$ and PR = 12 cm. Find the area of ΔPRQ.

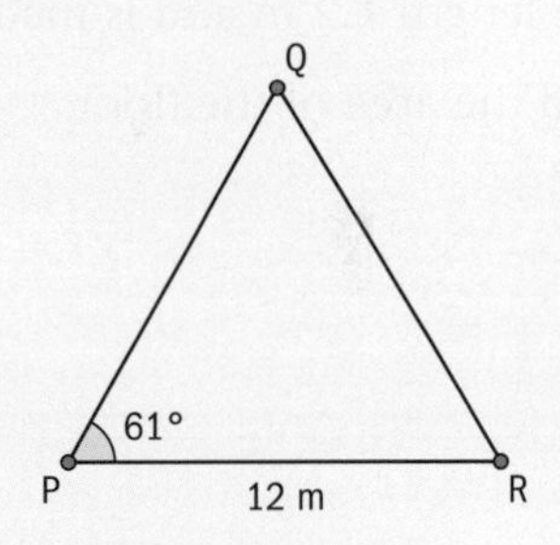

$\hat{P} = \hat{R} = 61°$	Since ΔPRQ is isosceles and PQ = RQ, $\hat{P} = \hat{R} = 61°$.
$\hat{Q} = 180° - 2 \times 61° = 58°$ $\frac{PQ}{\sin 61°} = \frac{12}{\sin 58°}$	Find the third angle by using the angle sum theorem.
$PQ = \sin 61° \times \frac{12}{\sin 58°} = 12.3759...$	Use the sine rule to find the length of [PQ].
Area $= \frac{1}{2} \times 12 \times 12.3759... \times \sin 61°$ Area = $64.9\ \text{cm}^2$ (3 sf)	Use the area formula to find the area of ΔPRQ.

To be able to determine the area of a triangle you need to know two sides and an inscribed angle, two angles and a side, or the three sides of the triangle.

Example 12

Find the area of a triangular vegetable garden with sides AB = 15 m, AC = 7 m and BC = 12 m.

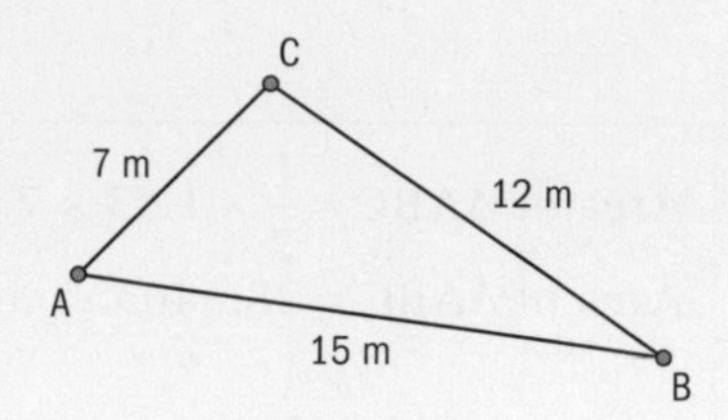

$\cos\hat{C} = \frac{12^2+7^2-15^2}{2\times12\times7}$	To use the area formula, you need to find one of the angles of the triangle.
$\hat{C} = \cos^{-1}\left(\frac{12^2+7^2-15^2}{2\times12\times7}\right) = 100.980...°$	You can use the cosine rule to find $\hat{C}$.
Area $= \frac{1}{2}\times12\times7\times\sin100.980...°$ Area $= 41.2\text{ m}^3$ (3 sf)	Now you can use the area formula to calculate the area of $A\hat{B}C$.

Example 13

The floor of a gazebo has the shape of a regular octagon with side length 1.2 m and is made of eight congruent triangles.

Find the area of the floor.

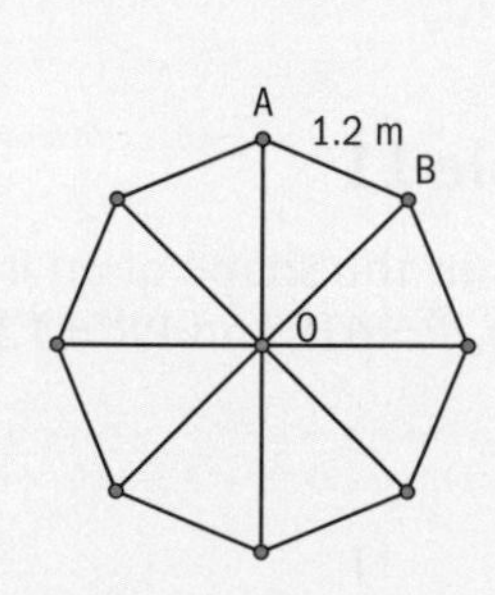

$A\hat{O}B = \frac{360°}{8} = 45°$	Area of the floor = 8 × area of ΔAOB. To find the area of ΔAOB, you first need to find $A\hat{O}B$.
$B\hat{A}O = A\hat{B}O = \frac{180°-45°}{2} = 67.5°$	To find the other two angles of ΔAOB you need to use the angle sum theorem and consider that ΔAOB is isosceles.
$\frac{AO}{\sin67.5°} = \frac{1.2}{\sin45°}$ $AO = \sin67.5°\times\frac{1.2}{\sin45°} = 1.56787...\text{ m}$	Then use the sine rule to find the length of [AO].
Area $= 8\times\frac{1}{2}\times1.56787...\times1.56787...\times\sin45°$ Area $= 6.95\text{ m}^2$ (3 sf)	To find the area of the octagonal floor you need to multiply the area of ΔAOB by 8.

Exercise 2C

1 Find the area of each of the following triangles:

a 16 cm, 28 cm, 23°

b 7.25 cm, 11 cm, 56°

c 10 cm, $7+3\sqrt{5}$ cm, 67°, 35°

2 Find the area of each of the following triangles:

a $a = 13.6$ cm, $b = 9.2$ cm and $\hat{C} = 49°$

b $c = 19$ m, $\hat{A} = 33°$ and $\hat{B} = 42.5°$

c $a = 25$ cm, $b = 14\sqrt{2}$ cm, $c = \sqrt{59}$ cm

3 A gardener leaves a triangular plot of his garden for a plant nursery. The two sides of the triangular plot are AC = 23.5 m and AD = 15 m.

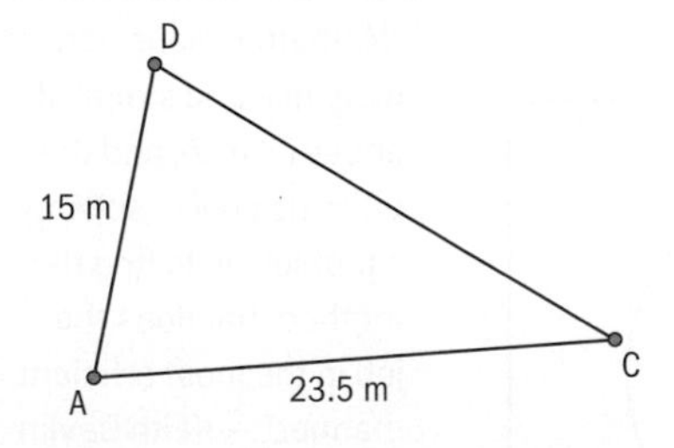

a If DC = 29 m, find the area of the plot designated for the plant nursery.

b 0.5 kg of herb seeds will cover an area of 150 m^2. Determine the mass of seeds, in kg, needed to plant the plot with herbs.

4 Find the area of a parallelogram ABCD with BC = 7.4 cm, $B\hat{A}C = 44.5°$ and $A\hat{C}B = 91°$.

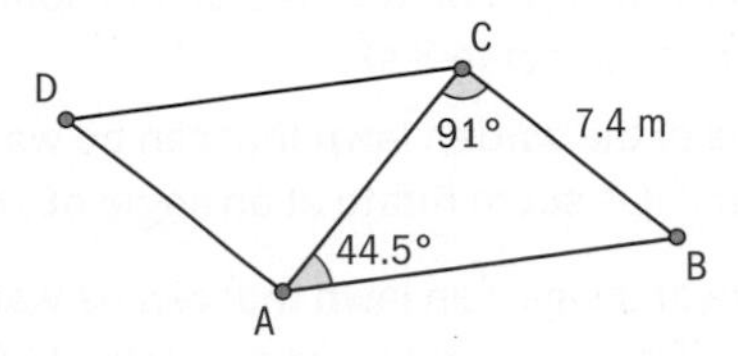

5 A regular hexagon is inscribed in a circle with radius 9 cm. Find the area of the hexagon.

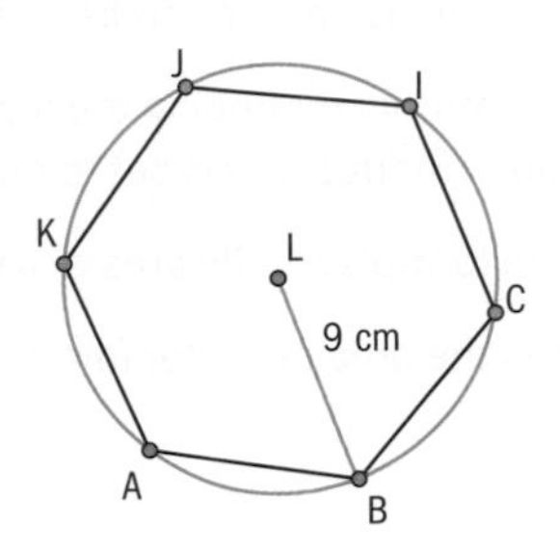

6 A garden is shaped in the form of a regular heptagon (seven-sided), MNSRQPO. A circle with centre T and radius 25 m circumscribes the heptagon as shown in the diagram below. The area of ΔMSQ is left for a children's playground, and the rest of the garden is planted with flowers. Find the area of the garden planted with flowers.

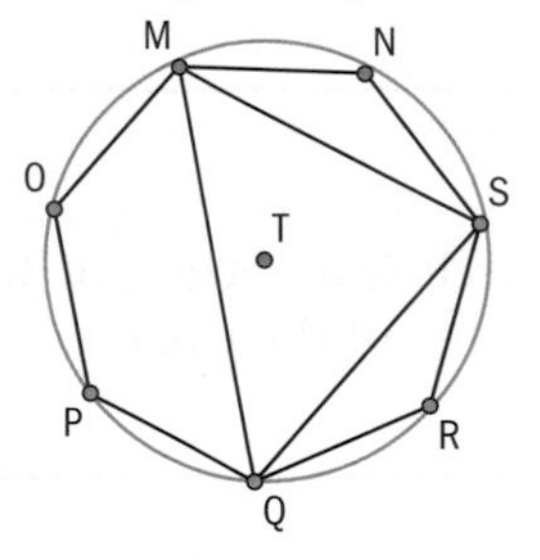

Investigation 7

A sprinkler located at the centre of a lawn sprays water a distance of 20 m as it rotates through a certain angle. What area of the garden lawn can be watered by the sprinkler? Write your answers correct to 3 sf.

1 Find the area of the garden lawn that can be watered by the sprinkler if it is set to rotate at an angle of 180°.

2 Find the area of the garden lawn that can be watered by the sprinkler if it is set to rotate at an angle of 60°.

3 Find the area of the garden lawn that can be watered by the sprinkler if it is set to rotate at an angle of 1°.

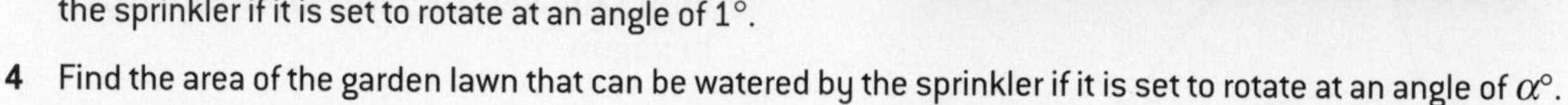

4 Find the area of the garden lawn that can be watered by the sprinkler if it is set to rotate at an angle of $\alpha°$.

5 Could you generate a formula for finding the area of the garden lawn that can be watered by a sprinkler that projects water out r metres and is set to rotate at an angle α?

6 **Factual** What is the formula for the area of a sector?

7 **Conceptual** How is the area of sector formula derived?

A sector is a portion of a circle lying between two radii and the subtended arc.

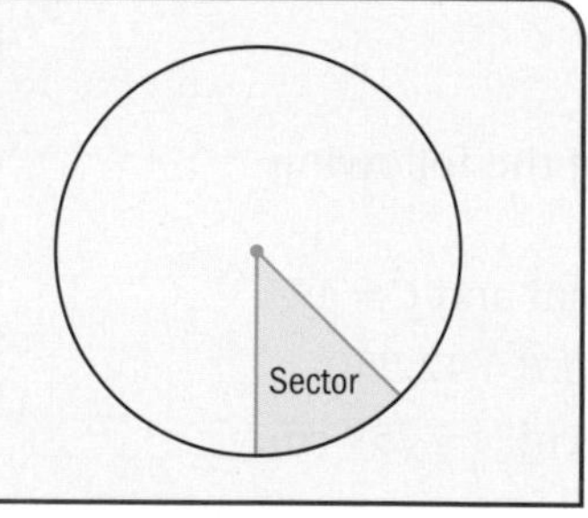

International-mindedness

Why are there 360 degrees in a complete turn? Why do we use minutes and seconds for time?

Any central angle in a circle is a fraction of 360°, so the area of a sector will be a fraction of the area of the circle.

$$\text{Area of sector} = \left(\frac{\theta}{360}\right) \times (\pi r^2) = \frac{\theta \pi r^2}{360}$$

where r is the radius of the circle and θ is the central angle in degrees.

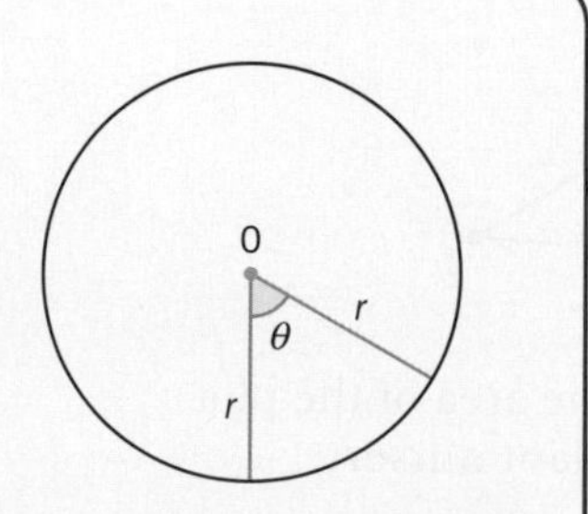

TOK

"Mathematicians admire elegance and simplicity above all else, and the ultimate goal in solving a problem is to find the method that does the job in the most efficient manner." —Keith Devlin

What is elegance and simplicity in mathematical proof?

Example 14

A crop sprinkler irrigates a circle with radius 5 m. Find the area of the sector BAC with radius 5 m and central angle 120° as shown in the diagram. Give your answer to 3 sf.

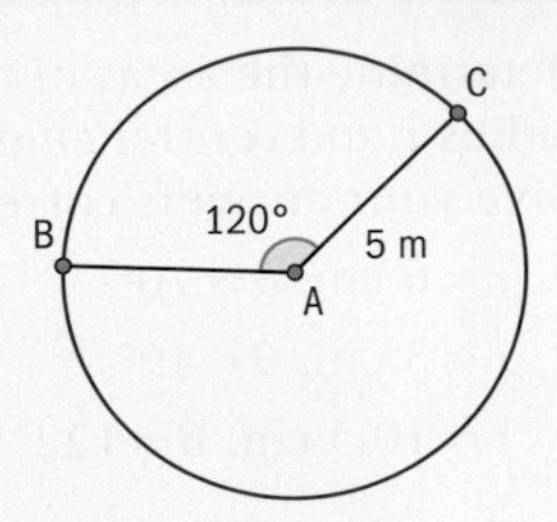

Area of sector $= \dfrac{120° \times \pi \times 5^2}{360°} = 26.2\ \text{m}^2$ (3 sf)	Substitute $r = 5$ and $\theta = 120°$ into the formula for the area of the sector.

Example 15

The area of a sector BAC with radius 13 cm is $30\pi\ \text{cm}^2$. Find x, the size of the central angle of sector BAC.

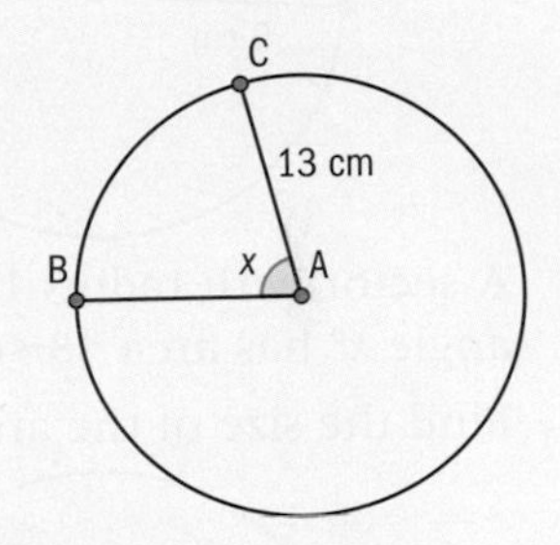

$30\pi = \dfrac{x\pi \times 13^2}{360}$	Use the formula for the area of a sector, and substitute the given values for the area and the radius.
$x = \dfrac{30 \times 360}{169} = 63.9°$ (3 sf)	Then use your GDC to calculate x.

Example 16

The sector AOB has radius 4 cm and central angle 75°.

Determine the area of the segment enclosed by chord [AB] and the subtended arc.

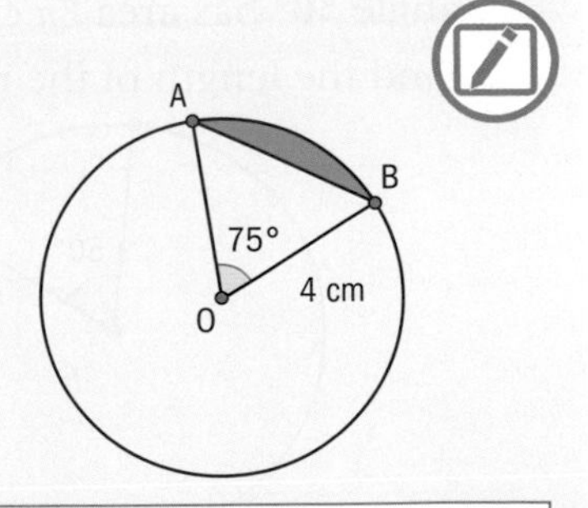

Area of sector $= \dfrac{75° \times \pi \times 4^2}{360°} = 10.4719\ldots\ \text{cm}^2$	Find the area of the sector AOB.
Area of triangle $= \dfrac{1}{2} \times 4 \times 4 \times \sin 75° = 7.7274\ldots\ \text{cm}^2$	Find the area of the triangle AOB.
Area enclosed $= 10.4719\ldots - 7.7274\ldots$ $= 2.74\ \text{cm}^2$ (3 sf)	

Exercise 2D

1 Determine the areas of the sectors with radius r and central angle θ given below. Give your answers correct to 3 sf.

a $r = 6$ cm, $\theta = 70°$

b $r = 3$ cm, $\theta = 49°$

c $r = 10.5$ cm, $\theta = 122°$

2 a A sector with central angle 140° and radius 9 cm is cut from a wooden circle.

Find the area of the sector.

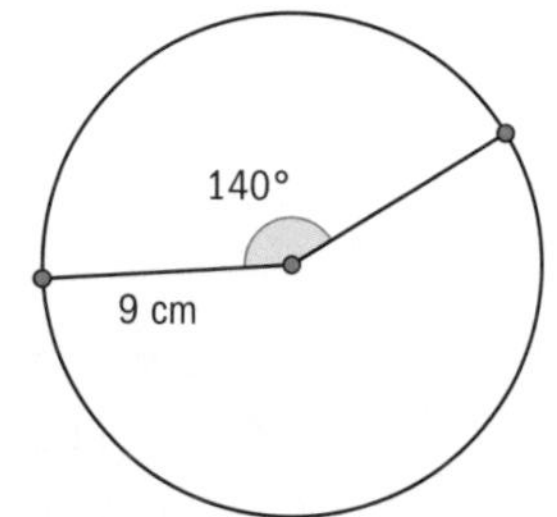

b A sector with radius 10 cm and central angle $x°$ has area 48π cm^2.

Find the size of the angle $x°$.

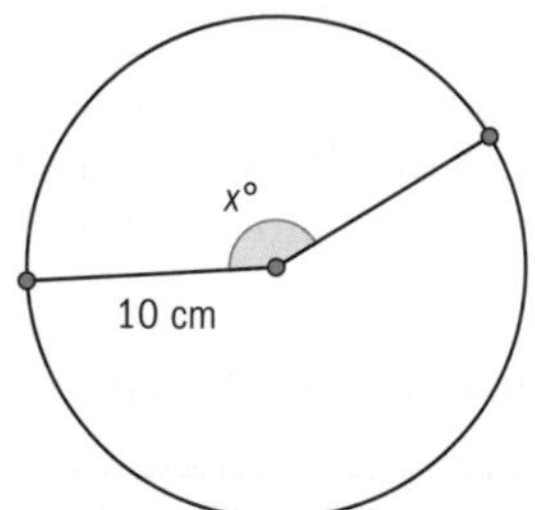

c A sector with radius x cm and central angle 50° has area 8π cm^2.

Find the length of the radius x.

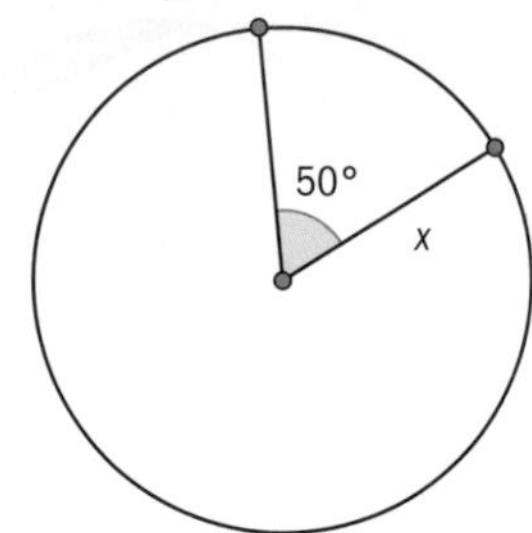

3 Determine the area of the shaded region of a wooded patch in a circular park if the radius of the circle is

a 8 cm **b** 12 cm.

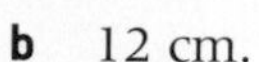

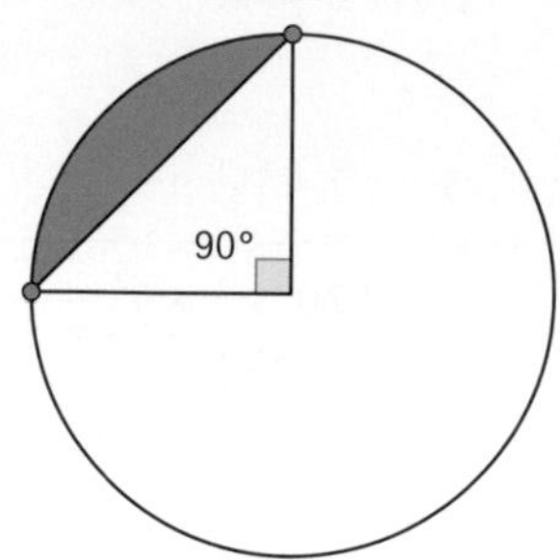

4 Find the area of the segment enclosed by chord [BC] and the subtended arc with radius 4 cm and central angle 55°.

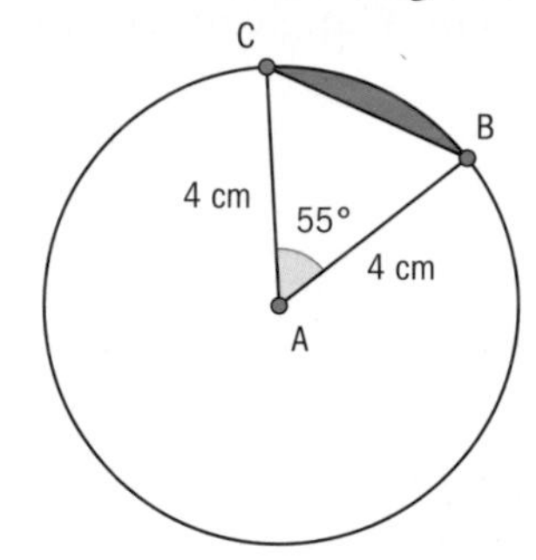

5 A landscaper builds a regular hexagonal patio in a circular garden. The area not covered by the patio will be covered in grass.

The radius of the garden is 1.5 m.

Find the area covered by grass in the garden.

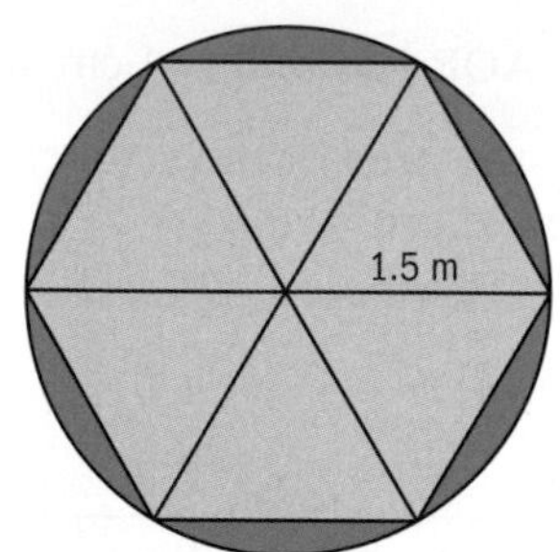

International-mindedness

Japanese mathematician Seki Takakazu calculated π to ten decimal places in the 17th century.

2.3 3D geometry: solids, surface area and volume

This cube is known as the Necker cube and shows the edges of the cube without any depth cues.	
When you stare at the cube frame for a while, you will be able to see two different images. When you draw a diagram of a solid you need to draw the edges that are not visible with dotted lines. Using square paper can help you draw the parallel and congruent edges.	
This cube is called the impossible cube and was invented by the Dutch artist M. C. Escher. Why do you think the cube is called impossible? What is impossible about it?	

Shapes with three dimensions (length, width and height) are called **solids**.

A **polyhedron** is a solid composed of polygonal **faces** that connect along line segments called **edges**. Edges meet at a point, called a **vertex** (plural **vertices**).

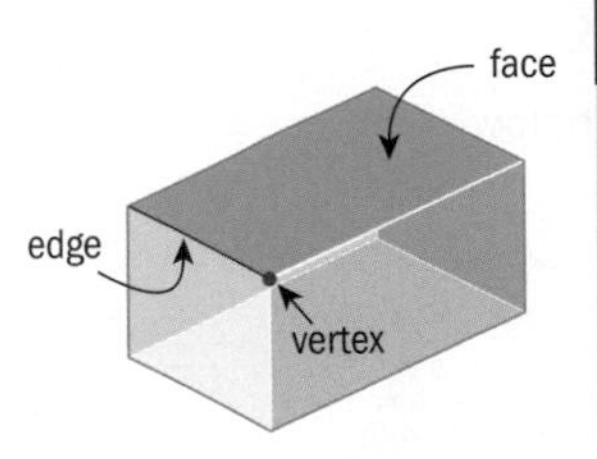

TOK

What are the platonic solids and why are they an important part of the language of mathematics?

Investigation 8

1 Six solids are given below.

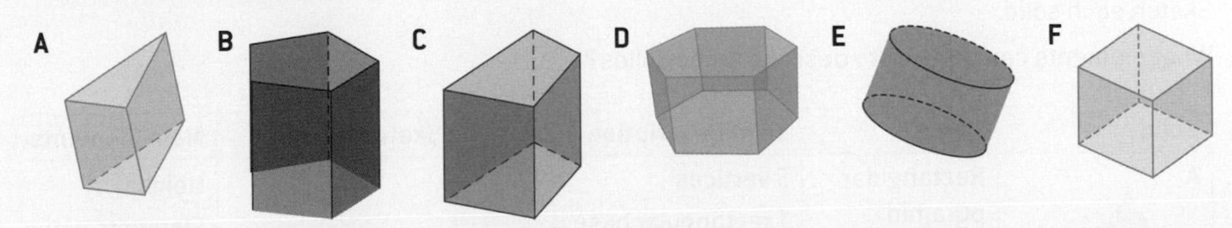

In your small group, compare the solids, discuss their similarities and differences and complete the table.

Sketch each solid.

What elements can you use to describe these solids?

Which of these solids are polyhedra? Are there solids that are not?

Continued on next page

Solid	Name	Solid description	Sketch of solid	Main elements
A	Triangular prism	6 vertices 2 equal triangular faces 3 rectangular faces 9 edges	C1, A1, B1, C, A, B	Height (h) Elements of the base: side and corresponding height, or two sides and an angle between

2 What are some real-life examples of the solids in the table?

A prism takes its name from the shape of its base.

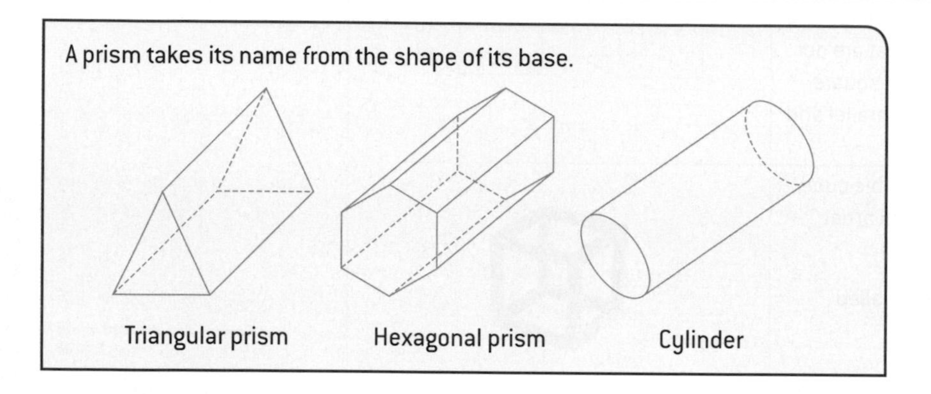

Triangular prism Hexagonal prism Cylinder

Investigation 9

1 Six solids are given below.

A 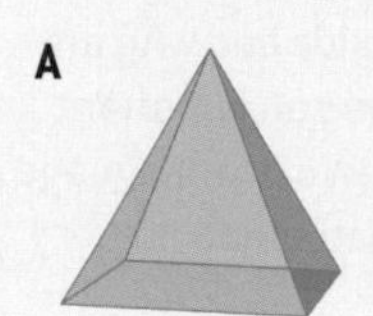 B 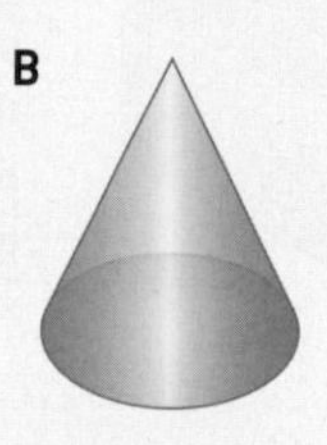 C 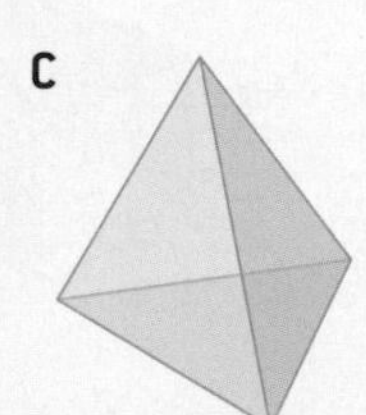 D 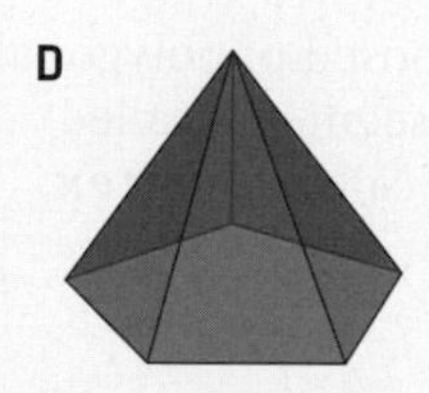 E 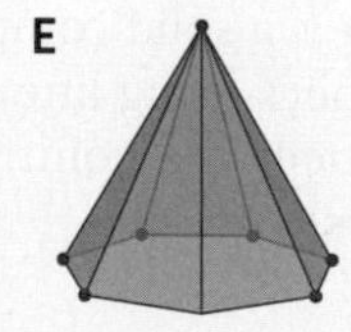 F 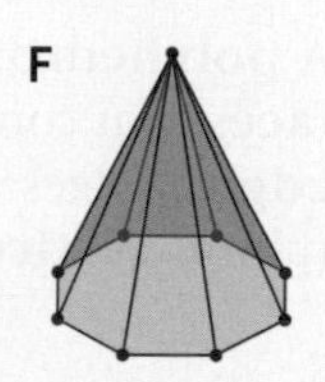

In your small group, compare the solids, discuss their similarities and differences and complete the table.

Sketch each solid.

What elements can you use to describe these solids?

Solid	Name	Solid Description	Sketch of Solid	Main Elements
A	Rectangular pyramid	5 vertices 1 rectangular base 5 triangular faces (the opposite faces congruent) 8 edges	E, D, C, A, B	Height (h) Elements of the base: length and width

2 What are some real-life examples of the solids in the table?

The base of a **pyramid** is a polygon, and the three or more triangular faces of the pyramid meet at a point known as the apex. In a right pyramid, the apex is vertically above the centre of the base.

The labelled figure below is a square-based pyramid, but there are also triangular-based pyramids (tetrahedrons), hexagonal-based pyramids and so on.

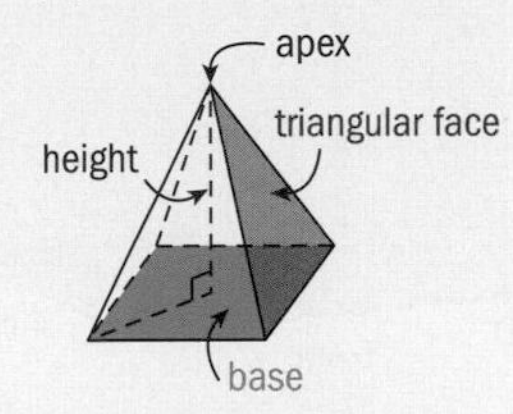

Square-based pyramid

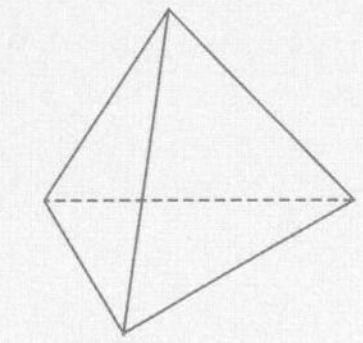

Tetrahedron

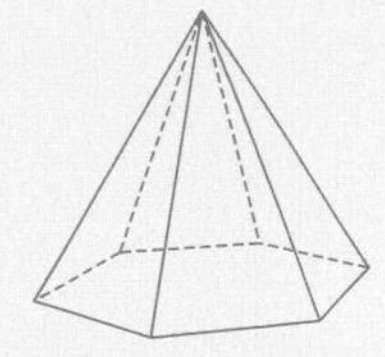

Hexagonal-based pyramid

There is a close relationship between pyramids and cones. The only difference is that the base of a cone is a circle.

A sphere is defined as the set of all points in three-dimensional space that are equidistant from a central point. Half of a sphere is called a hemisphere.

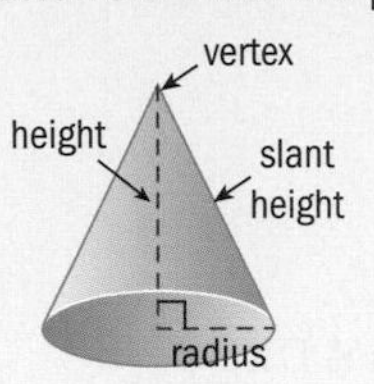

Cone

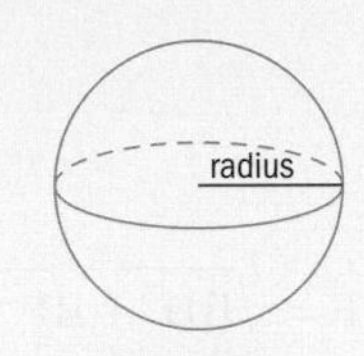

Sphere

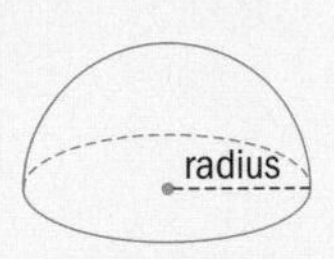

Hemisphere

International-mindedness

The stones used to make the pyramids in Egypt were constructed with a 13-knot rope replicating the shape of a 3-4-5 triangle to ensure sides meeting at right angles.

Example 17

Sketch a right rectangular pyramid.

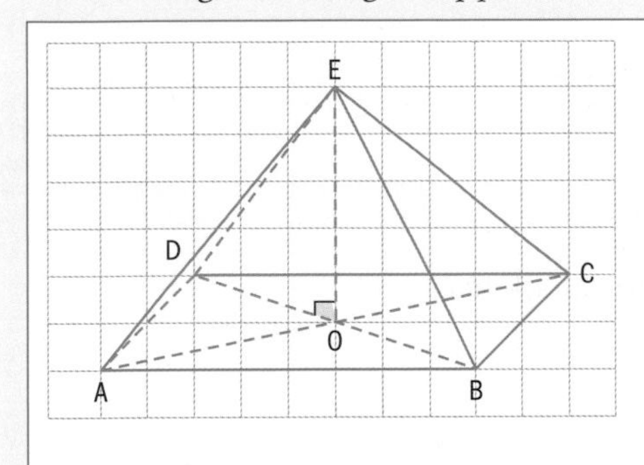

Use square paper if possible as it is easier to draw the parallel and congruent segments with equal length, such as [AB] and [DC], and [AD] and [BC].

Also note that all invisible edges, like [AD] and [DC], and other invisible segments, like [AC] and [BD], are drawn with dotted lines.

The height [EO] should be perpendicular to [AC] and [BD].

Label all vertices.

Exercise 2E

1 Sketch a cube with sides 3 cm.

2 Sketch a prism with length 5 cm and a regular hexagonal face with sides 3 cm and length 5 cm.

3 **a** Sketch a rectangular-based pyramid with base dimensions 3 cm by 4 cm and slant height 5 cm.

b Sketch a rectangular-based pyramid with base dimensions 3 cm by 4 cm and vertical height 5 cm.

Example 18

Cuboid ABCDEFGH has dimensions 4 cm, 5 cm and 6 cm as shown.

Find

a the length of the diagonal [DF]

b the size of $B\hat{D}F$.

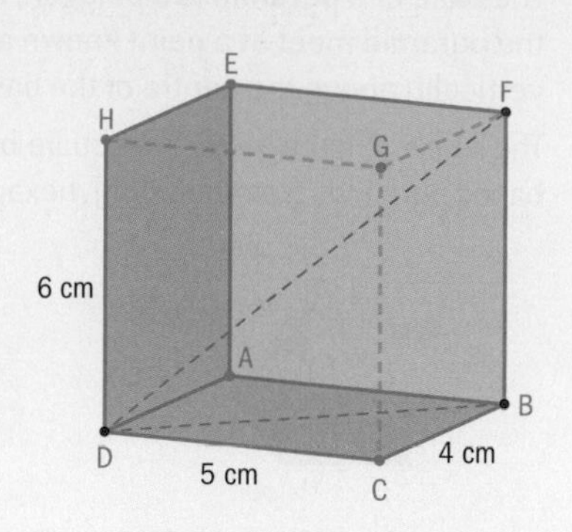

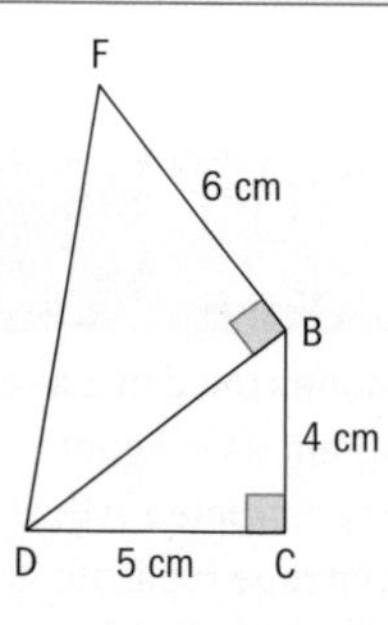

To find the length of [DF] you need to find right triangles that these segments are part of.

a $DF = \sqrt{CD^2 + BC^2 + BF^2}$

$= \sqrt{5^2 + 4^2 + 6^2}$

$= \sqrt{77}$

$= 8.77$ cm (3 sf)

$DF = \sqrt{BD^2 + BF^2}$ from ΔDBF

and $BD^2 = CD^2 + BC^2$ from ΔDBC

b $\sin B\hat{D}F = \frac{6}{\sqrt{77}}$

$B\hat{D}F = \sin^{-1}\left(\frac{6}{\sqrt{77}}\right) = 43.1°$ (3 sf)

$\sin B\hat{D}F = \frac{BF}{DF}$

You can use the Pythagorean theorem and trigonometry to calculate lengths and angles in 3D shapes.

Example 19

ABCDEFGH is a cuboid with dimensions 4 cm, 5 cm and 6 cm as shown.

M and N are the midpoints of [DH] and [BF] respectively.

Find $N\hat{E}M$.

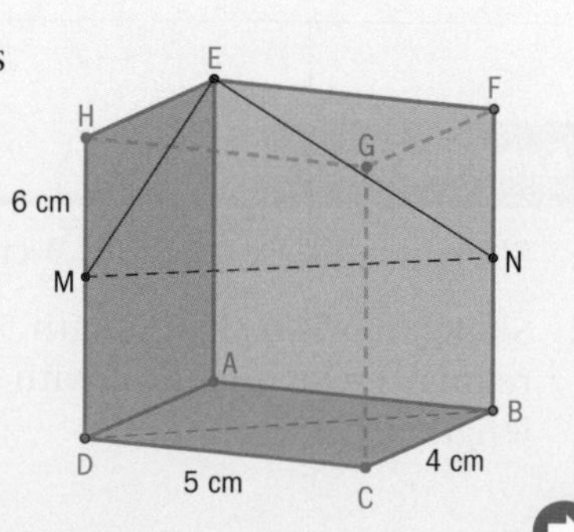

$\cos N\hat{E}M = \dfrac{EN^2 + EM^2 - MN^2}{2 \times EN \times EM}$	Consider ΔEMN. You can use the cosine rule to find $N\hat{E}M$.
$MN^2 = AB^2 + AD^2 = 41$	$MN = BD$
$EN^2 = EF^2 + FN^2 = 34$	[EN] is the hypothenuse of ΔEFN.
$EM^2 = EH^2 + HM^2 = 25$	[EM] is the hypothenuse of ΔEHM.
$N\hat{E}M = \cos^{-1}\left(\dfrac{34+25-41}{2\times\sqrt{34}\times\sqrt{25}}\right) = 72.0°$ (3 sf)	Substitute the lengths into the formula for the cosine rule.

Example 20

A rectangular pyramid ABCDE has a base with dimensions 8 cm and 5 cm and vertical height 10 cm. O is the centre of the base and is directly below the apex E.

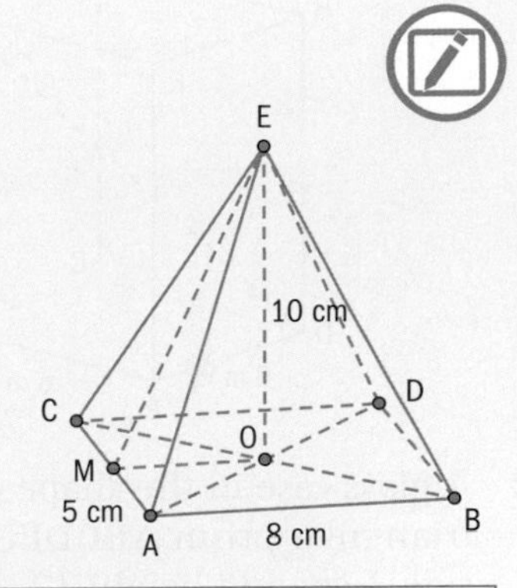

Find the size of the angle

a between the edge [BE] and the base ABDC of the pyramid

b between the edges [EB] and [EC]

c $E\hat{M}O$, where M is the midpoint of [AC].

a $\tan E\hat{B}O = \dfrac{OE}{OB}$	Since EO is the height of the pyramid, the angle between [EB] and the base is $E\hat{B}O$.... Consider the right triangle ΔEBO; you can use the tan ratio to find $E\hat{B}O$.
$OB = \dfrac{\sqrt{8^2+5^2}}{2} = \dfrac{\sqrt{89}}{2}$	O is the midpoint of [CB].
$E\hat{B}O = \tan^{-1}\left(\dfrac{10}{\frac{\sqrt{89}}{2}}\right) = 64.7°$ (3 sf)	
b $B\hat{E}C = 2\,B\hat{E}O$	The line [EO] bisects $B\hat{E}C$.
$B\hat{E}O = 180° - (90° + 64.7467...°) = 25.2532...°$	
$B\hat{E}C = 50.5°$ (3 sf)	
c $\tan E\hat{M}O = \dfrac{OE}{OM} = \dfrac{10}{4}$	M is the midpoint of [AC], so $OM = 4$ cm.
$E\hat{M}O = \tan^{-1}\left(\dfrac{10}{4}\right) = 68.2°$ (3 sf)	

Exercise 2F

1 A cuboid ABCDEFGH with dimensions 6 m by 6 m by 10.5 m is shown in the diagram below. Sketch the cuboid and mark the angles described below. Find:

a the length of [DF]

b the size of the angle between [DF] and base ABCD

c the size of the angle between [DF] and face DCGH.

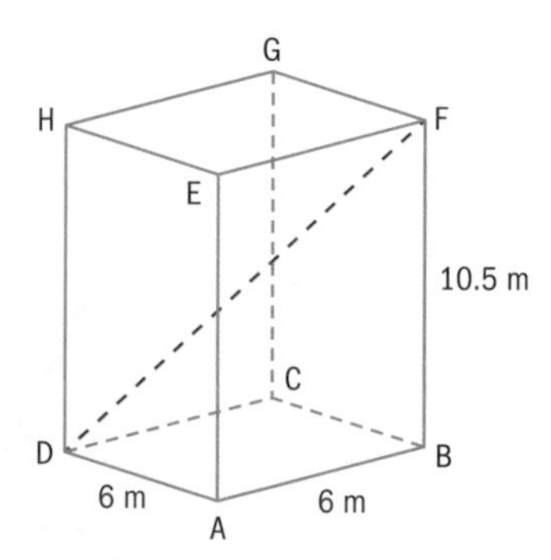

2 A glass case in the shape of a regular triangular prism ABCDEF has a base with side 3 cm and height 12 cm. H is the midpoint of [BC], G is the midpoint of [AB] and I is the midpoint of [EF]. Sketch the prism and mark the angles described below. The case is to enclose an art piece in the shape of a triangle attached at D, G and I.

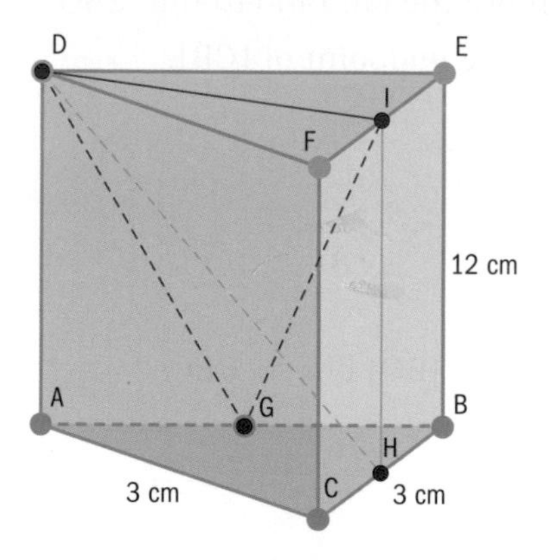

Find:

a the angle between [DH] and the base DEF

b the length of [GI]

c the length of [DG].

3 A heap of grain is shaped as a cone ADCF with height 5 m and base radius 2 m, as shown on the diagram. A and C are points on the circumference of the circular base of the cone and $A\hat{O}C = 120°$.

Sketch the cone and label the angles described below. Find:

a the angle between [AF] and the base of the cone

b the slant height of the heap

c the angle between [AF] and [CF].

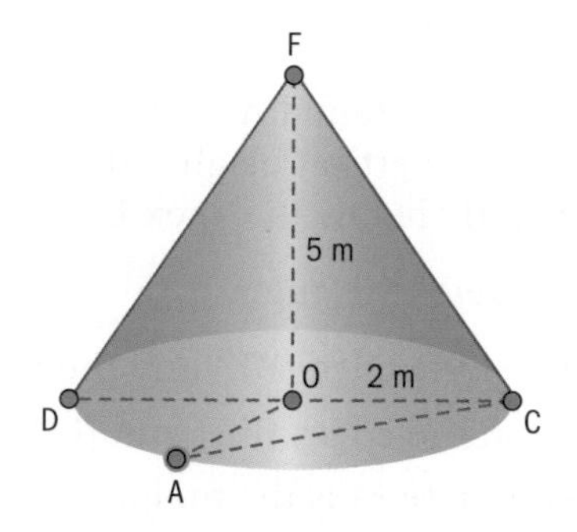

4 A cylinder with height 11 cm and radius 3 cm is shown in the diagram. D and O are the centres of the circular faces of the cylinder. A and C are two points on the circle with centre O, and $C\hat{A}O = 20°$. Point B is on the edge of the top face of the circle. The lines [AB] and [OD] are parallel.

Find:

a the length of [AC]

b the length of [BC]

c the angle between [BC] and the base with centre O.

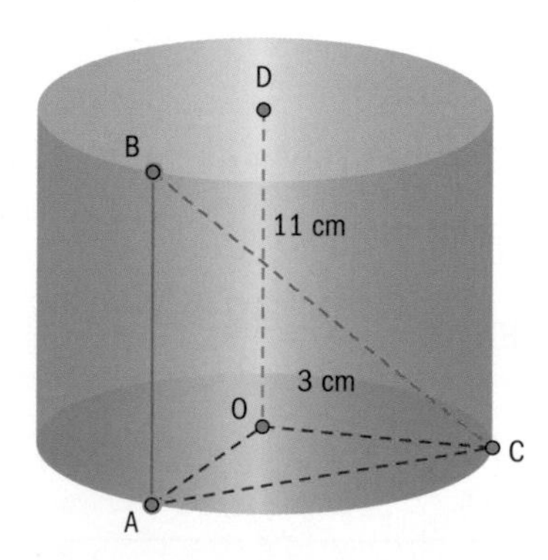

Nets of solids

Investigation 10

A net is a two-dimensional representation of a three-dimensional object.

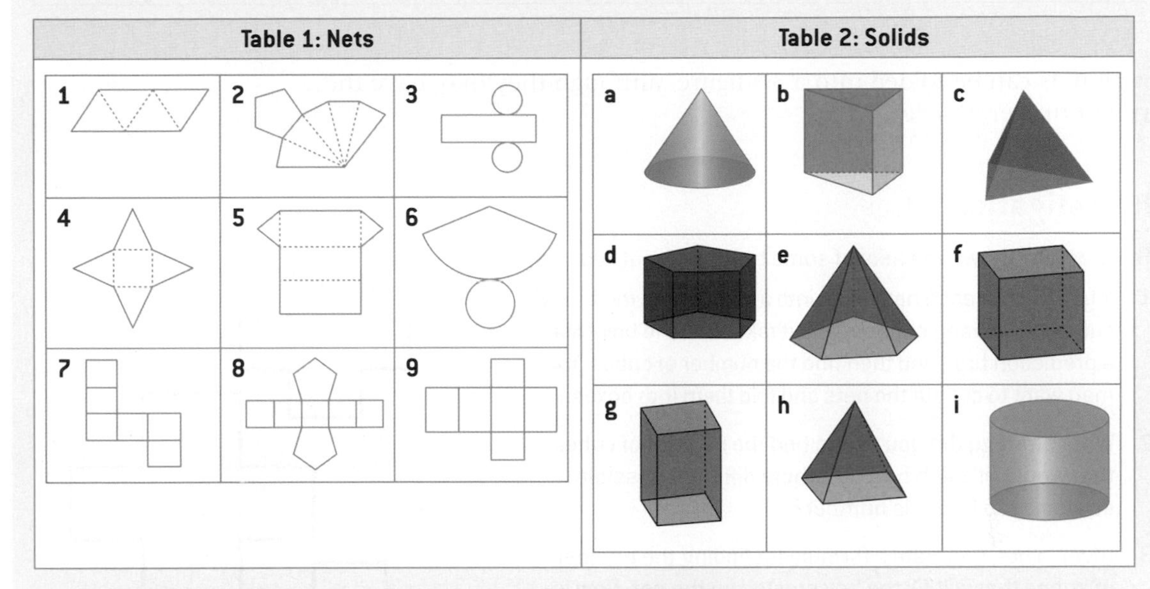

1. Decide which of the following nets in Table 1 are nets of 3D solids.
2. Match the nets with the corresponding solids from Table 2.
3. Name each solid.

Imagine making cuts along some edges of a solid and opening it up to form a flat figure. The flat (plane) figure is called a net of the solid. Each two-dimensional net of a solid can be folded into a three-dimensional solid, as in the diagram below.

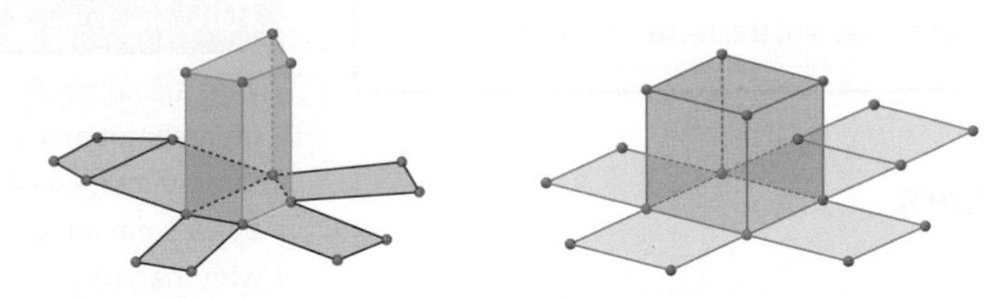

Example 21

State which of the following nets can make a cube.

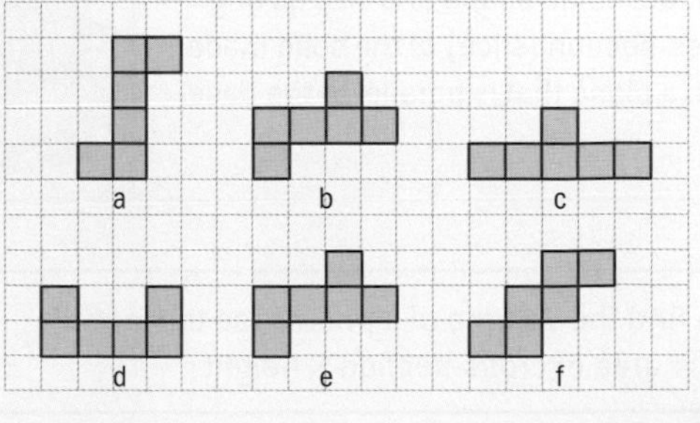

Continued on next page

The nets labelled a, b, e and f can be folded into a cube, whereas the ones labelled c and d cannot.	Check by cutting out the patterns and assembling them into a cube. Can you explain why the nets c and d cannot make a cube?

Not all nets can be folded into a 3D figure, although they may have the correct number of faces.

Investigation 11

The diagram shows the nets of some boxes without lids.

1. If you cut out each net, fold it into a box, and fill the box with cubes, how many cubes would it take to fill the box? Make a prediction first, and then find the number of cubes. You may want to cut out the nets and fold them into boxes.
2. What strategy did you use to find the number of cubes that would fill each box? Are there different possible strategies to find this number?
3. Given a net, generate a formula for finding the number of cubes that will fill the box created by the net. How is your generalization related to the volume formula for a rectangular prism (V = length × width × height)?
4. Imagine another box that holds twice as many cubes as Box A. What are the possible dimensions of this new box with a doubled volume?

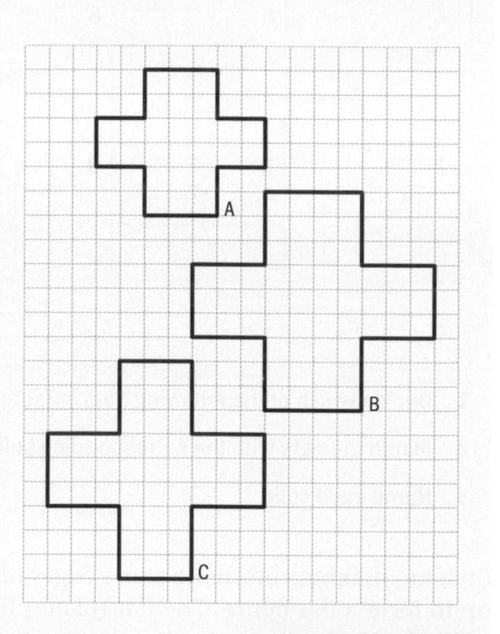

Volume of a cuboid = length × width × height = area of the base × height

Volume of prisms and cylinders

For some solids, called prisms, the base is of the same shape and size as any cross-section (slice) of the solid made with a plane that is parallel to the base.	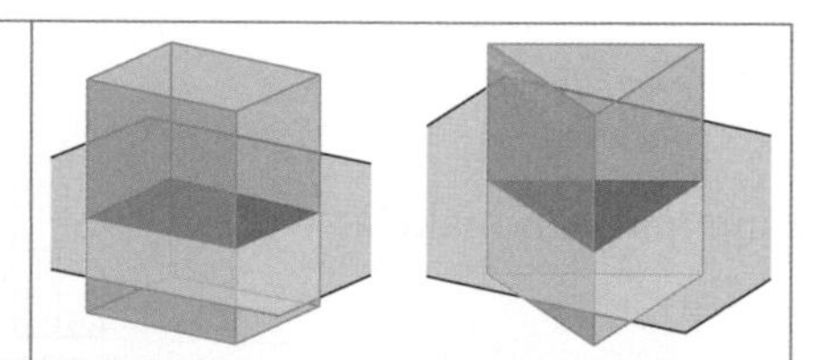

To find the volume of a prism, use the formula
V = area of cross-section × height

TOK

Why are symbolic representations of three-dimensional objects easier to deal with than physical models?

What does this tell us about our knowledge of mathematics in other dimensions?

This method can be used to find the volume of all prisms whether rectangular, triangular, hexagonal, octagonal or having irregularly shaped bases with congruent cross-sections.

Example 22

Babreka (The Kidney) is a lake in the Rila mountains, Bulgaria, with the shape of a kidney. Its area is 85 000 m^2, and the average depth is 28 m. Estimate the volume of water in the lake, in m^3.

V = area of cross-section × height $V = 85\,000 \times 28$ $V = 2\,380\,000$ m^3	We assume that the water in the lake has a 3D shape with uniform cross-sections that have the shown kidney shape. The depth of the lake is the same as the height of the water.

A cylinder is a special case of a prism, with a circular cross-section.

The volume of a cylinder where the radius of the circular cross-section is r and the height is h is $\pi r^2 h$.

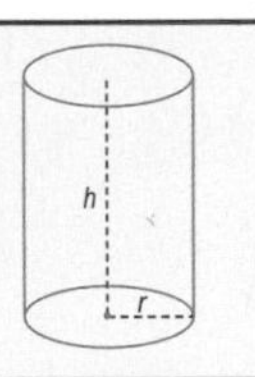

Example 23

Calculate the volume of the cylinder with radius 2 cm and height 7.5 cm.

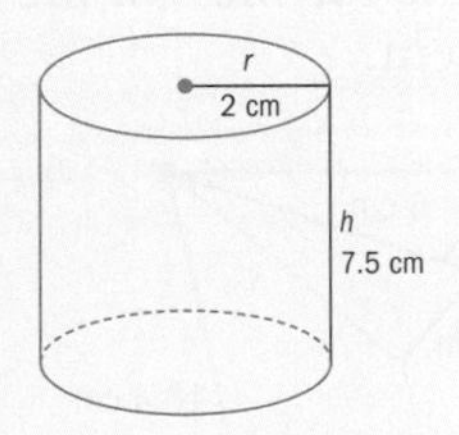

Volume = $\pi \times 2^2 \times 7.5 = 94.2$ cm^3 (3 sf)	Area of cross-section = πr^2.

Example 24

Find the volume of the triangular prism whose base is an isosceles triangle where the equal sides are 12 cm and the angle between them is 130°. The height of the prism is 15 cm.

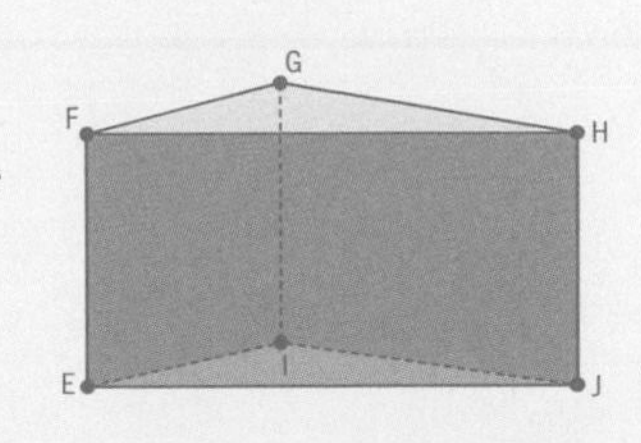

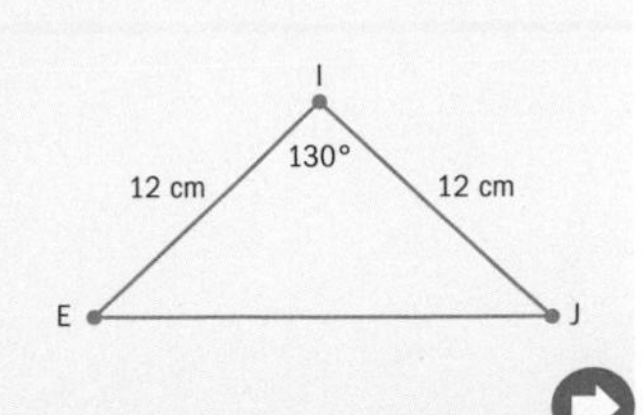

Continued on next page

$V = \frac{12 \times 12 \times \sin 130°}{2} \times 15$ $= 827 \text{ cm}^3$ (3 sf)	To find the volume you need to find the area of the triangular base, for which you can use the area formula. Remember that the volume is measured in cubic units.

Exercise 2G

1 A swimming pool with the dimensions shown is filled with water. The cost to fill the pool is \$1.50 per cubic metre of water. Find the cost of filling the pool.

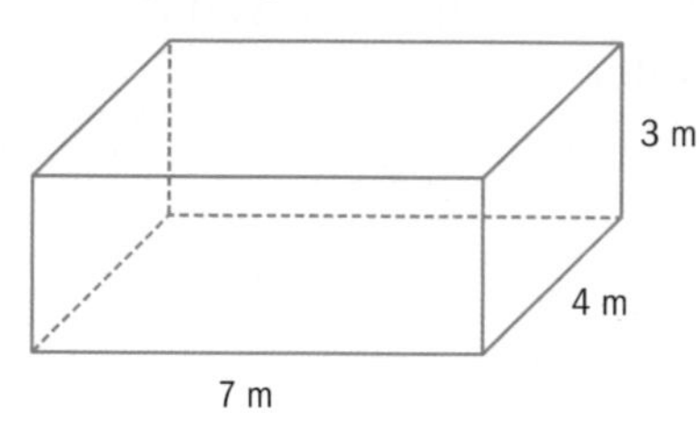

2 A ornament is made in the shape of a triangular prism. The ornament is made from mahogany, which has a density of 0.71 g/cm³. Calculate the mass of the ornament.

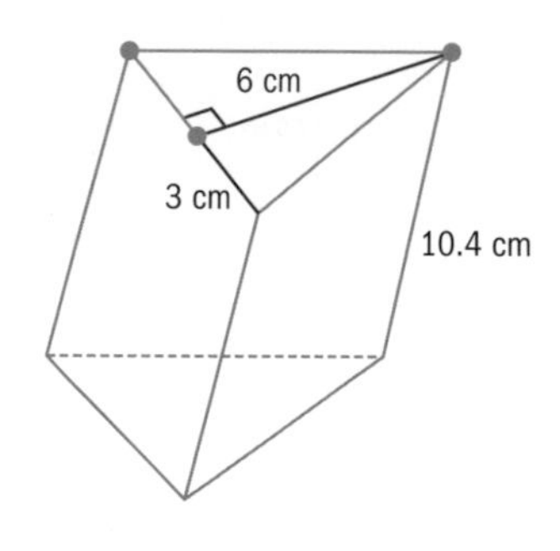

3 A solid metal cylinder has the following dimensions.

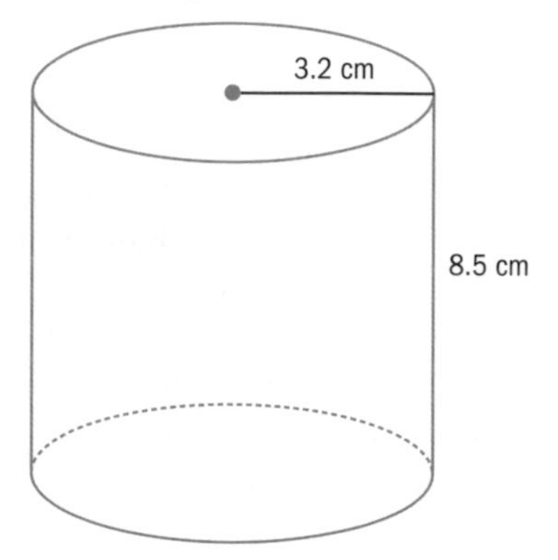

The cylinder is melted down into 2 cm cubes.

How many cubes can be made?

4 Nasim fills a measuring jug with 310 cm³ of water. She pours the water into a cylindrical vase with radius 4 cm. Find the height of water in cm.

5 The volume of a regular hexagonal prism is 2800 cm³. The height of the prism is 14 cm. Find the side of the hexagonal base in cm.

TOK

How is mathematical knowledge considered from a sociocultural perspective?

Volume of pyramids, cones and spheres

Investigation 12

1 Comparing the volume of prisms and pyramids

Take a plastic prism and a plastic pyramid with the same height and the same base. Fill the pyramid with water and pour into the prism. Repeat until the prism is filled to the top.

a What is the relationship between the volume of a prism and the volume of a pyramid with the same base area and height? Make a conjecture about the formula for the volume of a pyramid. Write it down.

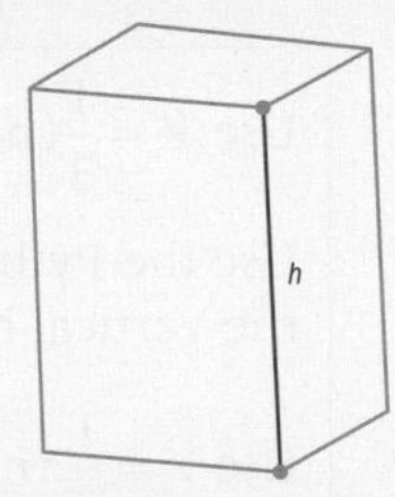

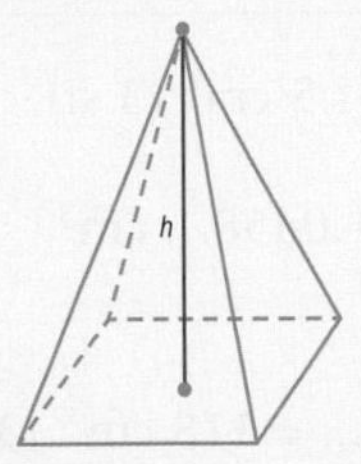

2 Comparing the volume of cylinders, cones and spheres

Take a plastic cone, a plastic sphere and a plastic cylinder with the same height and radius. If we take a close look at the sphere , we can see that $h = 2r$. Using water (or rice or popcorn), experiment with filling the 3D shapes to determine the relationships between their volumes.

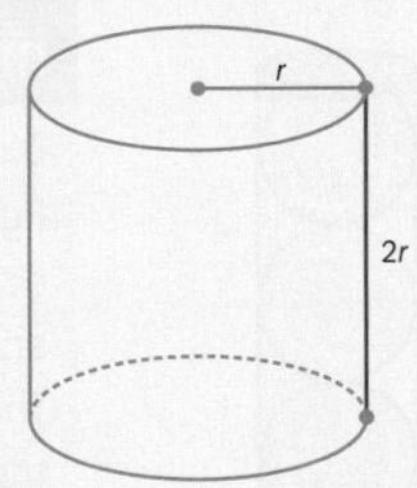

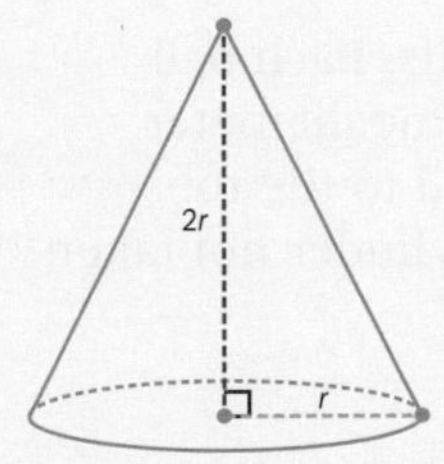

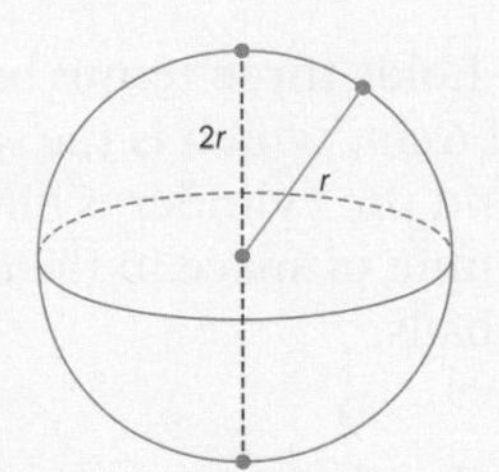

a What is the relationship between the volume of a cylinder and the volume of a cone with the same radius and height? Make a conjecture about the formula for the volume of a cone. Write it down.

b What is the relationship between the volume of a cylinder and the volume of a sphere with the same radius and height? Make a conjecture about the formula for the volume of a sphere. Write it down.

c Did you find another relationship between the volumes of two of the solids? Make a conjecture and write it down.

Shape	Pyramid	Prism	Cone	Cylinder	Sphere
Volume	$\frac{1}{3}(\text{base area} \times h)$	$\text{base area} \times h$	$\frac{1}{3}\pi r^2 h$	$\pi r^2 \times h$	$\frac{4}{3}\pi r^3$

- The volume of a pyramid is a third of the volume of a prism with the same base and height.
- The volume of a cone is a third of the volume of a cylinder with the same radius and height. The volume of the cone is also the same as a pyramid with the same base area and height.
- The volume of a sphere is four times the volume of a cone with the same radius and with height that is twice the radius.

Example 25

Calculate the volume of each solid.

a

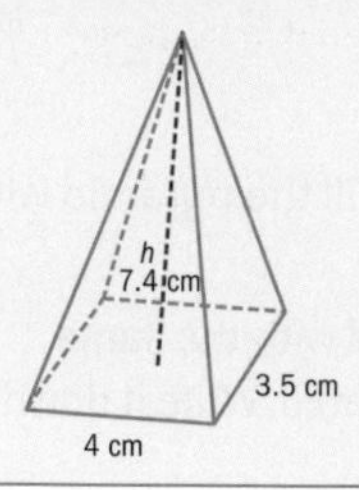

b

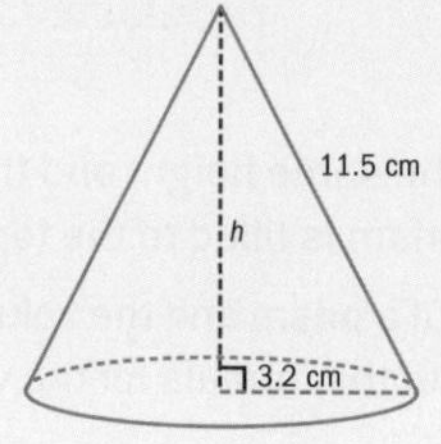

a $V = \frac{1}{3}(4 \times 3.5 \times 7.4) = 34.5\text{ cm}^3$ (3 sf)	Use $V = \frac{1}{3}(\text{base area} \times \text{height})$
b height = $11.5^2 - 3.2^2 = 11.0458\ldots\text{ cm}^3$	Use the Pythagorean theorem to find the vertical height of the cone.
$V = \frac{1}{3}\pi \times 3.2^2 \times 11.0458\ldots = 118\text{ cm}^3$ (3 sf)	Use $V = \frac{1}{3}\pi r^2 h$ to find the volume.

Example 26

A cylindrical can holds three tennis balls. Each ball has a diameter of 6 cm, which is the same diameter as the cylinder, and the cylinder is filled to the top. Calculate the volume of space in the cylinder not taken up by the tennis balls.

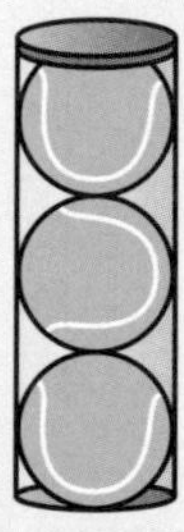

Volume of cylinder = $\pi \times 3^2 \times 18 = 508.9\text{ cm}^3$ Sphere: radius = 3 cm Volume of 1 ball = $\frac{4}{3}\pi \times 3^3 = 113.1\text{ cm}^3$ Volume of 3 balls = 339.3 cm^3 Space = $508.9 - 339.3 = 170\text{ cm}^3$ (3 sf)	The cylinder has radius 3 cm and height 18 cm.

International-mindedness

Diagrams of the Pythagorean theorem occur in early Chinese and Indian manuscripts. The earliest references to trigonometry are in Indian mathematics.

Exercise 2H

1 Calculate the volume of:

a a cylinder with radius 3 cm and height 23 cm

b a sphere with radius 2 cm

c a cone with radius 2.1 cm and height 7.3 cm

d a hemisphere with radius 3.1 cm

e a rectangular pyramid with length 8 cm, width 5 cm and vertical height 17 cm.

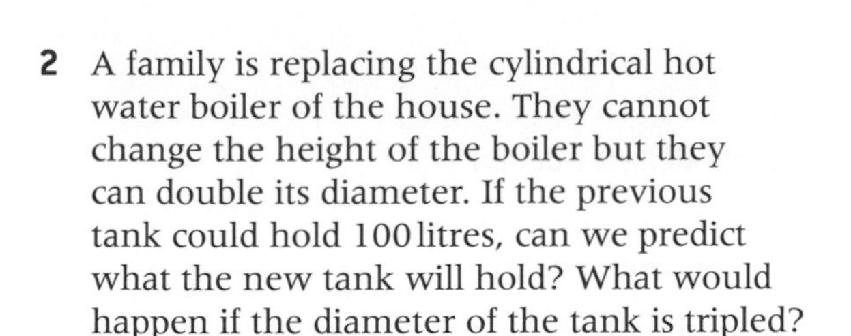

2 A family is replacing the cylindrical hot water boiler of the house. They cannot change the height of the boiler but they can double its diameter. If the previous tank could hold 100 litres, can we predict what the new tank will hold? What would happen if the diameter of the tank is tripled?

3 Palabora Mine, Phalaborwa, is a South African copper mine that is nearly 2000 m wide. A total of 4.1 million tonnes of copper have been extracted from this mine.

Imagine the total output of pure copper was compacted into the shape of a sphere. The size of this sphere would be much smaller than the total amount of material extracted from the mine. Find the radius of the copper sphere, given that 1 m^3 of copper weighs about 8930 kg.

Surface area of solids

Investigation 13

Surface area of a prism

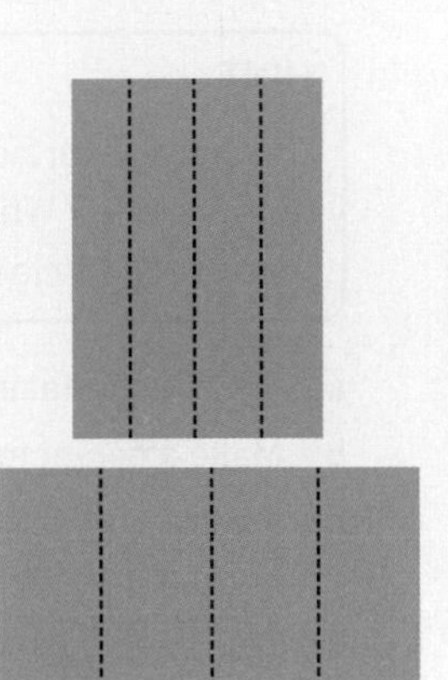

1 Take two standard sheets of paper for printing.

Fold one of then along the longer side so that you make four congruent rectangles, and fold the other one along the shorter side so that you make four congruent rectangles. Fold each sheet to make two open-ended prisms and use tape to connect the edges.

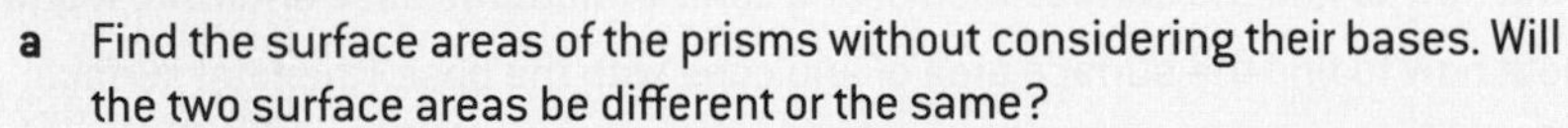

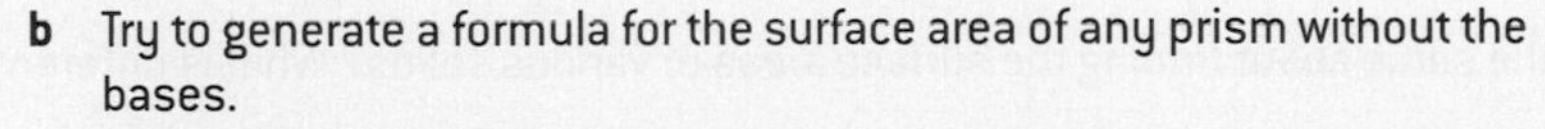

- **a** Find the surface areas of the prisms without considering their bases. Will the two surface areas be different or the same?
- **b** Try to generate a formula for the surface area of any prism without the bases.
- **c** What would be the formula for the surface area of any prism with the bases (the total surface area)?

Surface area of a cylinder

2 Take two standard sheets of paper for printing. Bend one of then along the longer side so that you make a cylinder, and bend the other one along the shorter side so that you make another cylinder. Use tape to connect the edges.

- **a** Find the surface areas of the cylinders without the areas of the bases. Are the curved surface areas the same or different? Make a conjecture about how the two surface areas compare and write it down.
- **b** Make a conjecture about how to find the surface area of any cylinder, with radius r and height h, without the bases.
- **c** Make a conjecture about how to find the surface area of any cylinder, with radius r and height h, with the bases.

Continued on next page

Surface area of a pyramid

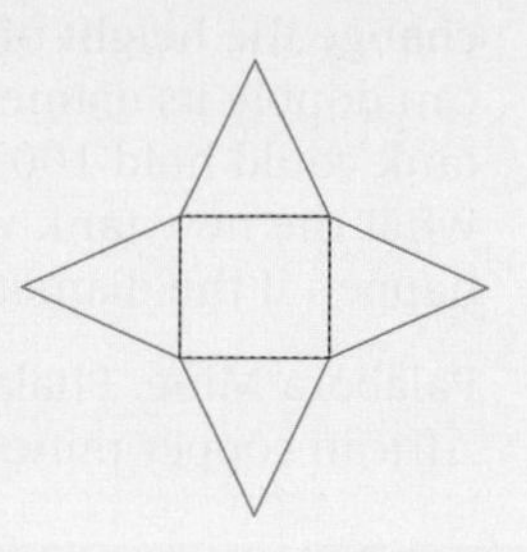

3 A net of a rectangular pyramid is given (with the base). Fold along the dotted lines and use tape to connect the edges.

- **a** Find the surface area of the pyramid without the base and write it down.
- **b** Make a conjecture about how to find the surface area of any pyramid without the base and write it down.
- **c** Make a conjecture about how to find the surface area of any pyramid with the base (the total surface area) and write it down.

Surface area of a cone

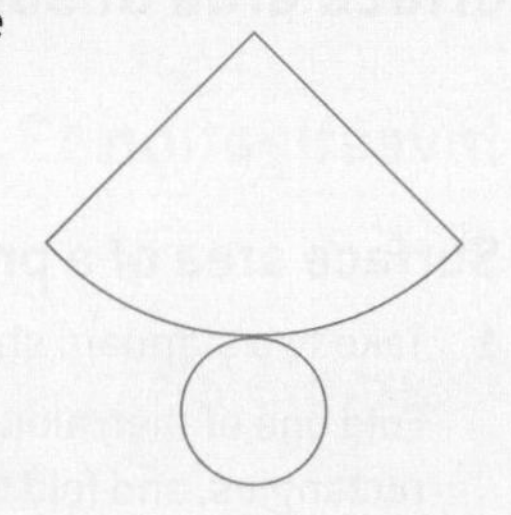

4 A net of a cone is given (with the base). Fold the net to make a cone and use tape to connect the edges.

> **HINT**
>
> The curved surface of a cone is a sector of a circle. What is the radius of this circle? What is the circumference of this circle? What is the area of the circle? Can we find the area of the circle sector?

- **a** Find the surface area of the cone without the base and write it down.
- **b** Make a conjecture about how to find the surface area of any cone without the base and write it down.
- **c** Make a conjecture about how to find the surface area of any cone with the base (the total surface area) and write it down.
- **d** **Conceptual** What is the same about finding the surface areas of various solids? What is different?

Surface area is measured in square units, eg cm^2 and m^2.

To calculate the surface area of a cylinder, open out the curved surface into a rectangle:

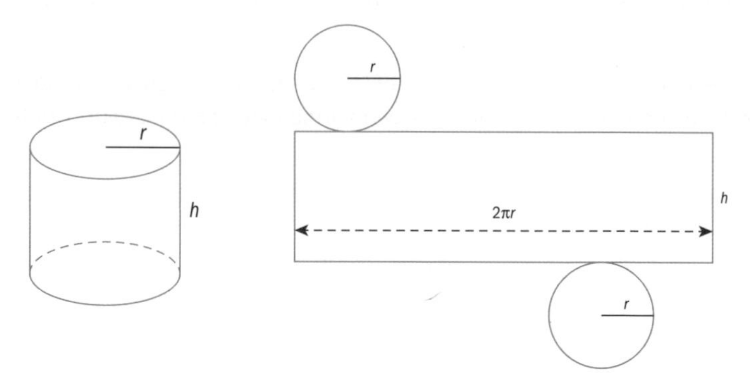

To find the curved surface area (CSA) of a cylinder use the formula $\text{CSA} = 2\pi rh$.

To find the total surface area of a cylinder, find the curved surface area and add on the areas of the two bases:

Total surface area $= 2\pi rh + 2\pi r^2$

The formula for the surface area (SA) of a sphere is

$SA = 4\pi r^2$

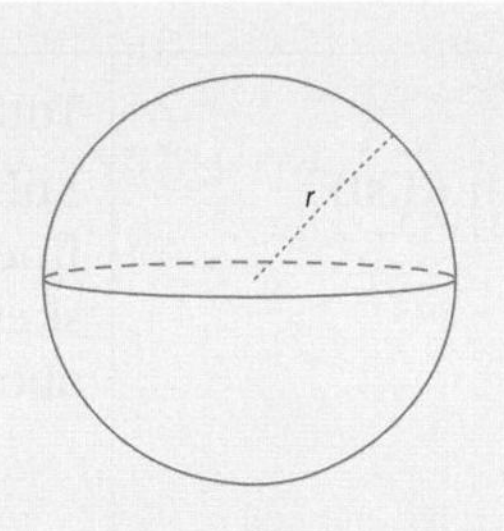

The curved surface area of a cone uses the length of the slanted height l:

$CSA = \pi rl$

To find the total surface area of the cone, add the area of the circular base:

$SA = \pi rl + \pi r^2$

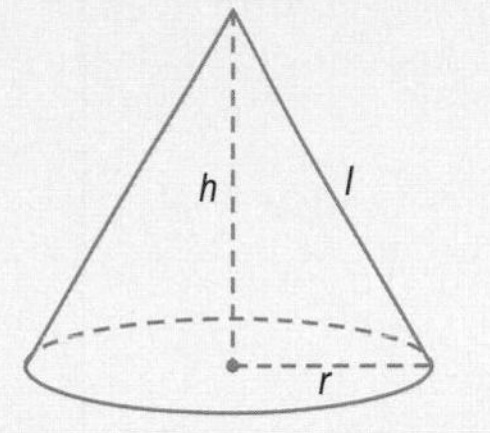

To find the total surface area of a pyramid, add together the areas of all the faces.

Example 27

Find the total surface area of a rectangular-based pyramid whose base has dimensions 10 m and 8 m, and where the slant heights for the faces ABE and BCE are 12 m and 13 m, respectively.

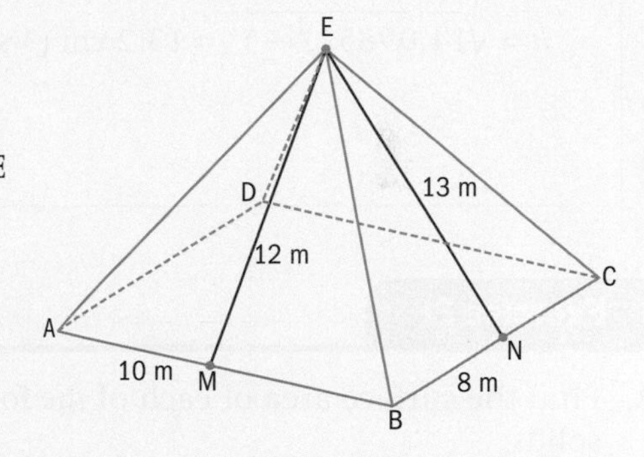

Base area $= 10 \times 8 = 80$ ABE area $= \dfrac{10 \times 12}{2} = 60$ BCE area $= \dfrac{8 \times 13}{2} = 52$ Total surface area $= 80 + 2 \times 60 + 2 \times 52$ $= 304\text{ m}^2$	To find the total surface area of the pyramid you need to add the areas of all faces. The pyramid has two pairs of congruent faces: ABE and DCE; BCE and ADE.

Example 28

A cone has radius 5 cm and a total surface area of 300 cm^2, rounded to the nearest integer.

Find:

a the slant height, l, of the cone

b the height, h, of the cone.

Continued on next page

a $300 = \pi \times 5 \times l + \pi \times 5^2$ $l = 14.0985... = 14.1$ cm (3 sf)	Total surface area $= 2\pi rh + 2\pi r^2$ Substitute the given values and then use your GDC to find the slant height, l. If you are not given a diagram, sketch the cone for yourself so that you can reason about what is given and what you need to find. 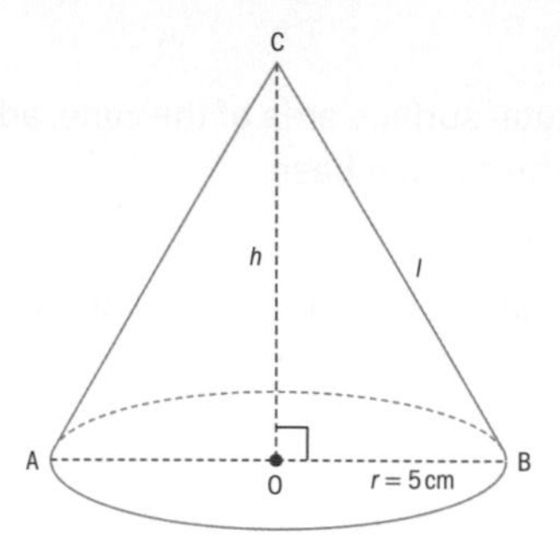
b $h^2 = l^2 - r^2$ $h = \sqrt{14.0985...^2 - 5^2} = 13.2$ cm (3 sf)	Now that you know the radius and the slant height you can find the cone height, h, from the triangle OBC by using the Pythagorean theorem. Remember to use the unrounded value for l when you calculate the value of h.

Exercise 2I

1 Find the surface area of each of the following solids:

a a cylinder with radius 2.5 cm and height 7.3 cm

b a cone with radius 3.5 cm and height 12 cm.

2 Find the surface area of the following solids:

a

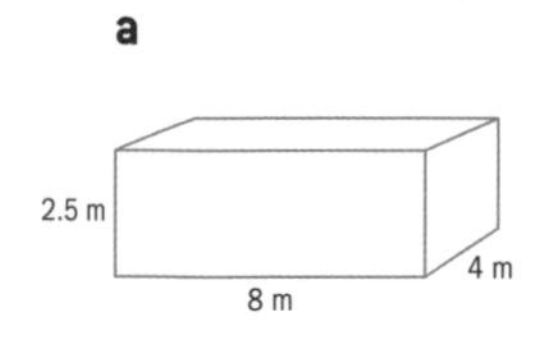

b

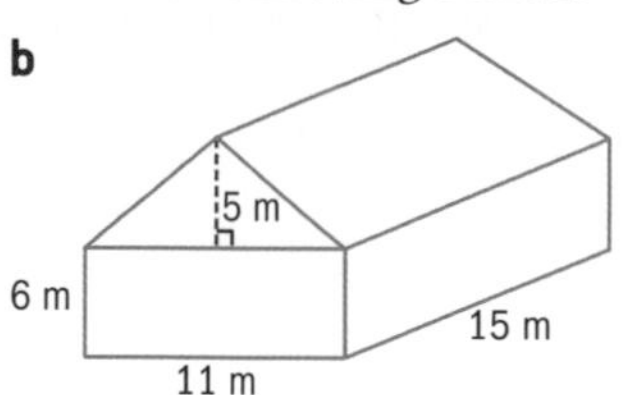

c

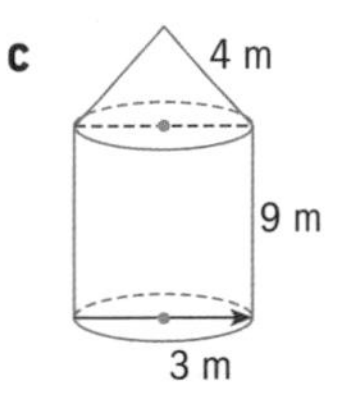

3 Find the surface area of a hemisphere with radius 4 cm.

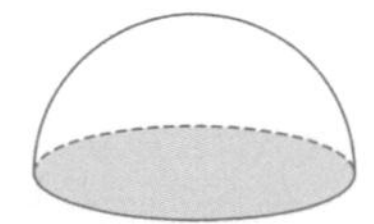

4 A silo has a cylindrical part and a roof that is a hemisphere. The radius of the cylinder is 3 m and its height is 12 m.

a Find the volume of the silo.

b The entire silo is to be painted. Find how much paint is needed if 1 L of paint covers 8.5 m² of surface.

Developing your toolkit

Now do the Modelling and investigation activity on page 92.

Chapter summary

The sine rule

- For any ΔABC, where a is the length of the side opposite $\hat{A}$, b is the length opposite $\hat{B}$ and c is the length of the side opposite $\hat{C}$:

  $\frac{a}{\sin \hat{A}} = \frac{b}{\sin \hat{B}} = \frac{c}{\sin \hat{C}}$ or $\frac{\sin \hat{A}}{a} = \frac{\sin \hat{B}}{b} = \frac{\sin \hat{C}}{c}$

Cosine rule

- For ΔABC, where a is the length of the side opposite $\hat{A}$, b is the length of the side opposite $\hat{B}$ and c is the length of the side opposite $\hat{C}$:

  $a^2 = b^2 + c^2 - 2bc \cos \hat{A}$

  $b^2 = a^2 + c^2 - 2ac \cos \hat{B}$

  $c^2 = a^2 + b^2 - 2ab \cos \hat{C}$

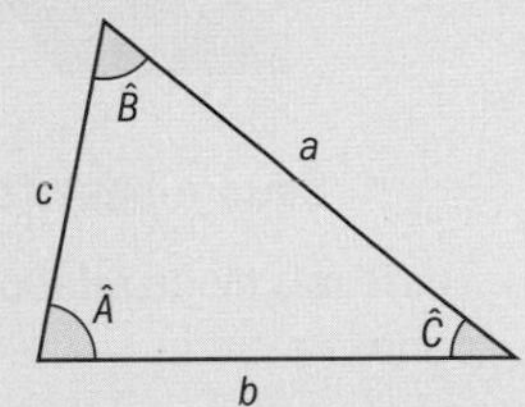

Triangle area formula

- The area of any triangle ΔABC is given by the formulae

  $\text{area} = \frac{1}{2} bc \sin \hat{A}$ or $\text{area} = \frac{1}{2} ac \sin \hat{B}$ or $\text{area} = \frac{1}{2} ab \sin \hat{C}$

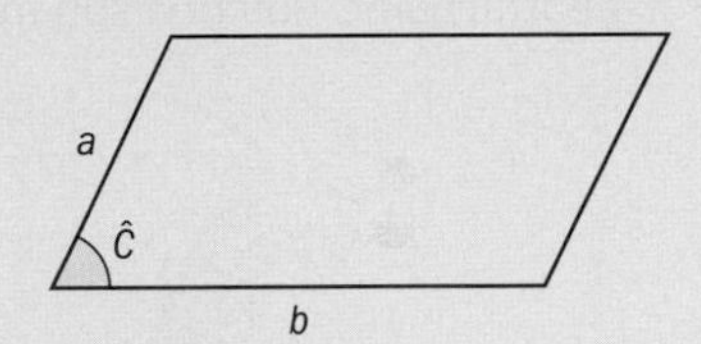

Parallelogram area formula

- The area of any parallelogram ABCD is given by the formula $\text{area} = ab \sin \hat{C}$

Sector

- A sector is a portion of a circle lying between two radii and the subtended arc.

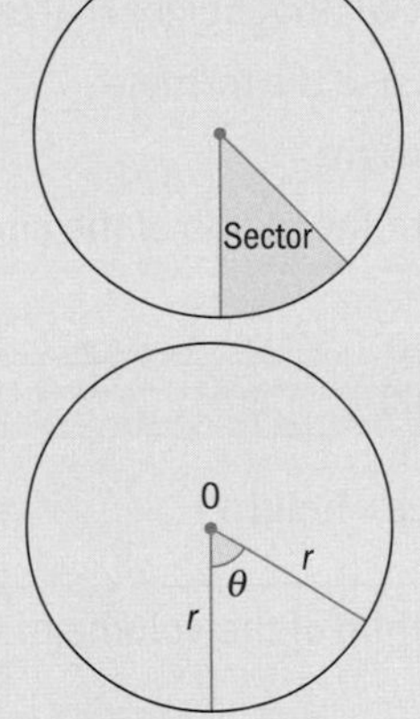

Sector area formula

- $\text{Area of sector} = \frac{\theta \pi r^2}{360}$

where r is the radius of the circle and θ is the central angle in degrees.

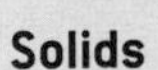

Solids

- A prism is a solid shape that has the same shape or cross-section all along its length.
- A prism takes its name from the shape of its cross-section

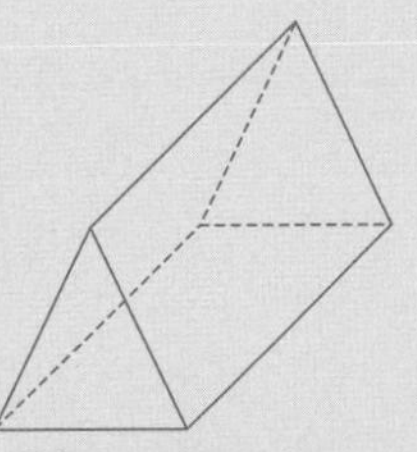
Triangular prism

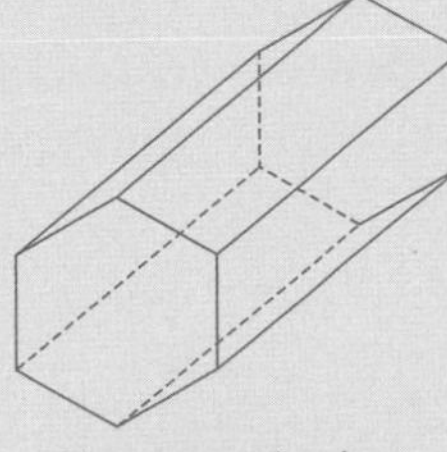
Hexagonal prism

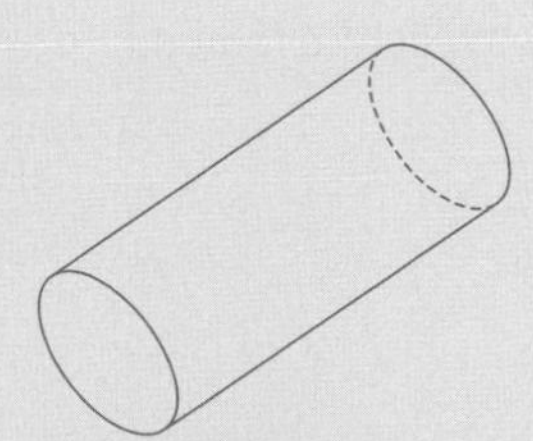
Cylinder

Continued on next page

- The base of a **pyramid** is a polygon, and the three or more triangular faces of the pyramid meet at a point known as the apex. In a right pyramid, the apex is vertically above the centre of the base.
- The figures below are examples of a square-based pyramid, a tetrahedron (triangular-based pyramid) and a hexagonal-based pyramid.

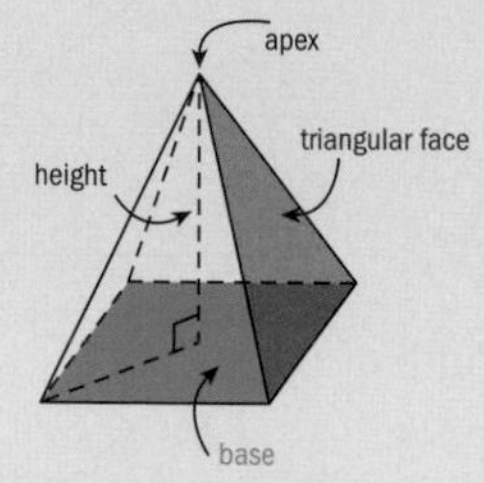

Square-based pyramid

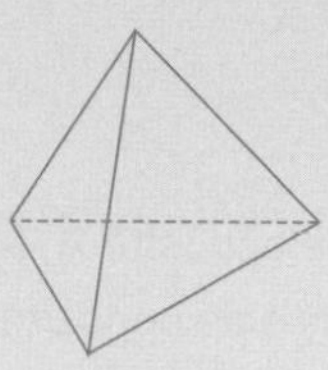

Tetrahedron

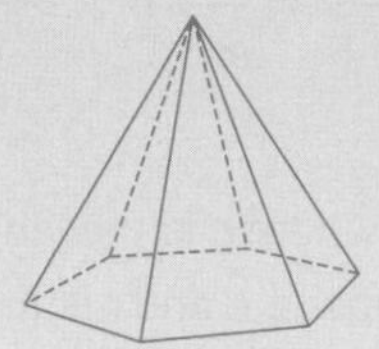

Hexagonal-based pyramid

- There is a close relationship between pyramids and cones. The only difference is that the base of a cone is a circle.
- A sphere is defined as the set of all points in three-dimensional space that are equidistant from a central point. Half of a sphere is called a hemisphere.

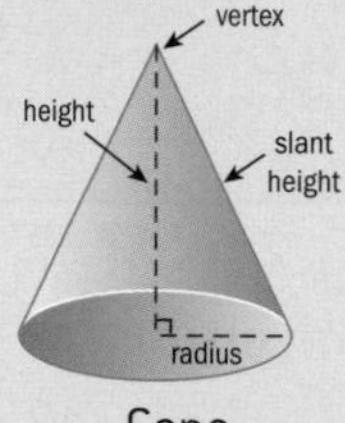

Cone

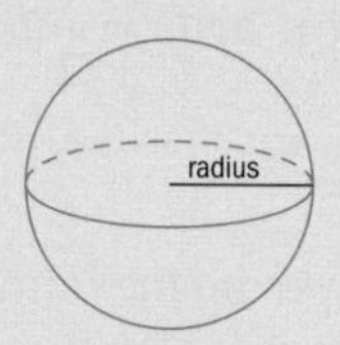

Sphere

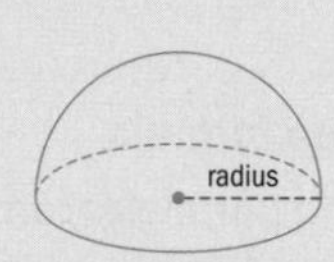

Hemisphere

- Volume of a cuboid = length × width × height = area of the base × height
- To find the volume of a prism, use the formula
- $V =$ area of cross-section × height
- The volume of a cylinder where the radius of the circular cross-section is r and the height is h is $\pi r^2 h$

Shape	Pyramid	Cone	Sphere
Volume	$V = \frac{1}{3}(\text{base area} \times \text{height})$	$V = \frac{1}{3}\pi r^2 h$	$V = \frac{4}{3}\pi r^3$

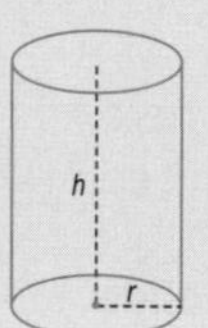

- The volume of a cone is a third of the volume of a cylinder with the same radius and height.
- The volume of a cone is the same as a pyramid with the same base area and height.
- The volume of a sphere is four times the volume of a cone with the same radius and with height that is twice the radius.

- To find the curved surface area of a cylinder use the formula $\text{CSA} = 2\pi rh$.
- To find the total surface area of a cylinder, find the curved surface area and add on the areas of the two circular ends:

  Total surface area $= 2\pi rh + 2\pi r^2$
- The formula for the surface area of a sphere is $\text{SA} = 4\pi r^2$

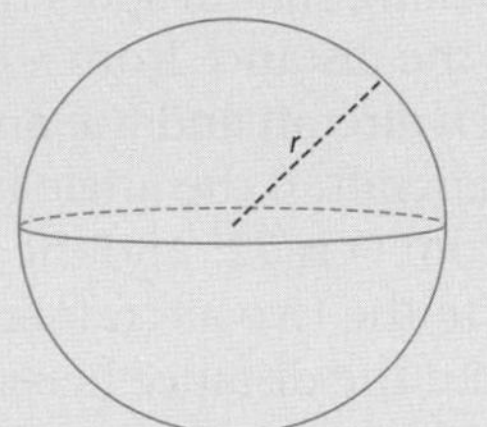

- The curved surface area of a cone uses the length of the slanted height l:

  $\text{CSA} = \pi rl$

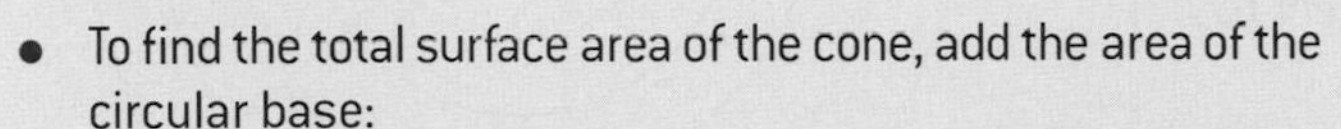

- To find the total surface area of the cone, add the area of the circular base:

  $\text{SA} = \pi rl + \pi r^2$
- To find the total surface area of a pyramid, add together the areas of all the faces.

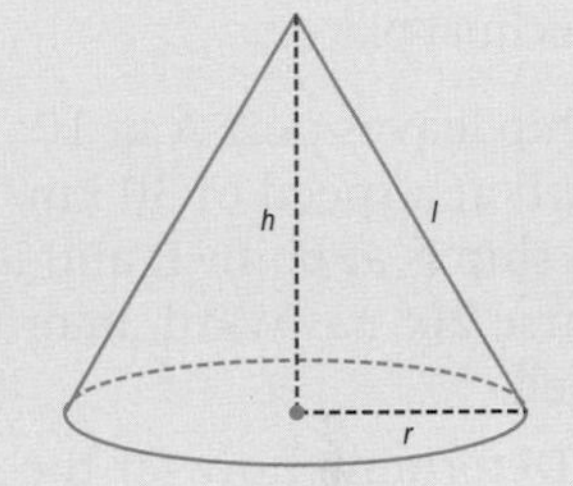

Developing inquiry skills

Look back at the opening problem. The radius of Mount Everest is approximately 16 km, and the average snow depth is approximately 4 m. Estimate the amount of snow on Mount Everest.

Chapter review

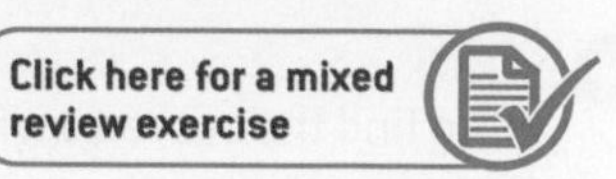

1 Find the area of a triangular area having two sides of lengths 90 m and 65 m and an included angle of 105°.

2 An aircraft tracking station spots two aircraft. It determines the distance from a common point O to each aircraft and the angle between the aircraft. If the angle between the two aircraft from O is 52° and the distances from point O to the two aircraft are 58 km and 75 km, find the distance between the two aircraft. Write down your answer correct to 1 decimal place.

3 A ship leaves port A at 10am travelling north at a speed of 30 km/hour. At 12pm, the ship is at point B and it adjusts its course 20° eastward, maintaining the same speed.

a Determine how far the ship is from the port at 1pm when it is at point C. Write down your answer correct to the nearest integer.

b Determine the angle CÂB.

4 A sector of a circle has a radius of 5 cm and central angle measuring 45°. Find the area of the sector and the length of the arc.

5 A circular sector has radius 4 cm and a corresponding arc length of 1.396 cm rounded to the nearest thousandth of a centimetre. Find the area of the sector.

6 Shown in the diagram are six concentric circles with centre O, where OA = 10 cm, OB = 20 cm, OC = 30 cm, OD = 40 cm, OE = 50 cm and OF = 60 cm.

Find:

a the length of arc AG

b the length of arc BH

c the length of arc CI

d Describe the relationship between the lengths of arcs AG, BH and CI, and find the lengths of arcs [DJ] and [EK] without calculating.

e Determine the area of sector OFL.

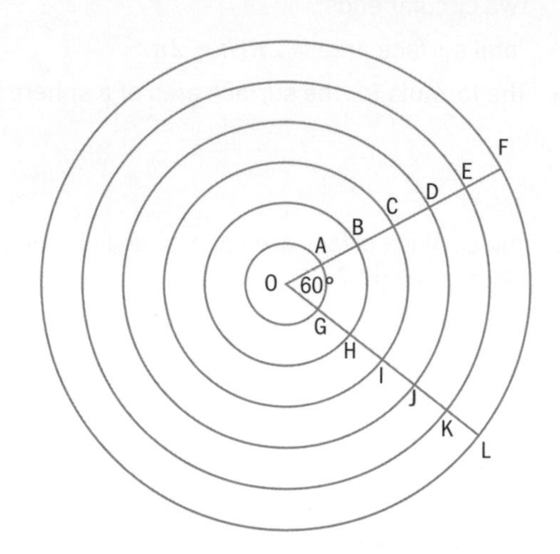

7 The surface area to volume ratio (SA:V) is an important measure in biology. Living cells can only get materials (like glucose and oxygen) in and can only get waste products out through the cell membrane. The larger the surface area of the cell membrane in relation to the cell volume the faster the cell is "serviced".

a Calculate the surface area, volume and surface area to volume ratio (SA:V) for each of the shapes below.

i Cubes

Side	1 cm	2 cm	4 cm
Surface area			
Volume			
Surface area to volume ratio, SA:V			

ii Spheres

Diameter	1 cm	2 cm	4 cm
Surface area			
Volume			
Surface area to volume ratio, SA:V			

iii Cylinders

Diameter × length	1 cm × 1 cm	1 cm × 2 cm	1 cm × 4 cm
Surface area			
Volume			
Surface area to volume ratio, SA:V			

iv Cuboids

Base side x length	1 cm × 1 cm	1 cm × 2 cm	1 cm × 4 cm
Surface area			
Volume			
Surface area to volume ratio, SA:V			

b State what conclusion you have reached about the best shapes (and size) for cells to take to achieve a higher surface area to volume ratio (SA:V).

8 A bolt is made of steel, and has dimensions shown in the diagram below.

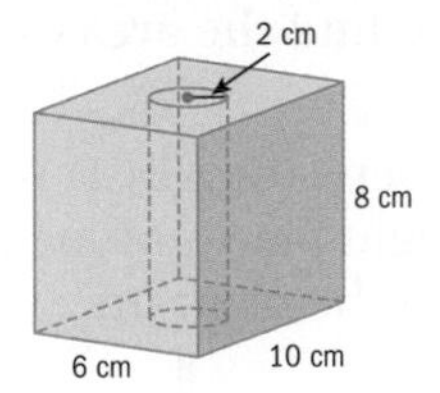

Find the:

a volume of the bolt

b surface area of the bolt.

9 Buildings A and B are rectangular prisms with height 8 m, as shown in the diagram. The length of building A is 10 m, and its width is 7.5 m. The length of building B is 22 m, and its width is 11 m. The two buildings intersect, and the intersection forms a prism with a square base of 5 m. Find the volume of the composite building.

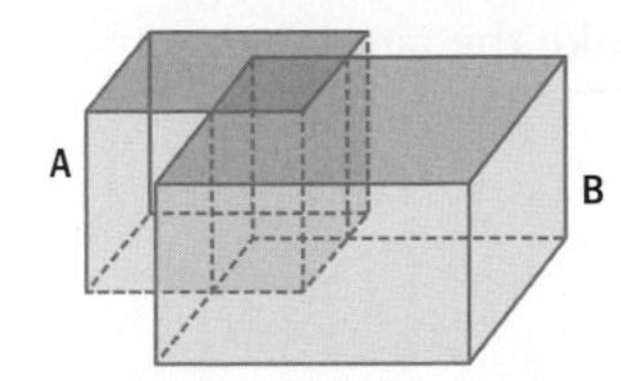

10 A triangular prism ABCDEF has edges AB = BC = 6 cm, AC = 4 cm and BD = 8 cm.

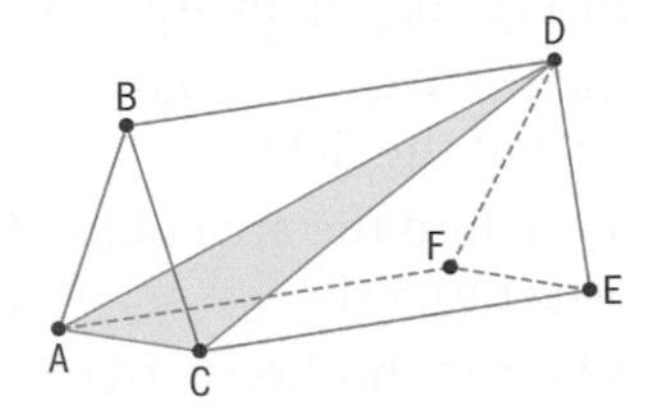

Calculate the area of ΔACD. Give your answer correct to 1 decimal place.

11 A regular dodecahedron has edges of 2 cm. Calculate the surface area.

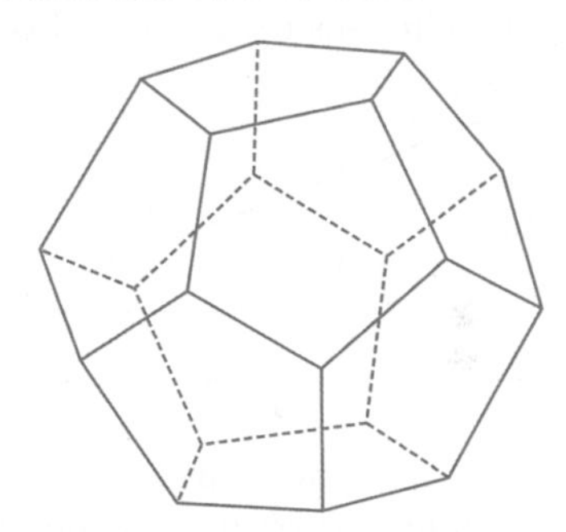

12 Cube ABCDEFGH has an edge length of 6 cm. I, J and K are the midpoints of edges [EF], [FG] and [FB] respectively. Vertex F has been cut off, as shown in the diagram, by removing pyramid IJKF. Find the remaining volume and surface area of the cube.

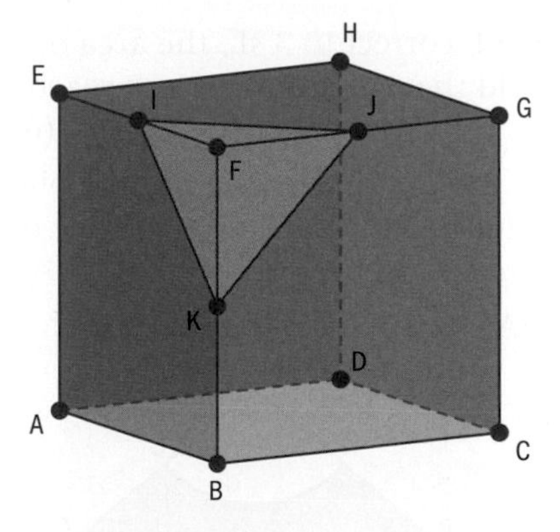

Exam-style questions

13 P1: A triangular field has boundaries of length 170 m, 195 m and 210 m.

a Find the size of the largest interior angle of the field. (3 marks)

b Hence find the area of the field, correct to 3 sf. (3 marks)

14 P1: To walk from his house (A) to his school (C), Oscar has a choice of walking along the straight roads AB and BC, or taking a direct shortcut (AC) across a field.

Find how much shorter his journey will be if he chooses to take the shortcut.

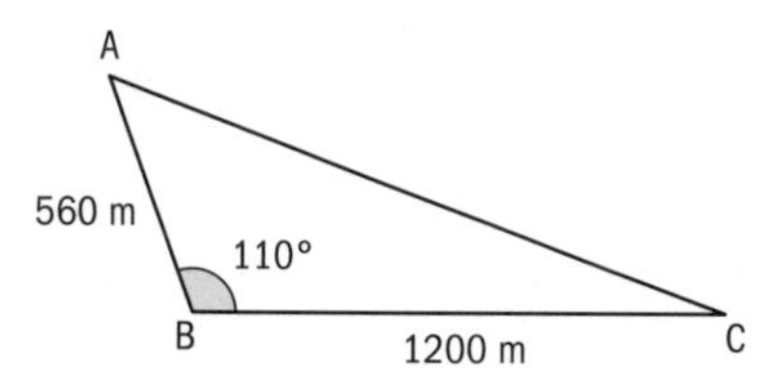

(5 marks)

15 P1: The diagram shows a triangular field ABC. A farmer tethers his horse at point A with a rope of length 85 m.

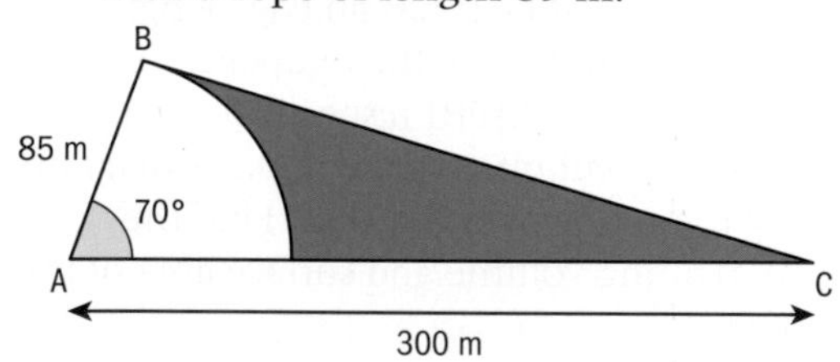

Find, correct to 3 sf., the area of the field that the horse cannot reach. (6 marks)

16 P1: In the following diagram, the plane figure ABCD is part of a sector of a circle.

OA = AB = 8 cm and $A\hat{O}D = \theta°$.

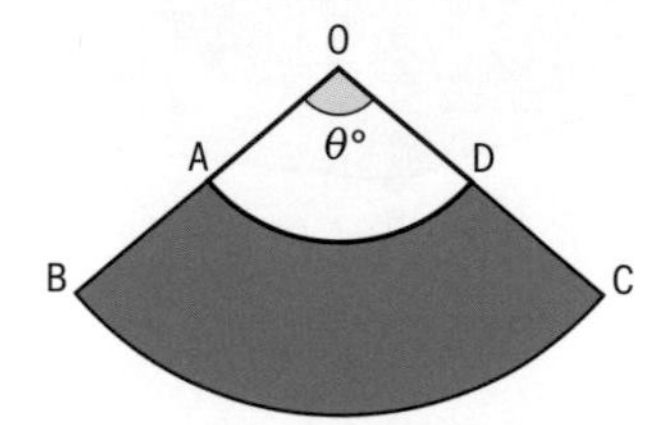

Given that the area of ABCD is 200 cm², determine the value of θ. Give your answer correct to the nearest degree. (5 marks)

17 P1: The diagram shows a quadrilateral ABCD.

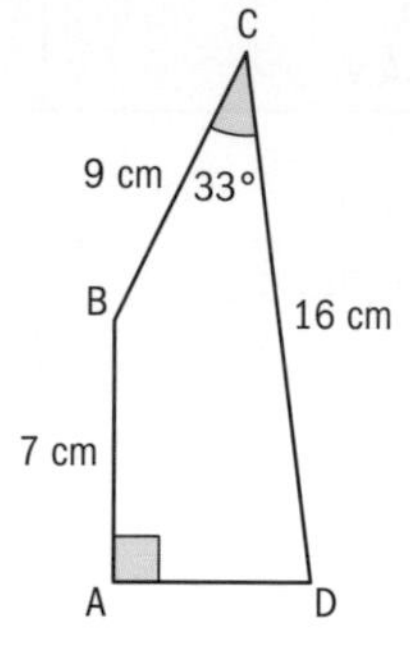

a Find the perimeter of the quadrilateral. (5 marks)

b Find the area of the quadrilateral. (3 marks)

18 P1: In ΔABC, $A\hat{C}B = 67°$, AB = 6.9 cm and BC = 5.7 cm.

a Calculate angle $B\hat{A}C$. (3 marks)

b Hence find the area of ΔABC. (3 marks)

19 P1: A small pendant ABCD is made from five straight pieces of metal wire as shown in the diagram.

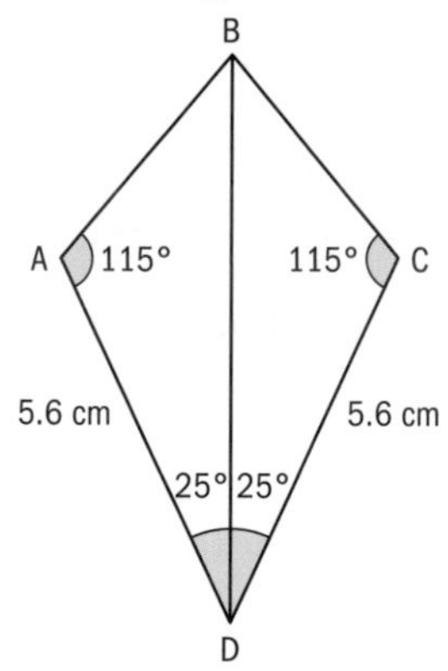

Calculate the total length of wire required to make the pendant. (7 marks)

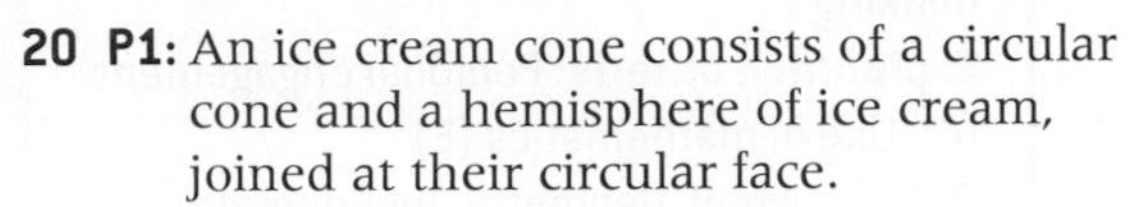

20 P1: An ice cream cone consists of a circular cone and a hemisphere of ice cream, joined at their circular face.

The hemisphere has radius 4 cm and the cone's perpendicular height is 9 cm.

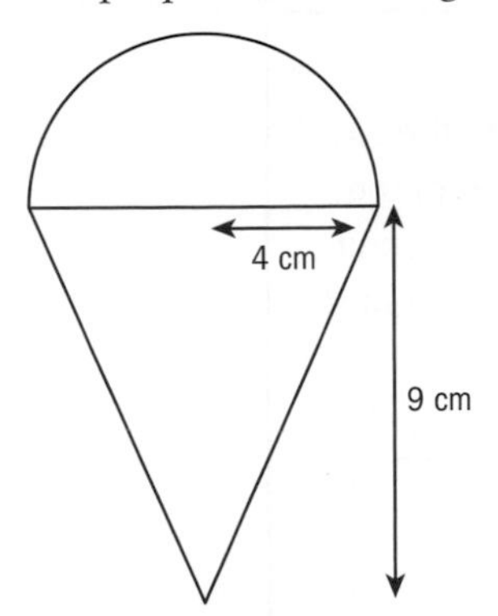

a Find the total volume of the ice cream and cone, giving your answer in terms of π. (4 marks)

b Find the total surface area, giving your answer to 3 sf. (5 marks)

21 P1: The diagram shows a hexagonal prism of height 12 cm.

The edges of each hexagonal face have length 4 cm.

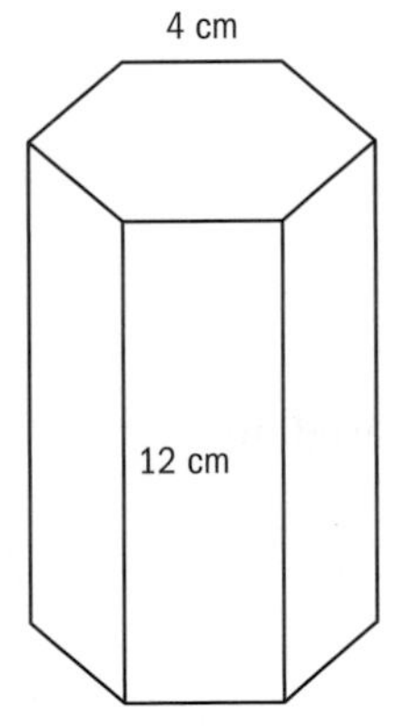

a Draw a net of the prism. (2 marks)

b Calculate the volume of the prism. (3 marks)

c Calculate the total surface area of the prism. (2 marks)

22 P1: Alan and Belinda stand on horizontal ground, 115 m apart.

Alan sees a bird in the sky at an angle of elevation of 27° from where he is standing.

Belinda sees the same bird at an angle of elevation of 42°.

Alan and Belinda are standing on the same side of the bird, and they both lie in the same vertical plane as the bird.

a Determine the direct distance of Alan to the bird. (4 marks)

b Determine the altitude at which the bird is flying. (2 marks)

23 P1: ABCDE is a square-based pyramid.

The vertex E is situated directly above the centre of the face ABCD.

AB = 16 cm and AE = 20 cm.

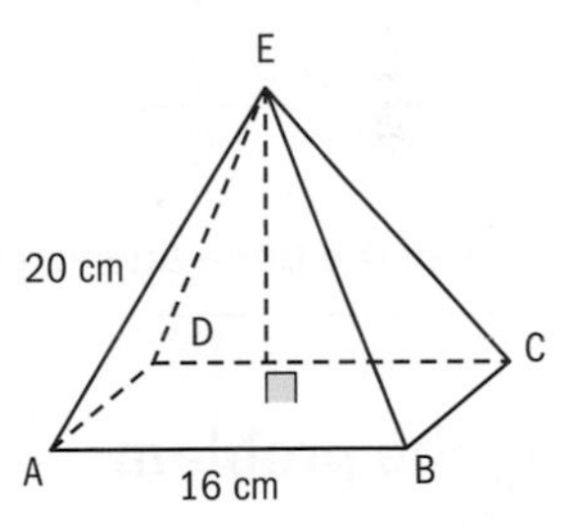

a Calculate the angle between the line [AE] and the plane face ABCD. (4 marks)

b Calculate the angle between the plane face BCE and the face ABCD. (4 marks)

c Calculate the angle between the line [AE] and the line [EC]. (2 marks)

d Find the volume of the pyramid. (2 marks)

e Find the total surface area of the pyramid. (3 marks)

Three squares

Approaches to learning: Research, Critical thinking
Exploration criteria: Personal engagement (C), Use of mathematics (E)
IB topic: Proof, Geometry, Trigonometry

The problem

Three identical squares with side length 1 are adjacent to one another. A line connects one corner of the first square to the opposite corner of the same square, another line connects to the opposite corner of the second square and a third line connects to the opposite corner of the third square, as shown in the diagram.

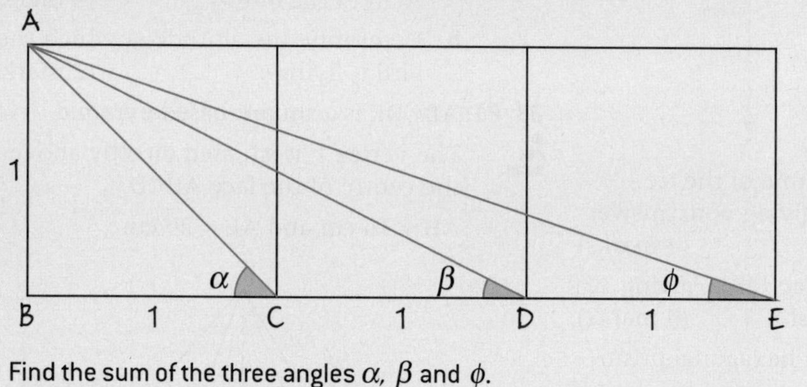

Find the sum of the three angles α, β and ϕ.

Exploring the problem

Look at the diagram.

What do you think the answer may be?

Use a protractor if that helps.

How did you come to this conjecture?

Is it convincing?

This is not an accepted mathematical truth. It is a conjecture, based on observation.

You now have the conjecture $\alpha + \beta + \phi = 90°$ to be proved mathematically.

Direct proof

What is the value of α?

Given that $\alpha + \beta + \phi = 90°$, what does this tell you about α and $\beta + \phi$?

What are the lengths of the three hypotenuses of ΔABC, ΔABD and ΔABE?

Hence explain how you know that ΔACD and ΔACE are similar.

What can you therefore conclude about $C\hat{A}D$ and $C\hat{E}A$?

Hence determine why $A\hat{C}B = C\hat{A}D + A\hat{D}C$ and conclude the proof.

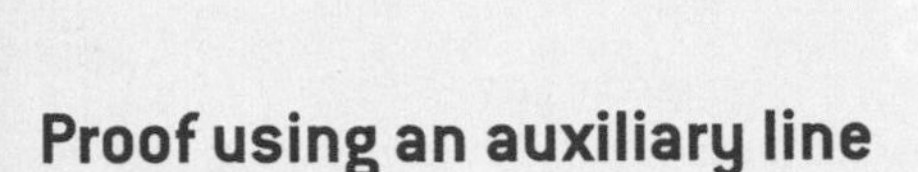

Proof using an auxiliary line

An additional diagonal line, [CF], is drawn in the second square, and the intersection point between [CF] and [AE] is labelled G:

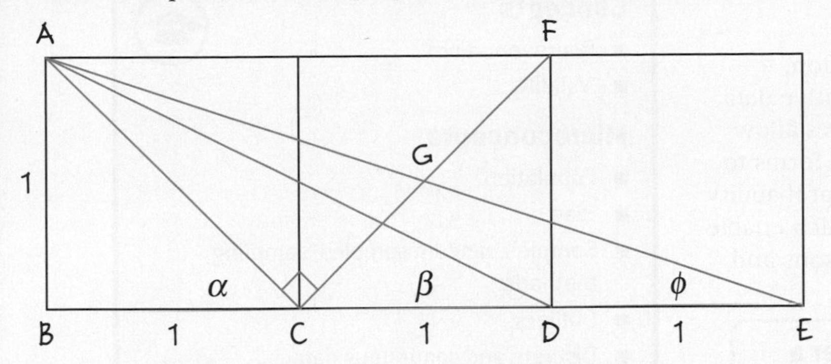

Explain why $B\hat{A}C = \alpha$.

Explain why $E\hat{A}F = \phi$.

If you show that $G\hat{A}C = \beta$, how will this complete the proof?

Explain how you know that ΔGAC and ΔABD are similar.

Hence explain how you know that $G\hat{A}C = B\hat{D}A = \beta$.

Hence complete the proof.

Proof using the cosine rule

The diagram is extended and the additional vertices of the large rectangle are labelled X and Y and the angle is labelled θ:

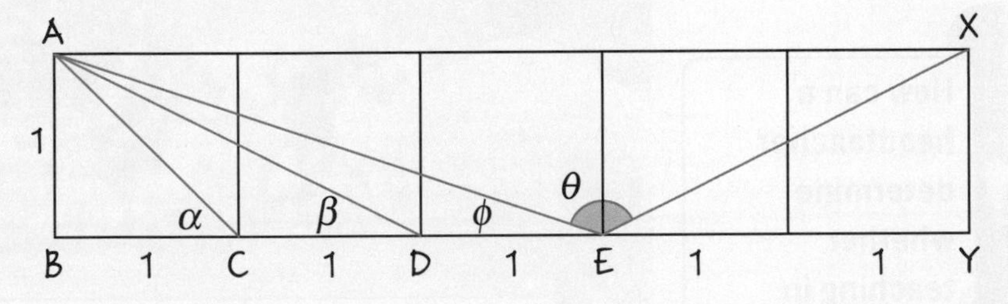

Explain why $X\hat{E}Y = \beta$.

Calculate the lengths of [AE] and [AY].

Now calculate $A\hat{E}Y$ (θ) using the cosine rule.

Hence explain how you know that $\beta + \phi = 45°$.

Hence complete the proof.

Extension

Research other proofs on the Internet.

You could also try to produce a proof yourself.

You do not have to stop working when you have the proof.

What could you do next?

In the task in Chapter 8 you will use Spearman's rank. As future extension work, you could use the methods in that task to rank these proofs and discuss the results.

Representing and describing data: descriptive statistics

Statistics is concerned with the collection, analysis and interpretation of quantitative data. Statistical representations and measures allow us to represent data in many different forms to aid interpretation. Both statistics and probability provide important representations which enable us to make predictions, valid comparisons and informed decisions.

Concepts

- Representations
- Validity

Microconcepts

- Population
- Bias
- Samples, random samples, sampling methods
- Outliers
- Discrete and continuous data
- Histograms
- Box-and-whisker plots
- Cumulative frequency graphs
- Measures of central tendency and dispersion
- Skewness
- Scatter graphs
- Correlation

How can scientists determine whether a new drug is likely to be a successful cure?

How can a headteacher determine whether teaching in the school has been effective?

How can a football coach determine whether a particular strategy is likely to be successful?

How can you persuade a potential customer that your product is better than the competition?

How can we tell if the oceans are warming?

Below is a graph of GDP per capita (gross domestic product per person) and life expectancy taken from Gapminder (www.gapminder.org). Click the icon to access the complete set of data.

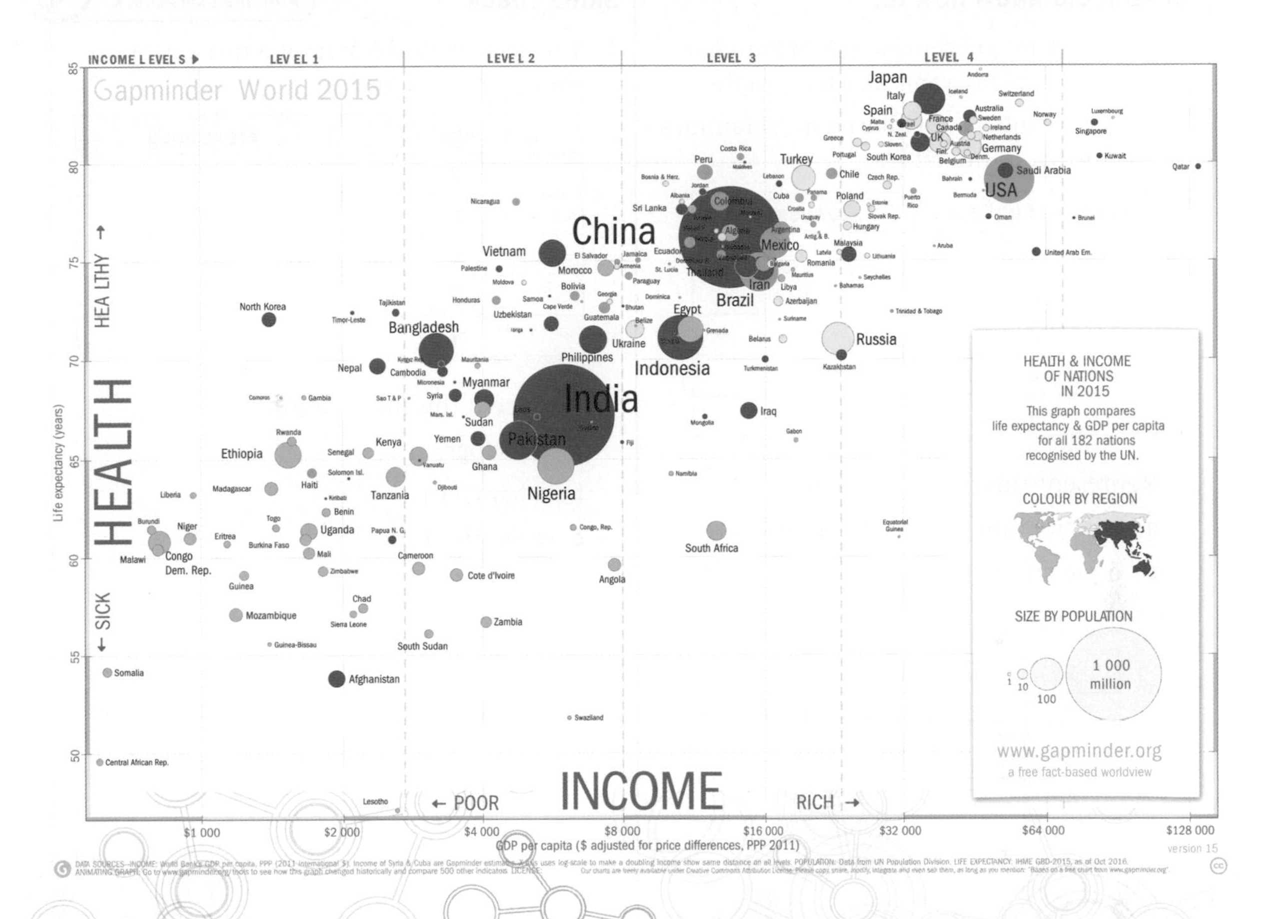

Name four pieces of information represented in this graph.

How do you think this data could have been collected? How exact do you think it might be?

Identify any relationships in the graph.

Do you find anything surprising in the graph?

Do you need to use all the data for analysis or can you just use a sample of the data?

Describe the scale on the x-axis. Why do you think it has been done like that?

Developing inquiry skills

Write down any similar inquiry questions you might ask to investigate the relationship between two different quantities, for example, GDP per capita and infant mortality or life expectancy and population.

How are these questions different from the ones used to investigate life expectancy and income?

Think about the questions in this opening problem and answer any you can. As you work through the chapter, you will gain mathematical knowledge and skills that will help you to answer them all.

Before you start

You should know how to:

1 Collect data and represent it in bar charts, pie charts, pictograms and line graphs.

eg The numbers of children in 20 families are shown in the table:

Number of children	Frequency
1	4
2	8
3	5
4	2
5	1

Represent this information in:

a a pictogram **b** a bar chart

c a pie chart.

a ☺ = 1 child

1	☺ ☺ ☺ ☺
2	☺ ☺ ☺ ☺ ☺ ☺ ☺ ☺
3	☺ ☺ ☺ ☺ ☺
4	☺ ☺
5	☺

b

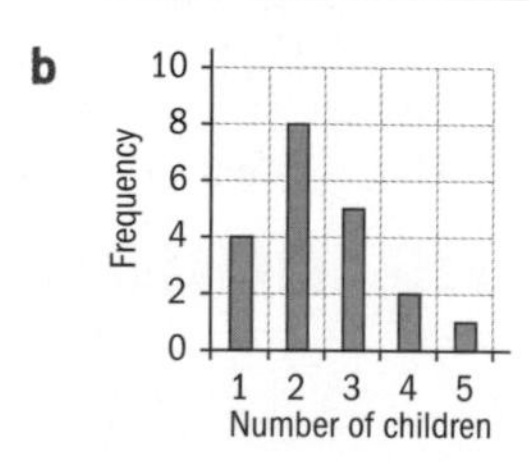

c

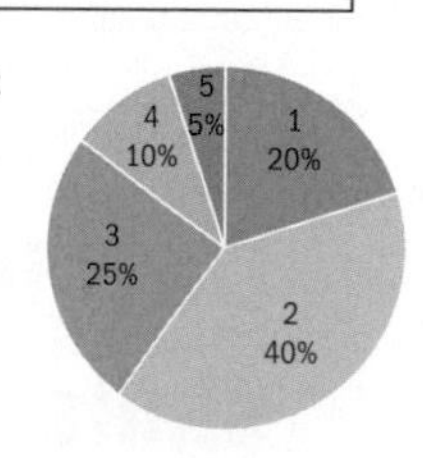

2 Obtain simple statistics from discrete data, including the mean, median, mode and range.

eg Find the mean, median, mode and range of the following data:

2, 2, 3, 4, 6, 6, 6, 7, 8, 9

Mean = 5.3 Median = 6

Mode = 6 Range = 7

3 Set up axes for graphs using a given scale.

Skills check

Click here for help with this skills check

1 The ages of 25 children at a nursery are shown in the table.

Age	Frequency
0	4
1	5
2	8
3	6
4	2

Represent this information in:

a a pictogram

b a bar chart

c a pie chart.

2 Find the mean, median, mode and range of the following data:

3, 3, 5, 7, 8, 8, 9, 9, 9, 9

3 Draw a set of axes such that 1 cm represents 5 units on the x-axis and 1 cm represents 2 units on the y-axis.

3.1 Collecting and organizing univariate data

Univariate data has only one variable.

Rosa works in a restaurant. The tips that the customers give to the waiters and waitresses are placed in a large jar. It is Rosa's job to count the tips every week, record the amount in a notebook and place the tips and notebook in the safe. At the end of each month, the tips are divided equally amongst the staff.

The manager is interested to see how the tips varied from week to week, and also how many of each type of note or coin there are (5¢, 10¢, 20¢, 50¢, €1, €5, €10, €20). He asks Rosa to make a presentation for the staff.

How can she best make this presentation? Which diagrams can she use? How can the staff use this information to improve service? How can the manager use the information to improve the restaurant?

International-mindedness

Ronald Fisher (1890–1962) lived in the UK and Australia and has been described as "a genius who almost single-handedly created the foundations for modern statistical science". He used statistics to analyse problems in medicine, agriculture and the social sciences.

There are two main types of data: qualitative and quantitative.
Qualitative data is data that is not given numerically, for example favourite ice cream flavour.
Quantitative data is numerical and can be classified as **discrete** or **continuous**.

Investigation 1

1 The numbers of cherries in 24 boxes are shown below.

44 43 42 42 43 41 45 42 40 43 44 41
42 41 43 42 42 40 45 43 43 41 45 43

This data is **discrete**.

Complete the frequency table for this data.

Number of cherries	Frequency
40	2
41	
42	
43	
44	
45	

Continued on next page

Statistics and probability

2 The lengths, in minutes, of 20 telephone calls are shown below.

4.2 6.8 10.4 8.2 11.5 1.6 5.9 7.6 3.1 21.5
13.5 5.8 4.1 22.8 13.6 11.2 9.5 1.8 12.4 4.9

This data is **continuous**.

Complete the frequency table for this data.

Length, t (minutes)	Frequency
$0 \le t < 5$	
$5 \le t < 10$	
$10 \le t < 15$	
$15 \le t < 20$	
$20 \le t < 25$	

3 Explain why the examples in parts **1** and **2** are different.

4 **Factual** What type of values can discrete data take? How is discrete data collected? What type of values can continuous data take? How is continuous data collected?

5 Men's jeans are sized by waist measurement, eg 28 inches, 30 inches, 32 inches and so on. Here are the jeans sizes of 10 men:

28 30 28 34 32 30 36 28 30 30

This is discrete data.

Do you need to change your answer to "What type of values can discrete data take?"

6 Here are the waist measurements, in cm, of 10 people:

24.3 27.2 22.1 28.3 27.0 29.6 32.4 23.8 21.7 35.2

Is this discrete or continuous data?

How does it differ from the previous example?

7 **Conceptual** What are the differences between discrete and continuous data?

Discrete data is either data that can be **counted**, for example the number of cars in a car park, or data that can only take specific values, for example shoe size.

Continuous data can be **measured**, for example height, weight and time.

Discrete and continuous data can be organized into a frequency table or a grouped frequency table.

For continuous data, the classes must cover the full range of the values and they must not overlap.

Example 1

The ages of boys in a football club are:

10 11 11 10 12 13 11 10 12 14 15 15 16 10 11 15 10 11 11 12
12 12 13 16 16 14 15 12 12 10 11 11 14 14 15 16 16 11 10 13

a State whether this data is discrete or continuous.

b Construct a grouped frequency table for this data. Let x represent age.

a The data is continuous.

Remember that you are 10 for a whole year!

b

Age, x	Frequency
$10 \leq x < 11$	7
$11 \leq x < 12$	9
$12 \leq x < 13$	7
$13 \leq x < 14$	3
$14 \leq x < 15$	4
$15 \leq x < 16$	5
$16 \leq x < 17$	5

Notice that all the possible ages are included in the classes.

International-mindedness

The 19th-century German psychologist Gustav Fechner popularized the median, although the French mathematician Pierre Laplace had used it earlier.

Exercise 3A

1 State whether the following data sets are discrete or continuous.

a the number of apples in a bag

b the weights of students in Grade 6

c the number of blue cars in a parking lot

d the football boot sizes of a football team

e the number of visitors to the Tower of London each week

f the weights of 20 puppies

g the depth of snow on a ski slope

h the number of sixes when you throw a die 25 times

i the time it takes to run 100 metres

j the lengths of 20 worms.

2 Construct a frequency table for this data.

The number of sweets in 25 packets:

21 23 22 24 21 22 23 25 24 24
22 23 25 21 23 23 24 26 25 25
21 22 22 24 22

3 Construct a grouped frequency table for the following data.

The heights, in metres, of 20 trees in a garden:

5.8 3.6 3.9 4.1 4.4 3.2 2.4 2.6 5.1 2.5
4.5 3.6 2.4 5.2 4.7 3.5 3.3 2.8 4.1 2.1

4 The following data shows the weights of 25 dogs, in kilograms.

2 5 31 22 16 7 12 35 9 18 5 11 15
6 3 14 8 10 12 25 27 34 7 1 5

Construct a suitable table for this data.

When faced with lots of numbers, how do you know which "average" is best to use?

Measures of central tendency (or averages)

- The most common measures of central tendency are the mean, median and mode.
- The **mode** of a data set is the value that occurs most frequently. There may be no mode or several modes.
- The **median** of a data set is the value that lies in the middle when the data is arranged in size. When there are two middle values, the median is the midpoint between the two values.
- The **mean** of a data set is the sum of all the values divided by the number of values. For a discrete data set of n values the formula is $\overline{x} = \frac{1}{n}\sum_{i=1}^{n} x_i$, where $\sum_{i=1}^{n} x_i = x_1 + x_2 + x_3 + \cdots + x_n$ and Σ means "the sum of". For example, the mean of the numbers 3, 4, 8, 12, 16 is the sum (Σ) of the numbers divided by 5.
- For a frequency data set, the formula is $\overline{x} = \frac{1}{\sum_{i=1}^{n} f_i}\sum_{i=1}^{n} f_i x_i$, where $\sum_{i=1}^{n} f_i x_i = f_1 x_1 + f_2 x_2 + f_3 x_3 + \cdots + f_n x_n$.

When there is a frequency table, you need to use the data values and the corresponding frequencies to calculate the mean.

Example 2

1 The grades in a history test for 14 students were as follows:

58 67 66 58 79 83 76 49 35 58 88 91 47 69

a Find the mode, median and mean.

When a 15th student took the test, the mean became 66.2.

b Calculate the grade for the 15th student.

2 Mindy opens some bags of candy and counts how many pieces are in each bag.

Her results are:

Number of pieces of candy	Frequency
23	2
24	3
25	9
26	5
27	1

Find the mean number of candies in a bag.

1 a Mode is 58	58 appears three times.
Median is 66.5	Arranging the data in order: 35 47 49 58 58 58 [66 67] 69 76 79 83 88 91. The middle number will be between 66 and 67.
Mean is 66	The mean is $\frac{58+67+66+58+79+83+76+49+35+58+88+91+47+69}{14}=66$
b The mark is 69	If the new mean is 66.2 then the total for all 15 students will be $66.2 \times 15 = 993$. Subtracting the total for the 14 students: $993 - 924 = 69$.
2 Mean is 25	$\text{Mean} = \frac{23\times2+24\times3+25\times9+26\times5+27\times1}{2+3+9+5+1}=25$

Example 3

Answer the following questions, and in each case interpret the meaning of the values calculated, and discuss why extreme values or an extreme mode affect the mean more than the median.

a The number of ice creams sold over a period of 13 weeks is as follows:

146 151 158 158 161 149 160 147 158 160 216 225 238

Write down the mode, and use technology to find the mean and median for this data set.

b Two dice are thrown 100 times and their total score is recorded in the table:

Score	2	3	4	5	6	7	8	9	10	11	12
Frequency	21	9	8	4	7	20	13	9	6	2	1

Write down the mode, and use technology to find the mean and median for this data set.

c The weights, w kg, of 50 cats are recorded in the table:

Weight (kg)	Frequency
$2 \le w < 3$	5
$3 \le w < 4$	19
$4 \le w < 5$	17
$5 \le w < 6$	5
$6 \le w < 7$	3
$7 \le w < 8$	1

Continued on next page

Statistics and probability

Find an approximation for the median and mean, and write down the modal class.

a Mean = 171.3; this is the average number of ice creams sold during the 13 weeks. Median = 158; this is the middle value. Half of the amounts are above this value and half are below it. Mode = 158; this is the number of ice creams that occurs the most frequently. The three large values have the effect of making the mean value larger, but the median is not affected since it is the middle value.	Put the numbers into a list on your GDC. Go to Statistics, Stat calculations, one-variable statistics, number of lists 1, enter the name of the list and press Enter. The GDC gives you a lot of data. The first is the mean. Then scroll down until you reach the median. Mode: 158 occurs the most often.
b Mean = 5.82; this is the average score for the 100 throws of the dice. Median = 7; this is the middle value. Half of the scores are above this value and half are below it. Mode = 2; this is the score that occurs the most often. Since the mode is 2, it makes the mean value smaller, but the median is not affected since half of the values are less than 7.	Put the scores into one list on your GDC and the frequencies into a second list. Go to Statistics, Stat calculations, one-variable statistics, number of lists 1, enter the names of the two lists and press Enter. The GDC gives you a lot of data. The first is the mean. Then scroll down until you reach the median. The mode is 2 since this occurs 21 times.
c Approximation for the mean = 4.2 kg; this is the approximate mean weight of the 50 cats. Approximation for the median = 4.5 kg; approximately half the weights are above and half below this middle value. Modal class = $3 \leq w < 4$; this group has more of the cats' weights than any of the other groups. The modal class is the second smallest group, but this does not have much effect on the mean in this example because the middle group is also quite large, whereas the top three groups are very small in comparison.	Using your GDC, enter the midpoints of the groups into one list and the frequencies into a second list and proceed as above. These values only give an estimate since we do not have the original data on the cats' weights and only know that five cats weigh between 2 kg and 3 kg, etc.

Investigation 2

1 Using technology, complete the table for the following data sets.

A The dress sizes of 15 females:

0 0 2 2 2 4 4 6 6 8 10 12 14 16 16

B The shoe sizes of 19 children:

23 23 23 23 26 28 35 35 36 36 36 37 39 41 43 40 38 37 41

TOK

Why have mathematics and statistics sometimes been treated as separate subjects?

C The number of times that 20 commuters travelled by train in one month:

40 50 41 28 51 52 49 50 51 28 48 33 35 28 45 40 51 62 28 49

D The ages of boys in a basketball club:

Age	10	11	12	13	14	15	16	17	18
Frequency	5	6	8	10	12	11	12	35	32

	Mode	**Median**	**Mean**
Data set A			
Data set B			
Data set C			
Data set D			

2 For each set of data, list the advantages and disadvantages of the mean, median and mode, and decide which best represents the data in each case. Explain your choice.

3 How do you decide which measure of central tendency best represents the data?

4 Why do we need more than one measure of central tendency?

This grouped frequency table shows data set E, the scores that 60 students gained in an entrance test:

Score	**Frequency**
$20 < x \le 30$	3
$30 < x \le 40$	5
$40 < x \le 50$	7
$50 < x \le 60$	8
$60 < x \le 70$	9
$70 < x \le 80$	12
$80 < x \le 90$	10
$90 < x \le 100$	6

5 What is the modal group (or class) for data set E?

6 Use your GDC to find approximations for the mean and median.

7 Why are these values only approximations?

8 Discuss which value is more appropriate to use in this case.

Data set F shows the weights, in kilograms, of 20 different breeds of dogs:

9 What is the modal group (or class) for data set F?

Weight (kg)	$0 < w \le 10$	$10 < w \le 20$	$20 < w \le 30$	$30 < w \le 40$	$40 < w \le 50$	$50 < w \le 60$
Frequency	6	5	4	3	2	1

10 Use your GDC to find approximations for the mean and median.

11 Why are these values only approximations?

12 Discuss which value is more appropriate to use in this case.

Exercise 3B

1 For the following sets of data find the mean, median and mode. State which of these measures is most appropriate to use in each case, giving a reason for your answer.

a The times, in minutes, to run 1500 metres:

7.2 7.3 7.5 7.8 8.0 8.3 8.6
8.6 8.6 9.0 9.2 9.5 10.0 10.5
10.6 11.1 15.3 16.8 17.2

b The weights, in kg, of 13 pumpkins:

2.6 2.9 4.7 6.8 6.9 7.2 8.5 8.9
10.1 11.5 12.5 14.7 15.0

c The monthly amounts of pocket money, in euros, for 21 Grade 6 students:

10 10 10 15 15 15 15 20 25
25 30 30 35 35 35 40 80 50
50 80 100

2 For the following sets of data, find

i the modal class

ii an approximation for the mean

iii an approximation for the median.

Comment on the meaning of these values and state which one is most appropriate to use in each case, giving a reason for your answer.

a

Number of cars (n)	Frequency
$0 \le n < 30$	12
$30 \le n < 60$	28
$60 \le n < 90$	39
$90 \le n < 120$	42
$120 \le n < 150$	54
$150 \le n < 180$	65

b

Speed of cars (s mph)	Frequency
$40 \le s < 45$	4
$45 \le s < 50$	8
$50 \le s < 55$	23
$55 \le s < 60$	15
$60 \le s < 65$	6
$65 \le s < 70$	4

c

Time to complete a puzzle (t minutes)	Frequency
$2 \le t < 3$	2
$3 \le t < 4$	5
$4 \le t < 5$	3
$5 \le t < 6$	7
$6 \le t < 7$	4
$7 \le t < 8$	9
$8 \le t < 9$	3

Example 4

The ages of 15 cats are:

10 10 11 11 11 12 12 12 12 13 13 14 14 24 25

Find the median, mean and mode for this data.

Comment on whether there are any data points that distort the calculation of the mean.

Remove these values and recalculate the mean. Discuss your answer.

The median is 12. The mean is 13.6. The mode is 12.	Enter the data into a list on the GDC. Go to Statistics, Stat calculations, one-variable statistics, number of lists 1, enter the name of the list and press Enter. The GDC gives you a lot of data. The first is the mean. Then scroll down until you reach the median. The mode is the number that appears the most.
24 and 25 are much larger than the other numbers. If they are removed, the mean becomes 11.9, which is much closer to the median and the mode.	24 and 25 are called **outliers**. Outliers are extreme data values that can distort the results of statistical processes.

Investigation 3

1 Find the mean, median and mode for the following sets of numbers:

a The monthly salaries, in Australian Dollars (AUD), of 13 employees in a factory:

4000 4200 4200 4250 4400 4400 4400 4450 4500
4550 4600 20 000 42 000

b The ages of students on a chemistry course at university:

19 18 18 21 22 19 20 17 20 21 22 19 19 19 20 17 55 63

c The lengths of time, in seconds, for which 15 people can hold their breath:

20 22 23 23 23 58 61 61 65 74 79 80 81 83 92

2 Which data entries do not appear to fit with the rest of the data?

3 Do you think that these entries are a result of an error in the recording of the data or not? Explain your answer.

4 Calculate the mean, median and mode of each data set without the entries you identified in part **2**. Do the values change?

Extreme data values that distort the mean are called **outliers**; they do not "fit" with the rest of the data.

5 **Conceptual** How can outliers affect measures of central tendency?

6 **Conceptual** How can identifying outliers help you decide which measure of central tendency to use to represent the data?

Outliers are extreme data values, or the result of errors in reading data, that can distort the results of statistical processes.

Outliers can affect the mean by making it larger or smaller, but most likely they will not affect the median or the mode.

TOK

Is there a difference between information and data?

Exercise 3C

1 Find the mean, median and mode for the following data sets and comment on any pieces of data that you think may be outliers.

a The times of 25 telephone calls in minutes:

1.0 1.5 2.3 2.6 2.8 3.0 3.4 3.8 4.1
4.5 4.6 4.8 5.2 5.3 5.5 5.8 6.0 6.3
6.6 7.3 7.5 7.5 7.5 17.8 25.0

b The heights, in metres, of 15 sunflowers:

1.1 2.2 2.5 2.5 2.5 3.1 3.5 3.6
3.9 4.0 4.1 4.4 4.6 4.9 6.1

c The results of a geography test:

22 39 45 46 46 52 54 58 62 62
62 67 70 75 78 82 89 91 95 98

Measures of dispersion

- Measures of dispersion measure how spread out a data set is.
- The simplest measure of dispersion is the **range**, which is found by subtracting the smallest number from the largest number.
- The standard deviation, σ_x, gives an idea of how the data values are spread in relation to the mean. The standard deviation is also known as the root-mean-squared deviation; its formula is

$$\sigma_x = \sqrt{\frac{1}{n}\sum_{i=1}^{n}(x_i - \bar{x})^2} = \sqrt{\frac{1}{n}\sum_{i=1}^{n}x_i^2 - \bar{x}^2}$$

In examinations you will use technology to find the standard deviation.

Investigation 4

In this investigation you will find the means and standard deviations for two sets of data and compare the results.

A quiz has 10 questions with one mark for each correct answer. Ten boys and ten girls took this test and their results are shown in the table:

Girls' scores	Boys' scores
2	4
3	5
4	5
5	6
5	6
6	6
8	7
8	7
9	7
10	7

Find the mean of the girls' scores and the mean of the boys' scores.

The standard deviation is called the **root-mean-squared deviation**, and to calculate this you work **backwards**.

First find the **deviation** of each score from the mean, then **square** these answers:

Girls' scores – girls' mean	Boys' scores – boys' mean	(Girls' scores – girls' mean)^2	(Boys' scores – boys' mean)^2
		$\Sigma =$	$\Sigma =$

Next you have to find the **mean** of (girls' scores – girls' mean)^2 and the **mean** of (boys' scores – boys' mean)^2.

Lastly you find the **square root** of each of these two values.

This gives you the **standard deviation**.

If you compare the value of the standard deviation to the mean in each case, what can you say about the spread of the data?

Conceptual What does the standard deviation represent?

Example 5

For each of the three data sets in Example 3, find the standard deviation and compare it with the mean.

a Standard deviation = 30.8; this would indicate that the data points are not all close to the mean.	Enter the data into a list on your GDC. Go to Statistics, Stat calculations, one-variable statistics, number of lists 1, enter the name of the list and press Enter. The GDC gives you a lot of data. The first is the mean. Then scroll down until you reach the standard deviation, given by the symbol σx.
b Standard deviation = 2.80; this is a small value and so most of the points will be close to the mean.	Enter the data into two lists on your GDC. Go to Statistics, Stat calculations, one-variable statistics, number of lists 1, enter the names of the two lists and press Enter.

Continued on next page

	The GDC gives you a lot of data. The first is the mean. Then scroll down until you reach the standard deviation, given by the symbol σx.
c The standard deviation is 1.1, which is quite small and would suggest that most of the weights are close to the mean. This is only an approximate value because the original data has not been given, only the grouped data.	Using your GDC, enter the midpoints of the groups into one list and the frequencies into a second list and proceed as above.

The **variance** is the standard deviation squared: $(\sigma_x)^2$.

TOK

Is standard deviation a mathematical discovery or a creation of the human mind?

While the standard deviation is useful for interpreting the spread of data about the mean, other statistical processes such as least squares regression, probability theory and investments use the variance.

The **interquartile range** (IQR) is the **upper quartile,** Q_3, minus the **lower quartile,** Q_1.
When the data values are arranged in order, the lower quartile is the data point at the 25th percentile and the upper quartile is the data point at the 75th percentile.

The interquartile range is another method of interpreting the spread of data. It is more reliable than the range because it is not affected by outliers.

Consider the following scores in a biology exam, arranged in order:

18, 22, 26, 39, 45, $\boxed{46}$, 46, 52, 54, 58, 62, $\boxed{62}$, 62, 67, 70, 71, 75, $\boxed{78}$, 82, 89, 91, 95, 98

The median is the middle value, 62, since half the numbers are above 62 and half the numbers are below 62.

To find Q_1, locate the number that is in the $\left(\frac{n+1}{4}\right)$th place. Here it is the number in the $\frac{23+1}{4} = 6$th place, so the lower quartile is 46, since one-quarter of the numbers are below 46 and three-quarters of the numbers are above 46.

To find Q_3, locate the number that is in the $\left(\frac{3(n+1)}{4}\right)$th place. Here it is the number in the $\frac{3(23+1)}{4} = 18$th place, so the upper quartile is 78, as three-quarters of the numbers are below 78 and one-quarter of the numbers are above 78.

The interquartile range is then $Q_3 - Q_1 = 78 - 46 = 32$.

Example 6

For the data sets in Example 3, find:

i the variance to 2 dp **ii** the range **iii** the IQR.

a **i** Variance $= (30.838...)^2 = 950.98$ **ii** Range $= 238 - 146 = 92$ **iii** IQR $= 188.5 - 150 = 38.5$	The variance is the square of the standard deviation. The maximum and minimum values as well as Q_1 and Q_3 are all found on the GDC as described above.
b **i** Variance $= (2.801...)^2 = 7.85$ **ii** Range $= 12 - 2 = 10$ **iii** IQR $= 8 - 3 = 5$ **c** **i** Variance $= (1.1)^2 = 1.21$ **ii** Range $= 7.5 - 2.5 = 5$ **iii** IQR $= 4.5 - 3.5 = 1$ Once again, these values are only approximate values since this is a grouped frequency table.	Note that different calculators use different methods for quartiles, so you may get slightly different answers from other students' and from the formula.

Investigation 5

1 For the data sets A–D in Investigation 2, complete the table:

	Mean	Standard deviation	Variance	Range	Lower quartile	Upper quartile	IQR
A							
B							
C							
D							

2 For each data set, compare the value of the standard deviation with that of the mean.

Discuss whether you think that the data is close to the mean or has a wide spread.

Discuss whether you think there are any outliers.

Continued on next page

3 Discuss the difference between the range and the interquartile range, and which one best represents the spread of the data.

4 **Factual** What is spread?

5 **Conceptual** Which value do you think gives a better representation of the spread: the range or the IQR? Why do you think this?

6 For data set D, do you think that the values from the GDC are exact or approximate? Explain.

7 **Conceptual** How does using technology save time and increase accuracy?

8 **Conceptual** What does the standard deviation represent?

Investigation 6

Complete the table for the following sets of numbers:

A: Find the mean and standard deviation of the numbers 3, 4, 6, 8, 9, 10, 15, 17.

B: Add 3 to each of the numbers in A and then find the mean and standard deviation.

C: Subtract 2 from each of the numbers in A and then find the mean and standard deviation.

D: Add 5 to each of the numbers in A and then find the mean and standard deviation.

E: Multiply the numbers in A by 3 and then find the mean and standard deviation.

F: Multiply the numbers in A by −2 and then find the mean and standard deviation.

G: Multiply the numbers in A by 0.5 and then find the mean and standard deviation.

	Mean	Standard deviation
A		
B		
C		
D		
E		
F		
G		

1 **Conceptual** What happens to the mean when you add or subtract a number from each data value?

2 **Conceptual** What happens to the standard deviation when you add or subtract a number from each data value?

3 **Conceptual** What happens to the mean when you multiply each data value by a constant?

4 **Conceptual** What happens to the standard deviation when you multiply each data value by a constant?

5 The mean of a set of numbers is 10 and the standard deviation is 1.5.

a If you add 3 to each number, write down the new mean and standard deviation.

b If you multiply each number by 4, write down the new mean and standard deviation.

The mean of a set of numbers is $\bar{x}$ and the standard deviation is σ_x.
If you add k to or subtract k from each of the numbers then the mean is $\bar{x} \pm k$ and the standard deviation is σ_x.
If you multiply each number by k then the mean is $k \times \bar{x}$ and the standard deviation is $|k| \times \sigma_x$.

TOK

Do different measures of central tendency express different properties of the data?

How reliable are mathematical measures?

Exercise 3D

1 Stan divided the lawn into 30 equal plots. He counted the number of daisies in each plot:

12 15 8 16 24 5 13 2 34 21 18 15 12 8 4

22 15 6 15 3 13 25 9 17 11 6 15 12 26 16

a State whether the data is discrete or continuous.

b Find the mean, the median and the mode, and comment on which is more appropriate to use.

c Find the standard deviation and comment on your result.

d Find the range and interquartile range.

2 Gal asked 60 people how much money they spent the last time they had eaten in a restaurant. The table shows his results.

Cost of dinner, UK£	Frequency
$10 \le c < 20$	6
$20 \le c < 30$	12
$30 \le c < 40$	28
$40 \le c < 50$	10
$50 \le c < 60$	4

a Write down the modal class.

b Find estimates for the mean and the median.

c Find an estimate for the standard deviation and comment on the result.

d Find estimates for the variance, the range and the interquartile range, and explain why these are all estimates.

3 The monthly salaries of the employees in a retail store had a mean value of US$3500 and standard deviation US$250. At the end of the year they all received an increase of US$100. Write down the new mean and the new standard deviation.

4 The table shows the number of orthodontist visits per year made by the students in Grade 10.

Number of visits	0	4	6	8	10	12	14
Frequency	3	2	8	4	2	12	5

a Find the mode, the median and the mean, and comment on which is the most appropriate to use.

b Find the standard deviation and comment on the result.

c Find the range and interquartile range, and comment on the spread of the data.

5 The number of sweets in 25 bags has a mean of 30 and a standard deviation of 3. In a special promotion, the manufacturer doubles the number of sweets in each bag. Write down the new mean and the new standard deviation of the number of sweets in a bag.

6 The table shows the heights of 50 wallabies.

Height (x cm)	Frequency
$150 \le x < 160$	3
$160 \le x < 170$	5
$170 \le x < 180$	13
$180 \le x < 190$	23
$190 \le x < 200$	4
$200 \le x < 210$	2

a Write down the modal class.

b Find estimates for the mean and standard deviation; comment on your results.

7 Mrs Ginger's Grade 8 class sat an English test. The grade was out of 40 marks. The mean grade was 32 marks and the standard deviation was 8 marks.

In order to change this to a mark out of 100, Mrs Ginger thinks that it would be acceptable to multiply all the grades by 2 and then add 20 to each one.

Mr Ginger thinks that it would be fairer to multiply all the grades by 2.5.

Miss Ginger suggests multiplying by 3 and subtracting 20 from each grade.

a Write down the new mean and the new standard deviation for each suggestion.

Matty had an original grade of 12, Zoe had an original grade of 25 and Ans had an original grade of 36.

b Find their new grades under all three suggested changes.

8 The heights in centimetres of 15 basketball players are:

175 183 191 196 198 201 203
203 204 206 207 209 211 212 213

The heights of 15 randomly chosen males are:

154 158 158 162 165 168 171
176 178 180 181 182 182 183 186

a Find the mean and standard deviation for each group.

b Compare your results and comment on any similarities or differences.

9 The table shows the monthly salaries of all the staff at Mount High College.

Monthly salary ($\$x$)	Number of males	Number of females
$1000 < x \le 1500$	4	9
$1500 < x \le 2000$	8	14
$2000 < x \le 2500$	14	11
$2500 < x \le 3000$	16	10
$3000 < x \le 3500$	7	3
$3500 < x \le 4000$	2	1
$4000 < x \le 4500$	3	0

a Estimate the mean and standard deviation for male staff and for female staff.

b Compare your results and comment on any similarities or differences.

Developing inquiry skills

Click the icon for the full set of life expectancy and GDP data. What can you say about the spread of data in both lists?

What are the standard deviations for life expectancy and GDP per capita?

Do these values imply that the points are all close to the mean values or not?

3.2 Sampling techniques

Here is some of the data on the number of airports in various countries from the CIA factbook for 2013. You can find the full table in the ebook.

Country	Number of airports	Country	Number of airports	Country	Number of airports
United States	13 513	Somalia	61	Lebanon	8
Brazil	4093	Chad	59	Turks and Caicos Islands	8
Mexico	1714	Ethiopia	57	Togo	8
Canada	1467	Yemen	57	Sierra Leone	8
Russia	1218	Suriname	55	Burundi	7
Argentina	1138	Morocco	55	Equatorial Guinea	7
Bolivia	855	French Polynesia	54	Rwanda	7
Colombia	836	Nigeria	54	Kuwait	7
Paraguay	799	Uzbekistan	53	Moldova	7
Indonesia	673	Austria	52	Falkland Islands (Islas Malvinas)	7
South Africa	566	Afghanistan	52	Benin	6
Papua New Guinea	561	Belize	47	Kosovo	6
Germany	539	Uganda	47	Micronesia, Federated States of	6
China	507	Israel	47	Western Sahara	6
...	...	...	...	...	...

It is possible to use all this data for analysis. However, it would be easier if you could just take a **sample** of the data instead.

A **population** is the whole group from which you may collect data.

A **sample** is a small group chosen from the population.

Simple random sampling is selecting a sample completely at random. For example, using a random number generator or picking numbers from a hat.

Systematic sampling is, for example, taking every fifth entry starting at a random place.

In the table of airports, all the data is from a website and is called the **population**. How can you take a **random** sample of this data to use for analysis?

Investigation 7

1 Using the data in the table, use technology to find the mean number of airports.

Using the spreadsheet of data in the ebook, click on the arrow next to the Σ AutoSum icon and select Average. Then highlight all the entries and press Enter. If you are using your GDC then you need to enter all the data into a list, select Statistics, start calculations, one-variable statistics, 1 list, enter the name you gave the list and press OK.

2 Which of the following methods do you think will give you a random sample? Using the spreadsheet or your GDC, find the mean in each case and give a reason why you think it does or does not give a random sample.

   a Take the first 50 countries. Which type of sample do you think this is? Discuss whether it will be a good representation for the mean.

   b Take the first 25 countries and the last 25 countries. Which type of sample do you think this is? Discuss whether it will be a good representation for the mean.

   c Take every fifth country. Which type of sample do you think this is? Discuss whether it will be a good representation for the mean.

   d Use the random number generator on your GDC to pick out 30 countries. Which type of sample do you think this is? Discuss whether it will be a good representation for the mean.

   To use the random number generator on the GDC, select Probability, Random, Integer(1, total number of countries, number of countries you want to select), eg Integer(1,450,9) will give you a list of nine random integers between 1 and 450. Be careful to check that none of the integers are repeated. If so, you will have to select some more numbers at random to make up your total.

   e Put the names of all the countries into a hat and pick out 50 countries at random.

   f Ask your friends which country they are from and use those countries.

3 Which methods in part **2** do you think will give the most reliable estimates of the population mean? Explain your answer.

4 How do you know whether the data is biased?

5 **Factual** What is biased data? What is a reliable result?

6 **Conceptual** How can you decide whether or not the results are biased or reliable?

Convenience sampling is getting data by selecting people who are easy to reach, for example people at a school, club, etc. It does not include a random sample of participants and so the results could be biased.

A **biased** sample is one that is not random—for example, researching spending habits on cars and only interviewing people exiting a garage.

Example 7

The following data shows the IQs of 200 people:

56	62	65	68	69	70	71	71	75	77	79	79	81	81	81	83	84	85	85	85
86	86	86	87	87	87	87	87	87	87	88	88	88	88	88	89	89	89	89	89
89	89	89	89	89	89	89	91	92	92	92	92	93	93	93	93	93	93	94	94
94	94	94	94	94	95	95	95	95	95	95	96	96	96	96	96	96	96	97	97
97	98	98	98	98	98	98	98	99	99	99	99	99	99	99	99	99	99	99	99
100	100	100	100	100	100	100	100	100	100	100	100	100	100	100	100	100	100	100	100
100	100	100	100	100	100	100	101	101	101	101	101	101	101	101	101	101	102	102	102
102	103	103	103	103	104	104	104	104	104	105	106	106	107	107	107	107	107	107	107
107	107	107	108	108	108	109	110	110	110	110	112	112	113	113	113	114	114	115	115
117	118	119	121	121	125	128	129	129	131	134	135	136	137	140	141	143	145	148	156

a Find the mean of the IQs.

b Find the mean of the first 20 numbers and the last 20 numbers.

Comment on the type of sample this is and any advantages and/or disadvantages it may have.

c Find the mean of the subset of the data consisting of every fifth IQ.

Comment on the type of sample this is and any advantages and/or disadvantages it may have.

d Find the mean of a random sample of 30 IQs.

Comment on the type of sample this is and any advantages and/or disadvantages it may have.

e Comment on which of these methods gives the best approximation to the mean of all 200 IQs.

Continued on next page

Statistics and probability

a 99.74	Put the data into a frequency table and find the mean.
b 104 This is a biased sample since it is not random. It is easy to use but unreliable.	
c 98.725 This is systematic sampling since every fifth entry is selected. It gives a good representation of the mean. It is easy to use and is not time-consuming.	You can start at any number and choose every fifth number until you have 40 numbers in total. Everyone will have a slightly different answer for this depending on the starting place.
d 101.7 This is a random sample. Each time a random sample is chosen, different numbers will be selected. The advantage is that the selection is truly random. The disadvantage is that it will select different numbers each time and can be time-consuming.	For example, using randomly generated integers, the selection that appeared was: 26, 194, 38, 77, 142, 174, 27, 153, 34, 176, 36, 40, 67, 122, 84, 148, 162, [38], 43, 132, 90, 98, 133, 166, 175, 103, 136, 185, 196, 64 38 is repeated, so another random number must be generated. This is 10. Find the corresponding number in the list and then find the mean of the selected numbers.
e In this case the systematic sample gave the closest value to the mean. It is also quite a simple method to use.	

Quota sampling is setting certain quotas for your sample, for example selecting a sample of eight boys and eight girls.

For example, the school canteen is considering introducing a new lunch menu and would like feedback from the students. The school has 250 boys and 300 girls and so the canteen manager decides to interview 25 boys and 30 girls to find out their opinion of the new menu. He stands at the entrance to the canteen and interviews the first 25 boys and 30 girls who come into the canteen.

This is called a quota sample. It is not random. It can be biased and unreliable. The advantage is that it is inexpensive, easy to perform and saves time.

However, it is more reliable than convenience sampling where people are selected based on availability and may not be representative of the population. This type of sampling produces a non-probability sample and can also be biased and unreliable.

Stratified sampling is selecting a random sample where numbers in certain categories are proportional to their numbers in the population.

For example, if 20% of students in a school were in Grade 7, then you would choose 20% of your sample from Grade 7. The 20% must be a random sample and not a convenience sample.

Example 8

Mandy asks all the students in her school to take a memory test. The students have to remember as many objects as they can from the 20 that Mandy shows them. The results are shown in the table.

Class 7 (20 students)	16, 15, 13, 15, 12, 8, 18, 16, 12, 11, 14, 17, 16, 9, 11, 10, 17, 13, 14, 13
Class 8 (27 students)	19, 15, 16, 14, 11, 16, 18, 15, 13, 12, 10, 8, 20, 14, 17, 12, 10, 7, 19, 20, 13, 17, 16, 16, 16, 15, 11
Class 9 (23 students)	17, 14, 15, 8, 7, 13, 15, 19, 16, 13, 11, 10, 17, 17, 20, 15, 11, 10, 7, 13, 16, 15, 15,
Class 10 (26 students)	9, 10, 10, 12, 18, 16, 17, 15, 11, 11, 14, 16, 19, 19, 11, 15, 17, 13, 13, 14, 13, 13, 9, 10, 8, 15
Class 11 (30 students)	16, 15, 15, 16, 16, 18, 11, 12, 13, 9, 10, 11, 16, 17, 15, 12, 12, 15, 15, 15, 18, 20, 16, 17, 17, 15, 14, 14, 14, 14
Class 12 (24 students)	9, 11, 16, 14, 13, 13, 18, 19, 12, 10, 11, 9, 16, 16, 18, 14, 15, 15, 16, 13, 13, 12, 18, 19

a In order to take a stratified sample of 40 students from the 150 in total, show that Mandy needs to select five students from Class 7.

b Determine how many students Mandy needs to select from each of the other classes.

a $\frac{20}{150} \times 40 = 5.333... \approx 5$ students	To select the 5 students, Mandy needs to use a random number generator to pick 5 numbers from the list for Class 7.
b From class 8 Mandy needs to select 7 students.	$\frac{27}{150} \times 40 = 7.2$, so 7 students from Class 8 Here again Mandy needs to select the 7 students using a random number generator.
From class 9 Mandy needs to select 6 students.	Similarly for the other classes.
From class 10 Mandy needs to select 7 students. From class 11 Mandy needs to select 8 students. From class 12 Mandy needs to select 6 students.	Note that due to rounding, the total is only 39.

Statistics and probability

Investigation 8

A dog kennel has 120 dogs. The ages of the dogs are:

1 1 1 1 1 1 1 1 1 1 2 2 2 2 2 2 2 2 3 3 3 3 3 3 3 3
3 3 3 3 3 3 3 3 4 4 4 4 4 4 4 4 4 4 4 4 4 5 5 5 5 5
5 5 5 5 5 5 5 5 5 5 5 5 5 5 5 5 6 6 6 6 6 6 6 6 6 7
7 7 7 7 7 7 7 7 7 8 8 8 8 8 8 8 8 8 9 9 9 9 9 10 10 10
10 11 11 11 11 12 12 12 13 13 13 14 14 15 16 16

1. How many dogs are in the population?
2. Represent the ages of the dogs and the frequencies in a table.
3. Which average can you read from the table?
4. Find the mean age of the dogs.
5. Discuss which method for finding the mean is easier: from the raw data or from the frequency table.
6. **Conceptual** Why can it be helpful to organize data in a table?
7. Describe how you would take a sample of 40 dogs.
8. **Conceptual** How can you decide whether your sample is unbiased?
9. Take a systematic sample of every five dogs, then find the mean of the sample.
10. Calculate the number of dogs of each age in a stratified sample of 40 dogs. Use the same method as in Example 8. Discuss why you do not get exactly 40 dogs using this method.
11. **Conceptual** How do you decide on the best sampling method to use?

Exercise 3E

1 The heights, to the nearest cm, of the students in a school are as follows:

Class 7 (28 students): 153, 149, 155, 148, 151, 150, 156, 154, 149, 152, 155, 154, 152, 156, 150, 151, 154, 155, 158, 147, 154, 155, 155, 156, 149, 151, 152, 153

Class 8 (30 students): 155, 154, 156, 158, 153, 155, 158, 157, 156, 155, 149, 151, 154, 153, 155, 154, 152, 159, 151, 149, 148, 153, 156, 155, 157, 155, 154, 157, 155, 156

Class 9 (26 students): 151, 158, 155, 156, 155, 158, 159, 160, 154, 153, 148, 156, 149, 150, 157, 156, 157, 156, 154, 155, 158, 153, 155, 150, 158, 160

Class 10 (24 students): 161, 158, 156, 148, 155, 156, 149, 159, 155, 156, 157, 158, 158, 161, 151, 159, 155, 156, 153, 160, 158, 155, 156, 158

Class 11 (25 students): 163, 160, 158, 149, 151, 159, 158, 162, 161, 156, 155, 154, 150, 151, 160, 159, 158, 156, 156, 155, 156, 157, 158, 158, 157

Class 12 (27 students): 151, 163, 165, 158, 155, 156, 159, 160, 161, 165, 159, 155, 156, 158, 158, 157, 155, 154, 152, 150, 163, 159, 158, 155, 156, 162, 158

a Find the mean height of the whole school.

b Use an appropriate sampling method to collect a sample of 50 students and find the mean of the sample. Comment on whether or not your sample is unbiased.

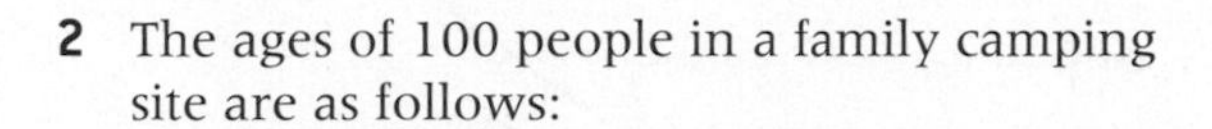

2 The ages of 100 people in a family camping site are as follows:

1	1	1	1	1	2	2	3	3	3
4	4	4	5	5	5	5	6	6	7
7	7	7	7	8	8	8	8	9	9
10	10	10	11	11	12	12	12	12	12
13	13	14	15	15	16	16	18	18	18
19	19	19	20	21	21	23	24	24	25
34	35	35	35	36	36	37	38	38	38
39	40	40	40	41	42	42	43	43	45
45	47	49	49	50	50	50	55	57	62
62	63	65	65	67	67	69	70	71	72

The manager decides to charge less for people over the age of 60.

a Find the mean age of the 100 people and decide whether or not the manager will lose much revenue due to this decision.

b Using an appropriate sampling method, pick a random sample of 35 people and find the mean age of the sample.

c Using a systematic sampling method of every third person, find the mean age of your sample.

d Comment on which method from parts **b** and **c** you think will give the better approximation to the population mean.

3 The number of goals scored in 50 hockey matches is as follows:

Girls: 0, 0, 1, 1, 1, 1, 1, 2, 2, 2, 2, 3, 3, 3, 4, 4, 5, 5, 5, 6, 6, 7, 7, 8, 9

Boys: 0, 1, 1, 1, 1, 1, 2, 2, 2, 3, 3, 3, 3, 4, 4, 5, 5, 5, 5, 6, 6, 6, 7, 7, 8

a Find the mean number of goals scored in all 50 matches.

b Taking a random sample of 12 girls and 12 boys, find the mean of these 24 matches.

c Comment on whether or not your sample gives a good approximation to the population mean.

Statistics and probability

3.3 Presentation of data

A zoo is open 360 days in the year. The number of visitors each day was recorded and displayed in several different types of graph.

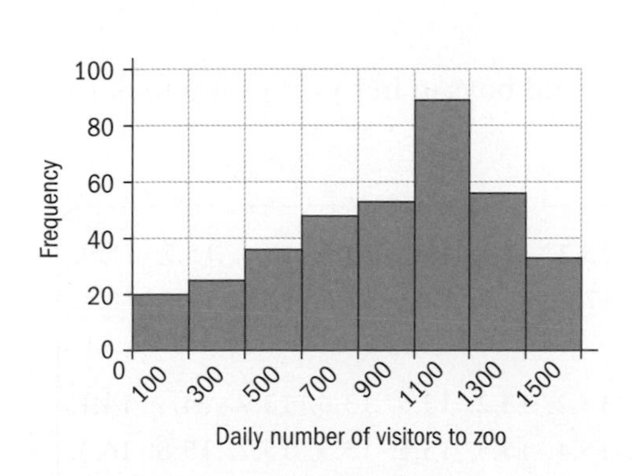

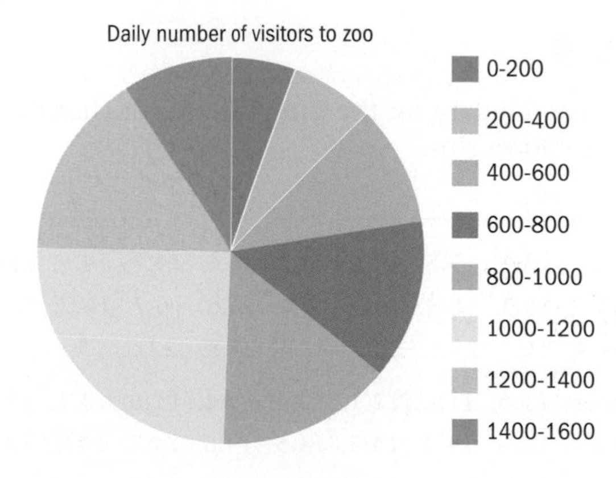

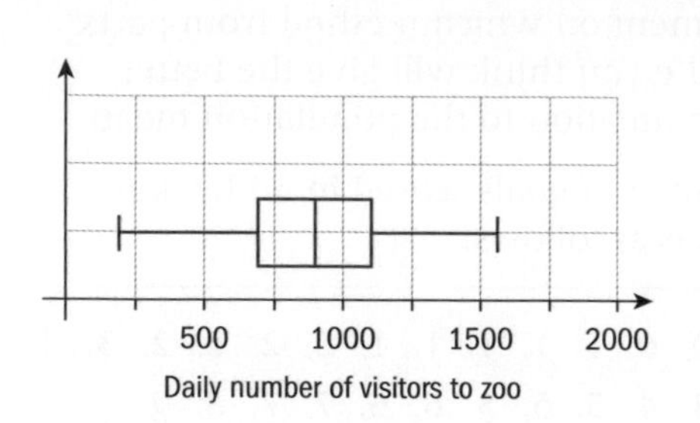

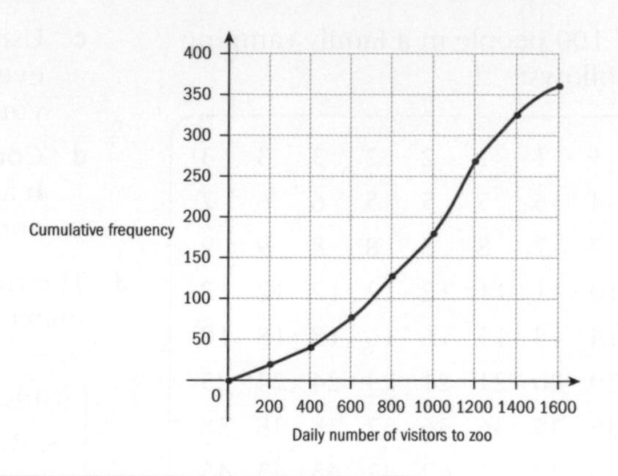

Reflect Which graph do you think is the most useful?

Discuss how the different types of graph might be interpreted.

Developing inquiry skills

Returning to the life expectancy and GDP data, what do you think would be the best method to use for taking a random sample of the data to use for finding an estimate of the mean?

Is the data discrete or continuous?

Frequency histograms

A frequency **histogram** is very similar to a **bar chart**. However, in a histogram there are no spaces between the bars.

Bar charts are useful for graphing **qualitative** data such as colour preference, whereas histograms are used to graph **quantitative** data.

In frequency histograms, as in bar charts, the vertical axis represents frequency.

To draw a frequency histogram, you need to find the lower and upper boundaries of the classes and draw the bars between these boundaries.

International-mindedness

What are the benefits of sharing and analysing data from different countries?

Example 9

Belinda collected data on the time in seconds that the girls and boys in her year group took to complete a 100 m run.

The results are:

Girls' times: 13.5, 13.8, 14.1, 14.3, 14.6, 14.7, 14.9, 15.2, 15.2, 15.3, 15.5, 15.5, 15.6, 15.7, 15.9, 16.1, 16.1, 16.3, 16.3, 16.3, 16.4, 16.6, 16.7, 16.7, 16.9, 17.2, 17.2, 17.5, 17.6, 17.8, 17.8, 18.4, 18.5, 18.9, 19, 20.1, 20,7, 21.4, 21.8, 22.5

Boys' times: 11.5, 11.8, 12.1, 12.4, 12.4, 12.6, 13.1, 13.2, 13.2, 13.2, 13.5, 13.6, 13.7, 14, 14.1, 14.1, 14.2, 14.3, 14.3, 14.3, 14.5, 14.6, 14.7, 14.9, 15.3, 15.4, 15.5, 15.5, 15.5, 15.7, 15.8, 16.3, 16.4, 16.6, 16.6, 16.7, 17.1, 17.4, 17.7, 20.5

a Complete the frequency table for this data.

Time (t seconds)	Frequency (girls)	Frequency (boys)
$11 < t \le 12$		
$12 < t \le 13$		
$13 < t \le 14$		
$14 < t \le 15$		
$15 < t \le 16$		
$16 < t \le 17$		
$17 < t \le 18$		
$18 < t \le 19$		
$19 < t \le 20$		
$20 < t \le 21$		
$21 < t \le 22$		
$22 < t \le 23$		

b Draw a frequency histogram for the girls and a frequency histogram for the boys to represent this data.

c The PE teacher was interested in comparing the times of the boys and the girls for the 100 m run. For both the girls and the boys, find the following information:

	Fastest time	Lower quartile	Median	Upper quartile	Slowest time
Girls					
Boys					

a

Time (t seconds)	Frequency (girls)	Frequency (boys)
$11 < t \le 12$	0	2
$12 < t \le 13$	0	4
$13 < t \le 14$	2	8
$14 < t \le 15$	5	10
$15 < t \le 16$	8	7
$16 < t \le 17$	10	5
$17 < t \le 18$	6	3
$18 < t \le 19$	4	0
$19 < t \le 20$	0	0
$20 < t \le 21$	2	1
$21 < t \le 22$	2	0
$22 < t \le 23$	1	0

Continued on next page

b

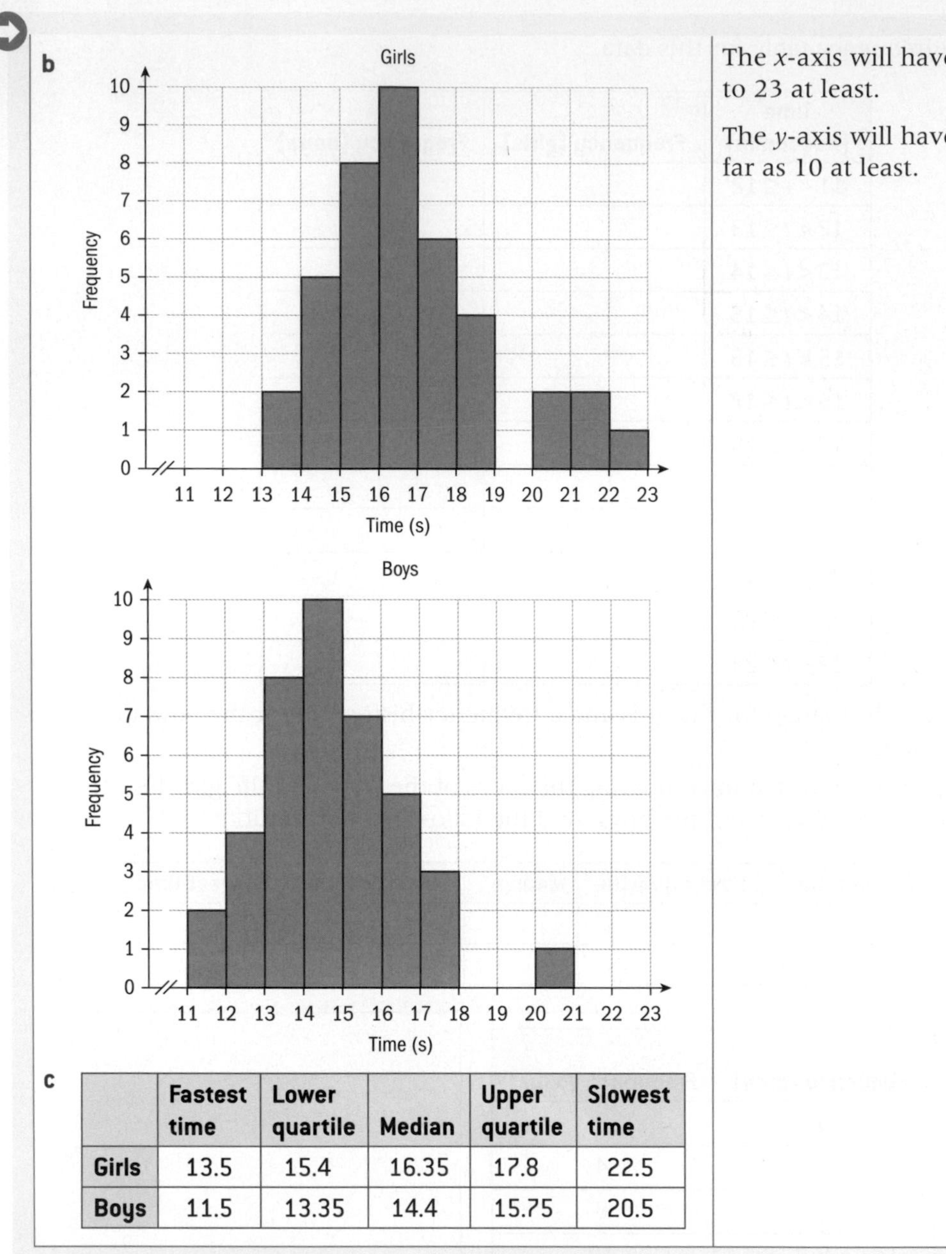

The x-axis will have to go from 11 to 23 at least.

The y-axis will have to go up as far as 10 at least.

c

	Fastest time	Lower quartile	Median	Upper quartile	Slowest time
Girls	13.5	15.4	16.35	17.8	22.5
Boys	11.5	13.35	14.4	15.75	20.5

You can use the five values from the example to draw a box-and-whisker plot to compare the data sets.

Box-and-whisker plots are very convenient for comparing sets of data. Here you can easily see that boys have a faster time than girls for all five values. However, the spread of data for girls and boys is fairly equal.

You can also see that in both cases the data is not symmetrical about the median. The data between the median and the slowest time is more spread out than the rest of the data.

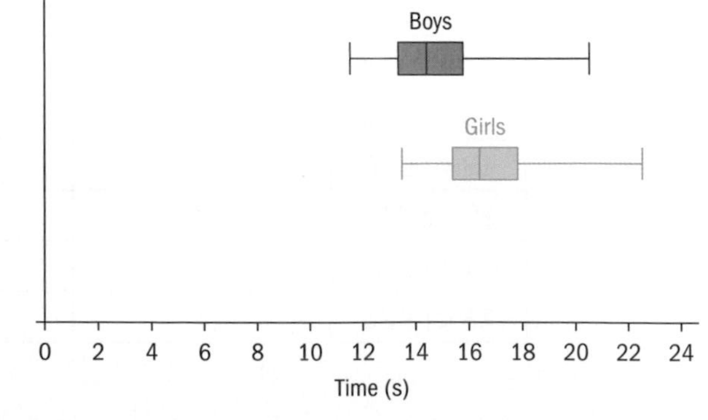

- To draw a box-and-whisker plot you need five pieces of information, called the five-number summary: the smallest value, the lower quartile (LQ), the median, the upper quartile (UQ) and the largest value.
- An outlier is a value that is much smaller or much larger than the other values. An outlier is a point less than the LQ $-1.5 \times$ IQR or greater than the UQ $+1.5 \times$ IQR.

Example 10

Data on the shoe sizes of a group of students is shown in the table.

Shoe size	34	36	37	38	39	40	41	42	49
Females	2	2	10	8	7	3	0	0	0
Males	0	0	3	7	12	9	3	1	1

Draw a box-and-whisker plot for the females and for the males and compare the two plots.

State whether the box plots are symmetrical.

Comment on whether there are any outliers.

Draw the box plots again showing any outliers clearly. Outliers are represented by crosses.

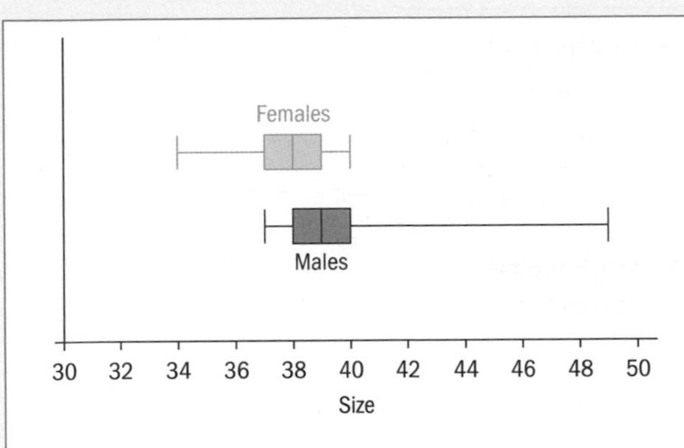

The five-number summary for the females is 34, 37, 38, 39, 40.

$IQR = 39 - 37 = 2$

$LQ - 1.5(2) = 37 - 3 = 34$

$UQ + 1.5(2) = 42$

So there are no outliers for the females.

The five-number summary for the males is 37, 38, 39, 40, 49.

$IQR = 40 - 38 = 2$

$LQ - 1.5(2) = 35$

$UQ + 1.5(2) = 43$

So 49 is an outlier for the males.

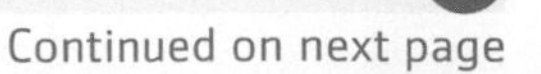
Continued on next page

Statistics and probability

The data for the females is more symmetrical than the data for the males.

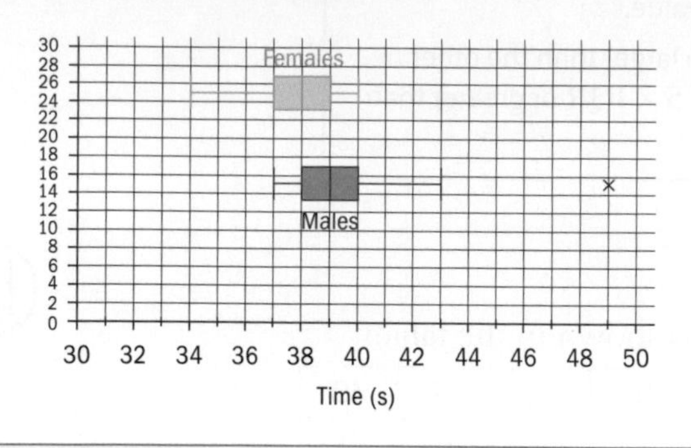

Investigation 9

The weights, w kg, of 30 new-born babies in the Town hospital are:

2.6 3.1 1.8 2.5 4.6 3.6 3.4 2.9 4.8 6.9 4.1 5 3.5 1.2 4.4
5.1 9.6 3.3 4.1 3.7 2.8 2.9 3.4 5.1 4.6 3.9 2.7 3.6 4.2 4.9

The weights, w kg, of 30 new-born babies in the Country hospital are:

2.9 4.1 2.6 3.2 5.1 4.7 3.9 3.8 5.6 5.9 4.8 4.5 2.9 2.6 4.8
6.8 9.2 8.3 5.7 6.3 3.8 2.9 4.4 1.8 4.3 4.9 3.5 6.6 3.7 4.6

1 Complete the grouped frequency table:

Town hospital frequencies	Weight (w kg)	Country hospital frequencies
	$1 \le w < 2$	
	$2 \le w < 3$	
	$3 \le w < 4$	
	$4 \le w < 5$	
	$5 \le w < 6$	
	$6 \le w < 7$	
	$7 \le w < 8$	
	$8 \le w < 9$	
	$9 \le w < 10$	

2 Draw a histogram to represent the data.

3 Using the **original data** find the five-number summary for each hospital and draw box-and-whisker plots. You use the original data because it will give you exact answers and not approximations.

4 Discuss whether you can tell whether there are any outliers from the histogram.

5 Compare the two box-and-whisker plots. Discuss whether either of them is skewed.

6 **Conceptual** How do box-and-whisker plots allow you to compare data visually?

7 **Conceptual** How do you know which diagram to use to represent a data set?

Interpreting a box-and-whisker plot:

- 25% of values are between the smallest value and the LQ.
- 25% of values are between the LQ and the median.
- 25% of values are between the median and the UQ.
- 25% of values are between the UQ and the largest value.

Investigation 10

The box-and-whisker plots show the weights, in kilograms, of 24 male poodles (upper) and 24 female poodles (lower).

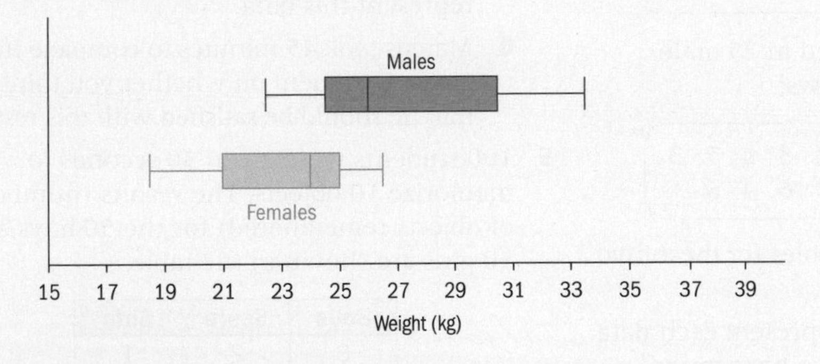

1 Write down the median for both groups.

2 Write down the IQR for both groups.

3 Write down the percentage of female poodles that weigh less than 24 kg.

4 Write down the percentage of male poodles that weigh between 26 and 30.5 kg.

5 Compare the two box plots and discuss the differences.

6 **Factual** Which measures of central tendency and dispersion can you read from a box-and-whisker plot and from a histogram?

7 **Conceptual** Is it useful to have more than one way of representing a univariate data set?

TOK

Can you justify using statistics to mislead others?

How easy is it to be misled by statistics?

Exercise 3F

1 Theo threw a die 40 times. The numbers that appeared were:

2 3 3 1 6 6 5 2 4 4 1 1 5 6 3 4 2 2 3 5
1 6 4 2 2 3 1 4 4 5 1 6 6 3 2 2 1 1 4 5

Millie also threw a die 40 times and the numbers that she threw are shown below.

6 5 5 6 1 1 3 4 5 4 3 2 2 2 4 5 4 6 6 1
1 2 2 1 3 3 3 6 5 5 4 1 2 2 3 3 6 5 1 3

a Construct frequency tables for these two sets of data.

b Draw a histogram to represent each set of data and compare the two histograms.

2 The number of goals scored in 25 female hockey matches is as follows:

0 3 1 4 2 3 4 0 1 0 5 2 6
3 1 3 3 2 4 2 5 1 0 2 1

The number of goals scored in 25 male hockey matches is as follows:

2 4 1 0 3 1 2 6 2 8 4 5 3
1 7 3 2 0 0 1 5 2 6 4 6

a Construct frequency tables for these two sets of data.

b Draw a histogram to represent each data set and compare the two histograms.

3 The heights, in cm, of 32 female gymnasts were recorded:

148 152 147 149 150 147 151
142 156 148 148 149 150 152
155 154 151 154 148 150 149
145 147 148 161 152 162 149
146 151 150 157

a Construct a grouped frequency table, using groups of 5 cm.

b Draw a histogram to represent this data.

c Draw a box-and-whisker plot to represent the data.

d State whether the data is symmetrical or not. Give a reason for your answer.

4 The times, in minutes, to complete 200 games of chess are shown in the table.

Time (x minutes)	Frequency
$20 \le x < 30$	36
$30 \le x < 40$	67
$40 \le x < 50$	48
$50 \le x < 60$	27
$60 \le x < 70$	10
$70 \le x < 80$	7
$80 \le x < 90$	5

a Draw a histogram to represent this data.

b Find the mean, median, LQ, UQ and range and determine whether there are any outliers.

c Given that the quickest time was 26 minutes and the longest time was 84 minutes, draw a box-and-whisker plot to represent this data.

d Marcus took 45 minutes to complete his game. Comment on whether you think that he should be satisfied with this result.

5 100 students were given 30 seconds to memorize 10 objects. The results (number of objects remembered) for the 50 boys and 50 girls are shown in the table.

Boys	Score	Girls
0	2	1
1	3	3
3	4	6
10	5	9
18	6	12
12	7	10
3	8	5
2	9	2
1	10	2

a Find the mean, median, LQ, UQ and range for the boys and for the girls, and determine whether there are any outliers.

b Draw box-and-whisker plots for the boys and the girls, and compare the results.

c Comment on whether the data sets are symmetrical.

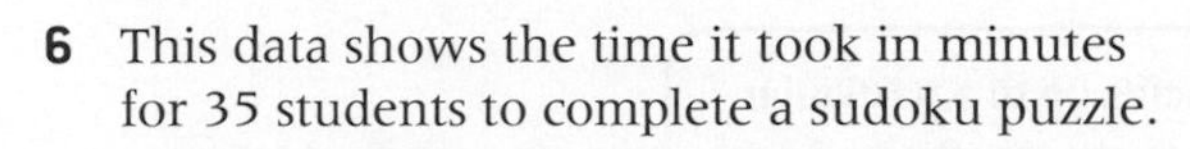

6 This data shows the time it took in minutes for 35 students to complete a sudoku puzzle.

9	12	21	15	15	8	12	11	22
24	17	10	15	6	12	18	35	12
8	19	22	26	24	17	18	21	20
16	9	43	12	16	15	12	18	

a Find the mean, median, LQ, UQ and range for this data and determine whether there are any outliers.

b Draw a box-and-whisker plot.

c If Jin took 16 minutes to complete the sudoku puzzle, find the number of students who took longer than him to complete the puzzle.

7 The box-and-whisker plot shows the results from tossing a die 60 times.

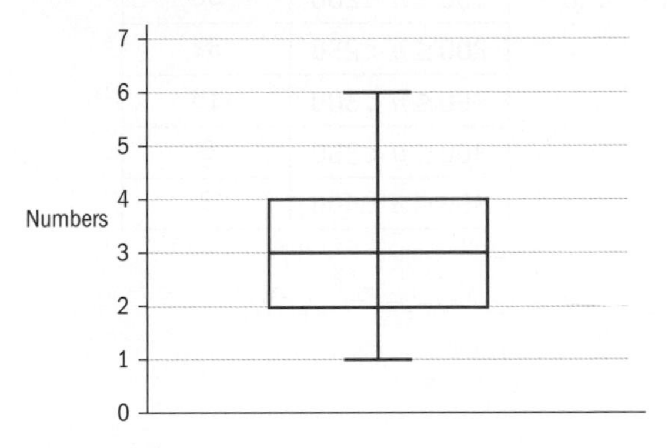

a Write down the median score.

b Find the interquartile range.

c Comment on whether the data is symmetrical.

8 The box-and-whisker plots show the scores in a mathematics test for 60 boys (top) and 60 girls (bottom).

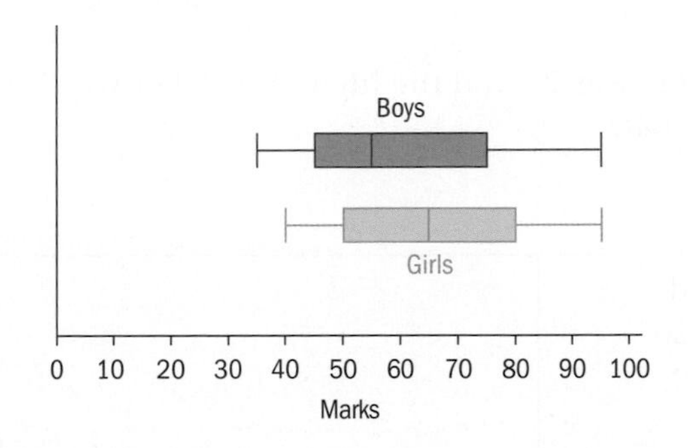

a Write down the median score for the boys and for the girls.

b Find the interquartile range for the boys and for the girls.

c Write down the percentage of boys who scored between 45 and 55.

d Write down the percentage of girls who scored between 65 and 95.

e Find the number of boys who scored less than 45.

f Find the number of girls who scored more than 50.

g Comment on whether the data is symmetrical.

9 The box-and-whisker plot shows the weights, in kilograms, of 40 pandas all of the same gender.

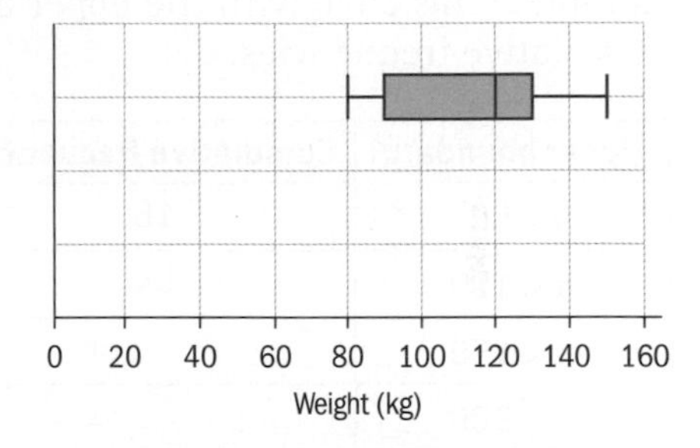

a Write down the median weight.

b Write down the range.

c Find the number of pandas that weigh less than 90 kg.

d Write down the percentage of pandas that weigh between 120 and 160 kg.

e Find the number of pandas that weigh between 90 and 130 kg.

f If the average weight of a panda is about 120 kg, state what information you can deduce from the 40 pandas in the sample.

Male pandas weigh, on average, between 80 kg and 140 kg, and females weigh, on average, between 70 kg and 120 kg.

g From the box plot, state the gender of the pandas in the sample.

The **cumulative frequency** is the sum of all the frequencies up to a particular value. To draw a cumulative frequency curve, you need to construct a cumulative frequency table, with the upper boundary of each class interval in one column and the corresponding cumulative frequency in another. Then plot the upper class boundary on the x-axis and the cumulative frequency on the y-axis.

Example 11

The number of visitors, n, to Hailes Castle was noted on 200 separate days of the year.

Number of visitors (n)	Frequency
$0 \le n < 50$	16
$50 \le n < 100$	38
$100 \le n < 150$	50
$150 \le n < 200$	36
$200 \le n < 250$	32
$250 \le n < 300$	19
$300 \le n < 350$	6
$350 \le n < 400$	3

a Explain how you can tell that there were fewer than 100 visitors on 54 of the days.

b Complete this table with the upper boundaries and cumulative frequencies.

Upper boundary	Cumulative frequency
$n < 50$	16
$n < 100$	54
$n < 150$	
$n < 200$	
$n < 250$	
$n < 300$	
$n < 350$	
$n < 400$	

c Draw a cumulative frequency curve for this data.

d Use your cumulative frequency curve to find an estimate for the median, the lower quartile and the upper quartile. (In other words, find the values on the x-axis corresponding to 100, 50 and 150 on the y-axis.)

e Find an estimate for the 85th percentile.

f If you are told that the lowest number of visitors was 25 and the highest number was 370, draw a box-and-whisker plot to represent this data.

g Determine whether there are any outliers.

a On 16 days there were $0 \le n < 50$ visitors and on 38 days there were $50 \le n < 100$ visitors, so in total there were fewer than 100 visitors on $16 + 38 = 54$ days.

b

Upper boundary	Cumulative frequency
$n < 50$	16
$n < 100$	54
$n < 150$	104
$n < 200$	140
$n < 250$	172
$n < 300$	191
$n < 350$	197
$n < 400$	200

Add 54 to 50 to get 104.

Add 104 to 36 to get 140 and so on.

c

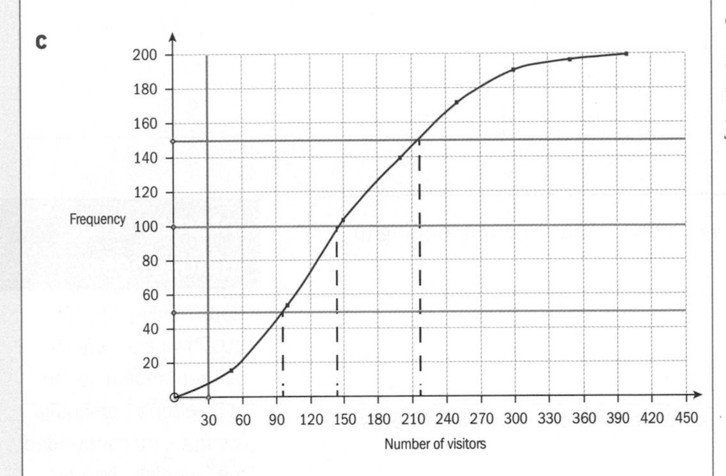

The upper boundary is plotted against the cumulative frequency, and the points are joined up with a smooth curve.

d Approximate values:

Median = 146

LQ = 95

UQ = 215

To find the median, draw a horizontal line from $100\left(=200\times\frac{1}{2}\right)$ on the y-axis to the curve.

From where this line meets the curve, draw a vertical line down to the x-axis and read off the answer.

Similarly for the LQ $\left(200\times\frac{1}{4}\right)$ and the UQ $\left(200\times\frac{3}{4}\right)$.

These are approximate values because you are not told the exact number of visitors each day.

Continued on next page

e Approximately 247	85% of 200 = 170. Draw a horizontal line from 170 on the y-axis until it meets the curve. Draw a vertical line from that point to the x-axis and read off the answer.

f

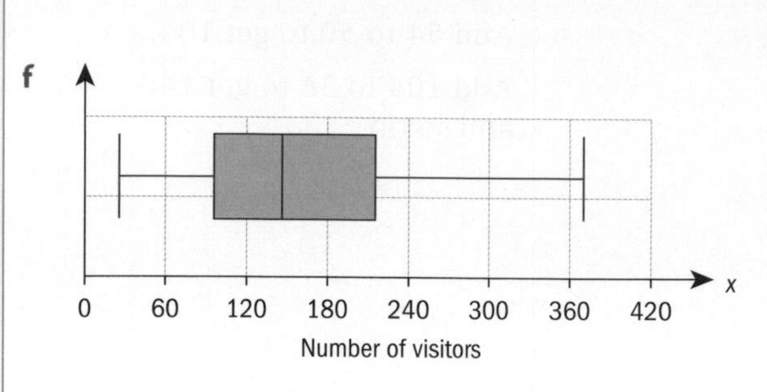

g IQR = 215 − 95 = 120

$95 - 1.5 \times 120 = -85$

$215 + 1.5 \times 120 = 395$

So there are no outliers.

To find any **percentile**, p%, you read the value on the curve corresponding to p% of the total frequency.

Investigation 11

This is the cumulative frequency curve of the weights of 100 male athletes.

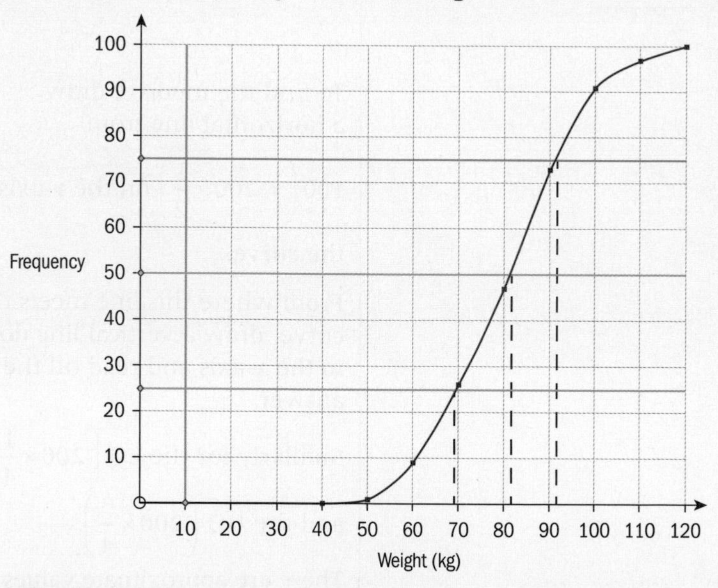

1 **Factual** What do the horizontal lines at 75, 50 and 25 on the vertical axis represent?

2 What percentage of the data values are below 70 kg?

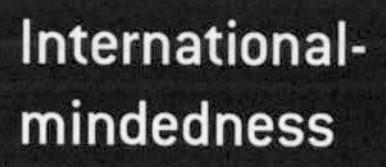

International-mindedness

Hans Rosling (1948 – 2017) was a professor of international health at Sweden's Karolinska Institute. He co-founded the Swedish chapter of Medécins Sans Frontières, and was able to clearly show the importance of collecting and analysing real data in order to understand situations and plan for the future.

3 If 90% of the athletes weigh more than x kg, how could you find the value of x?

4 How accurate are the values you have found?

5 **Factual** How can you find out whether there are any outliers?

6 **Conceptual** How does the cumulative frequency curve allow you to analyse the data?

Exercise 3G

1 The table shows the average times, in minutes, that 100 people waited for a train.

Time (x minutes)	Frequency
$0 \le x < 2$	5
$2 \le x < 4$	11
$4 \le x < 6$	23
$6 \le x < 8$	31
$8 \le x < 10$	19
$10 \le x < 12$	8
$12 \le x < 14$	3

a Construct a cumulative frequency table for this data.

b Sketch the cumulative frequency curve.

c Use your graph to find estimates for the median and interquartile range.

d Find the 10th percentile.

The train company will refund the fare if customers have to wait 11 minutes or more for a train.

e Determine how many customers can claim for a refund of their fare.

2 Nuria recorded the number of words in a sentence in one chapter of her favourite book. The results are shown in the table.

Number of words (x)	Frequency
$0 \le x < 4$	5
$4 \le x < 8$	32
$8 \le x < 12$	41
$12 \le x < 16$	28
$16 \le x < 20$	22
$20 \le x < 24$	12
$24 \le x < 28$	7
$28 \le x < 32$	3

a Construct a cumulative frequency table for this data.

b Sketch the cumulative frequency curve.

c Use your graph to find estimates for the median and interquartile range.

d Determine whether there are any outliers.

e Find the 90th percentile.

f The smallest sentence had 1 word and the longest sentence had 31 words. Draw a box-and-whisker plot to represent this data.

g A children's book has, on average, 8 words in a sentence and an adult book has, on average, 15 words in a sentence. State the type of book you think Nuria is reading, justifying your answer.

3 A tourist attraction is open 350 days of the year. The number of visitors each day for the 350 days was recorded and the results are shown in the table.

Number of visitors (n)	Frequency
$100 \le n < 200$	24
$200 \le n < 300$	36
$300 \le n < 400$	68
$400 \le n < 500$	95
$500 \le n < 600$	73
$600 \le n < 700$	38
$700 \le n < 800$	16

a Draw a suitable graph to represent this data.

b Use your graph or the data to find estimates for the median and interquartile range.

c Determine whether or not there are any outliers.

d The smallest number of visitors was 185 and the largest number was 792. Draw a box-and-whisker plot to represent this data.

If the number of tourists is fewer than 350 in a day, then the attraction loses revenue.

e Determine the number of days on which the attraction loses revenue.

4 The table shows the number of points that 120 students received on their IB diploma.

Number of points (n)	Boys	Girls
$21 \le x < 24$	2	1
$24 \le x < 27$	8	5
$27 \le x < 30$	10	8
$30 \le x < 33$	15	18
$33 \le x < 36$	9	12
$36 \le x < 39$	8	5
$39 \le x < 42$	4	8
$42 \le x < 45$	4	3

a Draw suitable graphs to represent this data.

b Use your graphs to compare the results for the boys and the girls.

c Mary and Martin both score 29 points. Compare their points with the other students of their gender.

5

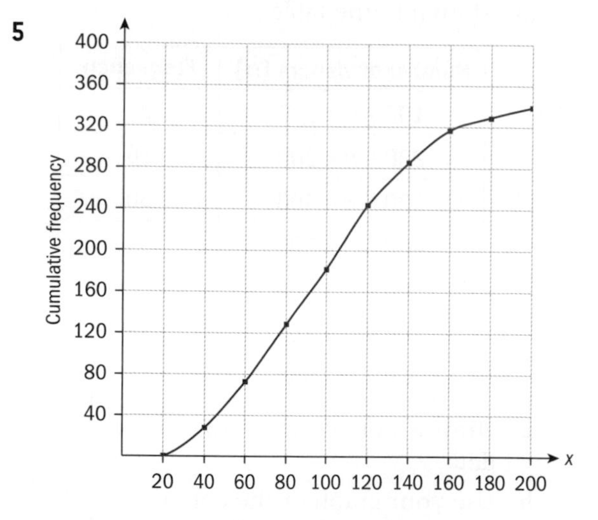

a Using the cumulative frequency curve, write down estimates for:

i the median

ii the interquartile range

iii the 90th percentile.

b Determine whether there are any outliers.

6

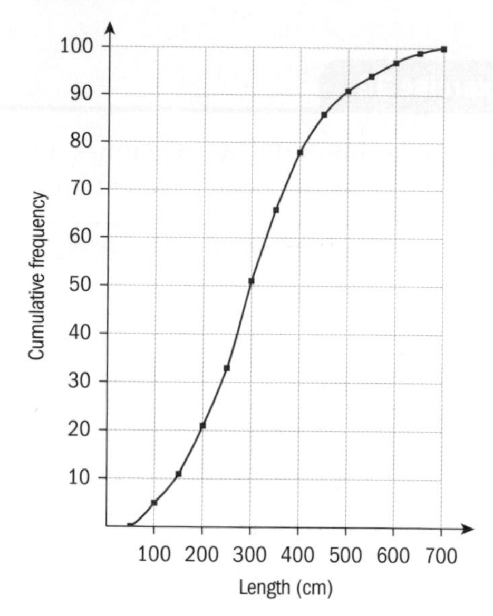

The cumulative frequency curve shows the lengths, in cm, of 100 snakes in a zoo.

a Write down estimates for the median, the lower quartile and the upper quartile.

b The smallest snake is 9 cm long and the longest is 650 cm long. Draw a box-and-whisker plot to represent this data.

c Construct a frequency table for the lengths of the snakes.

d Find estimates for the mean and standard deviation of the lengths of the snakes.

Developing inquiry skills

Return to the opening problem for the chapter. How can you best present the data on life expectancy and GDP?

3.4 Bivariate data

Monica and her friends are training for a charity run. She is interested to find out whether the height of the runners has any effect on the time taken to complete the race. You can see her data in Investigation 12.

TOK

Why are there different formulae for the same statistical measures, such as mean and standard deviation?

Bivariate data has **two** variables; **univariate** data has only one variable.

With bivariate data you have data on two different variables collected from the same individuals that you want to compare to see whether there is any **correlation** between the two variables.

Mr Price was interested to find out whether the number of past papers that his students completed had an effect on the grade they obtained in their final examination. The data he collected is shown below.

Number of past papers	2	6	5	1	4	8	3	12	7	4	2	8	10	9
Examination grade (%)	48	70	61	45	58	85	55	96	80	56	43	88	92	89

He plots all these points on a graph to see whether there is any correlation between the two variables. The number of past papers is the **independent** variable and this is plotted on the x-axis. The examination grade is the **dependent** variable and this is plotted on the y-axis.

The pattern of dots or crosses will give him an indication of how closely the variables are related.

Do you think that the two variables are related?

How closely do you think they are related?

What advice would you give to students who have to take examinations?

Types of correlation

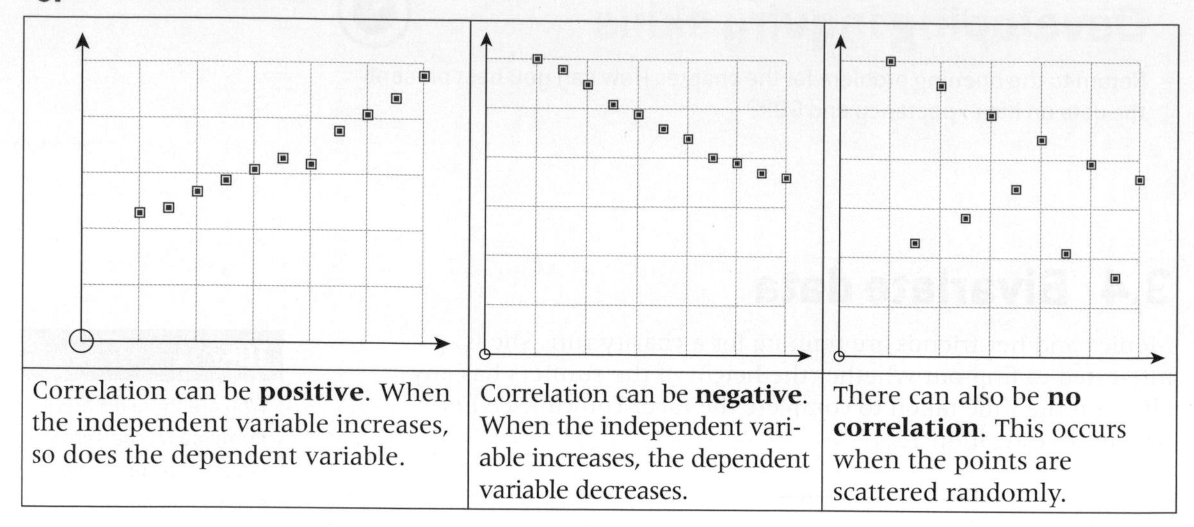

Correlation can be **positive**. When the independent variable increases, so does the dependent variable.	Correlation can be **negative**. When the independent variable increases, the dependent variable decreases.	There can also be **no correlation**. This occurs when the points are scattered randomly.

Correlation can also be described as strong, moderate or weak.

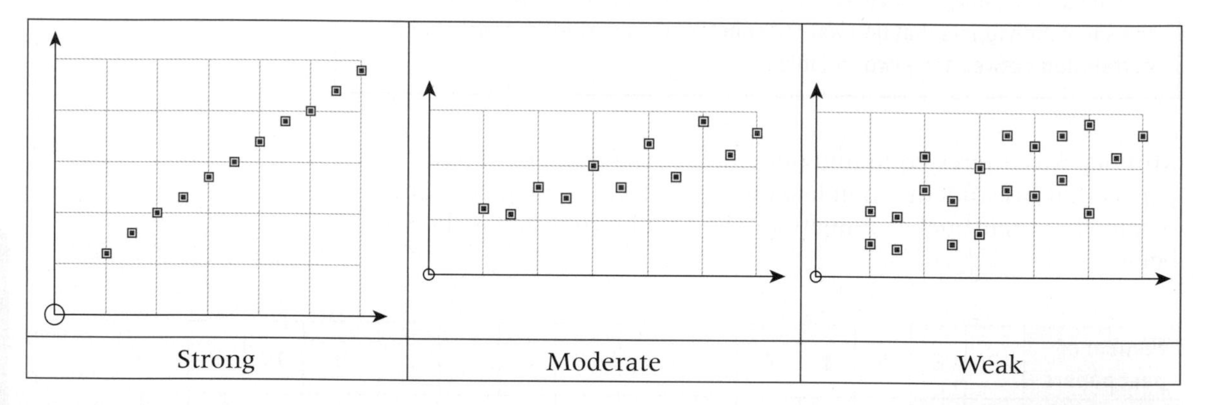

Strong	Moderate	Weak

Example 12

The table gives the heights and weights of 10 camels.

Weight (kg)	450	600	500	750	750	650	900	600	650	800
Height (m)	1.45	1.6	1.5	1.85	1.9	1.75	2.0	1.7	1.65	1.8

a Draw a scatter graph to represent this information.

b Comment on the relationship.

a

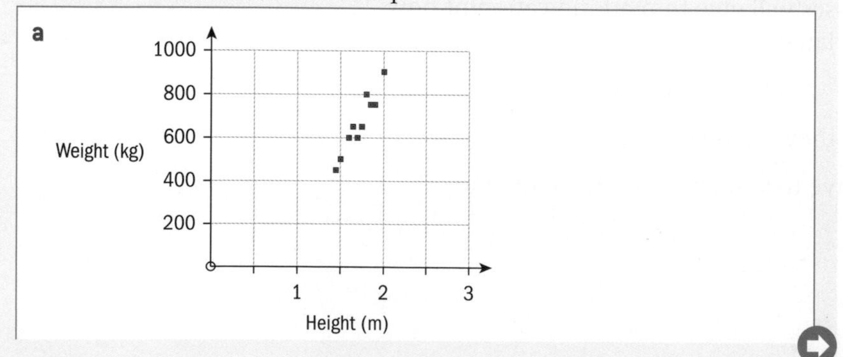

TOK

To what extent can we rely on technology to produce our results?

b There appears to be a strong, positive relationship between the height and the weight: the taller the camel, the more it weighs.

Investigation 12

Twelve students trained every week for a 5 km charity run. Their heights, h cm, and the times, t minutes, it took them to complete the run are shown in the table.

Height (h)	150	163	155	148	154	141	162	148	171	152	153	145
Time (t)	22	18	20	25	21	32	19	24	15	22	21	30

1 Draw a scatter graph to show this information. The height is the independent variable and goes on the horizontal axis; the time taken to complete the run is the dependent variable and goes on the vertical axis. Place a cross at each point, eg at (150, 22), (163, 18) etc.

2 Now that you have a visual picture, do you think that these variables are related?

3 If so, how would you describe the relationship?

4 **Conceptual** What are scatter diagrams useful for?

Example 13

The table shows the number of members in each of nine families and the number of pets the family has.

Number of members	2	3	4	4	5	5	6	7	8
Number of pets	1	0	3	2	5	4	4	7	5

a Draw a scatter graph to represent this data.

b Describe the correlation between the two variables.

c State, with a reason, whether you think that one variable "causes" the other.

a

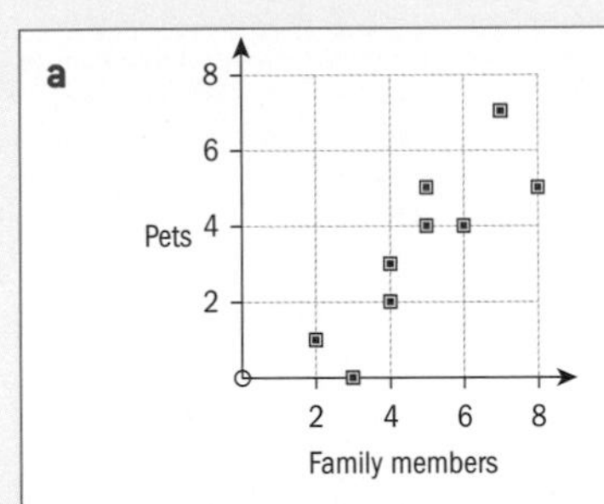

b The correlation is positive and moderate.

c No, the number of members in a family is not caused by the number of pets a family has, and vice versa.

Example 14

The table shows the number of schools and the number of restaurants in a town over a 40-year period.

Year	1980	1985	1990	1995	2000	2005	2010	2015	2020
Number of schools	12	13	15	15	16	17	17	18	19
Number of restaurants	28	30	33	34	36	36	38	39	40

a Draw a scatter graph of the number of schools and the number of restaurants.

b Describe the correlation between the two variables.

c State whether you think that one set of variable "causes" the other.

d State a possible reason why the number of schools and the number of restaurants increased over the 40-year period.

a

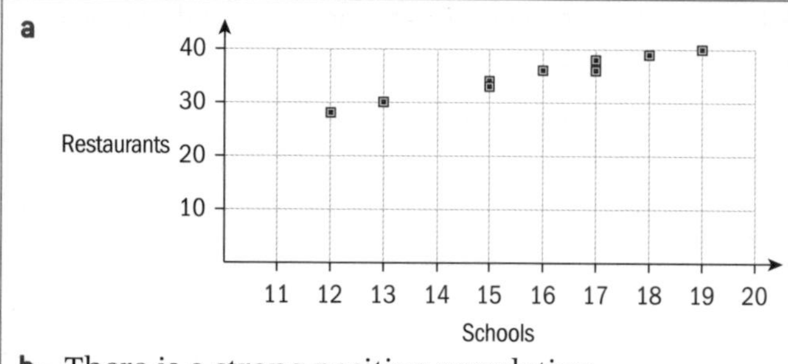

b There is a strong positive correlation.

c Not really.

d The population in the town could be increasing over the years, which could require more schools and more restaurants.

Example 15

The table shows the temperature in °C and the time in days taken for cream to turn sour.

Temperature (°C)	4	8	12	16	20	24	28
Time in days	15	13	9	6	4	2	1

a Draw a scatter graph to represent this data.

b Describe the correlation between the two variables.

c State whether one variable "causes" the other.

a

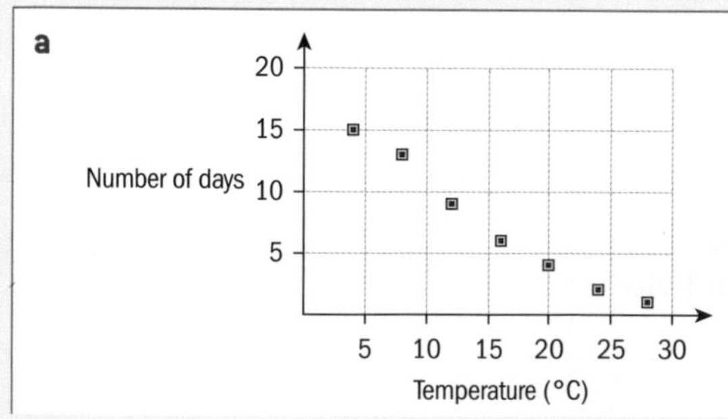

b There is a strong negative correlation: the higher the temperature, the lower the number of days for the cream to sour.	
c The increase in temperature could very well be the reason that the cream turns sour more quickly.	

Investigation 13

Consider the following data sets and the correlations that were found:

A: The number of hours spent training for a race and the time taken to complete the race has a strong negative correlation.

B: The age of participants in a race and the time taken to complete the race has a moderate positive correlation.

C: The number of fish in a garden pool and the number of trees in the garden has a strong positive correlation.

D: The speed of a car and its horsepower has a moderate positive correlation.

E: The temperature and the number of hats sold has a weak negative correlation.

1. In which sets do you think that one variable has an influence on or "causes" the other?
2. In which examples is there a moderate or strong correlation but one variable does not cause the other?
3. **Factual** What is causation?
4. **Conceptual** Does correlation imply causation?

Exercise 3H

1. For the following scatter graphs, describe the type of correlation and the strength of the relationship.

a

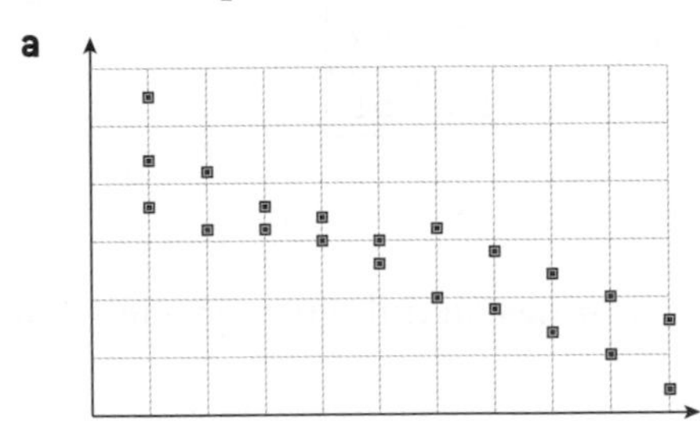

b

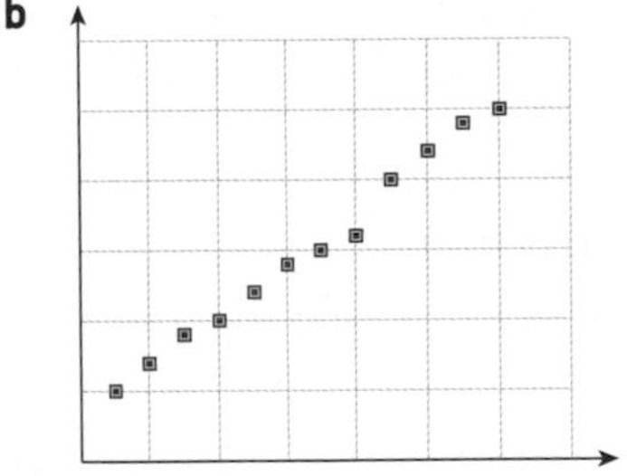

c

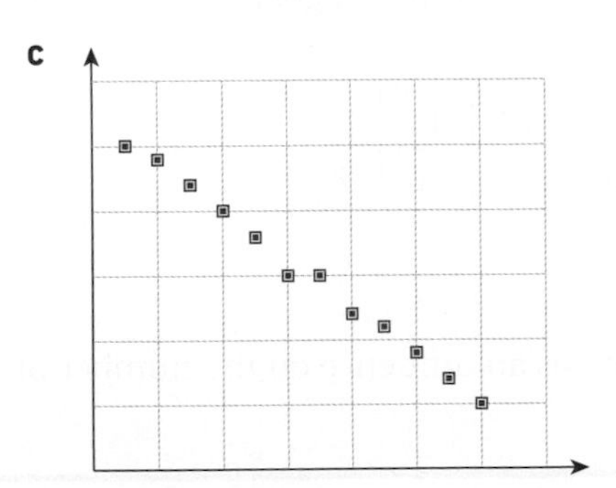

d

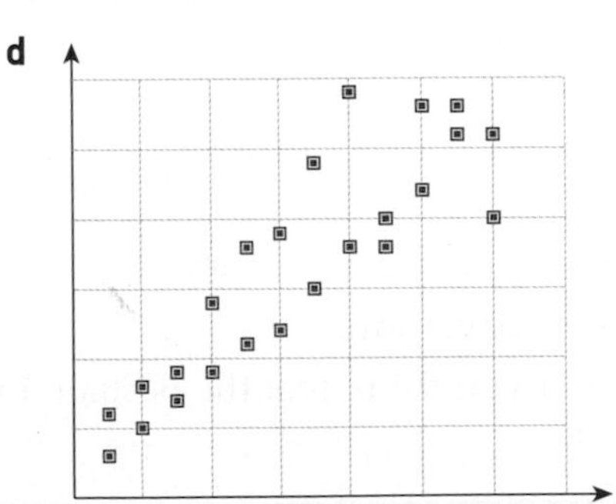

e

f

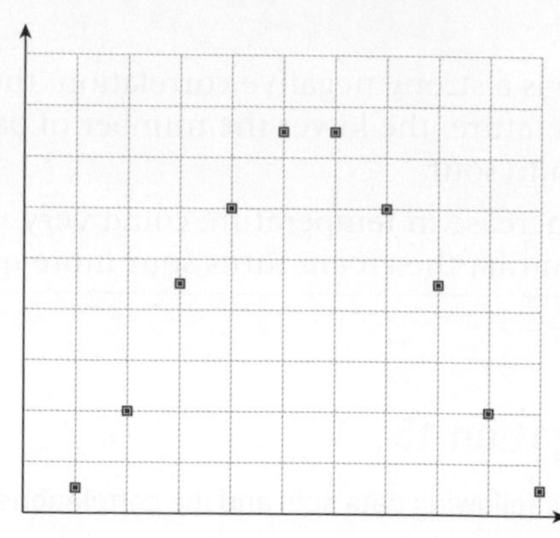

2 The table gives the heights, in cm, and weights, in kg, of 11 football players selected at random.

Height (h cm)	161	173	154	181	172	184	176	169	165	180	173
Weight (w kg)	74	76	61	80	76	88	79	76	75	83	75

a Plot the points on a scatter diagram.

b Comment on the type of correlation. Interpret what this means in terms of the football players.

c State whether the correlation might indicate a causation in this instance. Justify your answer.

3 The table shows the size, in inches, of 10 laptop screens and the cost, in euros, of the laptop.

Size (inches)	11.6	11.6	13.3	14	14	14	15	15.6	15.6	15.6
Cost (euros)	145	170	700	450	370	175	320	500	420	615

a Plot the points on a scatter diagram.

b Describe the correlation.

c State whether you think that the size has an influence on the cost.

4 Twelve students took tests in English and mathematics. The results are shown in the table.

English	44	66	71	33	87	90	55	76	65	95	40	58
Mathematics	71	75	58	63	55	87	54	58	77	54	56	51

a Plot the points on a scatter diagram.

b Describe the correlation.

c State whether you think that the grade for the English test has an influence on the grade for the mathematics test.

5 The data in the table shows the position in the league and the number of goals scored for each team in a hockey league.

Position	1	2	3	4	5	6	7	8	9	10	11	12
Goals scored	52	50	47	44	43	37	36	24	16	12	10	7

a Plot the points on a scatter diagram.

b Describe the correlation.

c State whether you think that the position in the league has an influence on the number of goals scored.

Developing inquiry skills

For the life expectancy and GDP data, use technology to draw a scatter graph of the data.

Is there any correlation between the two variables?

Is there any causal connection between life expectancy and GDP?

Developing your toolkit

Now do the Modelling and investigation activity on page 146.

Chapter summary

- **Univariate** data has only one variable.
- There are two main types of data: **qualitative** and **quantitative**.
- Qualitative data is data that is not given numerically, for example favourite ice cream flavour. Quantitative data is numerical and is classified as **discrete** or **continuous**.
- Discrete and continuous data can be organized into frequency tables or grouped frequency tables.
- For continuous data, the classes must cover the full range of the values and must not overlap.
- Discrete data is either data that can be counted, for example the number of cars in a car park, or data that can only take specific values, for example shoe size.
- Continuous data can be measured, for example height, weight and time.
- The most common measures of central tendency are the mean, median and mode.
- The **mode** of a data set is the value that occurs most frequently. There may be no mode or several modes.
- The **median** of a data set is the value that lies in the middle when the data values are arranged in size. When there are two middle values, the median is the midpoint between the two values.
- The **mean** of a data set is the sum of all the values divided by the number of values.

  For a discrete data set of n values the formula is $\overline{x} = \sum_{i=1}^{n} x_i$.

  For a frequency data set the formula is $\overline{x} = \dfrac{1}{\sum_{i=1}^{n} f_i} \sum_{i=1}^{n} f_i x_i$.
- **Outliers** are extreme data values that can distort the results of statistical processes.
- Measures of dispersion measure how spread out a data set is.
- The **range** is found by subtracting the smallest number from the largest number.
- The **standard deviation**, σ_x, gives an idea of how the data values are related to the mean. The standard deviation is also known as the root-mean-squared deviation and its formula is

$$\sigma_x = \sqrt{\frac{1}{n}\sum_{i=1}^{n}(x_i - \overline{x})^2} = \sqrt{\frac{1}{n}\sum_{i=1}^{n} x_i^2 - \overline{x}^2}$$

- The **variance** is the standard deviation squared, $(\sigma_x)^2$.

Continued on next page

- The **interquartile range** (IQR) is the upper quartile, Q_3, minus the lower quartile, Q_1.
- When the data is arranged in order, the **lower quartile** is the data point at the 25th percentile and the **upper quartile** is the data point at the 75th percentile.
- The mean of a set of numbers is $\bar{x}$ and the standard deviation is σ_x. If you add k to or subtract k from each of the numbers, then the mean becomes $\bar{x} \pm k$ and the standard deviation remains σ_x. If you multiply each number by k then the mean becomes $k \times \bar{x}$ and the standard deviation becomes $|k| \times \sigma_x$.
- The **population** is the whole group from which you can collect data.
- A **sample** is a small group chosen from the population.
- **Simple random sampling** is selecting a sample completely at random, for example by using a random number generator or picking numbers from a hat.
- **Systematic sampling** is, for example, taking every fifth entry starting at a random place.
- **Convenience sampling** is getting data from people who are easy to reach, for example the members of a school, club, etc. It does not select a random sample of participants and so the results could be biased.
- A **biased** sample is one that is not random, for example researching spending habits on cars and only interviewing people exiting a garage.
- **Quota sampling** is setting certain quotas for your sample, for example selecting a sample of eight boys and eight girls.
- **Stratified sampling** is selecting a sample where the numbers in certain categories are proportional to their numbers in the population. For example, if 20% of students in a school were in Grade 7, then you would choose 20% of your sample from Grade 7.
- To draw a box-and-whisker plot you need five pieces of information: the smallest value, the lower quartile (LQ), the median, the upper quartile (UQ) and the largest value.
- An outlier is a point less than the LQ $- 1.5 \times$ IQR or greater than the UQ $+ 1.5 \times$ IQR.
- Interpreting a box-and-whisker plot:
  - 25% of the values are between the smallest value and the LQ.
  - 25% of the values are between the LQ and the median.
  - 25% of the values are between the median and the UQ.
  - 25% of the values are between the UQ and the largest value.
- The **cumulative frequency** is the sum of all the frequencies up to a particular value. To draw a cumulative frequency curve, you need to construct a cumulative frequency table, with the upper boundary of each class interval in one column and the corresponding cumulative frequency in another. Then plot the upper boundary on the x-axis and the cumulative frequency on the y-axis.
- To find any **percentile**, p%, you read the value on the curve corresponding to p% of the total frequency.
- **Bivariate** data has two variables; univariate data has only one variable.
- With bivariate data you have paired data on two variables that you want to compare to see whether there is any **correlation** between the two variables.
- Correlation can **positive** or **negative**, or there may be **no correlation**, and correlation can also be described as **strong, moderate** or **weak**.

Developing inquiry skills

Return to the opening problem.

- Has what you have learned in this chapter helped you to answer the questions?
- What information did you manage to find?
- What assumptions did you make?
- How will you be able to construct a model?
- What other things did you wonder about?

Thinking about the inquiry questions from the beginning of this chapter:

- Has what you have learned in this chapter helped you to think about an answer to most of these questions?
- Are there any questions that you would like to explore further, perhaps for your internal assessment topic?

Chapter review

Click here for a mixed review exercise

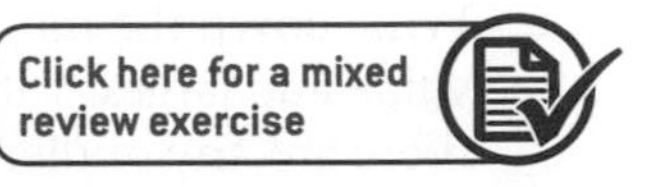

1 State whether the following sets of data are discrete or continuous, and, in each case, construct a frequency table.

a The number of apples in a 1 kg bag:

8 7 9 7 8 10 9 8 7
11 9 9 10 12 7 8 10

b The lengths of pencils, in cm:

7.4 8.5 9.6 7.1 14 13.5 8.8
7.4 11.2 13.6 12.8 14.2 9.8

c The shoe sizes of Grade 6:

34 35 34 33 36 37 36 38 35
36 37 38 35 37 37 38 35

2 Find the mean, median and mode for the following data sets. State which measure of central tendency is best to use in each case.

a The heights of 15 dogs, in cm:

7 23 32 41 32 56 64 67
88 91 110 78 56 45 32

b The price of a pair of shoes in dollars:

46 54 58 62 62 79 96
120 135 185 270 300

c The number of hours Grade 12 students sleep:

4 7 6 6 8 6 9 8 6 5
4 5 5 6 8 8 8 6 7

3 The data table shows the lengths of 120 pike fish.

Length of pike (l cm)	Frequency
$20 \le l < 30$	2
$30 \le l < 40$	12
$40 \le l < 50$	23
$50 \le l < 60$	46
$60 \le l < 70$	28
$70 \le l < 80$	9

a Write down the modal class.

b Find estimates for the median, mean and standard deviation.

c Draw a histogram to represent the data.

4 The marks, out of 50, for a history test have a mean of 38 and a standard deviation of 7. To get a percentage mark, Mr Thoughtful doubles all the marks. Write down the new mean and the new standard deviation.

Statistics and probability

5 Mr Pringle sells vegetables at the market. The number of tomatoes in a bag has a mean of 10 and a standard deviation of 1. On his birthday, Mr Pringle gives everyone who buys a bag of tomatoes three extra tomatoes. Write down the new mean and the new standard deviation on Mr Pringle's birthday.

6 Ursula measures the heights of 35 tulips in her garden. The data she gathered is:

20	20	21	22	22	22	24	25	27
28	28	29	30	31	32	33	33	34
34	34	35	35	36	37	39	39	39
40	41	41	42	43	43	44	45	

a Find the mean and standard deviation and comment on your answer.

b Find the range and interquartile range.

c Write down the median, LQ, UQ, smallest value and largest value and check whether there are any outliers.

d Draw a box-and-whisker plot to represent the data.

7 The grouped frequency table shows the number of hours of voluntary service completed by the 200 students at a community high school.

Number of hours (x)	Frequency
$0 \le x < 10$	8
$10 \le x < 20$	16
$20 \le x < 30$	41
$30 \le x < 40$	54
$40 \le x < 50$	36
$50 \le x < 60$	22
$60 \le x < 70$	17
$70 \le x < 80$	6

a Construct a cumulative frequency table for this data.

b Plot the points and draw the cumulative frequency curve.

c Use your curve to find approximate values for the median and the interquartile range.

The lowest number of hours was 8 and the greatest number was 76.

d Draw a box-and-whisker plot to represent the data.

8 Mr Farmer has 50 chickens. He collects data on the temperature and the average number of eggs that the chickens lay.

Temperature (°C)	Number of eggs
14	43
15	44
16	48
17	46
18	50
19	48
20	50
21	52
22	53
23	55

a Draw a scatter graph to represent this information.

b Describe the correlation.

c Comment on whether the temperature has an effect on the number of eggs laid.

Exam-style questions

9 **P1:** The grouped frequency table below shows the results of a statistics test taken by 70 students.

Test result x%	Frequency
$0 \le x < 20$	8
$20 \le x < 40$	17
$40 \le x < 60$	25
$60 \le x < 80$	13
$80 \le x < 100$	7

a State the modal class for the data. (1 mark)

b Find an estimate for the mean. (3 marks)

c Find an estimate for the standard deviation. (2 marks)

d A similar class took the same test. Their mean mark was 45% and the standard deviation was 19.5.

Compare the marks of the two classes, stating your conclusions (2 marks)

10 P2: The weights (in grams) of 25 mice were recorded as follows.

10, 11, 12, 12, 13, 14, 14, 14, 14, 15, 15, 15, 16, 16, 16, 17, 18, 18, 19, 19, 19, 20, 20, 20, 21

a Find the mean weight of the mice. (2 marks)

b Find the median weight of the mice. (2 marks)

c Find the interquartile range. (4 marks)

d The weight of another mouse was added to the data but found to be an outlier.

Find the least possible weight this new mouse could be, given that it is heavier than each of the others. (2 marks)

---

11 P2: The following tables show the mean daily temperatures, by month, in both Tenerife and Malta.

Tenerife	
Month	**Mean daily temperature (°C)**
January	19
February	20
March	21
April	21
May	23
June	25
July	28
August	29
September	28
October	26
November	23
December	20

Malta	
Month	**Mean daily temperature (°C)**
January	16
February	16
March	17
April	20
May	24
June	28
July	31
August	31
September	28
October	25
November	21
December	17

a Find the mean temperature over the course of the year for Tenerife. (2 marks)

b Find the standard deviation of temperatures in Tenerife. (2 marks)

c Find the mean temperature over the course of the year for Malta. (2 marks)

d Find the standard deviation of temperatures in Malta. (2 marks)

e By referring directly to your answers from parts **a–d**, make contextual comparisons about the temperatures in Tenerife and Malta throughout the year. (4 marks)

---

12 P1: A population of ferrets has mean age 5.25 years and standard deviation 1.2 years.

a Find the mean age of the same population of ferrets 3 years later. (2 marks)

b Find the standard deviation of the same population of ferrets 2 years later, justifying your answer. (2 marks)

13 P1: The following table shows the population sizes of England, Wales, Scotland and Northern Ireland.

Region	Population (millions)
England	54.8
Wales	3.10
Scotland	5.37
Northern Ireland	1.85

Statistics and probability

A polling company wishes to ask questions of a stratified sample of the UK population, and decides that a sample size of 5000 people would be appropriate.

Determine how many people from each region they should choose. (6 marks)

14 P1: The box-and-whisker diagram shows the average times taken for a class of students to walk to school.

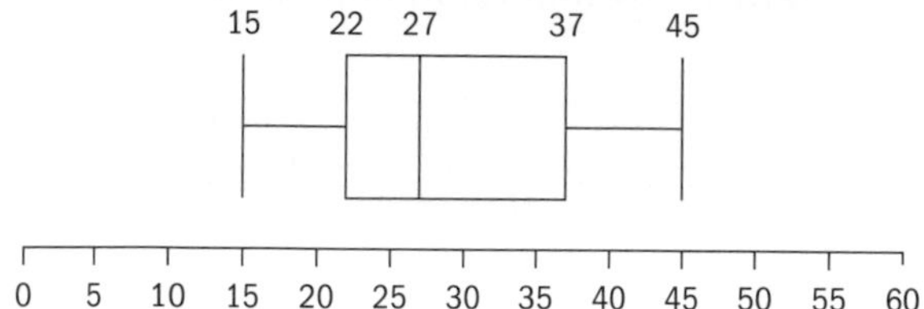

a Find the range. (2 marks)

b Find the interquartile range. (2 marks)

c Find the percentage of students who took between 15 and 37 minutes to walk to school. (1 mark)

d It was found that another student who took 60 minutes to walk to school.

Determine whether this time would be counted as an outlier. (3 marks)

15 P1: Ben practises playing the Oboe daily.

The time (in minutes) he spends on daily practice over 28 days is as follows:

10, 15, 30, 35, 40, 40, 45, 55, 60, 62, 64, 64, 66, 68, 70, 70, 72, 75, 75, 80, 82, 84, 90, 90, 105, 110, 120, 180

a Find the median time. (2 marks)

b Find the lower quartile. (2 marks)

c Find the upper quartile. (2 marks)

d Find the range. (2 marks)

e Determine whether there are any outliers in the data. (4 marks)

f Draw a box-and-whisker diagram for the above data. (3 marks)

16 P2: Ava practises the piano daily.

The time (in minutes) she spends on daily practice over 75 days is as follows.

Time (t minutes)	Frequency
$0 \le x < 15$	4
$15 \le x < 30$	5
$30 \le x < 45$	12
$45 \le x < 60$	24
$60 \le x < 75$	18
$75 \le x < 90$	7
$90 \le x < 100$	5

a State the modal class. (1 mark)

b Find the class in which the median time lies. (2 marks)

c Construct a cumulative frequency table for this data. (3 marks)

d Sketch the cumulative frequency curve. (2 marks)

e Use your curve to find estimates for the median and interquartile range. (4 marks)

17 P2: The following table shows the salaries of the members of a small private business.

Position	Salary ($)
Director	120 000
Line Manager 1	80 000
Line Manager 2	80 000
Analyst 1	25 000
Analyst 2	25 000
Analyst 3	25 000
Analyst 4	25 000
Analyst 5	25 000
Analyst 6	25 000
Analyst 7	25 000
Analyst 8	25 000

a Calculate the mean salary. (2 marks)

b Find the median salary. (2 marks)

c Calculate the interquartile range. (2 marks)

Analyst 8 decides to argue for a pay rise.

d Suggest which measure of average (mean, median or mode) Analyst 8 should use to support their case. Justify your answer. (2 marks)

e Suggest which measure of average (mean, median or mode) the managing director might use to counter the claim that Analyst 8 should be paid more. Justify your answer. (2 marks)

f Comment on which measure of average would be fairest as a representative salary for employees in this company. Justify your answer. (2 marks)

18 P1: a Define, as fully as you can, the terms random sampling, stratified sampling and systematic sampling. (5 marks)

b A researcher wishes to investigate the size of rats in a London Underground station.

i Suggest one reason why systematic sampling should not be used. (2 marks)

ii Determine whether random or stratified sampling would be more appropriate in this investigation. Justify your answer. (2 marks)

19 P2: The following raw data is a list of the height of flowers (in cm) in Eve's garden.

26.5, 53.2, 27.5, 33.6, 44.6, 39.5, 24.9, 45.1, 47.8, 39.3, 33.1, 38.7, 44.1, 22.3, 44.1, 30.5, 25.5, 35.9, 37.1, 40.2, 23.3, 36.2, 34.8, 37.3

a Copy and complete the following grouped frequency table.

Height (x cm)	Frequency
$20 \le x < 25$	
$25 \le x < 30$	
$30 \le x < 35$	
$35 \le x < 40$	
$40 \le x < 45$	
$45 \le x < 50$	
$50 \le x < 55$	

(3 marks)

b Find an estimate for the mean height, using the frequency table. (2 marks)

c Find an estimate for the variance, using the frequency table. (2 marks)

d Find an estimate for the standard deviation, using the frequency table. (2 marks)

e Eve's neighbour's garden was also surveyed.

It was found that the flowers in the neighbour's garden had a mean height of 32.1 cm and standard deviation 7.83 cm.

Compare the heights of the flowers in the two gardens, drawing specific conclusions. (3 marks)

20 P1 The population of Frankfurt in Germany was found to be 718 824.

A company chose a random sample of 1200 residents of Frankfurt to ask for comments on the city's proposed integrated transport system.

611 of the sample chosen were female.

a Calculate an estimate for the number of females in Frankfurt. (2 marks)

b The company decided to repeat their survey, but this time chose to use a stratified sample rather than a random sample.

Suggest two possible types of strata (apart from gender) that would be sensible for the company to use. (2 marks)

What's the difference?

Approaches to learning: Thinking skills, Communicating, Collaborating, Research
Exploration criteria: Presentation (A), Mathematical communication (B), Personal engagement (C), Reflection (D), Use of mathematics (E)
IB topic: Statistics, Mean, Median, Mode, Range, Standard deviation, Box plots, Histograms

Modelling and investigation activity

Example experiment

Raghu does an experiment with a group of 25 students.

Each member of the group does a reaction test and Raghu records their times.

Raghu wants to repeat the experiment, but with some change.

He then wants to compare the reaction times in the two experiments.

Discuss:

How could Raghu change his experiment when he does it again?

With each change, is the performance in the group likely to improve, stay the same or get worse?

Alternatively, Raghu could use a different group when he repeats the experiment.

What different group could he use?

With each different group, is the performance likely to improve, stay the same or get worse?

Your experiment

Your task is to devise an experiment to test your own hypothesis.

You will need to do your experiment two times and compare your results.

Step 1: What are you going to test? State your aim and hypothesis.

Write down the aim of your experiment and your hypothesis about the result.

Why do you think this is important?

What are the implications of the results that you may find?

Make sure it is clear what you are testing for.

Step 2: How are you going to collect the data? Write a plan.

- What resources/sites will you need to use?
- How many people/students will you be able to/need to collect data from to give statistically valid results?
- Exactly what data do you need to collect? How are you going to organize your data? Have you done a trial experiment?
- Are there any biases in the way you present the experiment? How can you ensure that everyone gets the same instructions?
- Is your experiment a justifiable way of testing your hypothesis? Justify this. What are the possible criticisms? Can you do anything about them?
- Is the experiment reliable? Is it likely that someone else would reach a similar conclusion to yours if they used the same method?

Step 3: Do the experiment and collect the data.

Construct a results sheet to collect the data.

Give clear, consistent instructions.

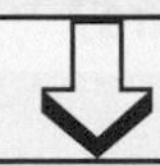

Step 4: Present the data for comparison and analysis.

How are you going to present the data so that the two sets can be easily compared?

How are you going to organize the summary statistics of the two data sets so that you can compare them?

Do you need to find all of the summary statistics covered in this chapter?

Step 5: Compare and analyse.

Describe the differences between your two sets of data.

Make sure that your conclusion is relevant to your aim and hypothesis stated at the beginning.

Step 6: Conclusions and implications.

What are the conclusions from the experiment?

Are they different from or the same as your hypothesis? To what extent? Why?

How confident are you in your results? How could you be more certain?

What is the scope of your conclusions?

How have your ideas changed since your original hypothesis?

Extension

- How could you test whether the spread (rather than the average) of the data has changed significantly?
- How could you analyse changes in individual results, rather than whole class changes?
- Investigate the "difference in means test".

4 Dividing up space: coordinate geometry, lines, Voronoi diagrams

This chapter explores two- and three-dimensional space and the relationship between points and lines in space. It includes finding the midpoint and distance between two points, intersection of lines and equations of lines, and explores their uses in air control to track aircraft, in technology, such as programming a robot cleaner to find the shortest distance from the line of its movement to the charging station, or in urban planning to find the best location for a train station. The chapter also looks at dividing space for the purposes of solving practical problems such as depositing toxic waste or the fair division of fishing territory among islands.

Concepts

- Space
- Relationships

Microconcepts

- Gradient
- Equations of straight lines
- Perpendicular lines
- Parallel lines
- Perpendicular bisectors
- Points of intersections of lines
- Coordinates of midpoint in 3D; distance in 3D
- Voronoi diagrams
- Area of Voronoi diagrams

How could you determine the incline of a road?

How could you determine the distance between stars?

How do you find the midpoint of a trail?

How could a town with four fire stations be divided into regions so that the nearest fire truck is dispatched to the fire location?

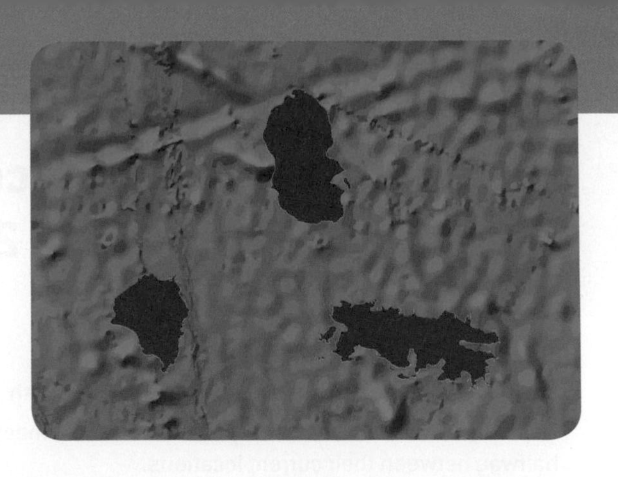

An island has control of all economic resources within its territorial waters. These resources may include fishing, mining, or offshore oil exploration.

The positions of three small islands are shown.

- How could you model the positions of the three islands?
- How could you use your model to decide how to divide the territorial waters between the islands?
- What information do you need to be able to answer this question?
- What assumptions would you need to make?
- What factors might influence how you would divide the waters between the islands?

Developing inquiry skills

Write down any similar inquiry questions you might ask if you were asked to divide an area between different landmarks—for example, deciding which fire station should assume responsibility for different areas of a town or deciding which hospital is closer to your home?

Are any of these questions similar to the ones used to investigate how territorial waters are being divided between three islands?

Think about the questions in this opening problem and answer any you can. As you work through the chapter, you will gain mathematical knowledge and skills that will help you to answer them all.

Before you start

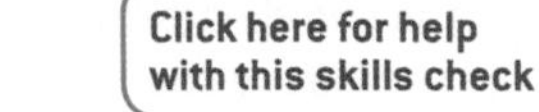

You should know how to:	Skills check
1 Solve equations of one variable. eg Solve for x: $\frac{x+(-1)}{2} = -3$ $x - 1 = -6$ $x = -5$	**1** Solve for x: **a** $2(x-5) = -3$ **b** $\frac{x-1}{2} = \frac{1}{3}$
2 Plot 2D coordinates of points. eg Place on a coordinate system the points $(3,-1)$, $(-2,-5)$, $(1,-4)$ and $(0,-2)$.	**2** Place the following points on a pair of coordinate axes: $A(1,3)$, $B(-2,-5)$, $C(1,-5)$, $D(-3,1)$, $E(0,4)$, $F(-4,0)$.
3 Substitute into an equation. eg $y = 2x - 1$. Find the value of y if $x = 3$. $y = 2(3) - 1$ $y = 5$	**3** **a** $y = -5x + 3$ Find the value of y if $x = -1$. **b** $y = \frac{1}{3}(x-1)$ Find the value of y if $x = -\frac{1}{2}$.
4 Use the Pythagorean theorem. eg Find the third side of a right-angled triangle with hypotenuse 7 and side 2. $\sqrt{72 - 22} = 6.71$	**4** A right-angled triangle has hypotenuse 10 cm and a side of length 6 cm. Find the length of the third side.

4.1 Coordinates, distance and the midpoint formula in 2D and 3D

Investigation 1

Two hikers walk towards each other on a straight path.

The hikers walk on a *horizontal plane* and want to meet halfway between their current locations.

The hikers locate their current positions in relation to the information centre O, shown on the coordinate system.

Hiker A is 6 metres north and 5 metres west, and hiker B is 19 metres east and 20 metres north.

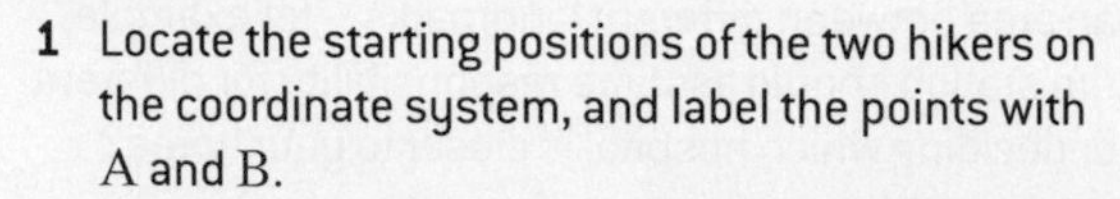

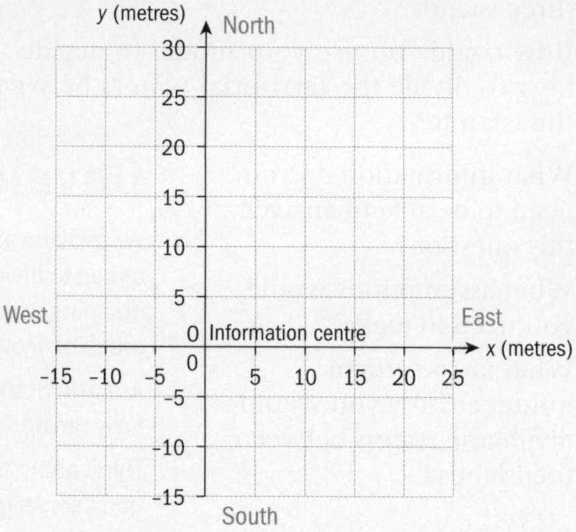

1. Locate the starting positions of the two hikers on the coordinate system, and label the points with A and B.
2. On the coordinate system, locate and label point C, such that ΔABC is a right-angled triangle and $\hat{C} = 90°$. Draw the line segments [AC], [BC] and [AB].

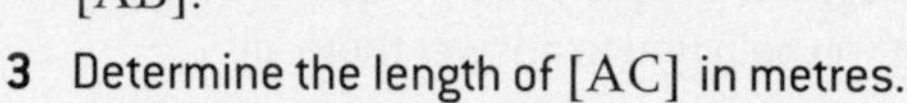

3. Determine the length of [AC] in metres.

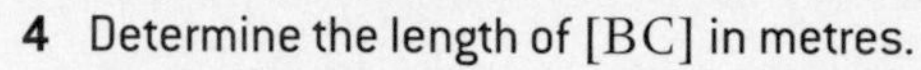

4. Determine the length of [BC] in metres.

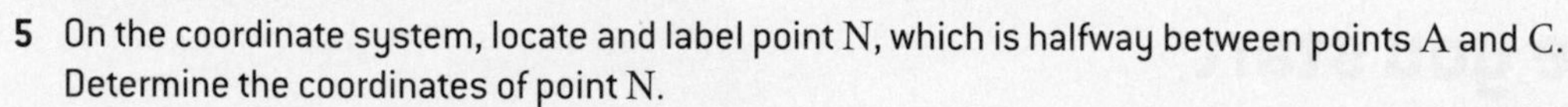

5. On the coordinate system, locate and label point N, which is halfway between points A and C. Determine the coordinates of point N.

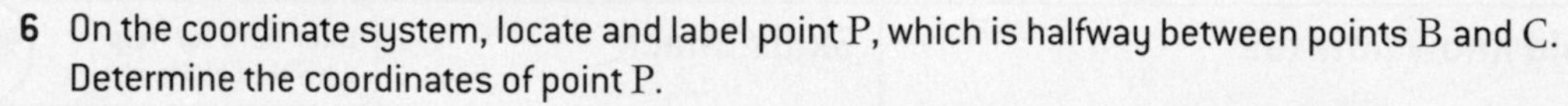

6. On the coordinate system, locate and label point P, which is halfway between points B and C. Determine the coordinates of point P.
7. From point N draw a perpendicular line i to [AC]. Determine the coordinates of the intersection point of line i with [AB].
8. From point P draw a perpendicular line j to [BC]. Determine the coordinates of the intersection point of line j with [AB].

The two lines j and i intersect at point M on [AB]. M is called the "midpoint" of [AB].

9. Explain why M is the point halfway between A and B.
10. Explain how you found the coordinates of the midpoint M.
11. **Conceptual** How would you find the midpoint of a line segment connecting any two points?

In a two-dimensional (2D) space, the midpoint, M, of the line segment joining points $A(x_1, y_1)$ and $B(x_2, y_2)$, is $M\left(\frac{x_1 + x_2}{2}, \frac{y_1 + y_2}{2}\right)$.

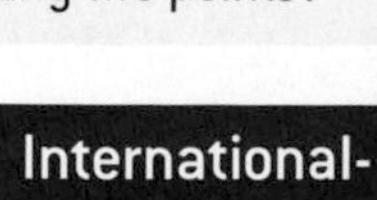

International-mindedness

Cartesian coordinates are named after the French mathematician René Descartes.

Example 1

The points A(−3, 5) and B(0, 7) are the endpoints of the diameter of a circle. Find the coordinates of the centre of the circle.

The coordinates of the centre will be $\left(\frac{-3+0}{2}, \frac{5+7}{2}\right) = (-1.5, 6)$.	Use the formula for finding the coordinates of the midpoint of the segment [AB], $\left(\frac{x_1+x_2}{2}, \frac{y_1+y_2}{2}\right)$

Example 2

Find the value of x if M(−3, 7.5) is the midpoint between A(x, 5) and B(−1, 10).

$\frac{x+(-1)}{2} = -3$ $x - 1 = -6$ $x = -5$	To find the x-coordinate of point A, use the formula for finding the coordinates of the midpoint of the segment [AB], $\left(\frac{x_1+x_2}{2}, \frac{y_1+y_2}{2}\right)$, and substitute the values of the known x-coordinates.

In a three-dimensional (3D) space, the location of any point A is determined by an ordered triple of coordinates (x, y, z), where each coordinate determines the location along the x-, y- and z-axis respectively.

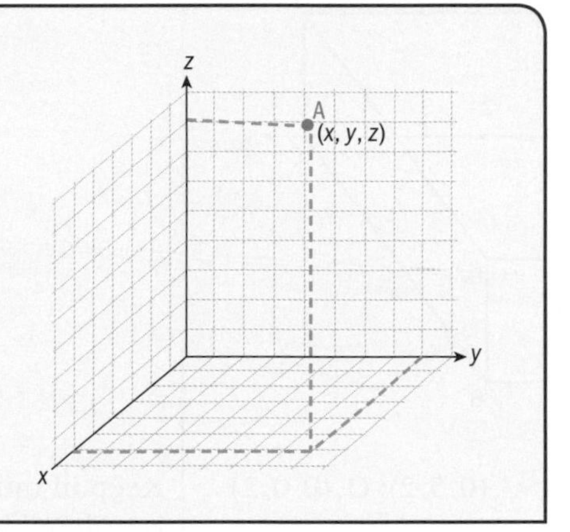

Investigation 2

A rock climber R is standing at a point with coordinates (60, 80, 0) and is about to start climbing a rock. Another rock climber S is standing at a point with coordinates (0, 0, 400). All coordinates are given in metres.

1. How many metres above the horizontal plane would the midpoint M between climbers R and S be?
2. Explain how you would find the coordinates of the midpoint M between points R and S.
3. **Factual** How could you find the coordinates of the midpoint of a segment connecting any two points $A(x_1, y_1, z_1)$ and $B(x_2, y_2, z_2)$?
4. **Conceptual** How does the formula for the coordinates of the midpoint of a segment in three dimensions relate to the same formula in two dimensions?

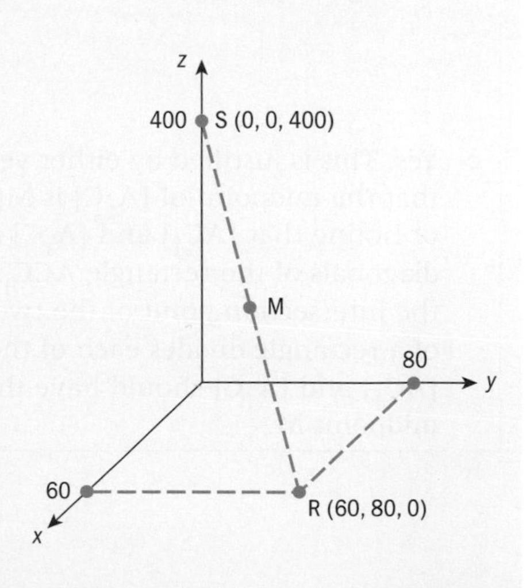

In 3D space, the midpoint, M, of the line segment joining points $A(x_1, y_1, z_1)$ and $B(x_2, y_2, z_2)$ is $M\left(\frac{x_1+x_2}{2}, \frac{y_1+y_2}{2}, \frac{z_1+z_2}{2}\right)$.

Example 3

A cuboid $ABCOA_1B_1C_1O_1$ has dimensions $4 \times 5 \times 2$. The vertices of the base have the following coordinates: $A(4, 0, 0)$, $B(4, 5, 0)$, $C(0, 5, 0)$ and $O(0, 0, 0)$.

a Sketch a coordinate system in 3D and the cuboid $ABCOA_1B_1C_1O_1$. Determine the coordinates of the vertices A_1, B_1, C_1 and O_1.

b Determine the coordinates of the midpoint M of diagonal $[AC_1]$.

c Explain whether or not you would expect $[AC_1]$ and $[A_1C]$ to have the same midpoint.

a

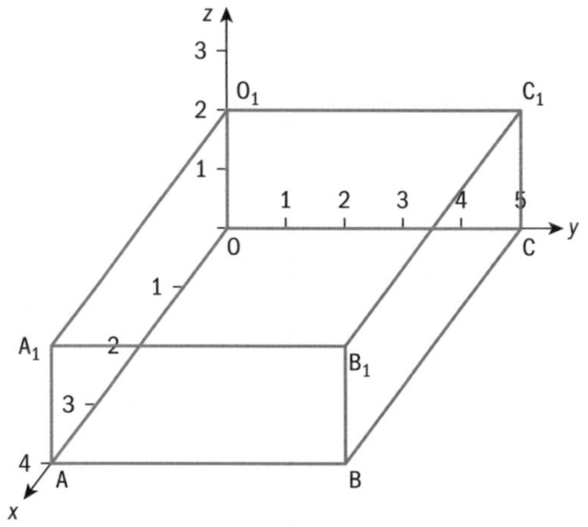

Since the given coordinates are in 3D, you will need to draw three perpendicular coordinate axes, Ox, Oy and Oz.

$A_1(4, 0, 2)$; $B_1(4, 5, 2)$; $C_1(0, 5, 2)$; $O_1(0, 0, 2)$

Keep in mind that each coordinate triple is ordered in the sequence (x, y, z).

b $M\left(\frac{4+0}{2}, \frac{0+5}{2}, \frac{0+2}{2}\right) = M(2, 2.5, 1)$

Use the 3D midpoint formula to determine the coordinates of point $M\left(\frac{x_1+x_2}{2}, \frac{y_1+y_2}{2}, \frac{z_1+z_2}{2}\right)$.

c Yes. This is justified by either verifying that the midpoint of $[A_1C]$ is $M(2, 2.5, 1)$ or noting that $[AC_1]$ and $[A_1C]$ are diagonals of the rectangle ACC_1A_1. Since the intersection point of the two diagonals of a rectangle divides each of them in half, $[AC_1]$ and $[A_1C]$ should have the same midpoint M.

Exercise 4A

1 Find the midpoint M of the given points:

a D$\left(2, \frac{1}{3}\right)$ and F$\left(\sqrt{2}, 3.5\right)$

b A$(-1, 5, -4.2)$ and B$(7, 8.5, -11)$

2 Determine the midpoint of the line segment connecting A and B shown below:

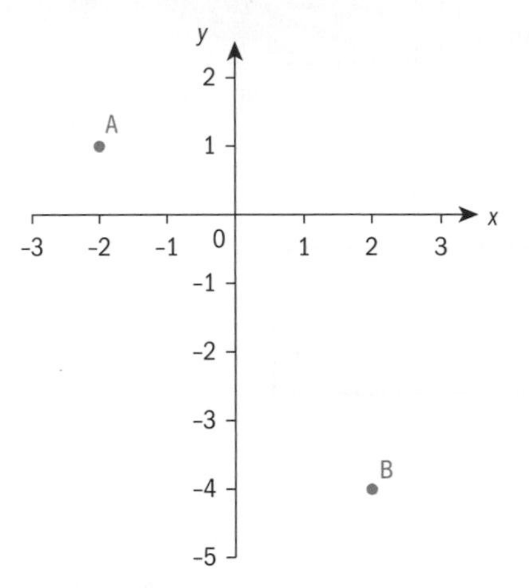

3 Find the coordinates of point C on segment [AB], shown below, such that the length of [AC] is $\frac{1}{4}$ of the length of [AB].

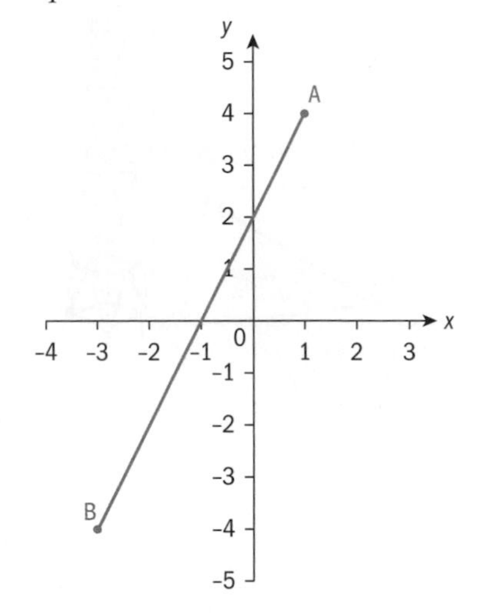

4 For the cuboid below:

a write down the coordinates of point B

b write down the coordinates of point A

c find the coordinates of the midpoint, M, of the diagonal [AO] of the cuboid.

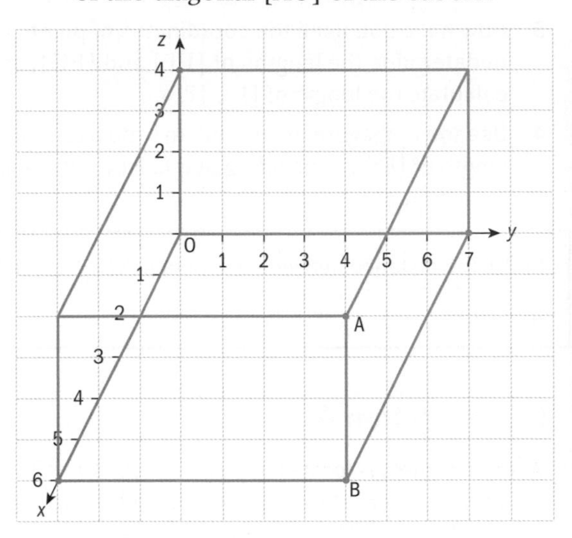

Coordinates of points are useful for finding distances. The two investigations on the next page provide two examples of finding distances in the 2D plane and in 3D space.

Investigation 3

Line segment [EG] represents a bridge over a river with endpoints located at E(10, 30) and G(60, 20).
On the coordinate system ΔEFG is such that $E\hat{F}G = 90°$.

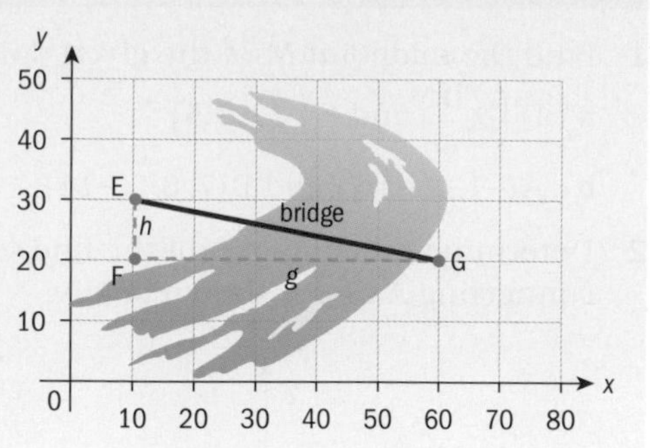

1. Find the lengths of segments [EF] and [FG].
2. Use the Pythagorean theorem to find the length of [EG].
3. How have you used the coordinates of points E and G to determine the lengths of [EF] and [FG], and then to calculate the length of [EG]?
4. Use the Pythagorean theorem to write a formula for the length of [EG] connecting points $E(x_1, y_1)$ and $G(x_2, y_2)$.

In 2D, the distance between two points $A(x_1, y_1)$ and $B(x_2, y_2)$ is
$d = \sqrt{(x_2 - x_1)^2 + (y_2 - y_1)^2}$.

Investigation 4

A traffic camera is located at point H(70, 80, 80) on a tower near the bridge [EG], where E(10, 30, 0) and G(60, 20, 0). The perpendicular, [HI], is drawn from point H to the ground, at point I.

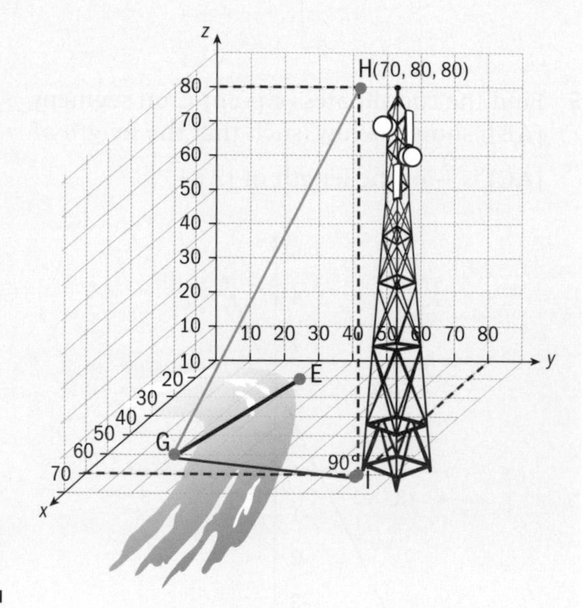

On the coordinate system ΔHIG is such that $H\hat{I}G = 90°$.

1. Find the x- and y-coordinates of point I.
2. Find the length of [HI].
3. Use the coordinates of G and I to find the distance [GI].
4. Find the distance [HG] of the traffic camera to the end of the bridge, G.
5. **Conceptual** How did you use the coordinates of points E and G to find HI, GI and HG?
6. **Factual** How could you extend the distance formula in 2D to three dimensions? How could you find the distance between any two points $A(x_1, y_1, z_1)$ and $B(x_2, y_2, z_2)$? Test whether your formula works in this real-life example.
7. **Conceptual** How does the distance formula in three dimensions relate to the same formula in two dimensions?

In 3D, the distance between two points $A(x_1, y_1, z_1)$ and $B(x_2, y_2, z_2)$ is
$AB = \sqrt{(x_2 - x_1)^2 + (y_2 - y_1)^2 + (z_2 - z_1)^2}$.

Example 4

A storage box for garden tools in the shape of a cuboid, OCDRFBAQ, is shown in the diagram. The vertex A has coordinates $(2, 3, 4)$. All coordinates are given in metres. Determine:

a the coordinates of vertices R, C, B and Q

b the maximum length of a garden tool that can fit in the storage box

c the height, RQ, of the storage box

d the angle between the diagonals [BR] and [BQ].

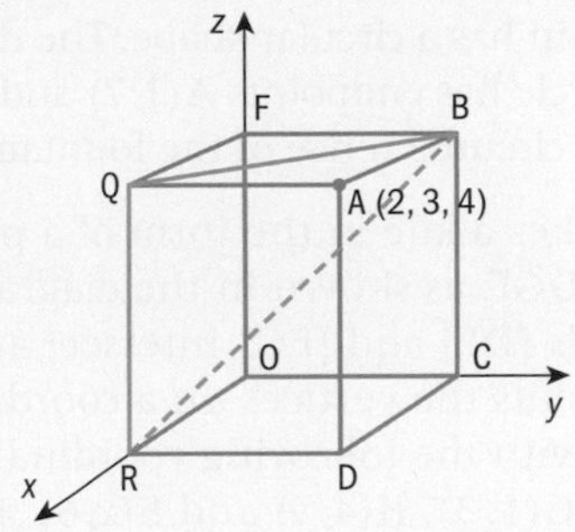

a $R(2, 0, 0)$	Point R has the same x-coordinate as A, and its y- and z-coordinates are 0.
$C(0, 3, 0)$	Point C is on the y-axis and its y-coordinate is the same as that of A.
$B(0, 3, 4)$	Point B has the same x- and y-coordinates as C and the same z-coordinate as A.
$Q(2, 0, 4)$	Point Q has the same x- and y-coordinates as R and the same z-coordinate as A.
b $BR = \sqrt{(2-0)^2 + (0-3)^2 + (0-4)^2}$ $BR = \sqrt{29} = 5.38516...$ Thus 5.39 metres is the maximum length of a tool that can be kept in the box.	The longest tool that can be kept in the box would be as long as the diagonal [BR]. Use the 3D distance formula to calculate BR and BQ: $d = \sqrt{(x_2 - x_1)^2 + (y_2 - y_1)^2 + (z_2 - z_1)^2}$
c $RQ = 4$ metres	
d $R\hat{B}Q = \sin^{-1}\left(\frac{RQ}{BR}\right)$ $R\hat{B}Q = \sin^{-1}\left(\frac{4}{\sqrt{29}}\right)$ $R\hat{B}Q = 48.0°(47.9688...)$	You could find the size of angle $R\hat{B}Q$ by using the right triangle ΔRBQ and the fact that $\sin R\hat{B}Q = \frac{RQ}{BR}$.

Exercise 4B

1 Find the distance between each pair of points given below:

a $A(4, -3)$ and $B(-1.5, 5)$

b $A(1, -2, 10)$ and $B(-4, 0, 3)$.

2 The distance between two towns P and Q is 2 km. On a coordinate system, town P has coordinates $(x, 7)$ and Q has coordinates $(3, 5)$. The values of the coordinates are given in kilometres. Find the value of x.

3 The vertices of a sail are plotted on a pair of coordinate axes and have the following coordinates: $A(-4,6)$, $B(6,2)$ and $C(-8,-4)$. Show that the triangular sail, ΔABC, is shaped as a right-angled triangle.

4 A fountain has a circular shape. The diameter of the circle has endpoints $A(1,7)$ and $B(-2,1)$. Find the circumference of the fountain.

5 Kari makes a kite in the form of a parallelogram, EDGF, as shown in the diagram. Its diagonals [DF] and [EG] intersect at point H. She plots the vertices on a coordinate system with the following coordinates: $D(3,6)$, $G(1,3)$, $H(4,y)$ and $E(x,6)$. Find the values of x and y.

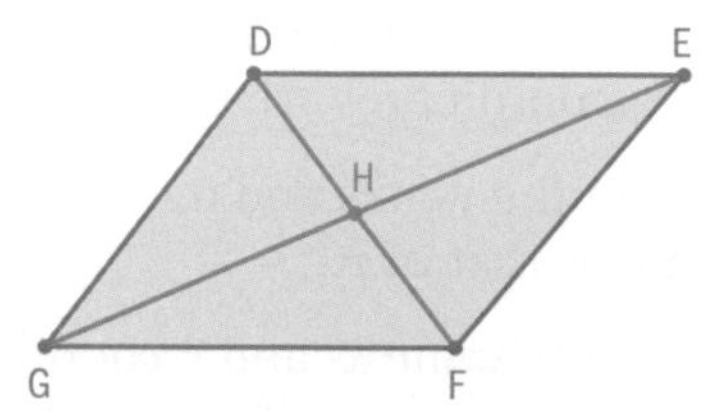

6 A tracking station lies at the origin of a coordinate system with the x-axis due east, the y-axis due north and the z-axis vertically upwards. Two aircraft have coordinates $(20,25,11)$ and $(26,31,12)$ relative to the tracking station. All coordinates are given in km.

a Find the distance between the two aircraft at this time.

b The radar at the tracking station has a range of 40 km. Determine whether it will be able to detect both aircraft.

Developing inquiry skills

Look back at the opening problem for the chapter. You were trying to divide the territorial waters between three islands.

The positions of the islands can be modelled on a graph grid.

The lines $x = 0$, $x = 100$, $y = 0$ and $y = 80$ mark the boundary of international waters and distances are given in kilometres. The islands are given exclusive fishing rights within these boundaries. The territory has to be divided between the islands so that each island is the closest one to any point in its given area.

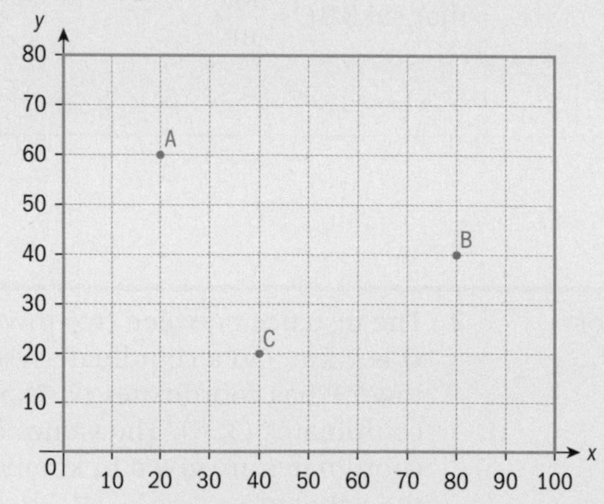

How can you find the distances between each pair of the islands?

4.2 Gradient of a line and its applications

Stairs are usually built of a number of equal steps, and each step has a horizontal part [DE], called a tread (run), and a vertical part [EF], called a riser (rise). The entire horizontal length of the stairs, [AB], is called a total run, and the entire height of the stairs, [BC], is called a total rise. The imaginary line that connects the tips of the steps, [AC], is called a pitch line.

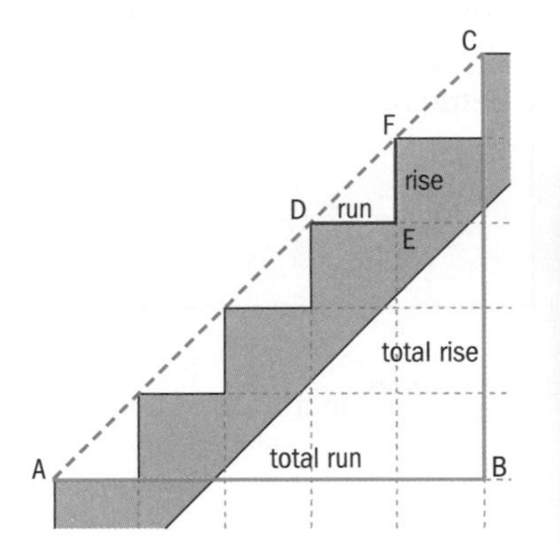

What do you think makes the pitch line of some stairs steeper than that of others?

What is the relationship between the run and rise of a step and the total run and total rise of the stair?

What is the relationship between the run and rise of a step and the pitch line?

Investigation 5

Some stairs are steeper than others. Four different stairs are given in the diagrams below. Their pitch lines are marked with dotted blue lines. One square on the grid has a side of 1 unit.

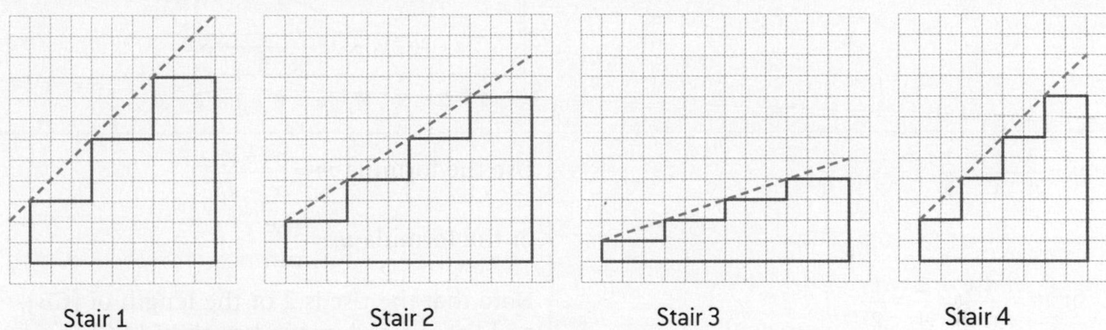

1 Complete the following table for each of the four stairs.

Stair #	Step rise (units)	Step run (units)	Ratio: $\frac{\text{rise}}{\text{run}}$	Pitch line steepness (Compare each pitch line to the one of stair 1: same, steeper or less steep?)
Stair 1				
Stair 2				
Stair 3				
Stair 4				

2 **Conceptual** Compare the ratios for the four stairs and the steepness of their pitch lines. What do you notice? Write down your observations.

3 **Conceptual** If there is a stair with rise 0, what would the pitch line be like? Could such a stair exist? Could a stair with run 0 exist?

4 **Conceptual** How could the steepness of any line be determined?

Gradient of a line

The **gradient of a line** is a measure of the steepness of the line. The bigger the gradient, the steeper the line. It is also known as the slope. Usually, the gradient (slope) of the line is denoted by the letter m.

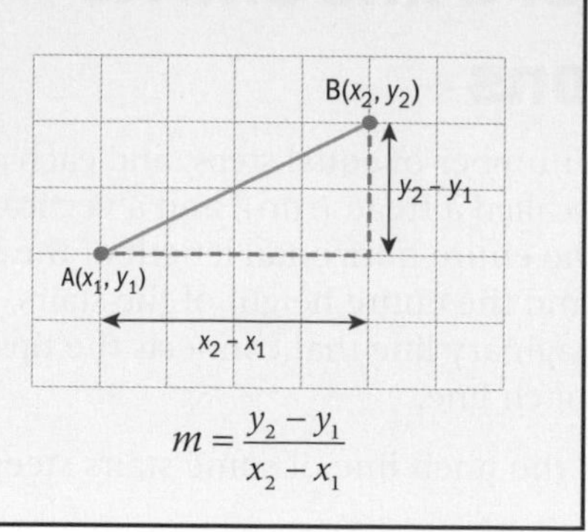

$$m = \frac{y_2 - y_1}{x_2 - x_1}$$

TOK

Around the world you will often encounter different words for the same item, such as gradient and slope, trapezium and trapezoid, or root and surd.

Sometimes more than one type of symbol might have the same meaning.

To what extent does the language we use shape the way we think?

Reflect In the formula $m = \frac{y_2 - y_1}{x_2 - x_1}$ which is the run and which is the rise?

Example 5

Determine the gradient of the line joining the points A(4, 1) and B(8, 3).

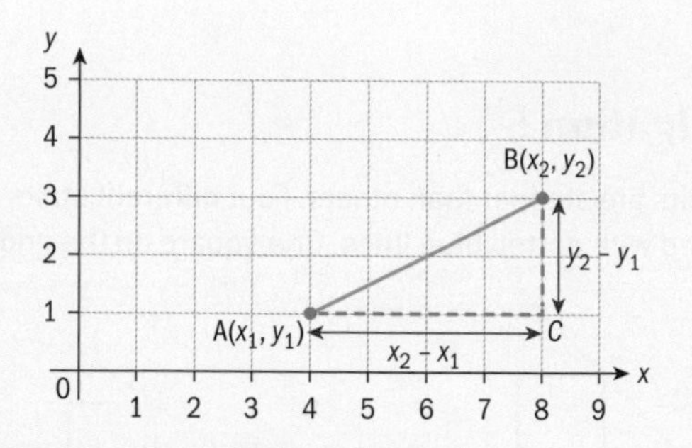

$m = \frac{3-1}{8-4} = \frac{1}{2}$ or $m = \frac{\text{rise}}{\text{run}} = \frac{2}{4} = \frac{1}{2}$	Use the formula $m = \frac{y_2 - y_1}{x_2 - x_1}$ or the formula $m = \frac{\text{rise}}{\text{run}}$ Note that the rise is 2 or the length of [CB], and the run is 4 or the length of [AC].

Reflect Does the gradient of a straight line depend on which two points A and B you use to calculate it?

Positive and negative gradients of straight lines

When the line goes upward towards the right the gradient is a **positive** number.

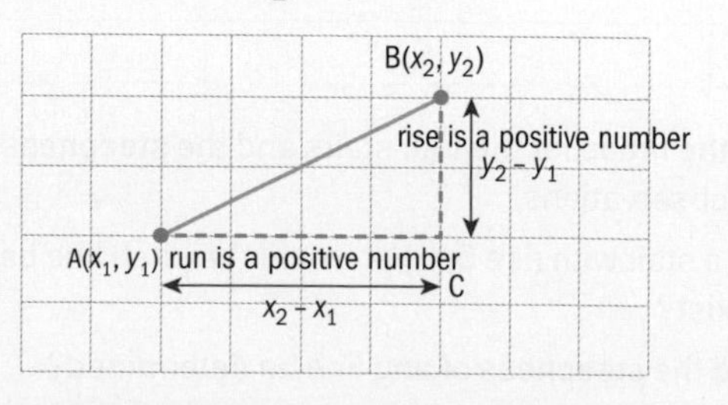

When the line goes downward towards the right the gradient is a negative number.

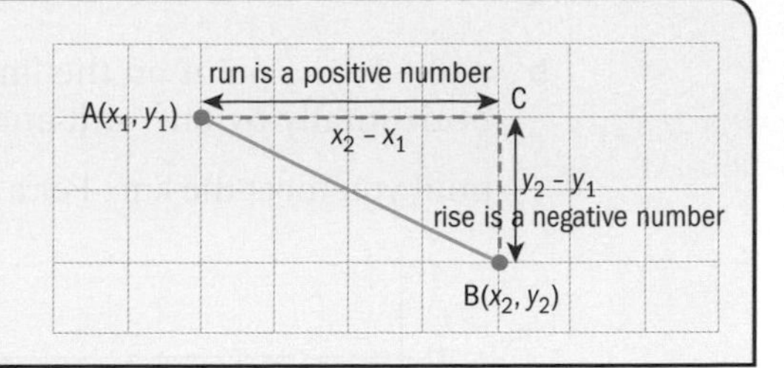

Zero gradient: When a line is horizontal its gradient is **zero.**

Undefined gradient: When a line is vertical it does not have a defined gradient.

Reflect What makes the gradient of a vertical line undefined?

What information does the gradient of a line give you?

The gradient of a line describes its steepness and direction.

Example 6

Determine the gradients of the lines shown below:

a

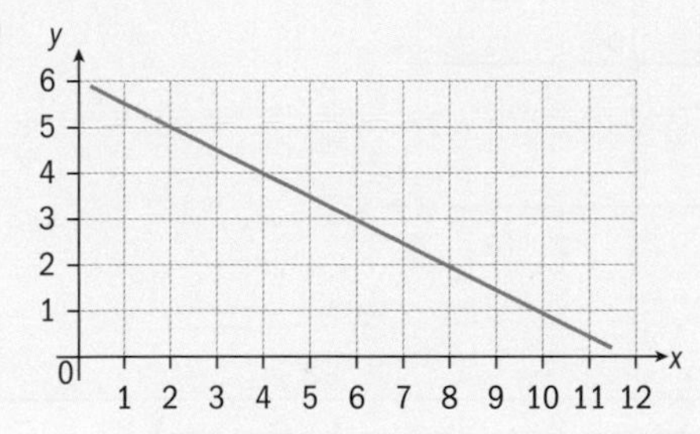

b

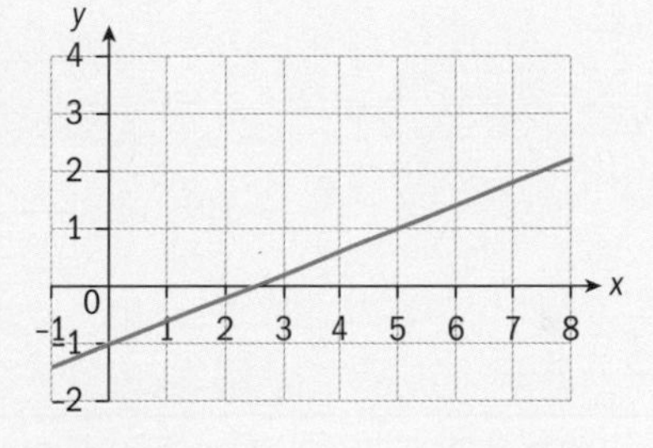

a $m = \dfrac{2-4}{8-4} = -0.5$	Find two points on the line and use them to find the gradient. Two easily identifiable points are A(4, 4) and B(8, 2). Substitute their coordinates into the gradient formula. 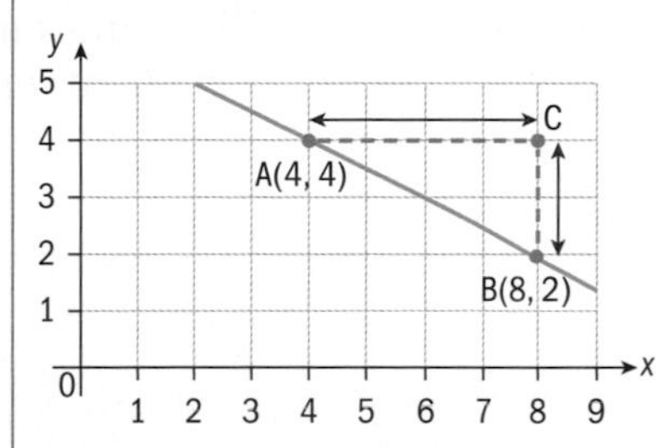 As the line goes downward to the right, the gradient is a negative number.

Continued on next page

b $m = 0.4$	**b** A(0,−1) is a point on the line. Imagine moving right horizontally by one unit and then moving vertically until you meet the line. For a run of 1 unit the rise is $\frac{4}{10}$. 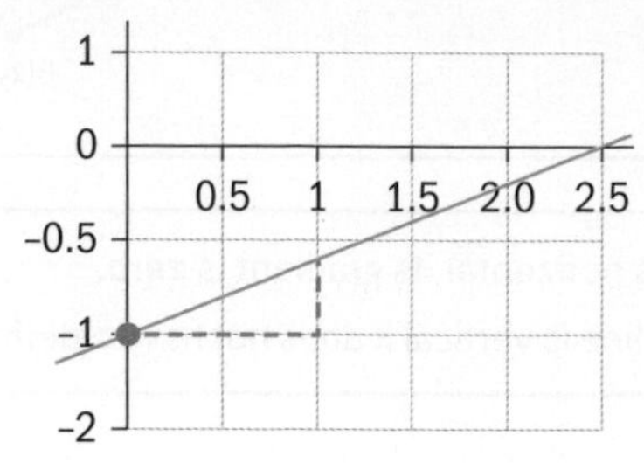

Example 7

a Determine the gradient of the horizontal line that passes through point D(1, 2).

b Show that the gradient of the vertical line that passes through point D(1, 2) is undefined.

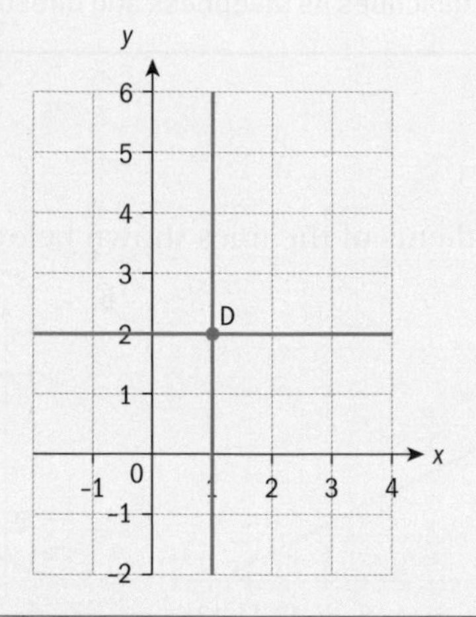

a Points A(0, 2) and D(1,2) are on the horizontal line passing through D. $m = \frac{2-2}{1-0} = \frac{0}{1} = 0$	Choose two points on the horizontal line and calculate the gradient using the formula.
b Points D(1,2) and F(1,5) are on the vertical line passing through D. $m = \frac{5-2}{1-1} = \frac{3}{0}$ The gradient is undefined.	Similarly, choose two points on the vertical line and substitute their coordinates into the gradient formula.

Example 8

Find the gradient of a line (AB) passing through A(−2,−11) and B(0,−5) using a GDC.

Enter the x-coordinates and the y-coordinates of the two points into your GDC. Then use the functions on your GDC to find the value of the gradient. The gradient of the line (AB) is 3.	The functions on your GDC can be used to enter coordinate values and to find the gradient of a line, given two points on that line.

Example 9

A funicular railway has several stations along its steep track. Station A with coordinates (30, 60) is 60 metres above Station Zero at ground level. Station B has coordinates $(230, h)$. All coordinates are given in metres. The gradient of the railway segment [AB] is 0.60. Find the height of station B above the ground.

$0.6 = \frac{h-60}{230-30}$ $0.6 = \frac{h-60}{200}$ $h-60 = 120$ $h = 180$ Station B is 180 metres above ground level.	Use the gradient formula and substitute the known coordinates of A and B into the formula.

Exercise 4C

1 Find, in each case, the set of possible coordinates of a point A which lies on a:

a horizontal line passing through B(−4, −8)

b vertical line passing through B(−3, 7).

2 Find the gradients (slopes) of the lines passing through the following pairs of points:

a A(2, 3) and B(−1, 4)

b A(−11, 2) and B(2, 7)

c A(−1.3, 2.2) and B $\left(1\frac{1}{5}, -0.3\right)$.

3 Find the gradients of the stairs for which information is given in the table below:

Stair #	Step rise (cm)	Step run (cm)	Gradient
Stair 1	12	25	
Stair 2	10	20	
Stair 3	15	18	

a State which stair has the greatest gradient.

b State which stair has the least gradient.

4 Draw the graph of the straight line:

a passing through (0, 0) and (−5, 6)

b passing through (−1, 2) and (−3, −5).

5 Find the gradient of the line:

a passing though (−1, −1) and (3, 2)

b passing through (0, 2.5) and $\left(\frac{-1}{3}, 5\right)$.

6 A skier travels downhill along two slopes: one with gradient −3 and the other with gradient $\frac{-1}{3}$. State which slope is the steeper, justifying your answer.

7 MegaLift plans ski slopes and lift stations along the slopes. Find the height of a ski lift station $B(490, h)$, which is on a slope with gradient 0.35 and which passes through a lower lift ski station A(90, 10). All coordinates are given in metres.

Gradient as unit rate

Today, 37 euros (EUR) is exchanged for 45 dollars (USD). If we plot the point (37,45) on a pair of coordinate axes and connect the point with the origin O(0,0), how can the graph help you find the exchange rate?

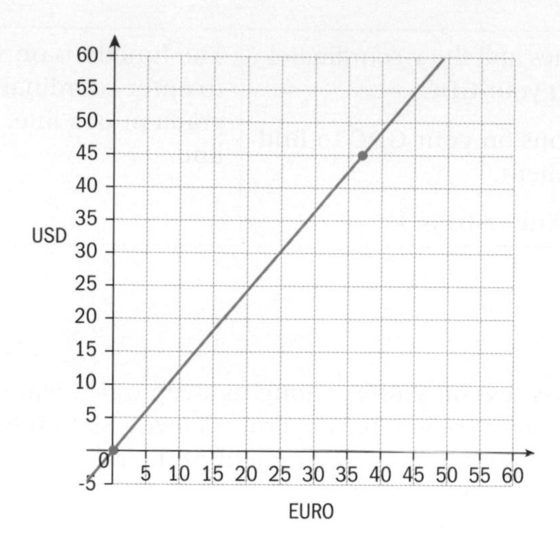

Investigation 6

Tomi is planning to rent a bicycle for up to a week. He finds two bicycle rental shops *Bikers' Village* and *Bike & Joy*, which rent bicycles per week. *Bikers' Village* charges 10 GBP per week plus 0.5 GBP per km cycled. *Bike & Joy* charges 20 GBP per week plus 0.1 GBP per km cycled.

1 Use the information to complete the table below.

Number of km cycled	Rent for a bicycle from *Bikers' Village* (GBP)	Rent for a bicycle from *Bike & Joy* (GBP)
10	15	21
15		
20		
25		
30		

2 Use technology to draw the lines representing the rent for a bicycle from each store for up to a week. Draw the lines on the same pair of axes, where x is the number of km cycled and y is the rent in GBP.

3 Which of the two options is cheaper for 20 km? And which is cheaper for 40 km? For what distance cycled is the rental cost the same?

4 Which shop should Tomi rent from? Would his decision depend on the number of kilometres he will ride the bicycle? Explain.

5 Determine the gradients of the two lines.

6 **Conceptual** What does the gradient of each line represent in the context of this question and why?

A third bicycle rental shop *Town on Bike* is about to open and wants to make an offer for renting a bicycle that is less expensive than the current two available offers.

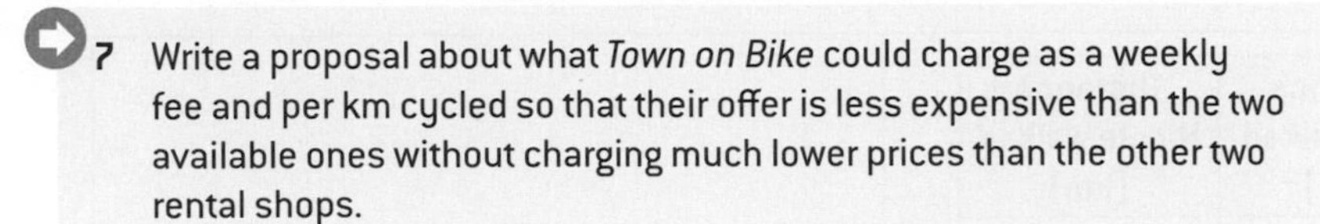

7 Write a proposal about what *Town on Bike* could charge as a weekly fee and per km cycled so that their offer is less expensive than the two available ones without charging much lower prices than the other two rental shops.

8 Use technology to draw the line representing the price scheme of your proposal and to justify it. Use the same pair of axes you used to draw the other two lines from part **2**.

Example 10

Sean and Nicole walk every day without taking any breaks during the walks. Sean walks 4 km per hour and Nicole walks 9 km in 2 hours.

a **i** Complete the table below.

Time (hours)	Distance Sean walked (km)	Distance Nicole walked (km)
1	4	
2		9
6		

ii Graph the lines representing Sean's and Nicole's daily walks on the same pair of axes, where x is the number of hours walked and y is the distance covered in km.

iii Determine the gradient of each line.

One Sunday, Nicole gives Sean a head start of 1.5 km, after which they time their walks.

b **i** Complete the table below, assuming that they both keep their rate of walking from part **a**.

Time (hours)	Distance Sean walked (km)	Distance Nicole walked (km)
1	5.5	
2		9
3		
4		
5		

ii Draw the lines representing Sean's and Nicole's walks on the same pair of axes, where x is the number of hours walked and y is the distance walked in km.

iii Determine the number of hours it will take Nicole to catch up with Sean after she starts her walk. How will you determine this from the graphs of the lines?

Continued on next page

a i

Time (hours)	Distance Sean walked (km)	Distance Nicole walked (km)
1	4	4.5
2	8	9
6	24	27

ii

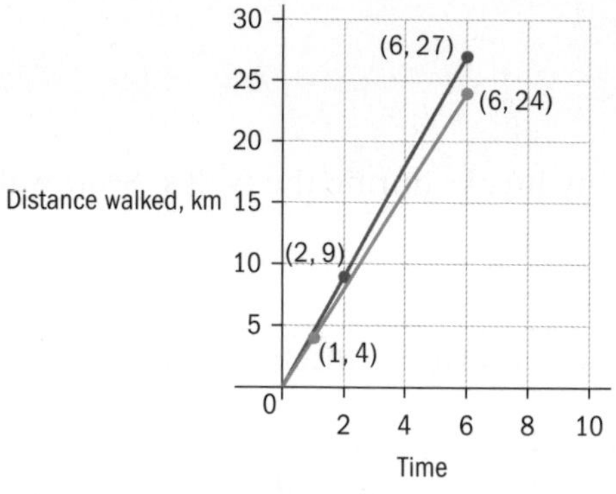

iii The line representing Sean's walk has a gradient of 4. The line representing Nicole's walk has a gradient of 4.5 since she walks 4.5 km per hour.

You can also determine the gradient for each line by calculating $\frac{\text{rise}}{\text{run}}$ for each line.

b i

Time (hours)	Distance Sean walked (km)	Distance Nicole walked (km)
1	5.5	4.5
2	9.5	9
3	13.5	13.5
4	17.5	18
5	21.5	22.5

ii

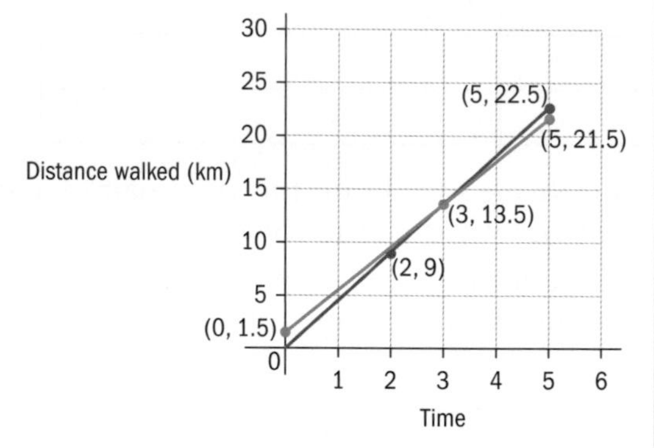

iii The two lines intersect at a point with x-coordinate 3, so Nicole will catch up with Sean after 3 hours. The y values are equal when $x = 3$ hours.

Exercise 4D

1 Safety regulations for wheelchair ramps mandate that the maximum gradient of the ramps should not exceed 0.08. Calculate the gradient of the triangular ramp ΔAOB ([AB] ⊥ [OB]) shown below, and determine whether it conforms to the safety regulations.

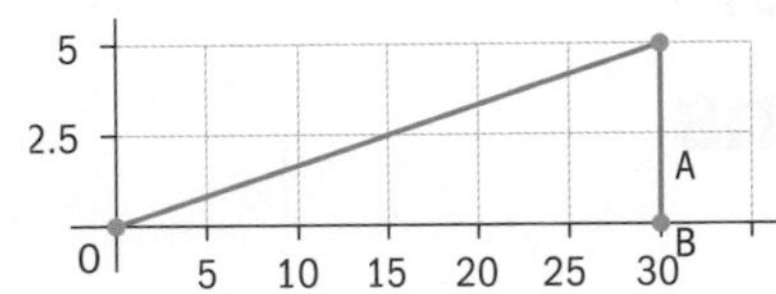

2 A town website advertises summer jobs for students. A student is trying to decide between a job as a city guide and a job at a flower shop. The job as a city guide pays $40 per day plus a monthly metro card with $120 value. The job at the flower shop pays $60 a day.

Compare the payments for the two jobs.

a Complete the table below:

Number of days worked	City guide (USD)	Flower shop (USD)
1	160	60
5		
10		

b Draw the lines representing the payments for each job. On the same pair of axes, where x is the number of days worked and y is the payment in dollars.

c Explain how the student should decide which job to take based on the number of days they plan to work during the month.

3 US roof regulations state that asphalt tiles can only be used for roofs with a maximum gradient of 0.17. The roof shown in the diagram has width 7 m and rise 1.6 m. Find:

a the roof run

b the roof gradient.

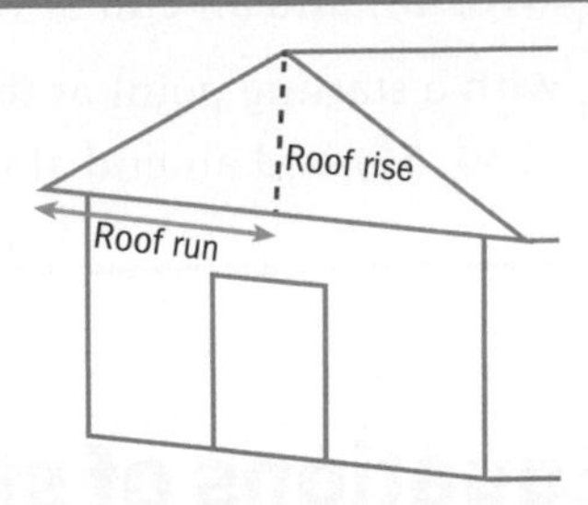

c Determine whether the roof satisfies the requirements for using asphalt tiles. Justify your conclusion.

4 Road signs showing the steepness of hills are often given as percentages, as in the one shown below, where the figure is derived using the formula

$$\frac{\text{rise (vertical change)}}{\text{run (horizontal change)}} \times 100\%$$

a For a certain road the vertical change is 5 m for a horizontal change of 20 m. Determine the percentage that would be written on the sign for this road.

b Which is the steeper road—the one with a sign indicating 10% incline or another with a sign indicating 15%?

c Determine the vertical change corresponding to a horizontal change of 5 km for a road with sign indicating 15% incline.

5 The ski slopes in the Scotch Pine Ski Resort are marked by colour as an indicator of the incline. The colours used are green, blue, red and black. Green is used for slopes with less than 20% incline, blue for 20–25%, red for 25–40% and black for above 40% incline.

a Classify the following slopes as blue, green, red or black:

i P_1 with a horizontal change of 450 m and a vertical change of 280 m that passes through O(0, 0)

ii P_2 with a horizontal change of 150 m and a vertical change of 50 m that passes through O(0, 0)

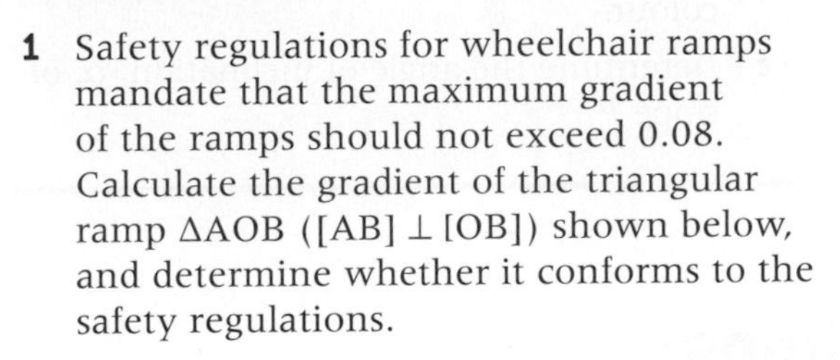

iii P_3 with a starting point at the top $B_1(310, 48)$ and an end at $O(0, 0)$

iv P_4 with a starting point at the top B_2 $(230, 56)$ and an end at $O(0, 0)$.

b On a pair of axes, graph each line, label it, and mark it with the respective colour.

c Determine the angle of inclination, α, of slope P_2.

4.3 Equations of straight lines: different forms of equations

Investigation 7

A ship is located at point $A(4, 6)$. The ship is moving along a straight line with gradient 0.5 as shown in the graph.

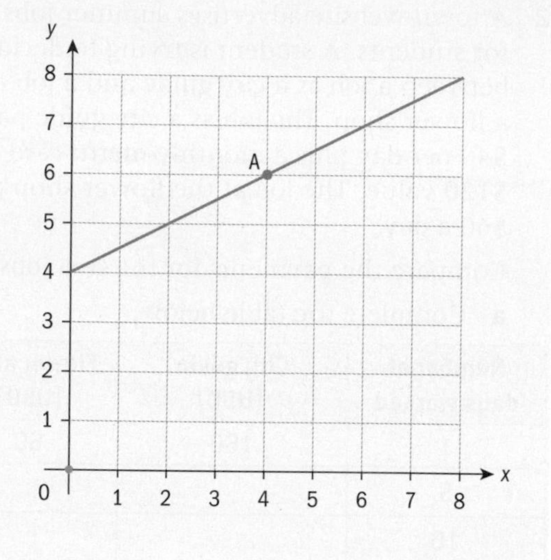

1 Use the graph of the line to determine the coordinates of three other points that the ship will pass through on its way.

2 Substitute point $A(4, 6)$ and the given gradient in the gradient formula to determine:

a the y-coordinate of the ship if its x-coordinate is 5

b the x-coordinate of the ship if its y- coordinate is −4.

3 How can you substitute a given point and a given gradient in the gradient formula to obtain an equation that describes the relationship between the x- and y-coordinates of a point on straight line with the given gradient that passes through the given point?

4 Substitute the point A(4, 6) and the given gradient in the gradient formula. Remove the parentheses and rewrite the equation in the form $0 = \ldots$

5 Using the equation in part **4**, find the coordinates of the point at which the line of the ship's movement intersects the y-axis.

6 How can you use the equation that you wrote in part **4** to check whether a given point lies on the line?

7 What would the equation of a line that passes through (4, 6) and (10, 6) look like? What do you notice about all points on a horizontal line?

8 What would the equation of a line that passes through (4, 6) and (4, 10) look like? What do you notice about all points on a vertical line?

Equation of a line

Point-gradient (point-slope) form

If a known gradient m of a line and a known point (x_1, y_1) on that line are substituted in the gradient formula, the relationship $m = \frac{y - y_1}{x - x_1}$ gives the **equation of the line.**

The form $y - y_1 = m(x - x_1)$ is called the **point-gradient (point-slope) form.**

Standard (general) form

The form $ax + by + c = 0$, where $a, b, c \in \mathbb{R}$, is called the **standard** or **general form**.

Equation of a horizontal line

All horizontal lines have equations $y = C$, where C is a constant.

The line in the diagram has equation $y = 2$, and each point on this line has y-coordinate 2.

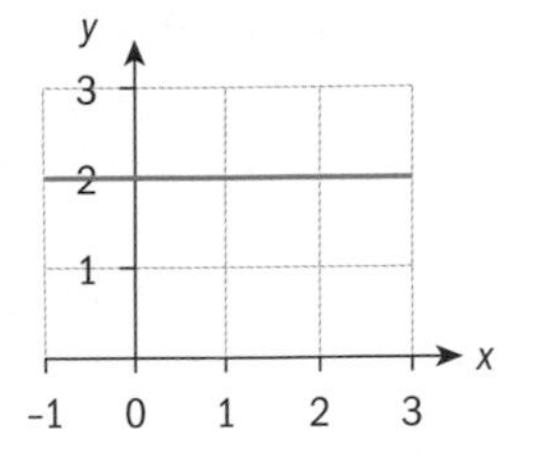

Equation of a vertical line

All vertical lines have equations $x = C$, where C is a constant.

The line in the diagram has equation $x = 3$, and each point on this line has x-coordinate 3.

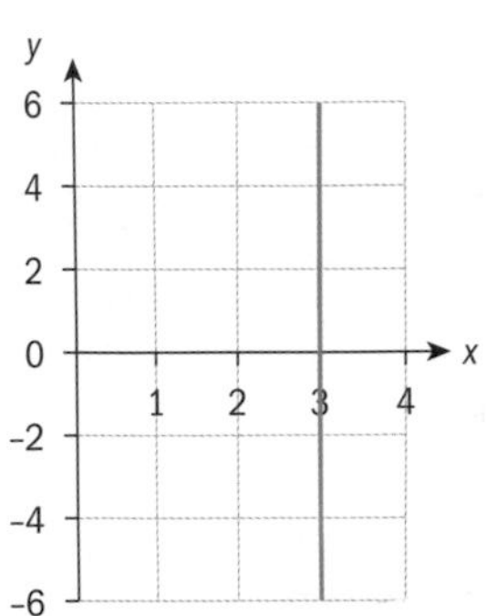

International-mindedness

The development of functions involved mathematicians from many countries, including France (René Descartes), Germany (Gottfried Wilhelm Leibniz) and Switzerland (Leonhard Euler).

Functions

Geometry and trigonometry

Example 11

Find the equation of the line:

a that has gradient 3 and passes through point A(1, 7)

b that passes though points A(1, 3) and B(2, 5).

Write down the equations in point-gradient form.

Draw the lines.

a The point-gradient form of the equation of the line is $y - 7 = 3(x - 1)$.	Substitute the given gradient and coordinates of the given point in the **point-gradient (point-slope) form** $y - y_1 = m(x - x_1)$
To draw the graph of line $y - 7 = 3x - 3$ on the GDC, rearrange the equation as $y = 3x + 4$.	Your GDC can be used to draw the graph of a line.

Continued on next page

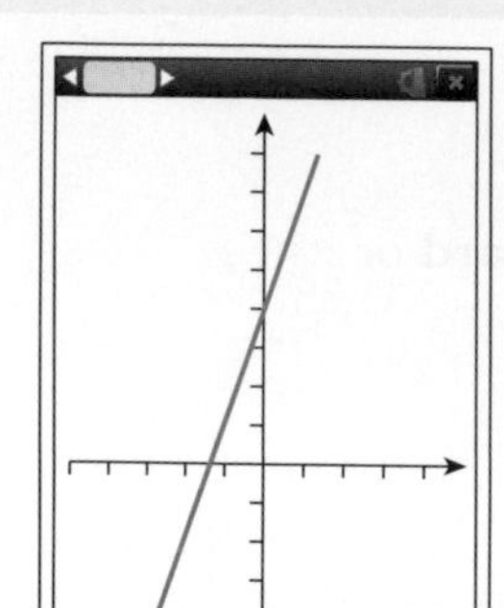

b $m = \dfrac{5-3}{2-1} = 2$

The point-gradient form of the equation of the line is $y - 3 = 2(x - 1)$.

The graph of the line is below:

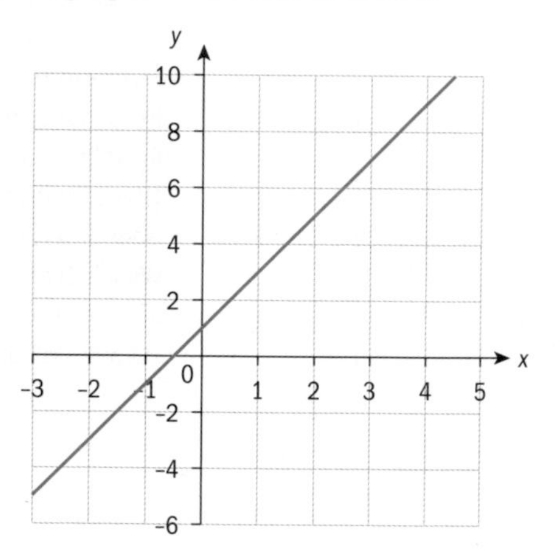

Similarly, substitute the coordinates of a point and the gradient in the **point-gradient (point-slope) form,** but before doing so you need to calculate the gradient of the line using the coordinates of the two given points.

When you substitute the gradient and one of the points in the point-gradient form, note that you could use the coordinates of either of the two given points.

Example 12

Draw the line $3x + y + 3 = 0$.

Identify and plot two points, A and B, that are on the line.

Point	x-coordinate	y-coordinate
A	1	$y = -3 - 3(1) = -6$
B	−2	$y = -3 - 3(-2) = 3$

To find a point on the line you could choose a value for one of the coordinates, substitute it in the equation and find the other coordinate.

A(1, −6) and B(−2, 3) are on the line.

(AB) is the graph of the line.

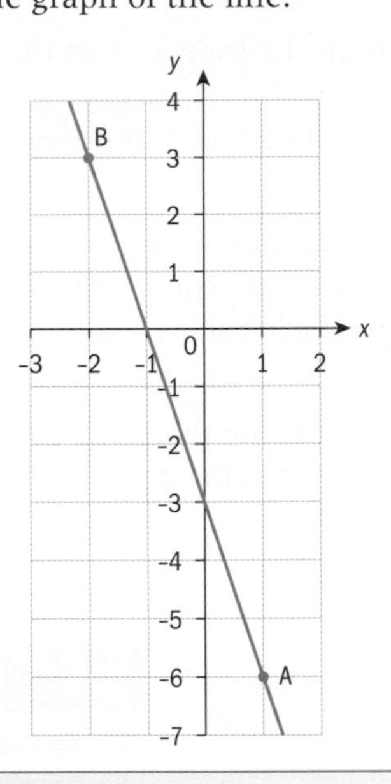

Then you plot these points and draw a line through them.

Example 13

Points A and B lie on the straight-line boundary between two plots of land. Write the equation of the line passing through points A and B. A tree is located at point C(25, 30).

Determine whether the tree is on the boundary line.

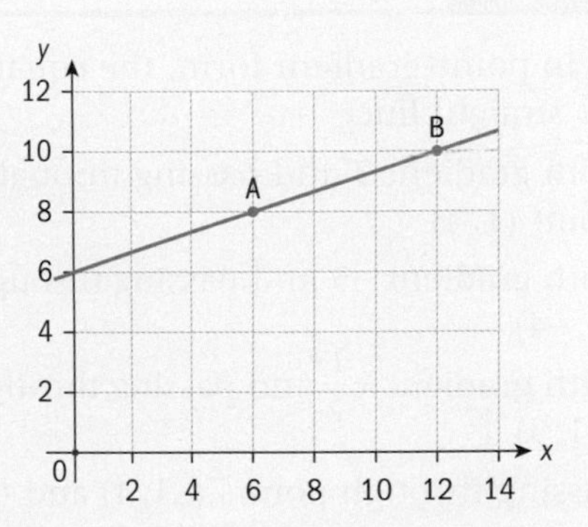

$m=\frac{10-8}{12-6}=\frac{1}{3}$

The equation of the line is $y-8=\frac{1}{3}(x-6)$ given in point-gradient form.

To determine whether point C is on the boundary line or not, check whether when we substitute the coordinates of point P in the equation, the left side of the equation will equal the right side.

$25-8\neq\frac{1}{3}(30-6)$ since $17\neq 8$

Therefore the tree is not on the boundary line.

From the graph you could determine the coordinates of the two points, A(6, 8) and B(12, 10), after which you can calculate the gradient and then substitute the gradient and coordinates of either A or B in the point-gradient form.

Example 14

A sailing boat is moving along a straight line with equation $-2.5x + y = 3$.

a Determine whether a tanker located at point A(3, −2) is on the trajectory of the sailing boat.

b Find the exact y-coordinate of the port of arrival for the sailing boat if the x-coordinate is 10.

a $-2.5 \times (3) + (-2) = -9.5$ Point A is not on the line as $-9.5 \neq 3$.	Substitute the coordinates of A(3, −2) in the left-hand side of the equation $-2.5x + y = 3$ and check whether the left side equals the right side.
b $-2.5\,(10) + y = 3$ $y = 3 + 25$ $y = 28$	If a point is on the line then its y-coordinate is found by substituting the given x-coordinate into the equation.

Reflect Could there be two different points with the same x-coordinate but different y-coordinates such that both points lie on a given line?

TOK

Are algebra and geometry two separate domains of knowledge?

Exercise 4E

1 Find, in point-gradient form, the equation of the straight line:

a with gradient 3 and passing through a point (1, 4)

b with gradient −5 and passing through (7, −4)

c with gradient $-\frac{1}{2}$ and passing through (−1, 3)

d passing through points (−1, 4) and (1, 8).

2 Draw the graph of each of the following lines:

a $y - 1 = -0.5(x + 2)$ **b** $y = -\frac{2}{5}x + 4$

c $y + 2x = \frac{1}{2}$

3 A ski resort planner works on designing ski slopes. Find the equations of the following slopes in the form of straight lines:

a a slope with gradient 4 and passing through (−2, 5)

b a slope with gradient $\frac{2}{7}$ and passing through (7, −2)

c a slope passing through (−3, 1) and (2, 0)

d a slope passing through points (2, 0) and (−1, −2).

4 Determine without graphing whether A(−4, −5) lies on the lines with equations:

a $y = -x - 9$ **b** $5x + 4y = 9$

5 Frances runs along a straight line with equation $y = \frac{1}{2}x + 4$. She starts at point C and finishes at point D. Find:

a the y-coordinate of point C if $x = 6$

b the x-coordinate of point D if $y = 9$.

Investigation 8

The graphs of lines L_1, L_2, L_3 and L_4 are given.

1 **a** Identify the points at which line L_1 intersects the x-axis and the y-axis.

b Determine the gradient of L_1.

c Find the equation of line L_1. Give the equation in point-gradient form.

d Rewrite the equation in the form $y = ax + b$, where $a \in \mathbb{Q}$ and $b \in \mathbb{Q}$.

2 **a** Identify the points at which line L_3 intersects the x-axis and the y-axis.

b Determine the gradient of L_3, and hence find the equation of line L_3. Give the equation in point-gradient form.

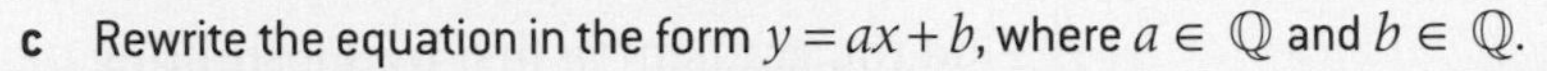

c Rewrite the equation in the form $y = ax + b$, where $a \in \mathbb{Q}$ and $b \in \mathbb{Q}$.

3 Examine each equation from parts **1d** and **2c** in relation to its gradient and axes intercepts.

4 **Conceptual** How do the values of a and b in your equations in the form $y = ax + b$ relate to the gradient and the axes intercepts?

Write down your conjecture about what a and b represent.

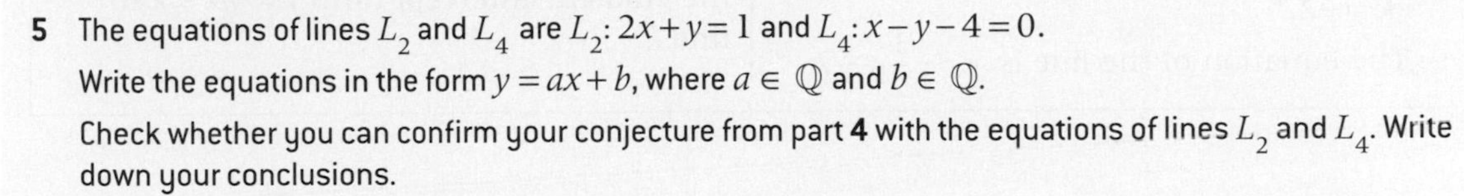

5 The equations of lines L_2 and L_4 are L_2: $2x + y = 1$ and L_4: $x - y - 4 = 0$.

Write the equations in the form $y = ax + b$, where $a \in \mathbb{Q}$ and $b \in \mathbb{Q}$.

Check whether you can confirm your conjecture from part **4** with the equations of lines L_2 and L_4. Write down your conclusions.

Other forms of an equation of a line: gradient-intercept form

Equations of lines given in point-gradient form or general form can be rearranged in the form below, which is called the **gradient-intercept form**:

$y = mx + k$, where m is the gradient and k is the value of y for $x = 0$

Intercepts: x-intercept and y-intercept

The point at which the graph intersects the x-axis is called the **x-intercept**, and the point at which the graph intersects the y-axis is called the **y-intercept**.

Note that the x-intercept has zero for its y-coordinate, and the y-intercept has zero for its x-coordinate.

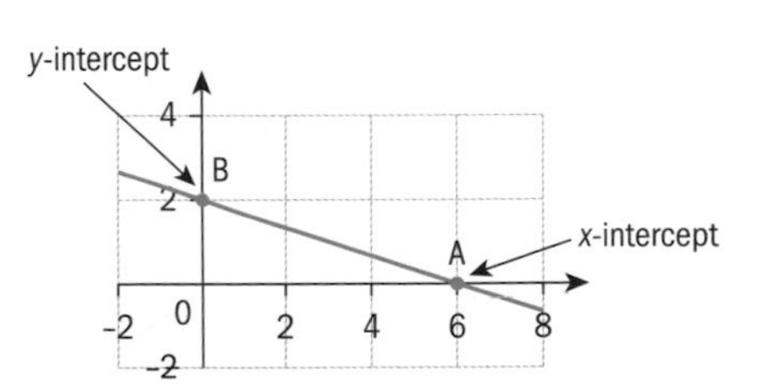

Example 15

Find the equation of the line:

a with gradient −2 and passing through (−4, 3)

b passing through (−2, −1) and (6, −5).

Write down the equations in the gradient-intercept form $y = mx + k$, where m is the gradient and k is the y-coordinate of the y-intercept.

a $y = -2x + k$ $3 = -2(-4) + k$ $k = -5$ The equation of the line is $y = -2x - 5$.	Substitute the gradient in the gradient-intercept form $y = mx + k$, where m is the gradient and k is the y-coordinate of the y-intercept. Then find k by substituting the coordinates of the point for x and y. Substitute the value of k in the gradient-intercept form.
b $m = \frac{-5-(-1)}{6-(-2)} = \frac{-4}{8} = -\frac{1}{2}$ $y = -\frac{1}{2}x + k$ $-1 = -\frac{1}{2}(-2) + k$ $k = -2$ The equation of the line is $y = -\frac{1}{2}x - 2$.	Find the gradient m first by substituting the coordinates of the two points in the gradient formula. Then substitute the value of the gradient and the coordinates of one of the points in the gradient-intercept form $y = mx + k$ to find k.

Example 16

A straight country road with equation $y = -0.75x + 11.5$ intersects two state highways with equations $x = 0$ and $y = 0$. A beverage shop owner wants to open two new shops exactly at the intersections of the country road with the state highways.

a Draw the graph of the line representing the country road.

b Find the exact locations of the two new beverage shops.

c Find the distance between these beverage shops.

a

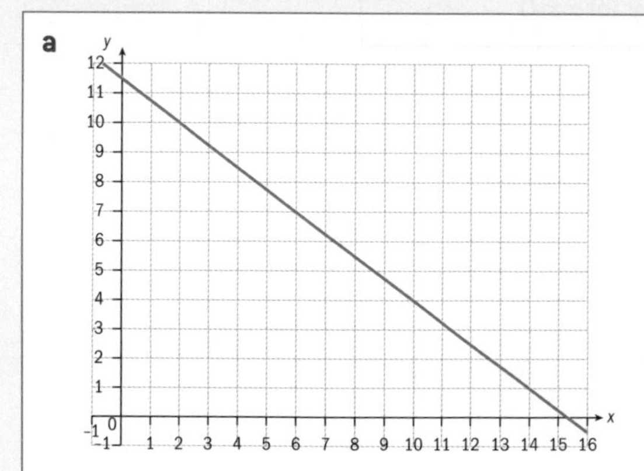

b When $x = 0$, $y = 11.5$ When $y = 0$, $x = 15.3$ (15.3333...) The locations of the two shops are $(0, 11.5)$ and $(15.3, 0)$. **c** $\sqrt{15.3333...^2 + 11.5^2}$ $= 19.2$ (19.1666...)	To find the locations of the two shops you need to find the x- and y-intercepts of the line. To find the y-intercept, substitute $x = 0$ in the equation of the line and solve for y. To find the x-intercept, substitute $y = 0$ in the equation of the line and solve for x.

Example 17

Draw the graphs of the following lines:

a $5x + 3y = 10$ **b** $y = \frac{1}{3}x + 4$

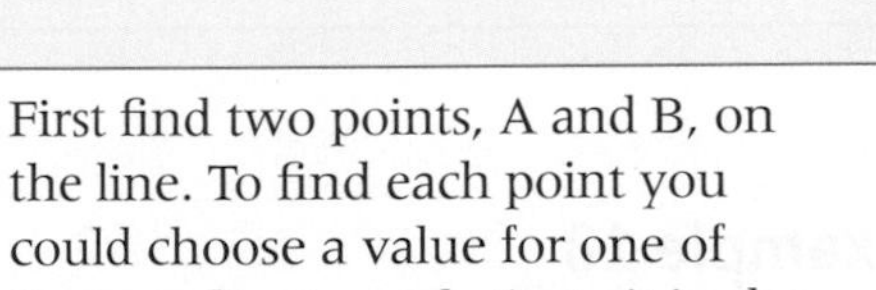

a For $x = 0$, $3y = 10$, so $y = \frac{10}{3}$ Thus the y-intercept is $\left(0, \frac{10}{3}\right)$. For $y = 0$, $5x = 10$ so $x = 2$. Thus the x-intercept is $(2, 0)$. A line drawn through the two intercepts is the graph.	First find two points, A and B, on the line. To find each point you could choose a value for one of the coordinates, substitute it in the equation and find the other coordinate. Such a point will be on the line. Then you plot A and B and draw a line through the points.

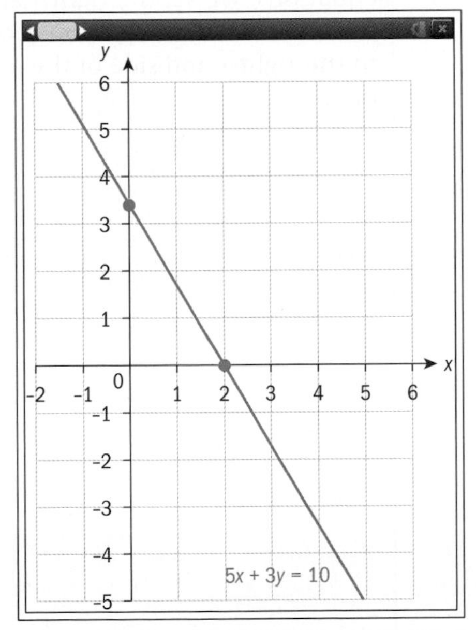

Continued on next page

b The y-intercept of the line with equation $y = -\frac{1}{3}x + 4$ is A(0, 4). The x-intercept is B(12, 0).

A line drawn through the two points is the graph.

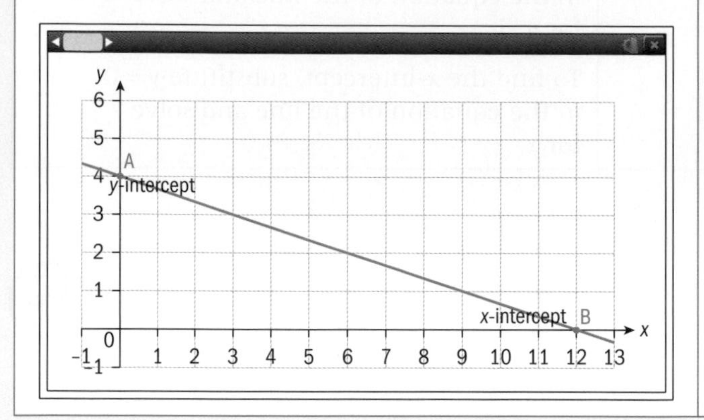

You could also use the gradient to find another point on the line. Start from (0, 4). Note that the gradient is $\frac{-1}{3}$, which means that for a run of 3 units the rise is −1 unit. Starting at A(0, 4) move one unit down to point C (rise −1) and then 3 units horizontally to point D (run 3). Draw a line through points A and D to obtain the graph.

Example 18

Two straight roads have equations $y = 3x + 15$ and $y = -2x + 5$. A traffic light has to be installed at their intersection point. Find the coordinates of the intersection point of the two roads.

One way to find the coordinates of the intersection point is algebraically:

$$3x + 15 = -2x + 5$$

$$5x = -10$$

$$x = -2$$

Then $y = 9$ and the intersection point is $(-2, 9)$.

Alternatively, plot both lines using your GDC.

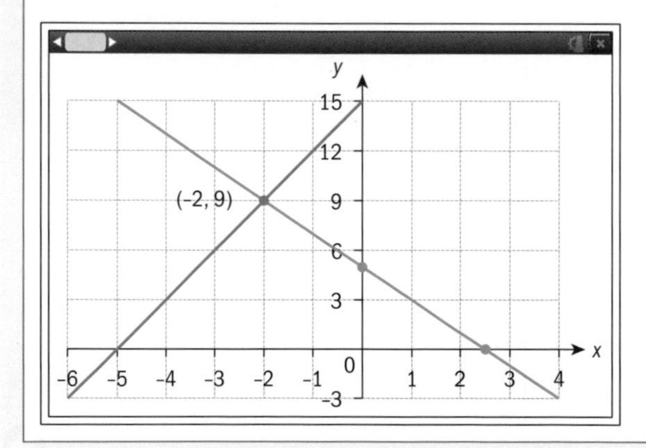

Since the intersection point lies on both lines and thus satisfies both equations, we set the right-hand side of one of the equations equal to the right-hand side of the other one.

Example 19

Bike & Bloom sells flower bouquets. The equation modelling the total cost, C, of making a number of bouquets, x, is $C = 6000 + 72x$, where C is given in Danish Krona, DKK.

a Find the gradient of the line representing the equation.

b Interpret the meaning of the gradient.

c Determine the y-intercept.

d Could the total cost per month be zero if no bouquets are made?

The total cost in a certain month is 112 720 DKK.

e Draw on the same axes the graph of:

i the line representing a cost of 112 720 DKK

ii the equation modelling the total cost, C.

f Find the number of bouquets made if the cost is 112 720 DKK.

a Gradient = 72

b It means that for any additional bouquet made the increase in cost is 72 DKK. This also means that 72 is the cost of producing one bouquet.

The assumption here is that the cost of each bouquet is the same.

c y-intercept is $(0, 6000)$

d No. Even if zero bouquets are made in the month, the total cost is 6000 (eg rent or other fixed expenses).

e **i** and **ii**

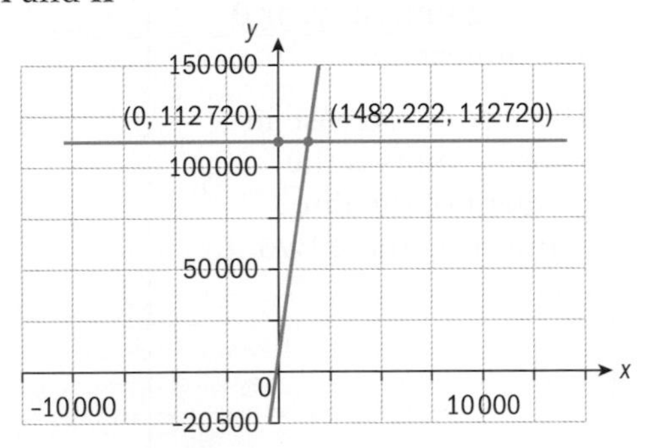

f $112720 = 6000 + 72x$

$x = 1482.222...$

1483 bouquets are made during the month.

Part **f** can be approached algebraically or graphically.

If you use graphs, the intersection point of the two graphs from part **e** will help you answer the question:

the x-coordinate gives you the number of bouquets.

Algebraically, equate $y = 6000 + 72x$ and $y = 112720$ and then solve for x.

Example 20

Brooklake town had only 50 social housing units when the new social programme started. Since then it has been adding new social housing units at a steady rate. Now, 17 years later, the town has a total of 322 social housing units.

a Find the rate of increase in social housing units per year.

b Write an equation for the total number of housing units, h, that the town had in relation to the number of years, x, that have passed since the social programme started.

c Draw the graph of the line representing this equation.

d Determine the gradient of the line. Interpret its meaning.

e Determine the y-intercept of the line. Interpret its meaning.

f Determine how many social housing units the town had 13 years after starting the programme.

g Predict the number of social housing units that the town will have 22 years after starting the programme. State your assumptions.

h Determine the year in which the town added its 200th social housing unit after the programme began.

a $m = \dfrac{322 - 50}{17 - 0} = 16$

b $h = mx + k$

$h = 16x + k$

$50 = 16 \times 0 + k$

$k = 50$

$h = 16x + 50$

c

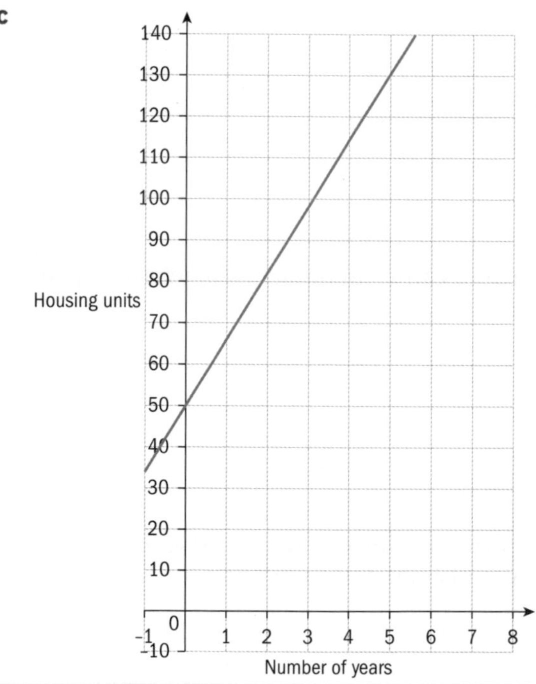

If we plot the given information, the y-coordinates of the points (0, 50) and (17, 322) represent the number of housing units at year 0 and at year 17.

The rate of increase in housing units is the gradient of the line connecting these two points.

TOK

"The object of mathematical rigour is to sanction and legitimize the conquests of intuition." Jacques Hadamard

Do you think that studying the graph of a function contains the same level of mathematical rigour as studying the function algebraically?

d Gradient is 16, which is the number of housing units added each year.	
e (0, 50), which means that there were 50 housing units when the town started the social programme.	
f $h = 16 \times 13 + 50 = 258$ housing units at the end of the 13th year.	
g $h = 16 \times 22 + 50 = 402$ housing units at the end of the 22nd year. We assume that they continue to add units at the same rate: 16 units a year.	
h $16 \times x + 50 = 200$ $x = 9.375$ The 200th unit will be added in the 10th year after the programme began.	We need to round up as more than 9 years were needed to build the 200th housing unit.

Exercise 4F

1 Find, in gradient-intercept form, the equation of the line:

a with gradient 3 and y-intercept $(0, -2)$

b with gradient -4 and passing through $(-1, 5)$

c passing through $(-1, -1)$ and $(2, -3)$.

2 Determine the x- and y-intercepts of the following lines:

a $y = 5x$ **b** $y = -0.4x + 3$

c $-0.2x + y = 1$

3 An air traffic controller examines four flight routes represented by straight lines given in A–D below.

A $y - 1 = -1\ (x + 4)$ **B** $y = 0.5x + 3$

C $2x + y - 5 = 0$

D line with a gradient of 2 and passing through $(-1, 3)$

a Draw the graphs of the four lines on the same pair of axes.

b Determine:

i the x- and y-intercepts of route A

ii the intersection point of routes A and B.

c Show that routes C and D have the same y-intercept.

4 Albena lives in Belgium. She travels to the USA where the temperature is measured in degrees Fahrenheit. She wants to know how hot it is in °C if the thermometer shows 83°F.

Use the fact that water boils at 100°C and 212°F, and that water freezes at 0°C and 32°F.

a i Plot the two points representing the given information on a pair of axes labelled °F (x-axis) and °C (y-axis).

ii Sketch the line passing through these two points.

b Write an equation for the relationship between C and F, where F represents degrees Fahrenheit and C represents degrees Celsius. Write your equation in the form $C = mF + k$.

c Write down the gradient of the equation.

d Interpret the meaning of the gradient.

e Write down the y-intercept of the line representing this equation.

f Interpret the meaning of the y-intercept.

g Find the temperature in degrees Celsius if the temperature is 83°F.

h Find the temperature in degrees Fahrenheit if the temperature is −10°C.

i Explain how you could rearrange the terms of the equation you obtained in part **b** so that it gives the relationship of C and F in the form $F = m_1C + k_1$.

5 A landscape architect has planned four new paths in a park. The paths are straight lines as shown on the diagram below. Find the equations of the paths L_1, L_2, L_3 and L_4.

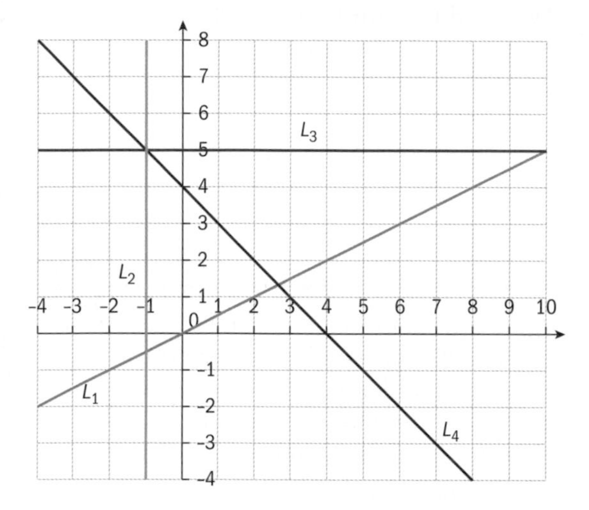

6 Maria works as a part-time waitress in the summer and earns \$30 per day plus 2% of all food and drink sales for the day.

a Write an equation to represent her daily earnings, A, in USD, in relation to the amount, x, in USD, of all sales of food and drinks during the day.

b Determine the gradient. Interpret its meaning.

c Determine the y-intercept. Interpret its meaning.

d Draw the graph of the line representing Maria's earnings A.

e Find what Maria's earnings would be if the amount of food and drinks sold during a certain day were \$2400.

Developing inquiry skills

Look back at the opening problem for the chapter. You were trying to divide the territorial waters between three islands.

How can you find equations for the lines that connect each pair of islands?

4.4 Parallel and perpendicular lines

A large farm is managing a crop dusting operation, which involves spraying crops with protection products from multiple agricultural aircraft. How could the crop dusting controller make sure that the agricultural planes do not collide?

Investigation 9

The graphs of the trajectory lines L_1, L_2, L_3, L_4, L_5 and L_6 of six agricultural planes are given in the diagram.

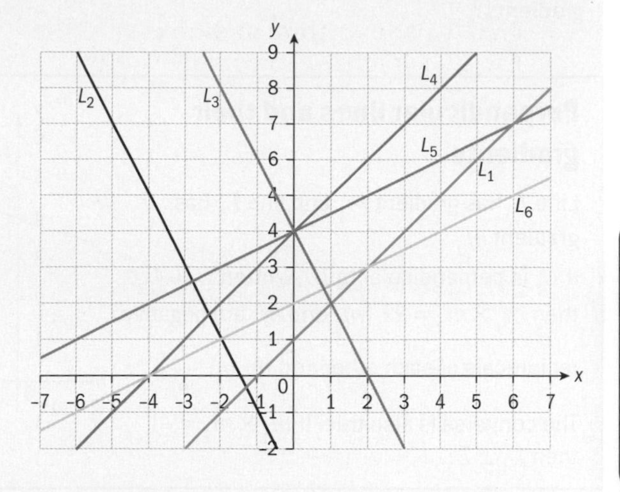

1 Examine the lines L_1, L_2, L_3, L_4, L_5 and L_6. List any common characteristics shared by two or more lines.

Lines L_1 (green), L_2 (black) and L_5 (blue) have the following equations:

$L_1: y = x + 1$

$L_2: y = -2x - 3$

$L_5: y = \frac{1}{2}x + 4$

2 Find the equations of the lines L_3 (purple), L_4 (orange) and L_6 (lime). Give the equations in the form $y = mx + k$, where m is the gradient and k is the y-intercept.

3 Use the equations of lines L_1, L_2, L_3, L_4, L_5 and L_6 to complete the table below.

Line	Gradient	y-intercept
L_1		
L_2		
L_3		
L_4		
L_5		
L_6		

Examine the gradients and the y-intercepts of the lines.

4 **Conceptual** If two lines are parallel, what is the relationship between their gradients? What about perpendicular lines and their gradients? What can you tell about the y-intercepts of parallel or perpendicular lines?

5 **Conceptual** Can you tell from the equations of two lines whether they are parallel, perpendicular or neither?

Parallel lines and their gradients

Line L_1 has gradient m_1 and line L_2 has gradient m_2.

If L_1 is parallel to L_2, written $L_1 \| L_2$, then $m_1 = m_2$. The converse is also true: If $m_1 = m_2$ then $L_1 \| L_2$.

For example, the lines shown in the diagram both have a gradient of −3, but they have different y-intercepts.

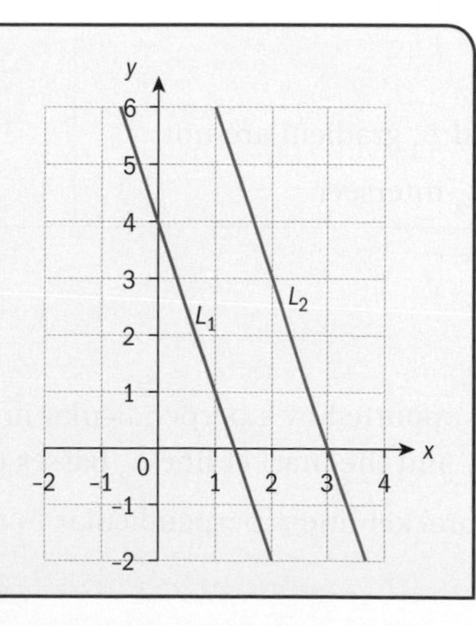

Reflect Why does it make sense that two parallel lines should have equal gradients?

Perpendicular lines and their gradients

Line L_1 has gradient m_1 and line L_2 has gradient m_2.

If L_1 is perpendicular to L_2, written $L_1 \perp L_2$, then $m_1 \times m_2 = -1$ (m_1 and m_2 are negative reciprocals of each other and $m_1 = -\frac{1}{m_2}$).

The converse is also true: If $m_1 \times m_2 = -1$ then $L_1 \perp L_2$.

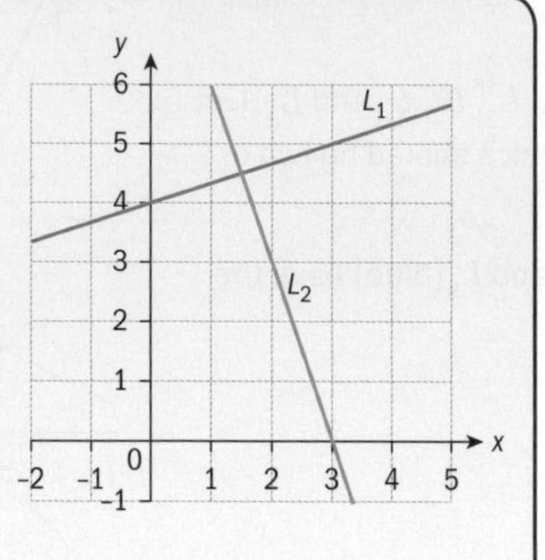

For example, the lines shown in the diagram above have negative reciprocal gradients of $\frac{1}{3}$ and -3, and they have different y-intercepts.

Example 21

The trajectory lines L_1 and L_2 of two cargo ships are given. In each case determine whether the lines are parallel or if they will intersect:

a L_1 passing through points $(2, 4)$ and $(-1, -2)$; and L_2 passing through points $(0, -4)$ and $(2, 0)$

b L_1 passing through points $(0.5, -1)$ and $(-7, -1.5)$; and L_2 passing through points $(-4, 0)$ and $(3, 2)$.

a L_1 gradient $= \frac{-2-4}{-1-2} = 2$ L_2 gradient $= \frac{0+4}{2-0} = 2$ Since L_1 gradient $= L_2$ gradient, $L_1 \parallel L_2$. **b** L_1 gradient $= \frac{-1.5+1}{-7-0.5} = \frac{1}{15}$ L_2 gradient $= \frac{2-0}{3+4} = \frac{2}{7}$ Since L_1 gradient and L_2 gradient are not equal, lines L_1 and L_2 intersect.	Find the gradients for each pair of lines and check whether they are equal.

Example 22

A roof structure is to be supported by a perpendicular bracket. The roof line L_1 passes through points $(-6, 2)$ and $(-1, 4)$; and the bracket line L_2 passes through points $(3, 5)$ and $(5, 0)$.

Determine whether the bracket line is perpendicular to the roof line.

L_1 gradient $= \frac{4-2}{-1+6} = \frac{2}{5}$ L_2 gradient $= \frac{0-5}{5-3} = -\frac{5}{2}$ Since L_1 gradient × L_2 gradient $= -1$, $L_1 \perp L_2$.	Find the gradients of the two lines and determine whether they are negative reciprocals.

Example 23

A tractor pulls a plough to make three parallel straight grooves for planting seeds. The graph of one of the grooves L_1 is given in the diagram.

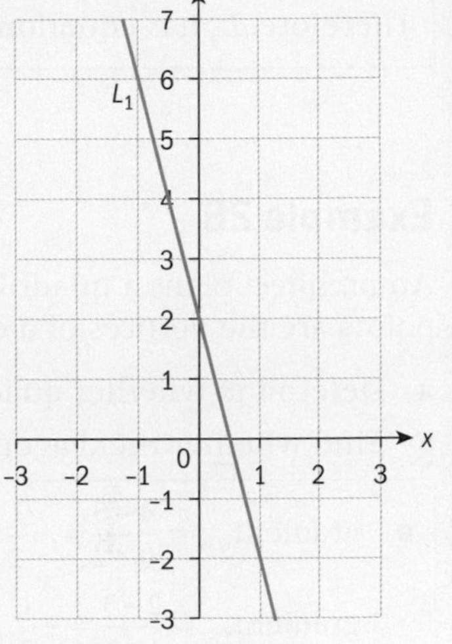

a Determine the equation of a groove L_2 parallel to L_1 such that it passes through the origin O.

b Determine the equation of a third groove that is equally distanced from L_1 or L_2.

a L_1 passes through $(0, 2)$ and $(1, -2)$ and therefore has run 1 and negative rise -4, and thus the gradient is -4. The y-intercept is $(0, 2)$, so the equation of L_1 is $y = -4x + 2$. A parallel groove L_2 passing through O will then have an equation $y = -4x$. **b** Another equally distanced groove would have an equation $y = -4x + 4$ or $y = -4x - 2$.	First determine the equation of the given line L_1.

Example 24

A machine is set to make parallel cuts on wooden boards. The cuts are made along two straight lines L_1 and L_2. Line L_2 has equation $y = 6$. Line L_1 has gradient $\frac{3-x}{4}$. Find x.

$\frac{3-x}{4} = 0$ $3 - x = 0$ $x = 3$	Equate the gradients of the parallel lines. Line L_2 is horizontal and its gradient is 0.

Example 25

A plane flies along a straight line L_1 with equation $y = 2x - 5$. Determine the equation of the line L_2 which represents the trajectory of another plane flying at the same altitude that should not intersect L_1 and should pass through point $(3, -1)$.

Line L_2 will have equation $y = 2x + k$.	The two trajectories must be parallel so that they do not intersect. Since parallel lines have equal gradients, L_2 will have the same gradient as L_1.
We substitute the coordinates of $(3, -1)$: $-1 = 2 \times 3 + k$ $k = -7$ Therefore, L_2 has equation $y = 2x - 7$.	To determine k, substitute the coordinates of point $(3, -1)$ in the equation of L_2.

Example 26

An architect plans a quadrilateral structure in the form of a parallelogram. The following four points are the vertices of a quadrilateral: $P(-1, 3)$, $Q(-2, 5)$, $M(0, 4)$ and $N(1, 2)$.

a Determine whether quadrilateral PQMN is a parallelogram.

b Find whether the diagonals are perpendicular.

a $\text{gradient}_{QM} = \frac{4-5}{0+2} = -\frac{1}{2}$ $\text{gradient}_{PN} = \frac{2-3}{1+1} = -\frac{1}{2}$ $\text{gradient}_{QP} = \frac{5-3}{-2+1} = -2$ $\text{gradient}_{MN} = \frac{2-4}{1-0} = -2$ Thus the opposite sides lie on lines with equal gradient, and therefore PQMN is a parallelogram.	To find out whether PQMN is a parallelogram, determine whether the opposite sides are parallel or not. It is always a good idea to sketch the quadrilateral to help you identify the opposite sides. Then check whether the pair [QM] and [PN] and the pair [QP] and [MN] are parallel. To do that, find the gradient of each line.
b $\text{gradient}_{QN} = \frac{2-5}{1+2} = -1$ $\text{gradient}_{MP} = \frac{4-3}{0+1} = 1$ The gradients are negative reciprocals $(-1 \times 1 = -1)$, so the diagonals are perpendicular.	To determine whether the diagonals [QN] and [MP] are perpendicular, calculate the gradients of the diagonals and determine whether they are negative reciprocals.

Reflect How do you determine whether two lines are parallel or perpendicular?

Exercise 4G

1 Find the gradient of a line which is parallel to each line with the given equation:

a $y = -x + 6$ **b** $x = 7$

c $x - 3y = 5$ **d** $y = -\frac{2}{5}(x - 1)$

2 Find the gradient of a line which is perpendicular to each of the following lines:

a $y = -3x + 11$ **b** $-\frac{x}{4} + 2y = 0$

c $y = -3$ **d** $y = \frac{2(x-1)}{3}$

3 L_1 and L_2 are the trajectories of two ships moving in straight lines. Determine whether the ships' trajectories are perpendicular, parallel or neither:

a Line L_1 passes through the points $(0, -4)$ and $(-1, -7)$; and line L_2 passes through the points $(3, 0)$ and $(-3, 2)$.

b Line L_1 has equation $2y - \frac{1}{2}x + 3 = 0$, and line L_2 has equation $y - 3 = 0.25(x - 1)$.

c Line L_1 has equation $y = -\frac{2}{5}x - 1$, and line L_2 has equation $x - \frac{y}{3} = 4$.

4 Determine whether the straight air routes with equations $x - \frac{y}{2} = -3$ and $x = -5$ are intersecting or not. If they are intersecting, find the point of intersection.

5 A ski resort is building two parallel straight ski slopes for children. One of them has a gradient of $\frac{1}{3}$. The other ski slope will pass through points $(2, -3)$ and $(s, -5)$. Find the value of s.

6 A straight connecting street segment is built perpendicular to an existing street with equation $y = \frac{2}{7}x + 3$. Determine the equation of the line of the new street segment, which passes through point $B(-1, -0.2)$.

7 A fish farm builds a breeding basin in the form of a quadrilateral ABCD, with $A(-3, -1)$, $B(2, 0)$, $C(5, 3)$ and $D(0, 2)$. Show that the quadrilateral ABCD is a parallelogram.

Investigation 10

Two towns, C and D, are to be serviced by a new railroad. Town C has coordinates $(2, 10)$, town D has coordinates $(10, 14)$ and the new railroad line L has equation $y = 2x + 3$, as shown in the diagram. A train station is to be built on the railroad line L to be used by both towns' inhabitants.

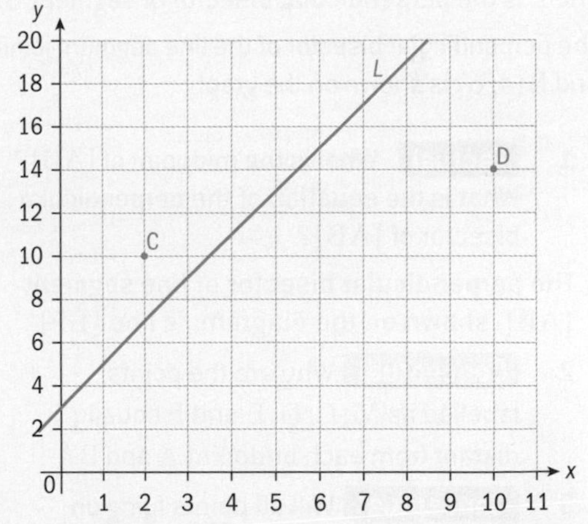

1 Connect points C and D. Write down the equation of line [CD].

2 [CD] intersects the railroad line L at point $E(x, y)$.

Point $E(x, y)$ lies on the line L. Using the equation of line L, write down an expression for the y-coordinate of the intersection point E.

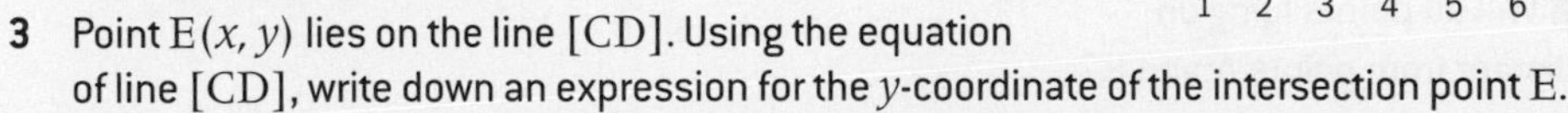

3 Point $E(x, y)$ lies on the line [CD]. Using the equation of line [CD], write down an expression for the y-coordinate of the intersection point E.

4 Find the value of the x-coordinate of point E by equating the two expressions from parts **2** and **3** for the y-coordinate of point E.

5 Find the value of the y-coordinate of point E.

Continued on next page

6 A committee compares point E and point F$(6, 15)$ to determine a location for the train station. Give one reason in support of and one reason against locating the train station at point E.

7 Next, the committee examines point M, the midpoint of [CD], as a possible location for the station. Find the coordinates of point M. Could point M be a suitable location for the train station? Explain why or why not.

8 The committee rejects points E, F and M as possible locations for the train station, and decides to find the exact point on the railroad line L that is equally distant from towns C and D.

Trace line segment [CD] and line L on a patty paper. Fold the patty paper so that point C overlaps with point D. Discuss with your partners how to find points that are equidistant from C and D. One such point is M, the midpoint of [CD]. What other points are also equally distant from C and from D?

9 All points that are equally distant from C and D lie on a line N. Determine the equation of this line.

10 Determine the intersection point T of line L and line N. Explain why point T is equally distant from town C and from town D and also lies on the railroad line.

11 Is T the only point satisfying the conditions for the train station location set by the committee? Or, are there other such points? How many such points are there?

Reflect Why are all points lying on the perpendicular bisector of a line segment equidistant from the two endpoints of the segment?

Investigation 11

On graph paper draw line L. Draw point B on L and point A, which is not on the line L. Fold the graph paper so that the parts of line L overlap and point A is exactly on the formed crease. Now fold along the new crease. Label the point where the crease intersects with line L as O. Identify point B_1 on the crease [BO], so that $B_1O = BO$. Explain why line L is the perpendicular bisector of segment BB_1.

The perpendicular bisector of the line segment joining A$(2, 2)$ and B$(4, 6)$ is shown on the graph.

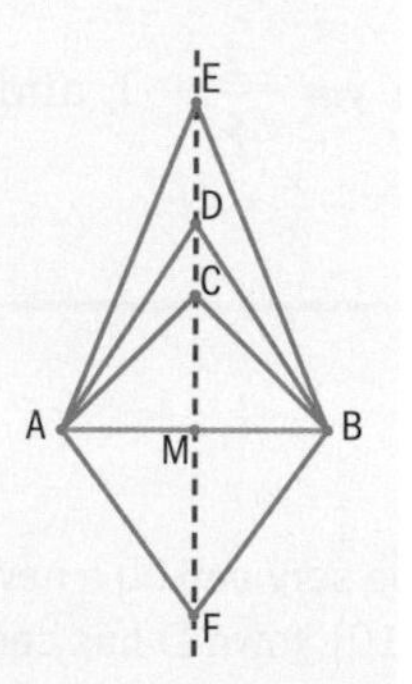

HINT

The perpendicular bisector of the line segment joining two points is the line that passes through the midpoint of the line segment and is perpendicular to it.

1 **Factual** What is the midpoint of [AB]? What is the equation of the perpendicular bisector of [AB]?

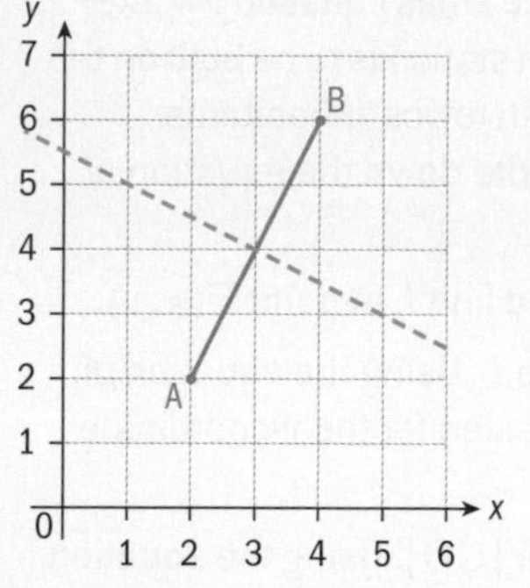

The perpendicular bisector of line segment [AB], shown on the diagram, is line [EF].

2 **Conceptual** Why are the points labelled as M, C, D, E and F equally distant from each endpoint A and B?

3 **Conceptual** Will all points lying on [EF] be equidistant from points A and B?

4 **Conceptual** Will a point equidistant from two points lie on the perpendicular bisector of the segment connecting the two points?

Example 27

Find the equation of the perpendicular bisector of the line segment whose endpoints are A(1, −1) and B(5, −3).

Midpoint $\text{M}\left(\frac{1+5}{2}, \frac{-1+(-3)}{2}\right) = \text{M}(3, -2)$	A perpendicular bisector always passes through the midpoint of the line segment. Find the midpoint, M, of the line segment [AB] using the formula $\text{M}\left(\frac{x_1+x_2}{2}, \frac{y_1+y_2}{2}\right)$.
	Then find the equation of a line which is perpendicular to [AB] and passes through point M.
$\text{gradient}_{\text{AB}} = \frac{-3-(-1)}{5-1} = -\frac{2}{4} = -\frac{1}{2}$	To find the equation, first find the gradient of line segment [AB].
$y = 2x + k$	The gradient of the perpendicular bisector is the negative reciprocal of $-\frac{1}{2}$, which is 2.
$-2 = 2 \times 3 + k$ $k = -8$	k is found by substituting the coordinates of point M in the equation.
The equation of the perpendicular bisector of [AB] is $y = 2x - 8$.	

Example 28

A border canal divides two countries. Bridge [AB] connects the two sides of the canal and is perpendicular to its banks. Point C(10, 17) is the midpoint of the bridge [AB], as shown in the diagram. The equation of the line [AB] is $y = 0.5x + 6$. Find the equation of the border line that equally divides the bridge and the canal between the two countries.

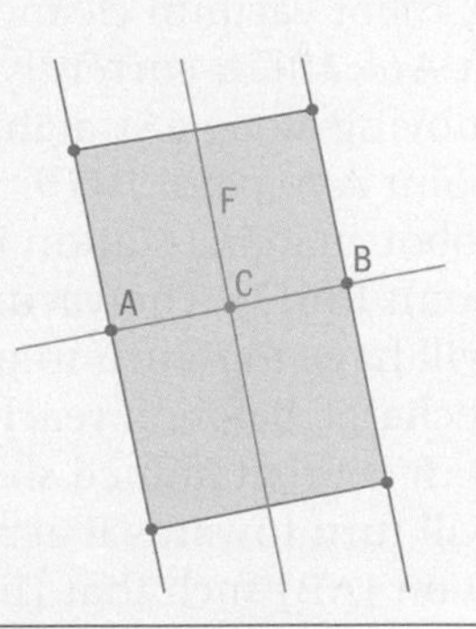

Line F is perpendicular to [AB] and passes through C. Then the equation of line F will be $y = -2x + b$.	The border line, F, that equally divides the bridge and the canal between the two countries is the perpendicular bisector of line [AB]. The perpendicular bisector F passes through point C, the midpoint of [AB].
$17 = -2 \times (10) + b$ $b = 37$	It will have a gradient that is the negative reciprocal of 0.5, which is −2.
Line F has equation $y = -2x + 37$.	b is found by substituting the coordinates of C(10, 17) in $y = -2x + b$.

Investigation 12

David is hiking. He is currently located at point $A(25, 90)$. He wants to take the shortest path to reach highway L, which follows a straight line with equation $y = \frac{1}{3}x + 5$.

He takes a straight perpendicular path to the highway.

1 Draw the perpendicular from A to line L.

2 Find the equation of the perpendicular path that David takes to the highway.

3 Find the coordinates of the point, B, at which the perpendicular path meets the highway.

4 Point $D(30, 15)$ is on the highway. Show that $AB < AD$.

5 Why is AB the shortest distance from A to the highway?

The shortest distance from a point A to a straight line (BC) is the segment [AD], where D lies on (BC) and $[AD] \perp (BC)$.

Reflect Why is the perpendicular passing through a given point to a given line the shortest distance from this point to the line?

TOK

Do you think that mathematics is just the manipulation of symbols under a set of rules?

Example 29

A robot vacuum cleaner is located at A(4, 10). It currently cleans by moving along a straight line from point A to point B(19, 5). The robot charging station is located at point E(6, 3). The vacuum cleaner will have to return to point E to recharge before it reaches point B. It is programmed so that it will turn towards E at a point D on [AB] such that [ED] is the shortest distance from E to [AB]. Find the coordinates of point D.

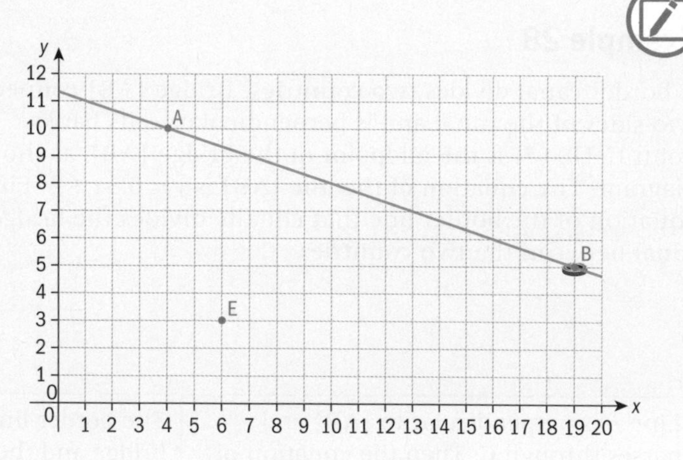

The shortest distance

$\text{gradient}_{AB} = \frac{5-10}{19-4} = -\frac{5}{15} = -\frac{1}{3}$ $y = -\frac{1}{3}x + k$ $10 = -\frac{1}{3} \times 4 + k$ $k = 11\frac{1}{3}$ Equation of [AB]: $y = -\frac{1}{3}x + 11\frac{1}{3}$	To identify point D on [AB] such that [ED] is the shortest distance from E to [AB], use the fact that the shortest distance from a point to a line is the perpendicular drawn from the point to the line. Thus you have to find the equation of the line perpendicular to [AB] and passing through E. Then find the intersection point of line [AB] and its perpendicular line through E.

To find the equation of the line [AB], find the gradient and substitute it and the coordinates of A or B in the point-gradient or gradient-intercept form.

The perpendicular to [AB] will have an equation of the form

$y = 3x + k$

$3 = 3 \times 6 + k$

$k = -15$

To determine k of the perpendicular to [AC] line, substitute the coordinates of point E.

The equation of the line perpendicular to [AB] is $y = 3x - 15$.

$3x - 15 = -\frac{1}{3}x + 11\frac{1}{3}$

$3\frac{1}{3}x = 26\frac{1}{3}$

$x = 7.9$

$y = 8.7$

Therefore point D has coordinates (7.9, 8.7).

Finally, determine the intersection point D of line [AB] and its perpendicular line passing through E. Equate the expressions for y from each of the two equations, and solve for the value x, which will be the x-coordinate of D. Use technology or an algebraic method. Substitute this x value in one of the two equations and find the corresponding y value.

As always, it is a good idea to draw a sketch that can help you reason and answer the question.

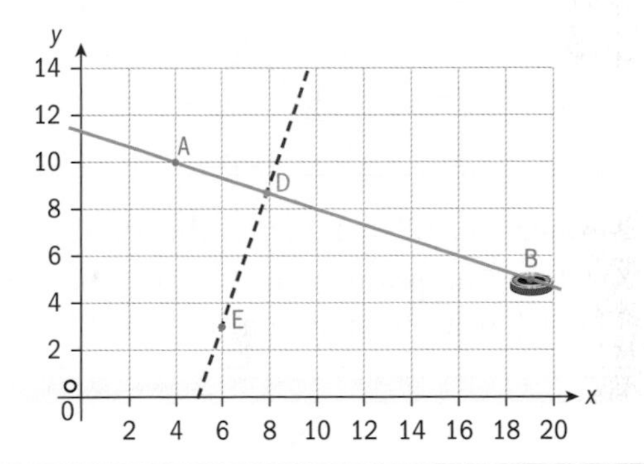

Reflect How do you find the shortest distance between a point and a line defined by two points?

Example 30

Two towns are located at point A(2, 20) and point B(14, 24), and a nearby road R is a straight line with equation $-x + y = 4$. A new taxi firm wants to find a location for a taxi station on the path R, so that the combined distance from the taxi station to the two towns A and B is shortest.

Continued on next page

a Draw a perpendicular from B to line R.

b Mark the intersection point C of road R and the perpendicular through B.

c Find E on [BC] such that BC = CE.

Point F(10, 14) lies on line R.

d State the relationship between the lines FC and BE.

e Show that AF + BF = AF + FE.

f From ΔAFE, explain why AE < AF + FE.

AE intersects road R at point D.

g Show that D is the point on road R such that the combined distance from it to the two towns A and B is shortest.

h Find the coordinates of D.

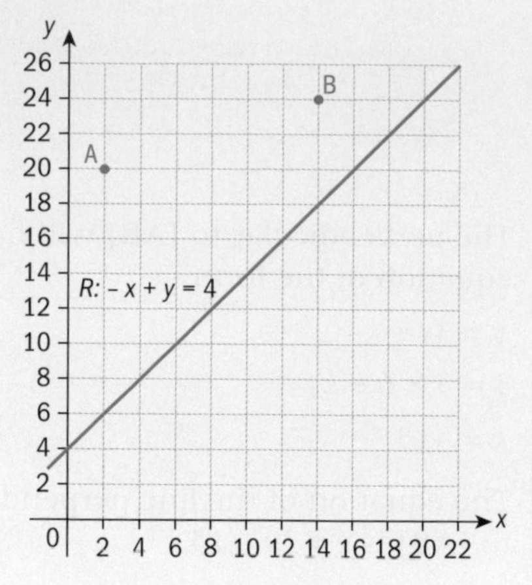

The shortest distance

a, b

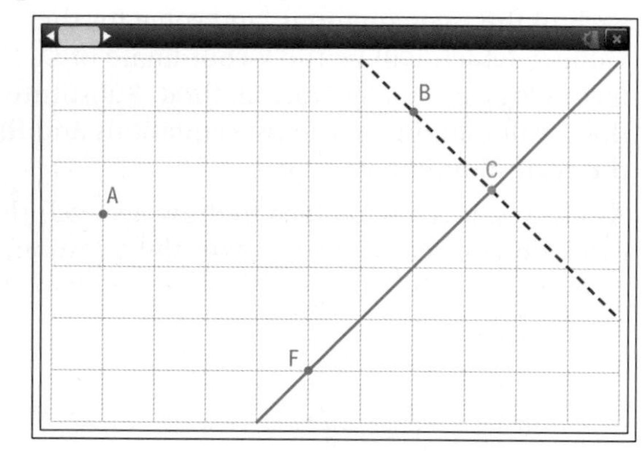

c We measure BC = 4.2 from the graph.

We then find point E on (BC), on the opposite side of R from B, such that EC = 4.2.

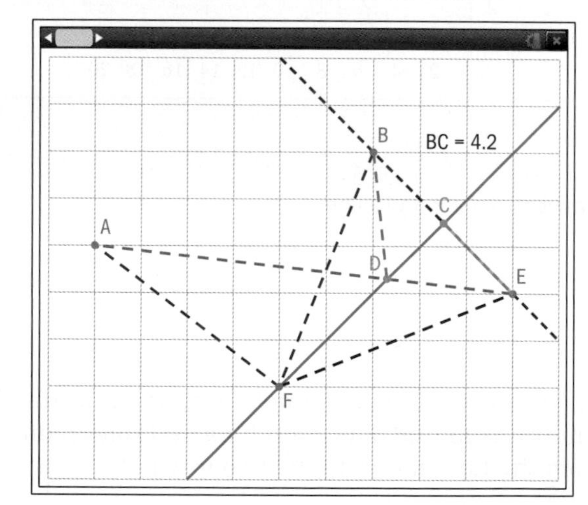

d (FC) is the perpendicular bisector of (BE) since (FC) $\perp$ (BE) and BC = CE.

e FB = FE since (FC) is the perpendicular bisector of (BE); thus AF + BF = AF + FE.

f From ΔAFE, AE < AF + FE because of the triangle inequality—the fact that one side of a triangle is always shorter than the sum of the other two sides.

g [AE] and line R intersect at point D. Since AE is shorter than AF + FE, for any F, and AE = AD + DB, AD + DB will be the shortest combined distance and D the location for the taxi station.

h From the graph: D(14.6, 18, 6)

Exercise 4H

1 Two friends, Alison and Bernard, are walking along two different roads. The roads can be represented by the lines with equations $y = -x + 410$ and $y = \frac{1}{2}x - 100$.

Alison is on the first road at the point with coordinates (0, 410) and Bernard is at the point with coordinates (50, −75).

a Verify that Bernard is on the road with equation $y = \frac{1}{2}x - 100$.

b Find the coordinates of the point of intersection of the two roads.

2 Find the equation of the perpendicular bisector of each of the following line segments:

a the line segment joining A(2, 2) and B(4, 6)

b [CD], if the equation of the line passing through C and D is $x + y = 3$ and the midpoint of [CD] is (2, 1).

3 Find the shortest distance from a hotel located at (2, 4) to a straight road with equation $3x + 5y + 8 = 0$.

4 Two towns are located at points A(2, −2) and B(8, 5). A new school is to be built on a straight road with equation $-x + 7y = -4$. Find the location of the school so that it is equidistant from the two towns.

5 A triangular park has vertices at A(−1, −1), B(4, 4) and C(7, −1). A fountain is to be built that is equidistant from all vertices. Find the location D of the fountain.

Developing inquiry skills

How could you use what you have learned in this section to find points that are an equal distance from two of the islands? What is significant about these points?

Can you find a point that is equally distant from all three of the islands?

4.5 Voronoi diagrams and the toxic waste dump problem

There are five airports in a state, represented by dots on the diagram. The coloured regions on the diagram indicate the points which are closest to a given airport. For example, Airport 5 is the closest one to all points in the purple region, and Airport 2 is the closest one to the points that are in the green region. What do you think this diagram can be used for? What kinds of questions can it help answer?

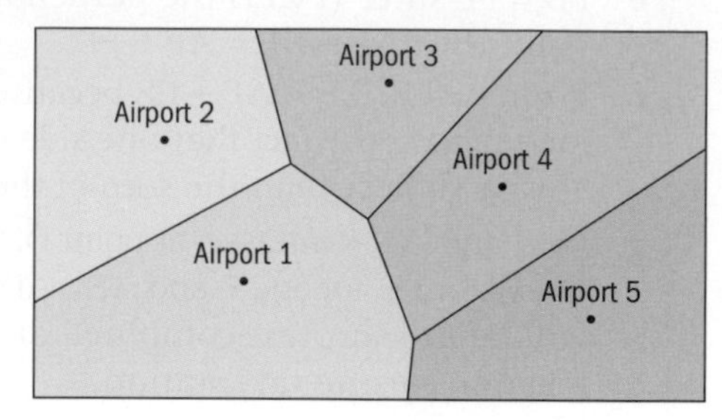

This is an example of a Voronoi diagram, named after the mathematician Georgy Voronoy (1868–1908). Voronoi diagrams have a large number of applications nowadays, such as in science, technology and visual art.

Reflect What does a Voronoi diagram show?

The regions that include the closest points to a given point are called **cells**, and the given points are called **sites**. The diagram above shows five sites—the five airports—and five cells—the five coloured regions. The lines separating the regions are called **boundaries** or **edges**.

Investigation 13

Three hospitals A, B and C lie inside a square of side length 10 units, which marks the boundaries of the town. The sides of the square are formed by the lines $x=0, x=10, y=0$ and $y=10$.

Use technology or graph paper to complete the investigation.

1 On a coordinate system, draw the lines with equations $x=0$, $x=10, y=0$ and $y=10$ to mark the boundaries of the town.

Hospitals A and B have coordinates (1, 8) and (8, 8).

2 Locate them on the coordinate system and label them with A and B.

3 Locate the midpoint of [AB] and label it with M1. Draw the perpendicular bisector of [AB].

4 Hospital C has coordinates (3 , 2). Locate the midpoint of [AC] and label it with M2. Draw the perpendicular bisector for A and C of [AC].

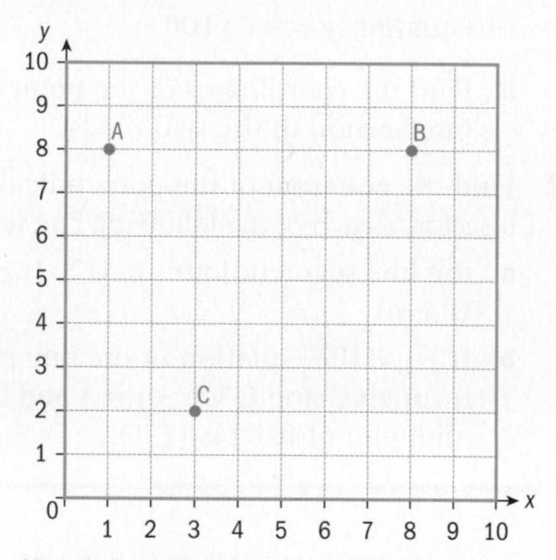

Hospitals A, B and C located within the town's boundaries

5 Locate the midpoint of [BC] and label it with M3. Draw the perpendicular bisector of [BC].

6 Would the perpendicular bisector of [AB] intersect the perpendicular bisector of [AC]? Explain why or why not.
Would the perpendicular bisector of [AB] intersect the perpendicular bisector of [BC]? Explain why or why not.
Would the perpendicular bisector of [AC] intersect the perpendicular bisector of [BC]? Explain why or why not.
Would all three perpendicular bisectors intersect in one point or not? Explain.

7 Where would the boundaries of the Voronoi diagram be?

8 **Conceptual** What does the boundary of a Voronoi diagram represent?

9 Use the pentagon tool to shade in a different colour each region separated by the Voronoi diagram boundaries.

10 Which point is equally distant from the three hospitals A, B and C? Is there more than one such point?

11 Samar's house is located at point (4, 5). Determine which hospital is closest to her house.

12 **Conceptual** What does the Voronoi diagram show?

13 **Conceptual** Why is a vertex always equidistant from three sites?

14 **Conceptual** How many edges meet at a vertex of a Voronoi diagram?

The boundaries of the **cells** in a Voronoi diagram are formed by the perpendicular bisectors of the line segments joining the sites. Usually all other lines that do not form boundaries to the cells are erased when the construction of the diagram is complete.

A point at which cell boundaries meet in a Voronoi diagram is called a **vertex**. A vertex is equidistant from the three surrounding sites.

TOK

John Snow used Voronoi diagrams to show that the Broad Street cholera outbreak that killed 616 people in London in 1852–60 was due to contaminated water and not air.

Investigation 14

There are three fire stations, A, B and C, in a town. The coordinates of the fire stations are $A(1, 4)$, $B(1, 0)$ and $C(4, 3)$.

In order to improve response times, the township has installed a new centralized fire response system, which allows a dispatcher to send a fire truck from the nearest fire station to the location of the fire. How should the township be divided into regions so that there is one fire station in each region and this fire station is the closest one to each house in the region?

Use technology or graph paper to complete the investigation.

1 Plot the fire stations A and B on coordinate axes, either by hand or using a software package.

2 Draw the perpendicular bisector of [AB] and gently shade those points nearest to A. How would you divide the township into two regions if fire station C is not yet in operation?

3 Add fire station C to the diagram, then find the equations of the perpendicular bisectors of [AC] and [BC] and add these to the diagram. How would you divide the township into three sections when the three fire stations A, B and C are fully operational?

4 The township is relocating fire station D within its territory. Station D will have location (5, 1). Divide the township into four regions for the four fire stations A, B, C and D.

5 Why does adding the fire stations one by one make sense for creating a town map?

6 What is the incremental algorithm?

HINT

Many software packages allow you to draw and obtain the equations of perpendicular bisectors directly. If not using software, a perpendicular bisector can be constructed using a compass and a ruler or by finding its equation and sketching it.

EXAM HINT

In an exam you will not be asked to draw a perpendicular bisector unless you already have its equation.

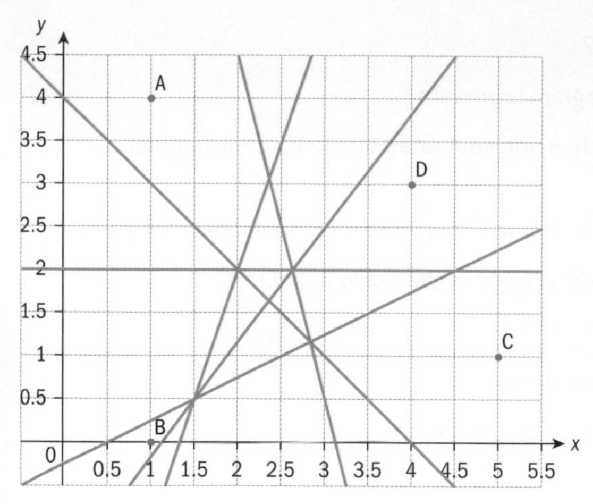

It can be difficult to draw all the perpendicular bisectors at the same time and then create a Voronoi diagram, especially if the number of sites is more than three. The diagram shows that even with just four points (sites) there are difficulties in deciding which perpendicular bisectors to use to form boundaries of the cells.

The **incremental algorithm** described below avoids this problem by adding each of the sites one at a time.

Incremental algorithm for constructing Voronoi diagrams

Method

1 Plot the points A and B on coordinate axes, either by hand or using a software package. Draw the perpendicular bisector of line segment [AB] and gently shade the region of points nearest to A.

2 Add point C to the diagram, then find the equations of the perpendicular bisectors of [AC] and [BC] and add these to the diagram.

3 The incremental algorithm:

 i Begin with the perpendicular bisector which lies between the new site C and the site in whose cell this vertex currently lies.

 ii Move along this line until you reach an intersection with another of the perpendicular bisectors between the new site and an existing one. (This will also be on a boundary of the previous Voronoi diagram.)

 iii Leave the intersection along this other perpendicular bisector in the direction that lies entirely in the cell surrounding another of the sites. (This will be the direction that creates a convex polygon around the new site.)

 iv The algorithm stops either when you return to your starting point (if the cell is bounded) or if there are no more intersections (if the cell is unbounded). In this case you may need to reverse the direction of the algorithm to ensure all sides have been found.

4 Shade the three regions making up the new Voronoi diagram.

5 Having completed the diagram for the three sites A, B and C, you can now add the next site, D. Repeat steps 2 and 3. This time there will be two intersection points with the perpendicular bisectors of [BD] and [AD].

TOK

Is mathematics independent of culture?

HINT

Usually a final version of the Voronoi diagram will have the perpendicular bisectors removed so that only the edges of the regions remain.

Reflect What is the incremental algorithm?

Exercise 4I

In these questions, unless told to calculate the equations of the perpendicular bisectors, you can construct the lines with a pair of compasses, by eye or by using software.

1 By finding the perpendicular bisectors between each pair of points, use the incremental algorithm to complete the Voronoi diagrams for the given sites.

a (1, 1), (3, 1), (2, 3)

b (1.1), (3, 1), (3, 5)

2 A neighbourhood has three bookstores located at A(1, 1), B(5, 1) and C(5, 5).

a Construct the Voronoi diagram for the three bookstores A, B and C either by hand or using a software package.

b Determine the closest bookstore to Kari, whose house is located at (1, 4).

A fourth bookstore was built recently at point D(3, 5).

c Construct the Voronoi diagram for the four bookstores A, B, C and D either by hand or using a software package.

d Determine whether there is another bookstore that is now closest to Kari's house. If so, determine which one.

The toxic waste dump problem

This problem is to find the point on the Voronoi diagram that is as far as possible from any of the sites. It is called the **toxic waste dump problem** because one application might be to find where waste can be deposited so that it is as far as possible from habitation.

Investigation 15

Toxic waste is to be deposited between three towns A, B and C. Find the point that is the farthest from all three towns.

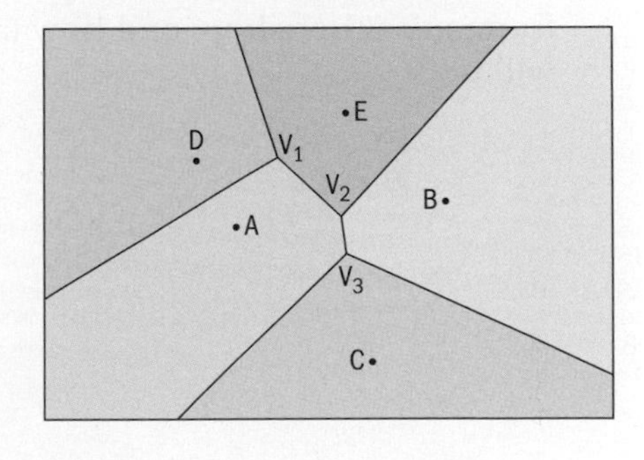

1 On a copy of the Voronoi diagram draw a circle that will pass through points A, E and D. Identify the centre of this circle.

2 Draw another circle through points A, B and E. Identify the centre of this circle.

3 Draw a third circle through points A, B and C. Identify the centre of this circle.

4 Write down what you notice.

5 Explain using the diagram below why there will always be three points (sites) equidistant from a vertex of a Voronoi diagram which has three edges incident to it.

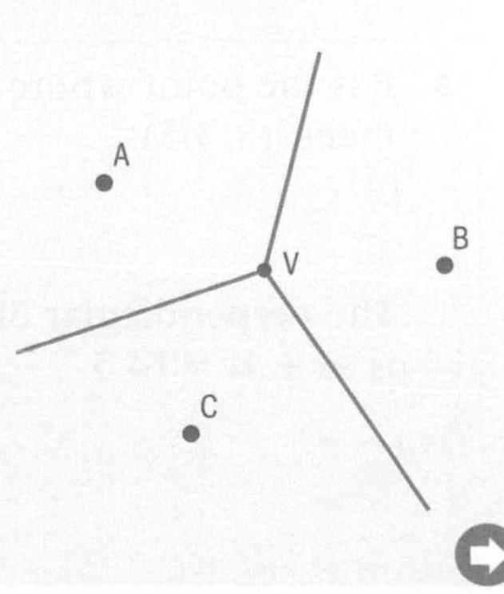

6 Explain why the circle with V_3 as a centre which passes through A will also pass through B and C.

7 Determine the radius of each of the three circles from parts **1**, **2** and **3**. Identify the circle with the largest radius.

8 Determine the location of the toxic waste deposit so that it is farthest from the surrounding towns A, B, C, D and E.

Continued on next page

9 Explain why another site (town) cannot be inside the largest circle, assuming that no new sites are added.

10 **Conceptual** Where is the point that gives the solution to the toxic waste problem?

11 **Conceptual** What applications are there for the solutions of the toxic waste problem?

12 **Conceptual** Why are vertices of a Voronoi diagram the only points that give the solution to the toxic waste problem (assuming a location between the towns is sought)?

EXAM HINT

Within a Voronoi diagram the solution to the toxic waste problem will be at an intersection of cell boundaries or on the boundary of the diagram. In exams the solution will always be one of the internal vertices rather than a boundary edge.

Example 31

A town has four coffee shops. An entrepreneur wishes to open a new shop in the town but would like it to be as far as possible from all the other four coffee shops. Where should he put it?

Consider the Voronoi diagram below showing the positions of the four coffee shops on a set of coordinate axes: A(1, 6), B(2, 2), C(8, 2) and D(8, 5), where one unit represents 1 km.

a Find the coordinates of the vertices P and Q in the Voronoi diagram.

b Determine where a fifth shop should be sited so as to be as far as possible from any other shop, and how far this will be.

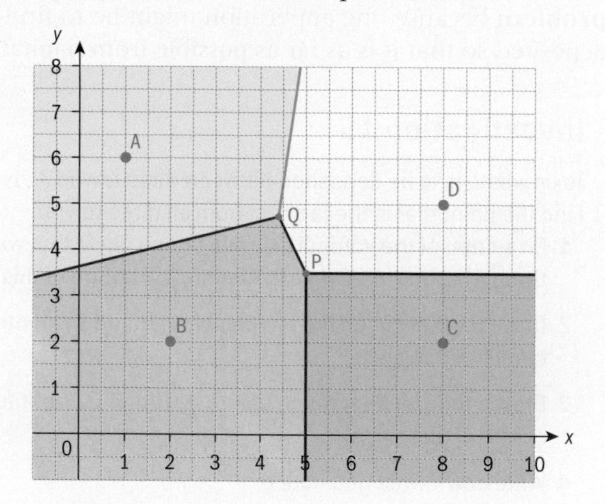

a P is the point where $x = 5$ and $y = 3.5$ meet: (5, 3.5)	Three perpendicular bisectors meet at each of the vertices. Finding the intersection of any two perpendicular bisectors will be sufficient to find the vertex.
The perpendicular bisector of [AB] is $-x + 4y = 14.5$.	Any two of these equations need to be calculated by finding the midpoint and gradient.

The perpendicular bisector of [AD] is $7x - y = 26$. The perpendicular bisector of [BD] is $2x + y = 13.5$. The coordinates of Q are (4.39, 4.72).	The coordinates can then be found algebraically or by using a GDC.
b Centred at P: $PD = \sqrt{(5-3.5)^2 + (8-5)^2}$ $= 3.35$	The solution will be at whichever of the points P and Q is farthest from the three sites nearest to them.
Centred at Q: $QA = \sqrt{(6-4.72)^2 + (4.39-1)^2} = 3.62$	Only one length for each vertex needs to be checked as the other two sites will be an equal distance from the vertex.
The new coffee shop should therefore be built at point Q.	

Reflect How do you find the distance from a site to a vertex?
Where is the point that gives the solution to the toxic waste problem?
How do you find the point that gives the solution to the toxic waste problem?

International-mindedness

Which do you think is superior: the Bourbaki group analytical approach or the Mandelbrot visual approach to mathematics?

Exercise 4J

1 In a town three schools A, B and C are located at the points with coordinates A(1, 3), B(6, 4) and C(6, 1).

It is decided that a new school should be built as close as possible to a point which is farthest from all three existing schools. All coordinates are in km.

a Explain why this point will be at the intersection of the perpendicular bisectors of line segments [AB], [BC] and [AC].

b Find the equations of the perpendicular bisectors of line segments [AB] and [BC].

c Hence find the coordinates of the point where the new school should be built.

d Determine how far the new school will be from each of the other schools.

2 At a fair there are three hamburger stands, A, B and C. The fairground is in the shape of a rectangle with dimensions 100 m by 50 m. The bottom-left corner of the field can be regarded as the origin of a coordinate system, with the diagonally opposite corner as (100, 50). The hamburger stands are at the points A(20, 30), B(80, 30) and C(40, 10).

a Find the equations of the perpendicular bisectors of:

i A and C **ii** B and C.

People will always go to the hamburger stand that is closest to them.

b Draw the Voronoi diagram that represents this situation.

c Find the area of the region of the fairground which is closest to:

i stand C **ii** stand A.

d A fourth hamburger stand is to be added to the fairground at a point as far as possible from the other three stands.

i State the coordinates of the position at which it should be built.

ii Determine how far the new stand will be from the other hamburger stands.

Chapter summary

- In two-dimensional (2D) space, the midpoint, M, of the line segment joining points $A(x_1, y_1)$ and $B(x_1, y_1)$ is $M\left(\frac{x_1 + x_2}{2}, \frac{y_1 + y_2}{2}\right)$.
- In three-dimensional (3D) space, the location of any point A is determined by an ordered triple of coordinates (x, y, z), where each coordinate determines the location along the x-, y- and z-axis respectively.
- In 3D space, the midpoint, M, of the line segment joining points $A(x_1, y_1, z_1)$ and $B(x_2, y_2, z_2)$ is $M\left(\frac{x_1 + x_2}{2}, \frac{y_1 + y_2}{2}, \frac{z_1 + z_2}{2}\right)$.
- In 2D, the distance between two points $A(x_1, y_1)$ and $B(x_2, y_2)$ is $d = \sqrt{(x_2 - x_1)^2 + (y_2 - y_1)^2}$.
- In 3D, the distance between two points $A(x_1, y_1, z_1)$ and

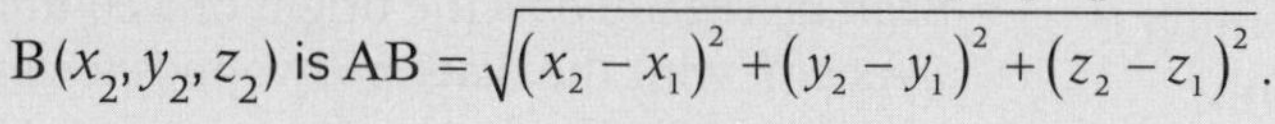

$B(x_2, y_2, z_2)$ is $AB = \sqrt{(x_2 - x_1)^2 + (y_2 - y_1)^2 + (z_2 - z_1)^2}$.

Developing your toolkit

Now do the Modelling and investigation activity on page 202.

- The **gradient** of a line is a measure of the steepness of the line. The bigger the gradient, the steeper the line. It is also known as the slope. Usually, the gradient (slope) of a line is denoted by the letter m.

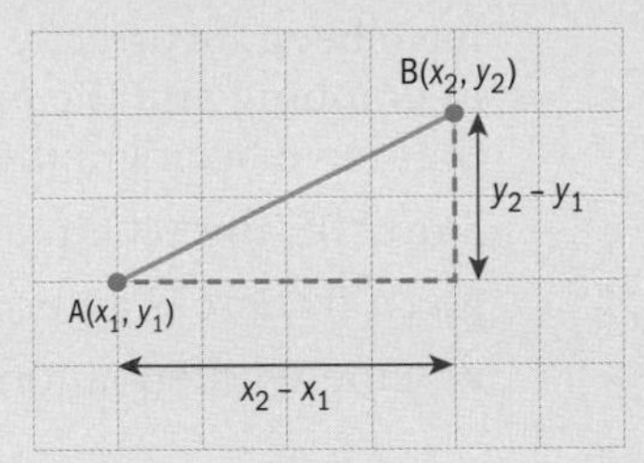

- $m = \dfrac{y_2 - y_1}{x_2 - x_1}$
- When the line goes upward towards the right the **gradient is a positive** number.
- When the line goes downward towards the right the **gradient is a negative** number.
- **Zero gradient:** When a line is horizontal its gradient is **zero.**
- **Undefined gradient:** When a line is vertical it does not have a defined gradient.
- The gradient of a line describes its steepness and direction.
- Equations of lines given in point-gradient form or general form can be rearranged in the form below, which is called the **gradient-intercept form**: $y = mx + k$, where m is the gradient and k is the value of y for $x = 0$.
- The shortest distance from a point A to a straight line (BC) is the length of the line segment [AD], where D lies on (AB) and $[\text{AD}] \perp (\text{AB})$.
- In a Voronoi diagram, the regions which include the closest points to a given point are called **cells**, and the given points are called **sites**. The lines separating the regions are called **boundaries** or **edges**.
- The boundaries of the **cells** in a Voronoi diagram are formed by the perpendicular bisectors of the line segments joining the sites. Usually all other lines that do not form boundaries of the cells are erased when the construction of the diagram is complete.
- A point at which cell boundaries meet in a Voronoi diagram is called a **vertex**. A vertex is equidistant from the three surrounding sites.

Developing inquiry skills

Look again at the problem at the beginning of the chapter. Can you use your knowledge of perpendicular bisectors and Voronoi diagrams to divide the territory into three regions so that each region contains all points that are closest to one island?

Can you now find the areas of these regions? You can use technology.

Is such an area allocation fair?

Chapter review

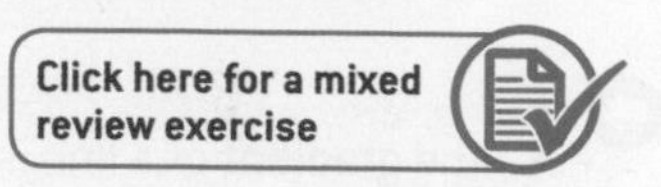

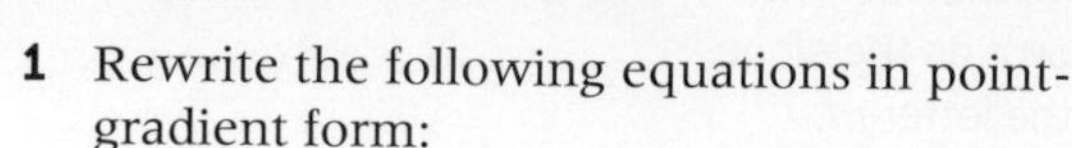

1 Rewrite the following equations in point-gradient form:

a $-x+y=2$ **b** $y-3=-2(x-8)$

c $2x+4y=-\frac{1}{3}$

2 Find the points of intersection of the lines with equations:

a $y=-\frac{1}{2}x-7$ and $y=-5x+10$

b $y-1=-(x-4)$ and $y=0$

c $-x+3y=-2$ and $-\frac{1}{4}x-y=2.$

3 Determine what kind of quadrilateral is formed by the following points:

M(−2,−1), L(1,3), T(5,0) and H(2,−4).

4 Three lines with equations $y=-x+3$, $y=\frac{1}{8}x-\frac{21}{8}$ and $y=\frac{5}{6}x-\frac{1}{2}$ intersect at three points and form a triangle. Find:

a the coordinates of these three points

b the lengths of the sides of the triangle

c the perimeter of the triangle.

5 Line [OB], shown in the diagram, represents Petya's walk to work.

a Find the number of km that Petya walks in 1 hour.

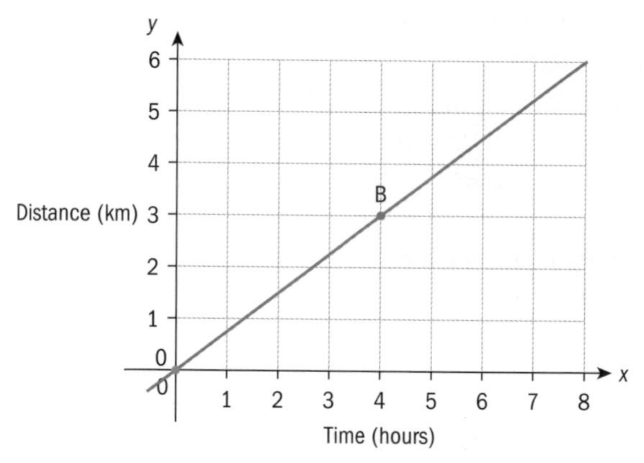

b Maria walks to the same workplace. She walks 5 km in 1.6 hours.

Determine who walks faster. Assume they each walk at a constant rate. Justify your conclusion.

c Write down equations of the lines describing Petya's and Maria's walks.

6 A student has graphed some equations incorrectly. Determine which equations have been incorrectly graphed, the error in graphing and the equations of the lines that have been graphed instead.

State the correct equation of the graph given if there is an error.

a Line with equation $y=5x-3$

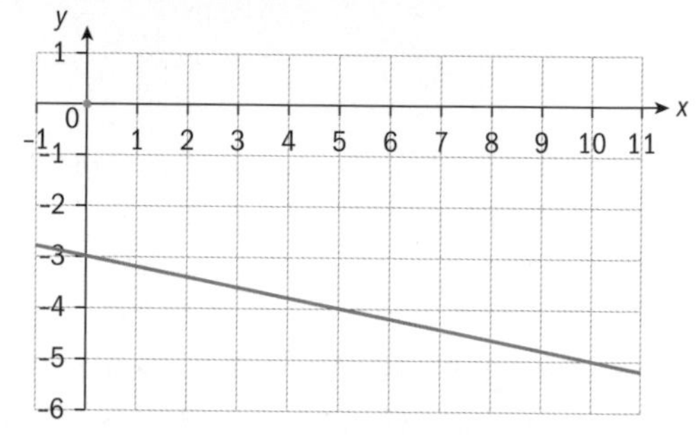

b Line with equation $x=5$

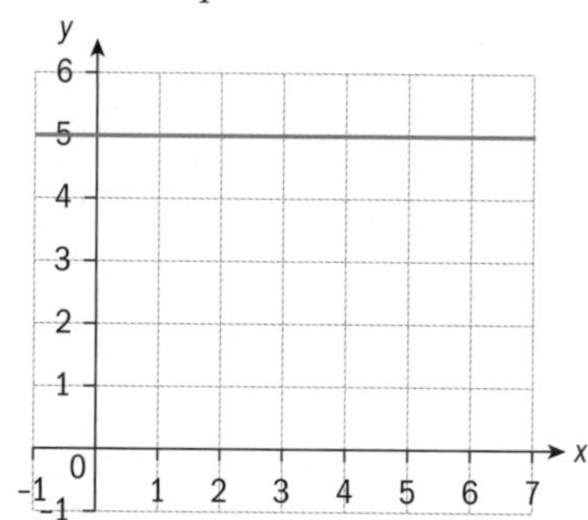

c Line with equation $-20x+55y=22$

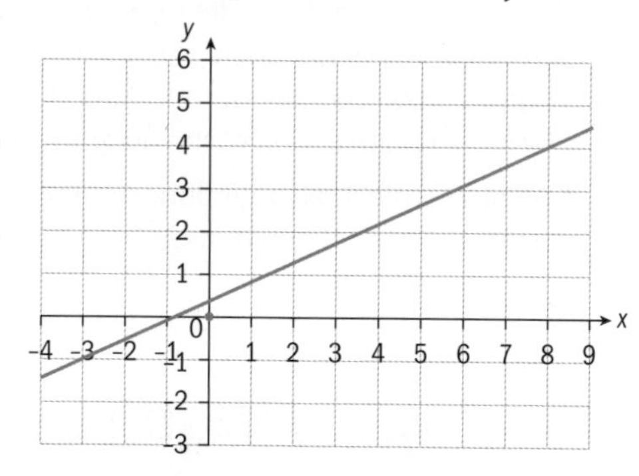

7 Find the equation of a line parallel to:

a the line $y=-\frac{1}{4}x+5$ and passing through (0, 0)

b the line $0.5x-3y=8$ and passing through (−3, −1).

8 Find the equation of a line perpendicular to:

a the line $x + 3y = 0$ and passing through point $(-3, 2)$

b the line $y = -3x + 0.75$ and passing through $(1, 1.5)$.

9 Three farmhouses are to be powered by wind. The farms are located at A(60, 20), B(220, 120) and C(240, 40), where the coordinates are in metres. A wind turbine is to be placed at the point equidistant from the farms A, B and C. Find:

a i the equation of the line (AB)

ii the midpoint of line segment [AB]

iii the equation of the perpendicular bisector of line segment [AB]

b i the equation of the line (AC)

ii the midpoint of line segment [AC]

iii the equation of the perpendicular bisector of line segment [AC]

c i the equation of the line (BC)

ii the midpoint of line segment [BC]

iii the equation of the perpendicular bisector of line segment [BC].

d Determine the coordinates of the point T that is equidistant from the farms A, B and C.

The blade of a wind turbine is 25 m long. Current regulations require that the distance between the turbine and a house be at least three times the length of the blade.

e Determine whether the wind turbine meets current regulations for installation at point T.

f Determine the area that one wind turbine needs to function. Give your answer to the nearest integer.

10 The map of a rectangular province is shown with the positions of four landing platforms for helicopters. The points A, B, C and D, which represent the landing platforms, are placed on a coordinate grid centred at one of the corners of the province. (Each unit along a coordinate axis is 100 miles.)

Emergency medical teams are flown to any point in the province as needed.

When an emergency occurs, the medical team that is based closest to the location of the emergency will fly out to the location.

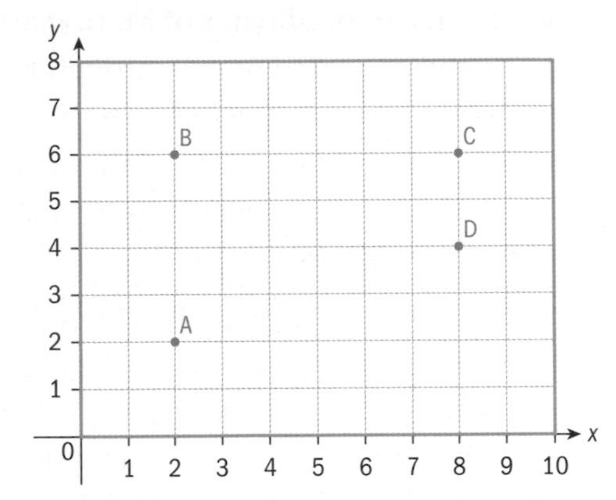

a Construct a Voronoi diagram for the four sites A, B, C and D.

A neighbouring province asks whether the medical team based at point B(2, 6) can also help out in their province. The director replies that it would be possible if the team is currently covering a smaller area than at least two of the other medical teams.

b Find where the perpendicular bisector of [AD] meets the line $y = 4$.

c Find the area of the cells surrounding:

i A ii B iii C iv D.

Hence state whether or not the medical team based at B will be able to support the other province.

Exam-style questions

11 **P1:** The point M is the midpoint of points A and B. If A $(4, -6, 10)$ and M $(7, 3, -5)$, find the coordinates of the point B.

(5 marks)

12 **P1:** The line L has equation $y = 3x - 5$. For the lines given below, state with reasons whether they are parallel to L, perpendicular to L, or neither. (6 marks)

i $y = \frac{1}{3}x - 7$ ii $-6x + 2y + 8 = 0$

iii $y - 5 = 2(x - 7)$ iv $y = \frac{-1}{3}x + 4$

v $x + 3y + 9 = 0$

13 P1: The coordinates of point P are (−3, 8) and the coordinates of point Q are (5, 3). M is the midpoint of [PQ].

a Find the coordinates of M. (2 marks)

L_1 is the line which passes through P and Q.

b Find the gradient of L_1. (2 marks)

The line L_2 is perpendicular to L_1 and passes through M.

c **i** Write down the gradient of L_2.

ii Write down, in the form $y = mx + c$, the equation of L_2. (3 marks)

14 P1: A ski resort is designing two new ski lifts. One lift connects station *B* at the base of a mountain and station *P* at the top of a ski run. The other lift connects station *P* with station *Q*, which is at the top of another ski run.

The three stations are placed on a three-dimensional coordinate system (measured in metres). The coordinates of the stations are $B = (0, 0, 0)$, $P = (500, 400, 300)$ and $Q = (900, 600, 700)$.

A skier wishes to reach the top of the run located at *P* from the base of the mountain.

a Determine the distance covered by the skier on the ski lift from the base of the mountain to *P*. (2 marks)

In order for a skier to reach the top of the ski run at *Q* they must take the lift from the base of the mountain to *P*, and then take a separate lift from *P* to *Q*.

b Determine the total distance covered by a skier on the ski lifts from the base of the mountain to *Q*. (3 marks)

15 P1: The swimming pool at West Park contains 500 000 litres of water when it is full. The pool can be drained of water at a constant rate of 50 000 litres per hour.

a Write an equation for the volume (*V* litres) of water remaining in the pool *t* hours since draining began. (1 mark)

At 8:00am the pool is full of water and the drain is opened.

b Determine the time when the pool is 25% full of water. (3 marks)

The swimming pool at East Park contains 800 000 litres of water when it is full, and can be drained at a rate of 100 000 litres per hour. At 8:00am (when the drain is opened at West Park pool), the drain is also opened at the East park pool.

c Calculate the time when the pool at West Park contains the same amount of water as the pool at East Park. Find also the volume of water in each pool at this time. (4 marks)

16 P1: Lines L_1 and L_2 are given by the equations L_1: $ax - 3y = 9$ and

L_2: $y = \frac{2}{3}x + 4$.

The two lines are perpendicular.

a Find the value of *a*. (3 marks)

b Hence, determine the coordinates of the intersection point of the lines. (2 marks)

17 P2: The cuboid ABCOFPDE has vertices with coordinates shown in the diagram.

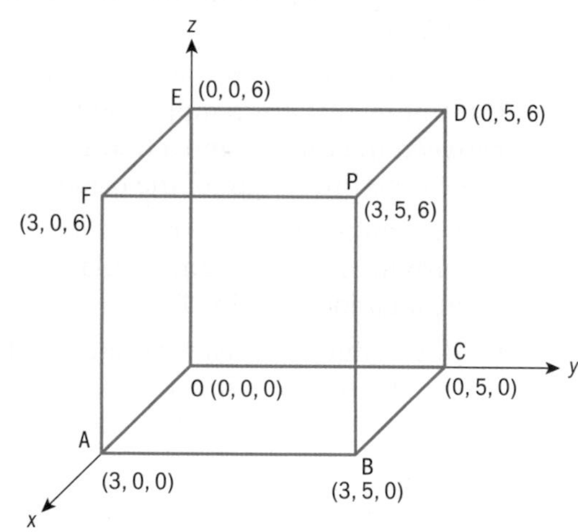

a Find the surface area of the cuboid. (2 marks)

b Find the length of the diagonal [BE]. (2 marks)

Diagonals [AD] and [BE] intersect at the point M.

c **i** Find the coordinates of M.

ii Find $A\hat{M}B$. (7 marks)

18 P1: A Canadian bank charges its customers a fixed commission fee of \$ a Canadian dollars (CAD) if they wish to exchange CAD into euros (€).

The bank then uses the exchange rate \$1:€r to convert the remaining amount.

Michael converts 1200 CAD and receives €765 from the bank.

Janet exchanges 500 CAD and receives €315 from the same bank.

a Find the values of **i** a **ii** r. (6 marks)

b If Michael and Janet had put their Canadian dollars together first and then exchanged it all in one transaction, calculate the amount in euros, to the nearest cent, that they would have received. (2 marks)

19 P1: A new airport, S, is to be constructed at some point along a straight road, R, such that its distance from a nearby town, T, is the shortest possible.

The town, T, and the road, R, are placed on a coordinate system where T has coordinates (80,140) and R has equation $y = x - 80$. All coordinates are given in kilometres.

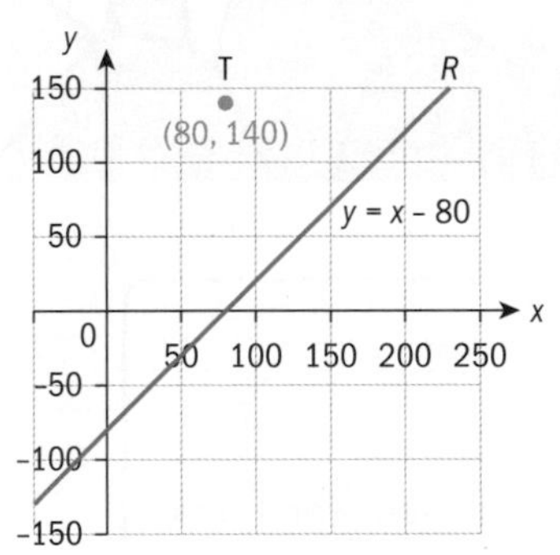

a Determine the coordinates of S, the new airport. (6 marks)

b Find the distance between the town and the new airport. (2 marks)

20 P2: Four mathematicians live on the bottom floor of a circular tower of radius 10 m. They sit 5 m from the centre of the circle equally spaced around it as shown in the diagram.

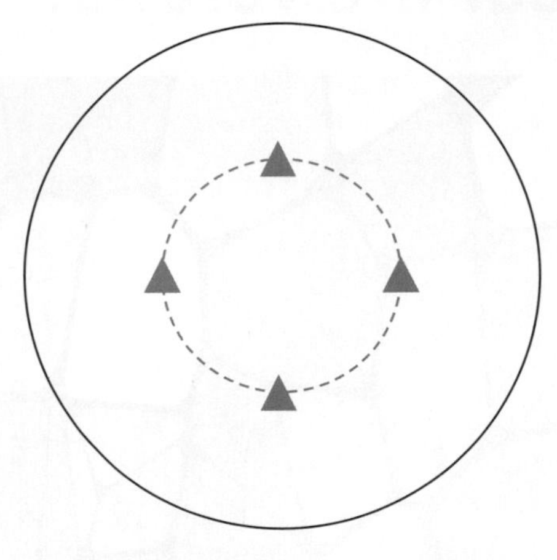

Each mathematician scatters papers, with equations written on them, on the floor around them but always ensures that their papers are nearer to themself than to another mathematician.

a Copy the above diagram and sketch a Voronoi diagram on it, showing where each mathematician's papers can be situated. (2 marks)

b Calculate the area of the floor that each mathematician uses. (2 marks)

Another mathematician joins the group and sits at the centre of the circle. All of the other mathematicians rearrange their papers according to the same rule as before.

c Make another copy of the original diagram and sketch on it a new Voronoi diagram to represent the new situation. (4 marks)

d Find the area of floor that the fifth mathematician ends up using. (2 marks)

e Calculate the area of floor that each of the original four mathematicians now uses. (2 marks)

f State how many points on the floor there are that are equidistant from exactly three mathematicians. (1 mark)

Click here for further exam practice

Real-life Voronoi

Approaches to learning: Thinking skills: Evaluating, Critiquing, Applying

Exploration criteria: Presentation (A), Personal engagement (C), Use of mathematics (E)

IB topic: Voronoi diagrams

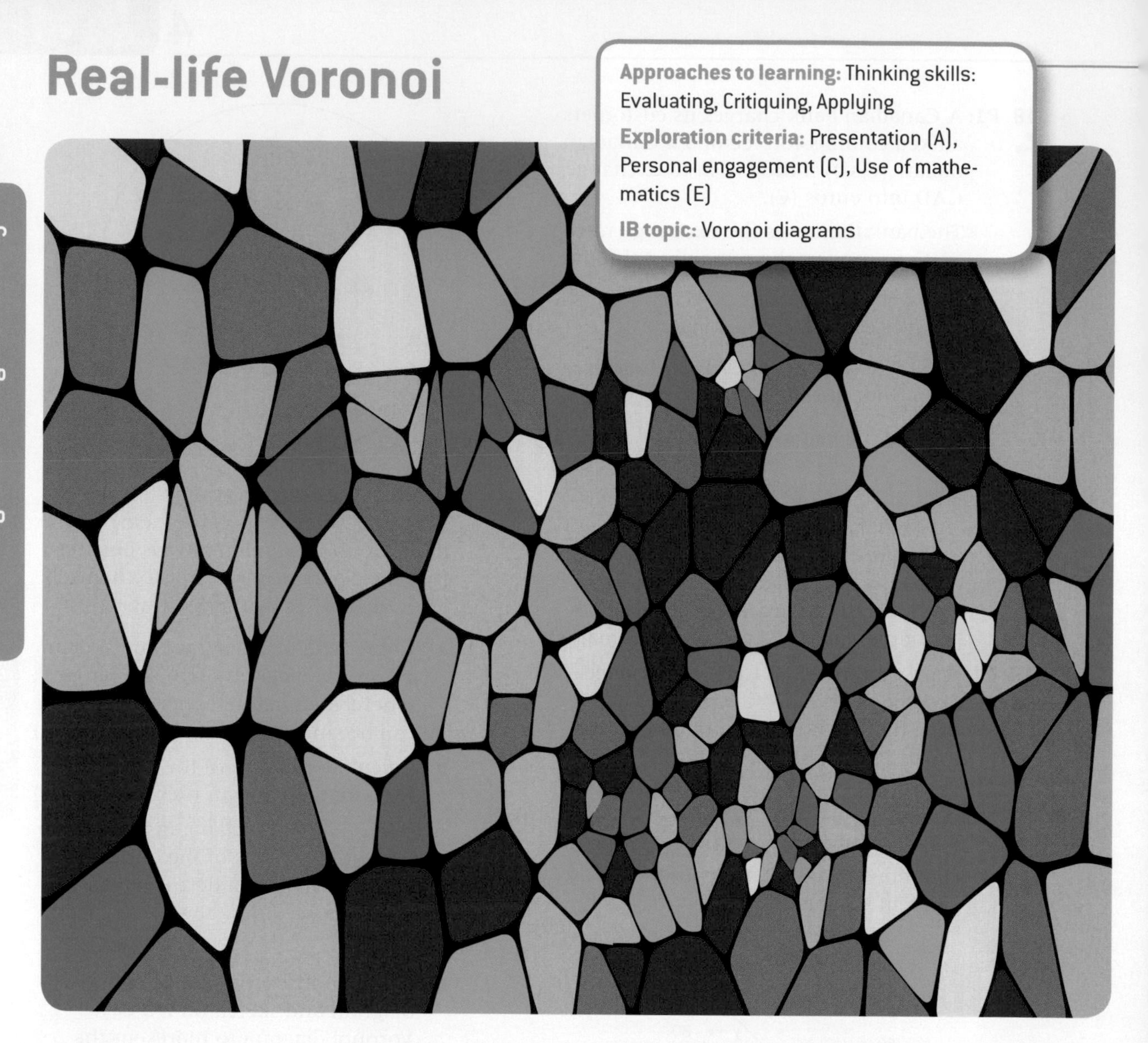

Voronoi diagrams

In this chapter you have been introduced to Voronoi diagrams.

Voronoi diagrams have been used here to answer questions about **distance** and **area**:

1. Distance: finding points that are equidistant to several sites **or** a path that stays far away from sites, **or** determining where something can be located so that it is as far as possible from existing sites.

   Which questions from the chapter could be listed in this group?

2. Area: considering the territories and regions of influence of animals or retail outlets or food places.

   Which questions from the chapter could be listed in this group?

You can use Voronoi diagrams to answer similar questions that you generate yourself in explorations.

Exploration

As part of your exploration you will need to provide an **aim** and a **context** for your choice of topic.

First write down **all** potentially interesting areas to consider for an exploration.

You are now going to try to create an exploration that you could use Voronoi diagrams to answer.

Consider the type of questions that are possible and combine this with something from your list of ideas.

Write the aim and context for your choice.

It may be helpful to think about how you would write the problem as a textbook question for someone else to answer.

Think about the following questions:

- Can you give a clear rationale for your choice of topic?
- Is the aim clearly stated?
- Do you have clearly articulated parameters for the problem?
- Is the aim manageable and completable within a sensible page limit and timeframe?
- Are you going to be able to find the relevant information?

You could now solve the problem you have created using the **incremental algorithm** introduced in this chapter.

Which method is best for answering your question given the available information and the form that this information is given in?

Could a computer software package (such as Geogebra) be used?

Or could the question be answered by hand using a compass and ruler for construction or by superimposing on a grid and using the equation of a perpendicular bisector?

Extension

Find out about "nearest neighbour interpolation".

This is an important application of Voronoi diagrams where the aim is to **interpolate values** of a function at points close to sites, given the value at those sites.

This could be used, for example, to estimate or predict pollution or precipitation levels when you know the pollution or precipitation readings only at particular sites in a region.

5 Modelling constant rates of change: linear functions

Conceptual lenses

- Change
- Modelling

Microconcepts

- Function
- Domain
- Range
- Graph of a function
- Linear models and their parameters
- Rate of change
- Direct variation
- Inverse function
- Arithmetic sequence, arithmetic series
- Common difference
- General term
- Sum of series
- Simple interest
- Prediction
- Extrapolation
- Interpolation

All of these questions are about situations that can be represented by a mathematical model. These models will allow you to study relationships between the variables and to make predictions.

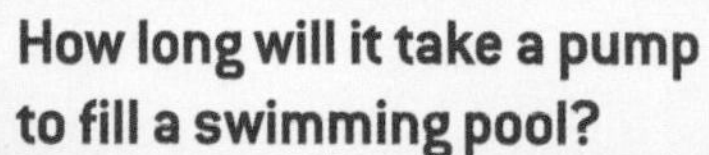

How long will it take a pump to fill a swimming pool?

How do you convert from one currency to another? And back again?

How much do you have to invest in your bank account to grow your savings so you can buy that yacht? And how much does the interest rate matter?

A rental car costs a fixed amount plus $\$t$ per kilometre. How can I find the maximum number of kilometres that can be driven with $\$A$?

Stefan drives taxis part-time, usually driving between 10 and 40 km per day. Currently, on each day that he drives he earns a fixed value of \$20 plus \$2.50 for every kilometre that the taxi's meter is on. Stefan wants to increase his annual profit so that it is equal to the annual average salary of \$44 000. He is thinking of increasing the profit per kilometre by \$0.30 per month for the next year.

What will his daily profit be if he drives 10, 20, 30 or 40 km?

How will his profits change if he increases the profit per kilometre by \$0.30 after one month?

How can you model the daily profit for different values of profit per kilometre?

Will he reach the average annual salary in one year's time?

If not, how long will it take?

What assumptions do you need to make to answer this question?

Developing inquiry skills

Write down any similar inquiry questions you could ask and investigate for another business or charging structure. What information would you need to find?

Think about the questions in this opening problem and answer any you can. As you work through the chapter, you will gain mathematical knowledge and skills that will help you to answer them all.

Before you start

You should know how to:

1 Solve linear equations and inequalities.

$6x \geq 2x - 4$

$4x \geq -4$ (subtract $2x$)

$x \geq -1$ (divide by 4)

2 Draw a line.

$y = -3x + 4$

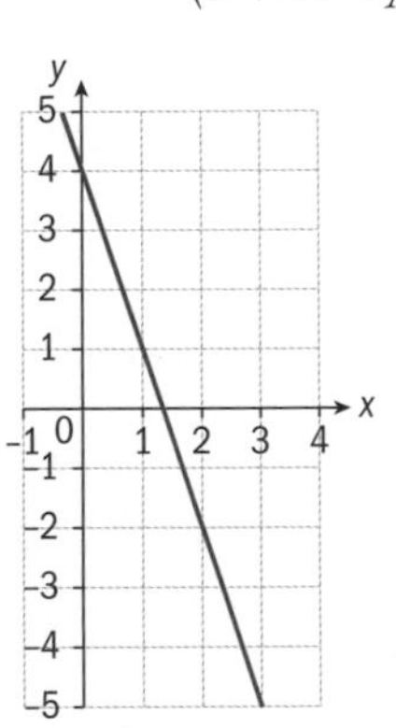

3 Find the equation of a line from a graph.

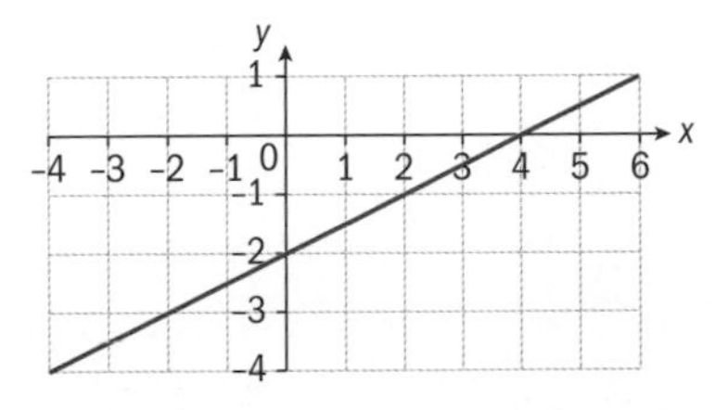

$y = \frac{1}{2}x - 2$

Skills check

Click here for help with this skills check

1 Solve the linear equation and inequality.

a $5 - x = \frac{x}{2}$

b $x \leq -x + 4$

2 Draw the line $y = \frac{3}{2}x + 1$.

3 Find the equation of this line:

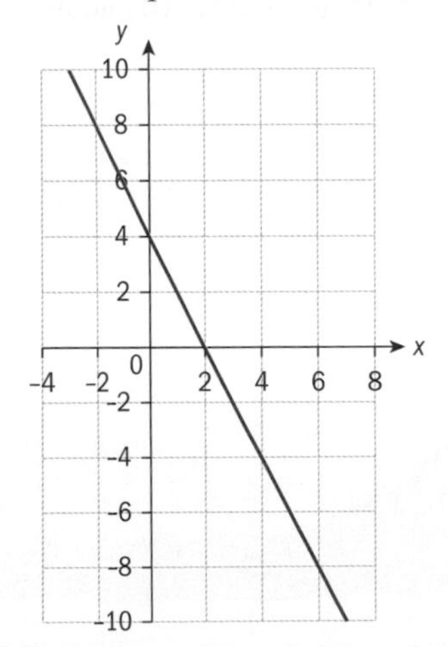

5.1 Functions

The table shows the number of assorted chocolates in a chocolate box and the price of the chocolate box.

Number of assorted chocolates	Price of chocolate box ($)
6	8
12	12
18	15
24	18

These "variables"—"number of assorted chocolates" and "price of chocolate box"—are clearly related.

Written as ordered pairs, the relationship can be described as: $(6, 8)$, $(12, 12)$, $(18, 15)$, $(24, 18)$.

> A **relation** is a set of ordered pairs.

The ordered pairs can also be represented on a graph. The first coordinate will be placed on the horizontal axis and the second coordinate on the vertical axis.

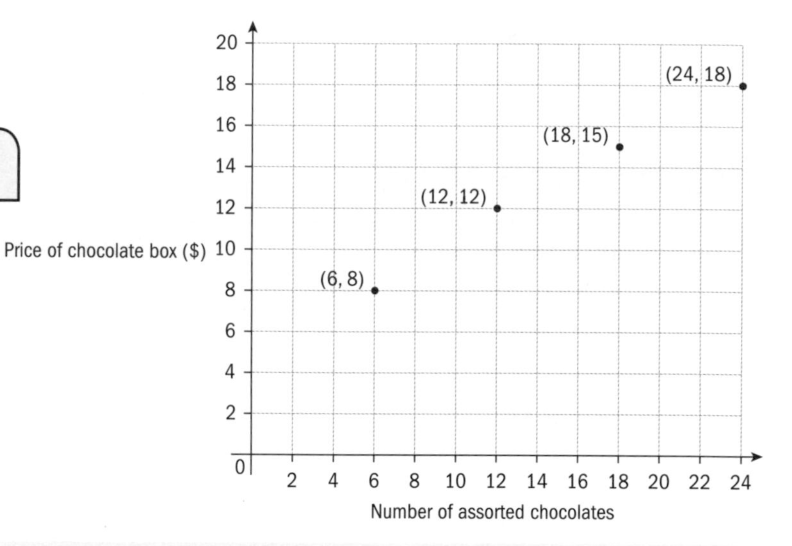

Investigation 1

Amir loves running. Here are the **contexts** of six of his recent runs, 1–6, and some graphs, A–G, that might describe the runs.

Run 1: Amir runs on flat ground for 10 minutes.

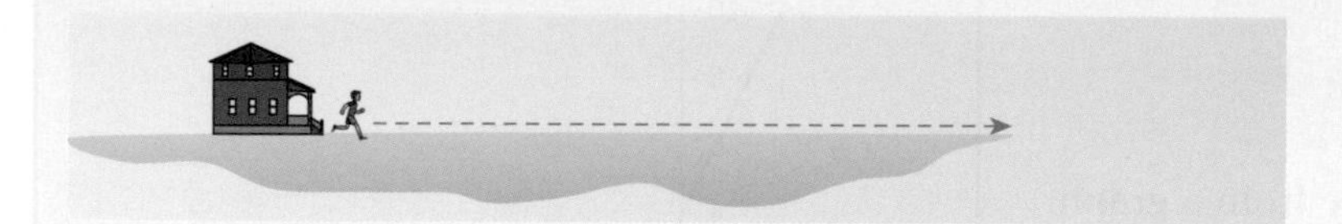

Run 2: He runs down a hill.

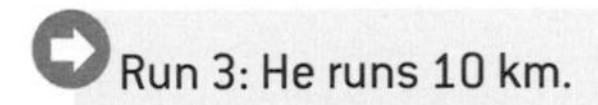

Run 3: He runs 10 km.

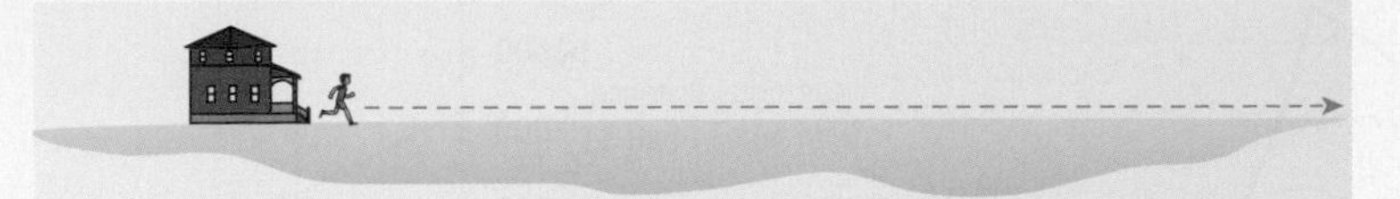

Run 4: He runs up a mountain and back.

Run 5: He stops to chat with a friend midway through his run.

Run 6: He encounters a rough trail that slows him down on his way home.

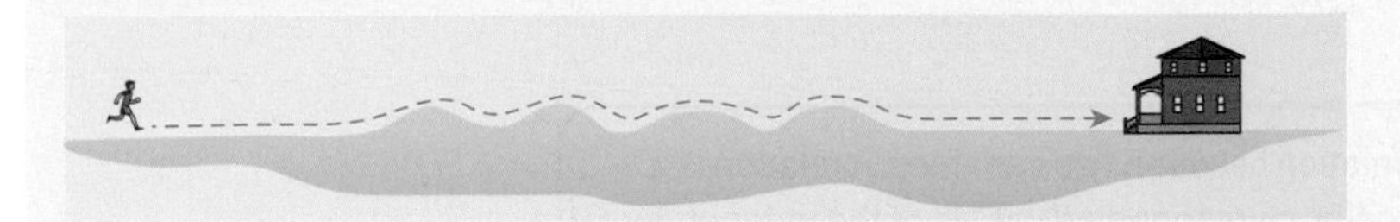

A

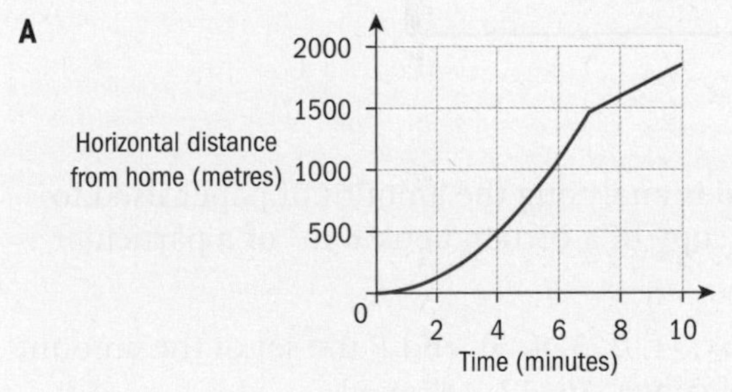

B

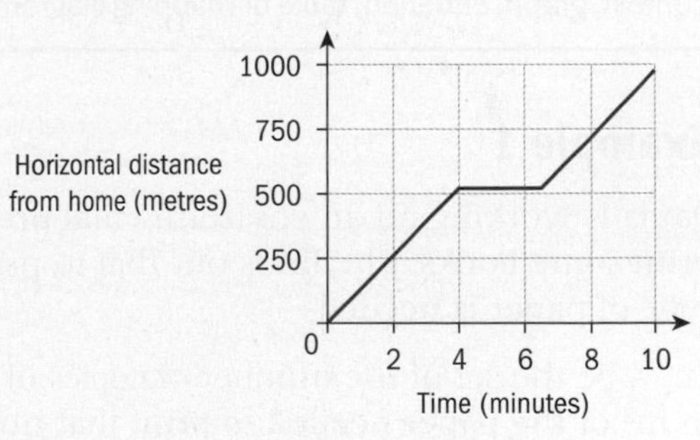

C

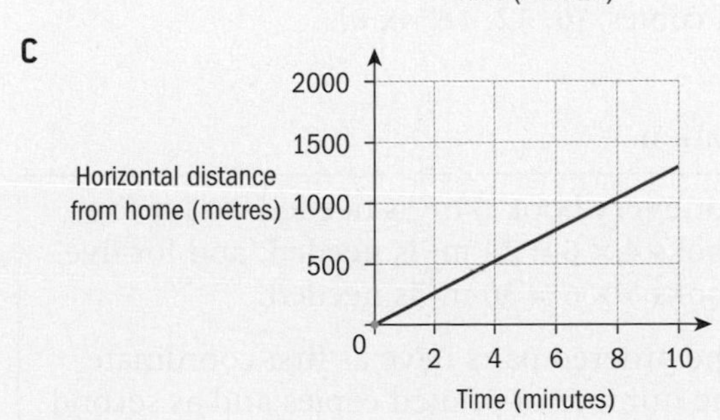

D

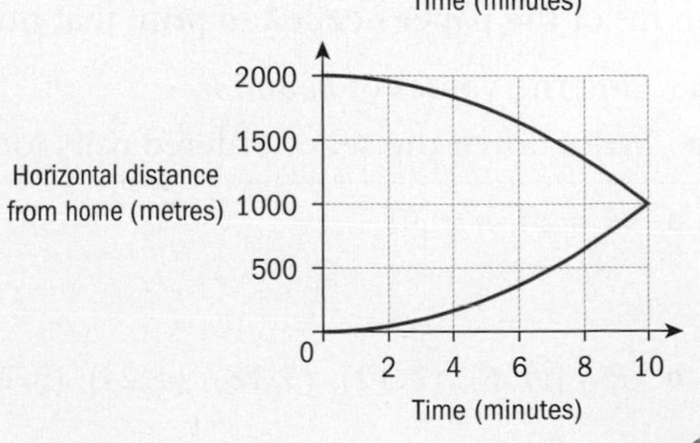

Continued on next page

E

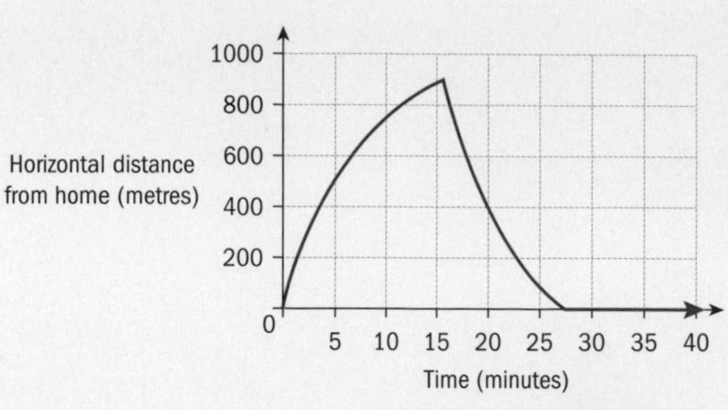

F

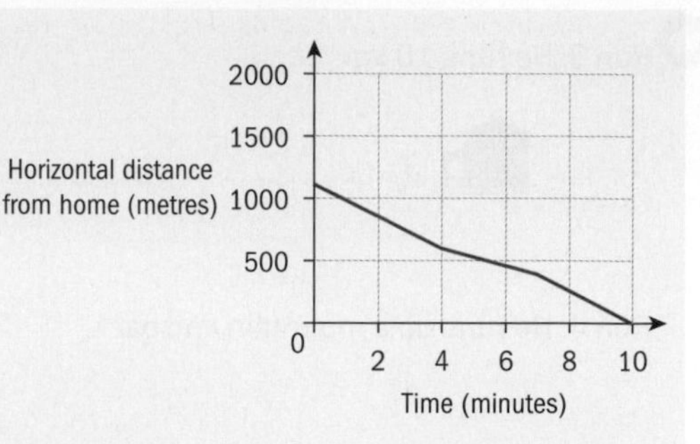

G

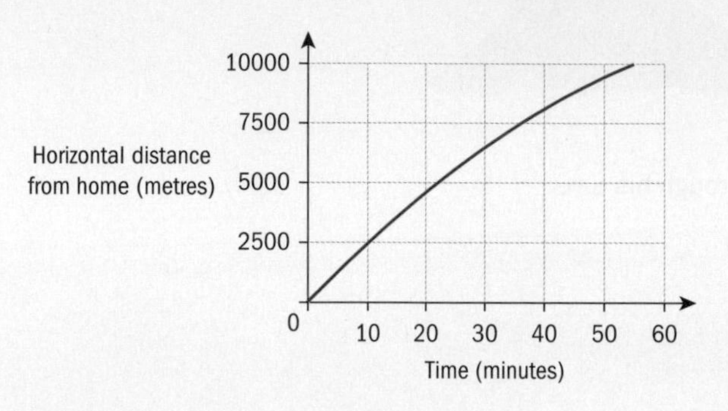

1 Match each run to a graph, giving two specific reasons why you know they match.

2 One graph does not match any of the runs. Explain why it is not possible for this graph to represent a person running over time.

A graph is one way to represent a relation between two variables. A relation is a set of ordered pairs (x, y) that specifies corresponding values of the independent and dependent variables. We can represent a relation between variables with a context, graph, equation, table or mapping diagram, as you will see.

Example 1

Paula is working on an environmental project and is analysing the amount of paper used to print some books. She finds out that to print one copy of a certain book 6 m^2 of a particular type of paper is needed.

Let A be the set of the number of copies of the book, {1, 2, 3, 4, 5}, and B the set of the amount in m^2 of the paper needed to print that number of copies, {6, 12, 18, m, n}.

a Find the values of m and n.

b Write down the set of ordered pairs for this relation.

a $m = 24$, $n = 30$	For every book 6 m^2 is needed, so for four books $4 \times 6 = 24$ m^2 is needed, and for five books $5 \times 6 = 30$ m^2 is needed.
b $R = \{(1, 6), (2, 12), (3, 18), (4, 24), (5, 30)\}$	The ordered pairs have as first coordinate the number of printed copies and as second coordinate the number of m^2 of paper used.

The relation from Example 1 can also be represented using a **mapping diagram**.

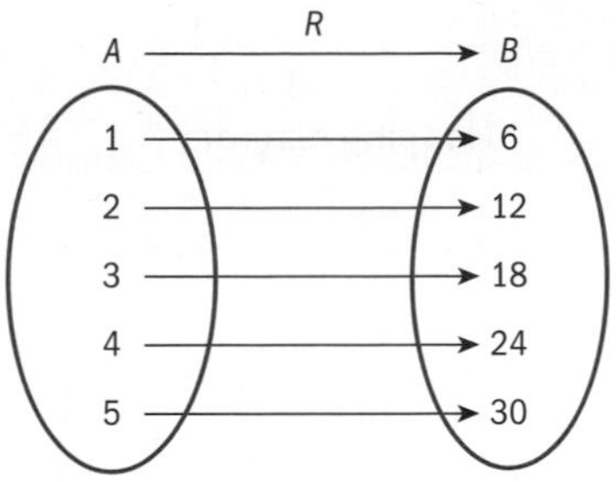

A **mapping diagram** is used to show how in a relation the elements of the first set (**inputs**) are mapped onto the elements of the second set (**outputs**).

Under the mapping R, 18, for example, is said to be the **image** of 3.

Reflect State the image of 4, and identify the element of A whose image is 12.

Investigation 2

1 Complete the second and third columns of the table. Where possible, complete the last column as well.

Mapping diagram	Set of ordered pairs	Graph of the relation	Relation in the form $x \rightarrow$
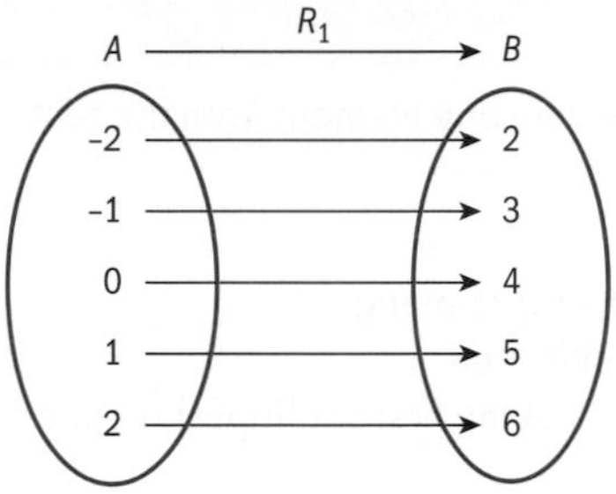 R_1: $A \rightarrow B$; $-2 \rightarrow 2$, $-1 \rightarrow 3$, $0 \rightarrow 4$, $1 \rightarrow 5$, $2 \rightarrow 6$			
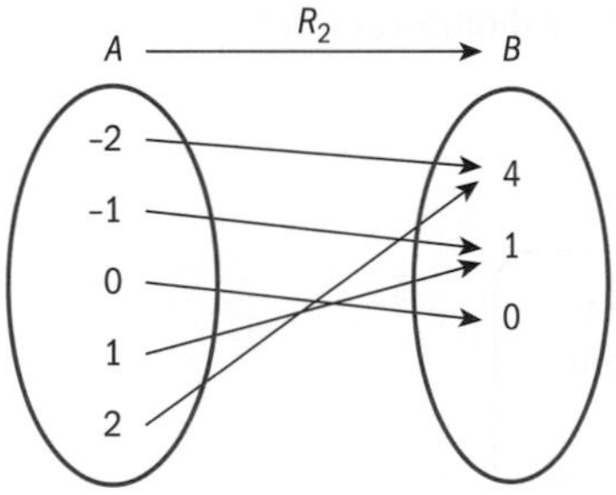 R_2: $A \rightarrow B$; $-2 \rightarrow 4$, $-1 \rightarrow 1$, $0 \rightarrow 0$, $1 \rightarrow 1$, $2 \rightarrow 4$			
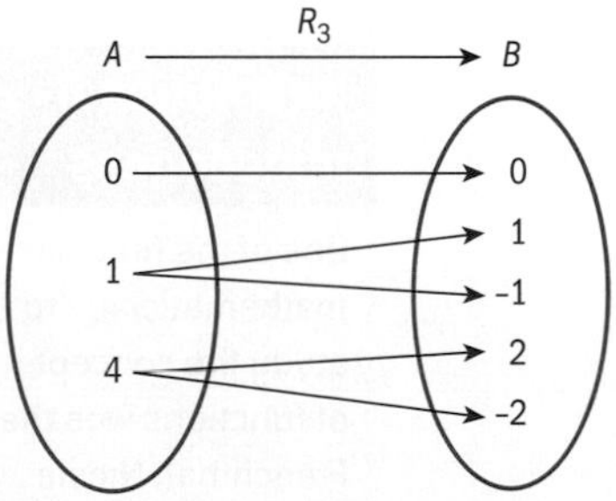 R_3: $A \rightarrow B$; $0 \rightarrow 0$, $1 \rightarrow 1$, $1 \rightarrow -1$, $4 \rightarrow 2$, $4 \rightarrow -2$			

Continued on next page

Mapping diagram	Set of ordered pairs	Graph of the relation	Relation in the form $x \rightarrow$
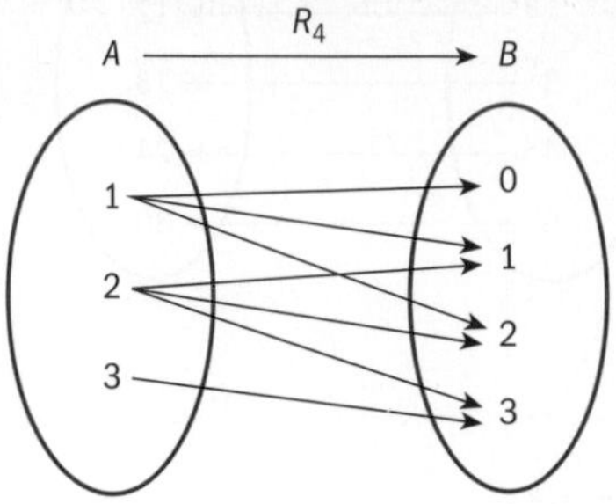 R_4: A → B; A: 1, 2, 3; B: 0, 1, 2, 3			
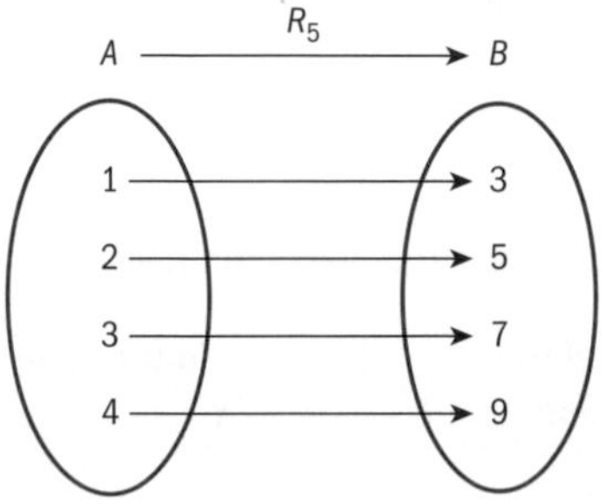 R_5: A → B; 1 → 3, 2 → 5, 3 → 7, 4 → 9			

2 R_2 from the table in part **1** is said to be a "many-to-one" mapping: more than one element from the first set is mapped onto the same element from the second set.

a How would you define these relations?

i one-to-one **ii** one-to-many **iii** many-to-many

b Classify the other relations in the table. Give a reason for each classification.

3 In which of the four different types of mappings in part **2** is each element of the first set (input) mapped onto **one and only one** element of the second set (output)?

4 **Factual** Which of the relations in part **1** are functions according to the definition below?

5 **Conceptual** How can you tell that a relation is a function?

A **function** is a relation between two sets in which every element of the first set (input, independent variable) is mapped onto **one and only one** element of the second set (output, dependent variable).

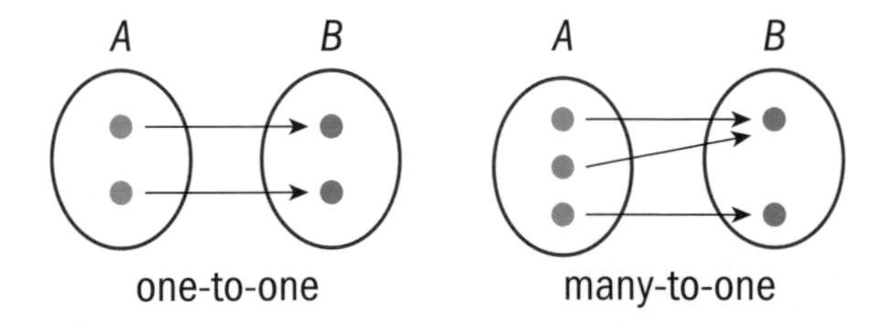

International-mindedness

One of the first mathematicians to study the concept of functions was the Frenchman Nicole Oresme in the 14th century.

The relations shown in the mapping diagrams above are functions: the first is a one-to-one function and the second is a many-to-one function.

Investigation 3

The graphs in Investigation 1 corresponding to the different contexts are examples of functions. They show Amir's distance from home as a function of time. The graph that did not match a context is an example of a non-function.

Here are some more examples of functions and non-functions and different ways that we can represent them.

1 For each pair of examples, describe how the relationship between the inputs (x-values) and outputs (y-values) differs between the function and the non-function.

<table>
<tr><th>Relations that are functions</th><th>Relations that are not functions</th></tr>
<tr><td>a x = length of foot in cm
y = shoe size</td><td>x = shoe size
y = length of foot in cm</td></tr>
<tr><td>b x = a student in your class
y = their birthday</td><td>x = a student in your class
y = the names of their siblings</td></tr>
<tr><td>c $y = x^2$
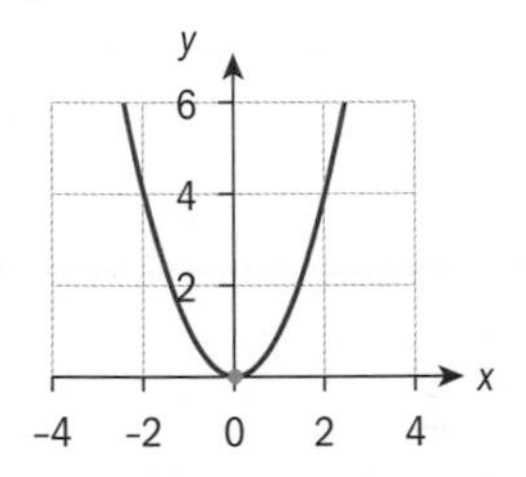
</td><td>$x^2 + y^2 = 1$
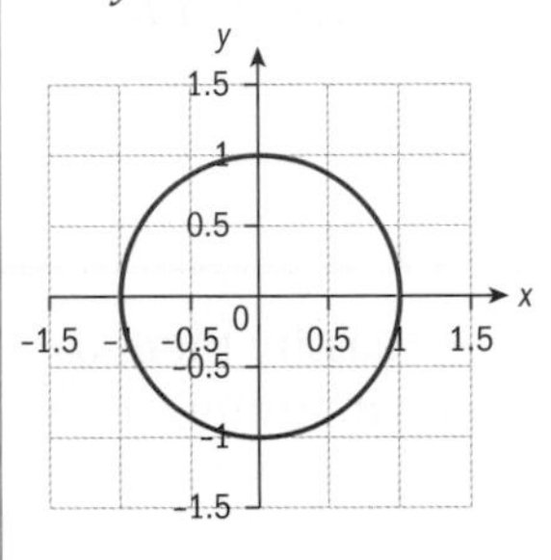
</td></tr>
<tr><td>d
<table>
<tr><th>x</th><th>y</th></tr>
<tr><td>−7</td><td>−22</td></tr>
<tr><td>−2</td><td>15</td></tr>
<tr><td>0</td><td>8</td></tr>
<tr><td>7</td><td>15</td></tr>
</table></td><td>
<table>
<tr><th>x</th><th>y</th></tr>
<tr><td>−2</td><td>3</td></tr>
<tr><td>0</td><td>0</td></tr>
<tr><td>7</td><td>−14</td></tr>
<tr><td>7</td><td>11</td></tr>
</table></td></tr>
<tr><td>e
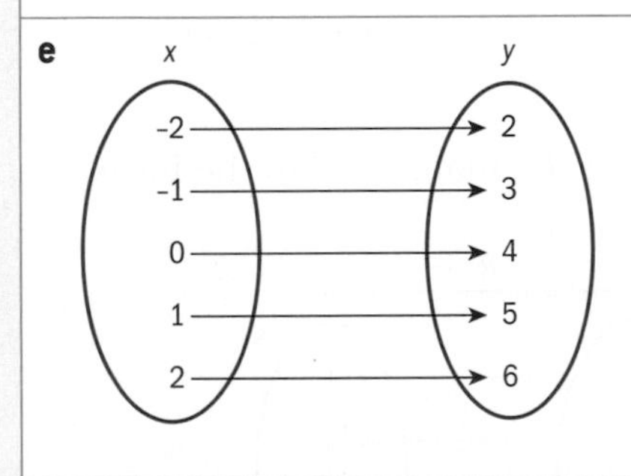
</td><td>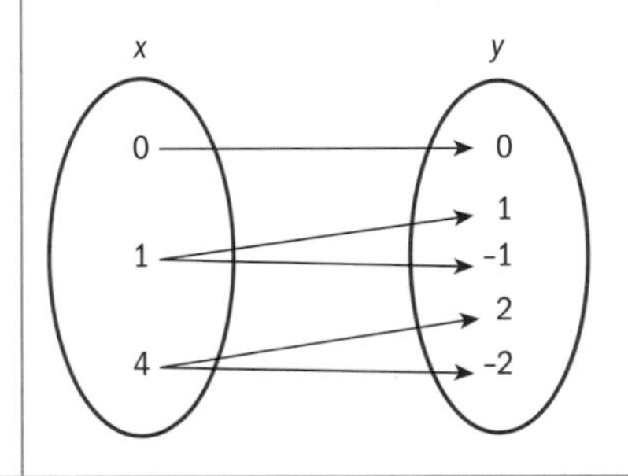
</td></tr>
</table>

2 Based on your descriptions, what generally is the difference between a relation that is a function and a relation that is not? Make sure that your generalization is consistent with all of the examples in part **1**.

3 **Factual** What types of relations are functions?

4 **Factual** What are the different ways that a function can be represented?

5 **Conceptual** How can you identify that a relation is a function from a graph, table, mapping diagram or context?

Example 2

Consider $A = \{-3, -2, 0, 2\}$ and the mapping $x \to 2x^2 - 1$.

a Find B, the set onto which A is mapped.

b Determine whether or not this mapping is a function from A to B.

a $B = \{-1, 7, 17\}$ **b** This mapping is a function (a many-to-one function). Every element of A is mapped onto one and only one element of B.	$-3 \to 2(-3)^2 - 1 = 17$ $-2 \to 2(-2)^2 - 1 = 7$ $0 \to 2 \times 0^2 - 1 = -1$ $2 \to 2 \times 2^2 - 1 = 7$

Exercise 5A

1 Consider these relations. Classify them as one-to-one, many-to-one, many-to-many or one-to-many.

a $\{(1,2), (2,3), (1,4)\}$

b $\{(1,2), (2,3), (3,6)\}$

c $\{(1,1), (2,1), (3,1)\}$

d $\{(1,1), (1,2), (1,3)\}$

e $\{(1,1), (1,2), (2,1)\}$

2 State which of the following mapping diagrams represent functions. In each case give a reason for your answer.

a

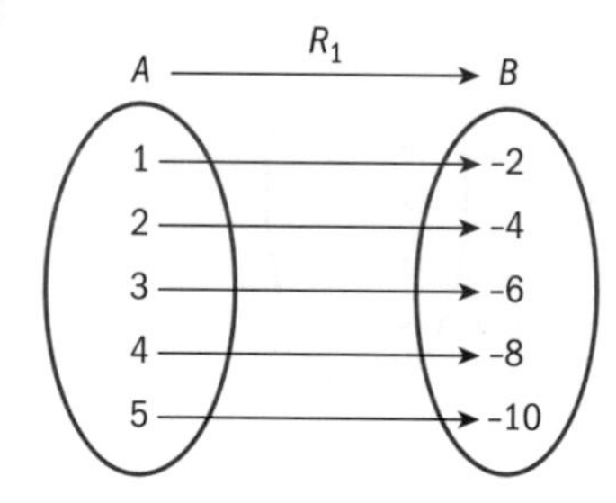

b

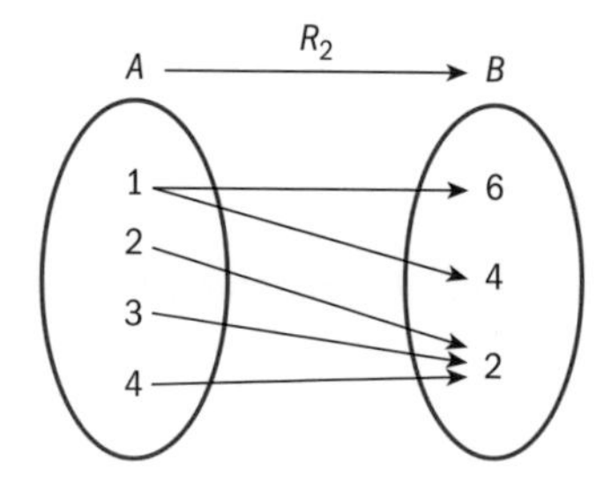

c

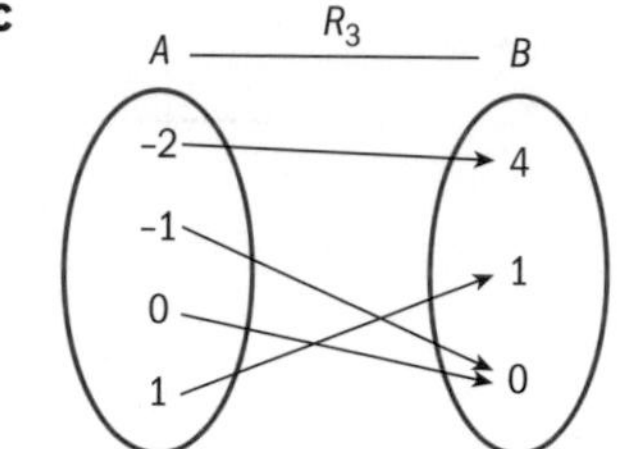

3 If $A = \{-1, 0, 0.5, 1, 2\}$, find B, the set to which the elements of A are mapped by the following relations:

a $x \to -x + 1$ **b** $x \to 2(1 + x)$

c $x \to x^2 + 1$

4 **a** Express the mapping R in the form $x \to \ldots$

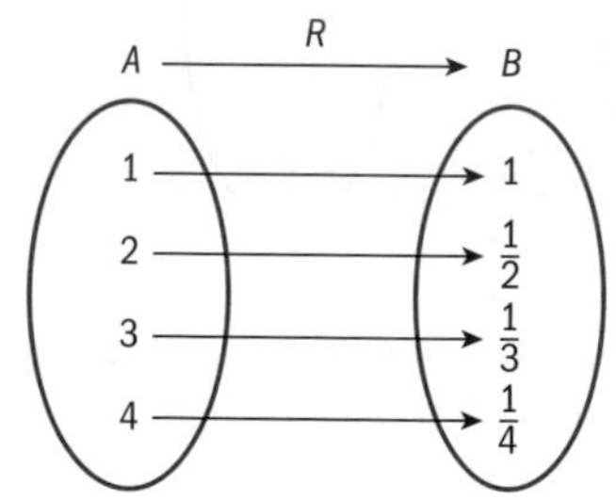

b State whether or not R is a function.

c Write down the image of 2 under R.

d The set A is extended to $\{1, 2, 3, 4, a\}$, and $a \to 3$. Find the value of a.

5 Consider the mapping $y = x^3$.

a Find the output when the input is $x = -1$.

b Find the input when $y = -64$.

c Let the set of inputs be $A = \{-1, 0, 1, 2, 3, 4\}$. Find B, the set of outputs.

d State whether or not the mapping $y = x^3$ from A to B is a function, giving a reason.

Investigation 4

1 Consider the function $x \to 2x$ with the input set $\{-2, -1, 0, 1, 2\}$.

a What is the set of images (outputs)?

b Represent this function on a set of axes.

2 Consider the function $x \to 2x$ where the input set is now $\{-2, -1.5, 0, 1, 1.5, 2\}$.

a What is the set of outputs?

b Represent this function on a new set of axes.

3 Is the function defined in part **1** the same one as defined in part **2**? Why, or why not?

4 a What is the set of images for $x \to 2x$ if the input set is $\{x \in \mathbb{R}: -2 \le x \le 2\}$.

b Represent this function on a new set of axes. What is different between this graph and those from parts **1** and **2**?

5 What is the set of outputs if the set of inputs is $\mathbb{R}$, all real numbers, and $x \to 2x$.

6 Now consider the mapping $f(x) = \frac{1}{x}$.

a There is one value of x for which there is no possible output. Write down this value of x and explain why it does not have an output.

b Consider the following sets. Analyse whether or not each could be the input set for this mapping rule.
$\{1, 2, 3\}, \{0, 1, 2, 3\}, \{-3, -2.5, 1, 1.2\}, \mathbb{N}, \{x \in \mathbb{R} : x \ge 0\}, \{x \in \mathbb{R} : x > 0\}, \mathbb{R}$

c What is the biggest set of possible inputs for the mapping $x \to \frac{1}{x}$? Why?

7 Think of a mapping rule in which the set of all possible inputs is $\{x \in \mathbb{R} : x \ge 0\}$.

8 **Conceptual** Why is a function not defined by the mapping rule only?

9 **Factual** How can we use a graph to decide whether the relation it represents is a function?

10 **Factual** How can we use the graph of a function to find its domain and range as defined below?

- The **domain** of a function is the set of all input values.
- The **range** of a function is the set of all output values.

$$\begin{array}{ccc} \text{input } (x) & \xrightarrow{\text{mapped onto}} & \text{output } (y) \\ \text{domain} & & \text{range} \end{array}$$

Number and algebra

Functions

Example 3

1 Consider the function $y = x + 1$, where $-2 \leq x \leq 1$.

a Find the image of $x = -0.5$.

b Write down the domain.

c Use your GDC to help you sketch the graph of this function.

d Hence, find the range of this function.

2 Consider this graph of a relation. Note that if a graph has a solid dot • then this means that the point is included in the graph. If there is a hollow dot ○ then this means that the point is not included in the graph.

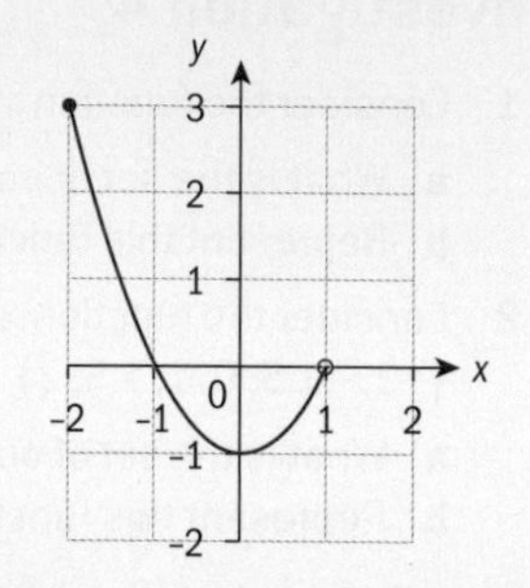

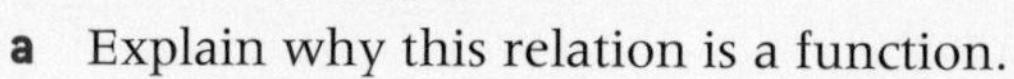

a Explain why this relation is a function.

b Determine its domain and its range.

1 a 0.5	Substitute $x = -0.5$ into the formula of the function: $y = -0.5 + 1 = 0.5$
b $\{x: -2 \leq x \leq 1\}$	In this case the domain is $-2 \leq x \leq 1$, as given next to the formula in the question.
c 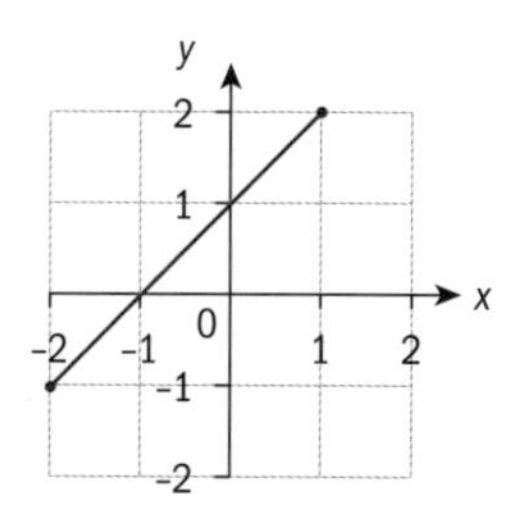	Use technology. Given that in this case the domain is $-2 \leq x \leq 1$, the graph of this function will be the line segment that starts when $x = -2$ and ends when $x = 1$. When $x = -2$, $y = -1$; the point is $(-2, -1)$. When $x = 1$, $y = 2$; the point is $(1, 2)$. You can then sketch the graph onto paper, remembering to mark the endpoints.
d The range is the set $\{y: -1 \leq y \leq 2\}$.	The endpoints for this graph are $(-2, -1)$ and $(1, 2)$. Once the graph is drawn, you can see its range by "squashing" or projecting it onto the y-axis. 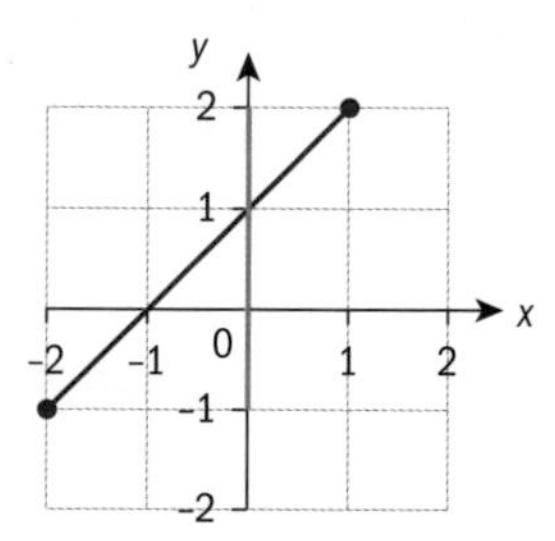 The blue line represents the range of this function, which is $-1 \leq y \leq 2$.

2 a For every input (x) there is only one output (y).	Any vertical line drawn between the endpoints will cut the graph exactly once. This means that every input, x, is mapped to one and only one output, y. 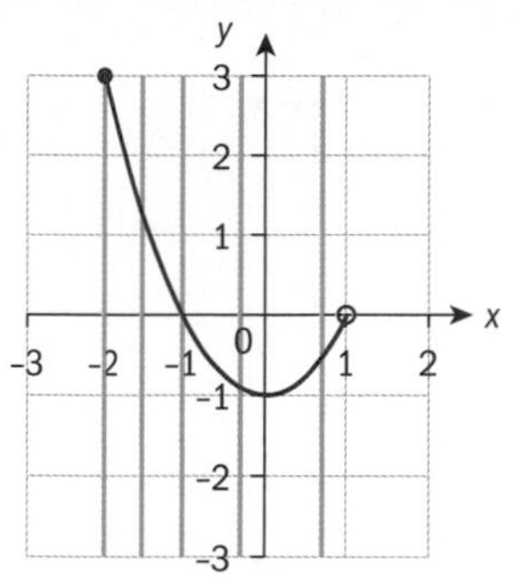 This is called the **vertical line test**.
b 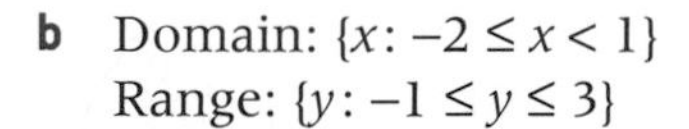 Domain: $\{x: -2 \le x < 1\}$ Range: $\{y: -1 \le y \le 3\}$	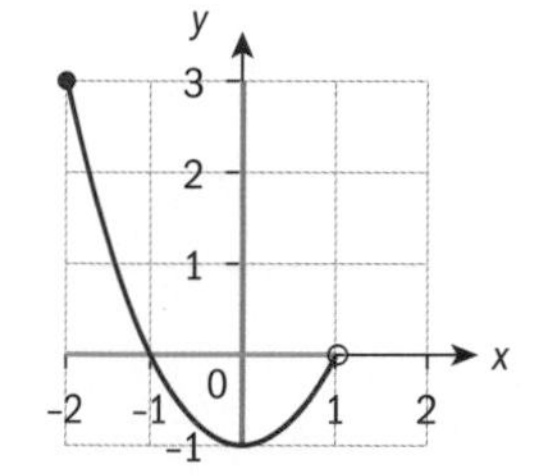 Project the graph onto the x-axis to see that the domain (the green line) is $-2 \le x < 1$. Observe that $x = 1$ is not included in the domain. Project the graph onto the y-axis to see that the range (the blue line) is $-1 \le y \le 3$.

If every vertical line drawn on the graph of a relation cuts the graph exactly once, then the relation is a function. This is the **vertical line test**.

Exercise 5B

1 Find the range of the function $y = x + 3$ when the domain is:

a $\{-1, 0, 1, 2\}$ **b** $\{x: -1 \le x \le 2\}$ **c** $\mathbb{R}$

2 Consider the function $y = -2x + 3$, where $-1 \le x \le 3$.

a Find each image:

i $x = -1$ **ii** $x = 3$

b Find the value of x when $y = 2$.

c Draw the graph of this function. Indicate clearly the **endpoints** of the graph.

d Hence, find the range of this function.

3 Determine which of the following relations are functions.

a

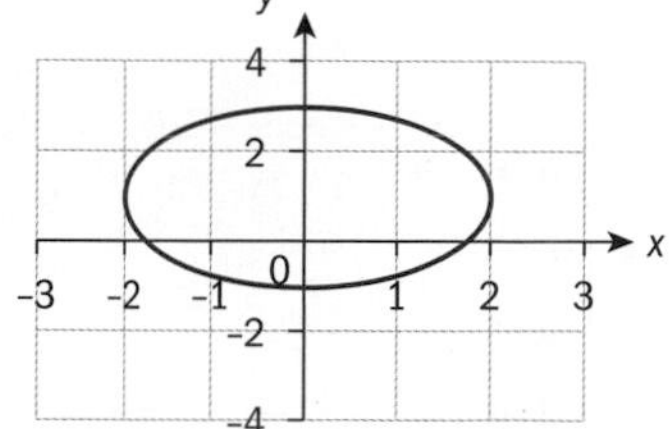

b

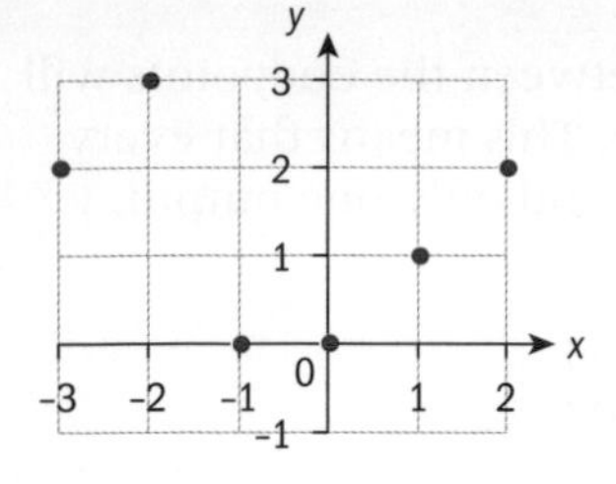

c

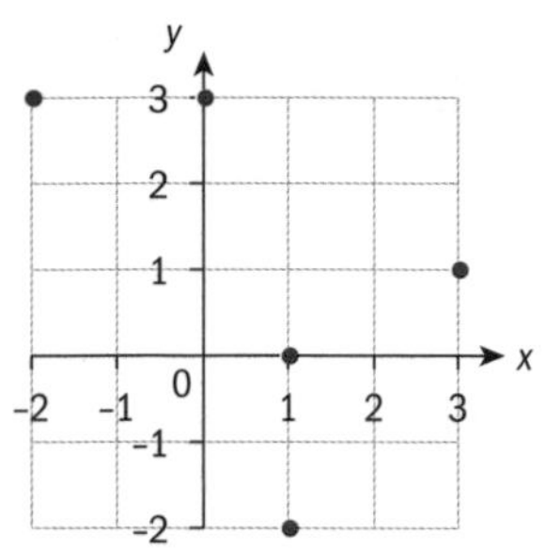

d

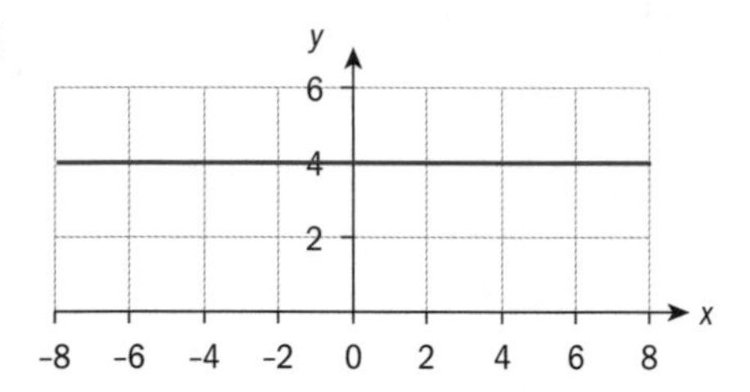

e

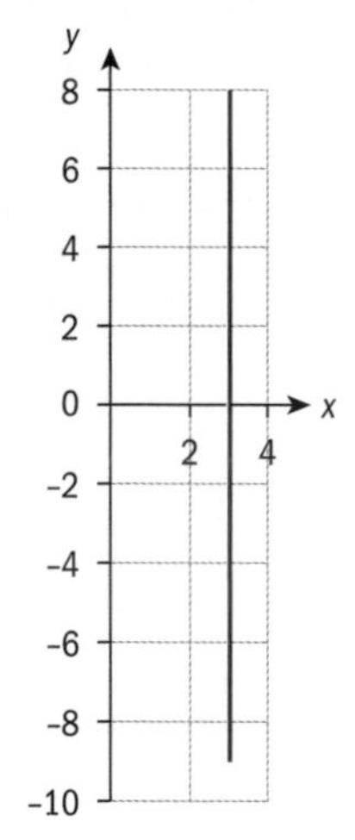

4 Find the domain and range of each of these functions.

a

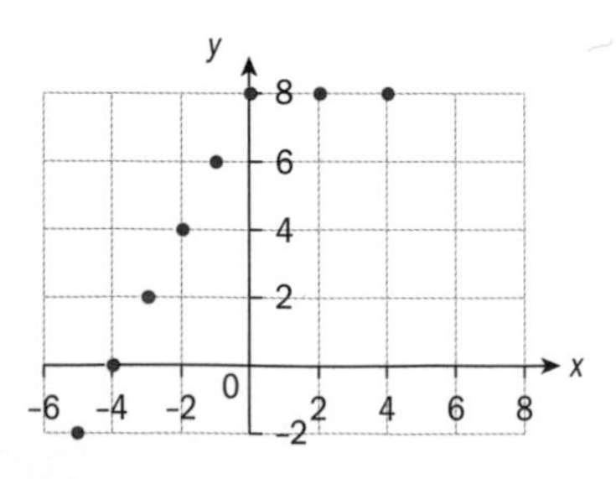

b

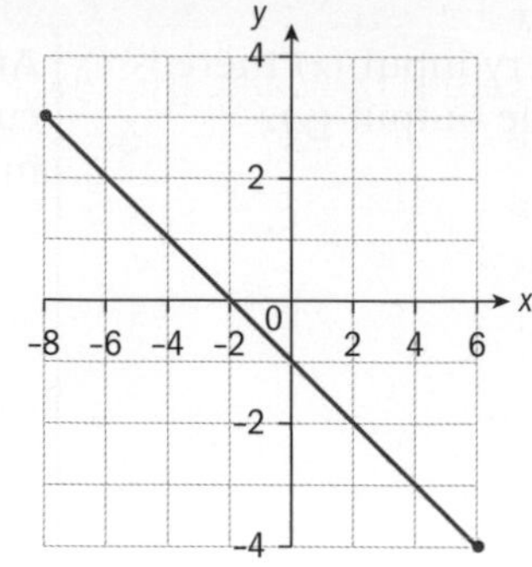

c

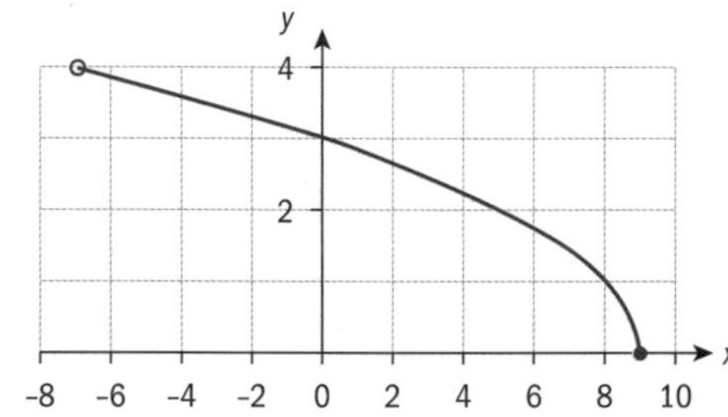

d

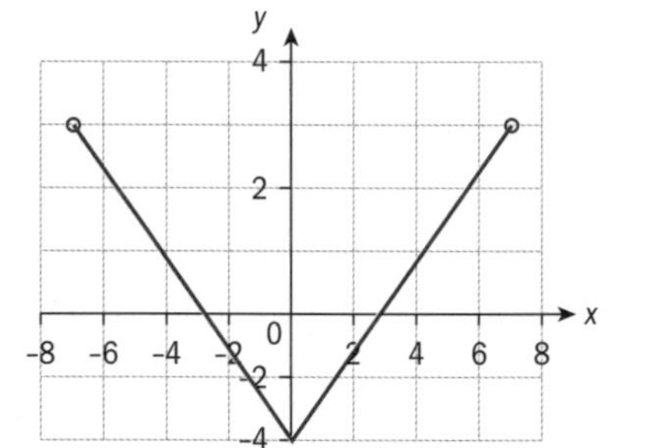

e

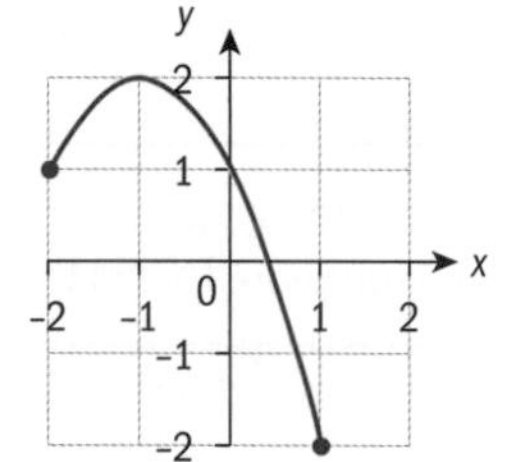

5 Determine whether these statements about the functions in question **4** are true or false.

a **i** The point $(-3, 2)$ lies on the graph of this function.

ii The image of $x = -2$ is $y = 3$.

iii When $x = -4$, $y = 0$.

b **i** The graph of this function crosses the x-axis at $(-2, 0)$.

ii The point $(0, -1)$ does not lie on the graph of this function.

iii There is only one value of x for which $y = -2$.

c **i** When $x = 9$, $y = 0$.

ii As the values of x increase, the corresponding values of y decrease.

iii The vertical line $x = -7$ intersects this graph once.

d **i** The line $y = 0$ cuts the graph of this function twice.

ii There are two values of x for which $y = 2$.

iii As the values of x increase, so do the corresponding values of y.

e **i** The line $y = 0.5$ intersects the graph of this function twice.

ii The line $y = 1.5$ intersects the graph of this function twice.

iii The value $x = 2$ has no image.

6 Find the domain of each of these functions if it is known that the range is $\{-3, -2, -1, 0, 1, 2\}$.

a $x \to 2x + 1$

b $x \to \frac{x}{2}$

c $y = -x + 2$

7 A special application measures the temperature of a person and records on a graph its variation from 37°C, considered to be the average optimum temperature. The application stops working when the temperature of the person is optimum.

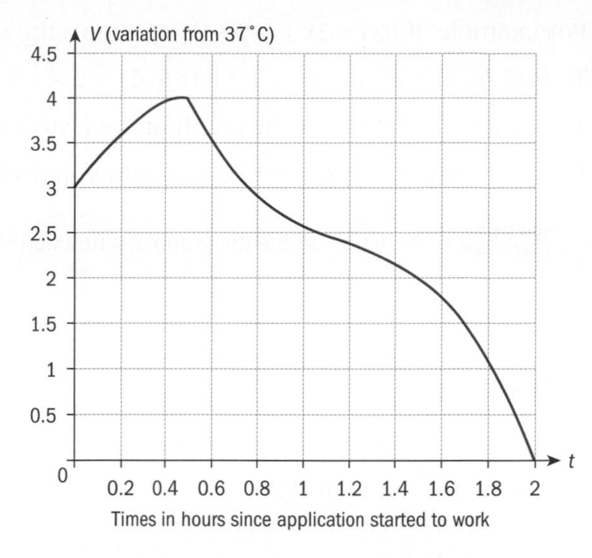

a State whether or not the relation represented in the graph is a function, giving the reason for your answer.

b State the domain and the range.

Explain their meaning in the context of the problem.

c Write down the missing coordinate for each ordered pair from the graph:

(0.2, __), (0.8, __), (__, 2), (2, __).

d State the temperature of the person at the start of the application.

e After how many hours was the temperature of this person equal to 37°C?

f During what period of time did the temperature of the person increase? Over what period did it decrease?

For a function f, the notation $y = f(x)$ means that x (an element of the domain, the input) is mapped through the function f to y (an element of the range, the output).

Note that:

- $f(x)$ is read "f of x".
- If $y = f(x)$ then x is said to be the independent variable and y is the dependent variable.
- Different variables and names can be used for functions, such as $d = v(t)$, $m = C(n)$, etc.

HINT

Be careful! $f(x)$ does **not** mean "f times x".

For example, if $f(x) = 2x + 5$, $f(-1)$ represents the value of y when $x = -1$.

To find $f(-1)$ we substitute $x = -1$: $f(-1) = 2 \times (-1) + 5 = 3$.

The point $(-1, 3)$ lies on the graph of the function f.

What is the value of $f(0.5)$? How would you find it?

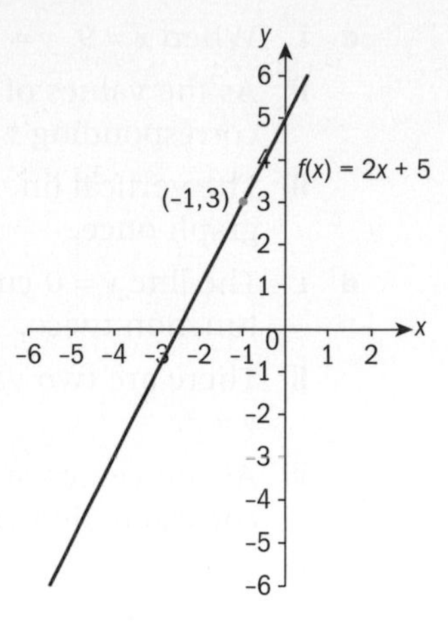

Reflect How do you define the graph of a function?

Example 4

Consider the function $f(x) = x^2 + 1$.

a Find the value of $f(2)$.

b Find the image of $x = -1$.

c Determine whether the point $(5, 26)$ lies on the graph of the function f.

a 5	To calculate $f(2)$, substitute $x = 2$ in the formula of the function by replacing x with 2: $f(2) = 2^2 + 1 = 5$
b 2	The image of $x = -1$ is found by replacing x with -1 in the formula of the function: $f(-1) = (-1)^2 + 1 = 2$ A table can be drawn in your GDC to find the answers to **a** and **b**.
c Yes, it does.	If the point $(5, 26)$ lies on the graph then when $x = 5$, $y = 26$. Replacing x in the formula of the function with 5: $f(5) = 5^2 + 1 = 26$ So $y = 26$, and therefore the point $(5, 26)$ lies on the graph of function f.

International-mindedness

The word "modelling" is derived from the Latin word "modellus", which means a human way of dealing with reality.

A **model** is a suitable mathematical representation of a particular situation. In the next example the model is given by a formula.

Example 5

The price, in dollars, of a motorcycle t years after purchase is modelled by $P(t) = 15\,000 - 2500t$, $0 \le t \le 6$.

a State the independent variable and the dependent variable.

b Find the value of $P(0)$. State the meaning of $P(0)$ in this context.

c Find the price of the motorcycle 1.5 years after purchase.

d Find the value of t for which $P(t) = 6250$. Interpret this value of t.

e Comment on the case $t = 6$.

f Sketch the graph of this function.

a The independent variable is t (time) and the dependent variable is P (the price of the motorcycle).	The coordinate pairs in this relation are of the form $(t, P(t))$. The first coordinate is the independent variable; the second is the dependent variable.
b $P(0) = 15\,000$ $P(0)$ represents the price of the motorcycle at purchase.	Substitute $t = 0$ into the formula for P: $P(0) = 15\,000 - 2500 \times 0 = 15\,000$
c \$11 250	Substitute $t = 1.5$ into the formula for P: $P(1.5) = 15\,000 - 2500 \times 1.5 = 11\,250$
d $t = 3.5$ After 3.5 years the value of the motorcycle is \$6250.	We need to find the value of t (time) for which the price is \$6250: $P(t) = 6250$ $15\,000 - 2500t = 6250$ Using algebra: $t = \frac{15\,000 - 6250}{2500} = 3.5$
e The price of the motorcycle is \$0.	$P(6) = 15\,000 - 2500 \times 6 = 0$
f 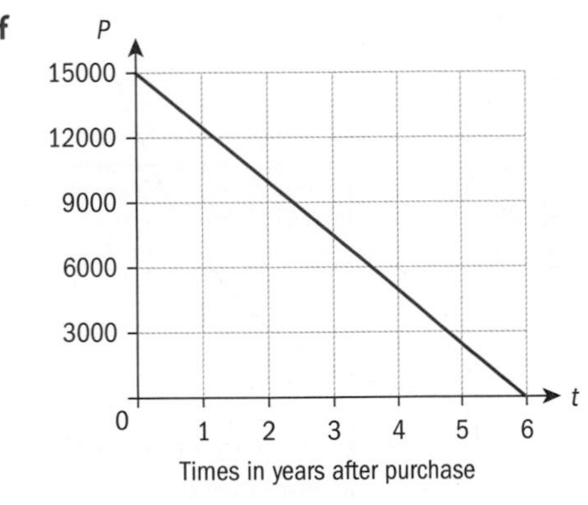	Plot the graph of $P = -2500t + 15\,000$ on your GDC. To do so, replace the independent variable by X and the dependent variable by Y. Recall that $0 \le t \le 6$. 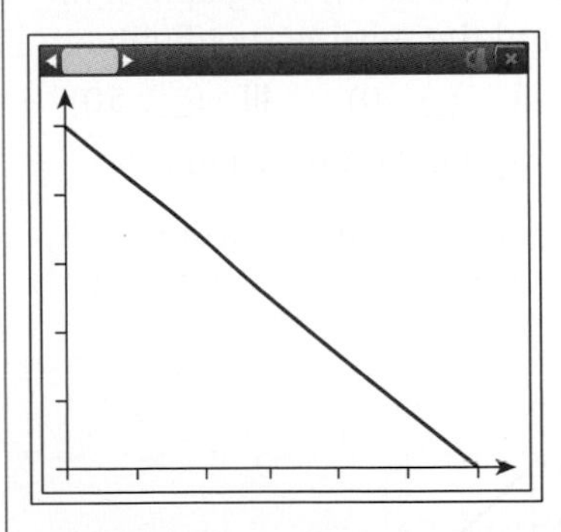 Label the axes and indicate some scale when you copy the graph. Mark the axis intercepts.

Reflect What notations do we use for functions?
What is an independent/dependent variable?
How can we use the formula of a function?
How can we represent functions?

Exercise 5C

1 A function $T(d)$ gives the average daily temperature T, in °C, of a certain city during last January in terms of d, the day of the month.

  a Explain the meaning of $T(2) = 25$.

  b State the domain of the function T.

  c Given the data in the table on $T(d)$, estimate a reasonable range for $T(d)$.

d	2	8	15	18	22	29
T	25	22	26	20	21	28

2 Consider the function $f(x) = 10 - 4x$.

  a Write down the independent variable for this function.

  b Calculate:

  i $f(2)$ **ii** $f\left(-\frac{1}{2}\right)$

  c Show that $f(2.5) = 0$.

  d There is an x for which $f(x) = -6$. Find this value of x.

3 Consider the function $g(x) = 100 + 2x$.

  a Calculate $g(-10)$.

  b These ordered pairs lie on the graph of the function g. Find the missing coordinate.

  i (0, __) **ii** (__, 0) **iii** (__, 50)

  c Sketch the graph of the function g.

4 The graph shows a particular function f.

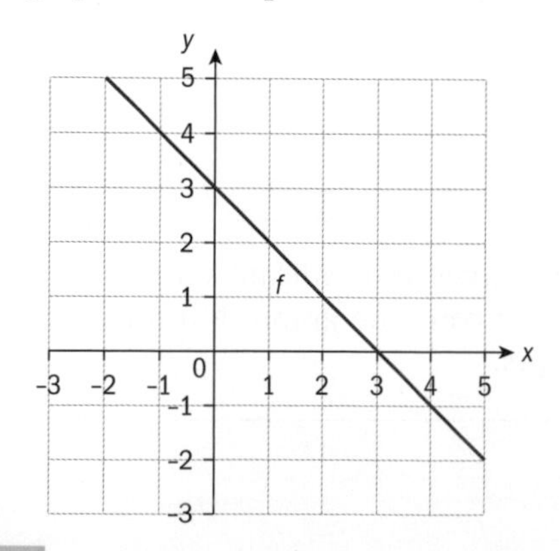

  a Write down the values of:

  i $f(-2)$ **ii** $f(6)$

  b Find the values of x that satisfy:

  i $f(x) = 3$ **ii** $f(x) = 0$

  c Find the set of values of x for which $f(x) < 0$.

5 **a** Use your GDC to sketch these straight lines:

  i $y = x + 1$ **ii** $y = x$ **iii** $y = -x$

  iv $x = -3$ **v** $y = 4$

  b State which of these graphs represent functions.

  c Describe those straight lines that can represent a function and those that cannot.

6 The number of cells in a culture, N, t hours after it has been established is shown in the graph for $0 \le t \le 20$.

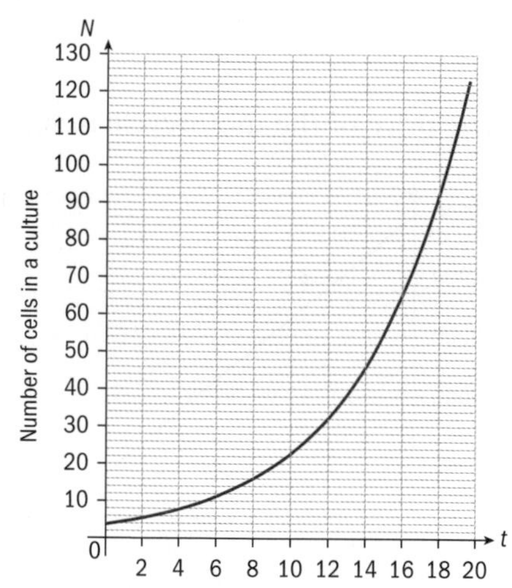

  a The relation represented in the graph is a function. Give reasons why this is the case.

  b State the independent variable and the dependent variable.

  c Estimate the initial number of cells in this culture.

  d Estimate the value of N when $t = 4$.

  e Estimate the value of t when $N = 32$.

  f Estimate the interval of time over which the number of cells varied from 64 to 128.

  g Find the range of this function.

7 The cost of renting a car is given by the formula $C(k) = 150 + 5k$, where C is the cost in dollars and k is the distance travelled in kilometres from the pick-up point.

a Find the cost of renting the car for 30 km.

b Find $C(72)$.

A group of friends rent a car and pay \$1275. They go to Reyes, a village which is k kilometres from where they pick up the car.

c Determine the distance between Reyes and the pick-up point.

8 The velocity $v\ \mathrm{m\,s^{-1}}$ of a moving body at time t seconds is given by $v(t) = 50 - 8t$.

a Calculate $v(0)$ and explain its meaning.

b Find the velocity of the body after 2 seconds.

c Find the time at which the velocity is $15.6\ \mathrm{m\,s^{-1}}$.

d Determine the time when the body comes to rest.

International-mindedness

What is the relationship between real-world problems and mathematical models?

5.2 Linear models

Investigation 5

1 In a chemistry experiment, a liquid is heated and its temperature observed. The graph shows the temperature of the liquid, T°C, t minutes after the experiment starts, up to the end of the experiment.

Describe how the temperature changes throughout the experiment. How much does T change for every one-minute increase in t?

This is the **rate of change** of T with respect to t.

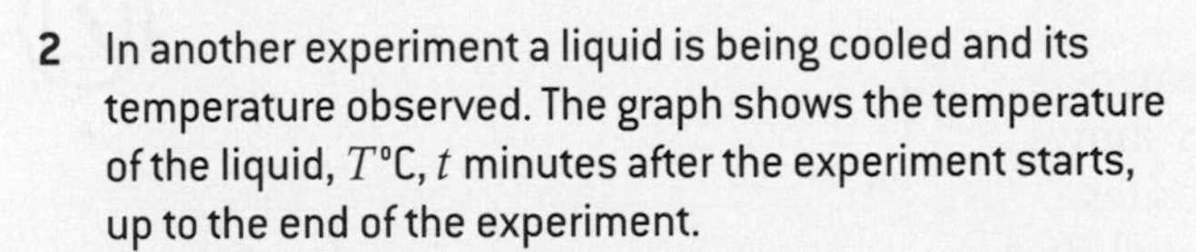

2 In another experiment a liquid is being cooled and its temperature observed. The graph shows the temperature of the liquid, T°C, t minutes after the experiment starts, up to the end of the experiment.

Describe how the temperature changes throughout the experiment. What is the rate of change of T with respect to t?

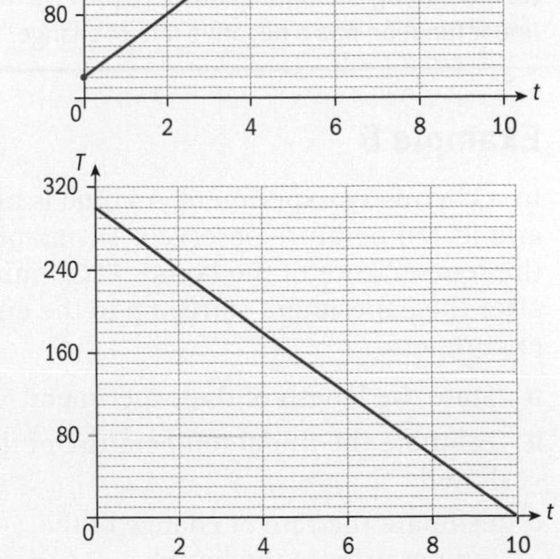

3 Consider these linear functions:

$f(x) = 2x + 1 \qquad g(x) = -3x + 2 \qquad h(x) = 3$

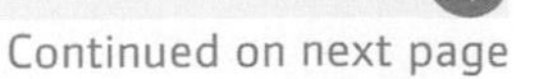
Continued on next page

For each function:

a What is the rate of change of the dependent variable with respect to the independent variable?

b When the independent variable increases, what is the behaviour of the dependent variable?

Now complete the table:

Linear function	Graph of the function	Rate of change of dependent variable with respect to independent variable	As the independent variable increases, the dependent variable ...
$f(x) = 2x + 1$			
$g(x) = -3x + 2$			
$h(x) = 3$			

4 **Conceptual** How is the rate of change of the dependent variable represented on the graph and in the function?

5 **Conceptual** What does the gradient of a linear function represent?

6 **Conceptual** When the rate of change is negative, what happens to the dependent variable as the independent variable increases?

A **linear model** has the general form $f(x) = mx + c$, where m and c are constants. The graph of a linear model is therefore a straight line. Recall that m is the gradient of the line and c is the y-intercept.

Note that not all processes have a constant rate of change, but when they do they can be modelled with a linear function.

The **rate of change** of a function is the increase/decrease in the dependent variable for every unit by which the independent variable increases.

An **increasing** linear function has a positive rate of change. A **decreasing** linear function has a negative rate of change.

Example 6

In a chemistry experiment, a liquid is heated and its temperature observed. The graph shows the temperature of the liquid, T°C, t minutes after the experiment starts, up to the end of the experiment.

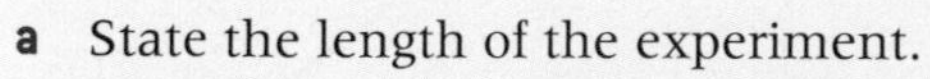

a State the length of the experiment.

b Estimate the initial temperature of the liquid.

c Estimate the rate of change of the temperature of the liquid.

d Suggest a formula to model the variable T in terms of t.

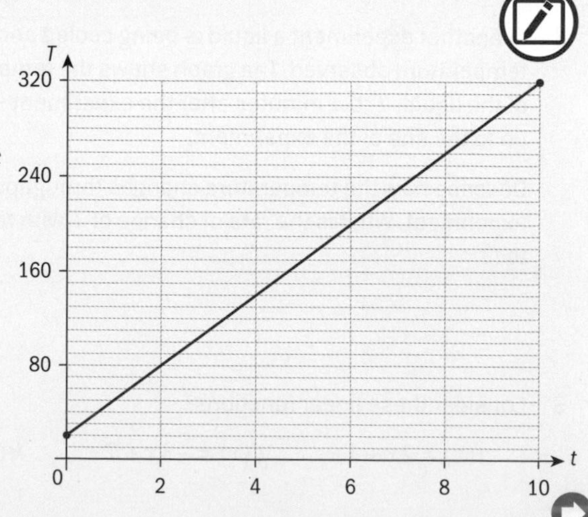

A different chemistry experiment lasts 8 minutes. A liquid is heated and its temperature rises at a constant rate. The initial temperature of the liquid is 40°C and the final temperature is 320°C.

e Find the formula for T in terms of t for this experiment.

a 10 minutes	The graph shows the temperature up to the end of the experiment, so the experiment lasts 10 minutes.
b 20°C	This can be read off the graph by looking at the point at which $t = 0$ (ie the start of the experiment). The point $(0, 20)$ is the T-intercept of the graph.
c 30°C per minute	The rate of change is the same as the gradient of the line. The line passes through the points $(0, 20)$ and $(2, 80)$, so the gradient is $m = \frac{80-20}{2-0} = \frac{60}{2} = 30$.
d $T = 30t + 20$	From the graph this is clearly a linear model, so it has the form $T = mt + c$. You found c in part **b** and m in part **c**.
e $T = 35t + 40$	If the temperature rises at a constant rate then the process can be represented by a linear model, so $T = mt + c$. We are told that when $t = 0$, $T = 40$, and also that when $t = 8$, $T = 320$. Therefore the points $(0, 40)$ and $(8, 320)$ both lie on the graph of this function. From the first point, $c = 40$. Method 1: Use the gradient formula. $m = \frac{320-40}{8-0} = 35$. Method 2: Substitute the point $(8, 320)$ into the equation $T = mt + 40$. $320 = 8m + 40$, so $m = 35$.

We have seen in Example 6 that the same function, $f(x) = mx + c$, has been used to model two different situations. When the experiment changes, the formula changes.

The **variables** in this formula are x, the independent variable, and $f(x)$ or y, the dependent variable.

In the equation, m and c are constants. However, for each different situation, the values of m and c may be different. They are called **parameters**.

To define a linear function you need to know the values of the two parameters m and c.

Reflect In Example 6, what is the value of c in the model in part **d**? What does this represent on the graph and in the context of the experiment?

In Example 6, what is the value of m in the model in part **d**? What does this represent on the graph and in the context of the experiment?

How do the parameters m and c of a linear model help us to interpret the model? How are variables different from parameters?

Example 7

A water tank drains at a constant rate. It contains 930 litres of water 3.5 minutes after it started to drain. It takes 50 minutes for the tank to empty. Let W be the amount of water in the tank (in litres) t minutes after it started to drain.

a Find a model for $W(t)$, the amount of water in the tank at any time after it started to drain. Write down the rate at which the water tank is emptying.

b Write down the amount of water in the tank when it starts to drain.

c Use your model to find the amount of water in the tank after 30 minutes.

a $W(t) = -20t + 1000$

The tank drains at 20 litres per minute.

If the tank drains at a constant rate then the amount of water can be modelled by a linear function:

$W(t) = mt + c \qquad (1)$

The information given can be displayed as follows:

Time (t minutes)	Amount of water (W litres)	Point on the graph of the function
3.5	930	(3.5, 930)
50	0	(50, 0)

Substituting each point into equation (1) gives:

$3.5m + c = 930$

$50m + c = 0$

Technology can be used to solve these simultaneous equations.

This gives $m = -20$, $c = 1000$.

The rate at which the tank drains is given by m.

b 1000 litres

When it starts to drain, $t = 0$.

c 400 litres

Substitute $t = 30$ into the formula:

$W(30) = -20 \times 30 + 1000 = 400$

Reflect Describe the steps involved in finding the equation of a linear function using technology.

When can we use a linear function to model a real-life situation?

Why is a model useful?

TOK

Does a graph without labels have meaning?

Exercise 5D

1 State which of the following are linear functions.

a $f(x) = 3$ **b** $g(x) = 5 - 2x$

c $h(x) = \frac{2}{x} + 3$ **d** $t(x) = 5(x - 2)$

2 For each context in **a–d** below:

i Identify the independent and dependent variables and their units, where appropriate.

ii Determine whether a linear function can be used to model the relationship between the two variables, and if so, state the rate of change.

a Parking at an airport is charged at \$18 for the first hour parked and \$12.50 for each subsequent hour.

b The population of fish in a lake declines by 7% each year.

c Italy charges a value-added tax (VAT) of 22% on all purchases. The amount of tax you pay depends on the amount of your purchase (in euros).

d Sal's Ski Resort notices the following trend in sales of daily passes:

Daily high temperature (°C)	−4	−1	3	7
Number of daily passes sold	430	406	374	342

3 **a** State whether these linear functions are increasing, decreasing or neither.

i $f(x) = 1 + 4x$ **ii** $g(x) = -2 + 0.5x$

iii $h(x) = 3 - x$ **iv** $s(x) = -1$

b Sketch the functions from part **a**. Show clearly their intersections with the axes. (The intersection with the x-axis occurs when $y = 0$; the intersection with the y-axis occurs when $x = 0$.)

4 The table shows values for a function $y = f(x)$.

x	1	3	5	7
$y=f(x)$	2	6	10	15

Determine whether or not the function defined in the table is linear.

5 The table shows some values for a linear function $f(x) = mx + c$.

x	1	3	5	7
$y = f(x)$	5	6.5	b	9.5

a Write down the value of b.

b Find the value of m.

c Find $f(0)$.

6 A school had 1000 students in 2016 and 1150 students in 2019. The number of students in the school has been increasing at a constant rate since 2015. Assuming that this continues to happen:

a Find a model for $N(t)$, the number of students at the school t years after 2015.

b Determine the number of students at the school in 2015.

7 A linear function f is such that $f(1) = 5$ and $f(5) = 1$.

a Find the formula for the function f.

b Find $f(-3)$.

8 The graph shows the velocity v of a moving body, in metres per second, at time t seconds.

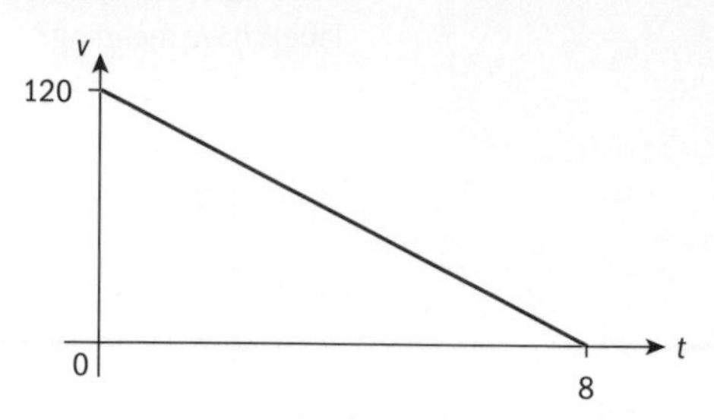

a Write down the initial velocity of the moving body.

b After how many seconds does the body stop moving?

c Find the rate of change of the velocity with respect to time.

d Find a model for $v(t)$.

In a relationship between two variables, if one variable is a constant multiple of the other then they are said to be **directly proportional**.

Investigation 6

1 For each of the following functions:

i Find its formula. **ii** Sketch its graph on an appropriate domain.

a The circumference C of a circle in terms of its radius r.

b The number of pounds in weight, p, relating to a weight in kilograms, k. Note that 1 kg is equivalent to 2.2 pounds.

c The price P you pay in Happy Shop in terms of the regular price R if there is a sign that says "Today you pay 15% less for every single item".

d The weight of a certain type of wood in terms of the volume:

Volume (V dm^3)	1	2	3
Weight (W kg)	0.9	1.8	2.7

2 What can you say about the relationship between the variables involved in each of the situations in part **1**?

3 What can you say about the graphs of these functions?

4 **Factual** What is the form of the formula for each graph?

5 **Conceptual** What type of graph represents two variables in direct proportion?

A linear model with formula $y = mx$ relates two variables, x and y, that are in **direct variation** (proportion).

- The graph has a y-intercept of 0.
- The gradient is the **constant of proportionality**, $m = \frac{y}{x}$, for any point (x, y) on the graph.

This type of linear model is used in conversion graphs.

Example 8

This is a conversion graph from pounds to kilograms.

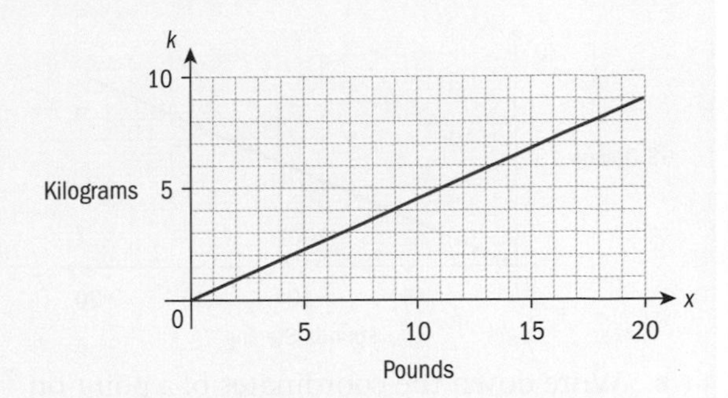

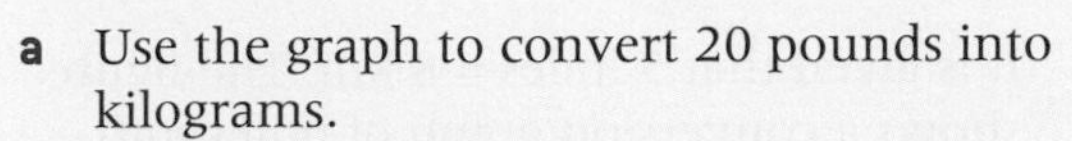

a Use the graph to convert 20 pounds into kilograms.

b Find the number of kilograms equivalent to 1 pound.

c Hence, write down the gradient of this line.

d Find a model for $k(x)$, where $k(x)$ is the number of kilograms and x is the number of pounds.

e Use your model to convert:

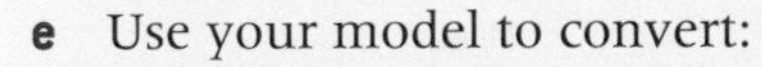

i 17 pounds into kilograms **ii** 25 kilograms into pounds.

a 9 kg	Reading off from the graph: when $x = 20$, $k = 9$. 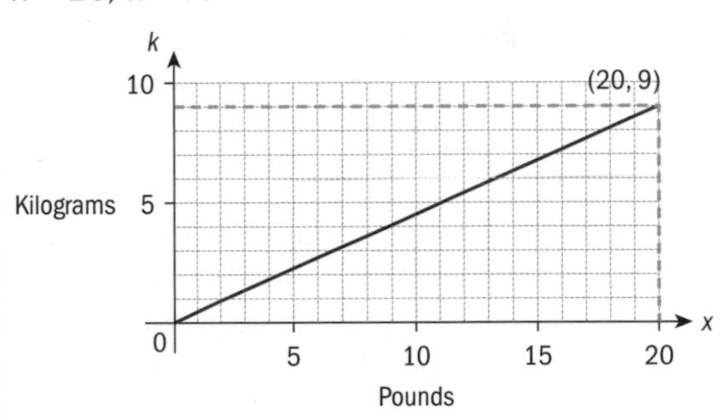
b 1 pound = 0.45 kg	Reading off from the graph, the answer would be a little less than 0.5. Looking for greater precision, it is clear that as the graph is a line that passes through the origin, the variables are directly proportional. Let p be the value we are looking for. Both $(20, 9)$ and $(1, p)$ lie on this graph, so $\frac{9}{20} = \frac{p}{1} = p$, and so $p = 0.45$.
c $m = 0.45$	The gradient of the line is 0.45 (the constant of proportionality).
d $k(x) = 0.45x$	The equation is of the form $k = mx$, and $m = 0.45$.
e **i** 7.65 kg	$k(17) = 0.45 \times 17 = 7.65$
ii 55.6 pounds	$25 = 0.45x$, so $x = 55.6$

Exercise 5E

1 The graph shows the equivalence between Pounds Sterling (UK£) and US Dollars (US$).

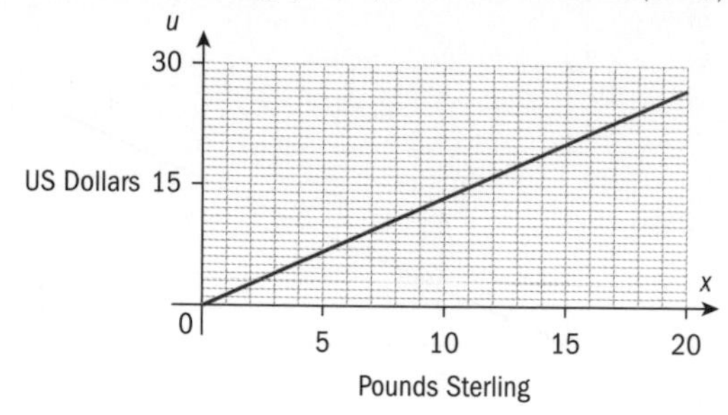

a Write down the coordinates of a point on the graph that gives you some accurate information about the relationship.

b Using your point from part **a**, find the gradient of the line.

c Find the formula for the model $u(x)$, where $u(x)$ is the number of US$ and x is the number of UK£.

d Use your model to find:

i the exchange rate from UK£ to US$

ii the amount in UK£ equivalent to US$100.

2 It is given that 5 miles = 8 km. The figure shows a conversion graph of miles into kilometres.

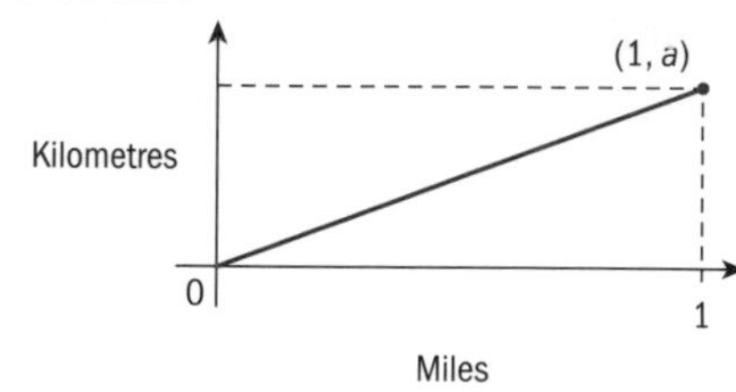

a Find the value of a.

b Write an equation to express the number of kilometres, k, in terms of the number of miles, m.

Investigation 7

1 a Consider all the points for which the second coordinate (y) is equal to 6 units less than the first coordinate (x). Write down this relation as a function, where y is written in terms of x.

b Consider all the points for which the sum of the two coordinates is equal to 4. Write down this relation as another function, where y is written in terms of x.

c Plot the functions from parts **a** and **b** on the same set of axes on your GDC.

d There is a point that satisfies both relations simultaneously.

i What is this point on the graph you have plotted?

ii What are the coordinates of this point?

iii If you are given **only** the point and the equations of both functions (but not their graphs), how would you check that this point lies on the graphs of both functions?

2 Now consider $f(x) = x + 5$ and $g(x) = 7x + 3$.

a Plot both functions on the same set of axes on your GDC.

b Is there any point that satisfies both equations simultaneously?

c Suppose the coordinates of this point P are (a, b). Complete the following:

If (a, b) lies on the graph of f then its coordinates satisfy the equation of f and $b =$ ________ (1)

If (a, b) lies on the graph of g then its coordinates satisfy the equation of g and $b =$ ________ (2)

As the b in equations (1) and (2) is the same, you can derive the value of a by equating (1) and (2).

i What is the value of a?

ii What is the value of b?

iii Write down the coordinates of P.

d How would you check that the point P lies on both graphs?

3 **Conceptual** What does the intersection of two lines represent?

Example 9

a Sketch the graphs of $f(x) = 2x + 3.5$ and $g(x) = 0.5x + 7$.

b Hence, find the intersection point of the two graphs.

a

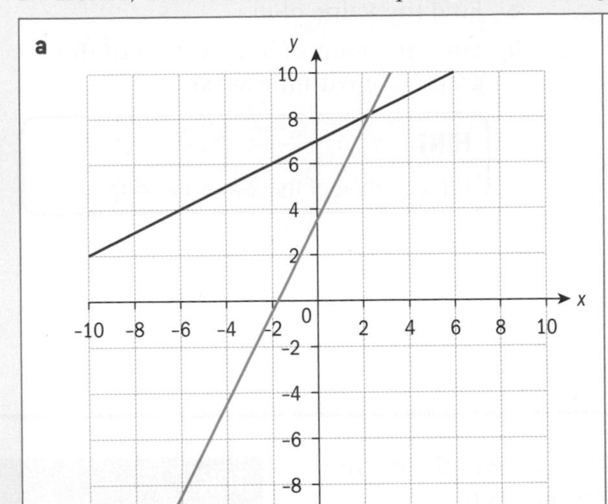

You can use technology to draw the graphs. You might need to change the level of zoom.

b (2.33, 8.17)

Method 1:

Technology can be used to find the intersection point graphically.

Method 2:

Given that both are linear functions, you can solve simultaneous equations in two unknowns algebraically.

$y = 2x + 3.5$ means $2x - y = -3.5$

$y = 0.5x + 7$ means $0.5x - y = -7$

Exercise 5F

1 Find the intersection point for each of the following pairs of functions.

  a $f(x) = x - 1$, $g(x) = 2x$

  b $f(x) = 2x + 3$, $g(x) = 1 - 4x$

2 The managers of a factory are trying to determine what to charge for an item they make. The demand for the item (a measure of how many are likely to sell) can be modelled by $D(p) = 800 - 2p$, where p is the selling price in dollars. The supply for the item (the number that can be made) can be modelled by $S(p) = 3p$.

  a Sketch both functions on the same set of axes.

  b Find the equilibrium point for this item. (The equilibrium point is where supply and demand are equal.)

  c Comment on the situation in which the selling price is \$100.

  d Contrast your answer in part **c** to what happens when the price is \$250.

3 Some friends are planning to rent a car during their holiday. They have two options:

Option A: US$5 per kilometre.

Option B: A fixed cost of US$150 plus US$2.30 per kilometre.

a Find the cost of renting the car if they travel 100 km under:

i Option A **ii** Option B.

b Write down, for each option, an expression for the cost of renting the car in terms of the number of kilometres travelled, x.

They know that above a certain number of kilometres, k, Option A is more economical than Option B.

c Find the value of k.

4 Consider the functions $f(x) = 2x + 1$ and $g(x) = mx + 3$, where m is a constant. The graphs of these functions intersect at the point $(-5, -9)$.

a Find the value of m.

b Find the point of intersection of the graph of g with the x-axis.

> **HINT**
>
> The equation of the x-axis is $y = 0$.

In Investigation 6 you found that the formula $C = 2\pi r$ gives the circumference, C, of a circle in terms of its radius, r. This means that given the radius, by using this formula you can find the circumference. The radius is the independent variable and the circumference the dependent variable.

If you were given that the circumference is equal to 10 cm, how could you find the radius?

And if you were given that the circumference is equal to 15 cm, what would be the radius?

There is only one circumference for a given radius. Is there only one radius for every circumference?

> **TOK**
>
> By 2000 BC, mathematicians in Babylon, Egypt and India had a good knowledge of mathematical models and used them to predict astronomical and environmental events.

> A one-to-one function f that maps A onto B will have an **inverse function** f^{-1} that maps B onto A. If $f(x) = y$ then $f^{-1}(y) = x$.

Consider this example:

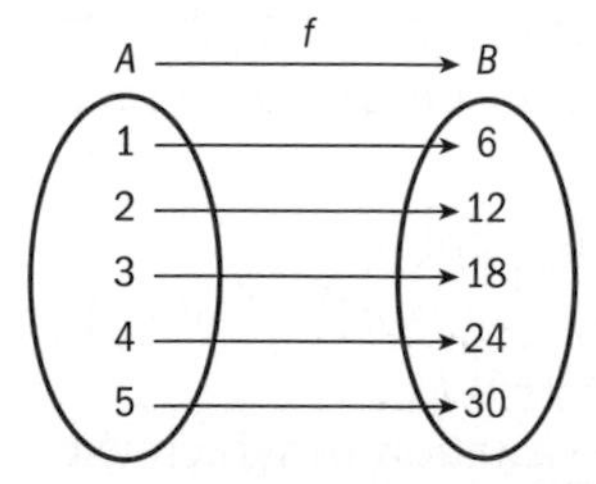

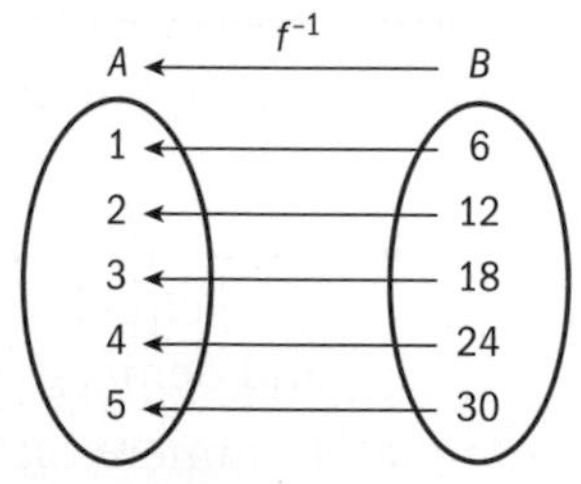

Here, f is a one-to-one function from A to B.

The inverse relation, f^{-1}, is a function that maps every element of B onto its corresponding element from A. For example, $f(1) = 6$, therefore $f^{-1}(6) = 1$.

What is the value of $f^{-1}(18)$? Can you explain its meaning?

If the image of x through f is y, what is the image of y through f^{-1}?

Example 10

The graph shows a function f.

a Use the graph to find:

i $f(2)$

ii $f^{-1}(9)$.

b Solve the equation $f^{-1}(a) = 1$.

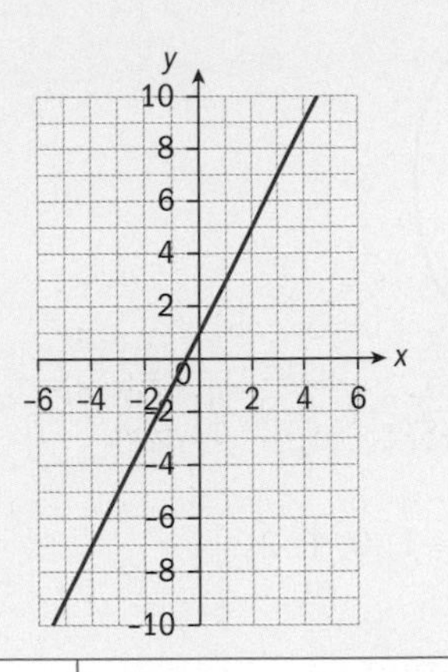

a **i** $f(2) = 5$	$x = 2$ is mapped onto $y = 5$.
ii $f^{-1}(9) = 4$	Let $f^{-1}(9) = b$. This is equivalent to $f(b) = 9$, so the point $(b, 9)$ lies on the graph of f. The point is $(4, 9)$, so $b = 4$. 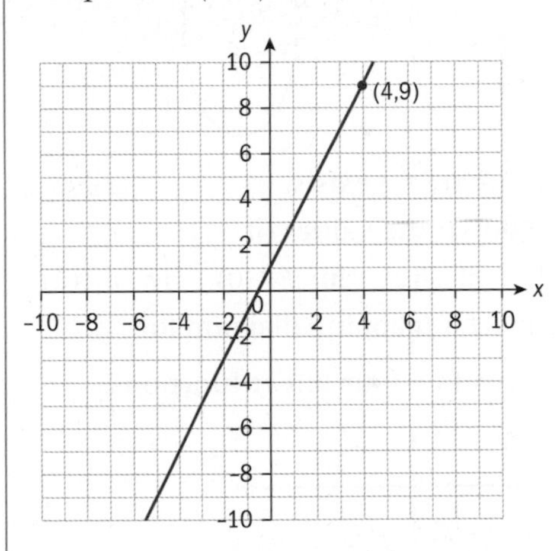
b $a = 3$	The equation $f^{-1}(a) = 1$ is equivalent to the equation $a = f(1)$, and is also equivalent to the point $(1, a)$ lying on the graph of f.

Exercise 5G

1 For the functions **b**, **c** and **d** from Investigation 6:

a Explain why each has an inverse function.

b Find the formula of each inverse function.

2 State which of the following functions has an inverse, giving a reason in each case.

a

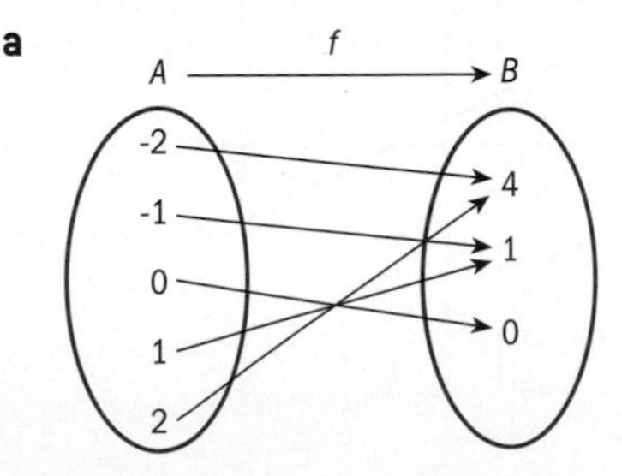

b

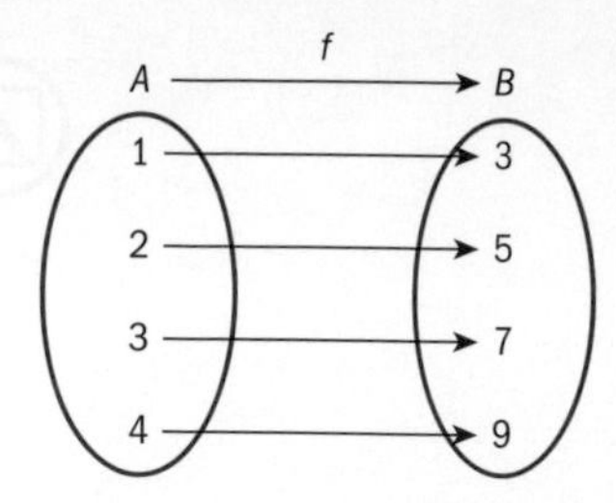

c $f(x) = x + 1$ maps $A = \{-2, -1, 0, 1, 2\}$ onto a set B.

d $f(x) = 1 + x^2$ maps $A = \{-1, 0, 1, 2\}$ onto a set B.

3 The diagram shows a function f.

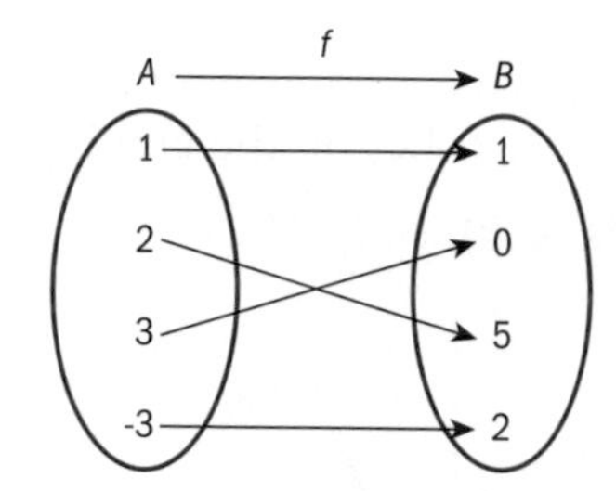

a Explain why f^{-1} exists.

b Write down the values of:

i $f(1)$ **ii** $f(2)$ **iii** $f^{-1}(0)$ **iv** $f^{-1}(2)$.

c Construct a mapping diagram for f^{-1}.

4 The graphs show four linear functions.

i

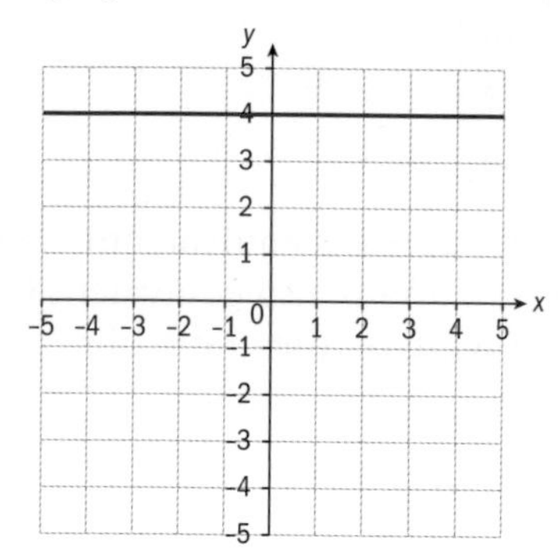

ii

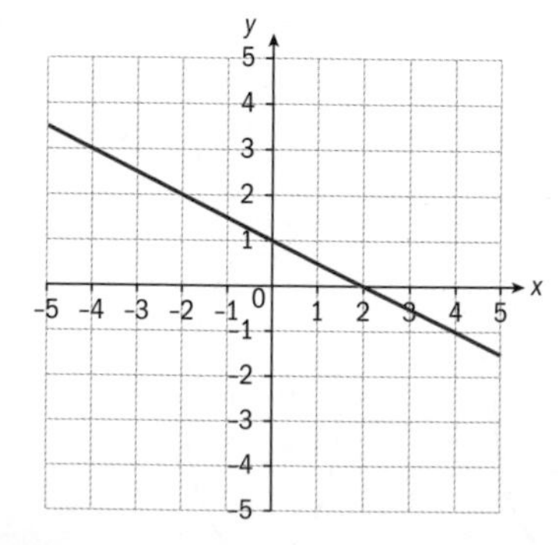

iii

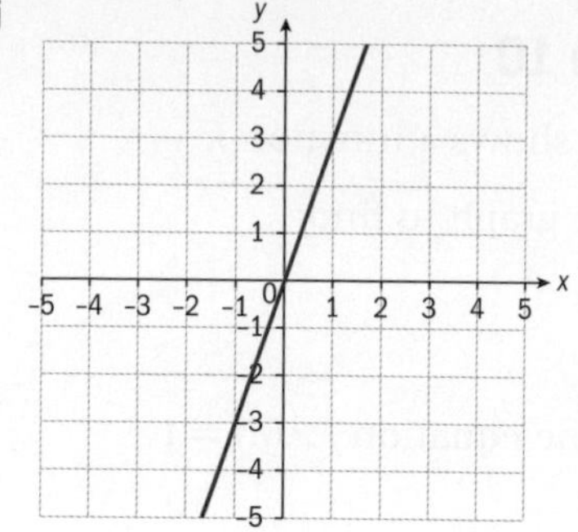

iv

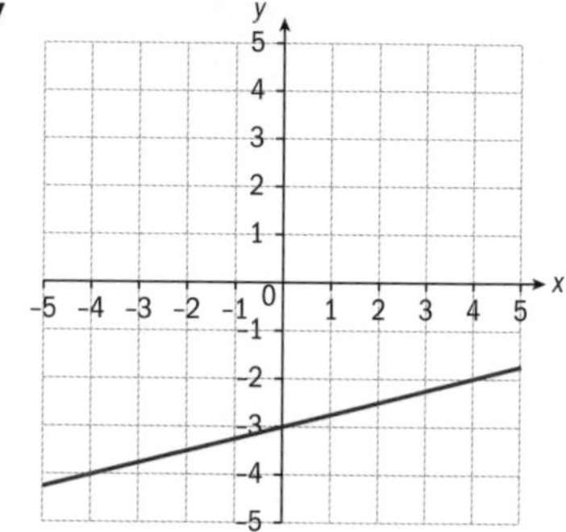

a State which of these are one-to-one functions, giving a reason in each case.

b Describe the difference between a linear function that has an inverse and one that does not.

5 The diagram shows the graph of a function f from A to B.

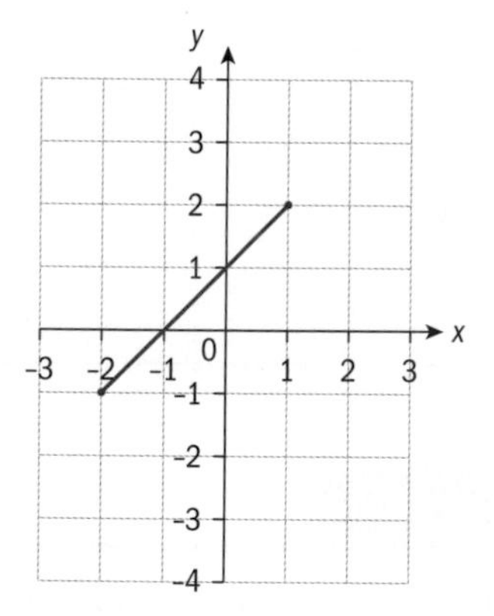

a Write down the set A.

b Write down the set B.

c State whether f is a one-to-one function, giving a reason for your answer.

d Find, if appropriate, the values of:

i $f(1)$ **ii** $f(-1)$

iii $f^{-1}(1)$ **iv** $f^{-1}(-1)$

6 Determine whether the following statements are true or false.

a A many-to-one function does not have an inverse function.

b Any linear function has an inverse function.

Investigation 8

1 Consider the function $f(x) = 2x + 4$.

a Explain why f has an inverse function f^{-1}.

b Complete the table for f and f^{-1}.

x	$y = f(x)$	$x \xrightarrow{f} y = f(x)$	$(x, f(x))$ lies on the graph of f	$y \xrightarrow{f^{-1}} x = f^{-1}(y)$	$(y, f^{-1}(y))$ lies on the graph of f^{-1}
−4			A(−4,_)		A′(_,−4)
	−2		B(_,−2)		B′(−2,_)
	0		C(_,0)		C′(0,_)
0			D(0,_)		D′(_,0)
1	6	$1 \xrightarrow{f} 6$	E(1, 6)	$6 \xrightarrow{f^{-1}} 1$	E′(6, 1)
3			F(3,_)		F′(_,3)

c The points E and E′ have already been plotted on the given set of axes.

i Draw the graph of $f(x) = 2x + 4$.

Also, clearly plot and label the other five points on the graph of f that were found in part **b**.

ii Plot and label the other five points on the graph of f^{-1} that were found in part **b**.

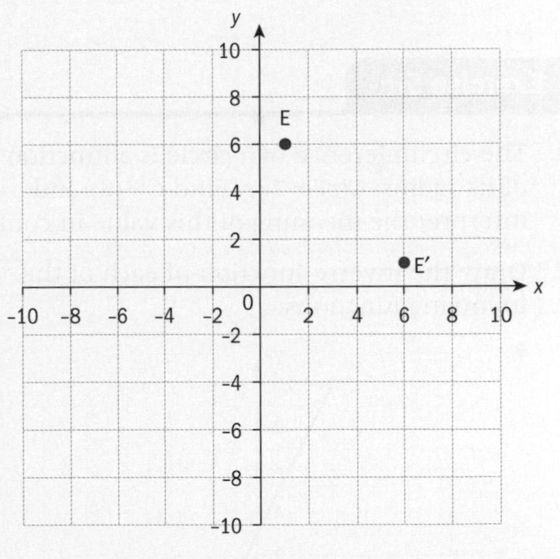

d What can you say about the type of graph f^{-1} is? How would you find more points on this graph? Write down at least three more points on the graph of f^{-1}, then complete the graph of f^{-1}.

e What can you say about the position of the point B′ with respect to the position of the point B? And what about the position of C′ with respect to C?

f There is one point that lies on both graphs. What is it? What do you notice about the coordinates of this point?

This point also lies on the line that bisects the first and third quadrants, the **identity line**. What is the equation of this line? Draw it with a dotted line on the set of axes. How would you describe the positions of the graphs of f and f^{-1} with respect to the identity line?

Continued on next page

g Now explain how you would draw the graph of f^{-1} if you are given the graph of f.

h What is the equation of the inverse function f^{-1}?

2 The diagram shows the graph of a function g. Some points on its graph have already been marked.

a Explain why g has an inverse function, g^{-1}.

b Complete the table:

Function	Domain	Range
g		
g^{-1}		

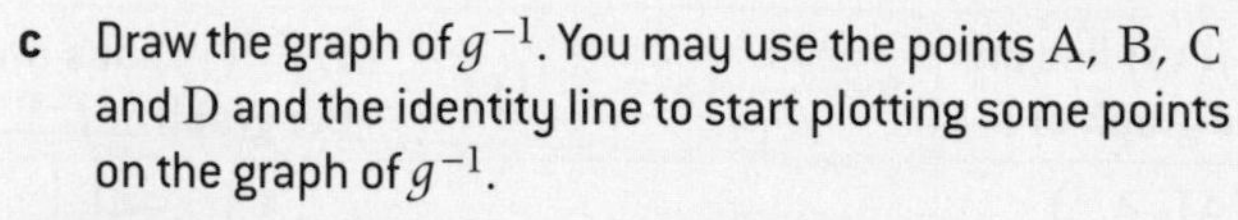

c Draw the graph of g^{-1}. You may use the points A, B, C and D and the identity line to start plotting some points on the graph of g^{-1}.

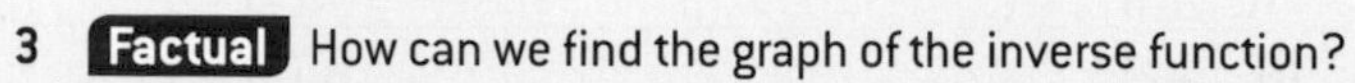

3 **Factual** How can we find the graph of the inverse function?

4 **Conceptual** How can the graph of the inverse function help us to find properties of the function?

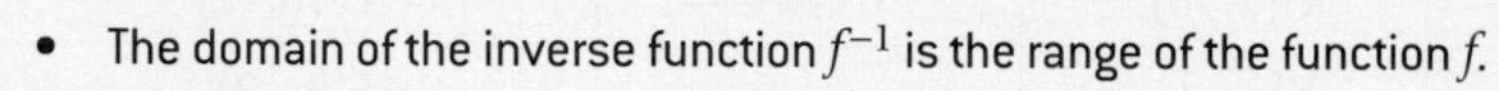

- The domain of the inverse function f^{-1} is the range of the function f.
- The range of the inverse function f^{-1} is the domain of the function f.

Exercise 5H

1 The circumference of a circle is a function of its radius: $C(r) = 2\pi r$. Find $C^{-1}(8)$ and interpret the meaning of this value in context.

2 Draw the inverse function of each of the following functions.

a

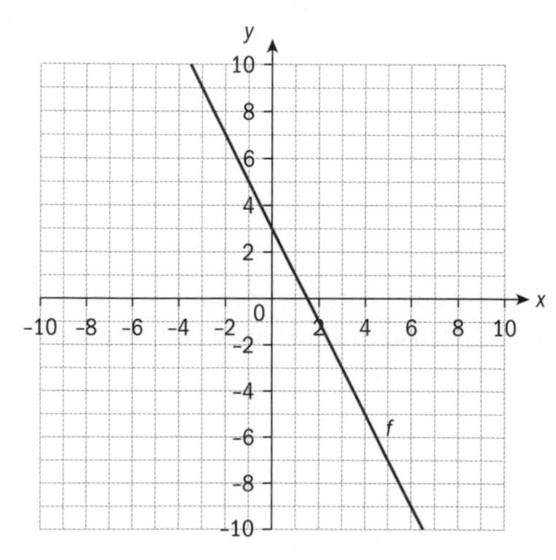

b

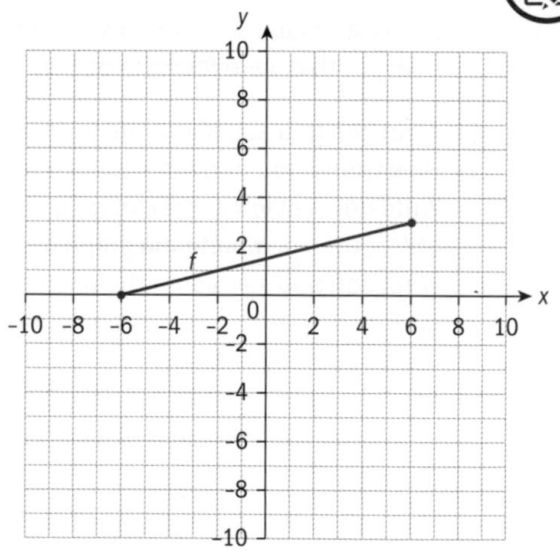

c

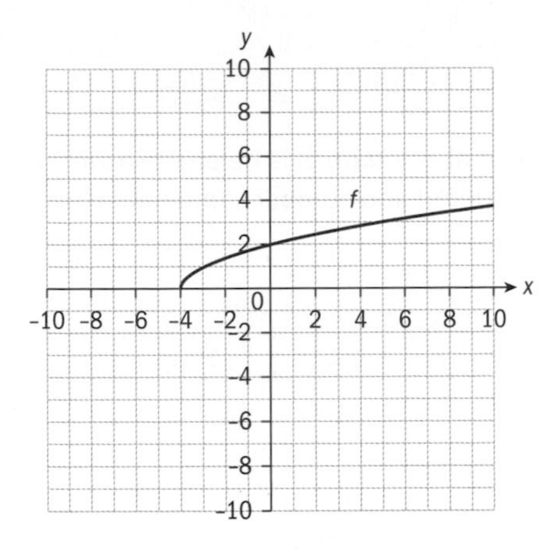

3 Consider $f(x) = -2.5x + 5$ for $0 \le x \le 3$.

a Draw the graph of f on a pair of axes. Use the same scale on both axes.

b Point A lies on the graph of f and has coordinates $(3, b)$. Find the value of b.

c Complete the table:

Function	Domain	Range
f	$0 \le x \le 3$	
f^{-1}		

d Draw the graph of f^{-1} on the same set of axes used in part **a**.

e Find the coordinates of the point that lies on the graph of f^{-1} and on the line $y = x$.

4 The one-to-one function f maps $A = \{x: x \ge 0\}$ onto $B = \{y: -1 \le y \le 2\}$; $f(1) = 0.5$ and $f(0) = 2$.

a Write down the domain and the range of f^{-1}, the inverse function of f.

b Write down the value of $f^{-1}(0.5)$.

c Determine whether f is an increasing or a decreasing function.

d Solve the equation $f^{-1}(x) = 0$.

Developing inquiry skills

How will the taxi driver's profit change if he increases the profit per kilometre by $0.30 after one month?

How can you model the daily profit for different values of profit per kilometre?

5.3 Arithmetic sequences

Investigation 9

Pablo starts his first full-time job at the age of 24. In his first year he earns $3250 per month after taxes, or $39 000 per year. He is given a salary schedule that shows how his salary will increase over time:

Years in job	1	2	3	4	5
Annual salary ($)	39 000	39 900	40 800	41 700	42 600

Continued on next page

We can also write the salaries for each year in a list, like so:

39 000, 39 900, 40 800, 41 700, 42 600

We call this a **sequence**, an ordered list of numbers. We call each number a **term**. It is helpful to have a way to write, for example, "the 13th number in this sequence". We use a letter and number combination for this, such as a_{13}. So, in the sequence above,

$a_1 = 39\,000$ (also called the **first term**)

$a_2 = 39\,000$ (the second term)

$a_3 = 40\,800$

and so on.

In general, a_n is the nth term. The entire sequence is denoted by $\{a_n\}$.

1 Write down the fifth term of the sequence above as a number and in the notation.

Sequences often have patterns that help us to predict them.

2 Use technology to graph the sequence $\{a_n\}$ as a set of points. Use the number of the term (1, 2, 3, ...) as the independent variable and the term in the sequence as the dependent variable. For example, the first point would be (1, 39 000).

What do you notice about the graph? Why do you think this is so?

3 A sequence that forms a straight-line graph is called **arithmetic**. How might we identify a sequence as arithmetic without graphing it? Make a conjecture.

4 Complete the table by calculating the difference between every pair of consecutive terms. The first is done as an example. What do you notice? Does this support your conjecture?

Years in job (term number)	1	2	3	4	5
Annual salary (term)	39 000	39 900	40 800	41 700	42 600
Difference between terms	—	39 900 − 39 000 = 900			

5 a What is the gradient of the line formed by plotting the points of the sequence? How can it be predicted from the terms of the sequence?

b What is the y-intercept of the line associated with the sequence? How can it be predicted from the terms of the sequence?

c Find the equation of the line associated with this sequence. Graph this line. Verify that it passes through the points that you plotted.

d **Conceptual** What advantages does representing Pablo's annual salary with a sequence have over representing it with a linear function?

6 Pablo decides to buy a car, but has to take out a bank loan to do so. Every month the bank charges him interest (a fee for the loan that he must pay). The table shows the amount that Pablo must pay each month.

Month	1	2	3	4	5
Payment ($)	55	52	49	46	43
Difference					

a Find the difference between consecutive terms as you did before, completing the third row of the table.

b Call the sequence of Pablo's payments $\{b_n\}$. What is b_1? What do you predict b_7 will be?

c What is different here from the previous situation? What is the same?

7 **Factual** How do you determine whether a sequence is arithmetic without graphing it?

8 **Factual** If a sequence is arithmetic, how are the parameters of the corresponding linear function related to the first term and the difference between terms of the sequence?

9 Use the pattern in the sequence $\{a_n\}$ to predict Pablo's annual salary if he stays in the same job and retires at the age of 65. You may find it helpful to complete the table:

Age	24	25	26	27	...	64	65
Number of term (n)	1	2	3	4	...		
Number of differences added ($d = 900$)	0	1	2		...		
Term (a_n)	39 000	39 000 + 900 = 39 900	39 000 + 2 × 900 = 40 800		...		

10 **Conceptual** How can we systematically calculate the nth term of an arithmetic sequence?

General sequences

A **sequence** of numbers is a list of numbers (of finite or infinite length) arranged in an order that obeys a certain rule.
Each number in the sequence is called a **term**. The ***n*th term**, where n is a positive integer, can be represented by the notation u_n.

HINT

Sometimes other letters are used to represent the nth term of a sequence, like a_n, b_n, ...

For example, 3, 6, 9, 12, 15, ... is a sequence. What is the rule?

The first term is 3, the second term is 6 and so on, so we can say $u_1 = 3$, $u_2 = 6$, and so on.

Note that a sequence can be thought of as a function from $\mathbb{Z}^+$ to $\mathbb{R}$, where the notation u_n is equivalent to $u(n)$.

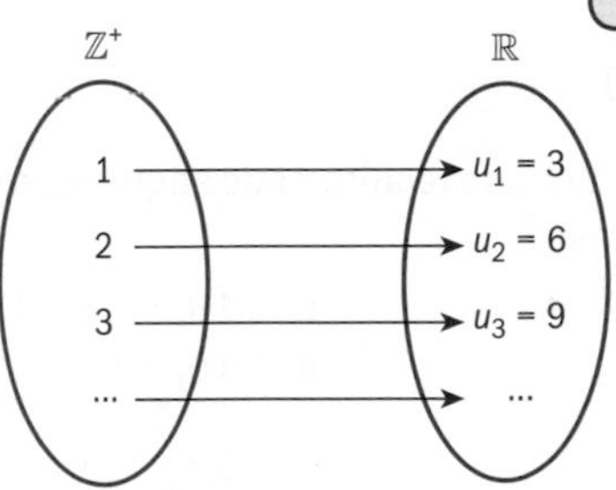

Example 11

For each of the sequences

i 2, 4, 6, 8, 10, ... ii $1, \frac{1}{2}, \frac{1}{3}, \frac{1}{4}, \ldots$

a Write down a general formula for the nth term. **b** Find the 12th term.

a i $u_n = 2n$	This is the sequence of positive even numbers. The terms can be written as $2 \times 1, 2 \times 2, 2 \times 3, \ldots$ Every term of this sequence is a positive integer multiplied by 2.
ii $u_n = \frac{1}{n}$	The denominators are the counting numbers.
b i 24	Substitute $n = 12$ into the formula of the sequence: $u_{12} = 2 \times 12 = 24$
ii $\frac{1}{12}$	$u_{12} = \frac{1}{12}$

Exercise 5I

1 Write down the next three terms in each sequence.

a 3, 8, 13, 18, ...
b 1, 2, 3, 5, 8, ...
c 2, 3, 5, 8, 12, ...
d 1, 2, 4, 8, ...
e 1, −1, 1, −1, 1, ...
f $\frac{1}{2}, \frac{2}{3}, \frac{3}{4}, \frac{4}{5}, \ldots$
g 100, 75, 50, 25, ...

2 Write down the first three terms of these sequences. Remember that $n \in \mathbb{Z}^+$.

a $u_n = n + 1$
b $a_n = 3n + 1$
c $b_n = 2^n$
d $t_n = 4 - 0.5n$

3 Write down a general formula for the nth term of each sequence.

a 1, 4, 9, 16, ...
b 1, 8, 27, 64, ...
c 1, 2, 3, 4, ...
d $2, 1, \frac{2}{3}, \frac{1}{2}, \frac{2}{5}, \ldots$

4 Consider the sequence $u_n = 3 + 4 \times (n - 1)$.

a Find u_6.
b Find the position n in which the number 207 appears in the sequence $\{u_n\}$.
c Find the value of n for which $u_n = 111$.
d Determine whether or not 400 is a term of this sequence.

Arithmetic sequences and series

A sequence in which the **difference** between each term and its previous one remains **constant** is called an **arithmetic sequence.**

This constant difference is called the **common difference** of the sequence.

Investigation 10

1 Which of these sequences are arithmetic sequences? For those that are arithmetic, what is the common difference in each case?

a 2, 4, 6, 8, 10, ...
b 10, 12, 14, 16, ...
c 1, 10, 100, 1000, ...
d 10, 8, 6, 4, ...
e −1, −0.5, 0, 0.5, ...
f $\frac{7}{4}, \frac{3}{2}, \frac{5}{4}, 1, \ldots$
g 1, 3, 4, 7, 11, ...
h $u_n = 5n + 2$
i $a_n = 3n + 1$
j $b_n = 2^n$
k $t_n = 4 - 0.5n$
l $r_n = n^2$

2 Write down:

a the first five terms of a sequence in which the difference between each term and the previous term is always equal to 4
b the first five terms of a sequence in which the common difference is equal to −3
c the first four terms of an arithmetic sequence in which the first term is 2 and the common difference is −0.5.

3 The sum of the terms of an arithmetic sequence is called an **arithmetic series.**

a Which of these sums are arithmetic series? How can you tell?
i $S_4 = 1 + 7 + 13 + 19$
ii $S_4 = 5 + 2.5 + 1.25 + 0.0625$
iii $S_5 = 9 - 7 - 5 - 3 - 1$

b Write down an arithmetic series with six terms in which the first term is equal to 4 and the common difference is equal to 0.20.

4 **Conceptual** How do you distinguish an arithmetic sequence or series from other sequences or series?

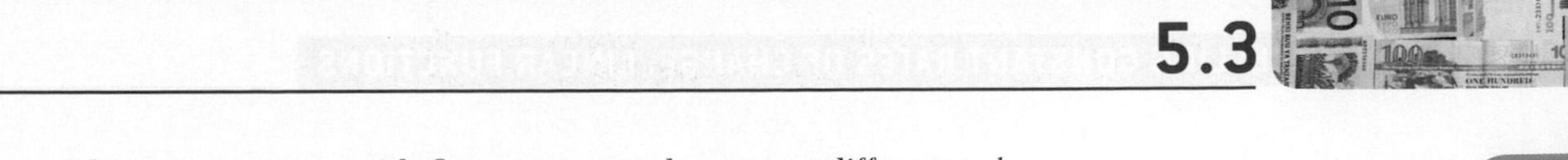

An arithmetic sequence with first term u_1 and common difference d can be generated as shown:

u_1

$u_2 = u_1 + d$

$u_3 = u_1 + d + d = u_1 + 2d$

$u_4 = u_1 + d + d + d = u_1 + 3d$

$u_5 = u_1 + d + d + d + d = u_1 + 4d$

What would the expression be for u_6? What do you notice about the number that multiplies d?

Following the pattern, $u_n = u_1 + (n - 1)d$.

> The general term (or nth term) of an arithmetic sequence with first term u_1 and common difference d is $u_n = u_1 + (n-1)d$, where $n \in \mathbb{Z}^+$.

Example 12

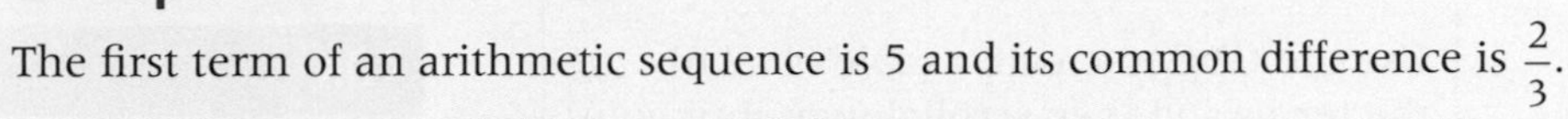

The first term of an arithmetic sequence is 5 and its common difference is $\frac{2}{3}$.

a Find the second and third terms of this sequence.

b Write down an expression for the nth term.

c Determine whether or not $\frac{49}{3}$ is a term of this sequence.

d Find the first term of this sequence that is greater than 25.

a $u_2 = \frac{17}{3}, \quad u_3 = \frac{19}{3}$	$u_1 = 5$ Every new term is found by adding $\frac{2}{3}$ to the previous term. Therefore, $u_2 = 5 + \frac{2}{3} = \frac{17}{3}$ and $u_3 = \frac{17}{3} + \frac{2}{3} = \frac{19}{3}$.
b $u_n = 5 + (n-1) \times \frac{2}{3}$	Substitute 5 for u_1 and $\frac{2}{3}$ for d in the formula for the general term.
c $\frac{49}{3}$ is the 18th term	If $\frac{49}{3}$ is a term of this sequence then there will be an n for which $u_n = \frac{49}{3}$: $5 + (n-1) \times \frac{2}{3} = \frac{49}{3}$ $(n-1) \times \frac{2}{3} = \frac{34}{3}$ $n - 1 = 17$ Since $n = 18$ is a positive integer, $\frac{49}{3}$ is in the sequence.
d 32	You can see this on your GDC.

Example 13

Consider the finite arithmetic sequence −3, 5, …, 1189.

a Write down the common difference, d.

b Find the number of terms in the sequence.

a $d = 8$	The common difference is the difference between any two consecutive terms: $d = 5 - (-3) = 8$
b $n = 150$	Method 1: The general term of this sequence is $u_n = -3 + (n - 1) \times 8$. We also know that $u_n = 1189$. So $-3 + (n - 1) \times 8 = 1189$. Use your GDC to solve for n. Method 2: Use the GDC to create a table of all of the terms and then scroll down until you see the value 1189. However, this may take you longer than Method 1 when the value of n is big (as in this example).

TOK

Is all knowledge concerned with identification and use of patterns?

Exercise 5J

1 The first term of an arithmetic sequence is −10 and the seventh term is −1.

a Find the value of the common difference.

b Find the 15th term of this sequence.

2 The nth term of a sequence is defined by $b_n = n(2n + 1)$.

a Find the values of b_1, b_2 and b_3.

b Show that this sequence is not arithmetic.

3 The first terms of an arithmetic sequence are 5, 9, 13, 17, …

a Write down the general term for this sequence.

b Determine whether or not 116 is a term in this sequence.

4 In an arithmetic sequence, $u_3 = 12$ and $u_{10} = 40$. The common difference is d.

a Write down two equations in u_1 and d to show this information.

b Find the values of u_1 and d.

c Find the 100th term.

5 When a company first started it had 85 employees. It was decided to increase the number of employees by 10 at the beginning of each year.

a Find the number of employees during the second year and during the third year.

b How many employees will this company have during the 10th year?

c After how many years will the company have 285 employees?

6 In Investigation 9, Pablo made monthly payments to his bank for a loan. The payments for the first five months were \$55, \$52, \$49, \$46, \$43.

- **a** Write down an expression for the nth term of this sequence.
- **b** Find the amount of Pablo's 12th payment.
- **c** Determine when Pablo will make his last payment.
- **d** Find the amount of the last payment.

7 A sequoia tree that was 2.6 m tall when it was planted in 1998 grows at a rate of 1.22 m per year.

- **a** Write down a formula to represent the height of the tree each year, with a_1 representing the height in 1998.
- **b** Find the height of the tree in 2025.
- **c** The tallest living sequoia tree, the General Sherman tree in the US Sequoia National Park, has a height of 84 m. In what year would the tree planted in 1998 reach this height if it continues to grow at the same rate?

8 Consider the finite arithmetic sequence 1000, 975, 950, ..., −225.

- **a** Write down the common difference.
- **b** Find the 10th term.
- **c** Find the number of terms in the sequence.

9 The diagram shows part of the graph of a sequence u_n.

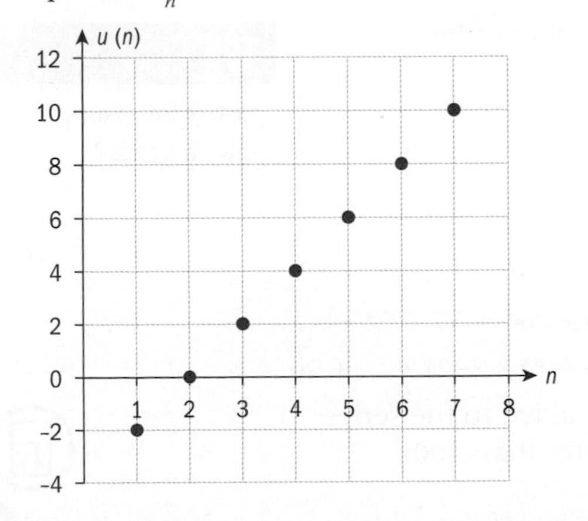

- **a** Explain why this sequence is arithmetic.
- **b** Write down the common difference of this sequence.
- **c** Write down the first term of the sequence.
- **d** Write down the general term of the sequence.
- **e** Determine whether or not the point $(20, 36)$ lies on the graph of this sequence.

10 Pedro is bored and is making patterns with sticks. Pattern 1 was made with four sticks, pattern 2 with seven sticks and so on.

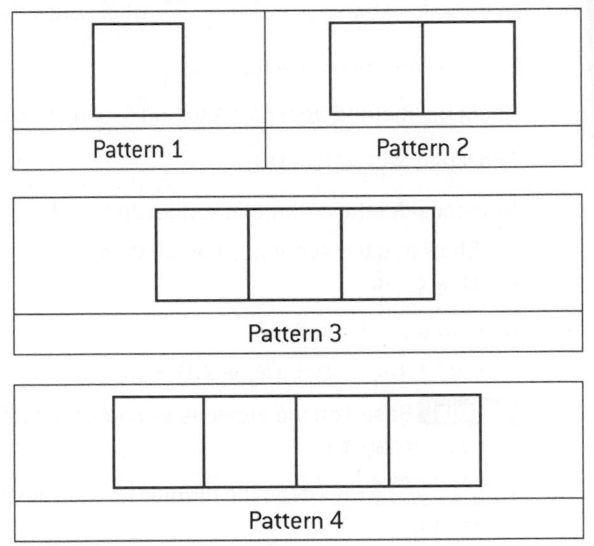

- **a** Write down the number of sticks needed to make pattern 5 and pattern 6.
- **b** Find the number of sticks needed to make pattern 20.
- **c** Find the pattern number for the pattern made with 127 sticks.

11 Consider the arithmetic series $S = 1 + 3 + 5 + 7 + \ldots + 61$.

Find the number of terms in this series.

Investigation 11

1 Consider the arithmetic sequence 1, 2, 3, 4, …, 100. We are going to find the sum of these 100 terms, S_{100}.

$S_{100} = 1 + 2 + 3 + 4 + \cdots + 97 + 98 + 99 + 100$, and also

$S_{100} = 100 + 99 + 98 + 97 + \cdots + 4 + 3 + 2 + 1$. Why?

Adding these two equations, we get

$$2S_{100} = \underbrace{101+101+101+101+\ldots+101+101+101+101}_{\text{This is } 100\times101}$$

Therefore $2S_{100} = 100\times101$, so $S_{100} = \frac{100\times101}{2} = 5050$.

In the calculation $S_{100} = \frac{100\times101}{2}$, where does the 100 come from? Where does the 101 come from?

Describe how you have found S_{100}.

2 Apply the method from part **1** to find S_{100} for the arithmetic sequence 3, 5, 7, 9, …

Show that $S_{100} = 10\,200$.

3 Now consider the arithmetic sequence 1, 4, 7, …, 148.

a Show that the sequence has 50 terms.

b Find S_{50}.

4 Now consider the series

$S_n = u_1 + (u_1 + d) + (u_1 + 2d) + (u_1 + 3d) + \ldots + (u_1 + (n-1)d)$

5 **Factual** Based on the previous examples, what formula do you suggest to find S_n? (Follow the same steps as in part **1**.)

6 **Conceptual** How does the formula for the sum of the first n terms of an arithmetic sequence use symmetry?

> The sum, S_n, of the first n terms of an arithmetic sequence $u_1, u_2, u_3, \ldots$ can be calculated using the formula $S_n = \frac{n}{2}(u_1 + u_n)$.

Note that if we replace u_n with $u_1 + (n-1)d$, the formula can be written as

$$S_n = \frac{n}{2}(u_1 + u_1 + (n-1)d) = \frac{n}{2}(2u_1 + (n-1)d).$$

TOK

How is intuition used in mathematics?

Example 14

1 Calculate the sum of the first 20 terms of the series 60 + 57 + 54 + …

2 a Find the sum of the arithmetic series (−10) + (−6) + (−2) + … + 90.

b Find the least number of terms that must be added to the series (−10) + (−6) + (−2) + … to obtain a sum greater than 100.

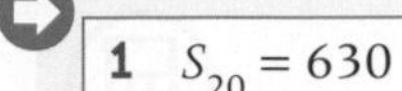

1 $S_{20} = 630$	The common difference is 57 – 60 = –3, and the first term is 60, so substitute $d = -3$, $n = 20$ and $u_1 = 60$ into the formula: $S_{20} = \frac{20}{2}\left(2\times 60 + (20-1)(-3)\right) = 630$
2 a $S_{26} = 1040$	This is an arithmetic series with common difference 4. The first and last terms are given, so we can use the formula $S_n = \frac{n}{2}(u_1 + u_n)$ if we can find n. The general term for the sequence is $u_n = -10 + (n-1)\times 4$. Setting $u_n = 90$ gives $n = 26$. So $S_{26} = \frac{26}{2}(-10+90) = 1040$.
b 11 terms	The sum of the terms (in terms of n) is $S_n = \frac{n}{2}\left(2\times(-10) + (n-1)\times 4\right) = \frac{n}{2}\left(-20 + (n-1)\times 4\right)$ and this must be greater than 100: $\frac{n}{2}\left(-20+(n-1)\times 4\right) > 100$ Use the GDC to make a table of values with the formula for S_n. Adding 10 terms gives a sum of 80, and adding 11 terms gives 110, so $n = 11$.

Sigma notation and series

The terms of a sequence can be added together. Adding the terms of a sequence gives a **series**.

If $u_1, u_2, u_3, \ldots, u_n$ is a sequence, then $u_1 + u_2 + u_3 + \ldots + u_n$ is a series.

S_n denotes the **sum** of the first n terms of a series: $S_n = u_1 + u_2 + u_3 + \ldots + u_n$.

The Greek letter Σ, called "sigma", is often used to represent a sum of values:

$$S_n = \sum_{i=1}^{n} u_i = u_1 + u_2 + u_3 + \ldots + u_n.$$

$\sum_{i=1}^{n} u_i$ is read as "the sum of all u_i from $i = 1$ to $i = n$".

For example, for the sequence 2, 4, 6, 8, 10, …, $S_1 = 2$, $S_2 = 2 + 4$, $S_3 = 2 + 4 + 6$, $S_4 = 2 + 4 + 6 + 8$ and so on, and $S_4 = 2 + 4 + 6 + 8$ can be written as $\sum_{i=1}^{4} 2i$.

TOK

Is mathematics a language?

Example 15

a Write the expression $\sum_{i=1}^{4}(2i+3)$ as a sum of terms.

b Calculate the sum of these terms.

a $\sum_{i=1}^{4}(2i+3)$ $= (2\times1+3)+(2\times2+3)+(2\times3+3)+(2\times4+3)$	It is important to remember that $\sum_{i=1}^{4}(2i+3)$ is a sum. There will be four terms in this sum. The first term is found by substituting $i=1$ in the expression $2i+3$, and the last term by substituting $i=4$.
b $5+7+9+11=32$	The sum can be found with the GDC.

Example 16

Write the series $1+3+5+7+9+11$ using sigma notation.

$\sum_{n=1}^{6}(2n-1)$	There are six terms in this series, which means that the index will vary from 1 to 6. This is the sequence of odd numbers: $u_n = 2n-1$, so that $u_1 = 1$, $u_2 = 3$, $u_3 = 5$, ...

Exercise 5K

1 Find the sum of the first 17 terms of the arithmetic series $15 + 15.5 + 16 + 16.5 + \ldots$

2 Find the sum of the first 20 terms of the arithmetic series $6 + 3 + 0 - 3 - 6 - \ldots$

3 Find the sum of the first 30 multiples of 8.

4 Consider the series $52 + 62 + 72 + \ldots + 462$.

a Find the number of terms.

b Find the sum of the terms.

5 Consider the arithmetic series $S_n = 1 + 8 + 15 + \ldots$

a Write down an expression for S_n in terms of n.

b Find the value of n for which $S_n = 5500$.

6 An arithmetic series has $S_1 = 4$ and $d = -3$.

a Write down an expression for S_n in terms of n.

b Find S_{10}.

c Find the smallest n for which $S_n < -250$.

7 Montserrat is training for her first race. In her first training week she runs 3 km, and in the second training week she runs 3.5 km. Every week she runs 0.5 km more than the previous one.

a Find how many kilometres Montserrat runs in her 10th training week.

b Calculate the total number of kilometres that Montserrat will have run by her 15th training week.

8 Consider the arithmetic sequence $5, a, 14, b, \ldots$

a Find the common difference.

b Find the values of a and b.

c Find S_{10}.

d Find the least value of n for which $S_n > 500$.

9 The first three terms of an arithmetic sequence are $2k + 1$, $-k + 10$ and $k - 1$.

a Show that $k = 4$.

b Find the values of the first three terms of this sequence.

c Write down the common difference of the sequence.

d Find the sum of the first 20 terms of the sequence.

10 Consider the sum $S = \sum_{i=3}^{x}(3i + 1)$, where x is a positive integer greater than 3.

a Write down the first three terms of the series.

b Write down the number of terms in the series, in terms of x.

c Given that $S = 520$, find the value of x.

Simple interest

Capital is the money that you put in a savings institution or a bank. If you put money in a bank, the bank will pay you **interest** on this money.

If you borrow money from a bank or savings institution, then you have to pay them **interest**.

The interest is the extra amount you pay or earn. Its value depends on the capital.

Simple interest is the simplest way to calculate interest. Of course, there are other ways.

As an example, Wang wants to borrow $1000 for three years at a rate of 4% per year.

The interest that Wang will have to pay for **1** year is 4% of 1000: 0.04×1000.

The interest that Wang will have to pay for **2** years is $(0.04 \times 1000) \times \mathbf{2}$.

The interest that Wang will have to pay for **3** years is $(0.04 \times 1000) \times \mathbf{3}$.

This is how the formula for simple interest is built.

The formula for simple interest is $I = C \times r \times n$, where
C is the capital (or principal)
r is the interest rate
n is the number of interest periods
I is the interest.

In the example above, how much money will Wang have to repay to the bank after three years?

Investigation 12

1 Capital of $1000 is invested in a bank account at a simple interest rate of 3% per annum.

a Complete the table for the amount of money in the bank account after n years (assuming no withdrawals).

Number of years	0	1	2	3	4	n
Amount in bank ($)						

Continued on next page

b What can you say about the difference in money between two consecutive years? What can you say about this sequence of numbers? What are the variables? What values does the independent variable take?

2 A capital amount C is invested in a bank account at a simple interest rate of r% per annum.

a Complete the table for the amount of money in the bank account after n years (assuming no withdrawals).

Number of years	0	1	2	3	4	n
Amount in bank ($\$A$)						

b What can you say about the difference in money between two consecutive years? What is the relationship between this value and the type of account? What can you say about this sequence of numbers? What are the variables? What values does the independent variable take?

3 The amount of money in a particular bank account that pays r% simple interest per annum can be calculated with the following formula:

$A_n = 1200 + 60n$

What does A_n mean? What is the capital in this bank account?
What is the value of r?

4 **Conceptual** What is a suitable model for simple interest?

Example 17

An amount of UK£5000 is invested at a simple interest rate of 3% per annum for a period of 8 years.

a Calculate the interest received after 8 years.

b Find the total amount in the account after the 8 years.

a UK£1200	In this example, $C = 5000$, $r = 0.03$ (3% is written as 0.03) and $n = 8$. Using the formula $I = C \times r \times n$, $I = 5000 \times 0.03 \times 8 = 1200$
b UK£6200	The total amount is found by adding the interest to the capital: $5000 + 1200 = 6200$

Example 18

Riddhi invests \$4000 in a bank account at a simple interest rate of 7.3% per year.
Find the number of years it takes for Riddhi's money to double.

14 years	$C = 4000$, and we are looking for it to double. We therefore need the interest earned to be $I = 8000 - 4000 = 4000$. Using the simple interest formula: $4000 = 4000 \times 0.073 \times n$, so $n = 13.69$. Since interest is only applied after each whole year, it will be 14 years before the capital doubles.

Exercise 5L

1 Calculate the simple interest on a loan of:

a $9000 at a rate of 5% pa over three years (pa means per annum)

b $10 000 at a rate of 8.5% pa over 18 months (18 months is 1.5 years)

c $6500 at a rate of 7% pa over 3 years and 5 months.

2 Find the amount of money borrowed if after seven years the simple interest charged is $9000 at a rate of 7.5% per annum.

3 Find the rate of simple interest charged if UK£1840 is earned after 5 years on a deposit of UK£8000.

4 Stephen deposits $8600 in a bank account that pays simple interest at a rate of 6.5% per annum. Find how many years it will take for Stephen's money to double.

5 True or false: The formula $A_n = 1000 + 30n$ can be used to calculate the final amount in an investment account that pays 0.3% simple interest pa on a capital sum of $1000.

Developing inquiry skills

Will the taxi driver reach the average annual salary in one year's time? If not, how long will it take?

What assumptions do you need to make to answer this question?

Developing your toolkit

Now do the Modelling and investigation activity on page 260.

5.4 Modelling

Investigation 13

A water tank is leaking. The amount of water in the tank was measured three times after it started to leak. The table shows the measurements.

Time (t minutes)	2	4	9
Amount of water (w litres)	1900	1800	1500

1 Using technology, plot the information from the table on a set of axes.

2 What can you say about the positions of the points? What do you think would be a good model to fit this data? Why?

Explain how you decided to choose your model. It is important to justify why you have chosen a particular model.

Find an equation for your model using technology. Plot your model on the same set of axes as in part **1**.

3 How could your model be useful? What results would it help to find?

Continued on next page

4 What is your estimate for the amount of water in the tank after 3 minutes? What is your estimate for the amount of water in the tank after 15 minutes?

Why are these estimates?

5 Do you think that your predictions are reliable? Are 3 and 15 in the domain of the given data?

Is it reasonable to use your model to estimate the amount of water in the tank after 40 minutes? Why or why not?

6 Is it reasonable to think that between the first and the third measurement times there was a moment at which there were 1300 litres of water in the tank? Explain.

Extrapolation means estimating a value **outside** the range of the given data.

Interpolation means estimating a value **within** the range of the given data.

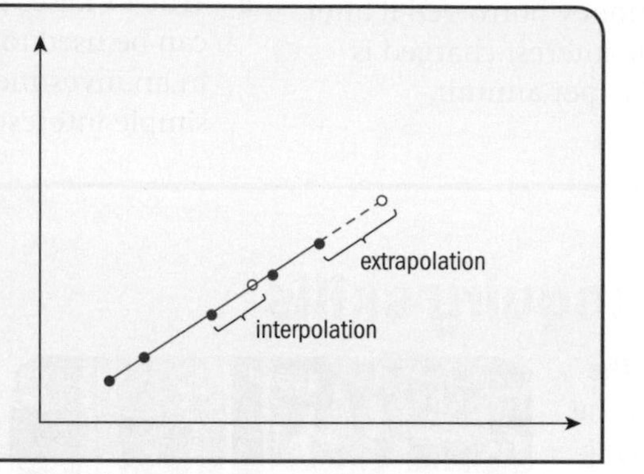

In Investigation 13, which predictions involved interpolation and which extrapolation? Which do you think were more reliable?

Reflect What is the difference between interpolation and extrapolation?

Why do we need to consider the reasonableness of results within the context?

Why is interpolation more reliable than extrapolation?

The diagram shows the mathematical modelling process.

Reject

Pose a real-world problem → Develop a model → Test the model → Accept → Reflect on and apply the model → Extend

There are three main stages in the mathematical modelling process that you need to apply:

- Given a real-world problem, develop and fit the model.
- Test and reflect upon the model.
- Use the model to analyse and interpret data and to make predictions.

Have you applied these stages when working on Investigation 13?

In which stage did you:

- determine a reasonable domain and range for the model?
- comment on the appropriateness and reasonableness of the model?
- recognize the dangers of extrapolation?

Example 19

In a chemistry experiment, a liquid is heated and the temperature is recorded at different times. Some results are shown in the table.

Time (x minutes)	4	6	8	10
Temperature (y °C)	130.0	209.8	290.3	369.2

a Plot a graph of this data on your GDC.

b Choose an appropriate model. Justify your decision.

c Determine a reasonable domain and range for your model.

d Find an equation for the model. Plot it on your graph.

e Comment on your model.

f Use your model to **estimate**:

i the temperature of the water 4.5 minutes after the experiment started

ii the temperature of the water 20 minutes after the experiment started.

g Comment on the predictions made in part **f**.

a

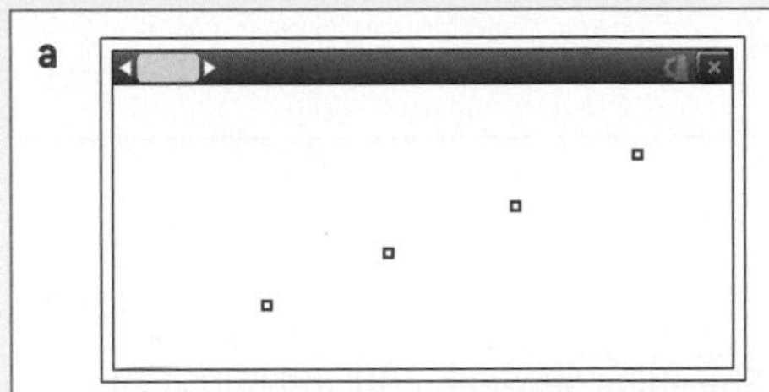

b The points seem to be aligned, so a linear model would be appropriate.

c Domain: $\{x: 4 \le x \le 10\}$

Range: $\{y: 130 \le y \le 369.2\}$

To determine a reasonable domain and range for your model, look at the table or graph on the GDC and identify the interval of values over which each of the two variables is defined:

Time is between 4 and 10, and temperature is between 130 and 369.2.

d

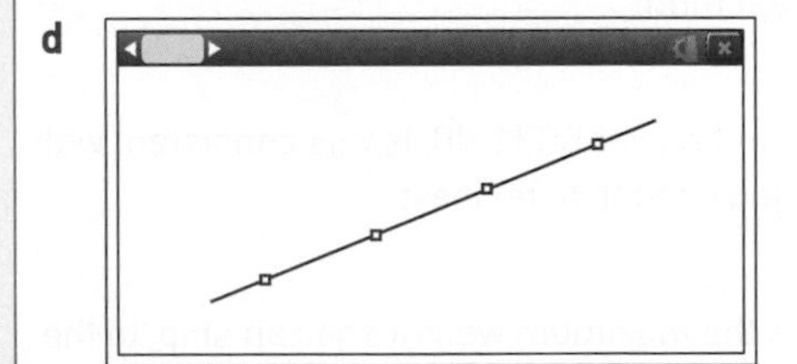

The equation of the model is

$y = 39.9x - 30.0$

(or $y = 39.905x - 29.51$).

Based on the position of the points, we may assume that the model is linear and of the form $y = mx + c$.

Your GDC gives the values of the gradient and the y-intercept.

Continued on next page

e The model fits the data points well.	
f **i** 150	Method 1: Your GDC can be used to estimate the temperature (y value) graphically. Method 2: Substitute $x = 4.5$ in the equation: $y = 39.905 \times 4.05 - 29.51 = 150$
ii 769	Similar methods can be used to estimate the temperature of the water 20 minutes after the experiment started.
g The answer to **i** is appropriate as $t = 4.5$ is within the domain on which the model has been defined, whereas in **ii** the model has been extrapolated and this may be risky as the linear tendency may not continue after 20 minutes; 20 is not within the domain on which the model has been defined.	

Investigation 14

Lucy is researching shipping companies for her business to use. She ships between 200 and 500 kg of products each week. Ted's Transport charges a flat rate of $15.99 per kg plus a flat fee. She knows that her friend used Ted's Transport and paid about $2800 to ship 170 kg of belongings.

Model this situation by following the diagram of the modelling process on page 248.

1. Pose a real-world problem:
   What question(s) might you ask here?
2. Develop a model:
   - **a** What are the independent and dependent variables in this situation?
   - **b** Explain why a linear model is appropriate for this situation.
   - **c** Find an equation for your linear model in gradient–intercept form.
3. Test the model:
   The following week, Lucy makes her first shipment of 310 kg and pays $5031.90. Is this consistent with your model? If not, revise the model, giving reasons for how you choose to revise it.
4. Apply your model:
   The following week, Lucy budgets $4500 for shipping. What is the maximum weight she can ship, to the nearest kilogram? (When applying a model, we **predict** the value of one variable given a known value for the other. It also allows us to make **decisions**, such as setting budgets.)
5. Reflect on your model:
   - **a** What is $C(0)$, and what is its meaning in the context of the problem?
   - **b** What is a reasonable practical domain and associated range if Lucy uses the model to predict weekly shipping costs?
6. **Conceptual** Why is modelling useful? What does a model allow us to do?

Developing inquiry skills

Returning to the opening problem, how have you used the modelling process discussed in this section to investigate it?

Exercise 5M

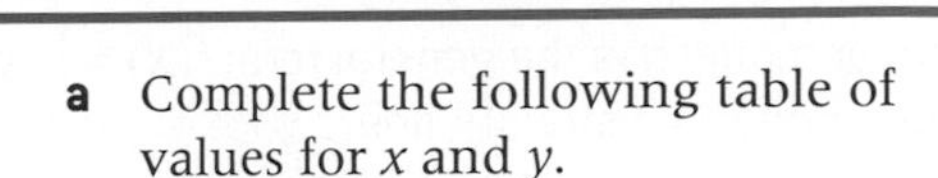

1 Consider all rectangles with perimeter equal to 20 cm. Let the base of the rectangle be x cm and the height of the rectangle be y cm.

a Complete the table with values for x and y.

x	9			0.5	0.1
y		5	8		

b Find a model to express the height of the rectangle (y) in terms of its base (x).

c **i** State the domain of the function defined in part **b**.

ii State the range of the function defined in part **b**.

d Sketch the function defined in part **b**.

e State the rate of change of y with respect to x, and interpret it in this context.

2 A fence is to be built around a rectangular garden against one wall of a farmhouse. The total length of fence is 70 metres, and it will enclose three sides of the garden as shown in the diagram.

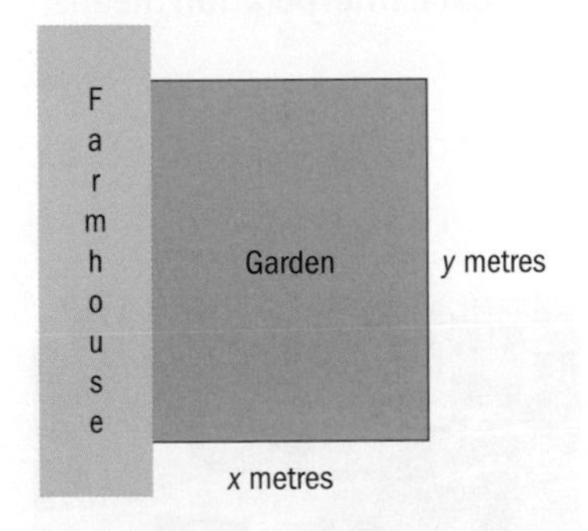

a Complete the following table of values for x and y.

Width (x m)	5	8	20	34
Length (y m)				

b Plot the points from the table on your GDC. State a good model for the length of the garden.

c Plot your model on the GDC.

d State whether it is reasonable to use your model to find the length when the width of the garden is 36 metres, explaining your answer.

e Find the domain and range for this model.

3 The sales of a company, S million dollars, x years after it opened are shown in the table.

x (years)	2	6	10
S (million dollars)	8	11.3	15.1

a Plot a graph of this data on your GDC.

b Choose an appropriate model for the sales of the company. Justify your decision.

c Find an equation for the model. Plot the model on the GDC.

d Comment on your model.

e Hence estimate:

i the sales of the company 7 years after it opened

ii the sales of the company 15 years after it opened.

TOK

What is the relationship between real-world problems and mathematical models?

Chapter summary

- A **relation** is a set of ordered pairs.
- A **function** is a relation between two sets in which every element of the first set (input, independent variable) is mapped onto **one and only one** element of the second set (output, dependent variable).
- The **domain** of a function is the set of all input values.
- The **range** of a function is the set of all output values.
- For a function f, the notation $y = f(x)$ means that x (an element of the domain, the input) is mapped by the function f to y (an element of the range, the output).
- A linear model has the general form $f(x) = mx + c$, where m and c are constants. The graph of a linear model is a straight line.
- A linear model with formula $y = mx$ relates two variables, x and y, that are in **direct variation** (proportion).
- A one-to-one function f that maps set A onto set B will have an **inverse function** f^{-1} that maps B onto A. If $f(x) = y$ then $f^{-1}(y) = x$.
- A **sequence** of numbers is a list of numbers (of finite or infinite length) arranged in an order that obeys a certain rule. Each number in the sequence is called a **term**.
- The terms of a sequence can be added together. Adding the terms of a sequence gives a **series**.
- The Greek letter Σ, called "sigma", is often used to represent a sum of values: $S_n = \sum_{i=1}^{n} u_i = u_1 + u_2 + u_3 + \ldots + u_n$
- A sequence in which the **difference** between each term and its previous one remains **constant** is called an **arithmetic sequence**. This constant value is called the **common difference** of the sequence.
- The general term (or nth term) of an arithmetic sequence with first term u_1 and common difference d is $u_n = u_1 + (n-1)d$, where $n \in \mathbb{Z}^+$.
- The sum, S_n, of the first n terms of an arithmetic sequence $u_1, u_2, u_3, \ldots$ can be calculated using the formula $S_n = \frac{n}{2}(u_1 + u_n)$.
- The formula for simple interest is $I = C \times r \times n$, where

  C is the capital (or principal)

  r is the interest rate

  n is the number of interest periods

  I is the interest.
- **Extrapolation** means estimating a value **outside** the range of the given data. **Interpolation** means estimating a value **within** the range of the given data

Developing inquiry skills

Has what you have learned in this chapter helped you to answer the questions from the beginning of the chapter about the taxi driver's profit? How could you apply your knowledge to investigate other business charging structures? What information would you need to find?

Thinking about the inquiry questions from the beginning of this chapter:

- Discuss whether what you have learned in this chapter has helped you to think about an answer to these questions.
- Consider whether there are any questions that you are interested in and would like to explore further, perhaps for your internal assessment topic.

Chapter review

Click here for a mixed review exercise

1 The diagram shows the graph of a relation R that maps set A onto set B.

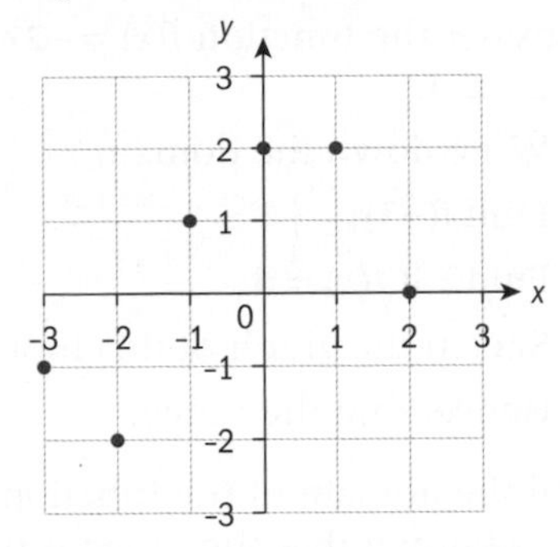

a Write down R as a set of ordered pairs.

b Explain why R is a function from A to B.

c Write down the domain of this function.

d Write down the range of this function.

2 State which of these graphs do not represent a function. Give a reason for your choice.

a

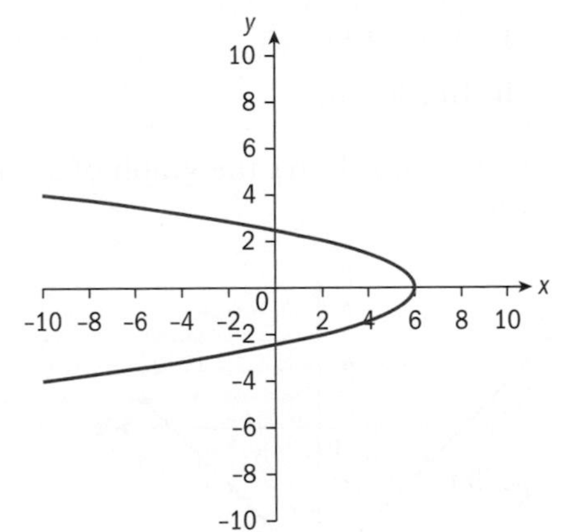

b

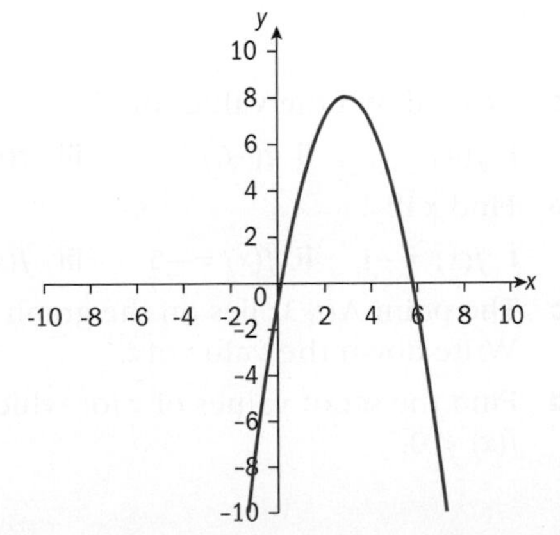

c

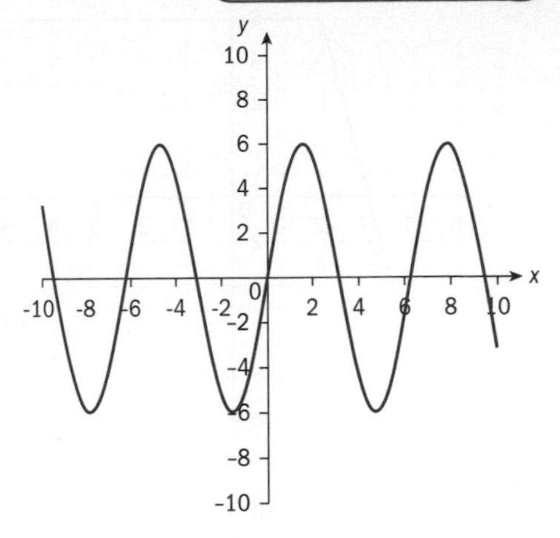

d

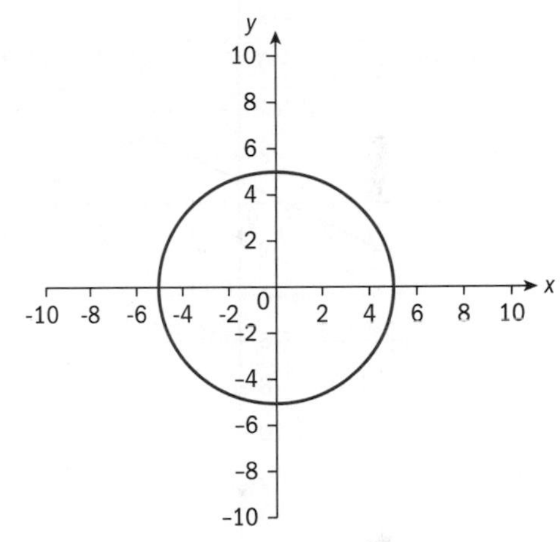

3 For each of these functions defined by their graphs:

i Determine whether it is many-to-one or one-to-one.

ii Write down the domain.

iii Write down the range.

a

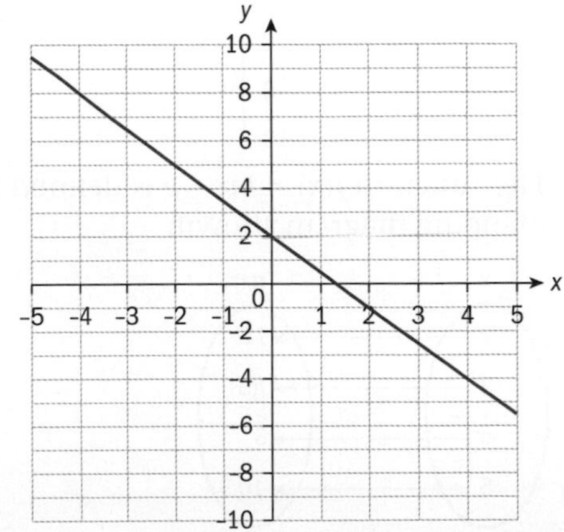

b

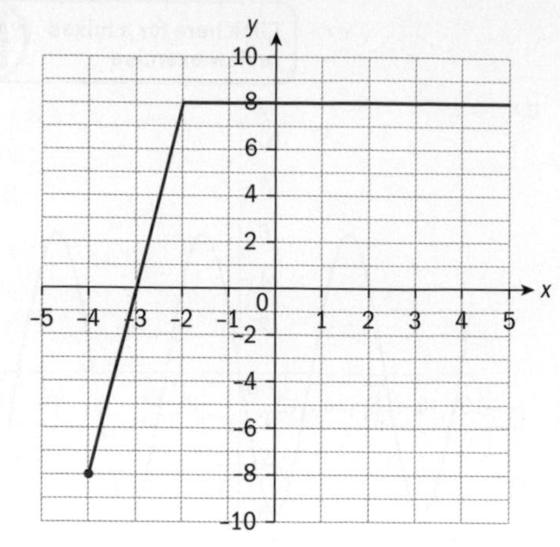

c

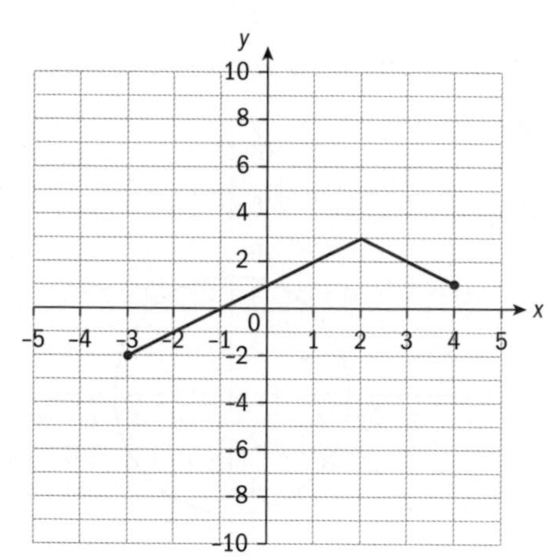

d

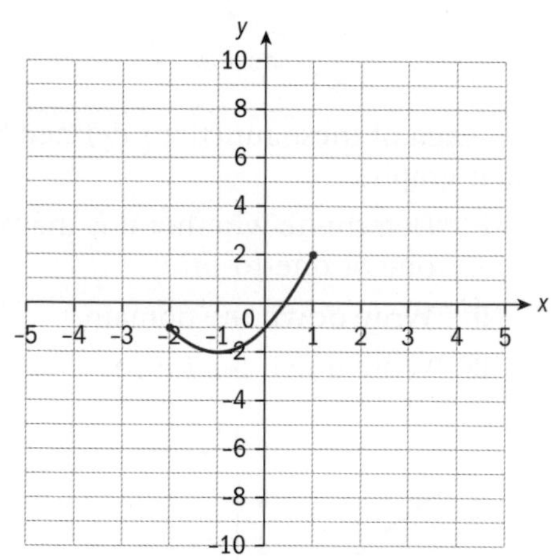

4 The function $f(x) = mx + b$ is defined by the mapping diagram shown.

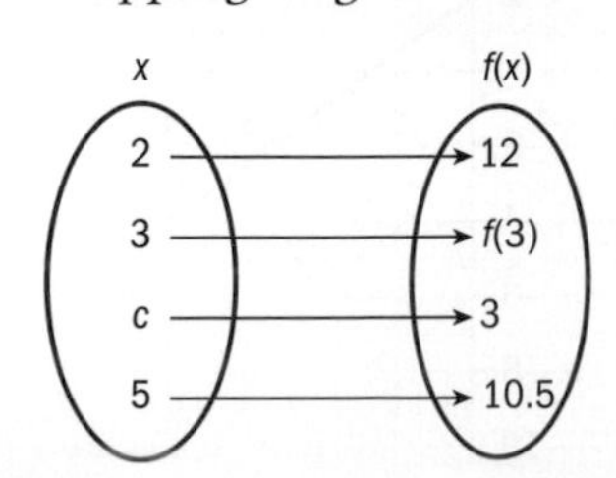

a Find the values of m and b.

b Find the value of $f(3)$.

c Find the value of c.

5 Consider the function $f(x) = -2x + 5$ where $-3 \leq x \leq 4$.

a Write down the domain.

b Find $f(-3)$.

c Find x if $f(x) = 4$.

d Sketch the graph of this function.

e Hence, find the range.

6 Find the domain of the function $y = 3 - x$ if it is known that the range is the set $\{-2, -1, 0, 0.5, 3\}$.

7 Consider the function $f(x) = 0.5x - 3$.

a Write down the domain of f.

b Sketch the graph of f. Mark clearly on your sketch the axis intercepts.

c Hence, find the range of f.

d Determine whether or not these points lie on the graph of f:

i $A(1, -2.5)$

ii $B(100, 46)$

8 The diagram shows the graph of a function $y = f(x)$.

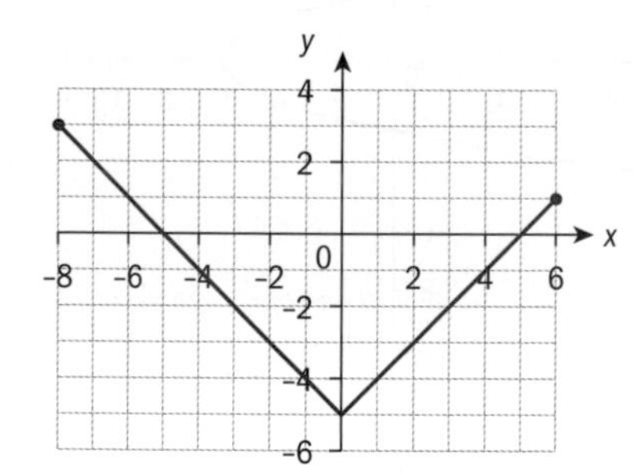

a Write down the values of:

i $f(4)$ **ii** $f(-6)$ **iii** $f(6)$

b Find x if:

i $f(x) = -1$ **ii** $f(x) = -5$ **iii** $f(x) = 2$

c The point $A(t, 3)$ lies on the graph of f. Write down the value of t.

d Find the set of values of x for which $f(x) < 0$.

9 In a certain city a travel card has a value of \$150. Every time the card is used, \$2.50 is deducted from the card.

a Write down a formula for $T(x)$, the amount of money remaining on the card after it has been used x times.

b Find $T(6)$ and explain its meaning.

c Find the number of times the card has been used when it has \$105 remaining.

d Find:

i the domain of the function T

ii the range of the function T.

10 Ezequiel walks from the park to home. After 7 minutes, he is 200 metres from home and after 10 minutes he is 80 metres from home. Assume that Ezequiel is walking at a constant speed.

a Find a model for the distance, d, from where Ezequiel is to home in terms of the time he has been walking, t.

b Determine how far it is between the park and Ezequiel's home.

c Determine how long it takes Ezequiel to get to home from the park.

d Sketch the graph of d against t for a suitable domain.

11 A linear function f is such that $f(2) = 0$ and $f(-2) = -1$.

a Find the formula for the function f.

b Find $f(5)$.

12 The exchange rate from Pounds Sterling (UK£) to Australian Dollars (AUD) is UK£1 = AUD1.73.

a Find the number of AUD equivalent to UK£70.

b Sketch a conversion graph for UK£ to AUD.

c Find a model $a(p)$, where a is the amount of money in AUD corresponding to UK£p.

d Find $a(100)$ and explain its meaning.

e Find $a^{-1}(50)$ and explain its meaning.

13 Find the intersection point of the graphs of the functions $f(x) = -x + 5$ and $g(x) = 0.5x + 4$.

14 Determine which of these functions have an inverse function. Give a reason.

a $f(x) = x^2$ **b** $f(x) = 5x + 1$

c $f(x) = 3$ **d** $f(x) = \sqrt{x}$

15 A car leaves Sun City and takes the road towards Mars City, which is 200 km away from Sun City. The car moves at a constant speed of 70 km per hour. At the same time a truck leaves Mars City on the same road but in the opposite direction. The truck moves at a constant speed of 60 km per hour.

a Find an equation for the distance from the car to Sun City, c km, in terms of the time x in hours since the car left Sun City.

b Find an equation for the distance from the truck to Sun City, t km, in terms of the time x in hours since the truck left Mars city.

c Find how long it takes for the car and the truck to be at the same distance from Sun City. Find this distance.

16 The diagram shows the graph of the function f.

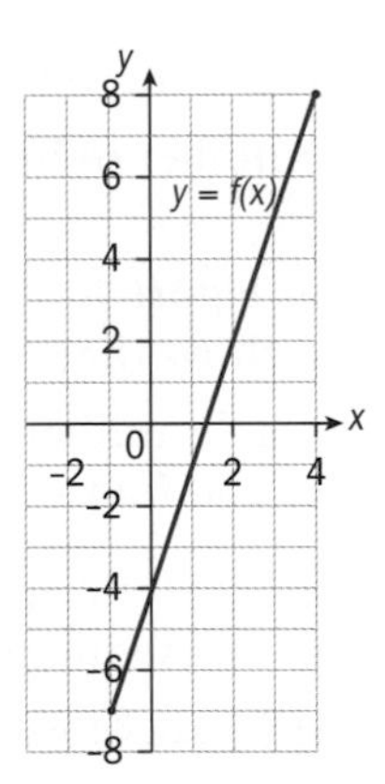

Use the graph to answer the following questions.

a Find the values of:

i $f(4)$ **ii** $f^{-1}(4)$

b Solve the equations:

i $f(x) = -7$ **ii** $f^{-1}(x) = 1$

c Write down the domain and the range of the function $f^{-1}(x)$.

d Draw the graph of $y = f^{-1}(x)$.

17 The first term of an arithmetic sequence is $u_1 = 5$ and the common difference is $d = 2.5$.

a Find u_2 and u_3.

b Write down the general term u_n of this sequence.

c Find the the smallest value of n for which $u_n > 377$.

18 Consider the arithmetic series $S = 20 + 20.4 + 20.8 + \ldots + 37.6$.

a Find the number of terms in this series.

b Find the value of S.

19 a Write down an expression for S_n, the sum of the first n terms of the arithmetic series

$105 + 100 + 95 + \ldots$

b Hence, find the value of n for which $S_n = 1140$.

20 Consider this sequence of figures.

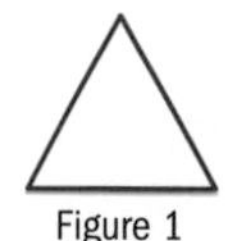
Figure 1

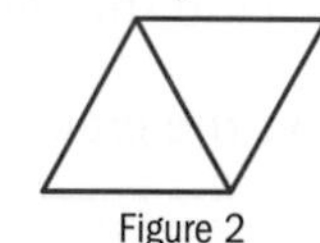
Figure 2

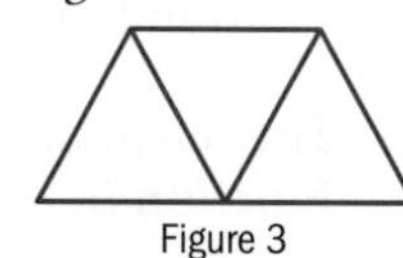
Figure 3

Figure 1 contains three line segments.

a Find the number of line segments in Figure 7.

b Given that Figure n contains 251 line segments, show that $n = 125$.

c Find the total number of line segments in the first 100 figures.

21 A sequence is given by $a_n = -10n + 3$.

a Show that this sequence is arithmetic.

b Find the smallest k for which $a_k < -1000$.

22 Deby borrows UK£1800 and wants to pay it back over three years. She has two options:

Option 1: Pay simple interest of r% pa.

Option 2: Pay interest at a flat rate of 18% of the borrowed amount.

a Determine how much interest Deby would have to pay back if she chose option 2.

b Find the value of r that makes Deby pay back the same amount for either option.

23 Transfers from the airport to a passenger's house have different prices. Here is a table with some journey lengths (in km) and their respective prices ($).

Journey length (km)	10	25	36
Price ($)	450	810	1200

a Plot a graph of this data on your GDC.

b Choose an appropriate model for the price of the journey.

c Hence find an equation for the model. Comment on your model.

d Use your model to estimate the price of the journey when the passenger lives

i 27 km from the airport

ii 40 km from the airport.

e Comment on the appropriateness of your predictions from part **d**.

Exam-style questions

24 P1: Find the range of each of the following functions.

a $f(x) = 5x + 1$, domain $\{x \in \mathbb{R} : -5 \le x \le 5\}$ (2 marks)

b $f(x) = 4 - 2x$, domain $\{x = -1, 0, 1, 2, 3, 4\}$ (2 marks)

c $f(x) = x^2$, domain $\{x \in \mathbb{R} : 0 \le x \le 10\}$ (2 marks)

d $f(x) = 250 - 12.5x$, domain $\{x \in \mathbb{R} : 0 \le x \le 10\}$ (2 marks)

25 P1: A taxi company's charge for a single taxi journey can be calculated using the formula $C = 2.8 + 1.6d$, where C is the cost (in pounds) and d is the distance travelled (in km).

a Draw a graph of distance against cost for any journey up to 8 km. (3 marks)

b State what the value 2.8 represents in the context of the question. (1 mark)

c State what the value 1.6 represents in the context of the question. (1 mark)

d State the range of the graph you have drawn. (2 marks)

e Using your graph, or otherwise, find how far a passenger could travel by spending £10. (2 marks)

26 P1: The nth term, u_n, of an arithmetic sequence is given by the formula $u_n = 23 + 7n$.

a State the first term of the sequence. (1 mark)

b Find the 50th term of the sequence. (2 marks)

c Explain why 1007 is not a term in this sequence. (3 marks)

27 P2: Denise organises a party for her work colleagues. The cost to rent a local hall is $430 for the evening. She then has to budget for food, which will cost approximately $14.50 per person.

a Write down a formula connecting the total cost of the evening ($$C$) with the number of people attending (P). (2 marks)

b Find the total cost for the evening if 25 people attend. (2 marks)

c Given that Denise has a maximum budget of $1000, find the greatest number of people she is able to invite. (3 marks)

d In the end, only 16 people will attend. Calculate how much each person should be charged so that Denise covers her costs. (3 marks)

28 P1: A function $f(x)$ is defined by $f(x) = 3x - 10$.

a Given that the range of $f(x)$ is $5 < f(x) < 50$, find the domain of $f(x)$. (4 marks)

b State the range of the inverse function. (2 marks)

29 P1: Zoran decides to go into business selling comic books. He decides to pay a local company to design a website for him at a one-off cost of $850. With this website, he hopes to earn approximately $1200 per month. Zoran has no other costs involved.

a Write down a formula connecting Zoran's total profits ($$P$) with the time ($T$ months) after he starts his business. (2 marks)

b Using this model, determine (to the nearest whole month) how long it will take for Zoran to make a total profit of $5000. (3 marks)

30 P1: One month, a currency exchange rate showed that €1 (euro) = $1.23 (US dollars).

a Draw a graph of the currency exchange rate, with euros along the x-axis and US dollars along the y-axis. The range of your graph should go up to approximately $100. (3 marks)

b Using the graph, find (approximately) how much $75 is worth in euros. (2 marks)

c Using the graph, find (approximately) how much €30 is worth in US dollars. (2 marks)

31 P1: State which of the following graphs represent functions, giving reasons for your answers.

a

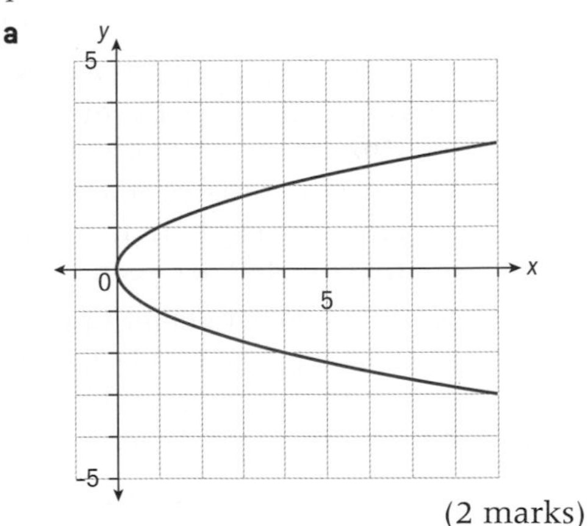

(2 marks)

b

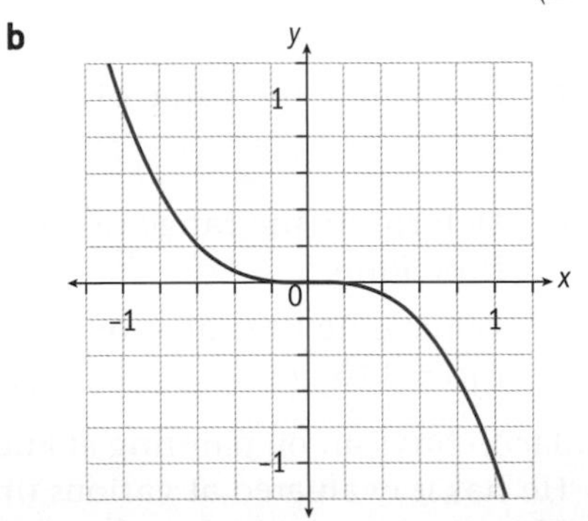

(2 marks)

c

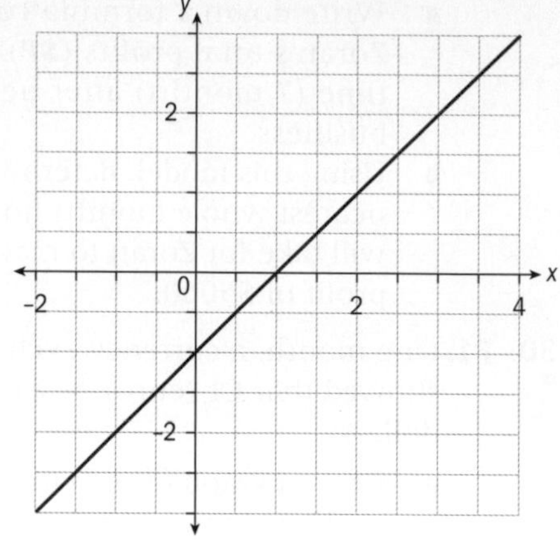

(2 marks)

d

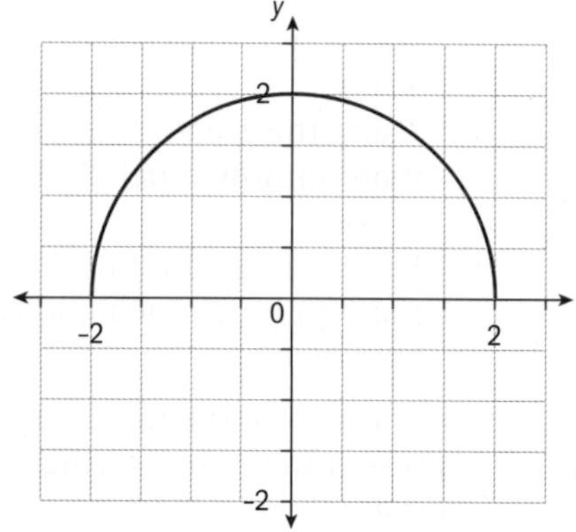

(2 marks)

32 P1: Ben buys a new car at the same time as Charlotte. Ben pays £25 600 for his car and Charlotte pays £18 000 for hers. Ben's car depreciates at a rate of £1150 per year, while Charlotte's depreciates at a rate of £480 per year.

Use technology to determine how long it will take for Ben and Charlotte's cars to have the same value. Find the value of their cars at this time. (7 marks)

33 P1: A function is given by $f(x) = 128x - 15$, $-3 < x < 15$.

a Determine the value of $f\left(\frac{3}{2}\right)$. (2 marks)

b Determine the range of the function f. (4 marks)

c Determine the value of a such that $f(a) = 1162.6$. (2 marks)

34 P1: Jacob buys an oil painting at auction. He has it evaluated at various times after the initial purchase.

After 5 years, the painting is worth £1575.

After 12 years, the painting is worth £2100.

After 20 years, the painting is worth £2700.

One model for the value (£V) of the painting after t years is given by $V = a + bt$, where a and b are constants.

a Use technology to determine the price Jacob paid for the painting. (2 marks)

b Determine the values of the constants a and b. (3 marks)

c Hence, calculate how much the painting should be worth after 50 years. (2 marks)

d Explain whether you think this model is realistic or not, giving your reasons. (2 marks)

35 P1: State:

i the domain, and **ii** the range for each of the following functions.

a

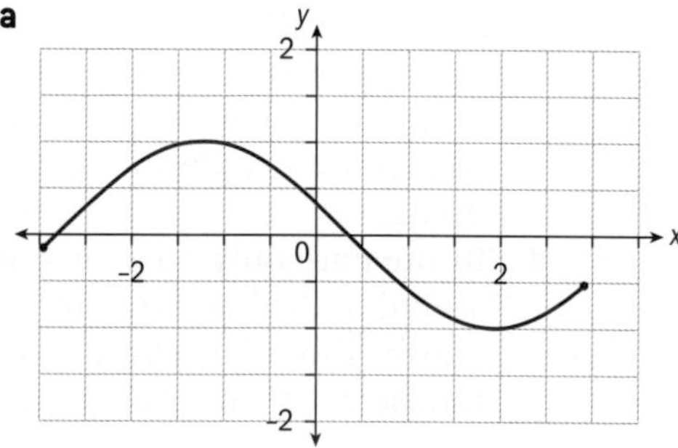

(2 marks)

b

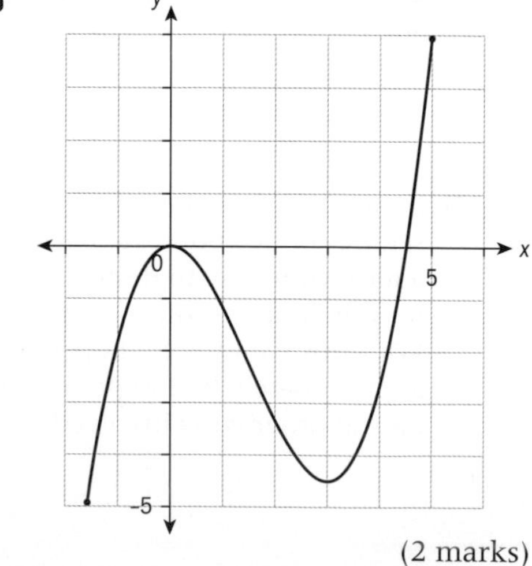

(2 marks)

c

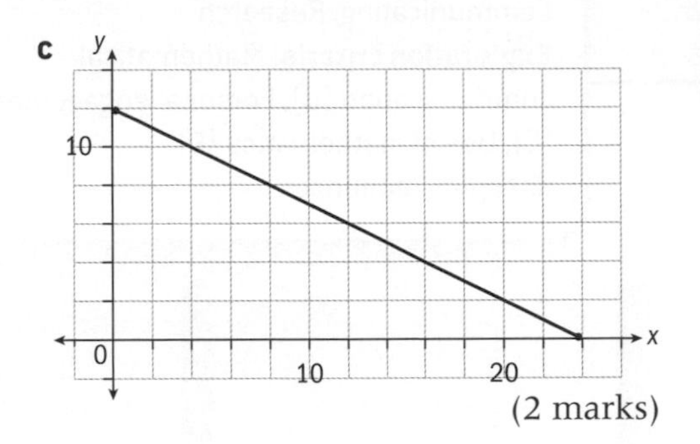

(2 marks)

d

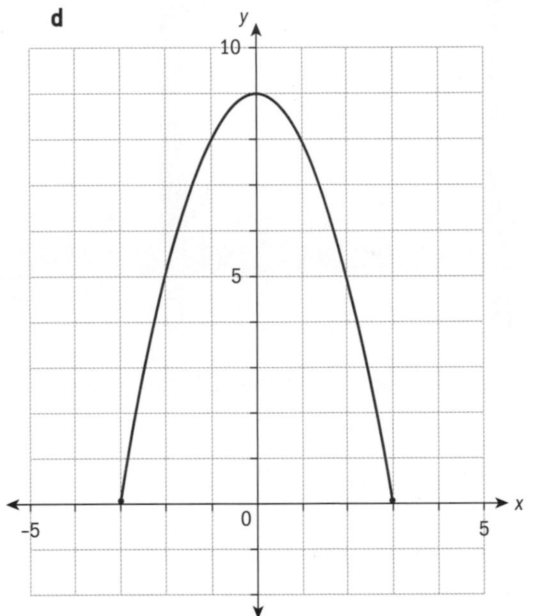

(2 marks)

36 P1: The dose D (in milligrams, mg) of a certain medicine for a child of mass m kg is determined by the equation $D = am + b$, where a and b are constants.

For a child of mass 20 kg the correct dose is 115 mg, and for a child of mass 25 kg the correct dose is 137.5 mg.

a Determine the value of a and the value of b. (3 marks)

b Find the correct dose that should be given to a child with mass 23 kg. (2 marks)

c Suggest reason(s) why the equation may not be appropriate for determining the dosage required for an adult with mass 100 kg. (2 marks)

37 P2: The temperature conversion formula to change degrees Celsius (C) to degrees Fahrenheit (F) is given by $F = \frac{9C}{5} + 32$.

a Find the temperature in °F that is equivalent to a temperature of 0°C. (1 mark)

b Determine which temperature is the same when measured in both degrees Fahrenheit and degrees Celsius. (3 marks)

c Find a formula that can be used to change degrees Fahrenheit to degrees Celsius. (3 marks)

d In a given year, the temperatures on the island of Menorca ranged from 46.4°F to 84.2°F. Find the equivalent range of temperatures in degrees Celsius. (4 marks)

38 P1: The price of renting a car (£C) from "Cars-R-Us" for d days is given by the formula $C = 30 + 12.5d$.

The price of renting a car (£C) from "Car-nage" for d days is given by the formula $C = 70 + 8.35d$.

Abel wishes to rent a car for the duration of his holiday.

He decides to rent from "Car-nage" as it will be cheaper for him.

a What is the minimum length of Abel's holiday? (3 marks)

b If Abel's holiday is between 14 and 21 days, find an inequality which shows the range in which Abel's car rental bill will lie. (4 marks)

The Towers of Hanoi

Approaches to learning: Thinking skills, Communicating, Research
Exploration criteria: Mathematical communication (B), Personal engagement (C), Use of mathematics (E)
IB topic: Sequences

The problem

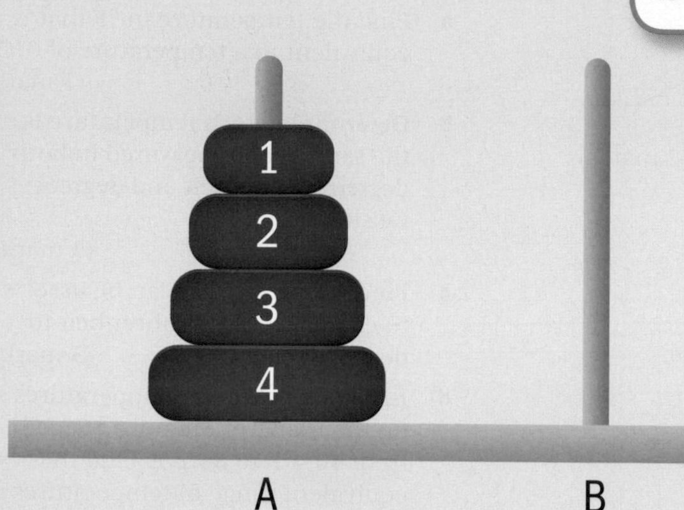

The aim of the **Towers of Hanoi problem** is to move all the disks from peg A to peg C following these rules:

1. Move only one disk at a time.
2. A larger disk may not be placed on top of a smaller disk.
3. All disks, except the one being moved, must be on a peg.

For 64 disks, what is the **minimum** number of moves needed to complete the problem?

Explore the problem

Use an online simulation to explore the Towers of Hanoi problem for 3 and 4 disks.

What is the minimum number of moves needed in each case?

Solving the problem for 64 disks would be very time-consuming, so you need to look for a rule for n disks that you can then apply to the problem with 64 disks.

Try and test a rule

Assume that the minimum number of moves follows an arithmetic sequence.

Use the minimum number of moves for 3 and 4 disks to predict the minimum number of moves for 5 disks.

Check your prediction using the simulator.

Does the minimum number of moves follow an arithmetic sequence?

Find more results

Use the simulator to write down the number of moves when $n = 1$ and $n = 2$.

Organize your results so far in a table.

Look for a pattern. If necessary, extend your table to more values of n.

Try a formula

Return to the problem with 4 disks.

Consider this image of a partial solution to the problem. The large disk on peg A has not yet been moved.

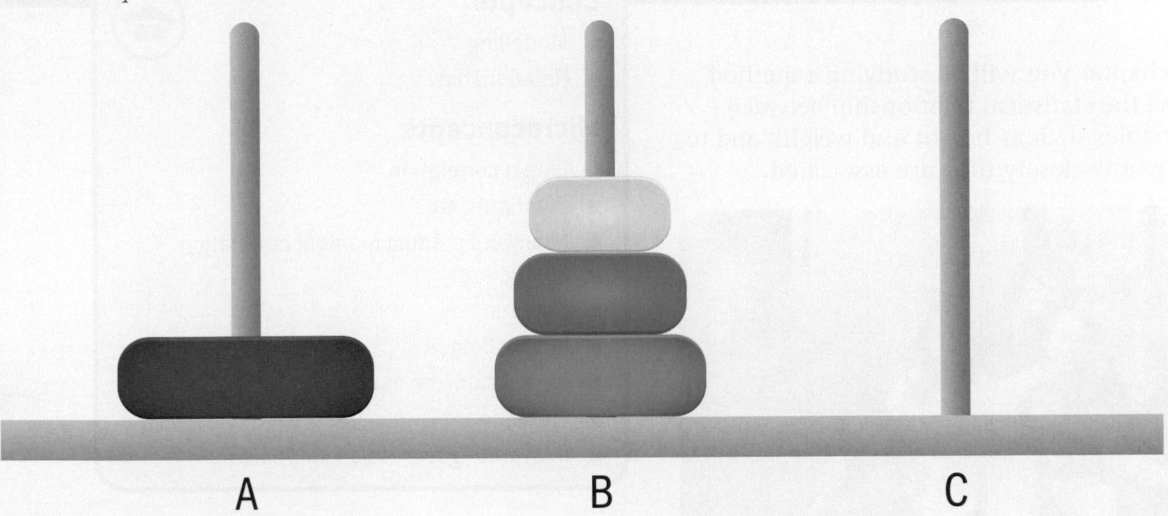

Consider your previous answers.

What is the minimum possible number of moves made so far?

How many moves would it then take to move the largest disk from peg A to peg C?

When the large disk is on peg C, how many moves would it then take to move the 3 smaller disks from peg B to peg C?

How many total moves are therefore needed to complete this puzzle?

Use your answers to these questions to write a formula for the minimum number of moves needed to complete this puzzle with n disks.

This is an example of a **recursive formula**. What does that mean?

How can you check whether your recursive formula works?

What is the problem with a recursive formula?

Try another formula

You can also try to solve the problem by finding an **explicit formula** that does not depend on you already knowing the previous minimum number.

You already know that the relationship is not arithmetic.

How can you tell that the relationship is not geometric?

Look for a pattern for the minimum number of moves in the table you constructed previously.

Hence write down a formula for the minimum number of moves in terms of n.

Use your explicit formula to solve the problem with 64 disks.

Extension

- What would a solution look like for 4 pegs? Does the problem become harder or easier?
- Research the "bicolor" and "magnetic" versions of the Towers of Hanoi puzzle.
- Can you find an explicit formula for other recursive formulae (eg Fibonacci)?

6 Modelling relationships: linear correlation of bivariate data

In this chapter you will be studying a method to model the statistical relationship between two variables such as height and weight, and to quantify how closely they are associated.

Concepts

- Modelling
- Relationship

Microconcepts

- Linear correlation
- Bivariate data
- Pearson's product moment correlation coefficient
- Outliers
- Line of best fit
- Regression line equation
- Prediction
- Interpreting regression coefficients

In general, is there a relationship between the number of children in a family and the family's income?

Do you have to be tall to be a good basketball player?

If a student does well in mathematics, does that mean they will do well in another subject?

In this city there is a strong correlation between the number of ice creams sold during a week and the amount of rubbish. Does this mean that the number of ice creams sold is the cause of the increased rubbish?

The table shows the total number of vaccine doses specified for infants aged less than one year along with the infant mortality rate (IMR; infant deaths per thousand) for a group of countries in 2009.

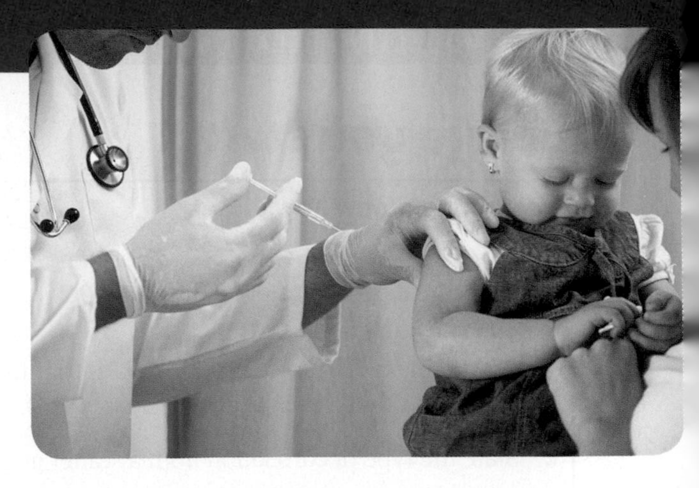

Country	Total number of vaccine doses	IMR	Country	Total number of vaccine doses	IMR
Sweden	12	2.75	France	19	3.33
Japan	12	2.79	Czech Republic	19	3.79
Iceland	12	3.23	Belgium	19	4.44
Norway	12	3.58	United Kingdom	19	4.85
Denmark	12	4.34	Spain	20	4.21
Finland	13	3.47	Portugal	21	4.78
Malta	15	3.75	Luxemburg	22	4.56
Slovenia	15	4.25	Cuba	22	5.82
South Korea	15	4.26	Austria	23	4.42
Singapore	17	2.31	Ireland	23	5.05
New Zealand	17	4.92	Greece	23	5.16
Germany	18	3.99	Netherlands	24	4.73
Switzerland	18	4.18	Canada	24	5.04
Israel	18	4.22	Australia	24	4.75
Italy	18	5.51	United States	26	6.22

Data from https://chemtrailsevilla.files.wordpress.com/2011/06/54772978-infant-mortality-rates-regress-w-higher-no-of-vaccine-doses.pdf

What do the figures in the second and fifth columns tell you?

What do the figures in the third and sixth columns tell you?

What can you say about the United States in particular?

What do you expect the relationship to be between the total number of vaccine doses specified and the IMR? Why?

Plot the points on a graph, with the number of vaccine doses specified on the horizontal axis and the IMR on the vertical axis. Comment on the correlation.

What do you wonder?

Developing inquiry skills

Write down any similar inquiry questions you could ask to investigate the relationship between another pair of variables, such as IMR and family income, or IMR and family size. What information would you need to find?

Think about the questions in this opening problem and answer any you can. As you work through the chapter, you will gain mathematical knowledge and skills that will help you to answer them all.

Before you start

You should know how to:

1 Interpret the gradient of a straight line.

What is the gradient of the line $y = -0.25x + 30$ and what does this mean?

The gradient is $m = -0.25$. This means that for every unit that x increases, y decreases by 0.25.

2 State the type of correlation from a scatter graph.

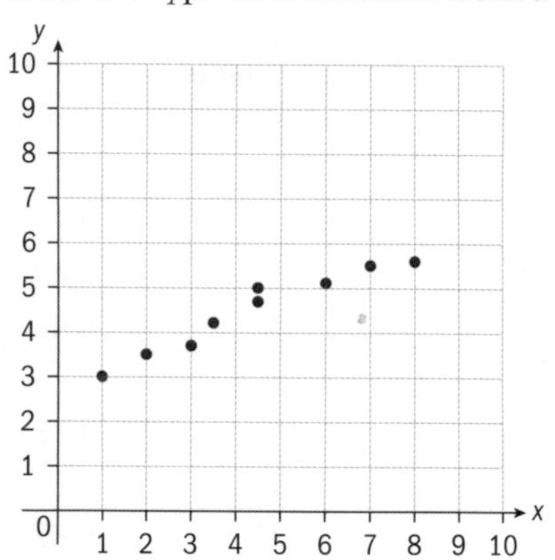

Strong and positive correlation.

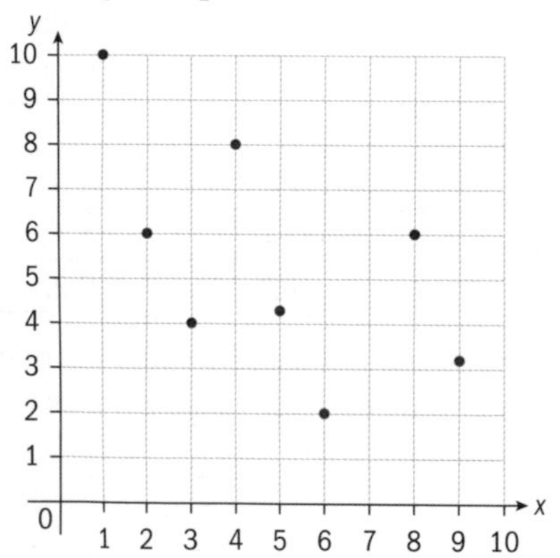

Weak and negative correlation.

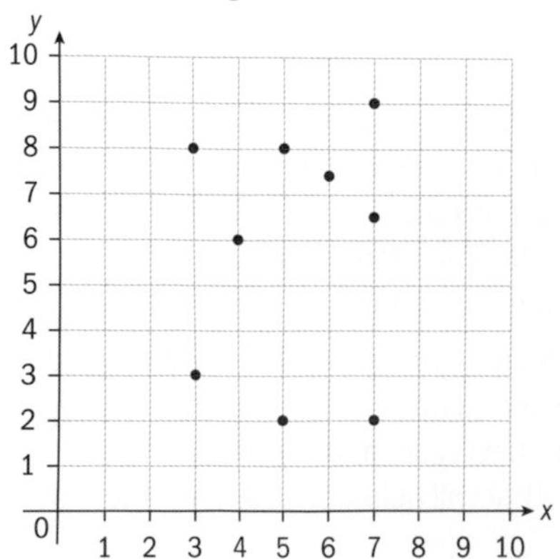

No correlation.

Skills check

Click here for help with this skills check

1 Interpret the gradient of each of these straight lines.

a $y = 3x - 20$

b $y = -\frac{1}{2}x + 100$

2 State the type of correlation shown in each of these scatter graphs.

a

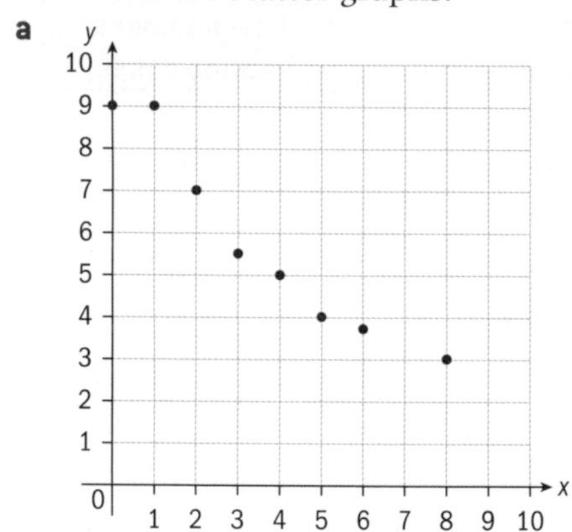

b

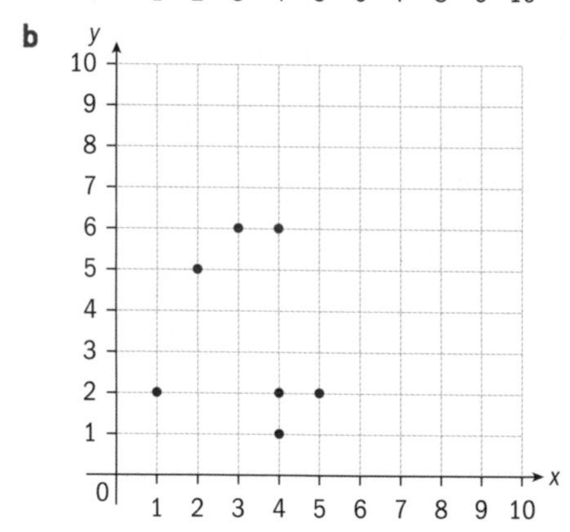

c

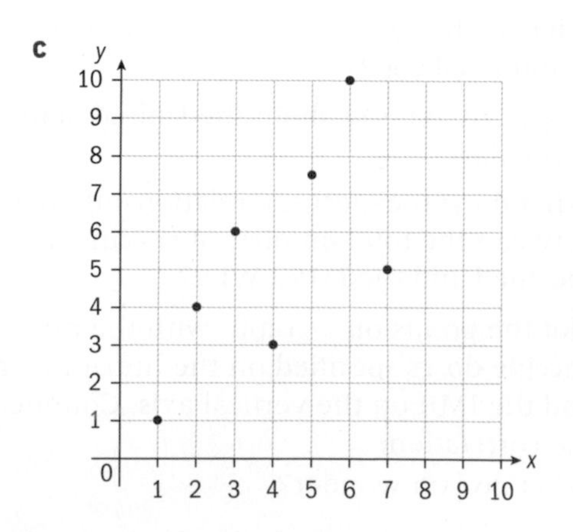

6.1 Measuring correlation

A **functional relationship** exists between variables when there is a formula (or function) that relates one variable to the other, so that all the data points lie on the same line or curve (the graph of the formula).

A **statistical relationship** means that there is not such a direct relationship, but the data points, when plotted, may indicate a tendency to be close to some line or curve.

International-mindedness

In 1956, Australian statistician, Oliver Lancaster made the first convincing case for a link between exposure to sunlight and skin cancer using statistical tools including correlation and regression.

Investigation 1

1 In each of the following situations, investigate whether there is a functional relationship or a statistical relationship between the pair of variables.

- **a** The gross domestic product (GDP) and the literacy rate in a country over a period of ten years.
- **b** The electricity consumed in a house and the cost of the electricity bill for the month of February.
- **c** The number of electrical appliances in a house and the cost of the electricity bill for a particular month.
- **d** The number of matches won in a football league and the number of points accumulated.
- **e** The number of matches played in a football league and number of points accumulated.
- **f** The side length of a square and area of the square.

2 **a** Write down two examples of a pair of variables that have a functional relationship.

b Write down two examples of a pair of variables that have a statistical relationship.

3 For each of the situations given in part **1**, if you were to plot a set of data points in a scatter graph, how would you expect the data points to be positioned if there was a functional relationship? And if there was a statistical relationship?

4 **Conceptual** How can real-life situations be modelled?

Reflect How is correlation different from a functional relationship?

HINT

Remember from Chapter 3 that correlation can be classified in terms of its direction (positive or negative) and its strength (zero, weak, moderate or strong).

It is possible to classify a correlation in terms of the form shown by the points in a scatter graph. If the data points all lie close to a line, then the correlation is said to be **linear**.

This is an example of **linear** correlation. It might, for example, represent the relationship between the ages of the runners in a race and their times for the race.

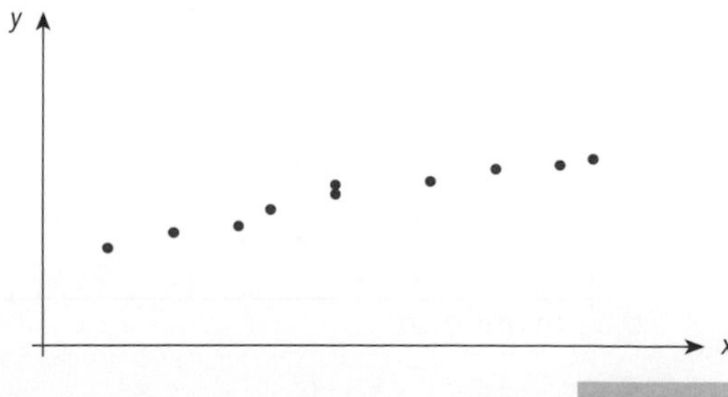

Statistics and probability

This is an example of **non-linear** correlation. It might, for example, show the number of germ cells in a dish in a laboratory as time passes.

If the data points all lie on a straight line, then the correlation is said to be a **perfect linear** correlation.

This is an example of a perfect linear correlation.

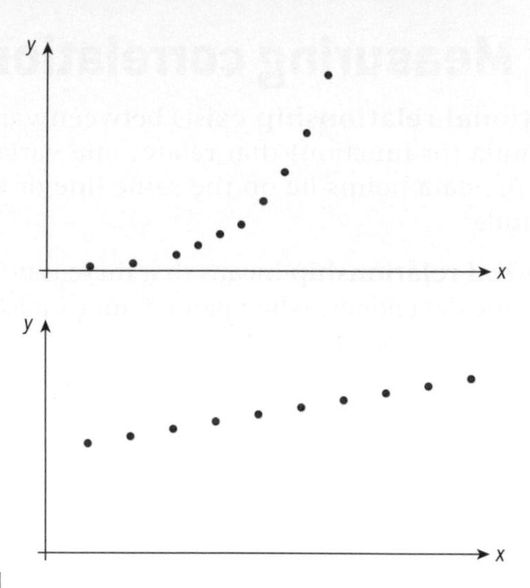

In perfect correlation, the data points lie on the graph of a function or their coordinates are linked through a formula.

This is an example of **perfect non-linear** correlation. It might, for example, show the relationship between horizontal distance and height when a ball is thrown in the air.

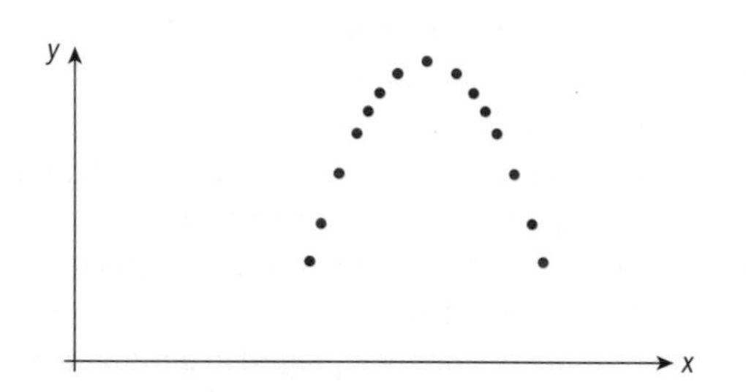

Reflect What is perfect correlation?

What is perfect linear correlation?

How does a scatter graph tell you the type of correlation?

Exercise 6A

1 For each of the situations described in Investigation 1 part **1**, state whether there is:

i correlation

ii linear correlation.

2 For each diagram, state (if possible) the direction (positive/negative), form (linear/non-linear) and strength (strong/moderate/weak/zero) of the relationship.

a

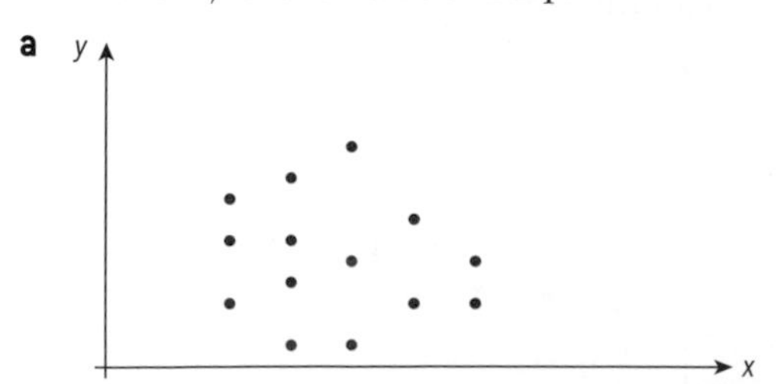

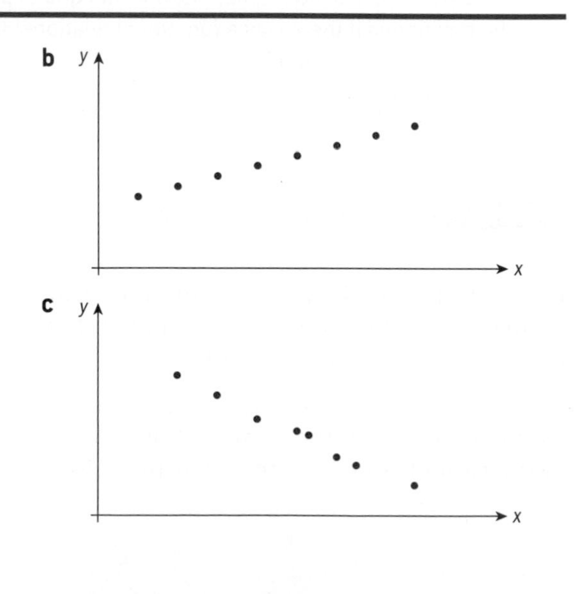

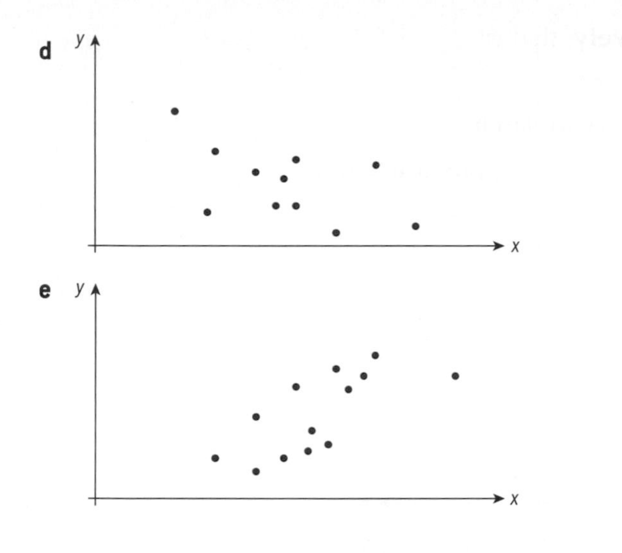

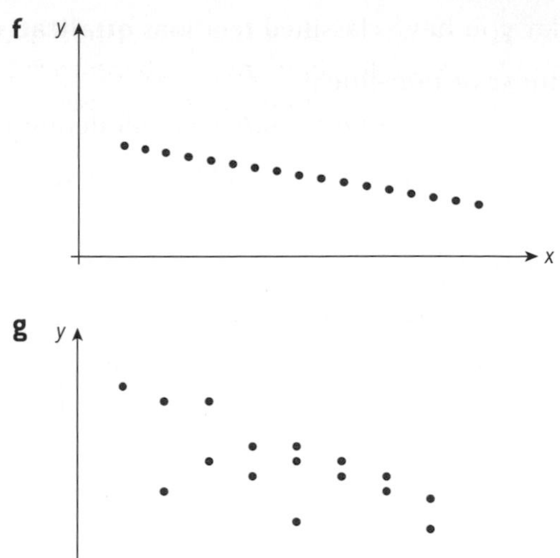

Remember from Chapter 3 that even when there appears to be a correlation between two variables, this does not necessarily mean that one variable **causes** the other. In other words, correlation does not imply causation.

Example 1

State whether the following statements are true or false.

a The IB grades achieved by a student are an effect of the predicted grades given by their teachers.

b The height of a person is a cause of their weight.

c The number of ice creams sold is a cause of high temperatures.

d Smoking is a cause of lung cancer.

a False	The predicted grades may well be correlated with the final grades, but that does not mean that the predicted grade causes the final grade.
b False	There are many other factors in addition to height that affect weight.
c False	The two are correlated, but the causation is likely to be in the other direction.
d True	This has been shown to be true independent of the correlation between the two variables.

TOK

What is the difference between correlation and causation?

To what extent do these different processes affect the validity of the knowledge obtained?

So far you have classified relations **qualitatively**, that is:

- linear or non-linear
- positive/negative/weak/strong/moderate/no correlation.

Now you will study a method to classify the correlation **quantitatively**.

Investigation 2

Consider these scatter graphs.

A

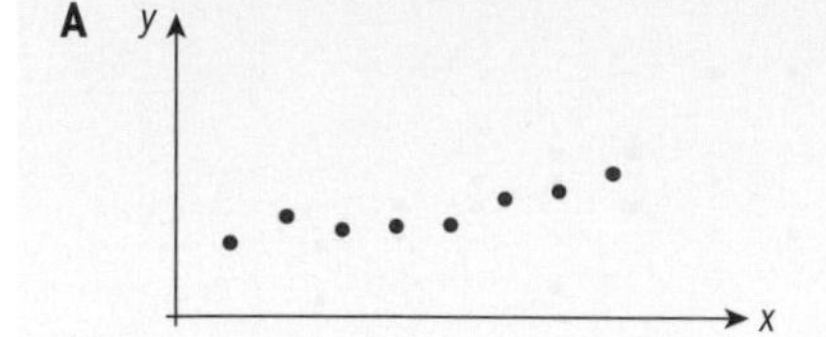

B

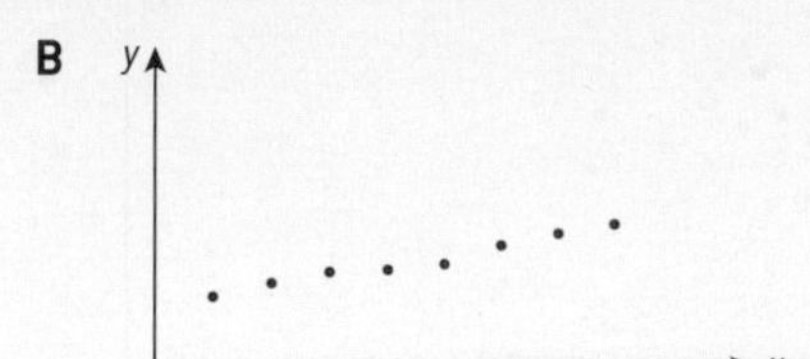

C

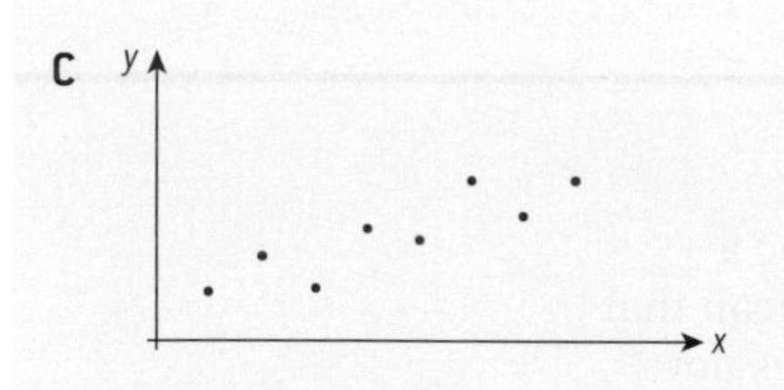

1 Explain why the correlation shown in each of these scatter graphs can be classified as linear.

2 Describe the strength of correlation in each scatter graph. What do you notice?

3 How would you distinguish the strength of the correlation in the three situations?

Suppose there was a number between −1 and 1 that describes the direction and strength of the correlation. It could be positive or negative according to the direction, and nearer to 1 or nearer to 0 according to the strength.

4 Match the following values of this number to the graphs at the start of the investigation using your judgment of the correlation. Note that one of the numbers does not match any of the graphs.

a 0.8 **b** −0.9 **c** 0.9 **d** 0.85

There are several ways to classify the strength of a correlation numerically. One of these is **Pearson's product moment correlation coefficient.**

Pearson's product moment correlation coefficient, r, is a measure of the **linear** correlation between two variables x and y, and can take a value between −1 and 1 inclusive.

Here are some examples of data sets and their r values:

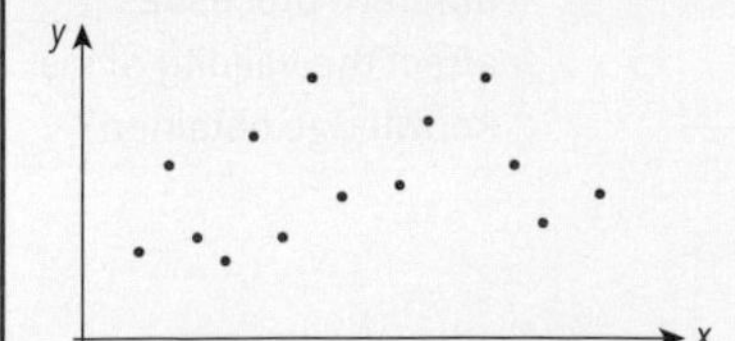

Weak positive linear correlation: $r = 0.3$

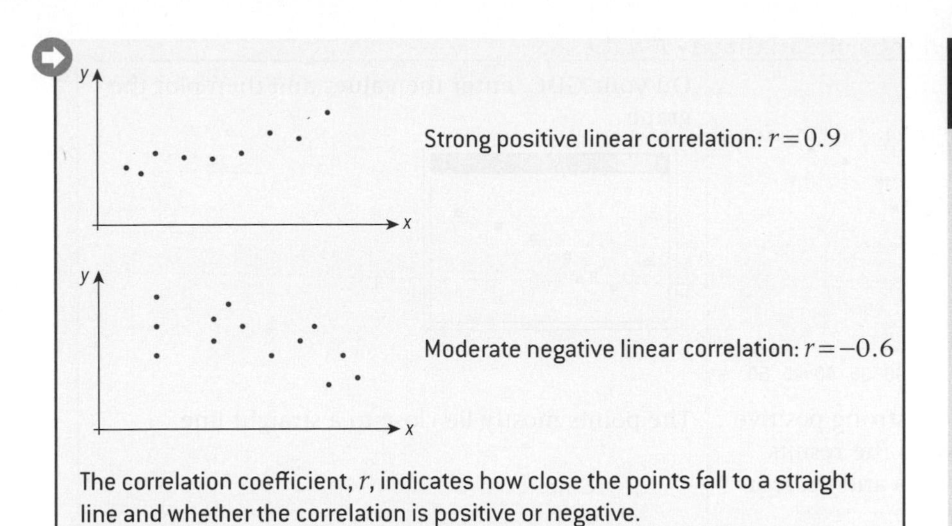

The correlation coefficient, r, indicates how close the points fall to a straight line and whether the correlation is positive or negative.

Reflect How can you quantitatively describe a relationship shown in a scatter plot?

International mindedness

Karl Pearson (1857-1936) was an English lawyer and mathematician. His contributions to statistics include the product-moment correlation coefficient and the chi-squared test.

He founded the world's first university statistics department at the University College of London in 1911

The formula to find Pearson's product moment correlation coefficient is

$r = \frac{s_{xy}}{s_x s_y}$, where s_{xy} is the covariance (a measure of how two variables vary together) and s_x and s_y are the standard deviations of x and y respectively.

$$s_{xy} = \sum xy - \frac{(\sum x)(\sum y)}{n}, \; s_x = \sqrt{\sum x^2 - \frac{(\sum x)^2}{n}} \text{ and } s_y = \sqrt{\sum y^2 - \frac{(\sum y)^2}{n}}.$$

EXAM HINT

In examinations you will only be expected to use technology to find the value of r. You do not need to learn these formulae; they are presented here to help you explore how things work.

Note that:

- When finding r, the order of the variables is not important.
- The correlation coefficient, r, has no units.

Example 2

The table gives test results in mathematics and biology for eight students.

Mathematics (x)	20	25	28	30	32	37	42	48
Biology (y)	24	20	22	21	25	28	30	32

a Plot the data points on a set of axes.

b Comment on any correlation you can see on the graph.

c Find Pearson's product moment correlation coefficient for this data and compare it with your answer to part **b**.

Continued on next page

Statistics and probability

a

On your GDC, enter the values and then plot the graph.

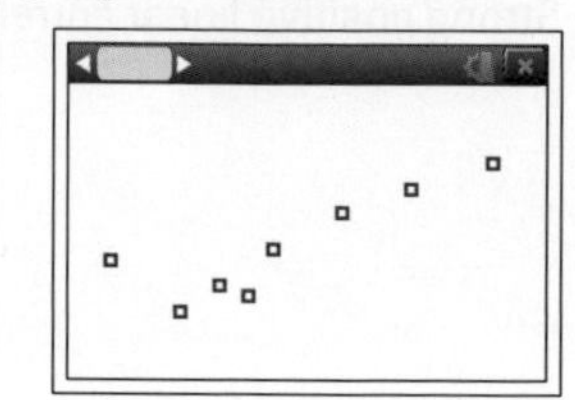

b The graph shows a strong positive correlation between the results in mathematics tests and those in biology.

The points mostly lie close to a straight line.

c $r = 0.864$ (3 sf)

This certainly shows a strong positive correlation.

Use your GDC to find the value of r.

Exercise 6B

1 For each scatter plot, state whether you would use Pearson's product moment correlation coefficient to measure the strength of the correlation.

a

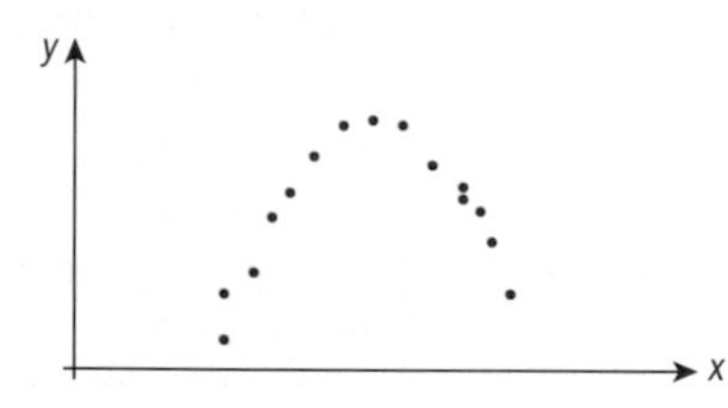

b

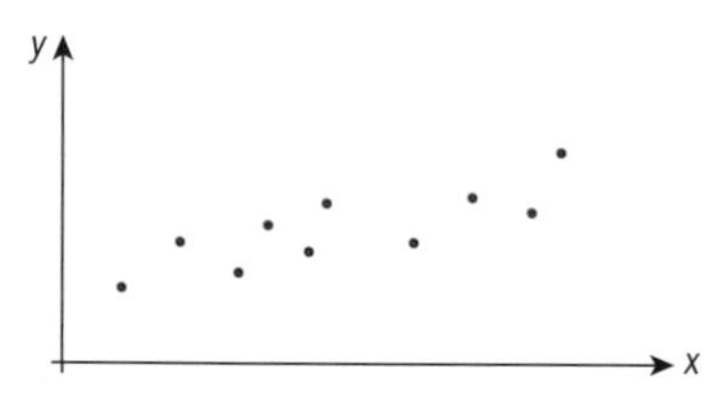

c

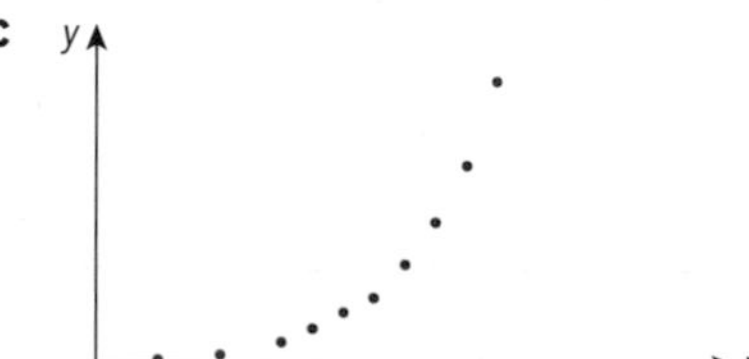

2 A small study involving 10 children is conducted to investigate the correlation between gestational age at birth (in weeks) and birth weight (in grams). The results are shown in the table.

Age at birth (weeks)	34.6	36	39.3	42.4	40.3	41.4	39.7	41.1	37	42.1
Birth weight (g)	1895	2028	2837	3826	3258	3660	3350	3300	3000	3900

a Plot the data points on a scatter graph. Label the axes.

b Find Pearson's product moment correlation coefficient, r.

c Comment on the correlation.

3 A biologist is studying the relationship between height above sea level and the numbers of certain species of plant at that particular height over an area of 100 m². The table shows the information collected.

Height (metres)	0	100	200	450	500	700	900	1000
Number of plants	1	2	5	8	8	10	12	13

a Plot the data points on a scatter graph. Label the axes.

b Find Pearson's product moment correlation coefficient, r.

c Comment on the correlation.

4 The heights (in metres) and the weights (in kilograms) of 10 basketball players are given in the table.

Height (m)	1.98	2.11	2.06	2.08	2.13	1.96	1.93	2.02	1.83	1.98
Weight (kg)	93	117.9	104.3	95.3	113.4	84	86.2	99.8	83.9	97.5

a Plot the data points on a scatter graph. Label the axes.

b Find Pearson's product moment correlation coefficient, r.

c Comment on the relationship between the coefficient and the graph.

Investigation 3

1 Consider these four data sets.

Set 1

x	2	3	4	5	6	7	8	9
y	3	6.5	4	7.5	5	5	7	13

Set 2

x	2	3	4	5	6	7	8	9
y	4	6	5	7	5.5	7	7.5	9

Set 3

x	2	3	4	5	6	7	8	9
y	4.5	5.8	5.5	6.8	6.5	7.2	7.7	8.6

Set 4

x	2	3	4	5	6	7	8	9
y	5	5.5	6	6.5	7	7.5	8	8.5

For each set:

a Plot the data points on a scatter graph.

b Comment on the correlation.

c Find the value of r, Pearson's product moment correlation coefficient.

d What do you notice? How would you relate the value of r to the strength of the correlation?

2 Consider another four data sets.

Set 1

x	2	3	4	5	6	7	8	9
y	13	8	12	7	10	11	6	5

Set 2

x	2	3	4	5	6	7	8	9
y	12.5	9	11	8	10	9	8	5

Continued on next page

Statistics and probability

Set 3

x	2	3	4	5	6	7	8	9
y	12	10	10.5	9	9.3	9	8.2	7

Set 4

x	2	3	4	5	6	7	8	9
y	11	10.5	10	9.5	9	8.5	8	7.5

For each set:

a Plot the data points on a scatter graph.

b Comment on the correlation.

c Find the value of r, Pearson's product moment correlation coefficient.

d What do you notice? How would you relate the value of r to the strength of the correlation?

3 **Conceptual** What does a value of r close to 1 or −1 tell you about the correlation?

Reflect What does a positive or negative value of r tell you about the correlation?

Here is a table to help you interpret a correlation from its r value.

r value	Correlation
$0 \le \lvert r\rvert \le 0.25$	Very weak
$0.25 < \lvert r\rvert \le 0.5$	Weak
$0.5 < \lvert r\rvert \le 0.75$	Moderate
$0.75 < \lvert r\rvert \le 1$	Strong

TOK

"Everything that can be counted does not count. Everything that counts cannot be counted" (Albert Einstein).

What counts as understanding in mathematics?

Exercise 6C

1 Label each of these scatter plots with one of the following values of r: −0.6, 1, 0.9.

a

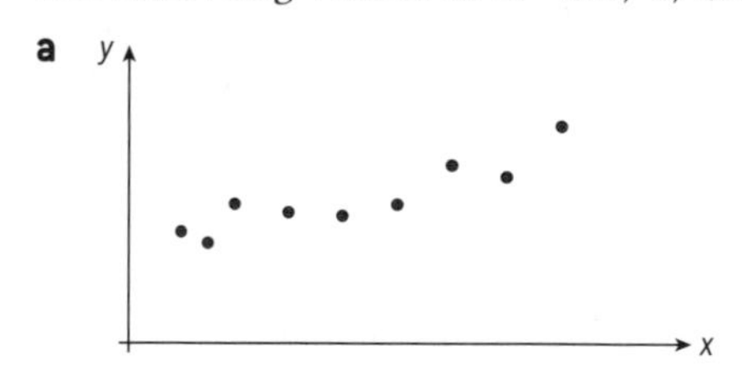

b

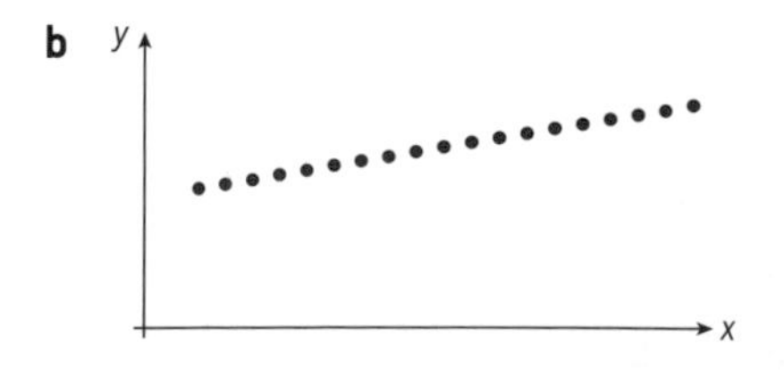

c y x

2 Label each of these scatter plots with one of the following values of r: 0, 0.7, −0.9.

a

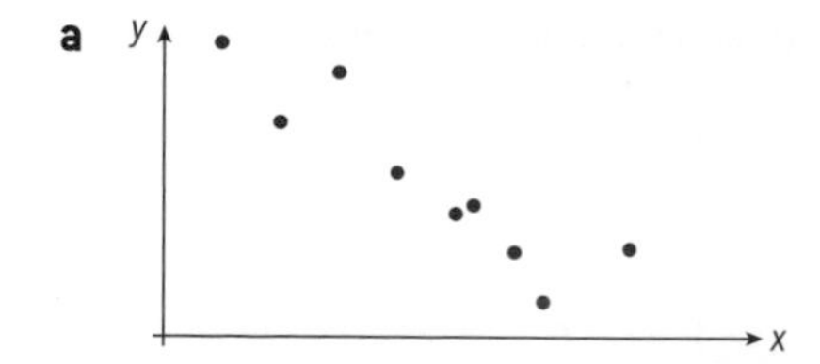

b

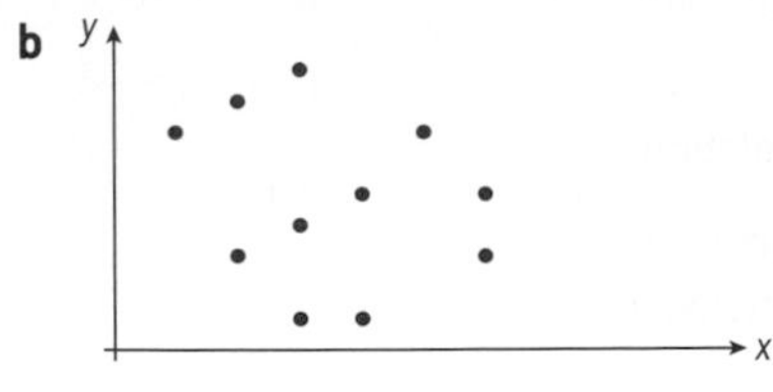

c

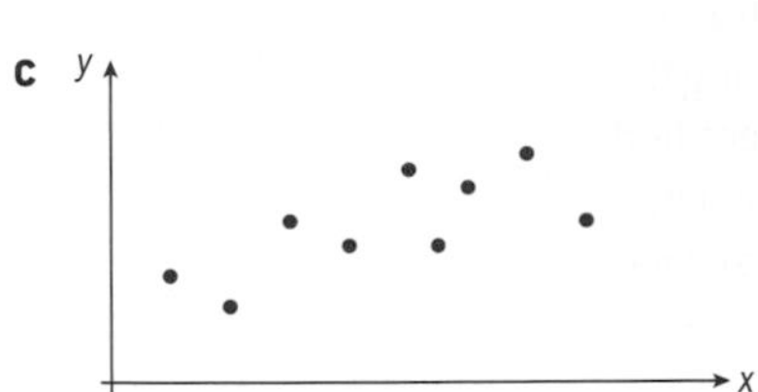

3 Label each of these scatter plots with one of the following values of r: -1, 0.5, -0.3.

a

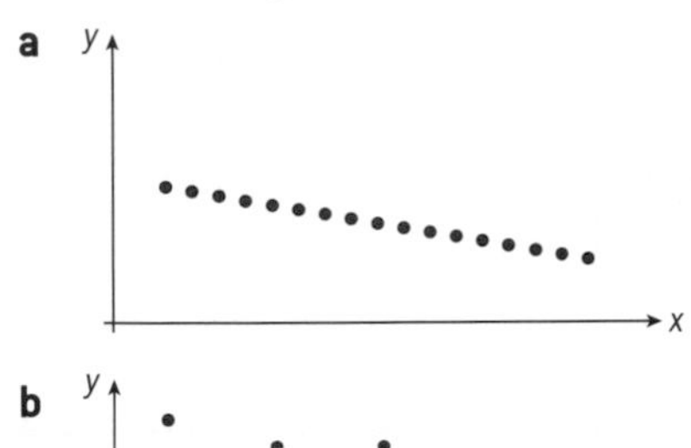

b

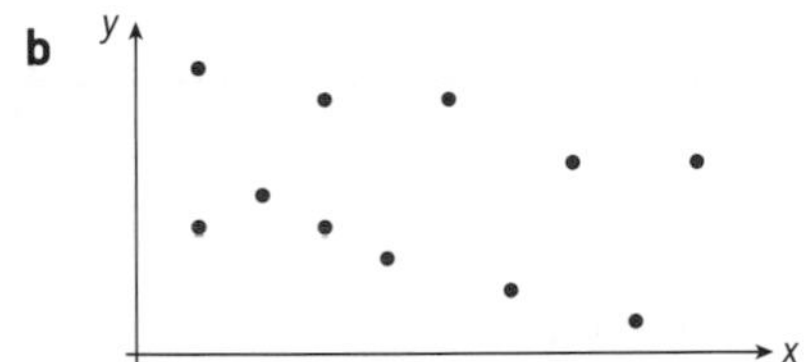

c

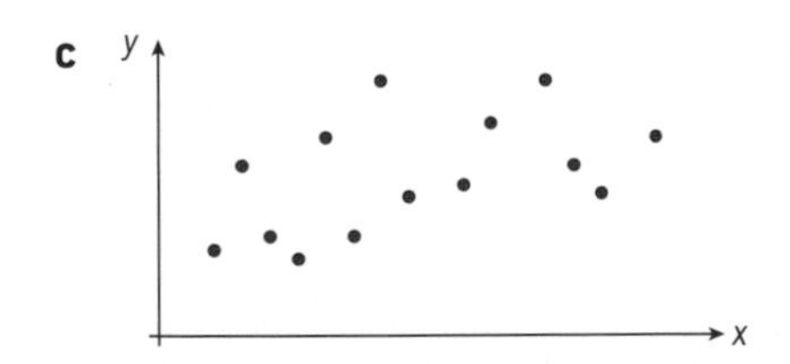

4 A school grading system ranges from 0 to 10. The table shows, for 10 students, the number of hours they study on average per week, s, and their grades, g.

s	32	30	40	48	35	50	52	60	43	50
g	4	4	6	7	6	7	8	9	6	10

a Find Pearson's product moment correlation coefficient, r, for this data.

b Describe the correlation.

c Comment on the effect on the grade of studying for more hours.

5 Different weights were attached to a vertical spring and the lengths of the spring measured. The table summarizes the results.

Weight (g)	0	10	30	50	90	110	150	200	220	250
Length (cm)	0	0.5	2	3	6	6.5	8	10	12	16

a Find Pearson's product moment correlation coefficient, r, for this data.

b Comment on the correlation.

Investigation 4

1 Consider this data set:

x	1	2	3	3	4	5	6	7	8	10
y	1	3	2	14	4	3	6	7	5	15

a Calculate Pearson's product moment correlation coefficient for this data. Comment on the correlation.

b Plot the data points on a scatter graph. What do you notice about the distribution of the points?

2 Now remove the point (3, 14) from the data set in part **1**. Find the correlation coefficient for the new data set and comment on the correlation.

Continued on next page

3 Now remove the point (10, 15) from the data set in part **1**. Find the correlation coefficient for the new data set and comment on the correlation.

4 Finally, remove both points (3, 14) and (10, 15) from the data set in part **1**. Find the correlation coefficient for the new data set and comment on the correlation.

The points (3, 14) and (10, 15) here are called outliers. **Outliers** were introduced in Chapter 3, and you learned how to identify them from a data set by using the upper and lower quartiles and the interquartile range. Once you have identified any outliers in a data set, you can then consider whether they might affect the correlation and whether they should be removed. From what we have just seen, an outlier that does not follow the general trend has more of an effect on the correlation coefficient than one that does.

5 **Conceptual** How does an outlier affect the correlation coefficient? How should you deal with them?

Reflect What is an outlier?

Why might there be outliers in a data set?

Exercise 6D

For each of the following data sets:

a Plot the points on a scatter diagram.

b Find Pearson's product moment correlation coefficient, r, for the data.

c Identify any outliers. Describe whether or not they follow the data trend.

d Calculate the change in the value of r if outliers are removed.

1

x	1	3	2	3	4	4	5	6	7	8	9
y	4	4	3	5	4	12	6	5	6	7	7

2

x	1	3	2	3	4	5	6	7	8	9	10
y	4	4	3	5	4	6	5	6	7	7	12

3

x	5	10	15	20	25	30	35	30	40	18	27	13
y	650	700	20	560	615	540	570	800	480	600	550	670

Developing inquiry skills

Thinking back to the problem at the start of the chapter, is it appropriate to find the correlation coefficient for this set of data points? Why? If so, find it.

Describe the correlation again. What does it mean in terms of the context?

Do you think that the IMR is an effect of the number of vaccine doses received?

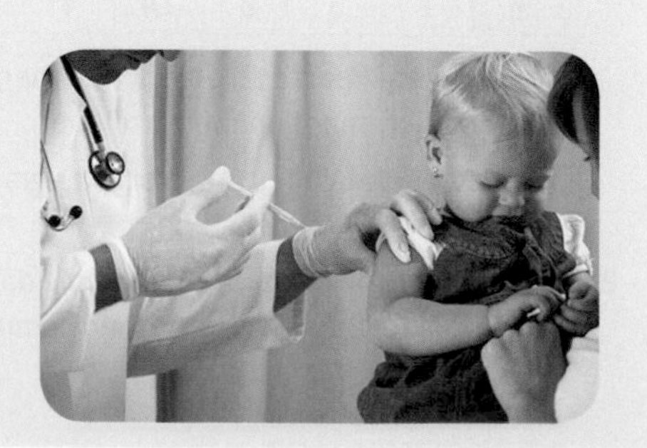

6.2 The line of best fit

Investigation 5

The table gives the age of a person, x, and the number of movies this person watched last year, y.

x	15	20	24	25	30	35	40	45	46	50
y	8	7	10	9	15	14	15	17	15	20

1 Plot the data points on a scatter diagram, using an appropriate scale on each of the axes.

2 Explain why it is appropriate to model the data points with a straight line.

3 Draw a line that you think would be a good model for this set of data points. This line will have to nearly fit all of the points. It is known as a **line of best fit**.

4 Compare your line with a line drawn by a classmate. How are they the same? How are they different?

5 Find the mean point of this data set, where the x-coordinate is the mean of all the x-values and the y-coordinate is the mean of all the y-values. Plot it on the scatter diagram and label it M.

6 Now draw a new line on the graph that passes through M and has as many of the data points above the line as below it. Does it match your line from part **3**?

7 Would you use this new line to estimate the number of movies that might be watched by a 48-year-old person? Why? How? Show your working on the scatter diagram.

8 **Conceptual** When and how can a line of best fit be used to model a set of data points?

A **line of best fit** is a line drawn on a scatter diagram that shows the trend followed by the data points. This line can then be used to make predictions.

To draw a line of best fit **by eye:**

a Find the **mean point**, whose coordinates are calculated by finding the mean of the x-values and the mean of the y-values; it is the point $(\bar{x}, \bar{y})$. Plot this point on the scatter diagram. It will be used as a reference point to draw the line.

b Draw a line through the mean point that balances the number of points above the line with the number of points below the line.

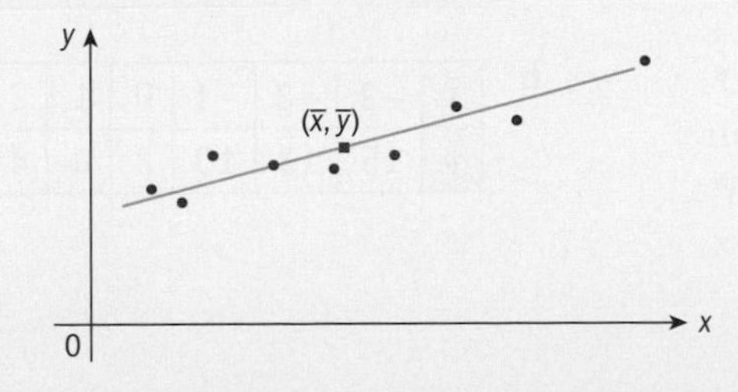

TOK

How can causal relationships be established in mathematics?

Example 3

Let us look again at the test results in mathematics and biology from Example 2.

Mathematics (x)	20	25	28	30	32	37	42	48
Biology (y)	24	20	22	21	25	28	30	33

a Find the mean of the mathematics test results.

b Find the mean of the biology test results.

c Plot and label the mean point on your scatter diagram and use it to draw a line of best fit by eye.

d Hence predict the biology test result for someone who got 40 in their mathematics test.

a $\bar{x} = 32.75$

b $\bar{y} = 25.25$

Calculate both means using your GDC.

c

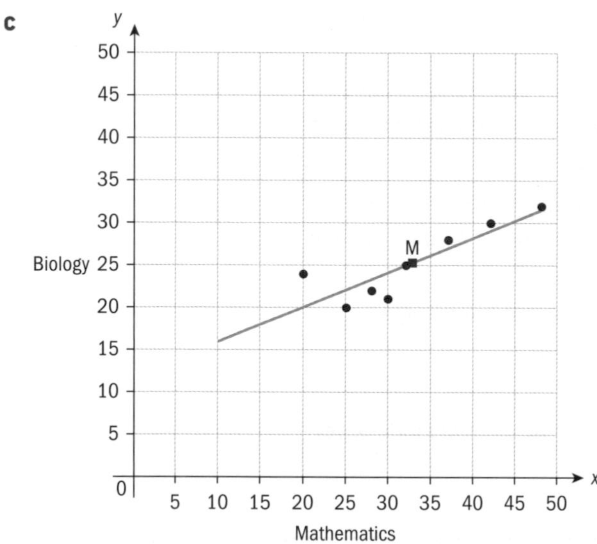

Plot the mean point (32.75, 25.25) on the scatter diagram. Draw a line of best fit through the mean point so that there is roughly an equal number of points above and below the line.

Do not forget to label the mean point, M.

d 28

Find the y-coordinate for the point at which the line of best fit crosses the line $x = 40$.

Exercise 6E

1 For each set of data:

i Plot the points on a scatter diagram.

ii Describe the type of correlation.

iii Find the mean of x and the mean of y.

iv Plot and label the mean point on your diagram and draw a line of best fit by eye through the mean point.

a

x	10	11	12	13	14	15	16	17	18	20
y	18	20	19	21	24	23	24	26	27	32

b

x	−3	−2	−1	0	1	2	3	4	5	6
y	15	13	10	7	6	4	5	2	3	0

2 The table shows the area (a, in millions of square feet) of ten shopping malls, along with the annual number of visitors (v, in millions).

a	2.7	2.2	1.8	2.6	1.8	2.2	2.7	2	1.4	1
v	28	27	26	25	23	22	22	21	20	18

a Plot a scatter diagram for this data.

b Describe the correlation.

c Find the mean area of the shopping malls.

d Find the mean annual number of visitors.

e Plot and label the mean point on the scatter diagram.

f Draw a line of best fit by eye through the mean point.

3 The table shows the gross domestic product (GDP) (x) and the average number of books read per person in a year (y) for 15 countries.

x	7904	10 326	7616	11 387	12 645	14 528	20 122	22 122	28 166	29 861	54 304	44 189	42 651	46 678	50 169
y	2.2	2.9	3.0	4.0	4.6	5.4	6.0	8.5	9.7	10.3	12.0	15.0	16.3	16.8	17.0

a Plot a scatter diagram for this data.

b Describe the correlation.

c Find the mean GDP.

d Find the mean number of books read per person in a year.

e Plot and label the mean point on the scatter diagram.

f Draw a line of best fit by eye through the mean point.

Investigation 6

In this investigation you will see that a line of best fit drawn by eye through the mean point can be improved.

1 What do you think "improve" means in this context?

2 The scatter graph shows a set of eight data points and also two lines of best fit. One of these lines models this set of points better than the other. Which line do you think is a better fit to the data? Why?

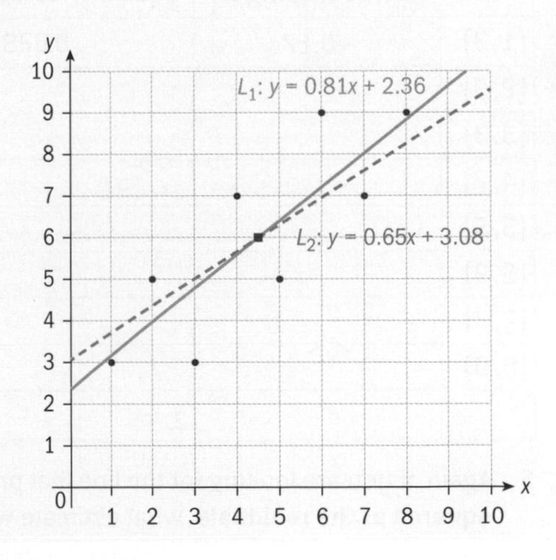

To find a better answer we need to introduce the concept of **residuals**.

A **residual** is the difference between the actual y-value and the predicted y-value. Therefore, residuals are the errors made when using lines of best fit to make predictions.

Statistics and probability

Investigation 6 (continued)

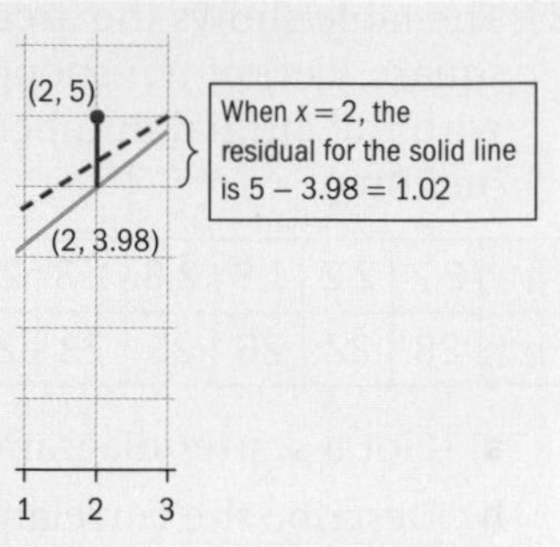

The line that will be the best model for this set of data points will be the one that makes these residuals as small as possible.

3 The row in the table corresponding to the first point has been completed for you. Complete the rows for the other seven points.

Point	x	y	Predicted y using L_1	Residual using L_1	Predicted y using L_2	Residual using L_2
(1, 3)	1	3	$0.81 \times 1 + 2.36 = 3.17$	$3 - 3.17 = -0.17$	$0.65 \times 1 + 3.08 = 3.73$	$3 - 3.73 = -0.73$
(2, 5)	2					
(3, 3)	3					
(4, 7)	4					
(5, 5)	5					
(6, 9)	6					
(7, 7)	7					
(8, 9)	8					

To find which of the two lines fits these data points better, the **sum of the squares** of the residuals must be **minimized**. The criterion used is called the **least squares** criterion.

4 Complete this table. Calculate the sum of the squares of the residuals where it says Σ.

Point	Residual using L_1	Square of residual using L_1	Residual using L_2	Square of residual using L_2
(1, 3)	−0.17	0.0289	−0.73	0.5329
(2, 5)				
(3, 3)				
(4, 7)				
(5, 5)				
(6, 9)				
(7, 7)				
(8, 9)				
		$\Sigma =$	$\Sigma =$	

5 Again, if you are looking for the line that produces the **smallest possible value for the sum of the squares of the residuals**, what estimate would you give for y when $x = 1.5$? And when $x = 7.5$? Which equation have you used to estimate these values? Why?

6 If you were given two lines of best fit, would you be able to choose which of the two lines is better for modelling the data? How would you make your choice? Explain briefly.

7 **Conceptual** How is the least squares regression line different from other lines of best fit?

The **least squares regression line** is the line of best fit that has the smallest possible value for the sum of the squares of the residuals.

For example, the line of best fit here is the least squares regression line when $a^2 + b^2 + c^2$ is a minimum.

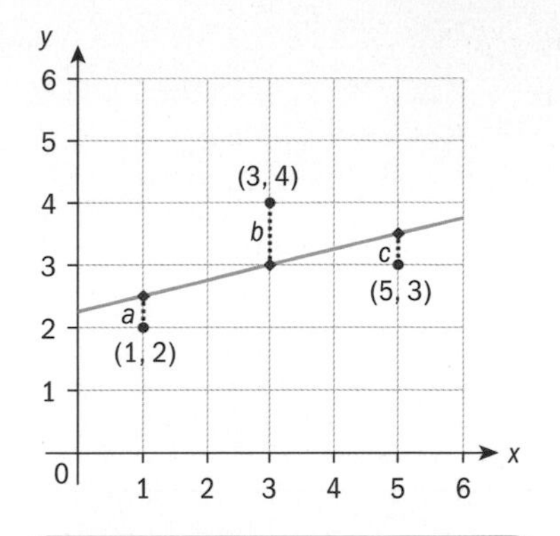

If x is the independent variable and y is the dependent variable then the least squares regression line is also called the regression line of y on x.

The equation of the regression line of y on x is

$$y - \bar{y} = \frac{s_{xy}}{(s_x)^2}(x - \bar{x})$$

where $\bar{x}$ and $\bar{y}$ are the means of the x and y data values respectively,

$$s_{xy} = \sum xy - \frac{(\sum x)(\sum y)}{n} \text{ and } (s_x)^2 = \sum x^2 - \frac{(\sum x)^2}{n}.$$

EXAM HINT

In examinations you will only be expected to use technology to find the equation of the regression line.

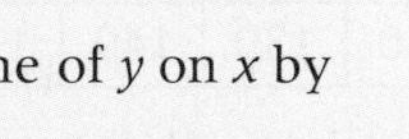

Example 4

For the set of data points $\{(1,2), (3,4), (5,3)\}$, find the regression line of y on x by using the formulae.

$y = 0.25x + 2.25$

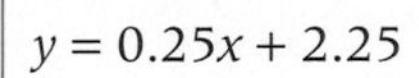

TOK

To what extent can you reliably use the equation of the regression line to make predictions?

Use a table like this to calculate the expressions needed for the formulae:

x	y	xy	x^2
1	2	2	1
3	4	12	9
5	3	15	25
$\sum x = 9$	$\sum y = 9$	$\sum xy = 29$	$\sum x^2 = 35$

$$s_{xy} = \sum xy - \frac{(\sum x)(\sum y)}{n}$$

$$= 29 - \frac{9 \times 9}{3} = 2$$

$$(s_x)^2 = \sum x^2 - \frac{(\sum x)^2}{n}$$

$$= 35 - \frac{9^2}{3} = 8$$

Continued on next page

Statistics and probability

	$\overline{x} = \frac{9}{3} = 3, \quad \overline{y} = \frac{9}{3} = 3$ $y - \overline{y} = \frac{s_{xy}}{(s_x)^2}(x - \overline{x})$ $y - 3 = \frac{2}{8}(x - 3)$, and by expanding we get $y = 0.25x + 2.25$

Example 5

At a coach station, the maximum temperature in °C (x) and the number of bottles of water sold (y) were recorded over 10 consecutive days. The collected data is summarized in the table.

Day	1	2	3	4	5	6	7	8	9	10
x	20	19	21	21.3	20.7	20.5	21	19.3	18.5	18
y	140	130	140	145	143	145	145	125	120	123

a Plot the data points on a scatter diagram.

b Comment on the correlation.

c Find the equation of the regression line of y on x.

d Draw the regression line on the scatter diagram.

e The temperature is predicted to be 19.5°C. Estimate how many bottles of water should be stocked according to this regression line.

a

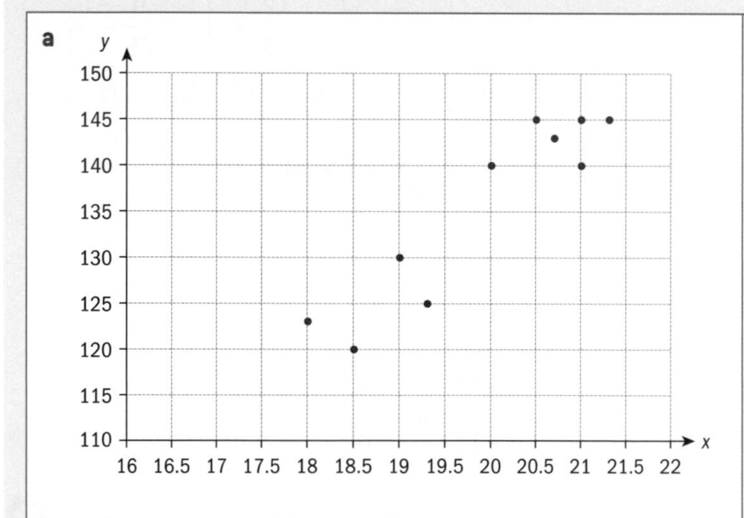

Your GDC can be used to plot the data points.

b It is a strong positive correlation.

c $y = 8.05x - 24.7$	You GDC can be used to find the equation of the regression line of y on x.
d 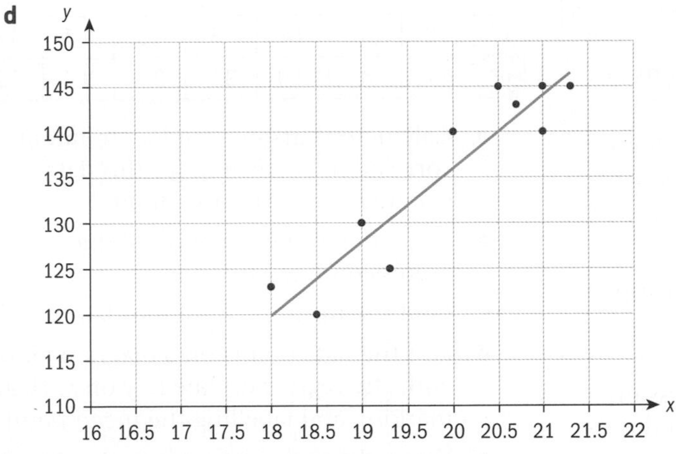	Your GDC can be used to draw the regression line.
e 133, because 132 would be insufficient.	For $x = 19.5$, $y = 132.2...$

Exercise 6F

1 The travel times in minutes (x) and the prices in euros (y) of ten different train journeys between various places in Spain are shown in the table.

x	128	150	102	140	140	98	75	130	80	132
y	25.95	40	24.85	31.8	30.2	28.95	21.85	34.5	23.25	26

a Calculate the value of r. Comment on the correlation.

b Write down the equation of the regression line of y on x.

c Find the mean travel time.

d Find the mean price.

e Plot the data points on a scatter diagram together with the regression line of y on x, clearly marking and labelling the mean point.

2 A memory test was given to a group of ten people. Each of them was shown images of 20 different objects, then after 5 minutes they were asked to name the objects they remembered. The times they took to remember those names (in minutes) were recorded. The results are shown in the table.

Objects remembered (x)	12	16	9	10	14	18	12	15	17	15
Time taken (y)	1.9	2.2	1.7	2	2	2.5	2.3	2.2	2.4	2.4

a Calculate the value of r. Comment on the correlation.

b Write down the equation of the regression line of y on x.

c Find the mean number of objects remembered.

d Find the mean time taken to remember the objects.

e Plot the data points on a scatter diagram along with the regression line of y on x, clearly marking and labelling the mean point.

3 The heights in metres (x) and weights in kilograms (y) of ten male gorillas are shown in the table.

x	1.9	1.83	1.81	1.79	1.74	1.91	1.93	1.86	1.81	1.95
y	275	267	260	257	258	272	273	268	261	273

a Calculate the value of r. Comment on the correlation.

b Write down the equation of the least squares regression line for this data.

c Find the mean height for this group of male gorillas.

d Find the mean weight for this group of male gorillas.

e Plot the data points on a scatter diagram together with the regression line of y on x, clearly marking and labelling the mean point.

4 The headmaster of a secondary school is investigating whether there is any relationship between a student's age (x) and the monthly average number of absences for that age group (y). The data collected is shown in the table.

x	12	13	14	15	16	17	18
y	4.2	4	3.9	3.5	3.4	3.4	3.2

a Calculate Pearson's product moment correlation coefficient for this data. Comment on the correlation.

b Write down the equation of the regression line of y on x.

c Find the mean point of this data.

d Plot the data points on a diagram along with the regression line of y on x, clearly marking and labelling the mean point.

e Hence determine how many absences per month you might expect from an 11-year-old, and comment on the reliability of the prediction.

Investigation 7

A provincial health department is studying whether there is any relationship between the number of beds in public hospitals and the daily cost per patient. It has collected data from eight hospitals in the province as shown in the table.

Number of beds (x)	300	200	180	250	450	330	500	510
Daily cost per patient (y)	40	45	50	43	38	40	37	35

1. Plot the points on a scatter graph. Label the axes.
2. Find Pearson's product moment correlation coefficient, r. Describe the relationship between these two variables.
3. Based on the value of r, is a linear model appropriate for the relationship between these two variables? Explain why it is appropriate to find the regression line of y on x.
4. Find the regression line of y on x and draw the line on the scatter plot.
5. Determine a reasonable domain and range for your model, the regression line of y on x.
6. What would be more reliable: using the regression line to estimate the daily cost per patient in a hospital in the same province that has 280 beds or in a hospital that has 600 beds? Why?
7. When is it safe to make predictions using the regression line of y on x? When is it risky to make predictions using the regression line of y on x?
8. **Conceptual** How can you use a line of best fit?

Reflect When is it appropriate to use a line of best fit?

Reflect Why might extrapolation be risky?

Example 6

A random sample of 12 students is taken to see whether there is a linear relationship between height in cm (x) and shoe size (y). The data collected is shown in the table.

Height (x cm)	160	187	175	180	186	170	185	172	174	180	165	170
Shoe size (y)	37	42	39	38	40	38	41	39	38	40	37	39

a Plot the data points on a scatter diagram. Comment on the correlation.

b Explain why it is appropriate to find the regression line of y on x.

c Find the regression line of y on x. Draw the line on the scatter diagram.

d Hence estimate the shoe size of a student who is 163 cm tall. Explain why this is a valid process.

e Explain why it is not reliable to use the regression line of y on x to estimate the shoe size of a student who is 150 cm tall.

f Explain why it is not appropriate to use the regression line of y on x to estimate the height of a student whose shoe size is 38.

a

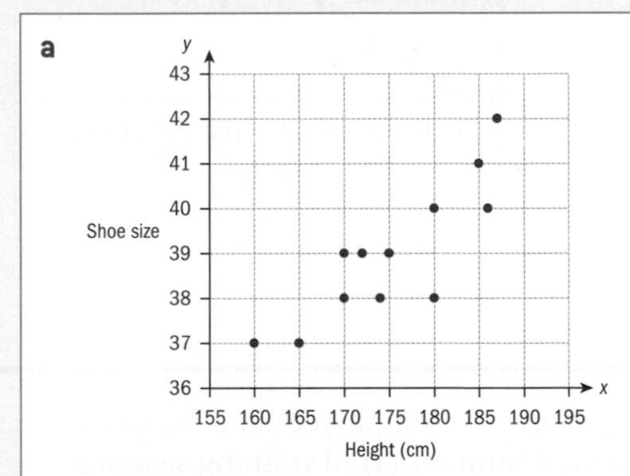

There is strong positive linear correlation.

Your GDC can be used to plot the data points.

b It is appropriate because the correlation is strong and linear. (It would also be appropriate if the correlation were moderate.)

c $y = 0.154x + 12.0$

Your GDC can be used to find the regression line of y on x.

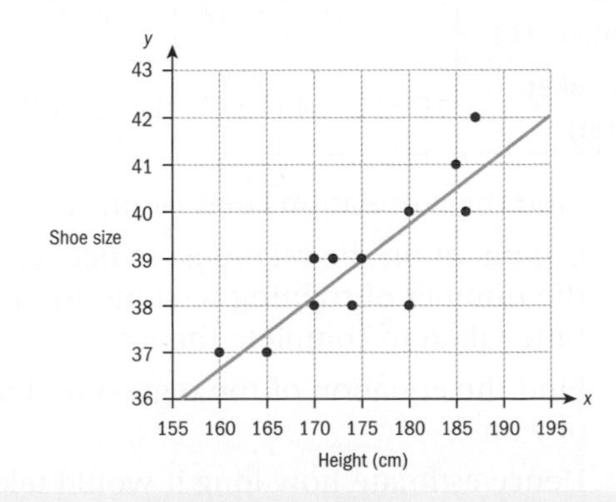

Continued on next page

Statistics and probability

d 36.9, which would round to a shoe size of 37.

The process is valid because 162 cm is within the data range $\{x: 160 \le x \le 187\}$.

Method 1:

Your GDC can be used to estimate the shoe size from the graph.

When $x = 162$, $y = 36.9$.

Method 2:

Substitute $x = 162$ in the equation:

$y = 0.154 \times 162 + 12.0 = 36.9$

Method 3:

Read off the graph, although this method is sometimes less accurate:

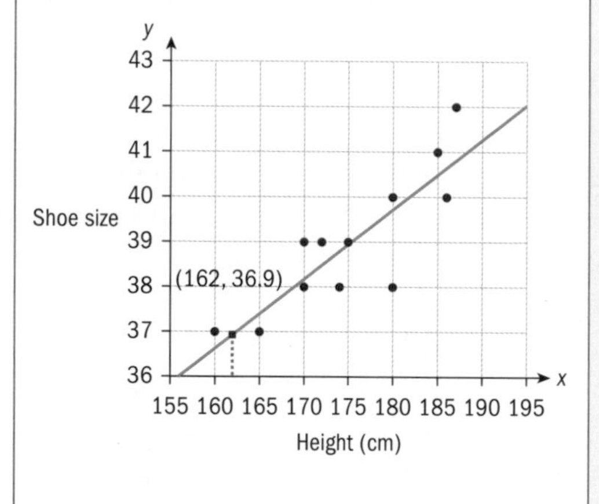

e 150 cm is beyond the data range. This would be an extrapolation.

f The regression line of y on x should only be used to estimate values of y when x is given, and not the other way round.

Exercise 6G

1 The CEO of a publishing company wants to know whether there is a linear association between the number of pages in a book (x) and the number of errors (y) found in the book. Ten books were chosen at random, and the information is shown in the table.

x	100	130	170	80	220	260	290	300	200	150
y	8	10	13	10	12	13	15	16	9	10

a Plot the data points on a scatter graph. Label the axes.

b Describe the correlation. Hence, explain why it is appropriate to find the regression line.

c Find the regression line of y on x.

d Hence estimate the number of errors in a book that has 280 pages.

e Comment on whether it would be reliable to use this equation to estimate the number of errors in a book that has 400 pages.

2 A group of 10 employees at a factory were given a number (x) of training sessions. They were then asked to complete a task. The times taken to complete this task (y) were recorded, measured in minutes. The results are shown in the table.

Number of sessions (x)	3	4	5	3	7	7	8	9	9	8
Time taken (y min)	10	15	14	12	7	12	6	5	6	4

a Find the correlation coefficient, r.

b Comment on the relationship between the number of training sessions and the time taken to complete the task.

c Find the equation of the regression line of y on x.

d Hence estimate how long it would take an employee to complete the task if they were given six training sessions.

3 There is a moderate positive linear correlation between variables x and y. The regression line for y on x is $y = 2x + 1$, and it was determined from a table of values where x is such that $3 \le x \le 12$.

a Explain why it is appropriate to find the equation of the regression line.

b Explain why it is not reliable to use the regression line to estimate the value of y when $x = 0$.

c Explain why it is reliable to use the regression line to estimate the value of y when $x = 10$.

d Explain why it is not valid to use the regression line to estimate the value of x when $y = 20$.

Piecewise linear functions

A real-world problem may not always be representable by a simple linear function. It is often possible, though, to break a problem down into elements that can each be modelled by a linear function.

> A function whose graph is made up of line segments is called a **piecewise linear function**.

TOK

We can often use mathematics to model everyday processes.

Do you think that this is because we create mathematics to emulate real life situations or because the world is fundamentally mathematical?

For example, if we think of someone walking at a constant speed and then stopping at a coffee shop for a drink, a graph of distance travelled (y) against time (x) would look like this.

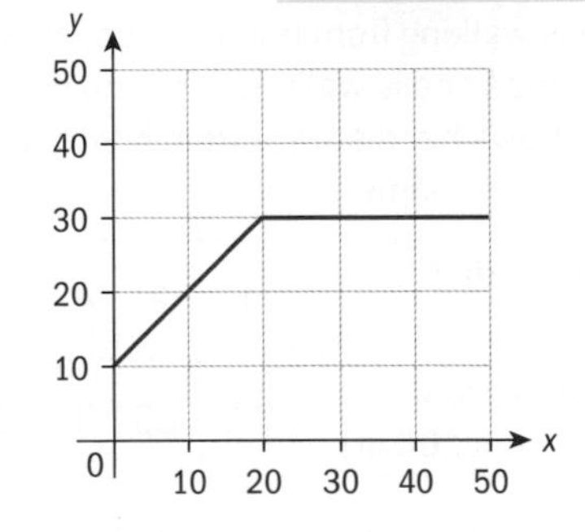

This function, f, is defined by two pieces:

- When $0 \le x \le 20$, the formula is $f(x) = x + 10$.
- When $x > 20$, the formula is $f(x) = 30$.

This can be summarized in the following way:

$$f(x) = \begin{cases} x + 10 & 0 \le x \le 20 \\ 30 & x > 20 \end{cases}$$

How would you use this formula to find $f(23)$ and $f(10.5)$?

Example 7

For the piecewise function

$$f(x) = \begin{cases} 4 & -2 \le x \le 3 \\ -x + 7 & 3 < x \le 7 \end{cases}$$

a Find:

  i $f(0)$ **ii** $f(6)$

b State the domain.

c Plot the graph.

Continued on next page

a i $f(0) = 4$	You first need to work out which part of the definition of f applies for $x = 0$. As $x = 0$ lies in $-2 \le x \le 3$, $f(0) = 4$.
ii $f(6) = 1$	$x = 6$ belongs to the interval $3 < x \le 7$, so substitute $x = 6$ into the formula $f(x) = -x + 7$.
b Domain is $\{x: -2 \le x \le 7\}$	Recall that the domain is the set of inputs for which the function is defined. In this case, the function is defined for $-2 \le x \le 3$ and $3 < x \le 7$.
c 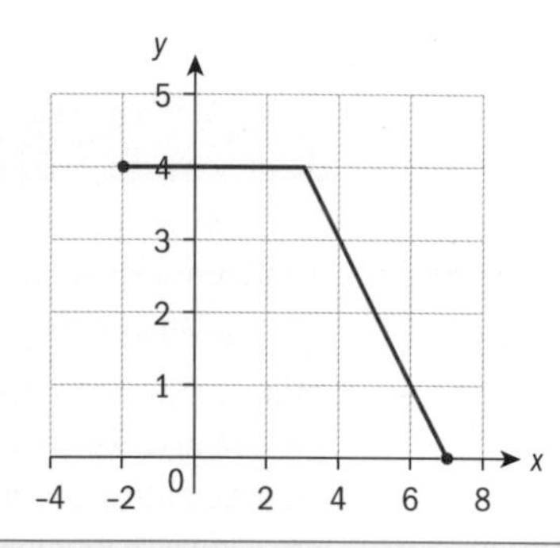	Your GDC can be used to plot the graph. You will need to enter the formula for each of the two pieces as different lines. You can start by setting the standard window, but you may need to adjust this.

Investigation 8

Nadia is walking from home to school. Sometimes she is on her own, and sometimes she is walking with one or more friends, so her pace varies. The table shows her distance from home (y, in hundreds of metres) x minutes after she left home.

Time walking (x min)	1	2	4	5	6	7	8	10	12	14	17	22
Distance from home ($y \times 100$ m)	0.5	1.5	3	4.5	6	7	8	8.5	9	9.5	10	11

1. Plot the data points on a scatter diagram using an appropriate scale. Label the axes.
2. How far is Nadia's school from home?
3. How long did she take to get from home to school?
4. Comment on Nadia's journey. Can you model the whole journey with one function?
5. Find a model to fit these data points. Explain why you have chosen this model. How did you deal with the change of pace?
6. Use your model to find Nadia's distance from home after 9 minutes and after 3 minutes.
7. **Conceptual** How do we decide whether a piecewise linear model is needed?

Exercise 6H

1 For each of the following piecewise linear functions:

  i State the domain.

  ii Plot the graph.

  iii Find $f(3)$ and $f(-3)$.

a $f(x)=\begin{cases}1+x & -3\le x\le 4\\ 9-x & 4<x\le 10\end{cases}$

b $f(x)=\begin{cases}x & x\ge 0\\ -x & x<0\end{cases}$

2 Determine the definition of the piecewise linear function shown in each of the graphs.

a

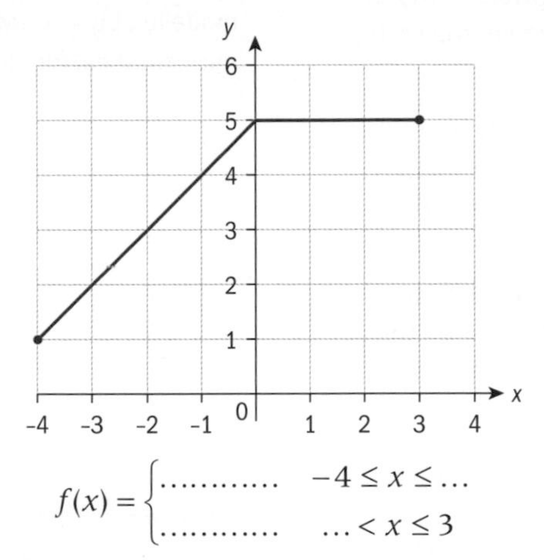

$f(x)=\begin{cases}\ldots\ldots\ldots\ldots & -4\le x\le \ldots\\ \ldots\ldots\ldots\ldots & \ldots<x\le 3\end{cases}$

b

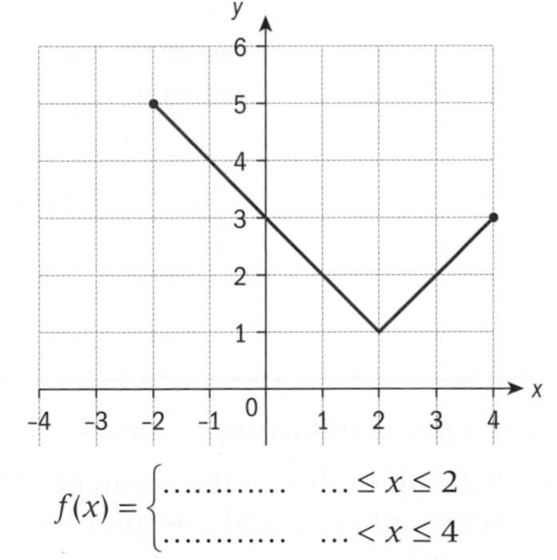

$f(x)=\begin{cases}\ldots\ldots\ldots\ldots & \ldots\le x\le 2\\ \ldots\ldots\ldots\ldots & \ldots<x\le 4\end{cases}$

3 Consider the following sets of data points. Decide in which cases you would choose a piecewise linear model to fit the data points.

a

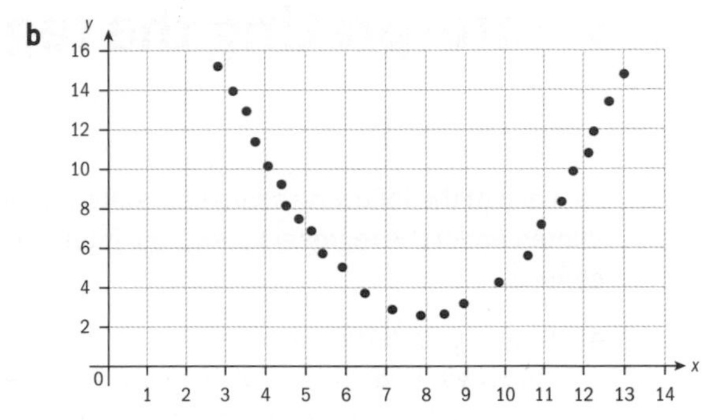

b

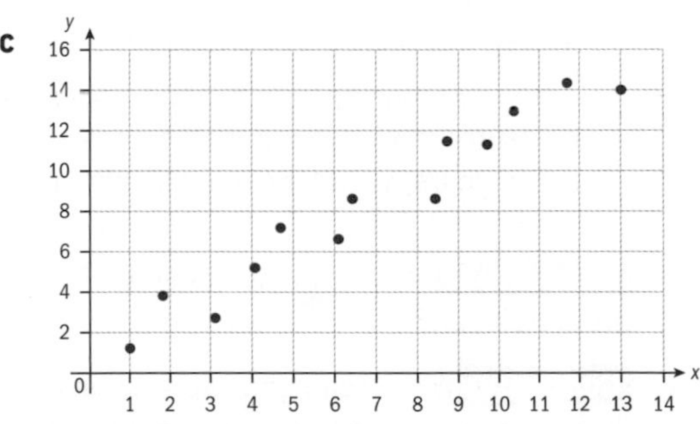

c

4 For any part of question **3** that was suitable for modelling with a piecewise linear function:

  a Draw this model "by eye".

  b Estimate the value of y when $x = 12$.

5 Consider the set of data points in the table.

x	1	3	4	6	7	9	10.5	12	13	15	14	17	16	18
y	5.3	4.8	4	4.2	3.4	3.1	3	2	4	7.3	8	11.5	11	15

  a Plot these points on a scatter diagram.

  b Find a piecewise linear model that best fits these data points.

  c Draw your model on the same set of axes used for part **a**.

  d Hence estimate the value of y when:

   i $x = 8$ **ii** $x = 15.5$

Developing inquiry skills

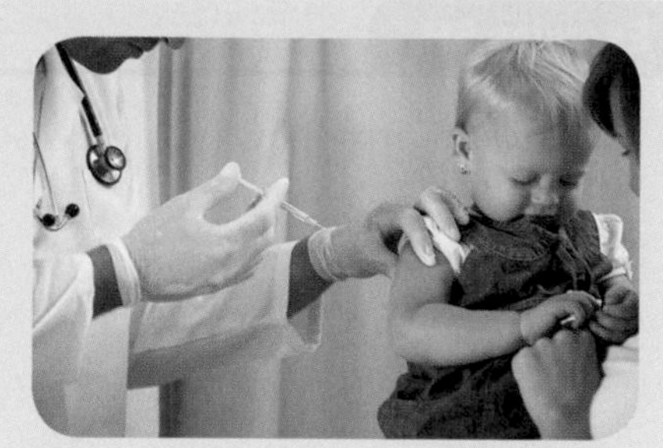

Looking back at the problem at the start of the chapter, find the mean point for this set of data points.

Find the equation of the least squares regression line. Draw the line on the scatter diagram, clearly labelling the mean point.

Is it reliable to use this line to estimate the IMR if a country had 9 vaccine doses specified? And if the country had 16 vaccine doses specified? Why? If you think the IMR value will be reliable, find it.

6.3 Interpreting the regression line

Investigation 9

1 For each of the following scenarios, state the value of the gradient and the y-intercept of the regression line, and interpret their meaning in the stated context.

  a A survey undertaken in a school asked how many hours per week a student spends on social media, x, and sleeping, y. It was found that there is a linear correlation between the variables, and the equation of the regression line of y on x is $y = 50 - 2x$.

  b A group of boys aged between 5 and 8 years had their hand lengths measured. It was found that there is a linear correlation between the age in years, x, and the hand length in centimetres, y. The regression line is $y = 1.7x + 1.5$.

2 **Conceptual** What can the gradient of the regression line tell us about the correlation?

3 **Conceptual** How can you explain that the y-intercept of the line of best fit may not be relevant in a given context?

TOK

Can all data be modelled by a known mathematical function?

Developing inquiry skills

Returning to the opening problem, what do the gradient and the y-intercept of the regression line mean in the context of the problem?

Developing your toolkit

Now do the Modelling and investigation activity on page 296.

Exercise 6I

1 Decide whether the following statements are true or false.

  a When the gradient of the regression line is positive, the correlation is strong.

  b A negative linear correlation will be modelled by a line with a negative gradient.

  c The independent variable, x, can never take the value zero because this would mean extrapolation.

2 For each of the following scenarios:

  i State the value of the gradient of the regression line and interpret its meaning.

  ii State the value of the y-intercept of the regression line and interpret its meaning if relevant, giving a reason if there is no meaning.

a A number of students were asked for their average grade at the end of the last year of high school, x, and their average grade at the end of their first year at university, y. On calculating the regression line for the resulting data, the result was $y = -2.50 + 1.04x$.

b It is found that the relationship between the height in centimetres, x, and the weight in kilograms, y, of a group of 15-year-old students can be modelled with the regression line $y = -70 + 0.87x$.

c A car salesman wants to study the relationship between the time in years after a particular type of car is bought, x, and the value of the car in US$, y. The regression line is found to be $y = -250x + 9000$.

3 Different weights are suspended from a spring and the length of the spring measured. The results are shown in the table.

Weight (x g)	100	150	200	250	300	350	400
Length of spring (y cm)	26	35	32	37	48	49	52

a Find the correlation coefficient, r.

b Comment on the correlation.

c The equation of the regression line of y on x is $y = ax + b$.

i Find the value of a. Comment on its meaning.

ii Find the value of b and interpret its meaning if relevant. If not relevant, explain why.

Chapter summary

- A correlation can be classified in terms of its **direction** (positive or negative) and its **strength** (zero, weak, moderate, strong).
- If the points on a scatter graph lie close to a line, then the correlation is said to be **linear**.
- **Causation** indicates that an event is the result of the occurrence of another event. A high correlation between two variables does not imply that one variable causes the other.
- **Pearson's product moment correlation coefficient**, r, is a measure of the linear correlation between two variables, giving a value between −1 and 1 inclusive. The table summarizes how to interpret the value of r.

r value	Correlation
$0 \le \lvert r \rvert \le 0.25$	Very weak
$0.25 < \lvert r \rvert \le 0.5$	Weak
$0.5 < \lvert r \rvert \le 0.75$	Moderate
$0.75 < \lvert r \rvert \le 1$	Strong

- Extreme or distant data points are called **outliers**.
- A **line of best fit** is a line drawn on a scatter diagram that shows the trend followed by the points. This line can then be used to make predictions.
- To draw a line of best fit **by eye**:
  - Find the **mean point**, whose coordinates are calculated by finding the mean of the x-values and the mean of the y-values. Plot this point on the scatter diagram. It will be used as a reference point to draw the line.
  - Draw a line through the mean point that balances the number of points above the line with the number of points below the line.

Continued on next page

- A **residual** is the difference between the actual y-value and the predicted y-value; residuals are the errors made when using the lines of best fit to make predictions.
- The **least squares regression line** is the line of best fit that has the smallest possible value for the sum of the squares of the residuals.
- If x is the independent variable and y is the dependent variable then the least squares regression line is also called the regression line of y on x.
- A function whose graph is made up of line segments is called a **piecewise linear function**.

Developing inquiry skills

Has what you have learned in this chapter helped you to answer the questions from the beginning of the chapter about the relationship between the total number of vaccine doses specified for infants aged less than one year and the IMR for a group of countries in 2009?

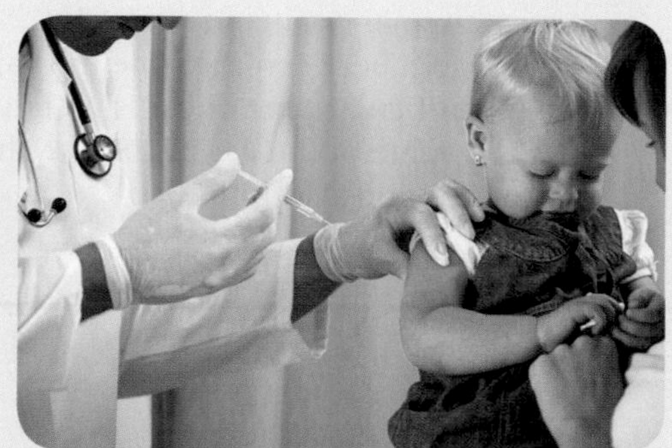

How could you apply your knowledge to investigate the relationships between other pairs of variables? Here are some you could think about. What information would you need to find?

- The relationship between per capita gross national income (in US$) and life expectancy (in years) in different countries.
- The relationship between the percentage of the female population aged 15 and older with no education and fertility rates.
- The relationship between the age of a student and the total amount of time spent on homework each week.

Thinking about the inquiry questions from the beginning of this chapter:

- Discuss whether what you have learned in this chapter has helped you to think about an answer to these questions.
- Consider whether there are any questions that you are interested in and would like to explore further, perhaps for your internal assessment topic.

Chapter review

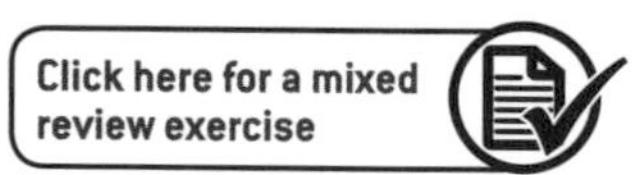

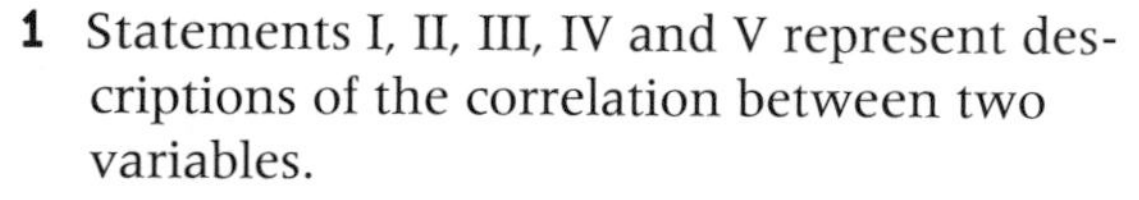

1 Statements I, II, III, IV and V represent descriptions of the correlation between two variables.

I Strong positive linear correlation

II Weak positive linear correlation

III No correlation

IV Weak negative linear correlation

V Strong negative linear correlation

Decide which of these statements **best** represents the relationship between the two variables shown in each of the following scatter diagrams.

a

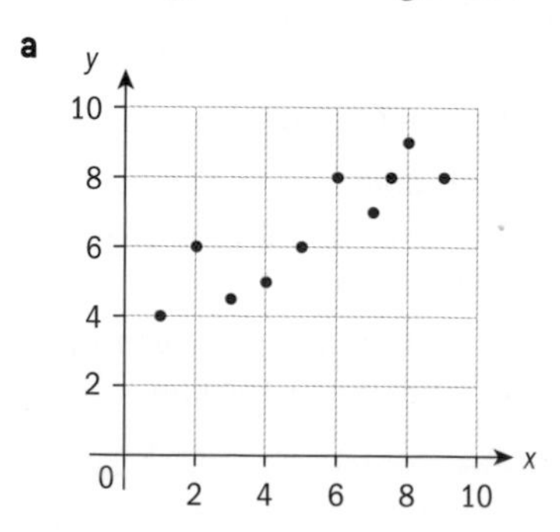

b

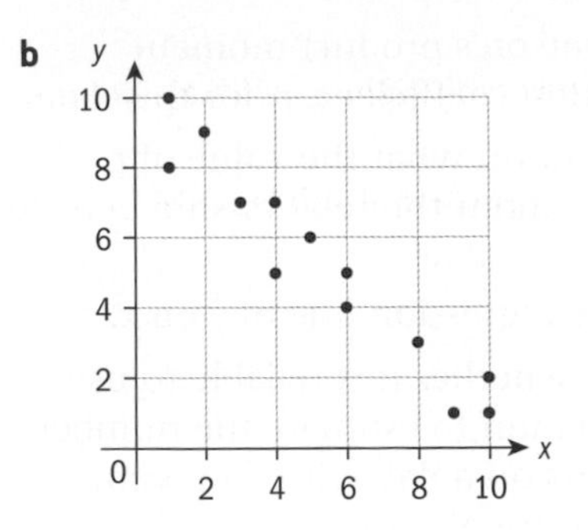

c

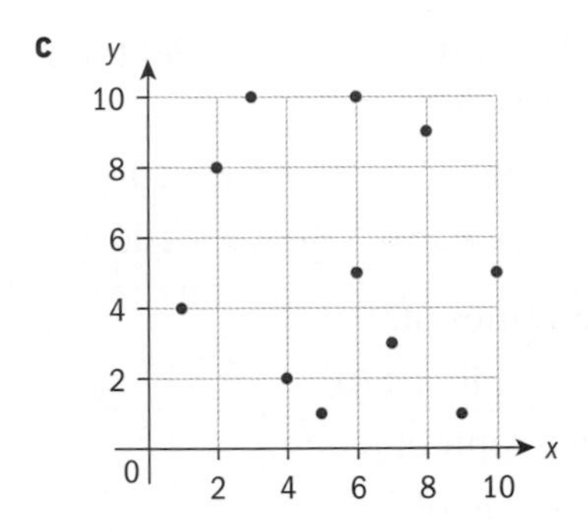

d

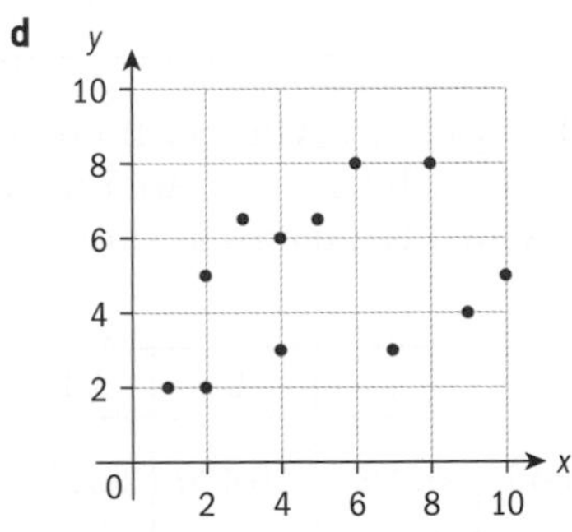

2 Label each of the following scatter diagrams with a value for the correlation coefficient, r, from this list: 0.86, −1, −0.9, 0, 1, −0.99.

a

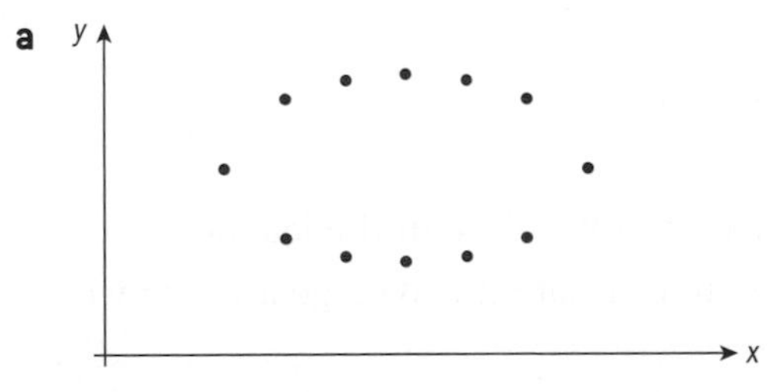

b

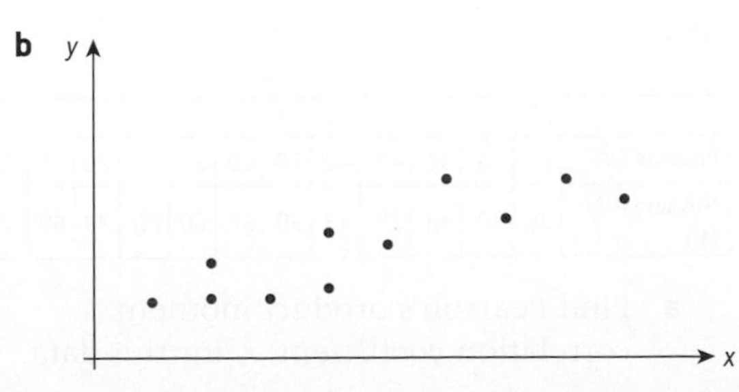

c

d

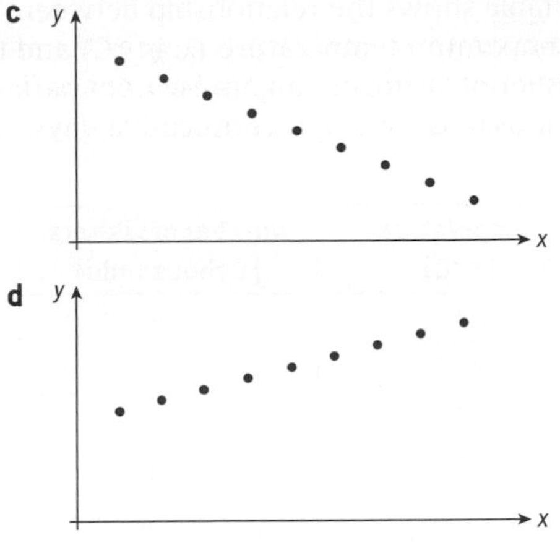

e

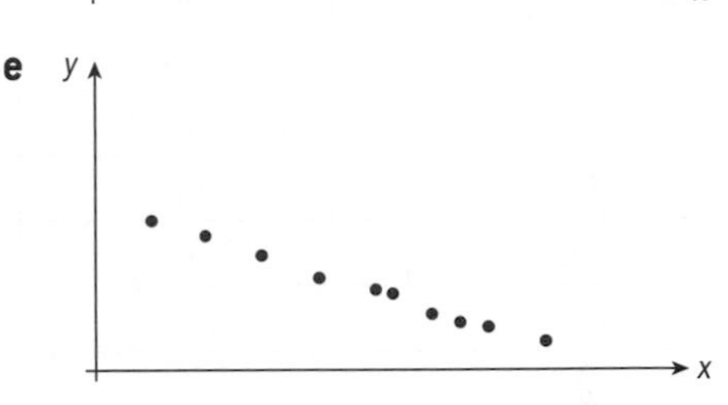

f

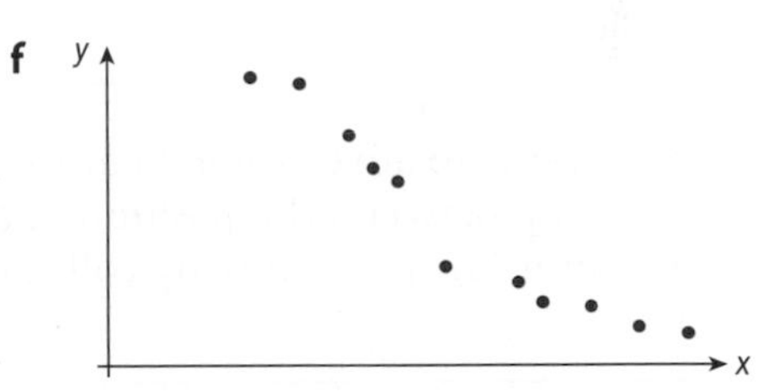

3 In a greenhouse, nine plants of the same species were chosen at random. Their heights and the times in months since they were planted are shown on the scatter diagram.

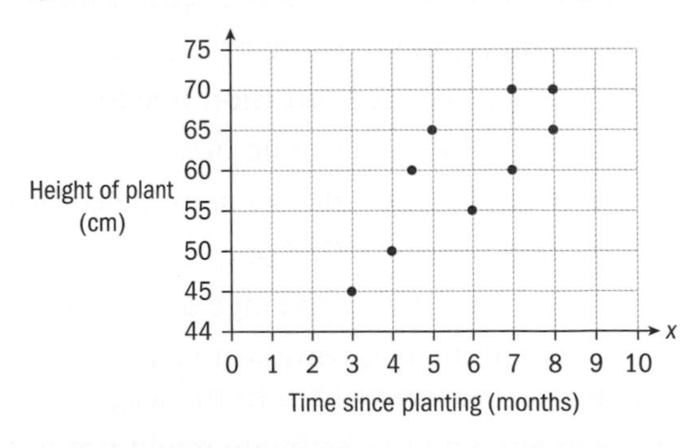

a Plot the point M(6, 60), which represents the mean time since planting and the mean height.

b Draw a suitable line of best fit.

c Hence estimate the height of a plant of this species that was planted 6.5 months ago.

d Explain why your answer to part **c** is reliable.

4 The table shows the relationship between the maximum temperature (x, in °C) and the number of visitors to an amusement park (y, in thousands) over ten consecutive days.

Temperature (x °C)	Number of visitors (y thousands)
24	3.5
23.6	3.4
25	3.6
26	3.44
25.6	3.52
27	3.35
26.5	3.42
23	3.67
23.5	3.6
28	3.2

a Find Pearson's product moment correlation coefficient, r, for this data.

b Comment on what the value of r suggests about the relationship between the two variables.

c Find the regression line of y on x.

d Explain whether it is reliable to use this equation to estimate the number of visitors on a day with a maximum temperature of 40°C.

5 It was found that the best-fit model for a set of data points is defined by

$$f(x) = \begin{cases} 100 + x & 0 \le x \le 50 \\ -2x + 250 & 50 < x \le 100 \end{cases}$$

a Find the values of:

i $f(30)$ **ii** $f(80)$

b Draw the graph of f.

c Find the range of f.

---

6 A company wants to advertise in Spanish football stadiums and is studying whether there is any relationship between the position of a team in the league (p) and the average attendance at their stadium (a). The company collected data from nine of the first ten teams as shown in the table.

Position (p)	1	2	3	4	6	7	8	9	10
Average attendance (a)	69170	77944	44719	32816	21447	39643	20150	16430	5363

a Find the correlation coefficient, r, for the variables p and a.

b Comment on the value of r.

The equation of the regression line for a on p is $a = mp + c$.

c **i** Write down the value of m.

ii Interpret the value of m in the context of the problem.

d Write down the value of c.

e Find the estimated average attendance for the team in fifth place in the league.

The head of the company wants to use this equation to estimate the average attendance for the team that comes 15th in the league.

f Explain why this estimate might not be reliable.

---

7 The physics and mathematics teachers are analysing whether there is any association between the percentage scores that the Y12 class obtained in last term's tests. The results of some of the students are shown in the table.

Physics (x)	27	35	39	40	42	52	60	67	72	75	78	83
Mathematics (y)	30	32	42	38	43	50	62	60	70	73	69	75

a Find Pearson's product moment correlation coefficient, r, for this data.

b Describe the correlation between the percentage scores in the physics and mathematics tests.

c Explain why it is appropriate to find the regression line of y on x.

d Find the regression line of y on x. State the domain on which it has been defined.

Minta scored 55% in the physics test but was absent for the mathematics test. The mathematics teacher uses the regression line to estimate the percentage score Minta would have obtained.

e Find the estimated percentage score for Minta in mathematics. Give your answer correct to 2 significant figures.

During the physics test Ernesto was not feeling well and scored 25% in the test. His score in mathematics was 93%.

f Comment on whether the teachers should leave this data point in their analysis. Explain how this data point would affect the correlation.

Exam-style questions

8 **P1:** Choose from the following list to accurately describe the correlation between two data sets which have each of the given Pearson's product moment correlation coefficients: perfect positive; strong positive; weak positive; zero; weak negative; strong negative; perfect negative.

i 1 ii −0.8 iii 0.4

iv −0.4 v 0 (5 marks)

9 **P2:** Eight pieces of paired bivariate data (x, y) are given in the table below.

x	1	2	3	4	5	6	7	9
y	9	8	7	6	5	5	5	2

Each data pair (x, y) represents the grade a student gained in maths (x) and the grade they gained in art (y).

a Draw, on graph paper, a scatter diagram to represent this data, using a scale of 1 cm for 1 unit on both the x- and the y-axis. (4 marks)

b By considering the scatter diagram, use two words to describe the linear correlation. (2 marks)

c i Calculate the mean of the x-values. (2 marks)

ii Calculate the mean of the y-values. (2 marks)

iii Mark the point $(\bar{x}, \bar{y})$ on the scatter diagram using the symbol ⊗. (1 mark)

d Draw the line of best fit "by eye" on the scatter diagram. (2 marks)

e If the x-value is known to be 8, use the diagram to estimate (to one decimal place) the y-value. Show on the diagram how this estimate was obtained. (2 marks)

10 **P1:** Paired bivariate data (x, y) that is strongly correlated has a y-on-x line of best fit $y = mx + c$. When $x = 70$, an estimate for y is 100. When $x = 100$, an estimate for y is 140. The data represents students' scores on a test in geography, x, and on a test in environmental systems, y.

a Find the values of:

i m ii c. (3 marks)

b State whether there is positive or negative correlation. (1 mark)

c The value of $\bar{x}$ is 90. Find the value of $\bar{y}$. (3 marks)

d When $x = 60$ find an estimate for the value of y, given that this is interpolation. (2 marks)

11 **P1:** The temperature, T°C, of liquid in a chemical reaction is a function of time t, where t is measured in seconds and is the time since the experiment started. The function is given by

$$T = \begin{cases} 40 + 2t & 0 \le t \le 30 \\ 130 - t & 30 \le t \le 60. \end{cases}$$

a Write down the initial temperature of the liquid. (1 mark)

b Find the temperature of the liquid after 60 seconds. (1 mark)

c Find the maximum temperature that the liquid reaches. (1 mark)

d i Sketch a graph of T as a function of t. (2 marks)

ii Find the time interval during which the temperature of the liquid is greater than or equal to 80°C. (4 marks)

12 P1: Ten teenagers were asked their age and the number of brothers and sisters they have. The data is shown in the scatter diagram below, where x represents the teenager's age and y represents the number of brothers and sisters that they have.

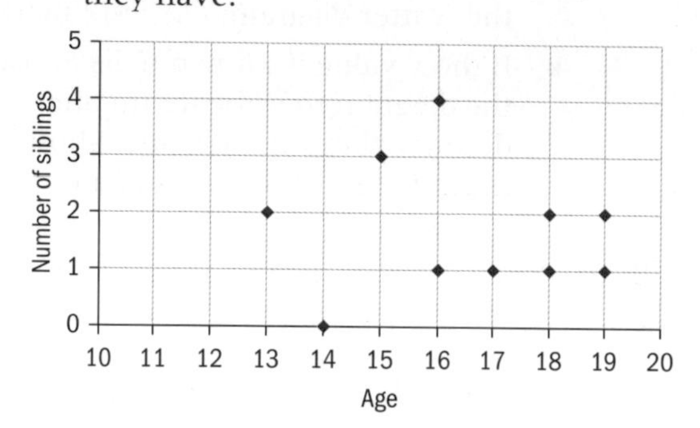

a Use the scatter diagram to complete the following table.

x	13	14	15	16	16	17	18	18	19	19
y					4			2	1	

(3 marks)

b Calculate Pearson's product moment correlation coefficient for this data. (2 marks)

c Give two reasons why it would not be valid to use this scatter diagram to estimate the number of brothers and sisters a 25-year-old might have. (2 marks)

13 P2: A ramp in a skate park can be modelled by a piecewise linear function. Each section of the ramp is a straight line, and the coordinates (measured in metres) of the endpoints of the sections are shown.

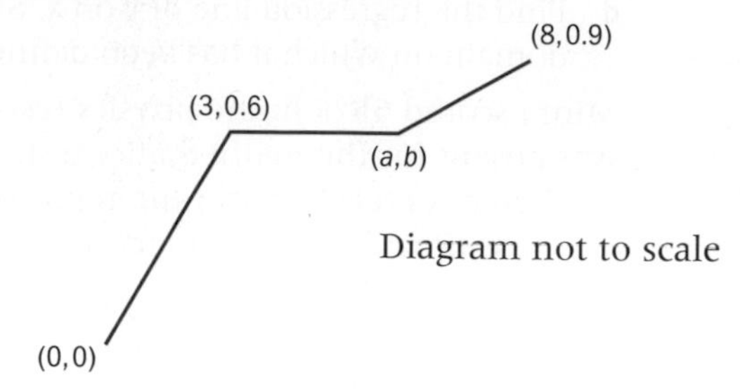

The middle section of the ramp is horizontal.

The formula for this piecewise linear function is given by

$$f(x) = \begin{cases} mx & 0 \leq x \leq k \\ l & k \leq x \leq 5 \\ px + q & 5 \leq x \leq r. \end{cases}$$

Find the values of:

i m **ii** l **iii** k

iv a **v** b **vi** p

vii q **viii** r. (11 marks)

14 P2: A school kept records of the winning times, t, in seconds, in the male 100 m sprint over the years. The data is given in the following table.

Year y	2005	2006	2007	2008	2009	2010	2011	2012	2013	2014	2015
Time t	12.8	12.7	12.7	13.0	12.6	10.8	12.4	12.5	12.3	12.2	12.3

a For this data, calculate Pearson's product moment correlation coefficient. (2 marks)

b For the data of the winning times, find the lower and upper quartiles and hence, with a reason, identify an outlier. (4 marks)

c Remove the outlier from the data and then calculate Pearson's product moment correlation coefficient for the remaining 10 pairs of data values. (2 marks)

d State how the description of the linear correlation has changed due to removing the outlier. (1 mark)

e For this new data, calculate the line of best fit for t on y. (2 marks)

f Hence, in the year of the removed outlier, predict what the winning time would have been if the extremely fast athlete (probably related to Usain Bolt) had not been at the school that year. (2 marks)

15 P1: For 20 paired pieces of bivariate data (x, y) with a Pearson's product moment correlation coefficient of $r = 0.93$, a line of best fit of y on x is calculated as $y = 0.51x + 7.5$. The data represents students' scores on a maths test, x, and on a physics test, y.

The mean of the x data is $\bar{x} = 100$.

a If an x-value of 120 is known, estimate the corresponding y-value, assuming that this is interpolation. (2 marks)

b Find the mean of the y-data, $\bar{y}$. (2 marks)

c Describe in two words the type of linear correlation that is shown by this example. (2 marks)

d A teacher knows a y-value and wants to find the corresponding x-value. State which linear regression line should be used to find an estimate for the x-value. (1 mark)

16 P2: Ten pairs of twins take an intelligence test. In each pair of twins, one twin is female and the other is male. The bivariate data on the test scores is given in the table below.

Female	100	110	95	90	103	120	97	105	89	111
Male	98	107	95	89	100	112	99	101	89	109

a Find Pearsons product moment correlation coefficient, r, for this data. (2 marks)

b State, in two words, a description of the linear correlation in this data. (2 marks)

c Letting the male score be represented by x and the female score by y, find the equation of the regression line of

i y on x

ii x on y. (4 marks)

d Another pair of female/male twins are discovered. The male scored 105 on the test but the female was too ill to take it. Estimate the score that she would have obtained, giving your answer to the nearest integer. (1 mark)

e Yet another pair of female/male twins are discovered. The female scored 95 on the test but the male refused to take it. Estimate the score that he would have obtained, giving your answer to the nearest integer. (1 mark)

f If, for a further pair of male/female twins, the male scored 140 on the test, explain why it would be unreliable to use a line of best fit to estimate the female's score. (1 mark)

17 P2: Paired bivariate data (x, y) is given in the table below. The data represents the heights (x metres) and lengths (y metres) of a rare type of animal found on a small island.

x	2.4	3.6	2.8	1.8	2.0	2.2	3.0	3.4
y	3.0	4.0	3.0	1.7	2.0	2.3	3.1	2.7

a **i** Calculate Pearson's product moment correlation coefficient for this data.

ii In two words, describe the linear correlation that is exhibited by this data.

iii Find the y-on-x line of best fit. (6 marks)

Another four specimens of this rare animal are found on a nearby smaller island. This extra data is given in the table below.

x	2.3	2.7	3.0	3.5
y	4.1	1.5	4.2	1.5

b **i** Calculate Pearson's product moment correlation coefficient for the combined data of all 12 animals.

ii In two words, describe the linear correlation that is exhibited by the combined data.

iii Suggest a reason why it would not be particularly valid to calculate the y-on-x line of best fit for the combined data. (5 marks)

Graphs of functions: describing the "what" and researching the "why"

Approaches to learning: Thinking skills, Communicating, Research
Exploration criteria: Presentation (A), Mathematical communication (B), Personal engagement (C)
IB topic: Graphs, Functions, Domain

Bulgaria population data

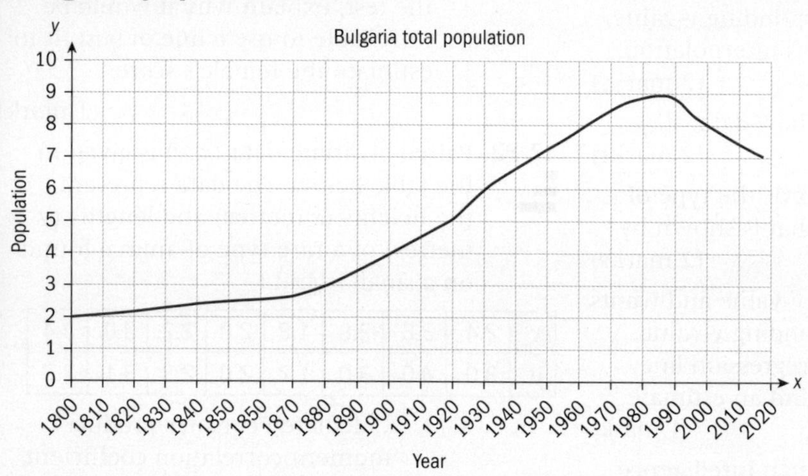

This graph includes two essential elements:

- a title
- x- and y-axis labels with units.

Without any research, write a paragraph to describe this graph.

Do not just describe the graph, but also explain why it might have this shape. Include any interesting points and regions on the graph "where things happen".

Using sources

You can use your general knowledge, printed sources and Internet sources to research data. Different sources can often give different explanations, and not all sources are valid, useful or accurate.

How do you know if a source is reliable?

Use Internet research to find out more precisely what happened at the key dates shown on the Bulgaria population graph.

Keep a record of any sources that you use.

Could you use the graph to predict what might happen to the population of Bulgaria in the future? Explain your answer.

Worldwide Wii console sales

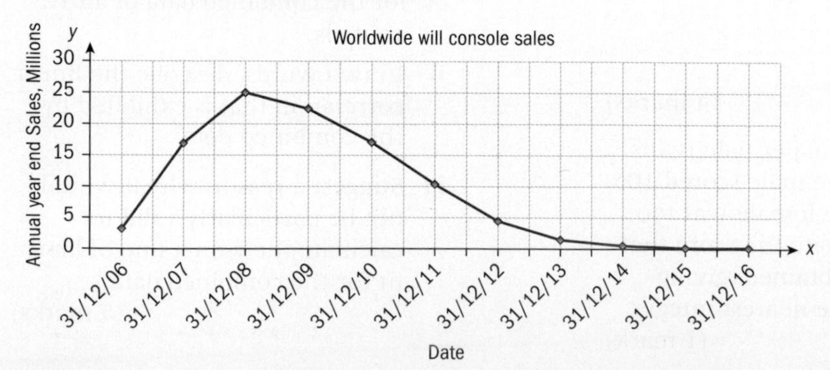

Initially **without research** write a paragraph about this graph, then research the reasons for the shape of this graph.

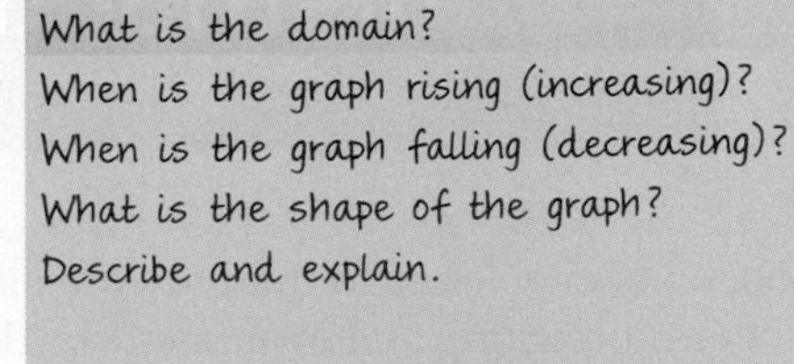

Global mean temperature anomaly

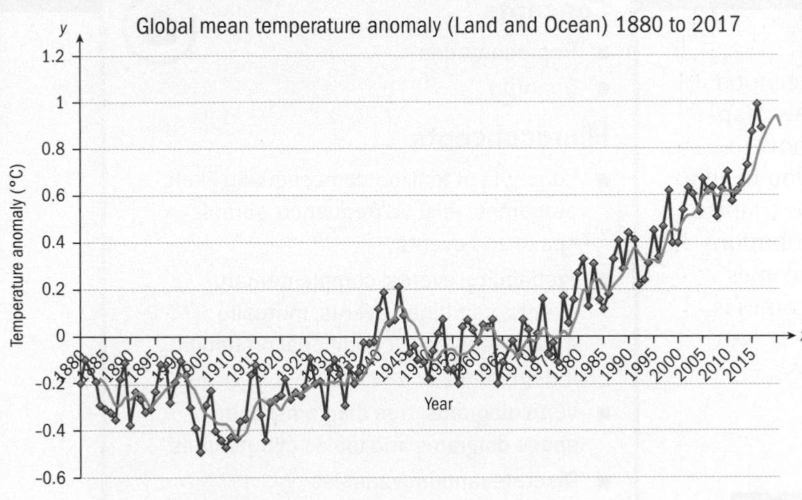

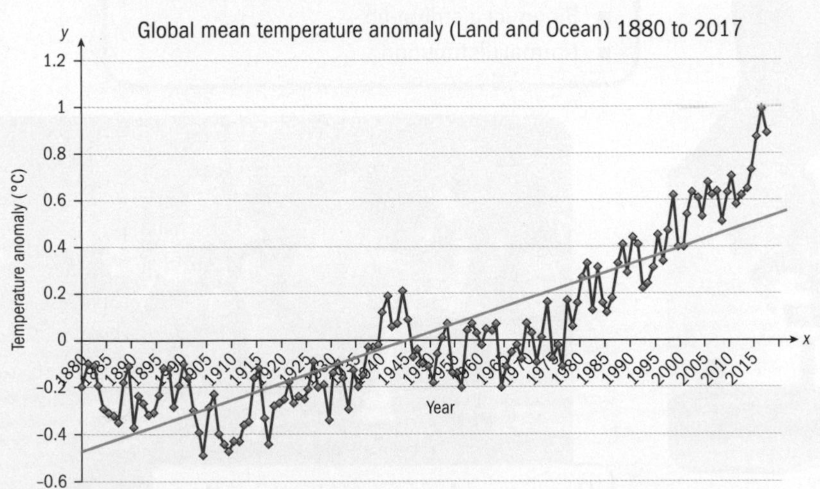

Research what is meant by "global mean temperature anomaly".

This data is based on deviations from the base average for 1951 to 1980.

On the graph in red, the 5-year moving average trend line is included. What is a 5-year moving average?

On the second graph is a linear trend line. What is a linear trend line?

What are the advantages and disadvantages of each of the representations of the data shown?

Describe the data, note any interesting points or trends and to try to explain and investigate why the trends may be as they are.

TOK

This is a potentially controversial topic with many opinions and theories. How can you protect against your own biases?

Extension

Find and research a graph from the news or an academic journal on one of your subjects or another source.

Describe and explain the trends and the reason for the shape of the graph.

Now display or print out the graph.

Write a series of questions for other students to answer that encourage them to describe and explain the graph.

7 Quantifying uncertainty: probability, binomial and normal distributions

Probability enables us to quantify the likelihood of events occurring and evaluate risk. This chapter looks at the language of probability, how to quantify probability and the basic tools you need to solve problems involving probability. This chapter also explores how to use probability distributions to model problems where individual outcomes are varied but the overall pattern of outcomes is predictable.

Concepts

- Representation
- Quantity

Microconcepts

- Concepts of trial, outcome, equally likely outcomes, relative frequency, sample space and events
- Probability: events, complementary events, combined events, mutually exclusive events, conditional probability, independent events
- Venn diagrams, tree diagrams, sample space diagrams and tables of outcomes
- Discrete random variables
- Binomial distribution
- Normal distribution

How can a geneticist quantify the chance that a child may inherit the same colour of eyes as his father?

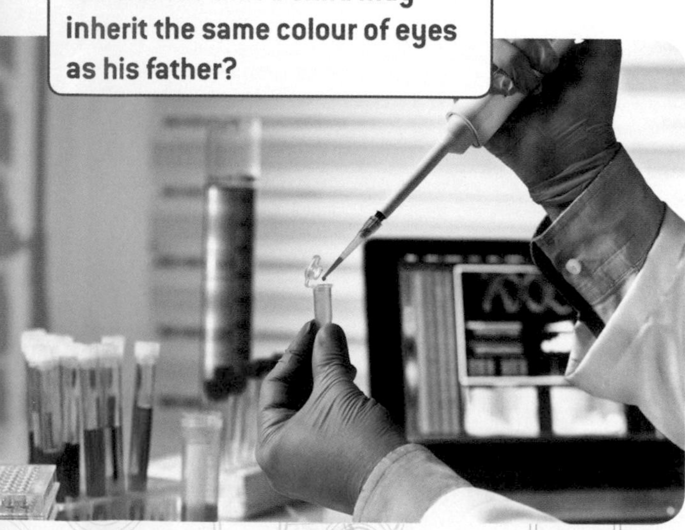

How can a lawyer make sure that a jury understands evidence based on probabilities?

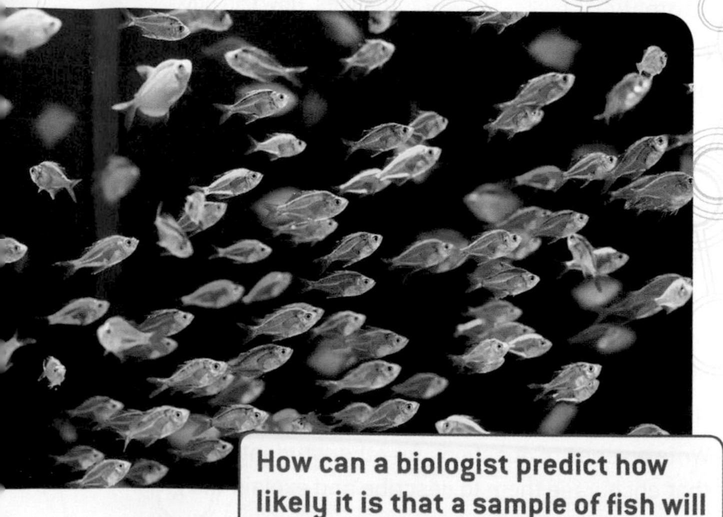

How can a biologist predict how likely it is that a sample of fish will have a certain average length?

How can an airline manage its booking system to allow overbooking for profitability while treating its customers fairly?

Daniel Kahneman (Nobel Prize winner in economic science) and Amos Tversky (cognitive and mathematical psychologist) spent decades collaborating and researching together. Below is an adaptation of one of the questions they set to a group of students:

Two taxi companies operate in Mathcity: Blackcabs and Yellowrides.

85% of the cabs in the city work for Blackcabs and are coloured black.

The rest of the cabs in the city work for Yellowrides and are coloured yellow.

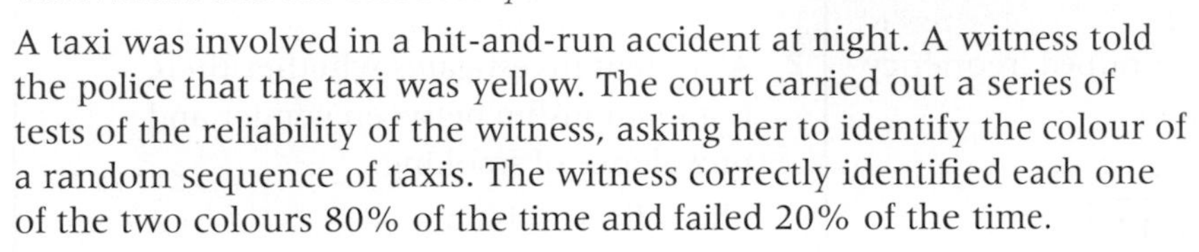

A taxi was involved in a hit-and-run accident at night. A witness told the police that the taxi was yellow. The court carried out a series of tests of the reliability of the witness, asking her to identify the colour of a random sequence of taxis. The witness correctly identified each one of the two colours 80% of the time and failed 20% of the time.

1. What is the probability that the taxi involved in the accident was yellow?
2. Karolina carries out a traffic survey in Mathcity. She sits at an interchange and notes the colour of the first six cabs that pass her. What number of yellow cabs is she most likely to observe?

- What types of diagram can be used to represent the problem?
- What assumptions did you make?

Developing inquiry skills

Write down any similar inquiry questions you might ask if you were asked to predict the reliability of the witness if 50% of the cars in the city were Yellowrides or if another taxi company, Blue Taxis, also operated in Mathcity.

What questions might you need to ask in these scenarios?

Think about the questions in this opening problem and answer any you can. As you work through the chapter, you will gain mathematical knowledge and skills that will help you to answer them all.

Before you start

You should know how to:

1 Find simple probabilities

A number is chosen at random from the set {1, 2, 3, ..., 100}.

Find the probability of choosing a cube number.

Cube numbers in the set are: 1, 8, 27, 64.

Probability of a cube number $= \frac{4}{100} = 0.04$

2 Find the mean from a grouped frequency table without technology

eg

x_i	4	7	8	10
f_i	2	1	5	7

$$\text{Mean} = \frac{\sum f_i x_i}{\sum f_i} = \frac{125}{15} \approx 8.33$$

Skills check

Click here for help with this skills check

1 A number is chosen at random from the set {1, 2, 4, 5, 9, 10, 11, 16, 17, 25, 26, 27}.

Find the probability that a number is:

a prime **b** odd **c** a square number.

2 A student investigates whether there is a relationship between gender and prevalence of smoking.

She collects this data:

	Smoker	Non-smoker
Male	12	47
Female	6	51

A person is chosen at random from the survey. Find the probability that they are:

a female **b** a male smoker

c a non-smoker.

3

x_i	1	2	3	6	11
f_i	9	7	3	2	1

Find the mean value of x.

7.1 Theoretical and experimental probability

Probability is synonymous with uncertainty, likelihood, chance and possibility. You can quantify probability through three main approaches: subjective, experimental and theoretical.

Subjective probability

You may judge that you are more likely to get to school on time if you take a particular route, based on your experience with traffic. Subjective probabilities are based on past experiences and opinions rather than formal calculations.

Investigation 1

1 **a** Discuss the likelihood of these outcomes.

A: The team winning the FIFA World Cup in 2030 will be from the Americas.	B: It will rain tomorrow.	C: Humans will reach Mars by 2050.
D: Choosing one digit at random from the decimal expansion of $\frac{1}{6}$, you get 6.	E: The world will be free of all dictators within the next 10 years.	F: The sequence 999999 is found somewhere in the first thousand digits of pi.

b Display your answers by plotting them on this probability scale:

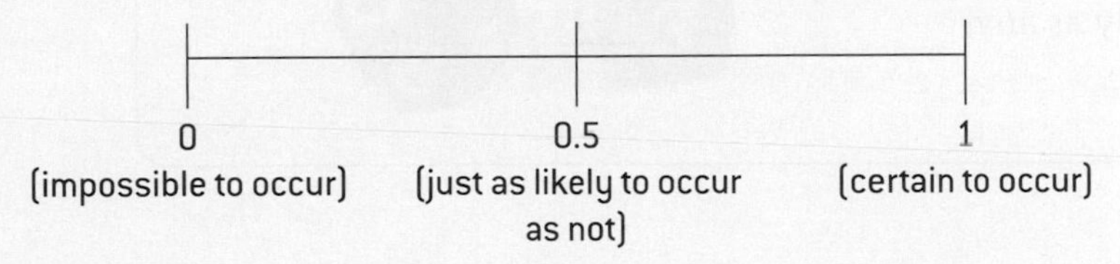

2 **Compare, contrast, discuss** and **justify** your answers within a small group.

You may find disagreements with others, based on your opinions, experience or beliefs.

3 When is it easier to reach a common agreement on the value of a subjective probability?

International-mindedness

Probability theory was first studied to increase the chances of winning when gambling. The earliest work on the subject was by Italian mathematician Girolamo Cardano in the 16th century.

Experimental probability

You should use these terms when discussing and quantifying probabilities:

Experiment: A process by which you obtain an observation

Trials: Repeating an experiment a number of times

Outcome: A possible result of an experiment

Event: An outcome or set of outcomes

Sample space: The set of all possible outcomes of an experiment, always denoted by U

These terms are illustrated in the following example.

Erin wants to explore the probability of throwing a prime number with an octahedral die. She designs an **experiment** that she feels is efficient and bias-free. Erin places the die in a cup, shakes it, turns the cup upside down, and reads and records the number thrown.

Erin repeats her experiment until she has completed 50 **trials**. She knows that the **outcome** of each trial can be any number from $U = \{1, 2, 3, 4, 5, 6, 7, 8\}$ and that the **event** she is exploring can be described as a statement: "throw a prime with an octahedral die" or a set of outcomes that make the statement true: $\{2, 3, 5, 7\}$.

Erin can either write **P(throw a prime)** to represent the probability of her event occurring or **P(A)** if A denotes the set $\{2, 3, 5, 7\}$.

A crucial assumption in many problems is that of **equally likely outcomes.**

A consequence of the geometry of the shapes shown here is that they form *fair dice*. Each outcome on a fair die is as equally likely as any other.

Investigation 2

1 Use card to build a triangular prism where the constant cross-sectional area is a right-angled triangle.

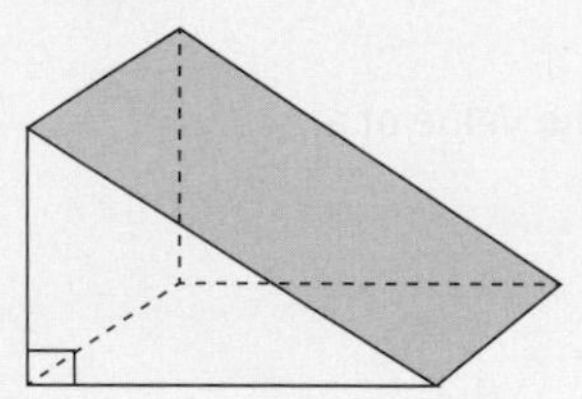

2 In your class, decide how to label the faces with A, B, C, D and E.

- Which face(s) do you predict are least likely to land face down when your prism is thrown?
- Which are most likely?
- **Discuss** in a group and try to rank the five outcomes in order of likelihood.

3 Throw your prism for 50 trials and record the frequency of each outcome. Store your data so that it can be used for further investigation in Section 7.3.

One way to quantify probability is with relative frequency, also known as experimental probability. The general formula for the relative frequency of an event A after n trials is:

Relative frequency of A

$$= \frac{\text{frequency of occurrence of event } A \text{ in } n \text{ trials}}{n}.$$

This is also known as the **experimental probability** of the event A.

4 Find the experimental probability of each outcome by completing a table:

Outcome	Frequency	Relative frequency (exact)	Relative frequency to 4 sf
A			
B			
C			
D			
E			
Totals	50 trials		

5 **Factual** What do the experimental probabilities add up to?

6 **Factual** What is the range of values that an experimental probability can have?

7 **Factual** What do you notice about how the smaller and larger experimental probabilities relate to each other in the context of the die?

8 **Collaborate** by adding your data to that of a friend to make a set of 100 trials. Do any patterns emerge? Then add up all the data in your class. Compare and contrast your experimental probabilities with your subjective probabilities from **2**.

9 **Conceptual** How can you have a more accurate estimate of the experimental probability?

Theoretical probability

Theoretical probability gives you a way to quantify probability that does not require carrying out a large number of trials.

The formula for the theoretical probability $P(A)$ of an event A is:

$P(A) = \frac{n(A)}{n(U)}$ where $n(A)$ is the number of outcomes that make A happen and $n(U)$ is the number of outcomes in the sample space.

Whenever $P(A)$ represents a subjective, experimental or theoretical probability, then $0 \leq P(A) \leq 1$.

Investigation 3

1 Imagine throwing a fair 12-sided die 15 times. Let A be the event "throw a prime number".

2 Follow the instructions on p.307 to build a spreadsheet or GDC document that shows the sequence of experimental probabilities of A after 1, 2, 3, ..., 100 throws. You should be able to create an image like one of these:

Continued on next page

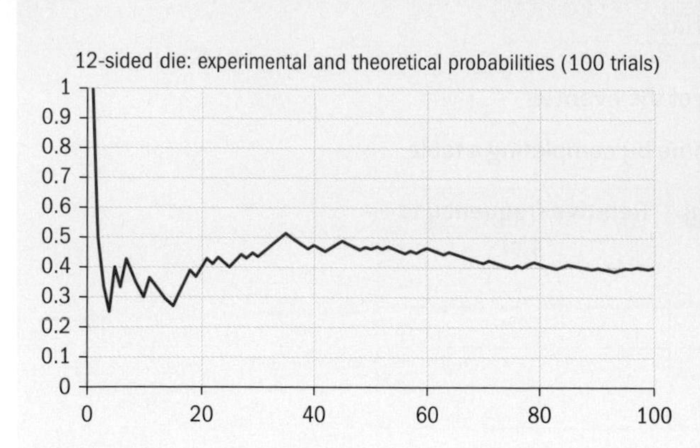

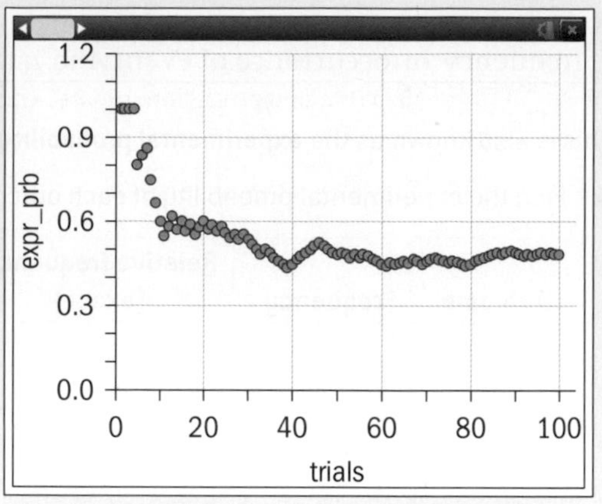

Number of trials (n)	Outcome	Event	Frequency of occurrence in n trials	Relative frequency in n trials
1	2	Prime	1	1
2	11	Prime	2	1
3	5	Prime	3	1
4	9	Not prime	3	0.75
5	4	Not prime	3	0.6
6	2	Prime	4	0.666667
7	4	Not prime	4	0.571429
8	8	Not prime	4	0.5
9	2	Prime	5	0.555556
10	12	Not prime	5	0.5
11	6	Not prime	5	0.454545
12	1	Not prime	5	0.416667
13	1	Not prime	5	0.416667
14	3	Prime	6	0.428571
15	11	Prime	7	0.466667

3 Show that for this experiment, $\mathrm{P}(A)=\frac{5}{12}\approx 0.417$. Add a horizontal line with equation $y=0.417$ to your graph. Carry out another 100 trials using your spreadsheet or GDC.

4 Repeat until you have seen each of these three scenarios:

- The experimental probability is always greater than the theoretical probability.
- The experimental probability is always less than the theoretical probability.
- The experimental probability is often equal to the theoretical probability.

You may adapt your spreadsheet so that it carries out 1000 trials. Examine the columns in your spreadsheet and the features on your graph.

5 **Factual** What is the set of **all** possible values of theoretical probabilities?

6 **Factual** What relationship does your graph have with the line $y = \frac{5}{12}$?

7 **Conceptual** What is the relationship between relative frequency and theoretical probability in the short term?

8 **Conceptual** What is the relationship between relative frequency and theoretical probability in the long term?

9 **Conceptual** Does random behaviour involve predictability in the short term or unpredictability?

10 **Conceptual** Does random behaviour involve predictability in the long term or unpredictability?

11 **Conceptual** How may we interpret and apply the number quantified by the formula for the theoretical probability of an event?

Example 1

Find the probability of each event and decide which event is least likely.

T: throw a factor of 24 on a four-sided die.

O: throw a prime on an eight-sided die.

D: throw at least 11 on a 12-sided die.

C: throw at most 3 on a six-sided die.

I: throw a multiple of 5 on a 20-sided die.

All the dice are fair.

$n(T) = n(\{1, 2, 3, 4\}) = 4 = n(U)$ so $\mathrm{P}(T) = 1$	Every element of $\{1, 2, 3, 4\}$ is a factor of 24. $\mathrm{P}(T) = \frac{n(T)}{n(U)} = \frac{4}{4} = 1$, so T is certain to happen.
$n(O) = n(\{2, 3, 5, 7\}) = 4,\ n(U) = 8$ so $\mathrm{P}(O) = \frac{4}{8} = 0.5$ $n(D) = n(\{11, 12\}) = 2,\ n(U) = 12$ so $\mathrm{P}(D) = \frac{2}{12} = \frac{1}{6} = 0.1\dot{6}$	"at least 11" means "11 or more".
$n(C) = n(\{1, 2, 3\}) = 3,\ n(U) = 6$ so $\mathrm{P}(C) = \frac{3}{6} = 0.5$ $n(I) = n(\{5, 10, 15, 20\}) = 4,\ n(I) = 20$ so $\mathrm{P}(I) = \frac{4}{20} = 0.25$ Hence D is the least likely event.	"at most 3" means "3 or fewer".

Statistics and probability

Exercise 7A

1 A letter is picked at random from the letters of RANDOM. Calculate the probability that it is a letter from MATHS.

2 A dartboard has 20 sectors of equal area.

If a dart lands in a numbered sector at random, find the probability that the number is:

a at least 4 **b** more than 6

c less than 30 **d** no more than 14

e prime **f** square

g a solution to the equation $x^2 = 3$.

3 Ann and Ruth are designing a game for a CAS project. The numbers 1, 2, 3, ..., 11 are written on identical tickets and one ticket is drawn at random from an envelope. Find the probability that the number on the ticket drawn is:

a odd **b** square

c prime **d** square and odd

e square and prime **f** prime and odd

g a prime and even.

4 A personal identification number (PIN) consists of four digits. Consider the PIN 0005 equal to the number 5 etc. Find the probability that a PIN is:

a equal to 0000

b less than 8000 and more than 7900

c divisible by 10 **d** at least 13.

Investigation 4

On 18 August 1913, at the casino in Monte Carlo, gamblers betting on whether the roulette ball would fall into a black or red slot witnessed a very unusual event. (For simplicity, assume that the roulette wheel only has red or black slots for the ball to fall into, and these are equal in number.)

The ball fell into the black slot 26 times in succession, a record. After 15 successive blacks, gamblers excitedly rushed to bet heavily on red, hugely increasing their bets with every time the ball ended up in black once more. Consequently, at the end of the sequence of 26 blacks, the casino made a huge profit because of the choices made by the gamblers.

Discuss in a group: Why were the gamblers panicking? What would the most rational response be to the 15 consecutive blacks? Have you ever experienced an unusual set of outcomes like the one in this story? Can we predict how long "unusual runs" are and when to expect them?

Ideas for further exploration: What is the maturity doctrine? How is it related to the representativeness fallacy? What about cognitive illusions and biases?

Discuss these questions in a group:

1 Discuss the validity of these three statements that could have been made after 15 blacks:

P: "It has to be black next—black is more likely today!"

Q: "It has to be red next—it's red's turn now!"

R: "There is clear evidence of cheating on the part of the casino."

T: "P and Q are just as poor examples of reasoning as each other."

2 How would you classify how we quantify probabilities? Assign each word from the top row to one or more of the three types and complete the table.

Rational, Emotional, Empirical, Subjective, Objective, Abstract, Concrete, Intuitive, Random, Predictive, Descriptive, Interpretive, Exact, Approximate, Data, Trials, Sample space, Equally likely outcomes, Personal knowledge, Shared knowledge		
Subjective probability	**Experimental probability**	**Theoretical probability**

Just as theoretical probability gives you a way to predict long-term behaviour of relative frequency, a simple rearrangement gives you a way to predict how many times an event is **likely** to occur in a given number of trials.

TOK

Do ethics play a role in the use of mathematics?

Investigation 5

You've established that P(throw a prime number on a fair 12-sided die) $= \frac{5}{12}$.

1 If we threw such a die 2512 times, how many times do you **expect** that a prime number will be thrown?

2 **Think** about and then **write down** your own intuitive answer to this question. **Discuss** your answer with a friend then **share** your ideas with your class. **Test** your prediction with a spreadsheet by following these steps:

- Enter "=RANDBETWEEN(1, 12)" in cell A1 of a new spreadsheet document.
- Copy this down to cell A2512 to create 2512 trials.
- Enter "=IF(OR(A1=2, A1=3, A1=5, A1=7, A1=11), 1, 0)" in cell B1 and copy down to cell B2512.
- Enter "=SUM(B1:B2512)" in cell C1. This is the frequency of primes occurring in your 2512 trials.
- Press F9 to refresh the trials and observe how the frequency compares to what you expected.
- Adapt your spreadsheet for dice with different numbers of sides and a different number of trials.

3 **Predict** other numbers of occurrences using a spreadsheet.

4 **Factual** For 128 trials, what is the expected number of occurrences of the event "throw a square number on a fair 12-sided die"?

5 Show that the formula

Expected number of occurrences of $A = nP(A)$ follows from the formula

$$\text{Relative frequency of } A = \frac{\text{frequency of occurrence of event } A \text{ in } n \text{ trials}}{n}$$

and the fact that you model relative frequency with theoretical probability in the long term.

6 **Conceptual** Why is the formula for an **expected** value and not just a value?

7 **Conceptual** How may we interpret and apply the number quantified by the formula for the expected number of occurrences?

Statistics and probability

Example 2

a A fair coin is flipped 14 times. Predict the number of times you expect a head to be face up.

b Statistical data built up over five years shows that the probability of a student being absent at a school is 0.05. There are 531 students in the school.

Predict the number of students you expect to be absent on any given day and interpret your answer.

c State the assumptions supporting your answer for part **b**.

a $14 \times 0.5 = 7$ Seven heads are expected.	The expected number of occurrences is $n\text{P}(A)$.
b $531 \times 0.05 = 26.55$. So, around 26 or 27 students are expected to be absent.	Note that 26.55 students cannot actually **be** absent.
c This assumes that absences on all days of the year are equally likely.	

Example 3

A coastal town carries out a survey of tourists in July 2018. The accommodation type chosen and the age of the tourist are recorded. The information is given in the table below:

		Accommodation type			
		Hotel	Campervan	Tent	Apartment
Age	18–30	67	81	125	32
	31–50	107	230	73	119
	51–70	87	76	34	89
	>70	109	32	15	54

a A tourist is chosen at random from the survey to receive a promotional offer. Determine what age group the tourist is most likely to belong to.

b Find the probability that the tourist chosen belongs to this age group, expressing your answer as a decimal.

c In July 2019, 16 000 tourists are predicted to visit the town. If the town's hotels have capacity for 5000 predict if the town has enough hotel capacity to meet the demand.

a The table shows that the total number of tourists in the 31–50 age group is $107 + 230 + 73 + 119 = 529$, and this is the most frequent age group.

Apply technology to manage tables of data efficiently. Cell E1 has the formula shown:

E1 =SUM(A1:D1)

	A	B	C	D	E
1	67	81	125	32	305
2	107	230	73	119	529
3	87	76	34	89	286
4	109	32	15	54	210
5	370	419	247	294	1330

b Finding the totals of all the other columns and adding them together gives 1330. Hence the probability required is $\frac{529}{1330} \approx 0.398$	Apply the formula for theoretical probability.
c The table shows that the total number of tourists choosing a hotel for their accommodation is $67 + 107 + 87 + 109 = 370$. Hence the expected number of hotel places is $\frac{370}{1330} \times 16\,000 \approx 4451$	Apply the formula for the expected value. Expressing the value to the nearest whole number is appropriate.
There is sufficient capacity.	State the conclusion.

Exercise 7B

1 A survey was carried out in a small city centre street one Saturday afternoon. Shoppers were asked about how they travelled that day. The results are shown in the table below.

Mode of transport	Car	Bus	Foot
Male	40	59	37
Female	33	41	29

One shopper was randomly selected.

a Find the probability that this shopper travelled by car.

One male shopper was randomly selected.

b Find the probability that this male shopper travelled on foot.

c 1300 shoppers visit the town in one week. Estimate the number of shoppers who travelled by bus.

2 In an experiment, a number a is chosen at random from the set {2, 3, 4, 5} and a number b is chosen at random from the set {3, 4, 5, 6}.

a Find the probability that $\frac{b}{a}$ is a natural number.

b The experiment is repeated 320 times. Find the expected number of times that $a - b$ will be positive.

3 A fair dodecahedral die has faces numbered 1, 2, 3, ..., 12. The die is thrown 154 times. Find the expected number of times that the die will show:

a a factor of 12 **b** a prime number

c a prime factor of 12.

4 The probabilities of each outcome of a biased die are modelled with the following theoretical probabilities:

Outcome	1	2	3	4	5	6
Probability	0.15	0.2	0.25	0.2	0.12	0.08

The die is thrown 207 times. Find the expected number of odd outcomes.

5 A large jar contains 347 marbles, 125 of which are red. A marble is chosen at random and replaced.

Find the expected number of times a non-red marble is chosen in 531 trials.

6 A letter is chosen at random from the word "ICOSAHEDRAL". Find the expected number of times a vowel is chosen in 79 trials.

Statistics and probability

7 Each day in June, Maged records types of cars passing a point on a road popular with tourists. The percentages of each type of car are given in the table below:

Type of car	Classic	Luxury	Compact	Family saloon	Estate	SUV	Other
Percentage	0.5%	1.2%	23.1%	30.9%	15.4%	19.8%	9.1%

a Use this information to predict the number of classic or luxury cars Maged would expect to observe in July given that he observes 573 cars in July.

b State the assumption made in your answer.

8 Quality control is being carried out in a clothing factory. 1.37% of garments produced by machine A have defects and 0.41% of garments produced by machine B have defects.

A quality control manager inspects 67 garments from machine A and 313 garments from machine B. Find the expected number of defects.

9 An artist chooses an area of her neighbourhood to photograph by throwing a dart each morning at a large map pinned to her studio wall.

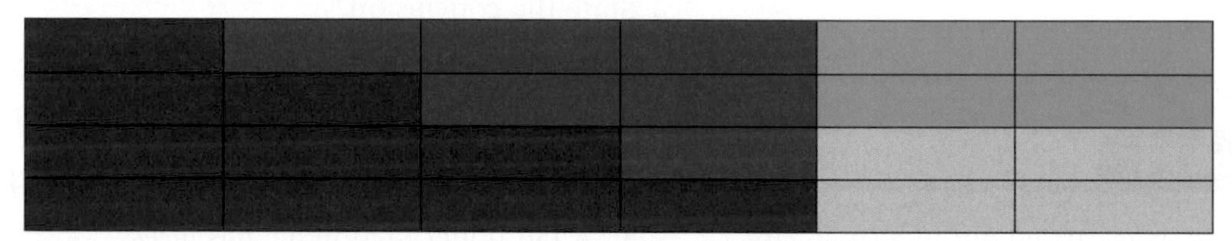

Key:

Red = central business district Blue = government buildings

Green = park Grey = housing

Find the expected number of times the artist will not photograph government buildings given that she takes photographs each day during August.

10 These dice compete in the "Dice world cup". The semifinals are A vs B and C vs D. The winners of each semifinal go in to the world cup final.

A

		4	
4	4	0	0
		4	

B

		3	
3	3	3	3
		3	

C

		2	
6	6	2	2
		2	

D

		1	
5	5	1	5
		1	

Write down your subjective judgment of which die you feel will be most likely to win the "Dice world cup".

Developing inquiry skills

There are four outcomes in the first opening scenario:

- A taxi is yellow and is identified as yellow.
- A taxi is yellow and is identified as black.
- A taxi is black and is identified as yellow.
- A taxi is black and is identified as black.

Are these equally likely outcomes?

In 1000 trials, how many occurrences of each outcome would you expect?

7.2 Representing combined probabilities with diagrams

You have taken the first steps in the quantification of probabilities, experienced random experiments and investigated how to make predictions in the world of chance by applying formulae.

Probability situations themselves have a structure that you can represent in different ways, for example in problems where two or more sets are combined in some way.

Investigation 6

For each situation, think about how best to represent the situation with a diagram.

Compare and contrast your diagrams with others in your class **then** solve the problems.

Situation 1

In a class survey on subject choices, Isabel, Clara, Coco, Anastasiia and Fangyu all state that they study biology. Isabel, Clara, Fangyu and Tomas all study chemistry whereas Barbora and Achille study neither biology nor chemistry.

1 **Factual** Find the probability that a student chosen randomly from this class studies both biology and chemistry.

2 Create your own probability question using your representation of the situation and have another student answer it.

Situation 2

One example of a Sicherman die is a fair cubical die with this net:

		6	
8	4	1	5
		3	

It is thrown together with a fair octahedral die whose faces are numbered 1, 2, 3, 4, 5, 6, 7 and 8.

3 **Factual** Find the probability that the number obtained by adding the two numbers thrown on each die is prime.

4 Find and describe a pattern in your representation of this situation and acquire some knowledge from your pattern.

5 **Conceptual** What advantages are there in using a diagram in problem-solving with combined probabilities?

Two frequently used representations of probability problems are Venn diagrams and sample space diagrams.

A **Venn diagram** represents the sample space in a rectangle. Within the rectangle, each event is represented by a set of outcomes in a circle or an oval shape and is labelled accordingly.

Example 4

In a class survey, Rikardo, Malena, Daniel, Maria, India and James reported that they study environmental systems and societies (ESS). India, Pietro, Mathea and Haneen said that they study geography. Rikardo and James were the only ones who reported that they studied Spanish whereas Sofia and Yulia studied none of the subjects mentioned in the survey. Represent the data in a Venn diagram.

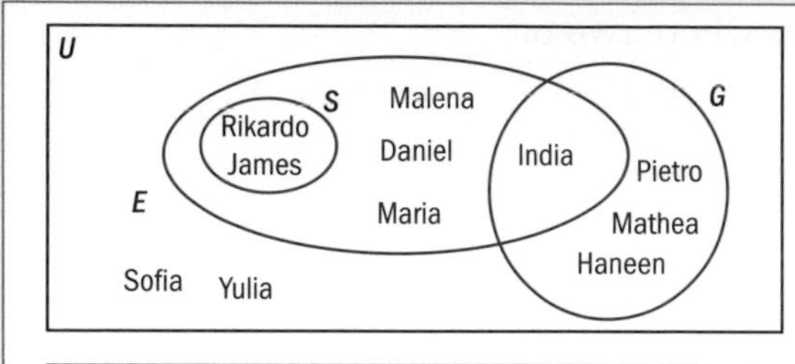

Each event is represented by an italic capital letter.

U represents the entire sample space. In set terminology, this is called the universal set.

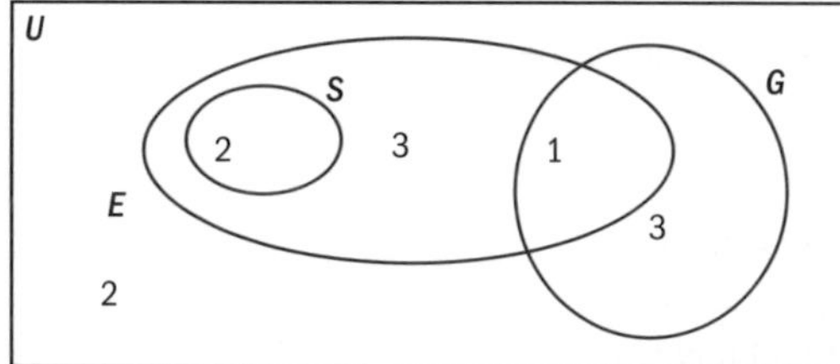

This diagram can be simplified to show the number of students in each region.

Example 5

Use the Venn diagram in Example 4 to find the probabilities that a student chosen randomly from this class:

a studies ESS

b studies ESS but not Spanish

c studies all three subjects

d studies exactly two of the subjects.

a $P(E) = \frac{n(E)}{n(U)} = \frac{2+3+1}{11} = \frac{6}{11}$

P(E) represents the probability of the event "the student chosen at random is from the set E".

b $\frac{4}{11}$

There are a total of four students within the ESS oval but outside the Spanish oval.

c 0

The diagram clearly shows that there are no students who study all three subjects.

d $\frac{3}{11}$

The diagram clearly shows that two students—Rikardo and James—study Spanish and ESS, whereas one student—India—studies both geography and ESS. These are the only three students who study exactly two of the subjects surveyed.

Example 6

In a class of 26 students, it is found that 10 study geography, 16 study history and 4 study neither history nor geography.

a Calculate the number of students who study both history and geography.

b Hence find the probability that a student selected at random from this class studies exactly one of these subjects.

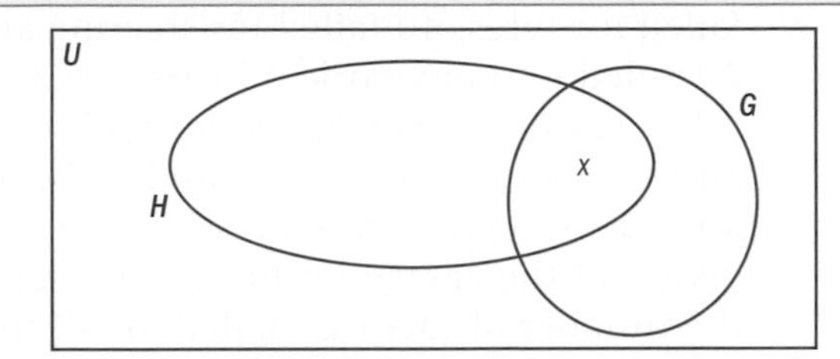

Since the information given asks for the number who study both subjects, you can draw a Venn diagram with two intersecting sets as shown.

Representing the unknown with x is an appropriate problem-solving strategy.

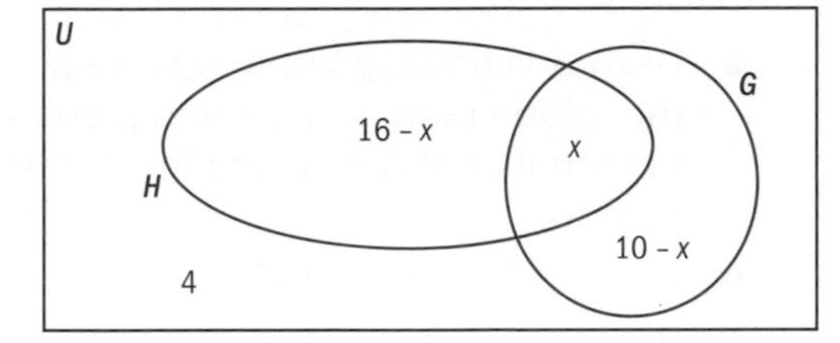

Completely filling in the Venn diagram with all the information given is another problem-solving strategy.

a $(16 - x) + x + (10 - x) = 22$

$\Rightarrow 26 - x = 22$

$\Rightarrow x = 4$

Since 4 study neither history nor geography the sum of these three regions must be 22.

Simplify and solve the equation.

b

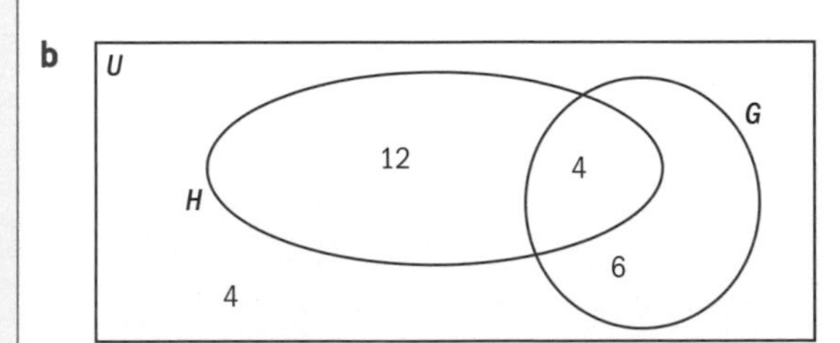

Modify the Venn diagram using your answer to part **a**.

The probability that a randomly selected student studies exactly one of the subjects is

$$\frac{12+6}{26} = \frac{18}{26} = \frac{9}{13}$$

From the Venn diagram read off the information needed to calculate the answer.

Exercise 7C

1 A survey of 127 consumers found that 81 had a tablet computer, 70 had a smartphone and 29 had both a smartphone and a tablet computer.

a Find the number of consumers surveyed who had neither a smartphone nor a tablet.

b Find the probability that when choosing one of the consumers surveyed at random, a consumer who only has a smartphone is chosen.

c In a population of 10 000 consumers, predict how many would have only a tablet computer.

2 In a class of 20 students, 12 study biology, 15 study history and 2 students study neither biology nor history.

a Find the probability that a student selected at random from this class studies both biology and history.

b Given that a randomly selected student studies biology, find the probability that this student also studies history.

c In an experiment, a student is selected at random from this class and the student's course choices are noted. If the experiment is repeated 60 times, find the expected number of times a student who studies both biology and history is chosen.

3 In a survey, 91 people were asked about what devices they use to listen to music. In total, 59 people used a streaming service to listen to music, 44 used a mobile device and 29 used vinyl. Also, 22 people used both a streaming service and a mobile device, 9 used both a mobile device and vinyl and 20 people used only a streaming service. Finally, 8 people said they used all three devices.

a Draw this Venn diagram with numbers assigned to the eight regions.

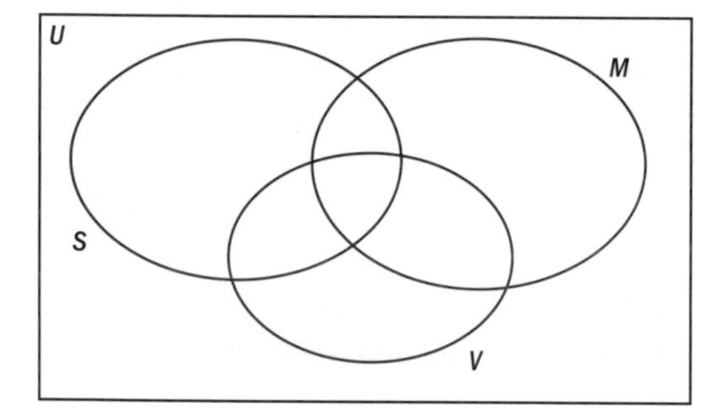

Hence find the probability that a person chosen at random from those surveyed:

b listens to music on exactly one device

c listens to music on exactly two devices.

4 A garage keeps records of the last 94 cars tested for roadworthiness. The main reasons for failing the test are faulty tyres, steering or bodywork. In total, 34 failed for tyres, 40 failed for steering and 29 failed for bodywork. 11 cars failed for other reasons. 7 cars failed for both tyres and steering, 6 for steering and bodywork and 11 for bodywork and tyres. The owner of the garage wishes to calculate the number of cars that failed for all three reasons.

a Draw a Venn diagram to represent the information, using x to represent the number of cars failing for all three reasons.

b Hence calculate the value of x.

c Hence find the probability that a car selected at random from this data set failed for at least two reasons.

A **sample space diagram** is a useful way to represent the whole sample space and often takes the form of a table.

Example 7

It is claimed that when this pair of Sicherman dice is thrown, and the two numbers obtained added together, the probability of each total is just the same as if the two dice were numbered with 1, 2, 3, 4, 5 and 6. Verify this claim.

		2	
4	2	1	3
		3	

		6	
8	4	1	5
		3	

Sample space diagram for the total of two dice numbered 1, 2, 3, 4, 5 and 6:	Form a sample space diagram for each experiment. Enter each total in the table as shown.

	1	2	3	4	5	6
1	2	3	4	5	6	7
2	3	4	5	6	7	8
3	4	5	6	7	8	9
4	5	6	7	8	9	10
5	6	7	8	9	10	11
6	7	8	9	10	11	12

Sample space diagram for the two Sicherman dice:

	1	2	2	3	3	4
1	2	3	3	4	4	5
3	4	5	5	6	6	7
4	5	6	6	7	7	8
5	6	7	7	8	8	9
6	7	8	8	9	9	10
8	9	10	10	11	11	12

Then find the probability of each outcome in the sample space, representing the total as T.

In both tables, $P(T = 2) = P(T = 12) = \frac{1}{36}$

$P(T = 3) = P(T = 11) = \frac{2}{36} = \frac{1}{18}$

$P(T = 4) = P(T = 10) = \frac{3}{36} = \frac{1}{12}$

$P(T = 5) = P(T = 9) = \frac{4}{36} = \frac{1}{9}$

$P(T = 6) = P(T = 8) = \frac{5}{36}$

and $P(T = 7) = \frac{6}{36} = \frac{1}{6}$

State your conclusion.

The probability of each total is the same for each pair of dice, so the claim is true.

Once time has been invested in drawing a diagram, it can be used to quantify many different probabilities.

Example 8

Use the sample space diagram in Example 7 to find the probability that the total found by throwing two Sicherman dice is:

a at most 4 **b** a factor of 24 **c** at least 8.

Continued on next page

TOK

How does a knowledge of probability theory affect decisions we make?

Statistics and probability

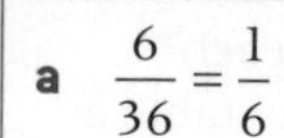

a $\frac{6}{36} = \frac{1}{6}$	The sample space diagram for the total of two dice numbered 1, 2, 3, 4, 5 and 6 is easier to use.
b $\frac{17}{36}$	
c $\frac{15}{36} = \frac{5}{12}$	Shade in or highlight the outcomes required to help count them.

	1	2	3	4	5	6
1	2	3	4	5	6	7
2	3	4	5	6	7	8
3	4	5	6	7	8	9
4	5	6	7	8	9	10
5	6	7	8	9	10	11
6	7	8	9	10	11	12

Exercise 7D

1 Jakub designs a fair cubical die numbered with the first six prime numbers. He throws his die and a fair tetrahedral (four-sided) die numbered with the first four square numbers and writes down the difference between the two numbers, D. Find the probability that D is:

a a prime number **b** a square number.

2 a Two fair cubical dice are rolled in a game. The score is the greater of the two numbers. If the same number appears on both dice, then the score is that number. Find the probability that the score is at most 4.

b In 945 trials of this game, find the expected frequency of the event "the score is at least 4".

3 Alex throws a fair tetrahedral die numbered 1, 2, 3 and 4 and an octahedral (eight-sided) die numbered 1, 2, 3, …, 8. He defines M as the product of his two numbers. Find:

a P(M is odd) **b** P(M is prime)

c P(M is both odd and prime).

Bethany has two fair six-sided dice that she throws. She defines N as the product of her numbers. Find:

d P(N is odd) **e** P(N is more than 13)

f P(N is a factor of 36).

g Bethany and Alex can see that the probability that M is odd equals the probability that N is odd and try to find more events that have the same probabilities in each of their experiments. Find at least one such event.

4 Genetic material contained in pairs of human chromosomes determine whether a child is male or female. Males have the pair XY chromosomes and females XX. Inheriting the XY combination causes male characteristics to develop and XX causes female characteristics to develop. Sperm contain an X or Y with equal probability and an egg always contains an X. An infant inherits one chromosome determining gender from each parent.

a Complete the following *Punnet square* to show the possible outcomes of pairs of chromosomes that can be inherited:

		Chromosome inherited from mother	
		X	X
Chromosome inherited from father	X	XX	
	Y	XY	

b Hence show that the probability a child is born female is 0.5.

Investigation 7

1 In a class of 11 students, students 1, 2, 3, 4, 5 and 6 all study art. Students 3, 4, 7, 8, 9 and 10 all study biology. Choose a diagram to represent this information and find the probability that a student selected randomly from this class studies biology but not art.

2 Two fair cubical dice A and B have faces numbered 1, 2, 3, 4, 5 and 6 and 3, 4, 7, 8, 9 and 10 respectively. The dice are thrown and the total noted. Choose a diagram to represent this information and find the probability that the total is a square number.

Ask yourself: from what I have learned in this section, what are all the similarities and all the differences between Venn diagram and sample space diagram representations of combined events involving sets?

Share your ideas in a group.

3 Construct and share a list of similarities and differences.

		Do you agree with the statement for each type of representation?	
Statement about representation		Venn diagram	Sample space diagram
	Involves construction of a rectangular figure and you filling in (calculated) numbers	Yes	Yes
	Represents how two or more sets relate to each other		
	Enables you to combine elements of two sets in many ways		
	Enables you to combine elements of three sets in many ways		
	...	...	...
	...	...	...

4 **Conceptual** How can we decide how best to represent events given in the context of combined events involving sets?

International-mindedness

A well-known French gambler, Chevalier de Méré consulted Blaise Pascal in Paris in 1650 with questions about some games of chance. Pascal began to correspond with his friend Pierre Fermat about these problems which began their study of probability.

Exercise 7E

1 Diego inspects the music collection in his study. He finds that he has albums by artists A, B, C, G, H, R, Q, L and M on CD and albums by artists B, C, V, D, E, H, I, J and M on vinyl.

Diego chooses an artist at random. Find the probability that he does not have a choice of formats on which to listen to his artist's music.

2 Two boxes A and B contain numbered cards as shown below:

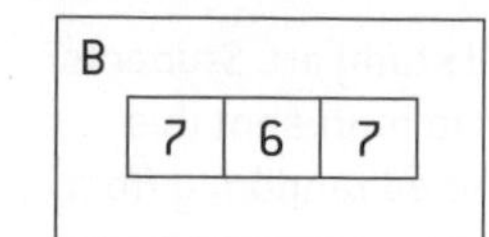

Two cards are chosen at random, one from each box and the number obtained from box B is raised to the power of the number obtained from box A. Find the probability that the number obtained is a factor of 5764801.

3 Assuming that the probabilities of a kitten being born male or female are the same, what is the probability that in a litter of four kittens there are two males and two females?

4 Pierfranco designs a game.

- He throws a fair coin with one side labelled X and the other labelled Y.
- If the outcome is X, a number is chosen at random from the set {5, 10, 15, 20, 25, 30, 45}.
- If the outcome is Y, the computer simulation chooses from the set {30, 40, 50, 60, 70, 80, 90}.

Pierfranco scores a point if the outcome is a multiple of 6.

Find the expected number of points Pierfranco would score if he played the game 54 times.

5 Rea throws two fair cubical dice numbered 1, 2, 3, ..., 6. Let R be the total of the numbers obtained by Rea.

Teodora throws a fair tetrahedral die numbered 1, 2, 4 and 8 and a fair five-sided die numbered 1, 2, 3, 4 and 5. Let the total of the numbers obtained by Teodora be T.

Determine which event is more likely: $R = 5$ or $T = 5$.

6 These dice compete in the "Dice world cup". A pair of dice is thrown and the highest number wins. The semifinals are A vs B and C vs D. The winners of each semifinal go in to the world cup final.

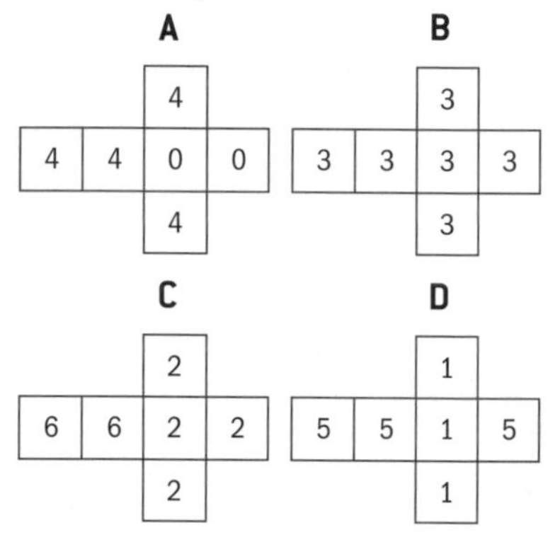

Construct sample space diagrams to find the probabilities of the outcomes of each semifinal.

Developing inquiry skills

In the first opening scenario, imagine 100 trials. How many outcomes would you expect in each area shown on this diagram?

		Cab yellow?	
		Yes	No
Witness correct?	Yes	??	??
	No	??	??

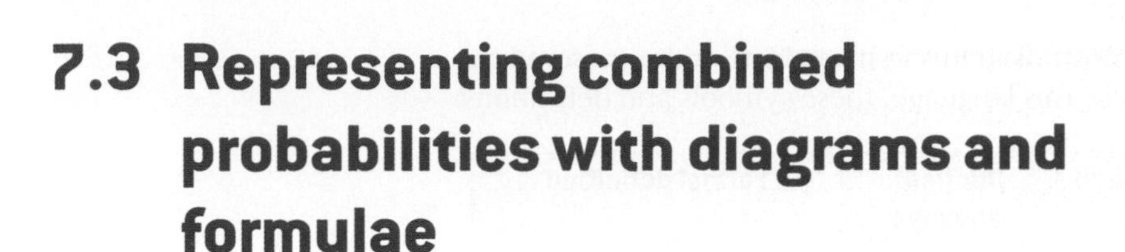

7.3 Representing combined probabilities with diagrams and formulae

In Section 7.2 you found probabilities by representing combined events in a sample space diagram or a Venn diagram. There are other ways to find probabilities of combined events, which can add to your problem-solving skills.

Investigation 8

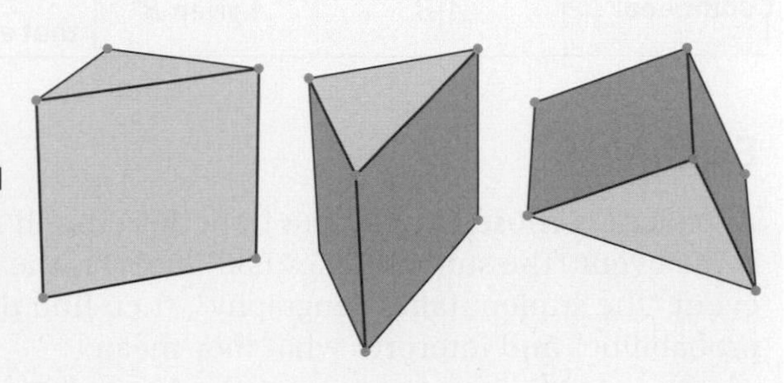

1 Retrieve your data from the triangular prism investigation in Section 7.1.

2 **Collaborate** in a pair to use your 100 trials to find experimental probabilities of the events A, B, C, D and E of the prism falling face down on these faces respectively.

3 In a pair, each put your own data for your 50 trials into one of the columns of a table:

First outcome	Second outcome	Combined outcome
B	C	BC
A	A	AA
E	C	EC
C	B	CB
...	...	...
...	...	...

4 Would the event AA be as likely as or less likely than the event A? Use subjective probability and reasoning to justify your answer.

5 **Factual** What is the experimental probability of the combined event AA?

6 **Factual** What is the relationship between your answer to **5** and the experimental probability of A? Is this relationship exact or approximate?

7 Repeat steps **4** and **5** for other combined events such as BD etc. If necessary, you may wish to collaborate by sharing data from the whole class on an online spreadsheet.

8 **Conceptual** Can you generalize your findings by completing: "The probability of event X and event Y **both** occurring is ..."

(Your generalization is true only for **independent** events, which are explored later in this section.)

9 **Conceptual** By what other methods, apart from diagrams, can combined probabilities be found?

Statistics and probability

In this section you will use Venn diagrams to investigate and represent laws of probability and you will use this language, these symbols and definitions:

Name	Symbol applied to event(s)	Informal language	Formal definition
Intersection	$A \cap B$	"A and B"	Events A and B **both** occur
Union	$A \cup B$	"A or B"	Events A **or** B **or both** occur
Complement	A'	"Not A"	Event A **does not** occur
Conditional	$A \mid B$	"A given B"	Event A occurs, **given that** event B **has** occurred

Example 9

A student is chosen at random from this class. If E is the event "the student takes ESS" and G is the event "the student takes geography", then find these probabilities and interpret what they mean:

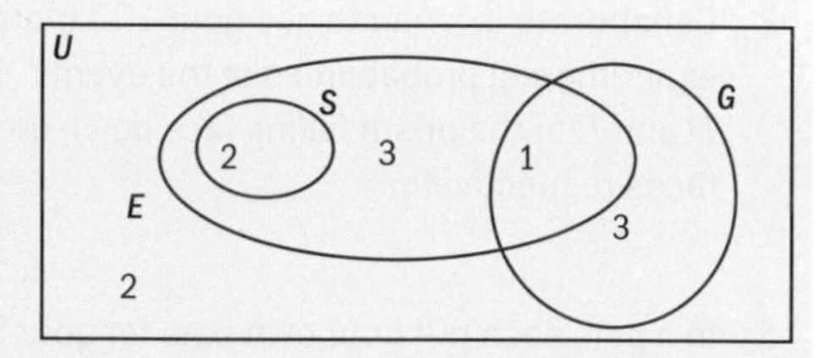

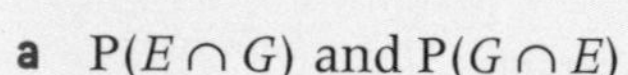

a $P(E \cap G)$ and $P(G \cap E)$ **b** $P(E \cup G)$ and $P(G \cup E)$

c $P(E')$ **d** $P(E|G)$ and $P(G|E)$.

a $P(E \cap G) = P(G \cap E) = \dfrac{1}{11}$ is the probability that a randomly chosen student studies **both** ESS and geography.

Only one student takes both ESS and geography. This example illustrates that $E \cap G$ means exactly the same as $G \cap E$. In general, such terms are always the same.

b $P(E \cup G) = P(G \cup E) = \dfrac{2+3+1+3}{11} = \dfrac{9}{11}$ is the probability that a randomly chosen student studies ESS **or** geography **or both**.

Similarly, $E \cup G$ means the same as $G \cup E$ hence $P(E \cup G) = P(G \cup E)$ is always true.

c $P(E') = \dfrac{5}{11}$ is the probability that a randomly chosen student **does not** study ESS.

There are 5 students outside the ESS oval. $P(E') = 1 - \dfrac{6}{11} = \dfrac{5}{11}$ is another way to find the probability required.

d $P(E|G) = \dfrac{1}{1+3} = \dfrac{1}{4}$, the probability that a randomly chosen students studies ESS **given** that he/she studies geography, whereas $P(G|E) = \dfrac{1}{2+3+1} = \dfrac{1}{6}$.

Since it is **given** that G has occurred, the sample space is now G, not U.

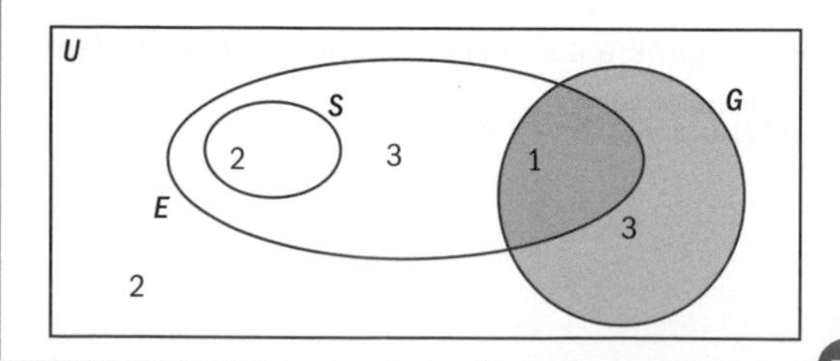

These are not equal since the information **given** changes the sample space. This example shows that the statement $P(E\|G) = P(G\|E)$ is not generally true.	Only 1 student studies ESS and geography hence $P(E\|G) = \frac{1}{4}$. Notice how this contrasts with $P(E) = \frac{6}{11}$.

Just as areas of mathematics like trigonometry or sequences have formulae, so does probability. In this investigation you will see some relationships that you can generalize as **laws of probability**.

Investigation 9

The following Venn diagrams represent how many students study art or biology in four different classes, using the sets A and B.

Fill in the probabilities for each Venn diagram and investigate your answers.

	1	2	3	4	5	6	7	8
Venn diagram	$P(A)$	$P(A')$	$P(A)+P(A')$	$P(B)$	$P(A)+P(B)$	$P(A\cup B)$	$P(A\cap B)$	$P(A)+P(B)-P(A\cap B)$
Class of 2019 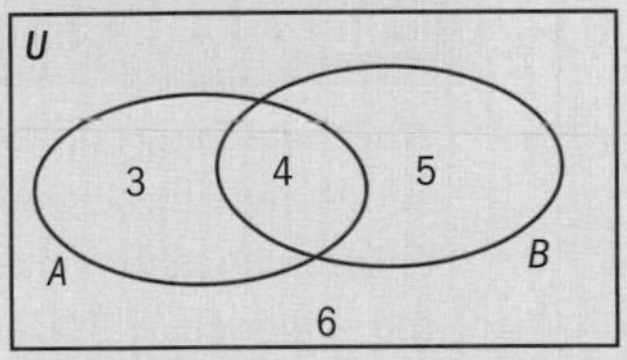 U; A only: 3; A∩B: 4; B only: 5; outside: 6								
Class of 2018 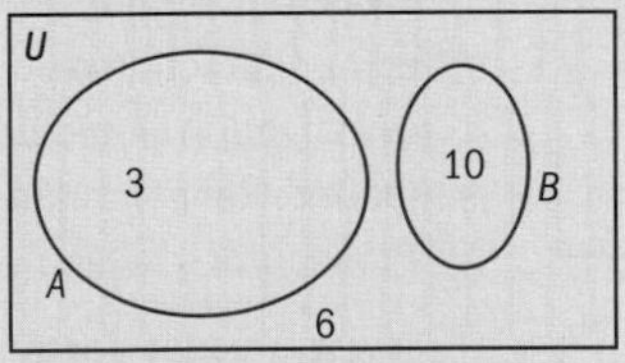 U; A: 3; B: 10; outside: 6								
Class of 2017 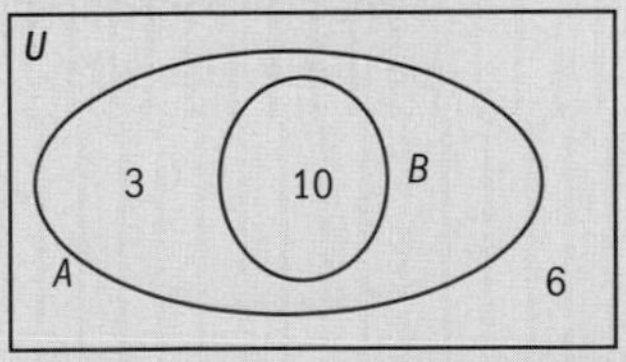 U; A only: 3; B (inside A): 10; outside: 6								
Class of 2016 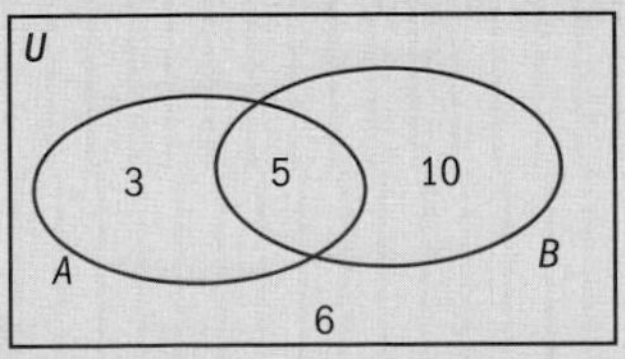 U; A only: 3; A∩B: 5; B only: 10; outside: 6								

Continued on next page

Answer these questions and discuss your answers in a group.

1. What relationship is shown in column 3?
2. This relationship is true in general. Why?
3. What relationship exists between the probabilities in columns 6 and 8?
4. This relationship is true in general. Why?
5. **Factual** How can you tell that two events cannot **both** occur?
6. Which class shows mutually exclusive events?
7. For which class can the relationship you wrote for question **3** be simplified?
8. Complete the table to summarize and justify what you have discovered.

Left-hand side of a law of probability	=	Right-hand side	When true?
$P(A \cup B)$	=	$P(A)+P(B)-$ ____	Always
$P(A \cap B)$	=	0	Only if A and B are ____ events.
Therefore, $P(A \cup B)$	=	$P(A)+P(B)$	

Two events A and B are **mutually exclusive** if they cannot **both** occur. Hence:

- Knowing that A has occurred means you know that B cannot, and that knowing that B occurs means you know that A cannot.
- The occurrence of each event **excludes** the possibility of the other.
- Consequently, $P(A \cap B)=P(B \cap A)=0$.

A and A' are **complementary** events. This means that

- A and A' are mutually exclusive.
- $P(A)+P(A')=1$

TOK

When watching crime series, reading a book or listening to the news, the evidence of DNA often closes a case. If it was that simple, no further detection or investigation would be needed. Research the "Prosecutor's fallacy".

How will reason contrast with emotion in making a decision based solely on DNA evidence?

Laws of probability can be used in the calculation of probabilities of combined events and also to make and justify statements about combined events.

Example 10

Catarina explores the names of the students in her class of 15 students.

She sorts through the data from all 15 students and determines the sets A and B:

A = {Students with exactly two vowels in their given name}
= {Clara, Tomas, Fanygu, Rea, James}

B = {Students with exactly three vowels in their given name}
= {Barbora, Achille, Malena, Daniel, Oliver, Rikardo}

a Represent Catarina's information on a Venn diagram.

A student from the class is chosen at random.

b Find P(A), P(B) and P($A \cap B$).

c State with a reason whether events A and B are mutually exclusive.

d Hence write down P($A \cup B$).

a

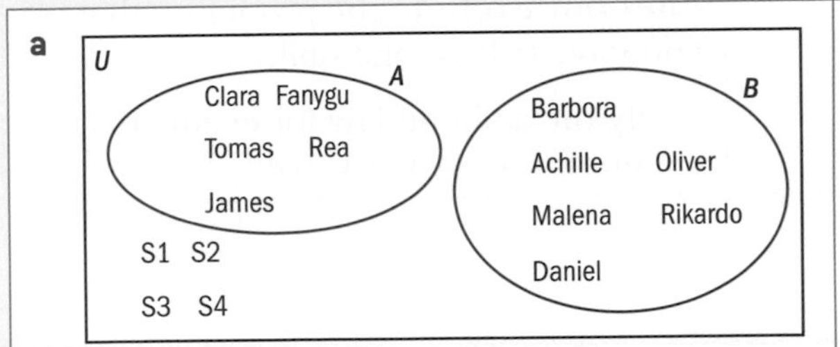

The sample space has 15 students; however, you are only given 11 names. Hence the remaining students have been denoted S1, S2, S3 and S4 for completeness.

b $\text{P}(A) = \frac{5}{15} = \frac{1}{3}$, $\text{P}(B) = \frac{6}{15} = \frac{2}{5}$ and $\text{P}(A \cap B) = 0$

c The events A and B are mutually exclusive since $\text{P}(A \cap B) = 0$.

Having exactly two vowels in a given name **excludes** the possibility that there are exactly three vowels.

d $\text{P}(A \cup B) = \frac{5}{15} + \frac{6}{15} = \frac{11}{15}$

Apply the addition law of probability for mutually exclusive events.

Example 11

Catarina further explores the number of siblings of the students in her class of 15 students.

She writes down another set in addition to those from Example 10:

C = {Students with at least one sibling} = {Alexandra, Isabella, Lukas, Paula, Oliver, Rikardo}

a Draw sets B and C on a Venn diagram.

A student from the class is chosen at random.

b Find P(B), P(C) and P($B \cap C$).

c State with a reason whether events B and C are mutually exclusive.

d Hence write down P($B \cup C$).

a

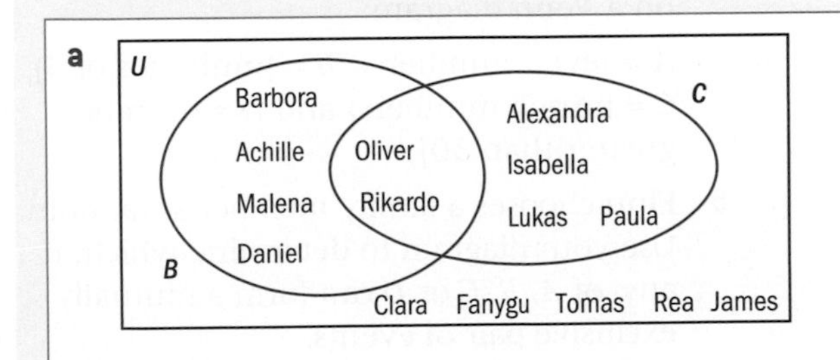

All 15 students can now be represented in the Venn diagram.

Continued on next page

b $P(B) = \frac{6}{15} = \frac{2}{5}$, $P(C) = \frac{6}{15} = \frac{2}{5}$ and $P(B \cap C) = \frac{2}{15}$	
c The events B and C are not mutually exclusive since $P(B \cap C) \neq 0$.	Having three vowels in your given name **does not exclude** the possibility that you have at least one sibling.
d $P(B \cup C) = \frac{6}{15} + \frac{6}{15} - \frac{2}{15} = \frac{10}{15} = \frac{2}{3}$	Apply the addition law for events that are not mutually exclusive.

Exercise 7F

1 A fair decahedral die numbered 1, 2, 3, ..., 10 is thrown and the number noted.

The events A, "throw a square number", and B, "throw a factor of six", are represented on the Venn diagram below:

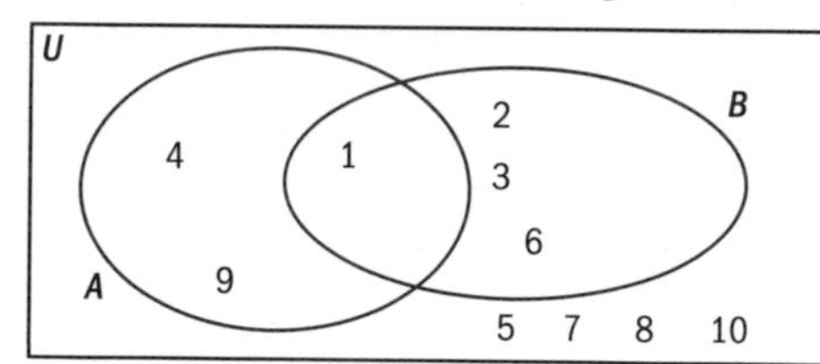

a Find $P(A)$, $P(B)$, $P(A \cap B)$ and $P(A \cup B)$.

b Hence show that $P(A \cup B) = P(A) + P(B) - P(A \cap B)$.

c State with a reason whether events A and B are mutually exclusive.

2 A fair dodecahedral die numbered 1, 2, 3, 4, 8, 9, 16, 27, 32, 81, 243 and 729 is thrown and the number noted.

The events C, "throw an odd number", and D, "throw an even number", are represented on the Venn diagram below:

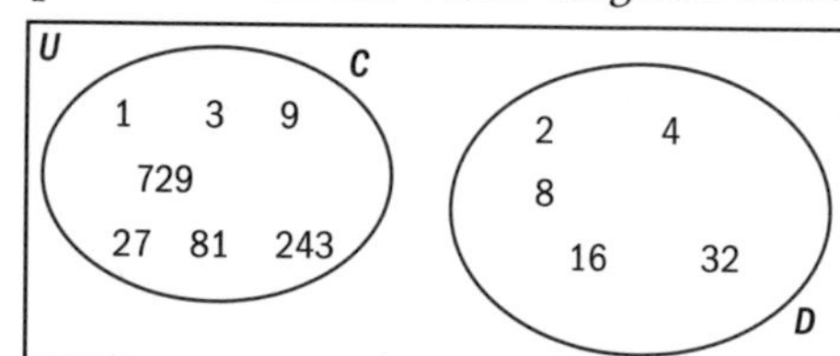

a Find $P(C)$, $P(D)$, $P(C \cap D)$ and $P(C \cup D)$.

b Hence show that $P(C \cup D) = P(C) + P(D)$.

c State with a reason whether events C and D are mutually exclusive.

3 A school is inspecting 24 student lockers before the start of the new academic year to see if they have been left tidy. It is found that some lockers have some food items left inside and some lockers have stationery items inside. Lockers 2, 5, 7, 8, 11, 17, 18 and 19 all have food items and lockers 1, 3, 4, 11, 13, 15, 17, 20 and 21 all have stationery items.

a Draw this information on a Venn diagram.

b State with a reason whether events "a randomly chosen locker contains foot items" and "a randomly chosen locker contains stationery items" are mutually exclusive.

c Find the probability that a locker selected at random has at least one type of item left inside.

4 Finn explores the ages of the people in his family. He represents his family's ages with the set $U = \{2, 3, 4, 6, 8, 9, 10, 12, 14, 15, 16, 25, 35, 55, 65\}$.

a Draw U and each of the following sets on a Venn diagram:

A = {even numbers}, B = {multiples of 3}, C = {prime numbers} and D = {numbers greater than 30}.

b Finn chooses a family member at random. Use your diagram to determine which, if any, of A, B, C or D can form a mutually exclusive pair of events.

You have learned the concept of **mutually exclusive events**. In the following investigation you learn about **independent events**. These terms are easy to confuse with each other.

> **TOK**
>
> What do we mean by a "fair" game? Is it fair that casinos should make a profit?

> Two events A and B are **independent** if the occurrence of each event does **not affect in any way** the occurrence of the other.
>
> Equivalently, **knowing** that A has occurred does not affect the probability that B does, and **knowing** that B occurs does not affect the probability that A does.
>
> Two events A and B are **dependent** if they are not independent.

Investigation 10

Fill in the probabilities for each Venn diagram and investigate your answers.

	1	2	3	4	5	6
Venn diagram	$P(A)$	$P(B)$	$P(A \cap B)$	$P(A\|B)$	$P(A\|B)P(B)$	$\frac{P(A \cap B)}{P(B)}$
Class of 2019 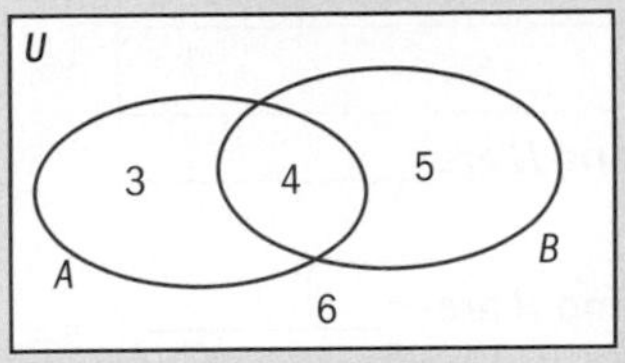						
Class of 2018 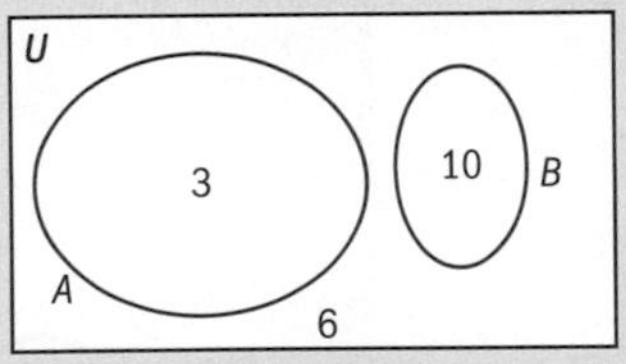						
Class of 2017 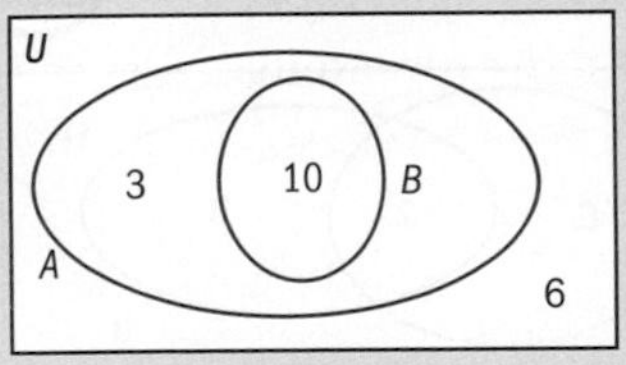						
Class of 2016 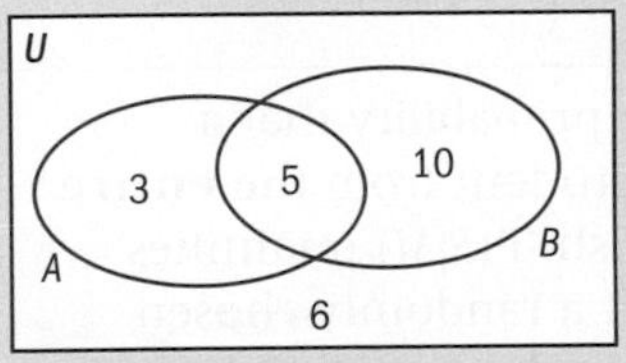						

Continued on next page

Statistics and probability

Answer these questions and discuss your answers in a group.

1 What relationship exists between the probabilities in columns 4 and 6? This relationship is true in general.

2 In which of the four cases above does **knowing** that a student studies biology **affect** the probability that they study art? How can you tell?

3 In which of the four cases above does **knowing** that a student studies biology **not** affect the probability that they study art? How can you tell?

Use your answers to **2** and **3** to find which of the following Venn diagrams represent pairs of independent events.

Class of 2015

U B A 6 2 3 9

Class of 2014

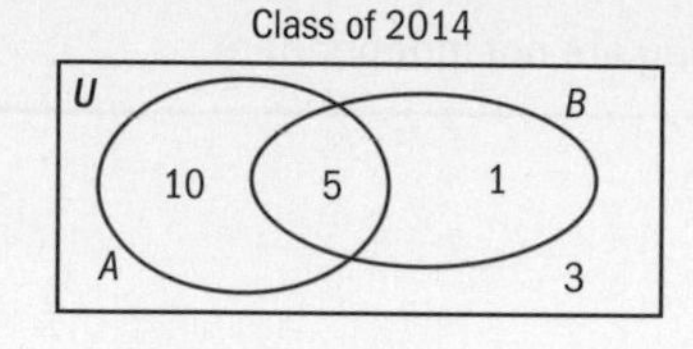

Class of 2013

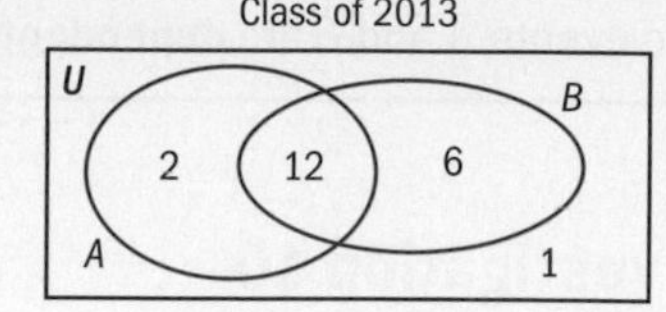

4 Complete the table to summarize and justify what you have discovered.

Left-hand side of a law of probability	=	Right-hand side	When true?
$P(A\|B)$	=	$\frac{P(A \cap B)}{\_\_}$	Always
Therefore, $P(A \cap B)$	=		Always
$P(A\|B)$	=	$P(A)$	Only if A and B are ________ events
Therefore, $P(A \cap B)$	=	$P(A) \times$ ______	Only if A and B are ________ events

These laws of probability can be used in making and justifying statements about combined events.

Example 12

This Venn diagram shows the number of students in a class who study Spanish and the number of students who study mathematics.

Determine whether S and M are independent events by:

a calculating and considering the values of $P(S)$ and $P(S|M)$

b calculating and considering the values of $P(S) \times P(M)$ and $P(S \cap M)$.

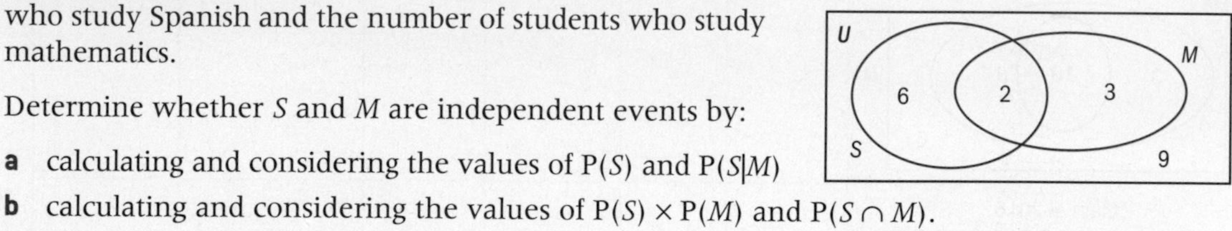

a $P(S) = \frac{6+2}{6+2+3+9} = \frac{8}{20} = \frac{2}{5}$ $P(S\|M) = \frac{2}{2+3} = \frac{2}{5}$	This quantifies the probability that a randomly chosen student from the **entire class** studies Spanish. $P(S\|M)$ quantifies the probability that a randomly chosen student from the **students who study mathematics** studies Spanish.

Since $P(S) = P(S \mid M)$, S and M are independent events.	Knowing that the randomly chosen student studies mathematics does not affect the probability that the student studies Spanish.
b Since $P(S) \times P(M) = \frac{2}{5} \times \frac{5}{20} = \frac{2}{20} = P(S \cap M)$, S and M are independent events.	This is an equivalent way to show the events are independent.

Now that you have learned about mutually exclusive events and independent events, you can gain knowledge and understanding about how these terms differ.

Investigation 11

At each stage, compare and contrast your ideas and results with a classmate.

1 Find the probabilities:

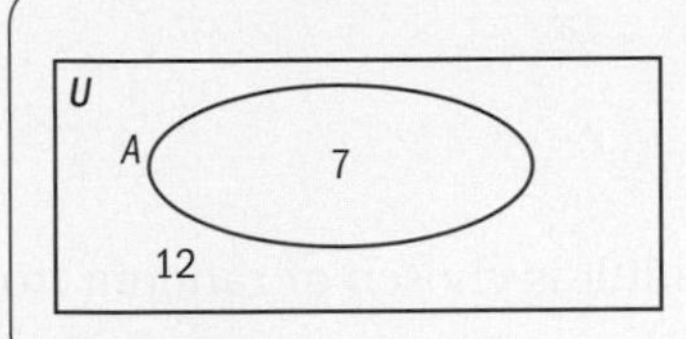

$P(A) =$		
$P(A') =$		
$P(A \cap A') =$		
$P(A \mid A') =$		

2 **Conceptual** Are complementary events always mutually exclusive? How do you know?

3 **Conceptual** Are complementary events always independent events? How do you know?

4 Identify a Venn diagram from the previous investigation that represents two mutually exclusive events that are not independent.

5 Identify a Venn diagram from the previous investigation that represents two independent events that are not mutually exclusive.

Write down your own three positive integers x, y and n and use them to fill in this Venn diagram, which represents two events A and B.

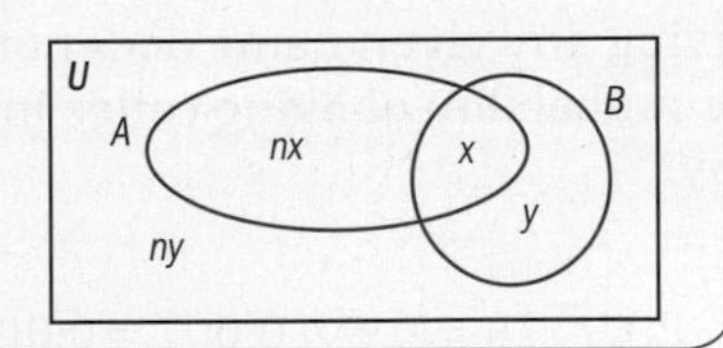

6 **Factual** Are A and B mutually exclusive? Are they independent?

7 **Conceptual** Can non-mutually exclusive events be independent?

Adapt your Venn diagram by replacing y with $y + 1$ but leaving ny, x and nx alone.

8 **Factual** Are A and B mutually exclusive? Are they independent?

9 **Conceptual** Can non-mutually exclusive events be dependent?

10 **Factual** If C represents the event "throw a 2 on a cubical die" and D the event "throw an odd number on a cubical die", are these events mutually exclusive? Are they independent?

Given the event C, write down your own event that excludes the possibility of C happening.

11 **Factual** Are your events independent?

Continued on next page

12 **Conceptual** Can mutually exclusive events be independent?

13 **Factual** If D represents the event "toss a head on a fair coin", would C and D be mutually exclusive? Would they be independent?

Given the event C, write down your own event that is independent of C happening.

14 **Factual** Are your events mutually exclusive?

15 **Conceptual** Must independent events be mutually exclusive?

16 **Factual** Choose the correct word to complete each sentence:

Complementary events are	sometimes / never / always	mutually exclusive events
Dependent events are		independent events
Mutually exclusive events		dependent events
Independent events are		mutually exclusive events

You can use the laws of probability to justify other statements.

Example 13

For a consumer survey, 2371 adults are asked questions. An adult is chosen at random from those taking part.

C is the event "the adult likes coffee".

D is the event "the adult is called David".

a You are given that C and D are independent events. Explain why this is the case.

b Given that $P(C) = 0.8$ and $P(D) = 0.007$, interpret your answers in context in each of the following. Find:

i $P(C \cap D)$ **ii** $P(C \cup D)$.

c Determine whether C and D are mutually exclusive. Justify your answer.

a Having any given name does not affect the probability of liking coffee in any way.	Recall and apply the definition of independent events. You could also say liking coffee does not affect the probability of having any particular name.
b i $P(C \cap D) = 0.8 \times 0.007 = 0.0056$ Choosing a coffee drinker called David is less likely than choosing a David and less likely than choosing a coffee drinker.	Since you are given that C and D are independent, you can use $P(C \cap D) = P(C) \times P(D)$.
ii $P(C \cup D) = 0.8 + 0.007 - 0.0056 = 0.8014$ Choosing a coffee drinker or someone called David is more likely than choosing a David and more likely than choosing a coffee drinker.	Apply the formula: $P(C \cup D) = P(C) + P(D) - P(C \cap D)$.
c C and D are not mutually exclusive because $P(C \cap D) \neq 0$.	Write a complete and clear reason.

Example 14

The Venn diagram shows the number of students in a class who can speak Spanish, Italian, both these languages and neither language.

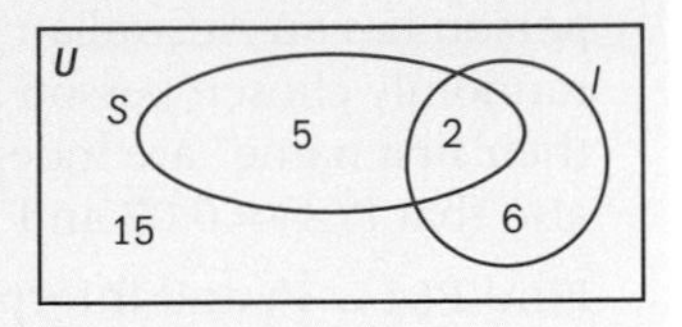

A student is chosen at random from the class.

Let S and I be the events "choose a Spanish speaker" and "choose an Italian speaker" respectively.

Find:

a $P(S)$ **b** $P(I)$ **c** $P(S|I)$ **d** $P(I|S)$ **e** $P(I \cap S)$ **f** $P(S) \times P(I)$.

Hence determine whether S and I are independent.

a $P(S) = \frac{n(S)}{n(U)} = \frac{5+2}{15+5+2+6} = \frac{7}{28} = \frac{1}{4}$ **b** $P(I) = \frac{n(I)}{n(U)} = \frac{2+6}{15+5+2+6} = \frac{8}{28} = \frac{2}{7}$	Write down the appropriate formula and fill in the numbers from the Venn diagram to show complete working out that is easy to check.
c $P(S\|I) = \frac{2}{2+6} = \frac{2}{8} = \frac{1}{4}$ **d** $P(I\|S) = \frac{2}{5+2} = \frac{2}{7}$ **e** $P(I \cap S) = \frac{2}{28} = \frac{1}{14}$	Recall that $P(S\|I)$ means you **know** the student chosen **does** speak Italian. So there are 8 students to consider, 2 of whom speak Spanish.
f $P(S) \times P(I) = \frac{1}{4} \times \frac{2}{7} = \frac{1}{14}$ Since $P(S\|I) = P(S)$, S and I are independent.	Write a complete and clear reason. $P(I\|S) = P(I)$ and $P(I \cap S) = P(I) \times P(S)$ are equivalent explanations.

Statistics and probability

Exercise 7G

1 For these pairs of events, state whether they are mutually exclusive, independent or neither.

a A = throw a head on a fair coin
B = throw a prime number on a fair die numbered 1, 2, 3, 4, 5, 6

b C = it will rain tomorrow
D = it is raining today

c D = throw a prime number on a fair die numbered 1, 2, 3, 4, 5, 6: E = throw an even number on the same die

d F = throw a prime number on a fair die numbered 1, 2, 3, 4, 5, 6: G = throw an even number on another die

e G = choose a number at random from {1, 2, 3, 4, 5, 6, 7, 8, 9, 10} that is at most 6, H = choose a number from the same set that is at least 7

f M = choose a number at random from {1, 2, 3, 4, 5, 6, 7, 8, 9, 10} that is no more than 5, H = choose a number from the same set that is 4 or more

g S = choose a Spanish speaker at random from a set of students represented by the set below, T = choose a Turkish speaker at random from this set

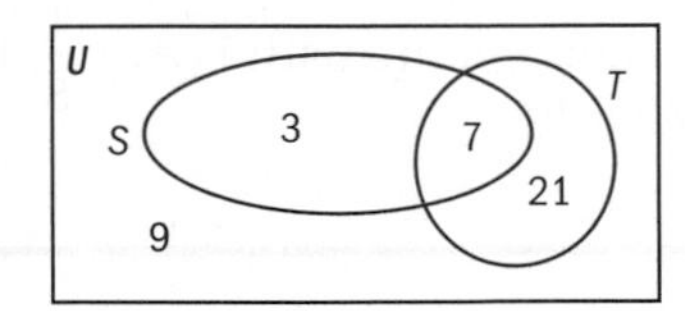

2 In a survey carried out in an airport, it is found that the events A: "a randomly chosen person has an Australian passport" and V: "a randomly chosen person has three vowels in their first name" are independent. It is found also that $P(A) = 0.07$ and $P(V) = 0.61$.

Find $P(A \cup V)$ and interpret its meaning.

3 A class of undergraduate students were asked in 2016 their major subject and whether they listen to music on their commute to university. S is the set of science majors and M is the set of students who listen to music on their commute.

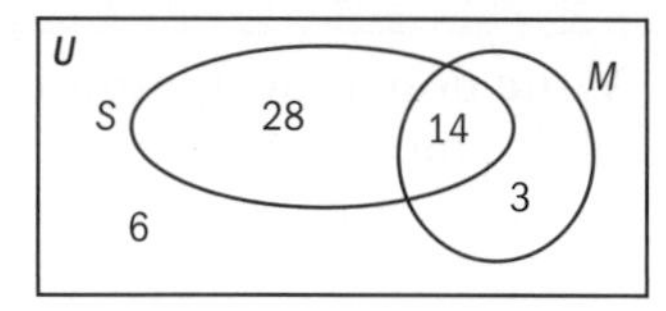

a Find $P(S) \times P(M)$ and $P(S \cap M)$, and hence determine whether S and M are independent events, stating a reason for your answer.

b The same questions were asked in a survey in 2017 with the results given in the Venn diagram below.
Find $P(S)$ and $P(S|M)$ and hence determine whether S and M are independent events, stating a reason for your answer.

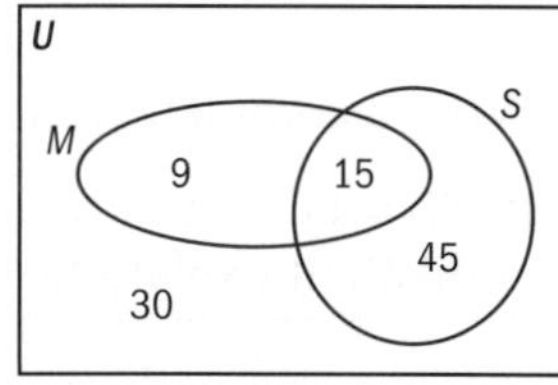

4 Students are surveyed about languages spoken. You are given the following Venn diagram that represents the events A: "a randomly chosen person can speak Arabic" and R: "a randomly chosen person can speak Russian".

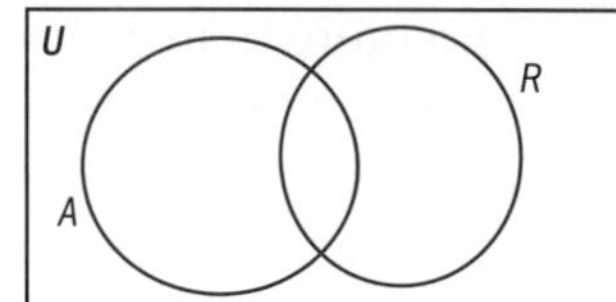

You are given also that A and R are independent events with $P(A) = \frac{1}{8}$, $P(R) = \frac{2}{7}$ and $n(U) = 56$.

a Draw the Venn diagram showing the numbers in each region.

b Hence find the probability that a student selected at random from this group speaks no more than one of the languages.

5 The letters of the word MATHEMATICS are written on 11 separate cards as shown below:

M	A	T	H	E	M
A	T	I	C	S	

a A card is drawn at random then replaced. Then another card is drawn.

Let A be the event the first card drawn is the letter A.

Let M be the event the second card drawn is the letter M. Find:

i $P(A)$ **ii** $P(M|A)$

iii $P(A \cap M)$.

b In a different experiment, a card is drawn at random and **not** replaced. Then another card is drawn. Re-calculate the probabilities that you found in part **a**.

6 A group of 50 investors own properties in north European cities. The following Venn diagram shows how many investors own properties in Amsterdam, Brussels or Cologne. One of the investors is chosen at random.

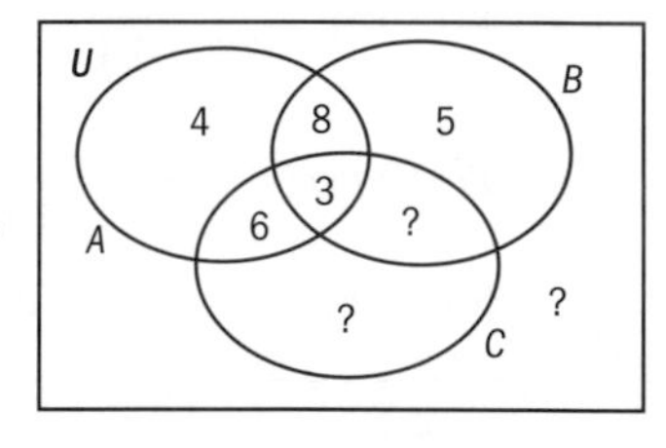

a Find $P(B \mid A)$.

b Find $P(C \mid A)$.

c You are given that $P(C \mid B) = \frac{10}{23}$ and $P\left(A|C\right) = \frac{1}{3}$. Calculate the remaining regions shown in the Venn diagram.

d Interpret your answers for **a** and **b**.

Developing inquiry skills

Which of the events in the first opening scenario are independent? Which are mutually exclusive?

7.4 Complete, concise and consistent representations

You can use diagrams as a rich source of information when solving problems. Choosing the correct way to represent a problem is a skill worth developing. For example, consider the following problem:

In a class of 15 students, 3 study art, 6 biology of whom 1 studies art. A student is chosen at random. How many simple probabilities can you find? How many combined probabilities can you find?

Let A represent the event "An art student is chosen at random from this group" and B "A biology student is chosen at random from this group".

If you represent the problem only as **text**, the simple probabilities $P(A) = \frac{1}{5}$ and $P(B) = \frac{2}{5}$ can be found easily but calculating these do not show you the whole picture of how the sets relate to each other.

TOK

In TOK it can be useful to draw a distinction between shared knowledge and personal knowledge. The IB use a Venn diagram to represent these two types of knowledge. If you are to think about mathematics (or any subject, in fact) what could go in the three regions illustrated in the diagram?

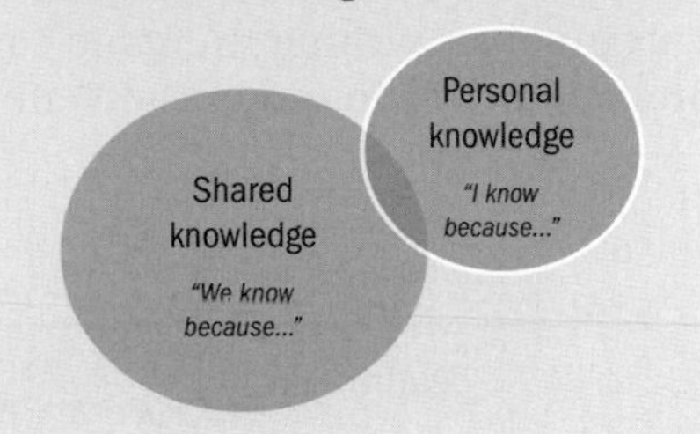

Represent this information as follows in a **Venn diagram** to see more detail:

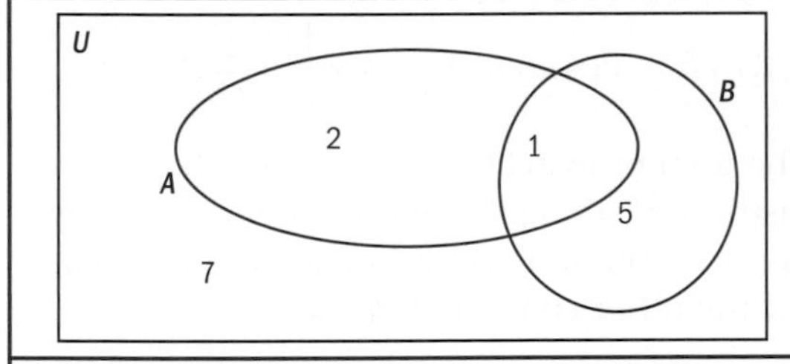	The rectangle represents the sample space U for which $P(U) = 1$, the total probability. The diagram allows us to find $P(B \mid A) = \frac{1}{3}$, $P(B' \mid A) = \frac{2}{3}$ etc easily.
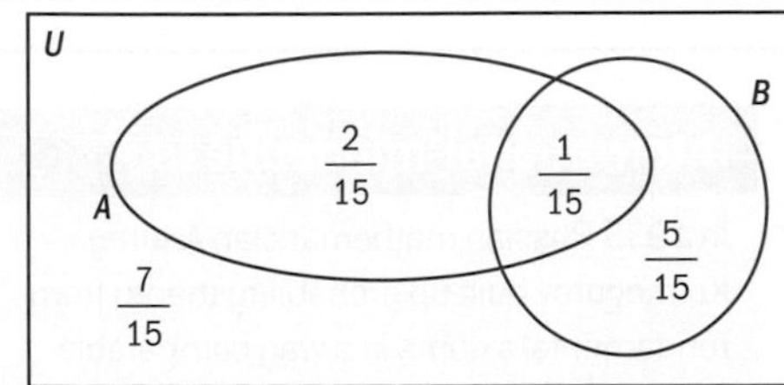	The Venn diagram can be adapted to show the distribution of the total probability in **four** regions that represent mutually exclusive events: $P(A \cap B) = \frac{1}{15}$, $P(A' \cap B) = \frac{5}{15}$, $P(A \cap B') = \frac{2}{15}$ and $P(A' \cap B') = \frac{7}{15}$.

Hence the probability that a randomly chosen student studies neither biology nor art is $P(A' \cap B') = \frac{7}{15}$. The simple probability

$\mathrm{P}(B) = \mathrm{P}(A \cap B) + \mathrm{P}(A' \cap B) = \frac{1}{15} + \frac{5}{15} = \frac{2}{5}$ is represented as a union of two mutually exclusive events in the Venn diagram.

The Venn diagram can be therefore be used to find all the simple, combined and conditional probabilities.

Represent the problem as a **tree diagram** by first choosing one student from the group and determining whether they study art or not.

The probabilities in this process can be represented as a tree with the two events A and A'.

A: $\mathrm{P}(A) = \frac{1}{5}$

A': $\mathrm{P}(A') = \frac{4}{5}$

To construct the next part of the tree, imagine a student who **does** study art and consider whether this student studies biology or not.

This involves writing the same conditional probabilities as found in the Venn diagram.

$\mathrm{P}(A) = \frac{1}{5}$, then from A: B with $\mathrm{P}(B|A) = \frac{1}{3}$; B' with $\mathrm{P}(B'|A) = \frac{2}{3}$

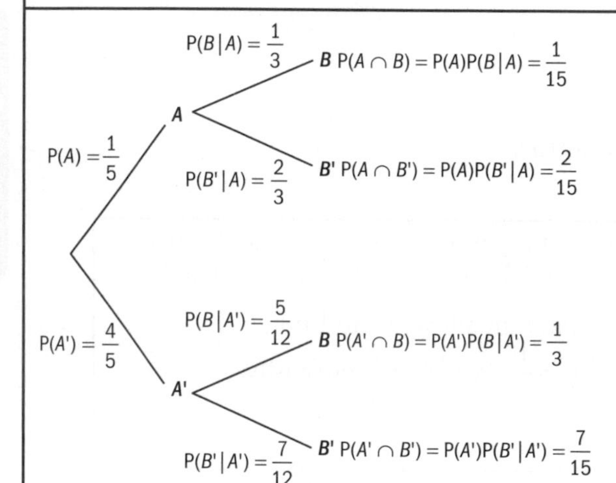

Similarly, complete the rest of the tree as shown.

Then apply the multiplication law of probability to find the probability represented at the end of each "branch" of the tree. For example,

$$\mathrm{P}(A \cap B) = \mathrm{P}(A)\mathrm{P}\left(B|A\right) = \frac{1}{5} \times \frac{1}{3} = \frac{1}{15}.$$

The total probability of 1 is seen to be distributed along the branches of the tree by applying the multiplication law of probability for the other combined events.

Notice that the simple probability $\mathrm{P}(B) = \mathrm{P}(A \cap B) + \mathrm{P}(A' \cap B) = \frac{1}{15} + \frac{1}{3} = \frac{2}{5}$ can be found from the probabilities at the end of two branches of the tree diagram.

A tree diagram is another way to represent all the possible outcomes of an event. The end of each branch represents a combined event.

International-mindedness

In 1933 Russian mathematician Andrey Kolmogorov built up probability theory from fundamental axioms in a way comparable with Euclid's treatment of geometry that forms the basis for the modern theory of probability. Kolmogorov's work is available in an English translation titled *The Foundations of Probability Theory*.

Investigation 12

Situation 1

A CAS project aims to raise funds by organizing a simple lottery. A number of coloured dice are placed in a box: 50 red dice, 30 blue and 20 green. To play the lottery, a die is chosen at random from the box and its colour noted. It is then replaced in the box and another die is chosen at random. If the two colours are the same, a small prize is awarded. The organizers of the project, Dani and Malena, want to know the probability of winning a prize in the most efficient way. They represent and interpret the information given by the lottery in two different ways.

Consider and reflect on the different approaches.

<table>
<tr><th>Dani's representation</th><th>Dani's application and interpretation</th></tr>
<tr><td>

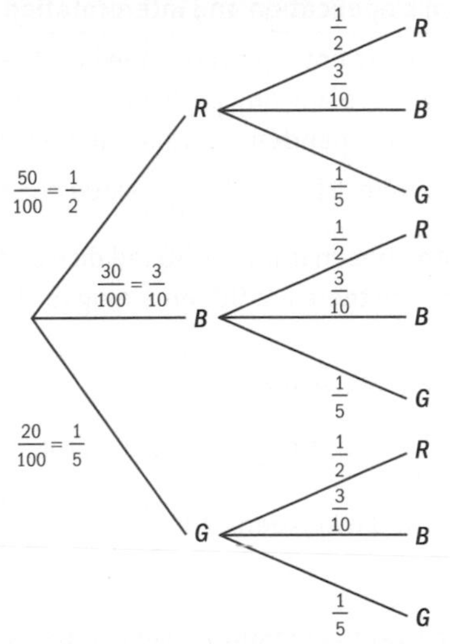

The first set of branches of the tree represent the colour possibilities of the first die chosen. The second set of branches represent the second, so there is an implicit time axis from left to right.

</td><td>

Since the dice are replaced in the box, the probability of choosing a red die on the second draw is the same as choosing a red on the first. The same argument applies to all the other pairs of combined events. Hence the probability of any particular choice on the second choice is **independent** of the first.

The probability of two reds is $\frac{1}{2}\times\frac{1}{2}=\frac{1}{4}$, similarly the probability of two blues and two greens are $\frac{9}{100}$ and $\frac{1}{25}$ respectively.

These three events are **mutually exclusive**, so you can add them to find the combined probability required.

P(the colours of the two dice are the same)

$=\frac{1}{4}+\frac{9}{100}+\frac{1}{25}=\frac{38}{100}=0.38$

</td></tr>
</table>

<table>
<tr><th>Malena's representation</th><th>Malena's application and interpretation</th></tr>
<tr><td>

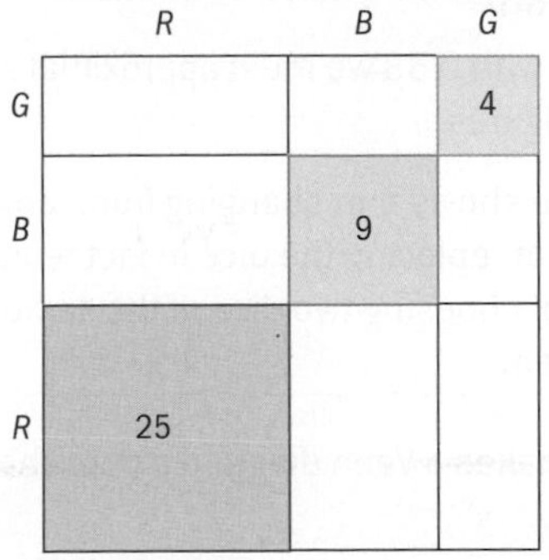

The horizontal *R*, *B*, *G* represent the colour possibilities of the first die chosen, the vertical letters the second.

</td><td>

Malena reasoned that she would expect 50 red dice on the first choice, and of these 50, 25 would be followed by a second red since the second choice is **independent** of the first. Similarly, she reasoned that out of an expected 30 blue die from the first throw, 9 would be followed by a second blue.

Malena reasoned that a total of 25 + 9 + 4 trials in which two identical colours were chosen would lead to a theoretical probability of 38 out of 100.

Dani was puzzled by this at first, but he could see after a while that Malena had created a slightly unusual looking Venn diagram. It was clear to them both from Malena's diagram that the sets are all **mutually exclusive.**

</td></tr>
</table>

Continued on next page

1 **Factual** Which representation involves less drawing and writing?

2 **Conceptual** Which diagram best represents how the stages of the game develop over time?

Situation 2

Then the CAS project organizers **changed the rules** as follows. A number of coloured dice are placed in a box: 50 red dice, 30 blue and 20 green. To play the lottery, a die is chosen at random from the box and its colour noted. It is **not** replaced in the box and another die is chosen at random. If the two colours are the same, a small prize is awarded.

Fill in the missing probabilities to find the probability of winning a prize.

Dani's representation	Dani's application and interpretation
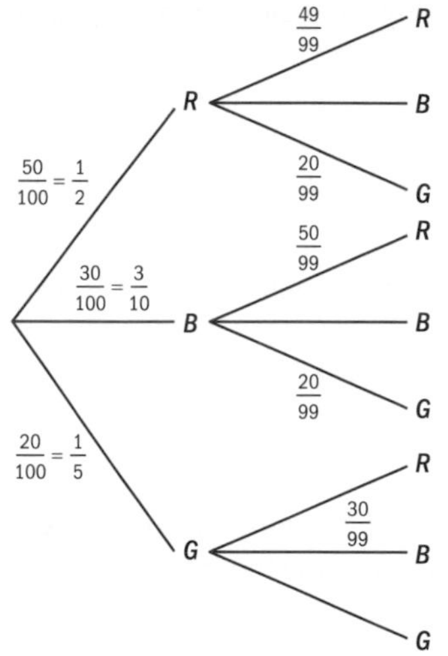 $\frac{50}{100}=\frac{1}{2}$ R: $\frac{49}{99}$ R, B, $\frac{20}{99}$ G $\frac{30}{100}=\frac{3}{10}$ B: $\frac{50}{99}$ R, B, $\frac{20}{99}$ G $\frac{20}{100}=\frac{1}{5}$ G: R, $\frac{30}{99}$ B, G The probabilities on the second set of branches represent the fact that since the die chosen in the first choice has not been replaced, the sample space has changed in size. Hence the probability of a red on the second choice given that a red was chosen at first can be represented as $P(R_2 \mid R_1)=\frac{49}{99}$ and so on.	Since the dice are not replaced in the box, the probability of choosing any particular colour on the second draw is **dependent** on what the first choice was. For example, $P\left(R_2 \mid R_1\right)=\frac{49}{99}$ since one red die has been chosen from the 50 red dice and **not** replaced, hence there are 99 remaining to choose from. The probability of two reds is $\frac{1}{2}\times\frac{49}{99}=\frac{49}{198}\simeq 0.247$, similarly the probability of two blues and two greens are $\frac{??}{330}$ and $\frac{19}{???}$ respectively. These three events are **mutually exclusive**, so we can add them to find the combined probability required. $\text{P(the colours of the two dice are the same)}$ $=\frac{49}{198}+\frac{??}{330}+\frac{19}{???}=\frac{??}{99}\simeq 0.???$ The answer $\frac{??}{99}$ is exact. However, to make the comparison with 0.38 we must approximate to three significant figures. This example shows that changing from replacing the dice to not replacing the dice in fact reduces the probability of choosing two dice of the same colour in this problem.

3 **Conceptual** What feature of choosing without replacement makes a Venn diagram a poor choice as a representation?

4 **Conceptual** What feature of choosing without replacement makes a tree diagram a good choice as a representation?

5 **Conceptual** How can we know when best to represent a problem with a tree diagram?

You can choose a diagram to make your calculation easier and more efficient. Also, sometimes you can choose how you calculate a probability to make it easier and more efficient, as shown in the following example:

Example 15

A box in an electronics store contains nine volt batteries of different colours. There are five red, six blue and seven orange. Two batteries are chosen to power up a radio that requires two batteries. Find the probability that both the batteries are different colours.

The batteries are selected **without** replacement since two batteries are needed to power up the radio.

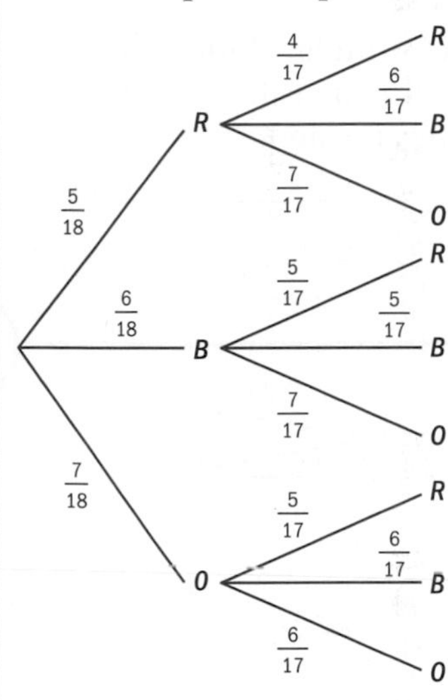

Interpret the context to make sure you are applying the maths correctly.

Draw a tree diagram, labelling all the branches. Take care to fill in the probabilities correctly to show that the batteries are not replaced.

The answer can be found in two ways:

P(Batteries are different colours)

$= \text{P}(RB \text{ or } RO \text{ or } BR \text{ or } BO \text{ or } OR \text{ or } OB)$

$$= \frac{5}{18} \times \frac{6}{17} + \frac{5}{18} \times \frac{7}{17} + \frac{6}{18} \times \frac{5}{17} +$$

$$\frac{6}{18} \times \frac{7}{17} + \frac{7}{18} \times \frac{5}{17} + \frac{7}{18} \times \frac{6}{17} = \frac{107}{153}$$

Or

P(Batteries are different colours)

= 1 − P(Batteries are identical colours)

$$= 1 - \frac{5}{18} \times \frac{4}{17} - \frac{6}{18} \times \frac{5}{17} - \frac{7}{18} \times \frac{6}{17} = \frac{107}{153}$$

Notice that the second method involves finding only **three** combined probabilities whereas the first method involves finding **six**.

The complementary probability law $\text{P}(A) = 1 - \text{P}(A')$ can give you a short way to solve problems.

Example 16

Mark buys his t-shirts from exactly two suppliers: 2aT and Netshirts, and he only buys two colours—grey and blue. He buys 73% of his t-shirts from 2aT. Moreover, 16% of his 2aT t-shirts are grey, and 80% of his Netshirts t-shirts are grey.

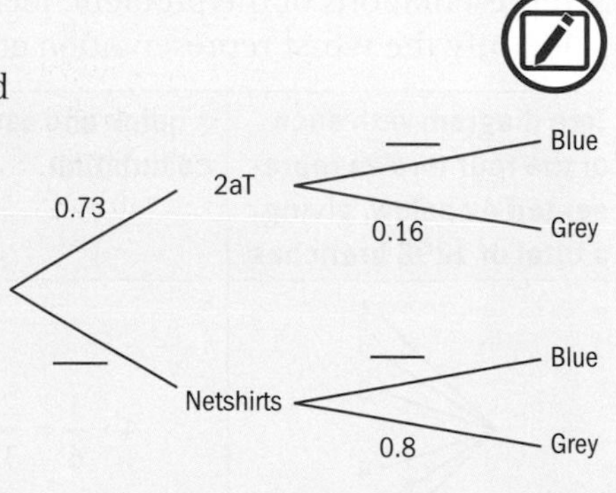

a Copy and complete the tree diagram.

b Calculate the probability that a t-shirt randomly chosen by Mark one morning is grey.

c Given that Mark chooses a grey t-shirt, determine the probability it was supplied by Netshirts.

Continued on next page

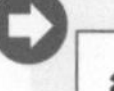

a

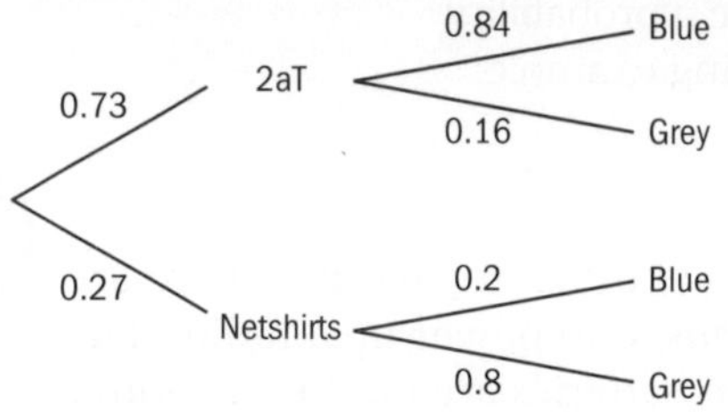

The question states that the events are complementary so the diagram can be filled in.

b $P(\text{grey}) = P(\text{2aT and grey}) + P(\text{Netshirts and grey})$
$= 0.73 \times 0.16 + 0.27 \times 0.8 = 0.3328$

Two mutually exclusive events are added to find the probability that Mark chooses a grey shirt.

c $P(\text{Netshirts} \mid \text{grey}) = \dfrac{P(\text{Netshirts and grey})}{P(\text{grey})}$

$= \dfrac{0.27 \times 0.8}{0.3328} \approx 0.649$

Apply the formula. Notice that you have already calculated the probabilities needed.

Exercise 7H

1 Denise can catch a local bus or an express bus to take her to work each day. The probability she catches the local bus is 0.8. If she catches the local bus, the probability that she is on time for work is 0.5. If Denise catches the express bus, the probability that she is on time for work is 0.95.

a Copy and complete the tree diagram below:

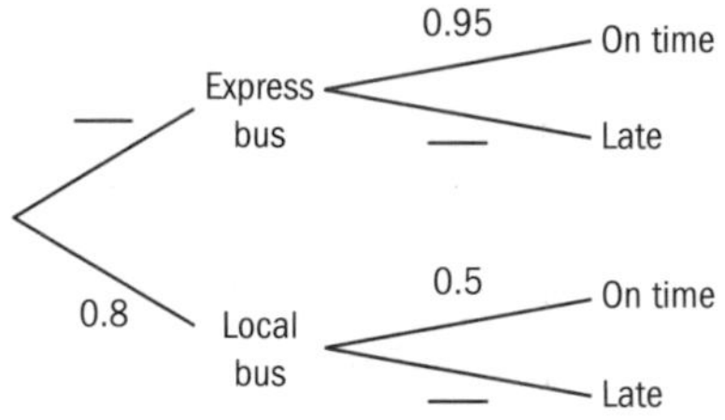

b Hence calculate the probability that Denise will be late for work.

2 A jewellery box contains 13 gold earrings, 10 silver earrings and 12 titanium earrings. Two earrings are drawn at random with replacement. Find the probability that they are made of different metals.

3 Chevy plays a game with four fair cubical dice numbered $\{1, 2, 3, 4, 5, 6\}$. She throws the four dice 160 times and finds that the event "throw at least one 6" occurs 77 times. She wonders what the theoretical probability is. She finds the answer after critically considering these four representations of the problem. Identify the best representation and justify your answer. Identify the worst representation and justify your answer.

Tree diagram with each of the four throws represented as below, giving a total of 1296 branches	A quick and easy calculation.	Tree diagram with each of the four throws represented as below, giving a total of 16 branches	A quick and easy application of probability laws.
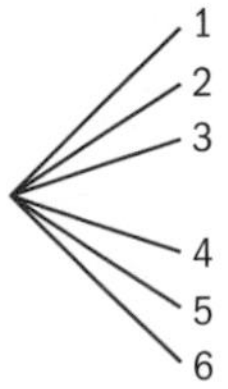	$4 \times \dfrac{1}{6} = \dfrac{2}{3}$	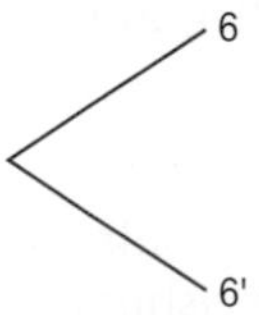	$1 - \left(\dfrac{5}{6}\right)^4$

4 A supermarket uses two suppliers, C and D, of strawberries. Supplier C supplies 70% of the supermarket's strawberries. Strawberries are examined in a quality control inspection (QCI); 90% of the strawberries supplied by C pass QCI and 95% of the strawberries from D pass QCI.

A strawberry is selected at random.

a Find the probability that the strawberry passes QCI.

b Given that a strawberry passes QCI, find the probability that it came from supplier D.

c In a sample of 2000 strawberries, find the expected number of strawberries that would fail QCI.

d The supermarket wants the probability that a strawberry passes QCI to be 0.93. Find the percentage of strawberries that should be supplied by D in order to achieve this.

5 Chevy plays a game in which she throws a pair of fair cubical dice numbered {1, 2, 3, 4, 5, 6} 24 times. Find the probability that she throws at least one double six.

6 A factory produces a large number of electric cars. A car is chosen at random from the production line as a prize in a competition. The probability that the car is blue is 0.5. The probability that the car has five doors is 0.3. The probability that the car is blue or has five doors is 0.6. Find the probability that the car chosen is not blue and does not have five doors.

Pietro solves this problem with a Venn diagram but Maria solves it with a tree diagram. They both get the correct answer. Solve the problem both ways. Discuss and then state which is the most efficient method.

7 A choir contains 15 girls and 9 boys. The choirmaster randomly selects four singers from the choir to be interviewed by a newspaper. Find the probability that at least one boy is in the four singers selected.

Developing inquiry skills

Apply what you have learned in this section to represent the first opening problem with a tree diagram.

Hence find the probability that a cab is **identified** as yellow.

Apply the formula for conditional probability to find the probability that the cab was yellow **given that** it was identified as yellow.

How does your answer compare to your original subjective judgment?

7.5 Modelling random behaviour: random variables and probability distributions

In the rest of this chapter you will apply your knowledge and understanding of combined events to model random behaviour in various processes that occur in real life.

Recall from Chapter 3 that **discrete** data is either data that can be **counted**, for example the number of cars in a car park, or data that can only take specific values, for example shoe size.

Continuous data can be **measured**, for example height, weight and time.

For example if a café has four coffee percolators, the **number** of percolators that fail on a given day, X, is a quantity that changes randomly according to the durability and reliability of the machinery. In contrast, the **weight** of a bag of coffee, Y, varies randomly according to the weighing and packaging processes in the factory. These are both examples of **random variables**: the value of these variables changes according to chance. The value of Y is a **measurement** that can take **any** real number in an interval.

TOK

A model is used to represent a mathematical situation in this case. In what ways may models help or hinder the search for knowledge?

We use the following terminology for random variables that are countable.

Example: Let X represent the number of percolators failing on a given day.

Terminology	Explanation
X is a **discrete random variable.**	Discrete: X can be found by **counting**.
	Random: X is the result of a **random** process.
	Variable: X can take **any value** in the domain $\{0, 1, 2, 3, 4\}$.

A first step in acquiring knowledge about X is to fill in a table:

Number of failing percolators (x)	0	1	2	3	4
$\mathrm{P}(X=x)$	$\mathrm{P}(X=0)$	$\mathrm{P}(X=1)$	$\mathrm{P}(X=2)$	$\mathrm{P}(X=3)$	$\mathrm{P}(X=4)$

The five probabilities must add to one, and they must each satisfy $0 \le \mathrm{P}(X=x) \le 1$.

Knowing them establishes the **probability distribution** of X since the table then shows how the entire probability of 1 is distributed to each value of the random variable in its domain.

A **discrete probability distribution** is the set of all possible values of a discrete random variable (a subset of $\mathbb{Z}$) together with their corresponding probabilities.

Investigation 13

One fair tetrahedral die with faces numbered 1, 2, 3 and 4 is thrown. Let A denote the value of the number thrown.

1 Complete the table to show the probability distribution of A:

a	1	2	3	4
$P(A = a)$				

2 Draw a bar chart to represent the probability distribution of A and make a general statement for $P(A = a)$.

Two fair tetrahedral dice with faces numbered 1, 2, 3 and 4 are thrown. Let B denote the sum of the two values thrown.

3 Use combined events to complete the table to show the probability distribution of B:

b	2	3	4	5	6	7	8
$P(B = b)$							

4 Draw a bar chart to represent the probability distribution of B and make a general statement for $P(B = b)$ using a piecewise function.

5 Compare and contrast the shapes of the bar charts for A and for B.

6 Compare and contrast the probability distributions of A and B.

7 **Factual** How could you use your answer to **3** to predict the number of times B is at least 6 in 200 trials?

8 **Conceptual** How can you represent a discrete probability distribution?

9 **Conceptual** How can you find a discrete probability distribution?

In the following investigation you will discover and explore an example of a discrete probability distribution with useful and surprising applications in real life.

Investigation 14

Imagine a list giving the populations of 267 countries. Let Z be the first digit in the number of people in a randomly chosen country.

1 **Factual** Make your own subjective judgment of the probability distribution of Z. What would the domain of Z and its shape be? Share this judgment in a pair and then with your class.

Now consider the data giving the population of 267 countries found here: https://www.cia.gov/library/publications/the-world-factbook/rankorder/2119rank.html

Collaborate with others to complete a frequency table as follows for the 267 populations:

First digit	1	2	...	9
Frequency				

Continued on next page

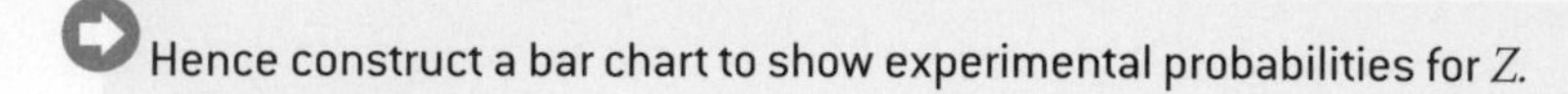

Hence construct a bar chart to show experimental probabilities for Z.

2 **Conceptual** How does your bar chart compare to your answer for **1**?

Let Y be the first digit of number in the measure of the **area** of a randomly chosen country.

Repeat **1** and **2** with the area of each country found here https://www.cia.gov/library/publications/the-world-factbook/rankorder/2147rank.html

3 **Factual** Use your knowledge of functions to propose a model for these sets of data.

4 **Conceptual** How may a discrete probability distribution be determined, apart from the method you gave in the previous investigation?

A **discrete probability distribution function** $f(t)$ assigns to each value of the random variable its corresponding probability: $f(t) = P(T = t)$.

$f(t)$ is commonly referred to by the abbreviation "pdf".

Example 17

a Show that the function $f(t) = \frac{t-2}{15}$, $t \in \mathbb{Z}$, $3 \le t \le 7$, defines a discrete probability distribution by constructing a table of values.

b If $P(T = t) = f(t)$, represent the distribution as a bar chart.

a

t	3	4	5	6	7
$f(t)$	$\frac{1}{15}$	$\frac{2}{15}$	$\frac{3}{15}$	$\frac{4}{15}$	$\frac{5}{15}$

Substitute each value of t from the domain given.

Since $\frac{1+2+3+4+5}{15} = 1$, $f(t) = \frac{t-2}{15}$ defines a probability distribution on the domain given.

Show that the values of the function add to 1.

b

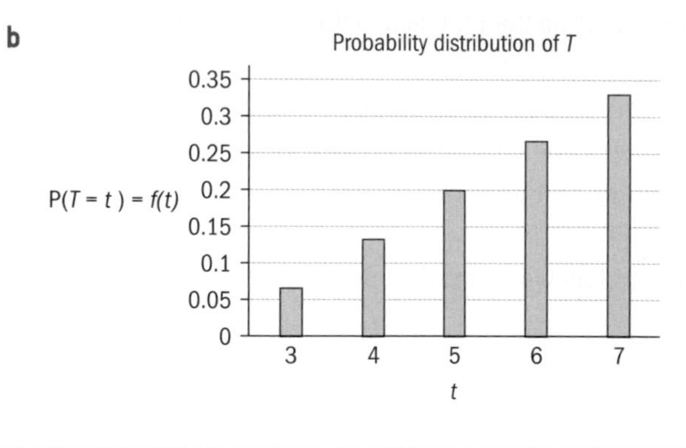

Label the graph completely in order to communicate what it represents.

Example 18

A fair cubical die and a fair tetrahedral die are thrown. The discrete random variable S is defined as the sum of the numbers on the two dice.

a Construct the probability distribution of S as:

i a table of values **ii** a bar chart **iii** a piecewise function.

b Hence find the probabilities:

i $P(S > 2)$ **ii** $P(S$ is at most $6)$ **iii** $P(S \leq 6 | S > 2)$.

a

	1	2	3	4	5	6
1	2	3	4	5	6	7
2	3	4	5	6	7	8
3	4	5	6	7	8	9
4	5	6	7	8	9	10

Draw a sample space diagram.

i

s	2	3	4	5	6	7	8	9	10
$P(S=s)$	$\frac{1}{24}$	$\frac{2}{24}$	$\frac{3}{24}$	$\frac{1}{6}$	$\frac{1}{6}$	$\frac{1}{6}$	$\frac{3}{24}$	$\frac{2}{24}$	$\frac{1}{24}$

Read off the probability of each event $P(S = s)$ in turn from the sample space diagram.

ii

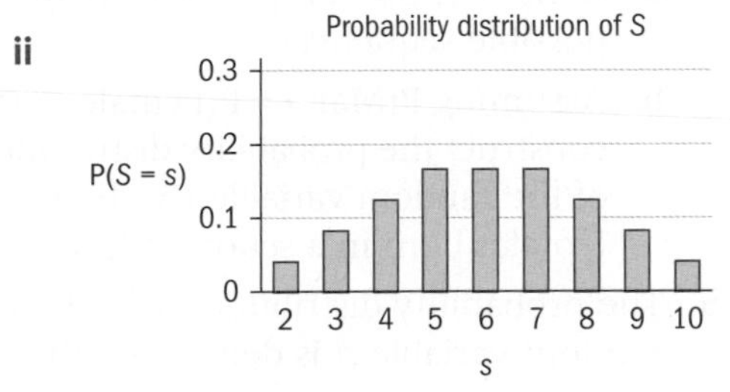

Do not forget to label both axes.

iii

$$f(s) = \begin{cases} \frac{s-1}{24} & 2 \leq s \leq 4 \\ \frac{1}{6} & 5 \leq s \leq 7 \quad \text{where } s \in \mathbb{N}. \\ \frac{11-s}{24} & 8 \leq s \leq 10 \end{cases}$$

Making a general statement helps represent the probability distribution function concisely.

b **i** $P(S > 2) = 1 - \frac{1}{24} = \frac{23}{24}$

ii $P(S \leq 6) = \frac{1}{24} + \frac{1}{12} + \frac{1}{8} + \frac{1}{6} + \frac{1}{6} = \frac{7}{12}$

Take care to interpret "at most 6" correctly.

iii $P(S \leq 6 | S > 2) = \frac{P(S \leq 6 \cap S > 2)}{P(S > 2)}$

$$= \frac{P(3 \leq S \leq 6)}{P(S > 2)} = \frac{\frac{13}{24}}{\frac{23}{24}} = \frac{13}{23}$$

Use the formula for conditional probability and find the intersection of the two sets.

Continued on next page

Statistics and probability

Or

s	2	3	4	5	6	7	8	9	10
P(S = s)	$\frac{1}{24}$	$\frac{2}{24}$	$\frac{3}{24}$	$\frac{1}{6}$	$\frac{1}{6}$	$\frac{1}{6}$	$\frac{3}{24}$	$\frac{2}{24}$	$\frac{1}{24}$

$P(3 \le S \le 6)$ can be found as $\frac{2}{24}+\frac{3}{24}+\frac{1}{6}+\frac{1}{6}=\frac{13}{24}$ from the diagram. The whole sample space of $P(S \le 6 | S > 2)$ is $P(S > 2) = \frac{23}{24}$, hence

$$P(S \le 6 | S > 2) = \frac{P(3 \le S \le 6)}{P(S > 2)} = \frac{\frac{13}{24}}{\frac{23}{24}} = \frac{13}{23}$$

You are not obliged to use the formula: a diagram can be used to give an equivalent solution. Many students find this method easier.

Exercise 7I

1 Consider these three tables. State which table(s) could not represent a discrete probability distribution, giving reasons.

a

a	1	2	3	7
P(A=a)	0.1	0.2	0.03	0.4

b

b	1	2	3	4
P(B=b)	0.1	−0.2	0.4	0.4

c

c	4.5	3	1	0
P(C=c)	0.2	0.2	0.5	0.1

2 Show that the function $f(t) = \frac{t-4}{21}$, $t \in \mathbb{Z}$, $5 \le t \le 10$, defines a discrete probability distribution by constructing a table of values.

3 $f(x) = \frac{x}{19}$, $x \in \{1, 5, 7, k\}$, defines a discrete probability distribution. Find the value of k.

4 Sarah researches multiple births in a clinic, where she keeps records over a period of years of the genders of triplets born there. There are eight possible sequences of genders in a set of triplets, for example MFM.

a Construct the sample space of all possible sequences.

b Assuming P(Male) = P(Female) = 0.5, construct the probability distribution table of the random variable F = the number of females born in a set of triplets.

5 The probability distribution of a discrete random variable A is defined by this table:

a	5	8	9	10	11	12
P(A=a)	0.5	0.05	0.04	0.1	0.2	P(A=12)

Find:

a $P(A = 12)$ **b** $P(8 < A \le 10)$

c P(A is no more than 9)

d P(A is at least 10) **e** $P(A > 8 | A \le 11)$.

6 Nico and Artem are designing a card game. In their pack of cards, each card has 2, 1, 0 or 5 printed on it. A card is selected from the pack and the number on the card defines a random variable X.

Nico states that the probability distribution of the game is as follows:

x	2	1	0	5
P(X=x)	0.3	0.27	0.25	0.1

a Explain why Nico is wrong.

b Artem correctly states that the probability distribution is

x	2	1	0	5
$P(X=x)$	0.28	0.2	p	$3p$

Find the value of p.

7 Part of the discrete probability distribution of the discrete random variable T with domain {1, 2, 3, 4, 5} is shown in the bar chart.

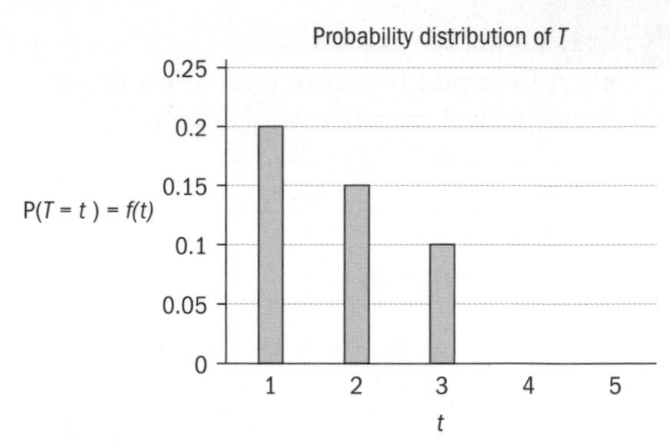

Given that $P(T=4) = 4P(T=5)$, construct the probability distribution table for T.

Example 19

The probability distribution of a discrete random variable U is defined by

$P(U=u) = k(u-3)(8-u),\ u \in \{4, 5, 6, 7\}$.

a Find the value of k and hence draw the table of the probability distribution of U.

b In 100 trials, calculate the expected value of each possible outcome of U.

c Find the mean of the values of U found in these 100 trials assuming the frequencies of each outcome of U are given by your expected values in **b**.

d Interpret your answer.

a

u	4	5	6	7
$P(U=u)$	$4k$	$6k$	$6k$	$4k$

Represent the probability distribution in a table.

$4k+6k+6k+4k=20k=1 \Rightarrow k=\frac{1}{20}$,

hence

u	4	5	6	7
$P(U=u)$	$\frac{1}{5}$	$\frac{3}{10}$	$\frac{3}{10}$	$\frac{1}{5}$

Use the fact that the probabilities must add to 1 on the domain of the pdf.

b The expected number of occurrences of $u=4$ is $100\times\frac{1}{5}=20$. Similarly, the expected number of occurrences of 5, 6 and 7 are 30, 30 and 20 respectively.

Use the formula: expected number of occurrences of $A = n\,P(A)$.

Continued on next page

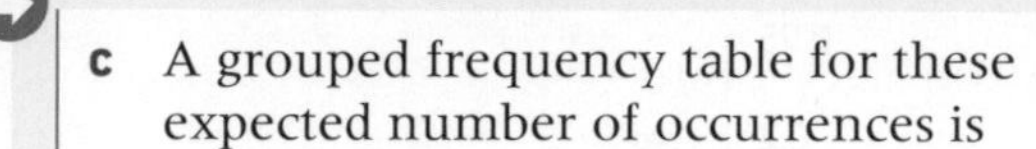

c A grouped frequency table for these expected number of occurrences is

u	4	5	6	7
Expected frequency of u	20	30	30	20

Hence the mean value of U with these frequencies is

$$\frac{20\times4+30\times5+30\times6+20\times7}{100}=\frac{550}{100}=5.5$$

Each number of occurrences is a frequency.

Use the formula for the population mean:

$$\mu=\frac{\sum_{i=1}^{k}f_i x_i}{n} \text{ where } n=\sum_{i=1}^{k}f_i$$

d 5.5 is not a number in the domain of the probability distribution function. Nevertheless, it models the central value of U expected in a data set of 100 trials.

Recall the definition of the mean as a measure of central tendency.

You can express the calculation in part **c** of the previous exercise more briefly as:

$$\frac{20\times4+30\times5+30\times6+20\times7}{100}=4\times\frac{20}{100}+5\times\frac{30}{100}+6\times\frac{30}{100}+7\times\frac{20}{100}$$

$$=4\times\frac{1}{5}+5\times\frac{3}{5}+6\times\frac{3}{5}+7\times\frac{1}{5}$$

Notice that this is the sum of the product of each value of the random variable with its corresponding probability, or $\sum_{u} u\mathrm{P}(U=u)$. This leads to a further generalization for all discrete random variables:

The **expected value** of a discrete random variable X is $\mathrm{E}(x)=\mu=\sum_{x} x\mathrm{P}(X=x)$

Example 20

A newsagent in Oxford takes delivery of six copies of a Scottish newspaper each Sunday. The newsagent has a regular order from her customers for three newspapers but sales vary according to current events, sport etc. The newsagent has collected data over several years to help predict her sales, creating a probability distribution table for the random variable S, the number of Scottish newspapers sold on Sunday.

s	2	3	4	5	6
P($S=s$)	0.05	0.39	0.29	0.22	0.05

Hence find the expected number of newspapers sold and interpret its meaning in context.

$E(S) = \sum_s sP(S = s) =$ $2 \times 0.05 + 3 \times 0.39 + 4 \times 0.29 + 5 \times 0.22 + 6 \times 0.05$ $= 3.83$	Apply the formula for the expected value.
On average, the newsagent should expect to sell more than 3 newspapers although 3 is the most likely outcome. There is more than a 50% chance that she will sell more than her regular order.	Draw conclusions from the given information.

Investigation 15

Consider the distribution in Example 18: "A fair cubical die and a fair tetrahedral die are thrown. The discrete random variable S is defined as the sum of the numbers on the two dice." Consider an experiment in which 100 values of S are calculated in 100 trials.

Predict the average of these 100 values of S by:

a deducing from the shape of the bar chart representation

b application of the formula for the expected value of a discrete random variable

c constructing a data set of 100 values of S with a spreadsheet and finding its mean by entering these formulae and dragging A1, B1 and C1 down to the 100th row.

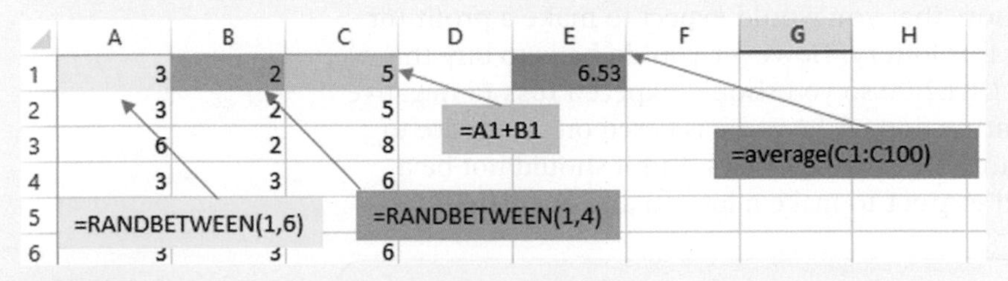

1 **Factual** What are the strengths and weaknesses of each approach? Discuss.

2 **Conceptual** What does the expected value of a discrete random variable predict about the outcomes of a number of trials?

Many countries organize national lotteries in which adults buy a ticket giving them a chance to win one of a range of cash prizes. Profits are often invested in "good causes". For example, the UK National Lottery has distributed over UK£37 billion through thousands of grants to good causes including sport, art and health projects since 1994.

It is possible to use the expected value formula to manage the prize structure of a lottery in order to maintain profitability.

Prize	Probability	Cash value per winner (UK £)
1st (Jackpot)	$\frac{1}{45\,057\,474}$	5 421 027
2nd	$\frac{1}{7\,509\,579}$	44 503
3rd	$\frac{1}{144\,415}$	1018
4th	$\frac{1}{2180}$	84
5th	$\frac{1}{97}$	25
6th	$\frac{1}{10.3}$	Free ticket

Looking at the cash prizes only for simplicity, we can find the expected winnings as follows:

Expected cash winnings =

$$5\,421\,027\times\frac{1}{45\,057\,474}+44\,503\times\frac{1}{7\,509\,579}+1018\times\frac{1}{144\,415}$$
$$+84\times\frac{1}{2180}+25\times\frac{1}{97}\approx 0.430$$

This would appear to show that you would expect to make a profit (or a positive gain) playing this lottery! However you do have to buy the ticket first, which costs UK£2.00 so you should expect a **loss** (a negative gain) of UK£1.57. The attraction of the game is based on the desire to win a large prize or contribute to good causes. But it should not be a surprise that you would **expect** to make a loss on any one game.

TOK

Do you rely on intuition to help you make decisions?

If X is a discrete random variable that represents the gain of a player, then if $\mathrm{E}(X)=0$, the game is **fair**.

Example 21

Some students have a meeting to design a dice game to raise funds for charity as part of a CAS project. Some of the decisions made in the meeting are lost.

This incomplete probability distribution table remains:

x (prize in US$)	1	2	4	6	7
$\mathrm{P}(X=x)$	$\frac{11}{40}$	$\frac{1}{4}$			$\frac{1}{8}$

The students also recall that $E(X) = \frac{67}{20}$ and that the probability distribution function generalizes to a linear model.

a Determine the missing entries in the table and hence find the probability distribution function.

b Find the smallest entry fee the students could set for playing the game in order to predict a profit. Comment on the advantages and disadvantages of a number of possible entry fees.

a Let the missing probabilities be represented by a and b.

Then $a+b=1-\frac{11}{40}-\frac{1}{4}-\frac{1}{8}=\frac{7}{20}$, also

$1 \times \frac{11}{40} + 2 \times \frac{1}{4} + 4a + 6b + 7 \times \frac{1}{8} = \frac{67}{20}$

Representing the unknown quantities with a variable and writing down true statements involving them is a problem-solving strategy.

The probabilities must add to 1. You can apply formula for the expected value.

Hence $a+b=\frac{7}{20}$ and $4a+6b=\frac{17}{10}$.

So $a=\frac{1}{5}$ and $b=\frac{3}{20}$.

Solve the system of simultaneous equations.

The completed table is :

x (prize in US$)	1	2	4	6	7
$P(X=x)$	$\frac{11}{40}$	$\frac{10}{40}$	$\frac{8}{40}$	$\frac{6}{40}$	$\frac{5}{40}$

Look for a pattern in the probability distribution table or alternatively apply the general equation for a linear function $y = mx + c$.

Hence $f(x) = P(X = x) = \frac{1}{40}(12 - x)$

b $E(X) = \frac{77}{20} = \text{US\$}3.85$ so charging a player US$3.85 would be a fair game. Therefore charging US$3.86 would predict a small profit, but this could easily give a loss. Perhaps charging US$4.00 would be more practical and it would predict a larger profit.

Apply the formula for the expected value and reflect critically.

Exercise 7J

1 The discrete random variable B has probability distribution function given by $P(B = b) = k(4 - b)$ for $b = 0, 1, 2, 3$. Find k and $E(B)$.

2 Apply your probability distribution table from Exercise 7I question 4 to find the expected number of male births in a set of triplets. Interpret your result.

3 A handbag contains seven coins and three keys. Two items are taken out of the handbag one after the other and not replaced. Find the expected number of keys taken out of the handbag.

4 A handbag contains five coins, four keys and eight mints. Two items are taken out of the handbag one after the other and not replaced. Find the expected number of mints taken out of the handbag.

5 Alexandre is designing a game. A spinning arrow rotates and stops on one of the regions A, B, C or D as shown in the diagram.

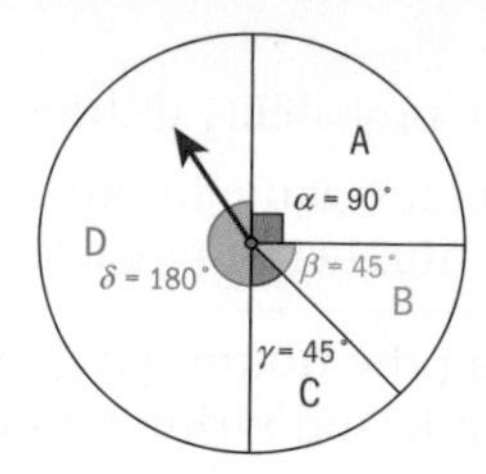

Alexandre proposes the prizes shown in the table and that the game should cost US$5 to play.

Letter	A	B	C	D
Prize	US$3	US$7	US$5	US$2

Determine whether Alexandre's game is fair and justify your answer.

6 Ten thousand US$10 lottery tickets are sold. One ticket wins a prize of US$5000, five tickets each win US$1000, and ten tickets each win US$200. Find:

a the probability of winning each prize in the lottery

b the expected gain from one ticket

c the price of a ticket to make the lottery a fair game.

7 Two fair tetrahedral dice with faces numbered 1, 2, 3 and 4 are thrown. The discrete random variable D is defined as the **product** of the two numbers thrown.

a Find the probability distribution table of D.

b Find $P(D \text{ is a square number} \mid D < 8)$.

c In a CAS fundraising game, a prize of US$12 is won if D is odd, and a prize of US$6 is won if D is even. Find the price of a ticket to ensure this is a fair game.

8 Xsquared Potato Crisps runs a promotion for a week. In 0.01% of the hundreds of thousands of bags produced there are gold tickets for a round-the-world trip. Let B represent the number of bags of crisps opened until a gold ticket is found.

a Find $P(B = 1)$, $P(B = 2)$ and $P(B = 3)$.

b Hence show that the probability distribution function of B is

$$f(b) = P(B = b) = 0.0001(0.9999)^{b-1}$$

c State the domain of $f(b)$.

d Determined to win a ticket, Yimo buys 10 bags of crisps. Find the probability that she finds a gold ticket after opening no more than 10 bags.

Developing inquiry skills

Read again the second question from the opening scenario.

Can this question be modelled by a discrete probability distribution?

How could you find its probability distribution? What assumptions would have to be made?

7.6 Modelling the number of successes in a fixed number of trials

You learned in Section 7.5 that a probability distribution function can be found as a generalization of a random process.

An example of the process you will learn about in this section is found in the work of cognitive psychologists Daniel Kahneman and Amos Tversky (1972).

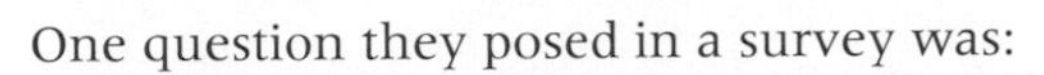

One question they posed in a survey was:

"All families of six children in a city were surveyed. In 72 families, the exact order of births of boys and girls was GBGBBG. What is your estimate of the number of families surveyed in which the exact order of births was BGBBBB?"

The median estimate was 30—suggesting the participants in the survey judged that GBGBBG was more than twice as likely an outcome as BGBBBB. Psychologists have studied this "representativeness fallacy" in research on subjective judgments and biases.

You will learn in this section the mathematics needed to model situations like this; a family of six children can be modelled as a sequence of six independent trials (births) in which the probability of a female birth is constant (0.5) over the six trials.

There are different ways to represent, experience and understand the processes behind quantifying the probabilities in this kind of experiment. Two examples are Pascal's triangle and the Galton board.

Investigation 16

In France in 1665, Blaise Pascal published the pattern partially shown.

```
        1
      1   1
    1   2   1
  1   3   3   1
1   4   6   4   1
```

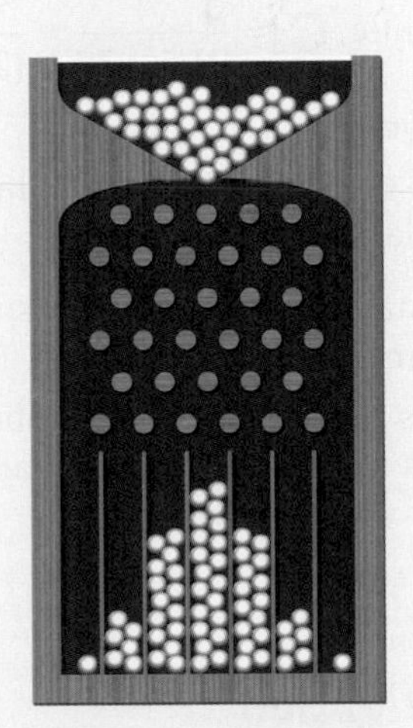

English polymath Sir Francis Galton invented a machine closely linked to Pascal's triangle.

In the **Galton board**, a ball is dropped into a vertical grid based on a tessellation of regular hexagons. Pegs force the ball to move down to the left or down to the right to reach the next level of the grid as shown.

At the end of the hexagonal grid the balls exit and stack up in vertical "bins". You may explore this process with technology here: https://www.mathsisfun.com/data/quincunx.html or here https://www.geogebra.org/m/ga8J6qDE.

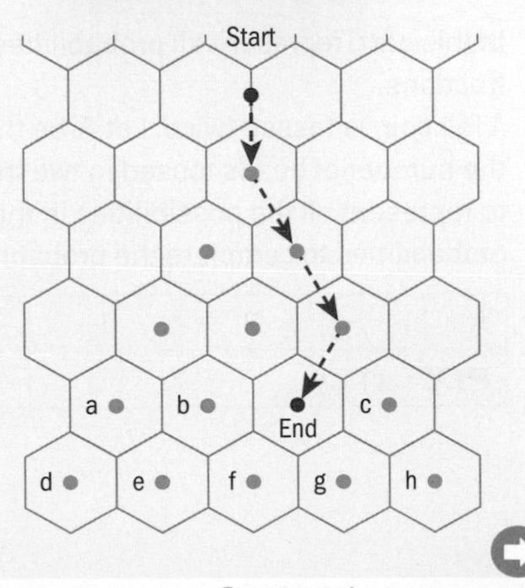

1 At each peg, how many choices are there for the direction of a ball?

2 How many different paths can the ball take to finish in the same place as that shown?

3 Show that the number of ways that the ball can finish in positions a, b, c, ..., h follow the same pattern as Pascal's triangle.

4 Which positions in the Galton board is the ball least likely to finish in?

Continued on next page

Statistics and probability

5 Which positions in the Galton board is the ball most likely to finish in?

6 What do the numbers in Pascal's triangle count?

7 Why is Pascal's triangle symmetrical?

A convenient way to experience the possible gender outcomes of a family of various sizes is by flipping a coin six times.

The numbers in Pascal's triangle are **binomial coefficients.** You can calculate them with technology.

Make a prediction of the next two rows and their sums and check your answers with technology.

The binomial coefficients in row $(n + 1)$ of the triangle are represented by the following notation: C_0^n, C_1^n, C_2^n, ...C_r^n, ..., C_{n-1}^n, C_n^n, where

$C_r^n = \frac{n!}{r!(n-r)!}$ and $n! = n \times (n-1) \times (n-2) \times ... \times 3 \times 2 \times 1$.

For example, $C_2^5 = \frac{5!}{2!(5-2)!} = \frac{5 \times 4 \times 3 \times 2 \times 1}{(2 \times 1) \times 3 \times 2 \times 1} = 10$.

Carry out 10 trials of flipping a coin six times, recording each trial as a sequence of H (heads) and T (tails). Compare your results with others in your class.

- How many of your trials resulted in outcomes like HTHTTH, which **appear** more random than THTTTT?
- Discuss this claim: "The probability of a total of three heads in six trials is more than the probability of a total of one head in six trials". Justify your answer.

International-mindedness

The Dutch scientist Christian Huygens, a teacher of Leibniz, published the first book on probability in 1657.

Investigation 17

In this part, represent all probabilities as fractions. Do not simplify the fractions.

A fair coin is tossed twice. Let X be the discrete random variable equal to the number of heads tossed in two trials of the fair coin. Use a tree diagram to represent all the possibilities in the sample space and hence find the probabilities to complete the probability distribution table.

x	0	1	2
$P(X=x)$			

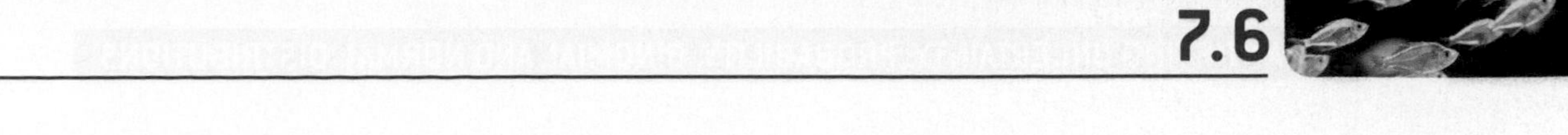

Repeat the above for three trials and then four trials.

1 **Factual** What is the connection between your results and Pascal's triangle?

2 Predict the probability distribution table for five trials.

3 **Conceptual** Would the probability distribution tables for 0 trials and 1 trial be consistent with the pattern in Pascal's triangle?

You can use your probability distributions from questions **1**, **2** and **3** and the binomial coefficients definition to make general statements for the probability distribution function.

4 Complete: "Let X be the discrete random variable equal to the number of heads tossed in n trials of a fair coin.

Then $\mathrm{P}(X=x) =$ ____________________ for $x \in \{0, 1, ...,\_\_\}$"

5 The experiment is changed so that the coin is not fair and it is thrown five times. $\mathrm{P}(H) = p$, $\mathrm{P}(T) = 1 - p$.

Complete the general statement "$\mathrm{P}(X=x) =$ ________ for $x \in \{0, 1, ...,\_\_\}$"

This investigation leads to the formal definition of the **binomial distribution:**

In a sequence of n independent trials of an experiment in which there are exactly two outcomes "success" and "failure" with constant probabilities $\mathrm{P}(\text{success}) = p$, $\mathrm{P}(\text{failure}) = 1 - p$, if X denotes the discrete random variable equal to the number of successes in n trials, then the probability distribution function of X is

$$\mathrm{P}(X=x) = \mathrm{C}_x^n\, p^x(1-p)^{n-x},\ x \in \{0, 1, 2, ..., n\}$$

These facts are summarized in words as "X is distributed binomially with parameters n and p" and in symbols as $X \sim \mathrm{B}(n, p)$.

You can use the binomial distribution to reflect on the questions at the start of this section.

Reflect Use the binomial distribution to find the probability of having exactly **three** boys in a family of six. Compare and contrast your answer with the probability of the outcome GBGBBG.

Which part of the formula for the binomial distribution counts the number of successes in n trials?

What is true about the number of trials and the probability of success in each binomial experiment?

In examinations, binomial probabilities will be found using technology.

Example 22

A multiple choice quiz has six questions each of which has four equally likely options A, B, C and D to choose from. If answers are guessed and if X represents the number of correct answers, find:

a $P(X = 2)$ **b** $P(X \le 2)$ **c** $P(X < 2)$ **d** $P(3 \le X < 6)$.

Assuming the guesses are equally likely and made independently of each other, X can be modelled by a binomial distribution with parameters 6 and $\frac{1}{4}$. Hence $X \sim B(6, \frac{1}{4})$.	Examine the context to see whether it fits the requirements of the binomial distribution. If it does, state the distribution and its parameters clearly before answering the question.
a $P(X = 2) = 0.297$	Use technology to find the probability required. $P(X = x) = C^6_x 0.25^x(0.75)^{6-x}$ Express your answer to three significant figures.
b $P(X \le 2) = P(X = 0) + P(X = 1) + P(X = 2)$ $P(X \le 2) = 0.831$	This event can be expressed as a sum of three mutually exclusive events. This sum can be found faster using technology.
c $P(X < 2) = P(X = 0) + P(X = 1)$ $P(X < 2) = 0.534$	Take care to apply the strict inequality correctly.
d $P(3 \le X < 6) = P(3 \le X \le 5)$ $P(3 \le X < 6) = 0.169$	Write the inequality in the form needed for applying technology.

It is essential to examine the context of a problem in order to understand whether the binomial distribution is an appropriate model to apply. To determine whether a context can be modelled by a binomial distribution, ask:

- What is the random variable **counting?**

  These **trials** should be independent of each other, which means that knowing what happens in one trial does not change the probabilities in any other trial.
- How many **trials** are there? This is the parameter $\boldsymbol{n}$.
- What is the **probability of success** in each trial? It should always be the same. This is the parameter $\boldsymbol{p}$.

TOK

Play the game of the St Petersburg Paradox and decide how much would you pay to play the game.

Example 23

For each situation state if the random variable is distributed binomially. If so, find the probability asked for.

a A coin is biased so that the probability of a head is 0.74. The coin is tossed seven times. A is the number of tails. Find $P(A = 5)$.

b A bag contains 12 white chocolates and 7 dark chocolates. A chocolate is selected at random and its type noted and then eaten. This is repeated five times. B is the number of dark chocolates eaten. Find $P(B = 4)$.

c A bag contains 10 red dice, 1 blue die and 7 yellow dice. A die is selected at random and its colour noted and replaced. This is repeated 12 times. C is the number of yellow dice recorded. Find $P(C \leq 6)$.

d Ciaran plays a lottery in which the probability of buying a winning ticket is 0.001. E is the number of tickets Ciaran buys until he wins a prize. Find $P(E < 7)$.

a Each toss of the coin is independent of the others. There are exactly two outcomes and a fixed number of trials. Therefore $A \sim B(7, 0.26)$. $P(A = 5) = 0.0137$	Write down the distribution. This clarifies your thoughts and you will receive method marks in the examination since you have demonstrated your knowledge and understanding. Use technology to find the binomial probability. Write the answer to three significant figures.
b Since the probability of selecting a dark chocolate is dependent on what was selected in previous trials, the trials are not independent. Hence the binomial distribution is not an appropriate model for B.	
c Since the die is replaced at each trial, the probability of success is constant and equal to $\frac{7}{18}$. Therefore $C \sim B(12, \frac{7}{18})$. $P(C \leq 6) = 0.861$	Use technology to find the binomial probability. Write the answer to three significant figures.
d There is not a fixed number of trials nor are the trials independent so the binomial distribution is not an appropriate model for E.	

The binomial distribution can be applied in problem-solving situations.

Example 24

Solve the problems, stating any assumptions and interpretations you make.

a In a family of six children, find

i the probability that there are exactly three girls

ii the probability that exactly three consecutive girls are born.

b A study shows that 0.9% of a population of over 4 000 000 carries a virus. Find the smallest sample from the population so that the probability of the sample having no carriers is less than 0.4.

Continued on next page

Statistics and probability

a i Assuming that boy and girl are the only two outcomes, then the probability of each is 0.5 if the gender of each child is independent of the other. Let G represent the number of girls, then $G \sim \mathrm{B}(6, 0.5)$ and	Understanding the context is the first step in solving the problem. State the distribution.
$\mathrm{P}(G = 3) = 0.3125$	Write down the event and find the probability with technology.
ii Three consecutive girls are born: GGGBBB, BGGGBB, BBGGGB, BBBGGG	Use a diagram to represent the entire sample space.
Each of these four outcomes has probability $\left(\frac{1}{2}\right)^6$, so	
$\mathrm{P}(\text{Three consecutive girls are born}) = 4 \times \left(\frac{1}{2}\right)^6 = \frac{1}{16} = 0.0625$	Both answers can be expressed exactly in this case.
This is considerably smaller than 0.3125 since there are many more combinations of three girls that are not consecutive than are.	Examine your result critically and check that it is feasible.
b In a population of this size the binomial distribution is an appropriate model since sampling without replacement does not alter the probability of choosing a carrier significantly.	State the assumptions.
If C is the number of carriers chosen in a sample of size n then $C \sim B(n, 0.009)$.	Write down the distribution.
Find the smallest value of n so that $\mathrm{P}(C = 0) < 0.4$.	Translate the problem into an inequality.
$n = 102$ is the minimum value required.	Solve using technology. Interpret the output from technology. The sequence is decreasing and 102 is the first value of n for which the probability is less than 0.4.

Exercise 7K

1 For each context, determine whether the binomial model can be applied. If it can, write down the distribution in the form $X \sim \mathrm{B}(n, p)$. If it cannot, state why.

a A fair die numbered 1, 2, 3 and 4 is thrown seven times. Find the probability that an even number is obtained exactly three times.

b A box contains six green dice, three red dice and three white dice.

A die is selected at random from the box.

The die's colour is noted and it is not replaced. This experiment is repeated four times. Find the probability that a green die is selected fewer than three times.

c A die is selected at random from the same box, its colour noted and the die replaced. This experiment is repeated four times. Find the probability that a green die is selected fewer than three times.

d Jasmine is packing her bag for a four-day music festival. The weather forecast for each of the days states that the probability of rain is a constant value of 30%. Find the probability that it will rain on at least two of the four days.

e In a soccer match, seven substitute players are available for selection during the match. FIFA rules state that a maximum of three substitutions can be made in a match. Find the probability that exactly three substitutions are made in a match.

2 A fair coin is flipped six times. Calculate the probability of obtaining:

a exactly three heads

b no more than four heads

c at least three heads and fewer than six.

3 Silk scarves are produced in a factory. Quality control investigations find that 0.5% of the scarves produced have flaws, which reduce their value. A sample of 30 scarves is selected from the factory. Calculate the probability that the sample has:

a exactly one flawed scarf

b no flawed scarves

c more than three flawed scarves.

4 Paul works Monday to Friday for a car recovery service in Mathcity. His records show that on any given day, the probability that he is called to recover a car in an area outside the city limits is 0.17. Calculate the probability that:

a In one working week, Paul has to recover a car outside the city limits more than three times.

b Paul has to recover a car outside the city limits more than three times on two consecutive weeks.

5 A weather station in a remote area is powered by 10 solar panels, which operate independently of each other. The manufacturers state that each solar panel has a probability of failure of 0.085 in any given week. The weather station needs at least six of the solar panels to be working in order to function.

a Find the probability that the weather station will not fail in a week.

b The weather station is inspected and any necessary maintenance is carried out every six weeks. Find the probability that the weather station will still be functioning when inspected.

6 Alexandre uses his spinning arrow to design another game. Only if the spinner stops in regions A or C is a prize is awarded. If the game is played eight times, find the probability that:

a exactly five prizes are awarded

b fewer than five prizes are awarded

c five prizes or fewer are awarded.

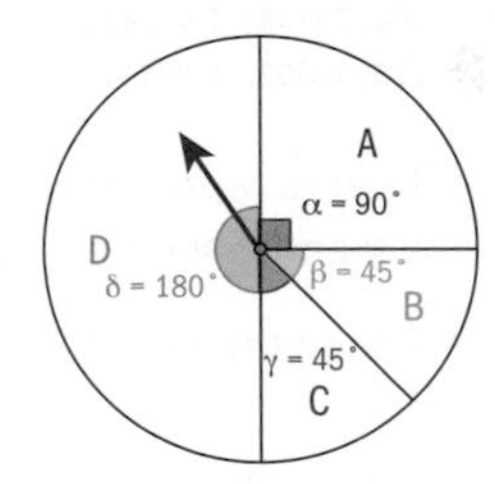

7 Zeke is exploring a biased coin. He tells Francesco that the probability of throwing a head on a coin he has designed is 0.964. However, the probability of throwing exactly four heads with this coin in five trials is approximately the same as the probability of the same event but with a fair coin. Francesco does not believe this is true. Demonstrate that Zeke is correct.

In Section 7.5 you learned about the expected value of a discrete random variable X: $\mathrm{E}(X) = \mu = \sum_{x} x\mathrm{P}(X = x)$.

You can use this to find the expected value when X is distributed binomially with parameters n and p. In the following investigation there are two other approaches.

Investigation 18

A A subjective perspective

Iana tosses a fair coin 20 times and counts the number of heads.

1 **Factual** On average, what is the number of heads she would expect? What if she tossed the coin 32 times? Discuss in a group.

2 **Factual** What if she tosses a biased coin with $\mathrm{P(Head)} = 0.7$ fifty times? What number of heads would she expect?

3 **Conceptual** Can you generalize your answers to make a conjecture on the expected number of heads for n tosses if $\mathrm{P(Head)} = p$? Discuss your justification in a group.

B An experimental perspective

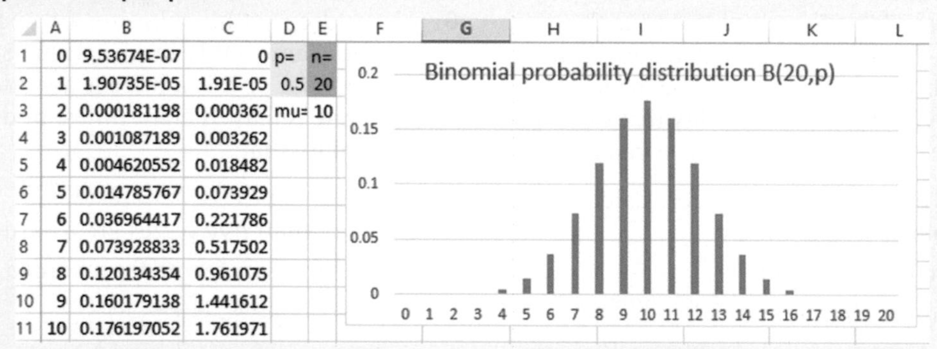

	A	B	C	D	E
1	0	9.53674E-07	0	p=	n=
2	1	1.90735E-05	1.91E-05	0.5	20
3	2	0.000181198	0.000362	mu=	10
4	3	0.001087189	0.003262		
5	4	0.004620552	0.018482		
6	5	0.014785767	0.073929		
7	6	0.036964417	0.221786		
8	7	0.073928833	0.517502		
9	8	0.120134354	0.961075		
10	9	0.160179138	1.441612		
11	10	0.176197052	1.761971		

4 Use a spreadsheet to represent $X \sim \mathrm{B}(n, p)$ as a probability distribution table and a bar chart by following these steps. Save your spreadsheet for use in the next investigation!

- In column A type the numbers 0, 1, 2, ..., 20 down to cell A21 to show the possible outcomes of the experiment.
- Fill in cells D1, D2, E1 and E2 as shown above to set up the parameters of the distribution.
- Type "= BINOM.DIST(A1,E2,D2,FALSE)" in cell B1. This is the probability of the event $\mathrm{P}(X=0)$ where $X \sim \mathrm{B}(20, p)$.
- Type "= A1*B1" in cell C1.
- Copy and drag cells B1 and C1 down to row 21 to complete the probability distribution table.
- Type "= sum(C1:C21)" in cell E3 and "mu =" in cell D3.
- Add a chart to display the domain of X on the x-axis and the corresponding probabilities on the y-axis.

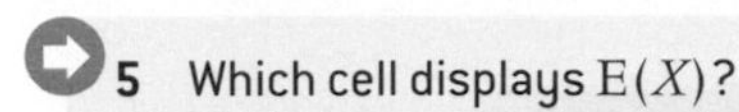

5 Which cell displays $\mathrm{E}(X)$?

6 Change the values of p and of n to test your conjecture from part (a) **a**.

7 **Factual** How do the parameters of the binomial distribution affect the expected value of the binomial random variable?

8 **Conceptual** How does the expected value of the binomial random variable relate to the formula for the expected number of occurrences?

In statistics you learned about the measures of central tendency (mean, median and mode), and measures of dispersion (range, interquartile range and standard deviation).

You have learned that the mean of the binomial distribution $X \sim \mathrm{B}(n, p)$ is $\mathrm{E}(X) = np$. Now you will explore its variance. The variance is the square of the standard deviation, and the standard deviation compares each data point with the mean.

TOK

During the mid-1600s, mathematicians Blaise Pascal, Pierre de Fermat and Antoine Gombaud puzzled over this simple gambling problem:

Which is more likely: rolling at least one six on four throws of one dice or rolling at least one double six on 24 throws with two dice?

Investigation 19

A Making visual comparisons of the spread: empirical evidence

With your spreadsheet from the previous investigation, alter the values of p and of n. Hence explore the effect these parameters have on the spread of the distribution. You may find it useful to collaborate with another student to compare and contrast, or just use Ctrl+Z when a parameter has been changed.

- Compare and contrast the spread of $X \sim \mathrm{B}(20, 0.15)$ with that of $X \sim \mathrm{B}(20, 0.5)$.
- Compare and contrast the spread of $X \sim \mathrm{B}(20, 0.15)$ with that of $X \sim \mathrm{B}(20, 0.85)$.
- Compare and contrast the spread of $X \sim \mathrm{B}(5, 0.85)$ with that of $X \sim \mathrm{B}(20, 0.85)$.

1 **Factual** By changing the probability of success, how can you maximize the spread for a fixed number of trials? Describe the reasons why your answer has this effect.

2 **Factual** By changing the number of trials, how can you make the spread greater for a fixed probability of success? Describe the reasons why your answer has this effect.

3 **Factual** Which parameters of $X \sim \mathrm{B}(n, p)$ affect variance?

B Looking for patterns and making a conjecture: inductive reasoning

You can investigate the variance with technology to see the effect of changing the probability and keeping the number of trials fixed as follows:

Continued on next page

Statistics and probability

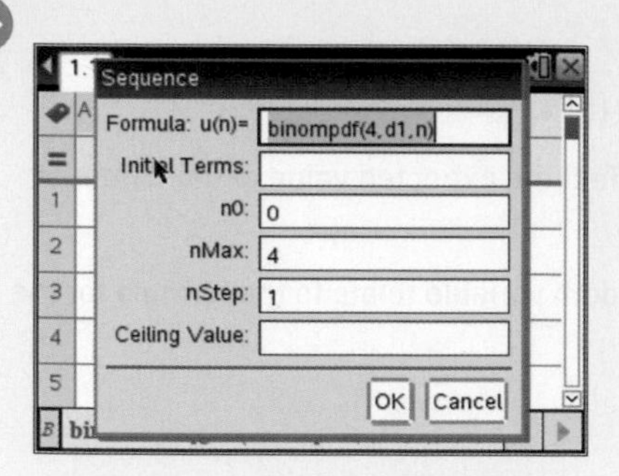

D1 is the cell where the probability parameter can be changed.

	A xval	B binom	C	D	E
=		=seqgen(binomp			
1	0	0.0256	p	0.6	
2	1	0.1536	var	0.96	
3	2	0.3456			
4	3	0.3456			
5	4	0.1296			

D2 $=g6^2$

Use the steps menu->statistics->stat calculations->one variable statistics. The cell reference G6 is where σx is found.

4 Use technology to collect data to complete the following table. You may repeat for other values of n.

5 Write down a conjecture for the variance of $X \sim \mathrm{B}(4, p)$.

6 Write down a conjecture for the variance of $X \sim \mathrm{B}(n, p)$.

7 **Factual** Is your conjecture consistent with your answers for **1**, **2** and **3**?

8 **Conceptual** How do the binomial distribution parameters model the spread and central tendency of the binomial distribution?

$X \sim \mathrm{B}(4, p)$	
p	Variance of X
0	
0.1	
0.2	
...	
0.9	
1	

If $X \sim \mathrm{B}(n, p)$ then $\mathrm{E}(X) = np$ and $\mathrm{Var}(X) = np(1-p)$.

Example 25

Two machines in a lightbulb factory are being inspected because quality control raised concerns. Managers have found that the probability that the first machine produces a defective lightbulb is 0.3, and that the probability that the second machine produces a defective bulb is 0.2.

Inspectors take a sample of six bulbs from the first machine and five from the second, and use $X \sim \mathrm{B}(6, 0.3)$ and $Y \sim \mathrm{B}(5, 0.2)$ to model the number of defective lightbulbs in the samples from the first and second machines respectively.

Compare and contrast the central tendency and spread of these binomially distributed random variables.

$\mathrm{E}(X) = 6 \times 0.3 = 1.8$ and $\mathrm{E}(Y) = 5 \times 0.2 = 1$	Apply the formula $\mathrm{E}(X) = np$.
X has the higher expected value: on average almost twice as many defective bulbs are predicted to appear in the sample of 6 than in the sample of 5.	Interpret your results.

Variance of $X = 6 \times 0.3 \times 0.7 = 1.26$ and Variance of $Y = 5 \times 0.2 \times 0.8 = 0.8$	Apply the formula variance of $X = np(1 - p)$.
There is more spread predicted in the values of the number of defective bulbs from the first machine.	Interpret your results.

Exercise 7L

1 Given $X \sim \mathrm{B}(6, 0.29)$ find the probabilities:

a $\mathrm{P}(X = 4)$ **b** $\mathrm{P}(X \le 4)$ **c** $\mathrm{P}(1 \le X < 4)$

d $\mathrm{P}(X \ge 2)$ **e** $\mathrm{P}(X \le 4 | X \ge 2)$.

f Use your answers to determine whether $X \le 4$ and $X \ge 2$ are independent events.

g Find $\mathrm{E}(X)$. **h** Find the variance of X.

2 A fair octahedral die numbered 1, 2, ..., 8 is thrown seven times.

Let Q denote the number of prime numbers thrown. Find:

a the probability that at least three prime numbers are thrown

b $\mathrm{E}(Q)$

c the variance of Q.

3 a A biased coin is coloured red on one side and black on the other. The probability of throwing red is 0.78. In 10 throws of the coin, find the probability that:

i exactly three blacks are thrown

ii the number of reds thrown is more than three but fewer than seven.

Let A represent the event "fewer than three blacks are thrown" and B "more than seven blacks are thrown".

b Find the probabilities $\mathrm{P}(A)$, $\mathrm{P}(B)$, $\mathrm{P}(A | B)$.

Hence determine if the events A and B are:

c independent

d mutually exclusive.

4 David plays a game at a fair. He throws a ball towards a pattern of 10 holes in this formation. The aim of the game is to have the ball fall into the red hole to win a point.

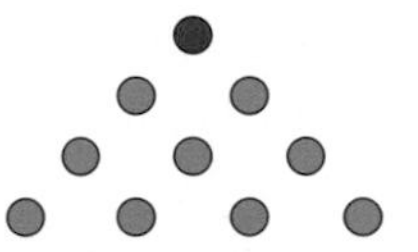

One game consists of throwing 10 balls. Assume David has no skill whatsoever at aiming and that a ball thrown must fall through one of the holes.

a Find the probability that David scores at least five points in a game.

b David plays six games. Find the probability that he scores no points in at least two games.

5 Zeke explores his biased coin in more detail. He designs a spreadsheet to simulate five throws of his biased (red) coin. The probability of a head is 0.964. He also throws a fair (black) coin five times. He collects data for 614 trials as shown below. The image also shows the outcome of the 614th trial.

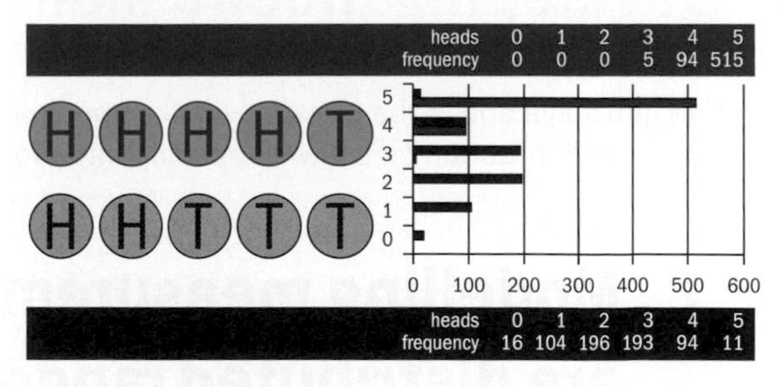

Let R and B represent the number of heads in five throws of the red coin and five throws of the black coin respectively.

a Find $\mathrm{E}(R)$ and $\mathrm{E}(B)$ and interpret your results.

b Find the variance of R and the variance of B and interpret your results.

6 In a mathematics competition, students try to find the correct answer from five options in a multiple choice exam of 25 questions. Alex decides his best strategy is to guess all the answers.

a State an appropriate model for the random variable A = the number of questions Alex gets correct.

Find the probability that the number of questions that Alex gets correct is:

b at most five **c** at least seven

d no more than three.

e Write down $E(A)$ and interpret this value.

f Find the probability that Alex scores more than expected.

g In the test, a correct answer is awarded 4 points. An incorrect answer incurs a penalty of 1 point. If Alex guesses all questions, find the expected value of his total points for the examination.

h Four students in total decide to guess all their answers. Find the probability that at least two of the four students will get seven or more questions correct.

7 Calcair buys a new passenger jet with 538 seats. For the first flight of the new jet all 538 tickets are sold. Assume that the probability that an individual passenger turns up to the airport in time to take their seat on the jet is 0.91.

a Write down the distribution of the random variable T = the number of passengers that arrive on time to take their seats, stating any assumptions you make.

b Find $P(T = 538)$ and interpret your answer.

c Find $P(T \geq 510)$ and interpret your answer.

d Calcair knows that it is highly likely that there will be some empty seats on any flight unless it sells more tickets than seats. Find the smallest possible number of tickets sold so that $P(T \geq 510)$ is at least 0.1.

e Determine the number of tickets Calcair should sell so that the expected number of passengers turning up on time is as close to 538 as possible.

f For this number of tickets sold, find $P(T = 538)$ and $P(T > 538)$. Interpret your answers.

8 You are given $X \sim B(n, p)$. Analyse the variance of X as a function of p to find the value of p that gives the most dispersion (spread) of the probability distribution.

Developing inquiry skills

Look back at the opening scenarios.

Can you solve one of the questions in the opening scenario with the binomial distribution? If so, what assumptions would you have to make?

7.7 Modelling measurements that are distributed randomly

In this section you will model an example of a **continuous** random variable.

For example, consider the height Y metres of an adult human chosen at random. Recall from Section 7.5 that Y is a measurement and would therefore be an example of **continuous** data.

We use the following terminology for random variables that are found by measuring:

Terminology	Explanation
Y is a **continuous random variable.**	Continuous: Y can be found by measuring and is therefore a real number.
	Random: Y is the result of a random process.
	Variable: Y can take any value in a domain that is a subset of R.
	(Y has domain $0.67\text{ m} \le Y \le 2.72\text{ m}$ according to the Guinness book of world records.)

Figure 1

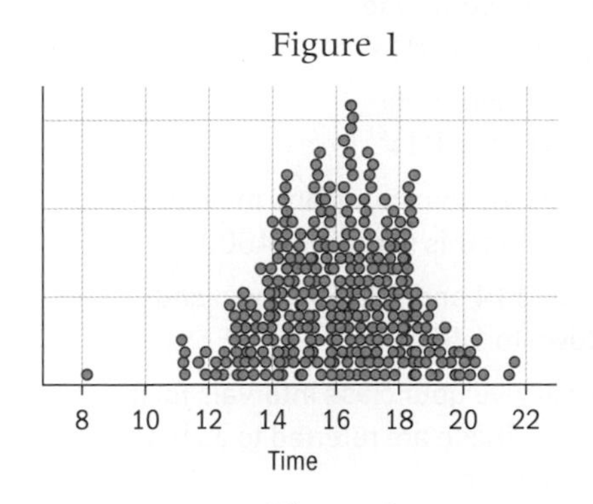

For example, 300 batteries are tested in a quality control exercise. The lifetime of each battery is measured to the nearest second.

The lifetime of a battery chosen at random, L, is a continuous random variable.

Figure 1 represents the 300 data points.

Figure 2

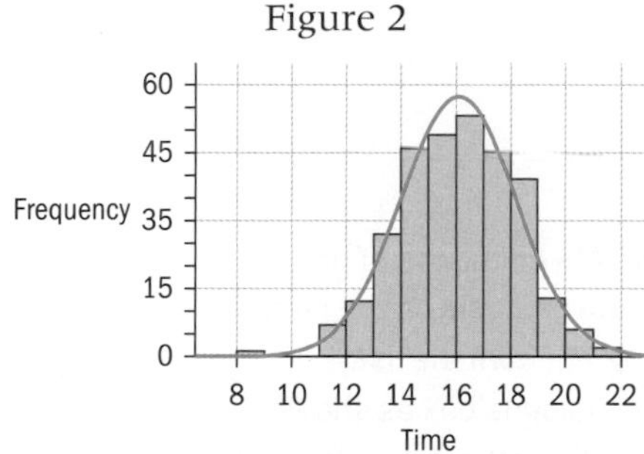

Figure 2 represents the same data set.

The frequency histogram is broadly symmetric and can be modelled by the "bell-shaped" curve shown.

Batteries lasting relatively long or short periods of time are rare.

If a set of continuous data can be modelled with this shape, we say that the data is **distributed normally** or that the data **follows a normal distribution.**

In this section you will make only **subjective** judgments of how well data follow the normal distribution. In the next chapter, you will learn techniques with which to make more objective judgments.

In the following investigation, you will explore whether some data sets are distributed normally or not.

International-mindedness

The normal curve is also known as the Gaussian curve, and is named after the German mathematician, Carl Friedrich Gauss (1777–1855), who used it to analyse astronomical data. This is seen on the old 10 Deutsche Mark notes.

Investigation 20

For this investigation, if a data set has a symmetric bell shape, consider that it follows a normal distribution. Click the icon to view the data set.

You are going to analyse W, the length of the wing from the carpal joint to the wing tip, of 4000 blackbirds. This data is in column E of the spreadsheet.

1 **Factual** Is W a discrete random variable?

Representing and exploring the data: procedure

E	F	G	H	I	J	K	L
Wing	Weight	Day	Month	Year	Time		
133	95	21	12	2006	14		114
134	106	25	11	2012	9		144
135	125	29	1	1994	9		112
135	113	5	2	1994	10		114
135	111	12	2	1994	8		116
134	105	15	2	1994	8		118
136	111	11	2	2004	8		120
127	103	23	2	2004	13		122
127	102	26	2	2004	9		124
126	104	25	2	2004	11		126
125	97	26	2	2004	9		128
135	102	27	2	2004	10		130
135	126	27	2	2004	15		132
135	121	2	3	2004	12		134
125	88	27	5	2004	14		136
123	90	8	6	2004	17		138
135	101	30	6	2004	13		140
129	100	7	9	2004	8		142
129	98	4	3	2005	16		144
129	97	6	3	2005	14		146
132	107	28	12	2005	10		

In cell L2 (green) type "$=\min(\text{E}2{:}\text{E}4001)$".

In cell L3 (blue) type "$=\max(\text{E}2{:}4001)$".

These give the maximum and minimum values of W in this sample of 4000.

Type 112, 114 and 116 as shown and drag down to 146 in cell L21.

These will give your class intervals for a histogram. These are referred to as bins by the software.

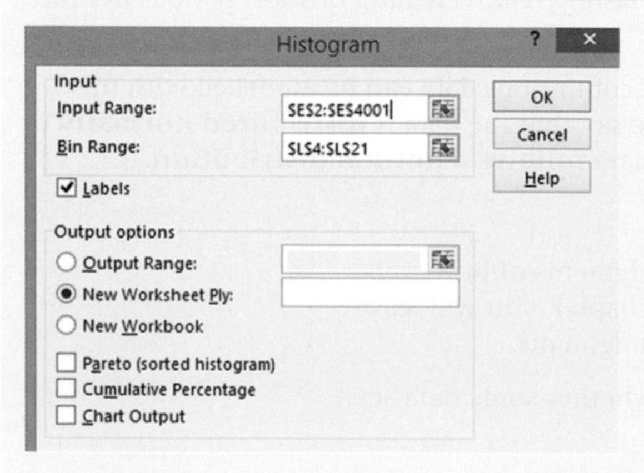

Now click on DATA on the top of your screen and Data Analysis on the far right. Choose Histogram from the list. Complete the dialogue box as shown. The software will put the data into class intervals. The output will begin:

114	1	This means thatthe frequency of W in the interval $(112, 114]$ is 1, the frequency in $(118, 120]$ is 28 etc.
116	3	
118	5	
120	28	
...	...	

Use the output to create a histogram of the 4000 values of W.

2 **Factual** Does the histogram for W show a symmetric bell-shaped curve?

3 **Factual** Can it be modelled by a normal distribution?

Repeat the procedure for E, life expectancy at birth. The data for 224 countries can be found here https://www.cia.gov/library/publications/the-world-factbook/rankorder/2102rank.html

4 **Factual** Is T a continuous random variable?

5 **Factual** Does the histogram for T show a symmetric bell-shaped curve?

6 **Factual** Can it be modelled by a normal distribution?

Repeat the procedure using other data sets that interest you.

7 **Factual** Which of these measurements do you feel could be modelled by a normal distribution?

- babies' birth weight
- ages of mother of new baby
- number of hairs on head of a new baby
- number of toes on a new baby
- annual salary of adults aged between 20 and 25 years
- life expectancy in Sweden
- the number of births on each day of January 1978 in the USA
- ages of people in Haiti
- journey time of a delivery van
- heights of sunflowers
- annual salary of professional footballers
- IQ scores of 2501 undergraduate students.

8 **Conceptual** In which contexts would you expect the normal distribution to be an appropriate model?

Now that you have experienced the bell-shaped curve used to model normally distributed data in different contexts, you can learn how it is applied to data with different measures of central tendency and spread.

Investigation 21

Notation: The data in Figure 1 is distributed normally with parameters 16.1 and 2.08^2.
You write this as $L \sim N(16.1, 2.08^2)$. The same data set is represented in the diagram below along with the curves that model three other data sets of 300 batteries sampled in the quality control exercise:
$P \sim N(16.1, 5^2)$, $Q \sim N(26.1, 5^2)$ and $R \sim N(36.1, 4^2)$.

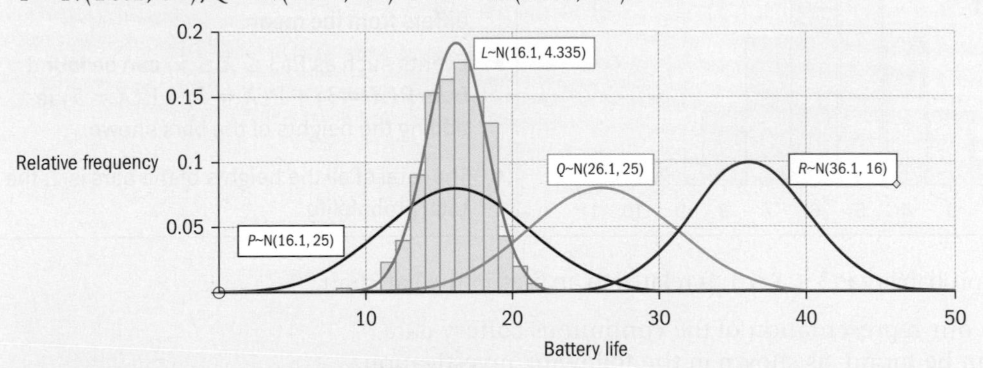

Reflect on how the parameters of these normal distributions affect the location and shape of the curves.
Relate the answers to the following four questions to the parameters of the distribution.

1 **Factual** Which of the data sets has the greatest mean?

2 **Factual** Which of the data sets has the lowest mean?

Continued on next page

Statistics and probability

3 **Factual** Which of the data sets has the widest spread?

4 **Factual** Which of the data sets has the narrowest spread?

5 **Conceptual** How can you predict the axis of symmetry from $X \sim \mathrm{N}(\mu, \sigma^2)$?

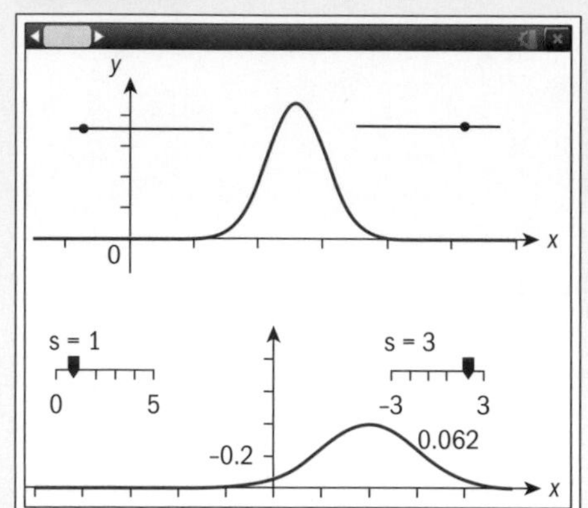

You can explore further either by using geometry software,

or by using a GDC.

In either case, alter the sliders and observe the effect that changing the parameters has on the shape of the curve.

6 **Conceptual** What do the parameters of $X \sim \mathrm{N}(\mu, \sigma^2)$ model?

You have learned and applied **discrete** probability distributions to contexts in Sections 7.5 and 7.6. Some facts relevant to this section are summarized here:

This bar chart represents $X \sim \mathbf{B}(11, 0.5)$	**Facts**
Figure 3 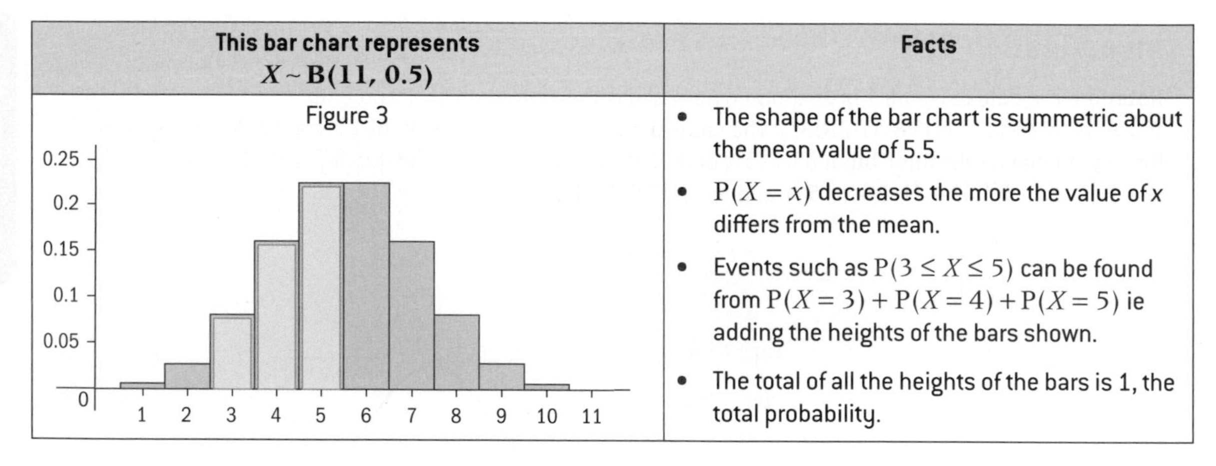	• The shape of the bar chart is symmetric about the mean value of 5.5. • $\mathrm{P}(X = x)$ decreases the more the value of x differs from the mean. • Events such as $\mathrm{P}(3 \le X \le 5)$ can be found from $\mathrm{P}(X = 3) + \mathrm{P}(X = 4) + \mathrm{P}(X = 5)$ ie adding the heights of the bars shown. • The total of all the heights of the bars is 1, the total probability.

Notice that the probability $\mathrm{P}(3 \le X \le 5)$ is related to an area in the bar chart.

Similarly, with our representation of the continuous battery data, probabilities can be found, as shown in the following investigation.

International-mindedness

French mathematicians Abraham De Moivre and Pierre Laplace were involved in the early work of the growth of the normal curve.

De Moivre developed the normal curve as an approximation of the binomial theorem in 1733 and Laplace used the normal curve to describe the distribution of errors on 1783 and in 1810 to prove the central limit theorem.

Investigation 22

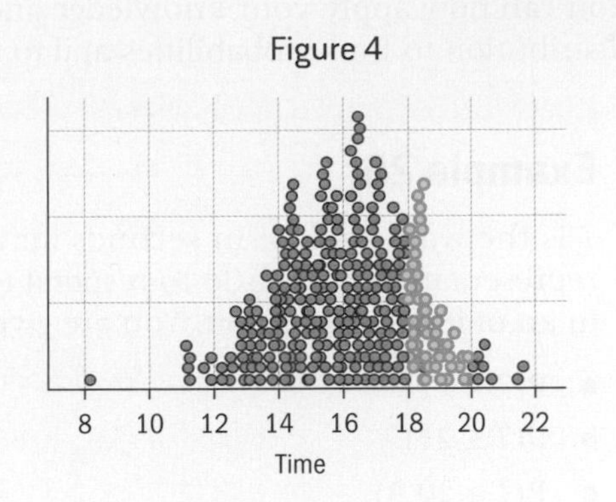

Recall that the lifetime of a battery chosen at random is L.

1 Write down an estimate of the mean battery lifetime.

In total, there are 300 data points, of which 60 are black.

2 Write down estimates for $P(18 \le L \le 20), P(12 \le L \le 14)$ and $P(14 \le L \le 18)$.

3 How does the shape of the dot plot help you make these estimates?

4 Is counting data points accurate and efficient in general?

Notice that Figures 3 and 4 use areas to find the probabilities. Probabilities from the normal distribution are **always** found by using technology or symmetry to find an **area** under the curve.

The total area under the normal distribution curve is 1.

For example,

Given $X \sim N(7,\ 1.5^2)$, you can find the probability $P(X \le 8) = 0.778$ as shown.

It is useful to bear in mind the symmetry of the curve in order to find related probabilities.

5 In the diagram above, why does the fact that 8 > 7 guarantee that $P(X \le 8) > 0.5$?

6 Use the graph to find $P(X \ge 8), P(X \le 6), P(X \ge 6)$ and $P(6 \le X \le 8)$.

Check your answers with technology.

7 Let $X \sim N(10,\ 1.7^2)$. Use technology to find the following probabilities:

$P(10 - 1.7 \le X \le 10 + 1.7), P(10 - 2 \times 1.7 \le X \le 10 + 2 \times 1.7), P(10 - 3 \times 1.7 \le X \le 10 + 3 \times 1.7)$.

Copy this diagram three times, complete the labelling of the regions and shade in the areas that represent your answers.

You are given that in $X \sim N(\mu, \sigma^2)$, μ is the mean and σ is the standard deviation.

8 Repeat **7** with your own values of μ and σ. What do you notice? Write your findings in a table and compare your results with those of others in your class.

If $X \sim N(\mu, \sigma^2)$, then:		
$P(\mu - \sigma \le X \le \mu + \sigma) =$		Approximately ____% of the data lies within _ standard deviations of the mean
$P(\mu - 2\sigma \le X \le \mu + 2\sigma) =$		Approximately ____% of the data lies within _ standard deviations of the mean
$P(\mu - 3\sigma \le X \le \mu + 3\sigma) =$		Approximately ____% of the data lies within ____ standard deviations of the mean

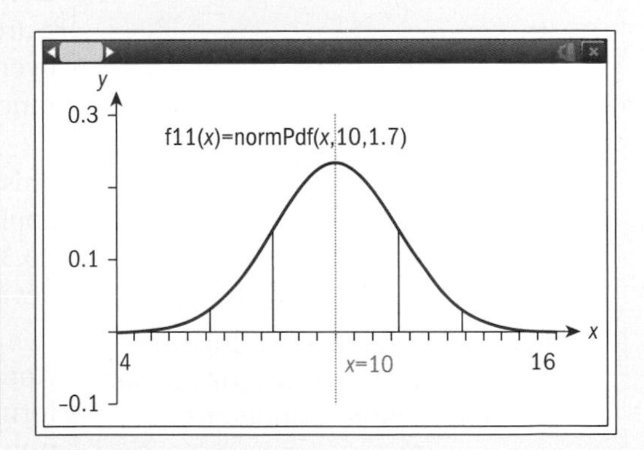

9 How do you know when you can find normal distribution probabilities without technology?

10 How do you know when you must find normal distribution probabilities with technology?

You can now apply your knowledge and understanding of the normal distribution to find probabilities and to solve problems.

TOK

Do you think that mathematics is a useful way to measure risks?

To what extent do emotion and faith play a part in taking risks?

Example 26

T is the waiting time in seconds for a customer care representative of TechCo to respond to a customer's first question in an online chat session. You are given that $T \sim N(19.1, 3^2)$. Find:

a $P(T < 17)$

b $P(T \leq 21)$

c $P(T \geq 20.3)$

d the expected number of times the waiting time is less than 21 seconds in a sample of 107 chats collected by TechCo management.

a $P(T < 17) = 0.0620$

b The answer to **a** can be represented on a sketch:

Noticing that 17 and 19 are both 2 units from the mean enables you to use the symmetry of the curve.

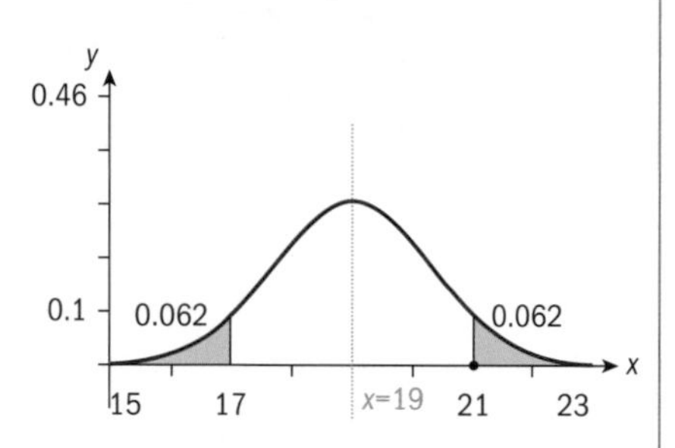

Hence $P(T > 21) = P(T < 17)$

So $P(T \leq 21) = 1 - P(T < 17)$
$= 0.938$

Apply the symmetry of the curve and complementary events. Note that $P(T \leq 21) = P(T < 21)$. In fact, $<$ and $\leq$ and $>$ and $\geq$ are interchangeable in events with continuous random variables.

c $P(T \geq 20.3) \approx \frac{1-0.68}{2} = 0.16$

This follows from applying the fact that 20.3 is one standard deviation from the mean.

d $107 \times 0.938 = 100.36$

Approximately 100 chats are predicted to have a waiting time of less than 21 seconds for the first question to be responded to.

Apply and interpret the formula for the expected number of occurrences.

Example 27

The lengths of trout in a fish farm are normally distributed with a mean of 39 cm and a standard deviation of 6.1 cm.

a Find the probability that a trout caught in the fish farm is less than 35 cm long.

b Cliff catches five trout in an afternoon. Find the probability that at least two of the trout are more than 35 cm long. State any assumptions you make.

c Find the probability that a trout caught is longer than 42 cm given that it is longer than 40 cm.

d Determine if the events $L > 42$ and $L > 40$ are independent.

a Let L represent the length of a randomly selected trout. Then $L \sim \text{N}(39, 6.1^2)$. $\text{P}(L < 35)$ can be found on your GDC: $\text{P}(L < 35) = 0.256$	Write down the random variable, the distribution and the event to clarify your thoughts and to demonstrate your knowledge and understanding. Use technology to find the probability and give the answer to three significant figures.
b Let C represent how many of the five fish are more than 35 cm long. Then $C \sim \text{B}(5, 0.744)$, assuming that the length of each fish caught is independent of the others. $\text{P}(C \geq 2)$ can be found on your GDC: $\text{P}(C \geq 2) = 0.983$	Five fish caught can be represented as five trials. Write down the distribution, using $1 - 0.256 = 0.744$ as the probability of success.
c $\text{P}(L > 42 \mid L > 40) = \dfrac{\text{P}(L > 42)}{\text{P}(L > 40)}$ $\text{P}(L > 42 \mid L > 40) = 0.716$	Apply the formula for conditional probability. $\text{P}(A \mid B) = \dfrac{\text{P}(A \cap B)}{\text{P}(B)}$
Since $\text{P}(L > 42 \mid L > 40) = 0.716$ and $\text{P}(L > 42) = 0.311$, the events are not independent.	Apply the definition of independent events. If $\text{P}(A \mid B) = A$ then A and B are independent.

Exercise 7M

1 The length of a local bus journey from Anrai's home to school is a normally distributed random variable T_L with mean 31 minutes and standard deviation 5 minutes. Sketch the following events on three separate diagrams:

a $\text{P}(T_L \geq 31)$ **b** $\text{P}(29 \leq T_L < 32)$

c $\text{P}(T_L < 36)$.

2 The length of an express bus journey from Anrai's home to school is a normally distributed random variable T_E with mean 25 minutes and standard deviation 3 minutes. Sketch the following events on three separate diagrams:

a $\text{P}(T_E \geq 31)$ **b** $\text{P}(29 \leq T_E < 32)$

c $\text{P}(T_E < 36)$.

3 A large sample of bottles of shampoo are inspected. The contents of the bottles S is distributed normally with mean 249 ml and standard deviation 3 ml.

a Sketch a diagram to represent this information.

b Hence estimate the probability that a randomly selected bottle of shampoo will contain less than 246 ml.

c Verify your answer using technology.

d The bottle is labelled "Contents 250 ml". Predict the number of bottles in a sample of 200 that will contain at least the amount claimed.

4 Data is collected from a large number of rush hour commuter trains.

The time T for all the passengers to board a train is distributed normally with mean 186 seconds and standard deviation 14 seconds.

a Sketch a diagram to represent this information.

b Hence estimate the probability that for a randomly selected commuter train it will take at least 214 seconds for all passengers to board.

c Verify your answer using technology.

d Hence predict the number of rush hour trains in a sample of 176 that will take longer than 200 seconds to be fully boarded.

5 $T \sim N(17.1, 3.1^2)$. Estimate these probabilities without technology:

a $P(T < 17.1)$ **b** $P(T < 14)$

c $P(T > 20.2)$ **d** $P(14 \le T < 23.3)$

e $P(T < 7.8)$ **f** $P(T < 23.3 | T > 20.2)$.

6 $Q \sim N(4.03, 0.7^2)$. Find these probabilities with technology:

a $P(Q < 4)$ **b** $P(Q < 3.4)$

c $P(Q > 5)$ **d** $P(3.5 \le Q < 4.5)$

e $P(Q < 4.9 | Q > 2.9)$.

7 **a** Match these five histograms with the correct box-and-whisker diagram.

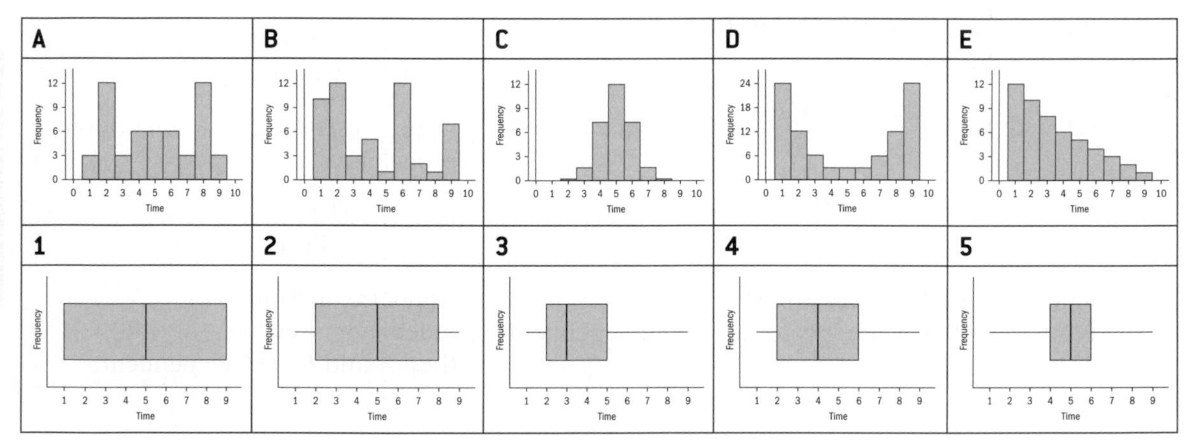

b Identify the one histogram of a data set that follows the normal distribution.

c State, with a reason, which statement is true:

p: A data set whose histogram is symmetric can be represented by a symmetric box-and-whisker diagram.

q: A data set that follows the normal distribution must have a symmetric box-and-whisker diagram.

r: A data set with a symmetric box-and-whisker diagram must be normally distributed.

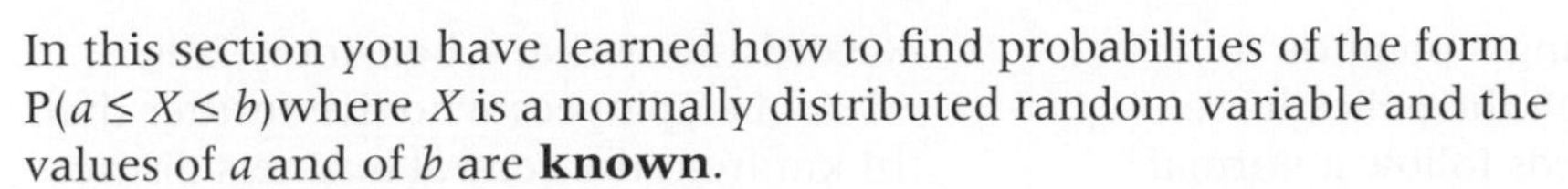

In this section you have learned how to find probabilities of the form $P(a \le X \le b)$ where X is a normally distributed random variable and the values of a and of b are **known**.

It is also possible to find the values of a and of b if, instead, $P(a \le X \le b)$ is known. This reverse process is illustrated in the next example.

International-mindedness

Belgian scientist Lambert Quetelet applied the normal distribution to human characteristics (l'homme moyen) in the 19th century. He noted that characteristics such as height, weight and strength were normally distributed.

Example 28

The weights W of cauliflowers purchased by a supermarket from their suppliers are distributed normally with mean 821 g and standard deviation 40 g. The heaviest 8% of cauliflowers are classified as oversized and re-packaged. Find the range of weights of cauliflowers classified as oversized. Express your answer correct to the nearest gram.

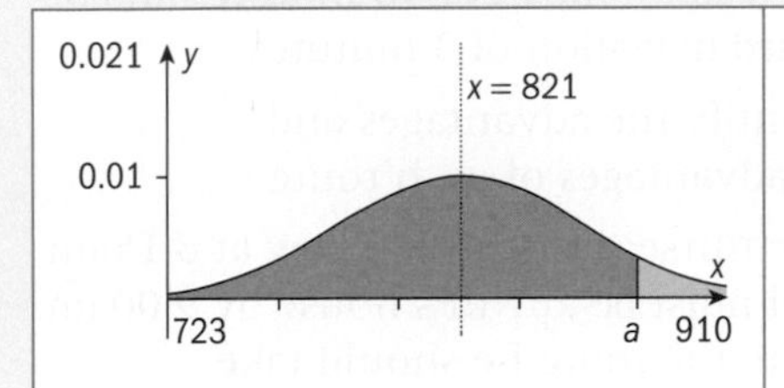	A sketch helps you orientate your reasoning in the correct place and is a good problem-solving strategy. You can see already that the lower limit for classification as oversized, labelled a, must be greater than 821.
$P(W < a) = 1 - 0.08 = 0.92$	Use complementary events to find $P(W < a)$.
Cauliflowers weighing at least 877 g will be classified as oversize.	This gives you the input necessary for the inverse normal function on your GDC. Interpret the result to state the range.

Exercise 7N

1 The weight of a bag of rice is distributed normally with mean 998 g and standard deviation 10 g. It is known that 20% of the packs weigh less than r g.

Find the value of r.

2 The weight of a pack of three bananas is distributed normally with mean 372 g and standard deviation 13 g. It is known that 17% of the packs weigh more than t g.

Find the value of t.

3 The weight of bags W measured in grams at an airport is distributed normally with mean 22 129 g and standard deviation 300 g.

a Identify the correct statement(s) about the upper quartile of the weights of the bags Q_3:

s : Q_3 can be found by solving $P(W > Q_3) = 0.75$.

t : Q_3 can be found by solving $P(W > Q_3) = 0.25$.

u : Q_3 can be found by solving $P(W < Q_3) = 0.75$.

b Hence calculate the interquartile range of W.

4 The speeds of cars passing a point on a highway are analysed by the police force. It is found that the speeds follow a normal distribution with mean 115.7 km/h and standard deviation 10 km/h.

   a Find the probability that a car chosen at random will be travelling between 110 km/h and 120 km/h.

   b A sample of eight cars is taken. Find the expected number of the sample that are travelling between 110 km/h and 120 km/h.

   c Find the probability that in the sample of eight, more than five cars are travelling between 110 km/h and 120 km/h. State the assumptions you must make.

5 An electronics company produces batteries with a lifespan that is normally distributed with a mean of 182 days and a standard deviation of 10 days.

   a Find the probability that a randomly selected battery lasts longer than 190 days.

   b In a sample of seven batteries chosen for a quality control inspection, find the probability that no more than three of them last longer than 190 days.

   c If a battery is guaranteed to last up to 165 days, find the probability that the battery will cease to function before the guarantee runs out.

   d Hence predict the number of batteries in a batch of 10 000 that would not last the duration of the guarantee.

6 The distance travelled to and from work each day by employees in a central business district is modelled by a normal distribution with mean 16 km and standard deviation 5 km.

   a Find the probability that a randomly chosen employee travels between 13 km and 15.3 km each day.

   b 13% of employees travel more than x km each day to and from work. Find the value of x.

   c Records show that when snow falls, 91% of employees who live further than 14 km from the central business district will fail to get to work. Predict how many of the 23 109 employees will fail to get to work on a snow day.

7 A nurse has a daily schedule of home visits to make. He has two possible routes suggested to him by an app on his phone for the journey to his first patient, Nur. Assume that the journey times are normally distributed in each case.

Route A has a mean of 42 minutes and a standard deviation of 8 minutes.

Route B has a mean of 50 minutes and a standard deviation of 3 minutes.

   a Identify the advantages and disadvantages of each route.

   b The nurse starts his journey at 8.15am and must be at Nur's house by 9.00am. State the route he should take.

   c If on five consecutive days, the nurse leaves home at 8.15am and takes route A, find the probability that he arrives at Nur's house:

      i by 9.00am on all five days

      ii by 9.00am on at least three of the five days

      iii by 9.00am on exactly three consecutive days.

8 Catarina finds a set of ages X measured in years that follow a normal distribution with mean 70 years and variance 25 years. She represents the data with a box-and-whisker diagram.

   a Calculate the upper quartile of X.

   b Hence determine whether the length of the box represents more than, less than or equal to 10 years.

Developing your toolkit

Now do the Modelling and investigation activity on page 376.

Chapter summary

- **Experiment**: A process by which you obtain an observation
- **Trials**: Repeating an experiment a number of times
- **Outcome**: A possible result of an experiment
- **Event**: An outcome or set of outcomes
- **Sample space:** The set of all possible outcomes of an experiment, always denoted by U
- The formula for the theoretical probability $\mathrm{P}(A)$ of an event A is:

  $\mathrm{P}(A)=\frac{n(A)}{n(U)}$ where $n(A)$ is the number of outcomes that make A happen and $n(U)$ is the number of outcomes in the sample space.

Whenever $\mathrm{P}(A)$ represents a subjective, experimental or theoretical probability, then $0 \le \mathrm{P}(A) \le 1$.

- A **Venn diagram** represents the sample space in a rectangle. Within the rectangle, each event is represented by a set of outcomes in a circle or an oval shape and is labelled accordingly.
- A **sample space diagram** is a useful way to represent the whole sample space and often takes the form of a table.
- Two events A and B are **mutually exclusive** if they cannot **both** occur. Hence:
  - Knowing that A has occurred means you know that B cannot, and that knowing that B occurs means you know that A cannot.
  - The occurrence of each event **excludes** the possibility of the other.
  - Consequently, $\mathrm{P}(A \cap B) = \mathrm{P}(B \cap A) = 0$.
- A and A' are **complementary** events. This means that
  - A and A' are mutually exclusive.
  - $\mathrm{P}(A) + \mathrm{P}(A') = 1$.
- Two events A and B are **independent** if the occurrence of each event does **not affect in any way** the occurrence of the other.
- Equivalently, **knowing** that A has occurred does not affect the probability that B does, and **knowing** that B occurs does not affect that probability that A does.
- Two events A and B are **dependent** if they are not independent.
- A tree diagram is another way to represent all the possible outcomes of an event. The end of each branch represents a combined event.
- The complementary probability law $\mathrm{P}(A) = 1 - \mathrm{P}(A')$ can give you a short way to solve problems.
- A **discrete probability distribution** is the set of all possible values of a discrete random variable (a subset of $\mathbb{Z}$) together with their corresponding probabilities.

Continued on next page

- A **discrete probability distribution function** $f(t)$ assigns to each value of the random variable its corresponding probability: $f(t) = \mathrm{P}(T=t)$.

$f(t)$ is commonly referred to by the abbreviation "pdf".

- The **expected value** of a discrete random variable X is $\mathrm{E}(X) = \mu = \sum_x x\mathrm{P}(X = x)$.
- If X is a discrete random variable that represents the gain of a player, then if $\mathrm{E}(X) = 0$, the game is **fair**.
- The binomial coefficients in row $(n+1)$ of the triangle are represented by the following notation:

  $\mathrm{C}_0^n, \mathrm{C}_1^n, \mathrm{C}_2^n, \ldots \mathrm{C}_r^n, \ldots \mathrm{C}_{n-1}^n, \mathrm{C}_n^n$, where $\mathrm{C}_r^n = \dfrac{n!}{r!(n-r)!}$ and

  $n! = n \times (n-1) \times (n-2) \times \ldots \times 3 \times 2 \times 1$.
- In a sequence of n independent trials of an experiment in which there are exactly two outcomes "success" and "failure" with constant probabilities $\mathrm{P}(\text{success}) = p$ and $\mathrm{P}(\text{failure}) = 1-p$, if X denotes the discrete random variable equal to the number of successes in n trials, then the probability distribution function of X is

$$\mathrm{P}(X=x) = \mathrm{C}_x^n p^x(1-p)^{n-x}, x \in \{0, 1, 2, \ldots, n\}$$

- These facts are summarized in words as "X is distributed binomially with parameters n and p" and in symbols as $X \sim \mathrm{B}(n, p)$.
- If $X \sim \mathrm{B}(n, p)$ then $\mathrm{E}(X) = np$ and $\mathrm{Var}(X) = np(1-p)$.
- The normal distribution enables you to model many types of data sets from science and society that involve measurements of **continuous** data.
- The parameters of the normal distribution are μ and σ^2. You write "X follows a normal distribution with parameters μ and σ^2" or $X \sim N(\mu, \sigma^2)$. Equivalently, you can state "X follows a normal distribution with mean μ and standard deviation σ."
- The graph of the normal distribution is a symmetric bell-shaped curve with these properties:
  - The axis of symmetry is $x = \mu$
  - The total area under the curve is 1
  - $P(\mu - \sigma \leq X \leq \mu + \sigma) \approx 0.68$
  - $P(\mu - 2\sigma \leq X \leq \mu + 2\sigma) \approx 0.95$
  - $P(\mu - 3\sigma \leq X \leq \mu + 3\sigma) \approx 0.997$
  - These properties enable you to find probabilities. In other cases, technology **must** be used.

Developing inquiry skills

Look back at the opening scenario, do any of the situations involve the normal distribution? If so, do you have enough information to solve them?

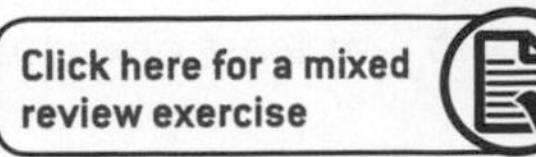

Chapter review

Click here for a mixed review exercise

1 To win a prize in a competition a door number in an apartment block is chosen at random from the numbers 1 to 150. Find the probability that the door number is:

a a square number

b at least 100

c divisible by 5

d no more than 11.

2 The discrete random variable K has probability distribution function given by $P(K = k) = \beta(3 - k)^2$ for $k = 0, 1, 2, 3, 4$. Find β and $E(K)$.

3 The discrete random variable D is distributed as shown in this table:

d	0	1	2	3	4
$P(D=d)$	0.3	$p+q$	0.15	$p-q$	$p+2q$

a Given that $E(D) = 1.7$, find p and q.

b Find $P(D = 3 \mid D \geq 1)$.

4 A packet of seeds contains 65% green seeds and 35% red seeds. The probability that a green seed grows is 0.85 and that a red seed grows is 0.74. A seed is chosen at random from the packet. Calculate the probability that the seed

a grows

b is green and grows

c is red or it grows.

5 **a** Find the probability of the outcome "throw at least one six" when a fair cubical die is thrown 1, 2, 3, 4, 5, ..., n times.

b Hence find the least value of n for which P(throw at least one six in n throws) > 99.5%.

6 B and C are independent events. $P(B \cap C) = 0.1$ and $P(B \cap C') = 0.4$. Find $P(B' \cup C)$.

7 The sides of a fair cubical die are numbered 2, 3, 3, 4, 4 and 5. The die is thrown twice and the outcomes are added to give a total T.

a Construct a table for the probability distribution of T.

b A game is designed so that if T is prime, a prize of US\$$T$ is won. If T is a square number, the player must pay US\$$x$. If T is any other number, no prize or payment is made. Find the value of US\$$x$ so that the game is fair.

8 Assume that the weights of 18-year-old males in a population follow a normal distribution with mean 65 kg and standard deviation 11 kg.

a Find the probability that a randomly chosen male weighs more than 70 kg.

b Find the interquartile range of the weights in the population.

c Find the weight exceeded by 7.3% of the population.

d Eight boys are chosen at random from the population. Find the probability that at most three of them weigh at least 70 kg.

e In a sample of 1000 taken from the population, estimate how many would weigh below 60 kg.

9 A box contains four blue balls and three green balls. Judith and Gilles play a game with each taking it in turn to take a ball from the box, without replacement. The first player to take a green ball is the winner.

a Judith plays first. Find the probability that she wins.

The game is now changed so that the ball chosen is replaced after each turn. Judith still plays first.

b Determine whether the probability of Judith winning has changed.

Exam-style questions

10 **P1:** A box contains 16 chocolates, of which three are known to contain nuts.

Two chocolates are selected at random.

Of the two chocolates selected, find the probability that

a exactly one chocolate contains nuts (3 marks)

Statistics and probability

b at least one chocolate contains nuts. (3 marks)

Give your answers as fractions.

11 P2: Hamid must drive through three sets of traffic lights in order to reach his place of work.

The probability that the first set of lights is green is 0.7.

The probability that the second set of lights is green is 0.4.

The probability that the third set of lights is green is 0.8.

It may be assumed that the probability of any set of lights being green is independent of the others.

a Find the probability that all three sets of lights are green. (2 marks)

b Find the probability that only one set of lights is green. (3 marks)

c Given that the first set of lights is red (ie *not* green), find the probability that the following two pairs of lights will be green. (2 marks)

d Find the probability that at least one set of lights will be green. (3 marks)

12 P1: Jake and Elisa are given a mathematics problem.

The probability that Jake can solve it is 0.35.

If Jake has solved it, the probability that Elisa can solve it is 0.6. Otherwise, the probability that Elisa can solve it is 0.45.

a Draw a tree diagram to illustrate the above question, showing clearly the probabilities on each branch. (3 marks)

b Find the probability that at least one of the students can solve the problem. (2 marks)

c Find the probability that Jake solves the problem, given than Elisa has. (4 marks)

13 P1: A and B are events such that $P(A) = 0.3$, $P(B) = 0.65$ and $P(A \cup B) = 0.7$.

By drawing a Venn diagram to illustrate these probabilities, find

a $P(A' \cap B)$ (2 marks)

b $P(A \cup B')$ (2 marks)

c $P(A \cap B)'$ (2 marks)

14 P1: In a survey, 48 people were asked about their holidays over the past year. It was found that 32 people had taken a holiday in Europe, and 25 people had taken a holiday in the USA.

Everyone surveyed had been to at least Europe or the USA.

a Determine how many people had taken a holiday in both Europe and the USA. (2 marks)

b Find the probability that a randomly selected person had been to Europe, but not the USA. (3 marks)

c Explain why the events "taking a holiday in Europe" and "taking a holiday in the USA" are not independent events. (3 marks)

15 P2: The Venn diagram illustrates the number of students taking each of the three sciences: physics, chemistry and biology.

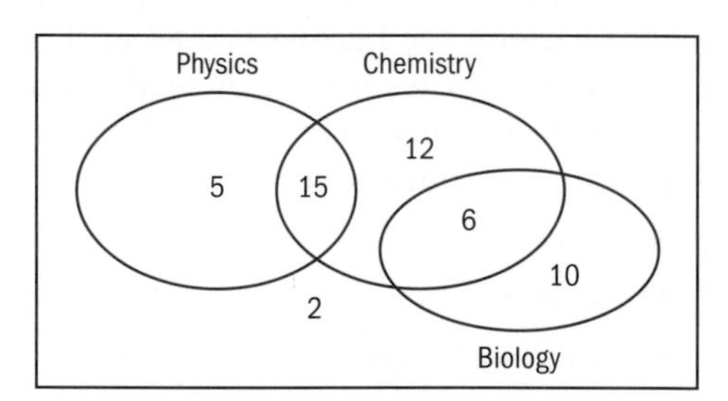

A student is randomly chosen from the group.

Find the probability that

a the student studies chemistry or biology (2 marks)

b the student studies neither physics nor biology (2 marks)

c the student studies physics, given that they study chemistry (2 marks)

d the student studies biology, given that they study physics (2 marks)

e the student studies physics, given that they do not study biology. (2 marks)

16 P1: A and B are independent events, such that $P(A) = 0.3$ and $P(B) = 0.5$.

Find the following probabilities.

a $P(A \cap B)$ (2 marks)

b $P(A \cup B)$ (2 marks)

c $P(B' \cap A)$ (2 marks)

d $P(B|A')$ (3 marks)

17 P1: The probability that Steve the striker scores in any given football match is 0.28.

Four matches over the course of a season are selected at random.

a Find the probability that Steve scores in at least one of the four matches. (3 marks)

b Find the probability that Steve scores in exactly one of the four matches. (2 marks)

c Find the probability that Steve scores in at least two of the four matches. (3 marks)

18 P1: In Victorian England, the probability of a child born being male was 0.512.

In a family of 10 children, find the probability that

a there were exactly 6 boys (3 marks)

b there were no boys (2 marks)

c there were more girls than boys. (2 marks)

19 P2: The discrete random variable X has the following probability distribution.

x	0	0.5	1	1.5	2
$P(X=x)$	$\frac{k}{2}$	k	k^2	$2k^2$	$\frac{k}{2}$

a Find the value of k. (4 marks)

b Determine the exact value of the mean of the distribution. (4 marks)

c Find the exact value of $P(X \geq 1.25)$. (3 marks)

20 P1: The times taken by Blossom to walk from her home to her local café are normally distributed, with a mean time of 35 minutes and a standard deviation of 3.4 minutes.

a Find the probability that, on a randomly selected day, Blossom's journey takes longer than 37 minutes. (2 marks)

b Find the probability that on a randomly selected day, Blossom's journey takes between 34 and 36.5 minutes. (3 marks)

Blossom walks to her café on 25 separate occasions.

c On how many occasions should she expect the journey to take less than 30 minutes? (3 marks)

21 P1: The masses of cans of baked beans are normally distributed, with a mean mass of 415 g and standard deviation 12 g.

a The probability that a randomly selected can of beans has a mass greater than m g is 0.65. Find the value of m. (3 marks)

b A can of beans is randomly selected. It is known that the mass of the can is more than 420 g. What is the probability that it weighs more than 422.5 g? (4 marks)

c Ashok buys 144 cans of beans from the supermarket.

What is the probability that at least 75 of them have a mass of less than 413.5 g? (4 marks)

Random walking!

Approaches to learning: Critical thinking
Exploration criteria: Mathematical communication (B), Personal engagement (C), Use of mathematics (E)
IB topic: Probability, Discrete distributions

The problem

A man walks down a long, straight road. With each step he either moves left or right with equal probability. He starts in the middle of the road. If he moves three steps to the left or three steps to the right, he will fall into a ditch on either side of the road. The aim is to find probabilities related to the man falling into the ditch, and in particular to **find the average number of steps he takes before inevitably falling into the ditch.**

Explore the problem

Use a counter to represent the man and a "board" to represent the scenario:

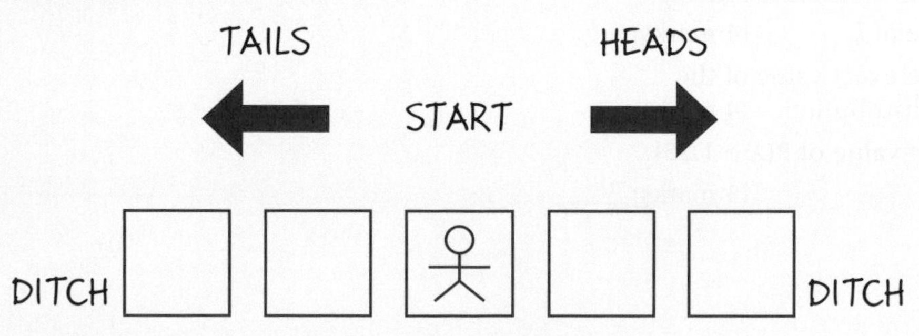

Toss a coin.

Let a tail (T) represent a left step and a head (H) represent a right step.

Write down the number of tosses/steps it takes for the man to fall into the ditch.

Do this a total of 10 times.

Calculate the average number of steps taken.

Construct a spreadsheet with the results from the whole class.

Calculate the average number of steps taken from these results.

How has this changed the result?

Do you know the actual average number of steps required?

How could you be certain what the average is?

Calculate probabilities

Construct a tree diagram that illustrates the probabilities of falling into the ditch within five steps.

Use your tree diagram to answer these questions:

What is the probability associated with each sequence in which the man falls into the ditch after a total of exactly five steps?

What is the probability that the man falls into the ditch after a total of exactly five steps?

What is the minimum number of steps to fall into the ditch?

What is the maximum number?

What is the probability that the man falls into the ditch after a total of exactly three steps?

Explain why all the paths have an odd number of steps.

Let x be the number of the steps taken to fall into the ditch.

Copy and complete this table of probabilities:

x	1	2	3	4	5	6	7	8	9	10	11	12	....
$P(X = x)$													

Look at the numbers in your table.

Can you see a pattern?

Could you predict the next few entries?

Simulation

You could use the table together with the formula $E(X) = \sum_{1}^{\infty} xP(X = x)$ to try to find $E(X)$, the exact theoretical answer to the problem posed.

However, since there is an infinite number of values of x, calculating the expected number of steps to fall into the ditch would be very complicated.

An alternative approach is to run a computer simulation to generate more results, and to calculate an average from these results.

You can write a code in any computer language available that will run this simulation as many times as needed.

This will allow you to improve on the average calculated individually and as a class.

Although this would not be a proof, it is convincing if enough simulations are recorded.

Extension

Once you have a code written you could easily vary the problem.

What variations of the problem can you think of?

You may also be able to devise your own probability question which you could answer using simulation.

8 Testing for validity: Spearman's, hypothesis testing and χ^2 test for independence

All of these questions require the collection of data. The data can then be analysed and presented in ways that allow you to understand and interpret the question more clearly. You might, for example, think that the data should fit a particular pattern; how could you test that?

Concepts

- Relationships
- Validity

Microconcepts

- Contingency tables
- Observed frequencies, expected frequencies
- Null hypothesis, alternative hypothesis
- Significance level
- Degrees of freedom
- probability values (p-values)
- χ^2 test for independence, goodness of fit
- t-test
- Spearman's rank correlation

Do the heights of the students at your school really follow a bell-shaped normal distribution?

A newspaper reports: "Scientific study shows people with more friends live longer." How do they figure this out?

Are your preferences for food positively or negatively correlated with their nutritional value? Or neither?

How can a teacher (or the IB) tell whether two different versions of a test were equally difficult?

Scientists are concerned with the effect of air pollution on the growth of trees. They measured the heights in metres of 24 young trees of the same species in each of two different forest areas. The results are shown in the table.

Area A		Area B	
5.30	5.17	5.64	4.73
5.26	4.97	3.90	4.21
3.74	4.87	4.38	5.07
5.55	5.17	4.91	4.82
4.77	5.85	4.87	4.84
6.00	5.48	3.89	5.14
4.44	5.24	4.61	4.95
4.53	4.96	4.88	3.12
4.04	5.61	4.47	4.02
4.73	6.14	5.12	4.12
4.83	5.10	4.46	5.23
5.12	5.75	4.64	5.06

It is claimed that the trees in area A are, on average, taller than those in area B. How could you test this claim?

How can the scientists ensure that their samples were not biased?

Which statistics can you calculate from the data?

How could you display the data to help test the claim?

Do you think there is enough evidence in these samples to make any claims about the tree heights in general?

Which tests can the scientists use to find out whether air pollution has had an effect on the growth of trees in either forest?

How valid will the results of these tests be?

Which test will be the most reliable?

Will scientists be able to use the results of these tests to give feedback on the effect of air pollution on the growth of the trees?

Developing inquiry skills

Write down any similar inquiry questions you might ask to decide whether a statement about two other data sets was true or not. Your variables might be, for example, the heights of children, or the sizes of pebbles on a beach, or the fuel consumption of cars. What would you need to think about in each case?

Think about the questions in this opening problem and answer any you can. As you work through the chapter, you will gain mathematical knowledge and skills that will help you to answer them all.

Before you start

You should know how to:	Skills check
1 Find the probability of independent events. eg A fair die and an unbiased coin are thrown. Show that the probability of getting a 6 on the die and a head on the coin are independent events.	**1** The numbers 1, 2, 3, 4, 5, 6 are written on cards. S is the event of picking a square number. E is the event of picking an even number. Show that E and S are independent.
2 Find the probabilities from a normal distribution. eg The heights of a sample of people are normally distributed with a mean of 156 cm and standard deviation of 7 cm. Find the probability that the height of a person chosen at random from the sample is between 152 cm and 161 cm.	**2** The diameters of washers are normally distributed with mean 35 mm and standard deviation 3 mm. Find the probability that a washer chosen at random has a diameter of less than 36 mm.
3 Calculate binomial probabilities. eg List the probabilities of all possible outcomes for a B(3, 0.7) distribution.	**3** List the probabilities of all possible outcomes for a B(4, 0.8) distribution.

Click here for help with this skills check

8.1 Spearman's rank correlation coefficient

Ranking data

For each of the following lists, rank the items in order of how much you like them, with a "1" being the one you like most and a "6" the one you like least.

You must rank them all, and ties are not allowed!

Subject	Rank
Maths	
English	
PE	
Art	
Science	
Drama	

Film genre	Rank
Sci-fi	
Romance	
Documentary	
Comedy	
Adventure	
Thriller	

Sport	Rank
Soccer	
Basketball	
Tennis	
Baseball	
American football	
Table tennis	

Now, find someone to compare your choices with.

In the following table, write down your ranks for the same items. Then, in the other columns, work out the difference between your ranks (d), and the square of the difference (d^2).

Finally, add up the totals in the last column (d^2).

Item	You	Them	d	d^2
			Total:	

Once you have this total, do the following:

- multiply it by 6
- divide the answer by 210
- subtract the answer from 1.

Your answer should be a number between −1 and +1.

Which other mathematical calculation in statistics has an answer between −1 and +1?

International-mindedness

Charles Pearson (1863–1945) was an English psychologist who developed rank correlation, usually shown as the Greek letter ρ (rho) or, r_s, as a tool for psychiatry.

What do you think an answer close to −1 means?

What do you think an answer close to +1 means?

What do you think an answer close to 0 means?

What can you say about your answer?

Consider the following sets of data:

Hours of study, x	0	1	2	3	4	5	6	7	8
Test results, y	43	50	52	70	68	75	81	78	92

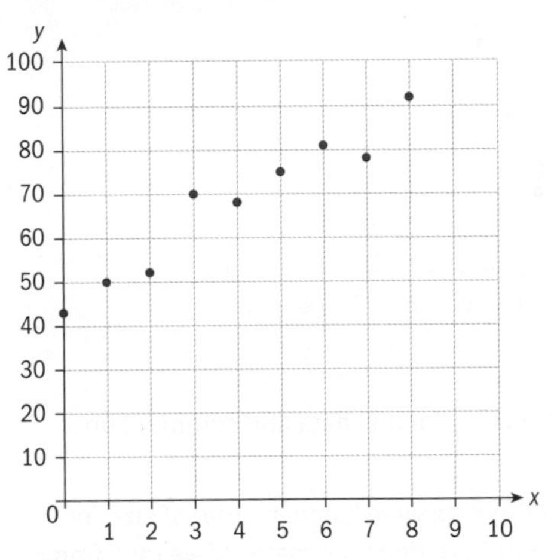

A scatter graph of the data is shown here, and Pearson's correlation coefficient is 0.97. So, there is a strong, positive relationship between the hours of study and the test results.

Now consider this data:

Number of minutes, x	0	5	10	15	20	25	30	35	40	45
Temperature of oven, y°C	0	140	165	175	180	180	180	180	180	180

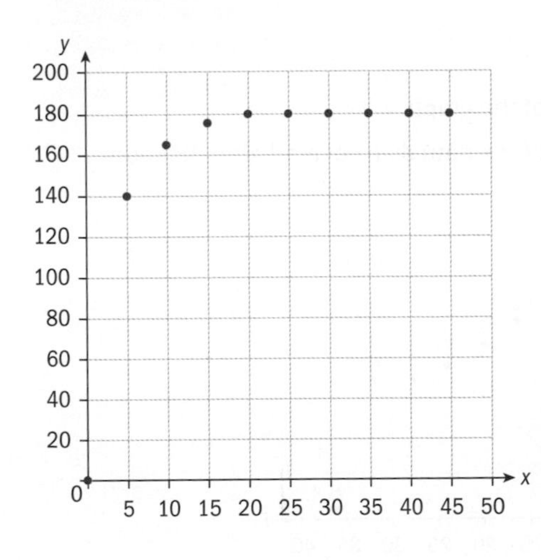

A scatter graph of this data is shown here, and Pearson's correlation coefficient is 0.73. So, there is only a moderate, positive relationship between the number of minutes and the temperature of the oven. However, the scatter graph indicates that there is an exponential relationship between the data points. Pearson's only measures for linear relationships. So, is there another test to find the strength of relationships that are not linear?

Investigation 1

Mould is grown in eight different petri dishes with different amounts of nutrients (X), and the area of the dish covered in mould after 48 hours (Y) is recorded. The results are given in the table and also shown on the graph.

X	Y
5.68	6.00
1.04	0.50
2.22	0.76
4.20	2.84
3.66	1.44
6.72	8.20
4.72	4.20
8.00	8.60

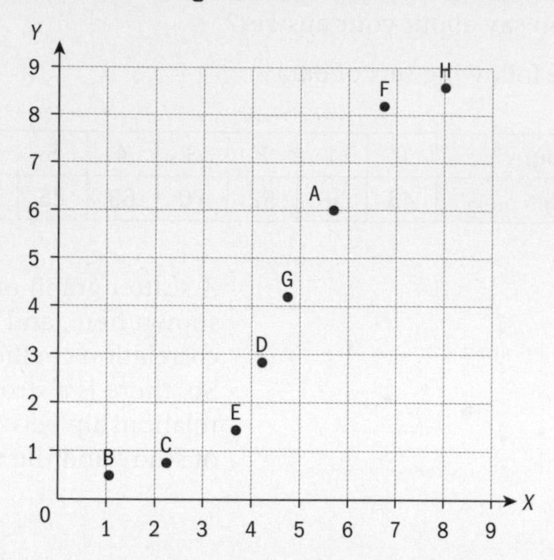

1 Use your GDC to graph these results.

2 Calculate the Pearson's product moment correlation coefficient (PMCC) for this data and comment on your results.

Now give each data point a rank, which is the position of the point if the data were listed in order of size for each of the variables. For example, H would be ranked 1 for both X and Y. (It does not matter if we rank from largest to smallest, like this, or from smallest to largest; the result will be the same.)

3 a Complete the following table showing the ranks for each of the data points.

	A	B	C	D	E	F	G	H
X rank								1
Y rank								1

b Use your GDC to graph these ranks.

c Calculate the value of PMCC for these ranks.

d Comment on your result, relating it to the particular shape of the graph.

In another experiment the temperature (T) is varied and the area of the petri dish covered after 48 hours (Y) is recorded.

T	Y
4.95	4.50
10.49	4.86
16.40	4.36
19.80	3.86
23.90	3.38
27.70	3.14
32.30	3.06
36.40	0.22

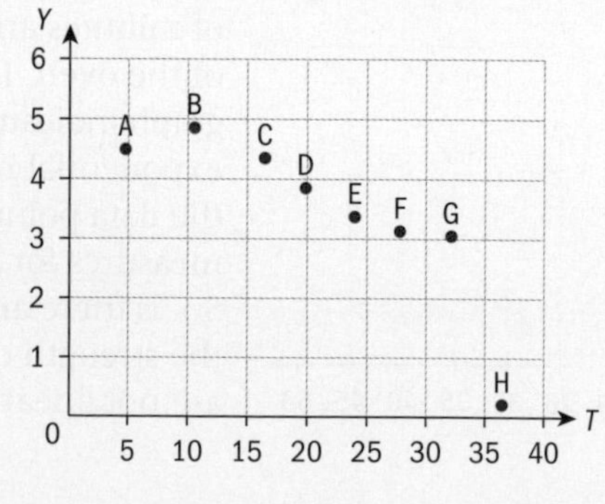

4 a Use your GDC to graph these results.

b Calculate the value of PMCC for this data and comment on your results.

c Complete the following table showing the ranks for each of the data points.

	A	B	C	D	E	F	G	H
T rank								1
Y rank								8

d Use your GDC to graph these results.

e Calculate the value of PMCC for this data and comment on your results.

f Discuss the features of the data that led to this value.

The PMCC of the rank values is called Spearman's rank correlation coefficient.

5 **Factual** What type of data is used for Spearman's?

6 **Factual** What type of data is used for Pearson's?

7 **Conceptual** What do correlation coefficients tell you about the relationship between two variables?

The product moment correlation coefficient of the ranks of a set of data is called Spearman's rank correlation coefficient. The IB notation is r_s.

Spearman's correlation coefficient shows the extent to which one variable increases or decreases as the other variable increases.

An r_s value of 1 means the set of data is strictly increasing, and a value of −1 means it is strictly decreasing. Data that is only increasing or only decreasing is known as **monotonic**.

A value close to 0 suggests that the data is not consistently increasing or decreasing.

Example 1

1 Find Spearman's rank correlation coefficient for the following sets of data.

a

Time spent training, x hours	23	34	17	23	29	45
Time to run 2 km, y min	12	10	14	11	11	8

b

Number of pets, x	1	2	3	4	5
Time spent each week caring for them, y hours	6	7	8	8	16

Continued on next page

2 A student was asked to rank nine different makes of burger in terms of which she liked best to which she liked least. She put 1 for the one she liked best and 9 for the one she liked least. These rankings and the costs of the burgers are given in the table.

Burger	A	B	C	D	E	F	G	H	I
Taste rank	7	3	4	6	1	9	2	5	8
Cost, US $	3.50	7.45	6.50	4.50	8.50	2.65	3.95	4.35	1.45

a Explain why you cannot use Pearson's in this example.

b Find Spearman's rank correlation coefficient for this data and comment on your answer.

1 a The ranks are

x	4.5	2	6	4.5	3	1
y	2	5	1	3.5	3.5	6

If we order the values of x by their size, we get their rank.

When more than one piece of data have the same value the rank given to each is the average of the ranks. For example, the two values of x equalling 23 here would have ranks 4 and 5; hence, each is given a rank of $\frac{4+5}{2} = 4.5$.

$r_s = -0.956$

The ranked data is put into a GDC and the PMCC obtained.

So, there is a strong, negative correlation. The more hours that you train the faster you can run the 2 km.

b The ranks are

x	5	4	3	2	1
y	5	4	2.5	2.5	1

$r_s = 0.975$

Often, when one of the variables increases at a fixed rate, for example measurements taken at one minute intervals, the order of the ranks will be the reverse of the order of the data.

So, there is a strong, positive correlation. The more pets you have the more hours it takes each week to look after them.

2 a Because the ranks are given rather than quantifiable data.

b Ranking the costs, you get:

Taste	7	3	4	6	1	9	2	5	8
Cost	7	2	3	4	1	8	6	5	9

and $r_s = 0.8$. So, there is a moderately strong relationship between the taste and the cost of the burgers.

Spearman's rank correlation coefficient is only valid for the data given in the question. If some data points are similar then any small changes could affect the value of r_s.

TOK

What practical problems can or does mathematics try to solve?

Exercise 8A

1 Write down the value of Spearman's rank correlation coefficient for each of the sets of data shown.

a

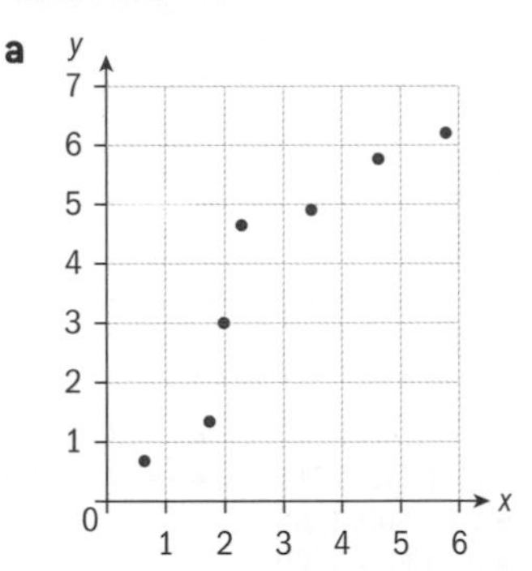

b

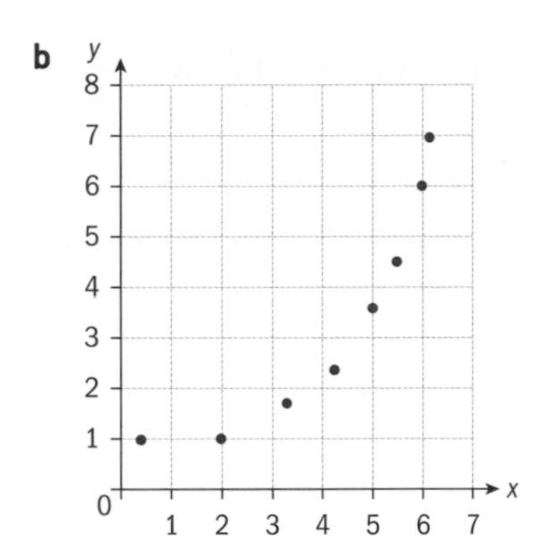

c

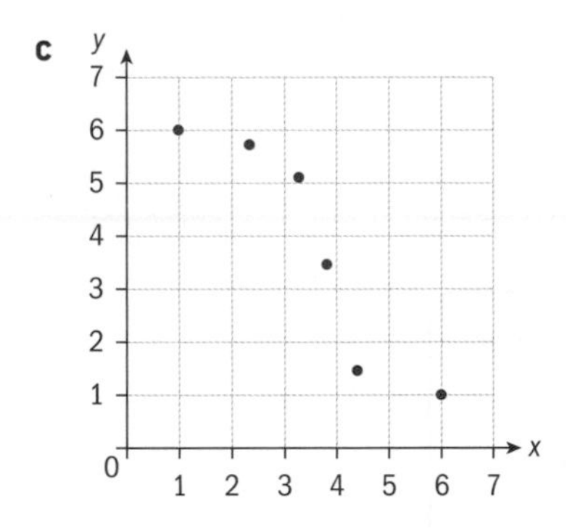

d

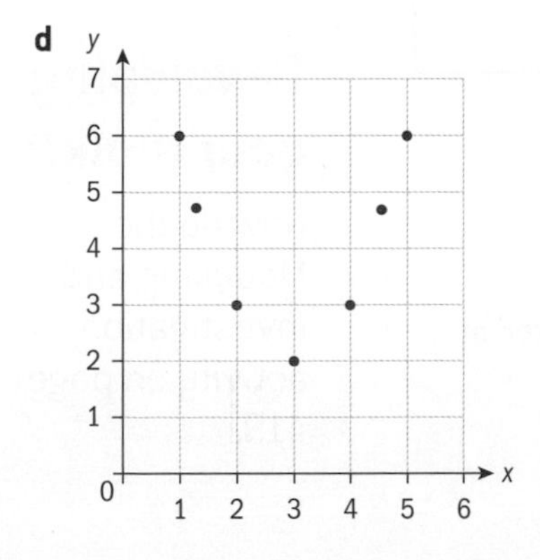

2 A group of students is asked to rank six snack foods by taste and value for money. The ranks are averaged and recorded in the following table.

Calculate Spearman's rank correlation coefficient for the data and comment on your results.

	Pop-corn	Crisps	Choco-late bar	Chews	Chocolate-chip cookie
Taste	2	4	1	5	3
Value	5	3	2	4	1

3 Find Spearman's rank correlation coefficient for the following data sets.

a

x	0	5	10	15	20	25	30
y	23	18	10	9	7	7	7

b

x	10	12	9	6	3	14	8
y	12	11	8	5	7	14	9

4 A sports scientist is testing the relationship between the speed of muscle movement and the force produced. In 10 tests the following data is collected.

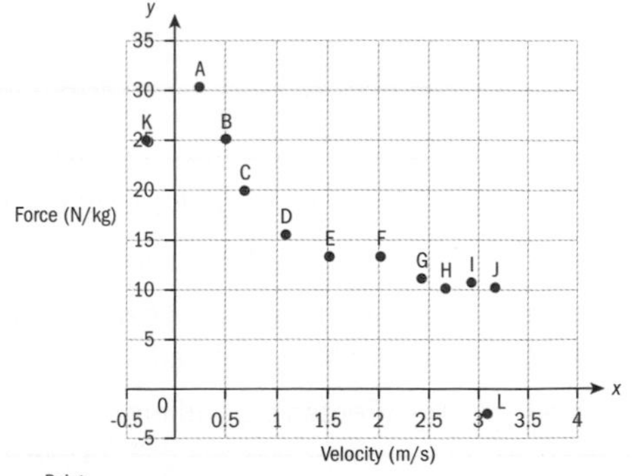

a Explain why it might not be appropriate to use the PMCC in this case.

b Calculate Spearman's rank correlation coefficient (r_s) for this data.

c Interpret the value of r_s and comment on its validity.

5 A class took a mathematics test (marked out of 80) and an English test (marked out of 100), and the results are given in the following table.

Maths	15	25	37	45	60	72	74	78	78	79	79
English	44	47	42	49	52	44	54	59	69	78	89

a Calculate the PMCC for this data and comment on the result.

b Use graphing software to plot these points on a scatter diagram and comment on your result from **a**.

c Calculate Spearman's rank correlation coefficient for this data and comment on your result.

d State which is the more valid measure of correlation, and give a reason.

6 In a blind tasting, customers are asked to rank six different brands of coffee in terms of taste. These rankings and the costs of the different brands are given in the following table.

Brand	A	B	C	D	E	F
Taste rank	1	2	3	4	5	6
Cost	450	360	390	320	350	300

a Explain why you cannot use PMCC in this case.

b Find Spearman's rank correlation coefficient for this data and comment on your answer.

7 Consider the following data set:

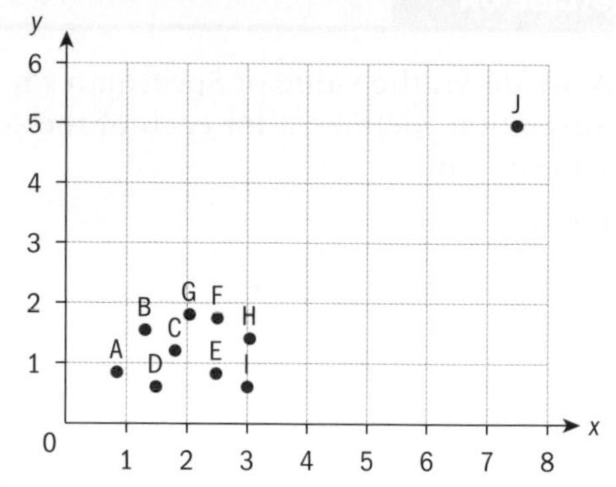

Point
- A = (0.82, 0.86)
- B = (1.28, 1.56)
- C = (1.78, 1.22)
- D = (1.46, 0.62)
- E = (2.46, 0.84)
- F = (2.48, 1.76)
- G = (2.02, 1.82)
- H = (3.02, 1.42)
- I = (2.98, 0.62)
- J = (7.46, 4.98)

a For this data, calculate the PMCC:

i with the outlier J

ii without the outlier J.

b Calculate Spearman's rank correlation coefficient:

i with the outlier J

ii without the outlier J.

c Comment on the results.

The advantages of Spearman's rank correlation coefficient over the PMCC are:

- It can be used on data that is not linear.
- It can be used on data that has been ranked even if the original data is unknown or cannot be quantified.
- It is not greatly affected by outliers.

Developing inquiry skills

Can you use the PMCC or Spearman's rank correlation coefficient to compare the data in the opening scenario of this chapter, which looked at tree heights in different forest areas?

Why, or why not?

Developing your toolkit

Now do the Modelling and investigation activity on page 418.

8.2 χ^2 test for independence

So far you have been looking at samples of data and working out summary statistics related to the sample.

For example, if a scientist takes 20 plants from a field and measures their heights, he can use this data to find the mean or standard deviation of this particular sample of the plants.

What does this tell us about the mean or standard deviation of all the plants in the field?

What might the accuracy of the prediction depend on?

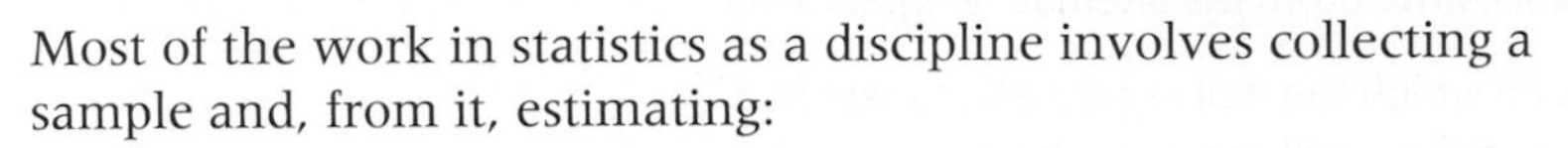

Most of the work in statistics as a discipline involves collecting a sample and, from it, estimating:

- **parameters**; for example, the mean or correlation coefficient of a whole population, when all you have are these values for a sample of the population
- the **distribution** of the population; for example, whether or not the population is normally distributed (see Chapter 7).

Hypothesis testing

In statistics a **hypothesis** is a statement about the unknown parameters or features of a data set.

The aim of a **statistical test** is to try to find out whether the data supports your hypothesis. If it does, this is an example of **statistical inference**: you have inferred something from the statistics of the sample you are considering.

Our initial hypothesis is called the **null hypothesis** and is written as H_0.

Every hypothesis has an **alternative hypothesis** that will be accepted if H_0 is rejected; we write this as H_1.

For example, if we were interested in whether a population mean is 20 cm or more than 20 cm we could write:

$H_0 : \mu = 20, \ H_1 : \mu > 20$

Investigation 2

You need to test whether a coin is fair or biased in favour of either heads or tails. You decide to do an experiment in which you toss the coin 10 times to see what happens.

Continued on next page

Statistics and probability

1 a Without doing any calculations, write down a range of values for the number of heads to appear in the 10 tosses that will make you reject the null hypothesis that the coin is fair.

b Also without doing any calculations, write down a range of values for the number of heads to appear in the 10 tosses for which you would not reject the null hypothesis but still might be suspicious that the coin is not fair. If this is the case, how might you be able to reduce your suspicions?

2 a Use the binomial distribution to work out the probability of the events listed in **1a** happening if the coin is fair.

b Do you feel that this probability is small enough that you could reject the null hypothesis if one of the events listed occurred?

3 In hypothesis testing it is important that you can test your null hypothesis mathematically. Suppose the probability of getting a head is p.

a Explain why you would choose your null hypothesis to be $H_0: p = 0.5$ rather than $H_0: p \neq 0.5$.

b For your chosen null hypothesis, write down the alternative hypothesis.

4 The events, such as those in **1a**, for which the null hypothesis is rejected are called the **critical region**. Statisticians will not normally reject a null hypothesis unless the probability of an outcome occurring in the critical region is less than 0.05. This means that data would appear by chance in the critical region even when H_0 is true less than 5% of the time. This is referred to as a 5% **significance level**.

a Write down the significance level of your test.

b Find the significance level for a different choice of critical region. Would you prefer this one over the one chosen previously? Justify your answer.

c Comment on the symmetry of your critical regions and discuss whether they have to be symmetrical.

5 a Deduce whether you can ever be sure that the coin is biased; explain your answer.

b Suppose the result of your experiment fell just outside the critical region. Explain whether or not this means that the null hypothesis is true.

6 **Conceptual** What is meant by the significance level of a hypothesis test?

7 **Conceptual** What is the purpose of hypothesis testing?

Saanvi is a member of a sports club. She has noticed that more males play squash than females, and is interested to find out whether there is any relationship between gender and favourite racket game. She sent around a survey to the other members in the club to find out which game they prefer: tennis, badminton or squash. The results of the survey are shown in the table.

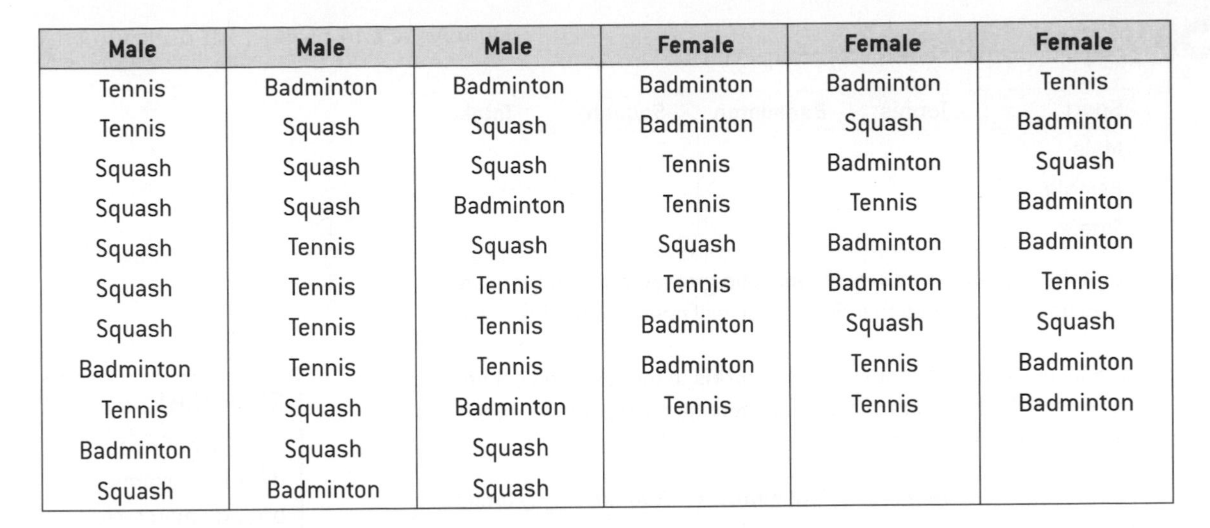

Male	Male	Male	Female	Female	Female
Tennis	Badminton	Badminton	Badminton	Badminton	Tennis
Tennis	Squash	Squash	Badminton	Squash	Badminton
Squash	Squash	Squash	Tennis	Badminton	Squash
Squash	Squash	Badminton	Tennis	Tennis	Badminton
Squash	Tennis	Squash	Squash	Badminton	Badminton
Squash	Tennis	Tennis	Tennis	Badminton	Tennis
Squash	Tennis	Tennis	Badminton	Squash	Squash
Badminton	Tennis	Tennis	Badminton	Tennis	Badminton
Tennis	Squash	Badminton	Tennis	Tennis	Badminton
Badminton	Squash	Squash			
Squash	Badminton	Squash			

Saanvi wants to know whether or not the choice of game is independent of gender.

A χ^2 **test for independence** shows whether two data sets are independent of each other or not. It can be performed at various significance levels.

Saanvi decides to perform a χ^2 test for independence at the 5% significance level to find out whether the preferred game is independent of gender or not.

Her null hypothesis, H_0, is:

H_0 : Preferred racket game is independent of gender.

Her alternative hypothesis is:

H_1 : Preferred racket game is not independent of gender.

EXAM HINT

In the examination it will only be tested at the 1%, 5% or 10% significance level.

Investigation 3

Complete the following table. This is the table of **observed frequencies,** f_o, and is called a contingency table.

Sport	Tennis	Badminton	Squash	Totals
Male	10			
Female				
Totals				60

1 Calculate the probability that a person chosen at random is male.

2 Calculate the probability that a person chosen at random likes tennis best.

3 If these two probabilities are independent, find the probability that a person chosen at random is male and likes tennis best.

4 There are 60 people in total. If the events are independent, find the expected number of males who like tennis best.

These are called the **expected frequencies,** f_e.

Continued on next page

TOK

The use of *p*-values and null hypothesis testing is "surely the most bone-headedly misguided procedure ever institutionalized in the rote training of science students."

— William Rozeboom

In practical terms, is saying that a result is significant the same as saying it is true?

5 Complete the table of expected frequencies.

Sport	Tennis	Badminton	Squash	Totals
Male				
Female				
Totals				

Note that the expected frequencies must be greater than 5. If there are expected frequencies less than 5 then you will need to combine rows or columns.
In the table of expected frequencies, the **totals** of the rows and columns are fixed to match the numbers of males and females and players of each sport in the sample. In this example:

Sport	Tennis	Badminton	Squash	Totals
Male				33
Female				27
Totals	19	20	21	60

EXAM HINT

You will not need to do this in exams, but you may need to in your internal assessments.

6 Find the smallest number of entries that you need to calculate by multiplying probabilities before you can fill in the rest of the table from the numbers already there.

7 If your table had three rows and three columns, find the smallest number of entries that you would need to calculate by multiplying probabilities.

8 If your table had three rows and four columns, find the smallest number of entries that you would need to calculate by multiplying probabilities.

9 Find the smallest number of entries if the table had n rows and m columns.

This number is called the **degrees of freedom**, often written as ν. This is because you only have a "free" choice for the numbers that go into that many cells. After that, the remaining numbers are fixed by the need to keep the totals the same.

10 **Conceptual** What does the number of degrees of freedom represent?

The formula for the degrees of freedom is

$$\nu = (\text{rows} - 1)(\text{columns} - 1)$$

In examinations, ν will always be greater than 1.

Investigation 3 (continued)

To decide whether two variables are likely to be independent it is necessary to compare the observed values with those expected. If the observed values are a long way from the expected values then you can deduce that the two variables are unlikely to be independent and reject the null hypothesis. But

how do you measure how far away they are, and, if you have a measure, how do you decide when the difference is large enough to reject the null hypothesis?

11 Looking back at the results, which categories are furthest from the expected values? Which are closest?

12 Find the sum of the differences between the observed and expected values in the tables above and comment on how suitable this would be as a measure of how far apart they are.

13 Comment on an advantage of squaring the differences before adding them.

14 Comment on a disadvantage of using this sum as a measure of the distance between the observed and expected values.

In order to make sure that differences are in proportion, it would be better to divide each difference squared by the expected value (as long as the expected value is not too small).

This calculation will give you the χ^2 **test statistic**.

EXAM HINT

In examinations you will use your GDC to find this value. Go to Matrix, enter the number of rows and the number of columns (not including the totals, if included in the question). Enter the data and store the matrix. Next go to Stat tests, chi-squared 2-way test, enter the letter that you stored the matrix under and press OK. You will see the chi-squared valuc, the p-value and the number of degrees of freedom.

The χ^2 test statistic is

$$\chi^2_{calc} = \sum \frac{(f_o - f_e)^2}{f_e}$$

where f_o are the observed values and f_e are the expected values.

If this number is larger than a **critical value** then reject the null hypothesis. If it is smaller than the critical value then accept the null hypothesis.

Exercise 8B

1 Misty was interested to find out whether preference for car colour was dependent on gender. She asked 80 of her friends and the results are shown in the table.

Colour of car	White	Black	Red	Blue	Totals
Male	6	14	10	8	38
Female	12	8	9	13	42
Totals	18	22	19	21	80

a Show that the expected number of males who prefer black cars is 10.45.

b Show that the expected number of females who prefer white cars is 9.45.

c Find the χ^2 value.

2 Max was watching people coming out of a take-away coffee shop. He was noting down the gender and whether they had bought a small, medium or large coffee. The results were:

Coffee size	Small	Medium	Large	Totals
Female	22	16	18	56
Male	12	14	28	54
Totals	34	30	46	110

a Show that the expected number of males who buy a small coffee is 16.7.

b Show that the expected number of females who buy a large coffee is 23.4.

c Find the χ^2 value.

3 Brenda works in a pet shop. She watches the rabbits, guinea pigs and hamsters to see if they eat the lettuce or carrots first. Her results are:

Pet	Rabbits	Guinea pigs	Hamsters
Lettuce	16	16	28
Carrots	34	18	18

a Show that the expected number of rabbits who eat carrots first is 26.9.

b Show that the expected number of hamsters who eat lettuce first is 21.2.

c Find the χ^2 value.

4 Ziyue asked her year group how they travelled to school. Her results are shown in the table.

Transport	Car	Bus	Bicycle	Walk
Male	12	12	28	8
Female	21	13	15	11

a Show that the expected number of males who come by bicycle is 21.5.

b Show that the expected number of females who come by car is 16.5.

c Use your GDC to find the χ^2 value.

Investigation 3 (continued)

15 Use the entries in the tables from earlier in the investigation for the observed and expected frequencies to find the χ^2 test statistic. Complete the following:

f_o	f_e	$(f_o - f_e)$	$(f_o - f_e)^2$	$\frac{(f_o - f_e)^2}{f_e}$
10	10.45	−0.45	0.2025	0.1937...
			$\sum \frac{(f_o - f_e)^2}{f_e}$	

For a 5% significance level the critical value is chosen so that the probability of the test statistic being greater than this value if the two variables are independent is 0.05.

16 Will the critical value be larger or smaller for a 1% significance level than for a 5% significance level?

The size of the critical value also depends on the number of degrees of freedom. With a larger table, and hence more degrees of freedom, more numbers are being added to create the test statistic. In examinations you will always be given the critical value if you need to use it.

The critical value for two degrees of freedom at the 5% significance level is $\chi^2_{5\%} = 5.991$.

17 Use this value and your test statistic to decide whether or not to accept the null hypothesis.

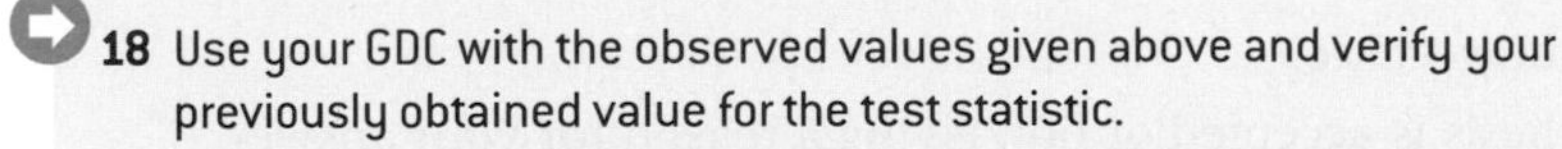

18 Use your GDC with the observed values given above and verify your previously obtained value for the test statistic.

Your GDC also gives you a p-value. This is the probability value. If the p-value is smaller than the level of significance then you do not accept the null hypothesis.

19 Use the p-value that you found on your GDC and the significance level of 5% to reach a conclusion on Saanvi's test.

20 **Conceptual** How would you use the result of the chi squared test to determine whether two variable are independent or not?

21 **Conceptual** What is the purpose of the χ^2 test?

- If the χ^2 test statistic is less than the critical value then you accept the null hypothesis. If it is greater than the critical value then you do not accept the null hypothesis.

$$\chi^2_{\text{calc}} < \text{critical value} \Rightarrow \text{accept } H_0$$
$$\chi^2_{\text{calc}} > \text{critical value} \Rightarrow \text{reject } H_0$$

- If the p-value is greater than the significance level (0.01, 0.05 or 0.10) then you accept the null hypothesis. If it is less than the significance level then you do not accept the null hypothesis.

$$p\text{-value} > 0.05 \Rightarrow \text{accept } H_0$$
$$p\text{-value} < 0.05 \Rightarrow \text{reject } H_0$$

TOK

How can a mathematical model give us knowledge even if it does not yield accurate predictions?

You can use **either** the test statistic and the critical value **or** the p-value and the significance level to reach a conclusion. You may only be given one of these values (the critical value or the significance level), so you should know how to use both.

Statistics and probability

Example 2

Eighty people were asked for their favourite genre of music: pop, classical, folk or jazz. The results are in the following table.

Genre	Pop	Classical	Folk	Jazz	Totals
Male	18	9	4	7	38
Female	22	6	7	7	42
Totals	40	15	11	14	80

A χ^2 test was carried out at the 1% significance level. The critical value for this test is 11.345.

a Write down the null and alternative hypotheses.

b Show that the expected value for a female liking pop is 21.

c Write down the number of degrees of freedom.

Continued on next page

d Find the χ^2 test statistic and the p-value.

e State whether the null hypothesis is accepted or not, giving a reason for your answer.

a H_0 : Favourite music genre is independent of gender

H_1 : Favourite music genre is not independent of gender

b E(female liking pop) = P(female) × P(likes pop) × total $= \frac{42}{80} \times \frac{40}{80} \times 80 = 21$

c $\nu = (2-1)\times(4-1) = 3$

Remember that you do not count the Totals row or column.

d $\chi^2 = 1.622...$ and $p = 0.654...$

e $0.654 > 0.01$ or $1.622 < 11.345$ and so you accept the null hypothesis: favourite music genre is independent of gender.

Example 3

American bulldogs are classified by height, h, as Pocket, Standard or XL. Pockets have $h < 42$ cm high, Standards have $42 \le h < 50$ and XLs have $50 \le h < 58$. At a dog show, Marius measures and weighs 50 dogs. He is interested to find out whether class of dog is independent of weight and decides to perform a χ^2 test at the 5% significance level. The results are shown in the table.

Height	Weight	Height	Weight	Height	Weight
36	30	42	38	50	39
37	33	42	39	51	41
37	36	43	36	51	42
38	31	43	44	52	45
38	38	44	42	52	45
39	32	44	48	52	51
39	39	45	46	53	53
39	42	46	49	54	55
40	41	46	38	54	48
40	43	46	42	54	56
40	38	47	46	55	58
41	38	47	50	55	51
41	44	47	52	56	54
41	46	48	49	56	53
41	45	48	48	56	55
41	47	48	42	57	58
		49	53	57	59

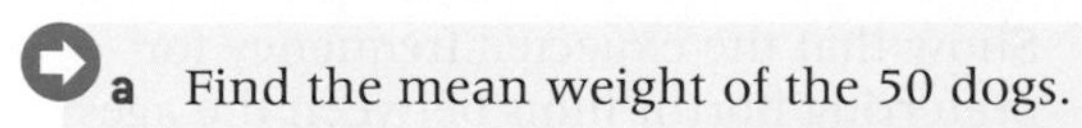

a Find the mean weight of the 50 dogs.

b Complete the following contingency table:

Class	Pocket	Standard	XL
< mean			
≥ mean			

c Write down the null and alternative hypotheses.

d Write down the number of degrees of freedom.

e Show that the expected number of XL dogs that weigh less than the mean is 8.16.

f Find the χ^2 test statistic and the p-value.

g Comment on your answer.

a Mean weight = 44.96

b

Class	Pocket	Standard	XL
< mean	13	8	3
≥ mean	3	9	14

c H_0 : Class of dog is independent of weight.
H_1 : Class of dog is not independent of weight.

d $v = 2$

e $\frac{17}{50} \times \frac{24}{50} = 8.16$

f $\chi^2 = 13.4$ and p-value $= 0.001\,25$

g $0.001\,25 < 0.05$, therefore do not accept the null hypothesis.

Exercise 8C

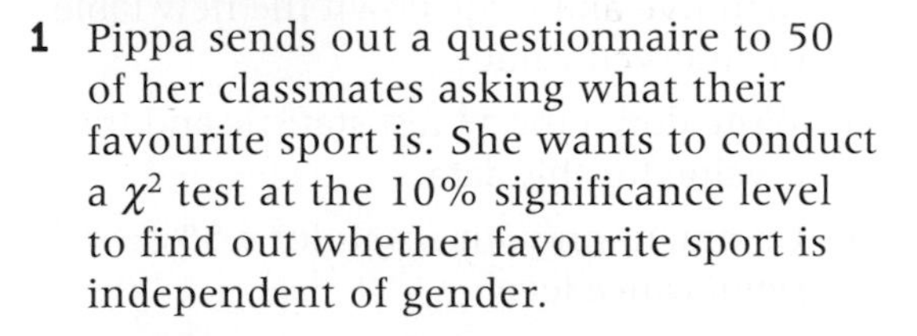

1 Pippa sends out a questionnaire to 50 of her classmates asking what their favourite sport is. She wants to conduct a χ^2 test at the 10% significance level to find out whether favourite sport is independent of gender.

The results are as follows:

Males: cycling, cycling, basketball, football, football, football, basketball, basketball, basketball, basketball, football, football, football, cycling, cycling, basketball, basketball, basketball, basketball, basketball, cycling, cycling, cycling.

Females: football, football, football, basketball, basketball, cycling, basketball, basketball, basketball, cycling, cycling, cycling, football, football, football, cycling, basketball, basketball, cycling, football, football, football, football, cycling, cycling, cycling, basketball.

a Set up a contingency table to display the results.

b Write down the null and alternative hypotheses.

c Write down the number of degrees of freedom.

d Find the χ^2 value and the p-value.

e Check that expected values are greater than 5.

The critical value is 4.605.

f Write down the conclusion of the test.

g Comment on whether the p-value supports this conclusion.

2 A survey was conducted to find out which type of bread males and females prefer. Eighty people were interviewed outside a baker's shop and the results are shown below.

Bread	White	Brown	Corn	Multi-grain	Totals
Male	14	10	7	8	39
Female	17	6	6	12	41
Totals	31	16	13	20	80

Using the χ^2 test at the 5% significance level, determine whether the favourite type of bread is independent of gender.

a State the null hypothesis and the alternative hypothesis.

b Show that the expected frequency for female and white bread is approximately 15.9.

c Write down the number of degrees of freedom.

d Write down the χ^2 test statistic and the p-value for this data.

The critical value is 7.815.

e Comment on your result.

3 Three hundred people of different ages were interviewed and asked which genre of film they mostly watched (thriller, comedy or horror). The results are shown below.

Film type	Thriller	Comedy	Horror	Totals
0–20 years	13	26	41	80
20–50 years	54	48	28	130
51+ years	39	43	8	90
Totals	106	117	77	300

Using the χ^2 test at the 10% significance level, determine whether the genre of film watched is independent of age.

a State the null hypothesis and the alternative hypothesis.

b Show that the expected frequency for preferring horror films between the ages of 20 to 50 years is 33.4.

c Write down the number of degrees of freedom.

d Write down the χ^2 test statistic and the p-value for this data.

e Comment on your result.

4 Three different flavours of dog food were tested on different breeds of dogs to see whether there was any connection between favourite flavour and breed. The results are shown in the table below.

Flavour	Beef	Chicken	Lamb	Totals
Boxer	14	6	8	28
Labrador	17	11	10	38
Poodle	13	8	14	35
Collie	6	5	8	19
Totals	50	30	40	120

Perform a χ^2 test at the 5% significance level to test whether favourite flavour is independent of breed of dog.

a State the null hypothesis and the alternative hypothesis.

b Write down the table of expected frequencies.

c Combine the results for collie and poodle so that all expected values are greater than five and write down the new table of observed values.

d Write down the χ^2 test statistic and the p-value for this data.

e Comment on your result for a 5% significance level.

5 Sixty people were asked what their favourite flavour of chocolate was (milk, pure, white). The results are shown in the table below.

Flavour	Milk	Pure	White	Totals
Male	10	17	5	32
Female	8	6	14	28
Totals	18	23	19	60

A χ^2 test at the 1% significance level is set up.

a State the null hypothesis and the alternative hypothesis.

b Write down the number of degrees of freedom.

c Write down the χ^2 test statistic and the p-value for this data.

The critical value is 9.210.

d Comment on your result.

6 Nandan wanted to know whether or not the number of hours spent on social media had an influence on average grades (GPA). He collected the following information:

Grade	Low GPA	Average GPA	High GPA	Totals
0–9 hours	4	23	58	85
10–19 hours	23	45	32	100
≥ 20 hours	43	33	9	85
Totals	70	101	99	270

He decided to perform a χ^2 test at the 10% significance level to find out whether there is a connection between GPA and number of hours spent on social media.

a State the null hypothesis and the alternative hypothesis.

b Show that the expected frequency for 0–9 hours and a high GPA is 31.2.

c Write down the number of degrees of freedom.

d Write down the χ^2 test statistic and the p-value for this data.

The critical value is 7.779.

e Comment on your result.

7 Hubert wanted to find out whether the number of people walking their dog was related to the time of day. He kept a record covering 120 days and the results are shown in the following table.

Time of day	Morning	Afternoon	Evening	Totals
0–5 people	8	6	18	32
6–10 people	13	8	23	44
> 10 people	21	7	16	44
Totals	42	21	57	120

Test, at the 5% significance level, whether there is a connection between time of day and number of people walking their dog. The critical value for this test is 9.488.

8 Carle has a part-time job working at a corner shop. He decides to see whether there is a connection between the temperature and the number of bottles of water sold. His observations are in the table below.

Temperature	< 21°C	21°C–30°C	> 30°C	Totals
< 30 bottles	14	23	12	49
30–60 bottles	10	31	17	58
> 60 bottles	8	26	9	43
Totals	32	80	38	150

Test, at the 1% significance level, whether there is a connection between temperature and the number of bottles of water sold.

a State the null hypothesis and the alternative hypothesis.

b Write down the number of degrees of freedom.

c Write down the χ^2 test statistic and the p-value for this data.

The critical value is 13.277.

d Comment on your result.

9 Samantha wanted to find out whether there was a connection between the type of degree that a person had and their annual salary in dollars. She interviewed 120 professionals and her observed results are shown in the following table.

Degree	BA	MA	PhD	Totals
< US $60 000	17	8	4	29
US $60 000–US $120 000	14	19	9	42
> US $120 000	8	14	27	49
Totals	39	41	40	120

Test, at the 5% significance level, whether there is a connection between degree and salary.

a State the null hypothesis and the alternative hypothesis.

b Write down the number of degrees of freedom.

c Write down the χ^2 test statistic and the p-value for this data.

The critical value is 9.488.

d Comment on your result.

Developing inquiry skills

Let the measurement of the largest height of the trees in the opening problem be h. Divide the trees into small, medium and large, where small trees have $h \leq 4.5$ m, medium, $4.5 < h \leq 5.0$ m, and large $h > 5.0$ m.

Use these categories to form a contingency table for the trees in areas A and B, and test at the 5% significance level whether the heights of the trees are independent of the forest area they were taken from.

Does the conclusion of the test support the hypothesis that the trees from area A, on average, are taller than those from area B? Justify your answer.

8.3 χ^2 goodness of fit test

Investigation 4

Jiang wonders whether the die he was given is fair. He rolls it 300 times. His results are shown in the table.

Number	Frequency
1	35
2	52
3	47
4	71
5	62
6	33

TOK

How have technological innovations affected the nature and practice of mathematics?

1 Write down the probability of throwing a 1 on a fair die.

2 If you throw a fair die 300 times, how many times would you expect to throw a 1?

3 Write down the expected frequencies for throwing a fair die 300 times.

Number	Expected frequency
1	
2	
3	
4	
5	
6	

4 **Factual** Do you need to combine any rows on your table?

Since all the expected frequencies are the same, this is known as a **uniform distribution.**

5 **Factual** Is the formula for the χ^2 test suitable to test whether Jiang's results fit this uniform distribution?

The null hypothesis is H_0: Jiang's die satisfies a uniform distribution.

6 Write down the alternative hypothesis.

7 Given that the critical value at the 5% significance level for this test is 11.07, use the formula for the χ^2 test, $\chi^2_{calc} = \sum \frac{(f_o - f_e)^2}{f_e}$, to find out whether Jiang's results could be taken from a uniform distribution.

Normally you would solve this using your GDC, which may ask you to enter the degrees of freedom.

8 **Factual** What is the number of degrees of freedom in a χ^2 goodness of fit test? (Consider in how many cells you have free choices when completing the expected values table.)

9 Write down the number of degrees of freedom for this test.

10 Using your GDC, find the test statistic and the p-value.

11 What is your conclusion from this test?

12 **Conceptual** What is the purpose of the χ^2 goodness of fit test?

These types of test are called "goodness of fit" tests as they are measuring how closely the observed data fits with the expected data for a particular distribution. The test for independence using contingency tables is an example of a goodness of fit test, but you can test for the goodness of fit for any distribution.

In a χ^2 goodness of fit test, the number of degrees of freedom is $v = (n - 1)$.

Example 4

The students in Year 8 are asked what day of the week their birthdays are on this year. The table shows the results.

Day	Sunday	Monday	Tuesday	Wednesday	Thursday	Friday	Saturday
Frequency	12	14	18	17	15	15	14

a Write down the table of expected values, given that each day is equally likely.

b Conduct a χ^2 goodness of fit test at the 5% significance level for this data.

c The critical value is 12.592. Write down the conclusion for the test.

a

Day	Sunday	Monday	Tuesday	Wednesday	Thursday	Friday	Saturday
Frequency	15	15	15	15	15	15	15

b H_0 : The data satisfies a uniform distribution.

H_1 : The data does not satisfy a uniform distribution.

$v = (7 - 1) = 6$

Using GDC, $\chi^2 = 1.60$ and p-value $= 0.953$.

c $0.95 > 0.05$ or $1.60 < 12.592$, so you can accept the null hypothesis: the data does satisfy a uniform distribution.

Exercise 8D

1 Terri buys 10 packets of Skittles and counts how many of each colour (yellow, orange, red, purple and green) there are. In total she has 600 sweets.

According to the Skittles website, the colours should be evenly distributed with 20% of each colour in a bag.

The results for Terri's 10 bags are:

Colour	Frequency
Yellow	104
Orange	132
Red	98
Purple	129
Green	137

a Find the expected frequencies.

b Write down the number of degrees of freedom.

c Determine the results of a goodness of fit test at the 5% significance level to find out whether Terri's data fits a uniform distribution. Remember to write down the null and alternative hypotheses.

The critical value for this test is 9.488.

d State the conclusion for the test and give a reason for your answer.

2 There are 60 students in Grade 12. Mr Stewart asks them which month their birthdays are in, and the results are shown in the table.

Month	Jan	Feb	Mar	Apr	May	Jun	Jul	Aug	Sep	Oct	Nov	Dec
Frequency	3	5	4	6	5	6	4	7	8	6	3	3

The months in which people have birthdays are uniformly distributed.

a Write down the table of expected values.

b Write down the number of degrees of freedom.

c Determine the results of a goodness of fit test at the 10% significance level to find out whether the data fits a uniform distribution. Remember to write down the null and alternative hypotheses.

The critical value for this test is 17.275.

d State the conclusion for the test and give a reason for your answer.

3 Sergei works in a call centre. One week he answers 840 calls. The number of calls that he answers each day are shown in the table.

Day	Mon	Tues	Wed	Thurs	Fri	Sat	Sun
Frequency	146	98	103	106	93	204	90

The calls are uniformly distributed.

a Show that the expected value of the number of calls each day is 120.

b Write down the number of degrees of freedom.

c Determine the results of a goodness of fit test at the 5% significance level to find out whether the data fits a uniform distribution. Remember to write down the null and alternative hypotheses.

The critical value for this test is 12.592.

d State the conclusion for the test and give a reason for your answer.

4 The last digit on 490 winning lottery tickets is recorded in the table.

Last digit	0	1	2	3	4	5	6	7	8	9
Frequency	44	53	49	61	47	52	39	58	42	45

a Each number should be equally likely to occur. Write down the table of expected values.

b Write down the number of degrees of freedom.

c Determine the results of a goodness of fit test at the 10% significance level to find out whether the data fits a uniform distribution. Remember to write down the null and alternative hypotheses.

The critical value for this test is 14.684.

d State the conclusion for the test and give a reason for your answer.

Statistics and probability

Example 5

The scores for IQ tests are normally distributed with a mean of 100 and standard deviation of 10. Cinzia gives an IQ test to all 200 IB Diploma Programme students in the school. Her results are shown in the table.

Score, x	Frequency
$x < 90$	5
$90 \le x < 100$	14
$100 \le x < 110$	74
$110 \le x < 120$	58
$120 \le x < 130$	34
$130 \le x$	15

Cinzia wants to test whether these results are also normally distributed and performs a χ^2 goodness of fit test at the 10% significance level.

a Write down her null and alternative hypotheses.

b Find the expected values.

c If any expected values are less than 5 then rewrite both tables.

d Write down the number of degrees of freedom.

The critical value is 6.251.

e Find the χ^2 test statistic and the p-value, and state the conclusion for the test.

a H_0 : The scores are normally distributed with mean of 100 and standard deviation of 10.

H_1 : The scores are not normally distributed with mean of 100 and standard deviation of 10.

Continued on next page

b

Score, x	Probability	Expected score
$x < 90$	0.1587	31.7
$90 \le x < 100$	0.3413	68.3
$100 \le x < 110$	0.3413	68.3
$110 \le x < 120$	0.1359	27.2
$120 \le x < 130$	0.0214	4.28
$130 \le x$	0.00135	0.270

Use the normal cdf function on your GDC to find the probability and then multiply your answer by 200.

c The last two scores are both less than 5, even when added together, and so you will have to combine them with the one above.

Observed and expected values are now:

Score, x	Observed score	Expected score
$x < 90$	5	31.7
$90 \le x < 100$	14	68.3
$100 \le x < 110$	74	68.3
$110 \le x$	107	31.7

$4.28 + 0.270$ is still less than 5.

d $v = 3$

$v = (4 - 1)$

e χ^2 value is 245 and the p-value is 7.89×10^{-53}. The p-value is less than 0.10 and so the null hypothesis is rejected: the scores are not normally distributed with mean of 100 and standard deviation of 10.

Also, the critical value for $v = 3$ at the 10% significance level is 6.251. $245 > 6.251$, so the null hypothesis is rejected.

Exercise 8E

1 Marius works in a fish shop. One week he measures 250 fish before selling them. His results are shown in the table.

Length of fish, x cm	Frequency
$9 \le x < 12$	5
$12 \le x < 15$	22
$15 \le x < 18$	71
$18 \le x < 21$	88
$21 \le x < 24$	52
$24 \le x < 27$	10
$27 \le x < 30$	2

Marius is told that the lengths of the fish should be normally distributed with a mean of 19 cm and standard deviation of 3 cm, so he decides to perform a χ^2 goodness of fit test at the 5% significance level to find out whether the fish that he measured could have come from a population with this distribution.

a Write down his null and alternative hypotheses.

b Find the probability that a fish is between 9 cm and 12 cm.

c In total, 250 fish were measured. Calculate how many fish you expect to be between 9 cm and 12 cm.

d Complete the table of expected values for 250 normally distributed fish with a mean of 19 cm and standard deviation of 3 cm.

Length of fish, x cm	Probability	Expected frequency
$9 \le x < 12$	0.009 386	2.35
$12 \le x < 15$	0.081 396	20.35
$15 \le x < 18$		
$18 \le x < 21$		
$21 \le x < 24$		
$24 \le x < 27$	0.043 960	10.99
$27 \le x < 30$	0.003 708	0.927

Two of the expected frequencies are less than 5.

e Discuss what you have to do in this case. Remember that this may be encountered in internal assessments, but will not be in examinations.

f Rewrite the original table and the table of expected values so that all the expected values are greater than 5.

g Write down the number of degrees of freedom.

h Find the χ^2 value and the p-value.

The critical value is 9.488.

i Write down your conclusion for this test.

2 The weights of sample group are normally distributed with a mean of 52 kg and standard deviation of 3 kg.

The district nurse weighs 200 people and her results are shown in the table.

Weight, w kg	$w < 45$	$45 \le w < 50$	$50 \le w < 55$	$55 \le w < 60$	$w \ge 60$
Frequency	12	44	82	53	9

a Complete the expected frequency table.

Weight, w kg	$w < 45$	$45 \le w < 50$	$50 \le w < 55$	$55 \le w < 60$	$w \ge 60$
Expected frequency	1.96	48.54			0.77

b Rewrite the table of observed frequencies so that all the expected frequencies are greater than 5 and find the corresponding expected frequencies.

c Write down the number of degrees of freedom.

d Determine the results of a goodness of fit test at the 5% significance level to find out whether the data fits a normal distribution. Remember to write down the null and alternative hypotheses.

The critical value for this test is 5.991.

e State the conclusion for the test and give a reason for your answer.

3 The grades for an economics exam for 300 university students are as follows.

Grade, x%	$x < 50$	$50 \le x < 60$	$60 \le x < 70$	$70 \le x < 80$	$x \ge 80$
Frequency	8	72	143	71	6

The grades are normally distributed with a mean of 65% and standard deviation of 7.5%.

a Complete the expected frequency table.

Grade, x%	$x < 50$	$50 \le x < 60$	$60 \le x < 70$	$70 \le x < 80$	$x \ge 80$
Expected frequency	6.83			68.92	6.83

b Write down the number of degrees of freedom.

c Determine the results of a goodness of fit test at the 10% significance level to find out whether the data fits a normal distribution. Remember to write down the null and alternative hypotheses.

The critical value for this test is 7.779.

d State the conclusion for the test and give a reason for your answer.

4 The heights of elephants are normally distributed with a mean of 250 cm and standard deviation of 11 cm. Two hundred and fifty elephants are measured; the results are shown in the table.

Height, h cm	$h < 235$	$235 \le h < 245$	$245 \le h < 255$	$255 \le h < 265$	$h \ge 265$
Frequency	10	69	88	63	20

a Complete the expected frequency table.

Height, h cm	$h < 235$	$235 \le h < 245$	$245 \le h < 255$	$255 \le h < 265$	$h \ge 265$
Expected frequency	21.59				

b Write down the number of degrees of freedom.

c Determine the results of a goodness of fit test at the 5% significance level to find out whether the data fits a normal distribution. Remember to write down the null and alternative hypotheses.

The critical value for this test is 9.488.

d State the conclusion for the test and give a reason for your answer.

5 The lifespan of light bulbs is normally distributed with a mean lifespan of 1200 hours and standard deviation of 100 hours. Four hundred light bulbs are tested and the results are shown in the table.

Lifespan, h hours	$h < 1000$	$1000 \le h < 1100$	$1100 \le h < 1200$	$1200 \le h < 1300$	$1300 \le h < 1400$	$h \ge 1400$
Frequency	24	52	92	164	42	26

a Complete the expected frequency table.

Lifespan, h hours	$h < 1000$	$1000 \le h < 1100$	$1100 \le h < 1200$	$1200 \le h < 1300$	$1300 \le h < 1400$	$h \ge 1400$
Expected frequency	9.1					

b Write down the number of degrees of freedom.

c Determine the results of a goodness of fit test at the 5% significance level to find out whether the data fits a normal distribution. Remember to write down the null and alternative hypotheses.

The critical value for this test is 11.070.

d State the conclusion for the test and give a reason for your answer.

Example 6

Using what you learned in Chapter 7, find the probability when you toss three coins of obtaining: 0 heads, exactly 1 head, exactly 2 heads, 3 heads.

Hagar tosses three coins 200 times and makes a note of the number of heads each time. Her results are as follows.

Number of heads	Frequency
0	28
1	67
2	83
3	22

She is interested in finding out whether her results follow a binomial distribution and performs a χ^2 goodness of fit test at the 5% significance level.

a Using the terms of B(3, 0.5) and the fact that Hagar tossed the coins 200 times, find the expected values for the number of heads.

b Comment on whether any of these values are less than 5.

c Write down the null and alternative hypotheses and the degrees of freedom.

The critical value is 7.815.

d Find the χ^2 value and the p-value.

e Write down the conclusion for this test.

TOK

To what extent can shared knowledge be distorted and misleading?

a

Number of heads	Expected frequency
0	$200 \times 0.125 = 25$
1	$200 \times 0.375 = 75$
2	$200 \times 0.375 = 75$
3	$200 \times 0.125 = 25$

b They are all greater than 5.

c H_0 : The number of heads follows a binomial distribution.

H_1 : The number of heads does not follow a binomial distribution.

d The χ^2 value is 2.426… and the p-value is 0.4886…

e 2.426… < 7.815, so the null hypothesis is accepted. Or, 0.4886… > 0.05, so the null hypothesis is accepted. The number of heads follows a binomial distribution.

International-mindedness

The physicist Frank Oppenheimer wrote:

"Prediction is dependent only on the assumption that observed patterns will be repeated."

This is the danger of extrapolation. There are many examples of its failure in the past: for example share prices, the spread of disease and climate change.

Exercise 8F

1 Percy sews three seeds in each of 50 different pots. The probability that a seed will germinate is 0.75. The number of seeds that germinated in each pot is shown in the table.

Number of seeds germinating	0	1	2	3
Frequency	5	10	15	20

a Using the binomial expansion B(3, 0.75), find the expected probabilities of 0, 1, 2 or 3 seeds germinating.

b Write down the table of expected frequencies.

c State whether or not there are any expected values less than 5 and combine results if required.

d Write down the number of degrees of freedom.

e Determine the results of a goodness of fit test at the 5% significance level to find out whether the data fits a binomial distribution. Remember to write down the null and alternative hypotheses.

The critical value for this test is 5.991.

f State the conclusion for the test and give a reason for your answer.

2 The number of boys in 90 families with three children is shown in the table.

Number of boys	0	1	2	3
Frequency	16	23	32	19

a If the probability of having a boy is 0.5, use the binomial expansion B(3, 0.5) to find the expected probabilities.

b State whether or not there are any expected values less than 5 and combine results if required.

c Write down the number of degrees of freedom.

d Determine the results of a goodness of fit test at the 1% significance level to find out whether the data fits a binomial distribution. Remember to write down the null and alternative hypotheses.

The critical value for this test is 11.345.

e State the conclusion for the test and give a reason for your answer.

3 Esmerelda tosses two unbiased dice 250 times. She records the number of 6s that she tosses.

Number of 6s	0	1	2
Frequency	135	105	10

a Use the binomial expansion $B(2, \frac{1}{6})$ to complete the table of expected values.

Number of 6s	0	1	2
Expected frequencies			6.94

b State whether there are any expected values less than 5 and combine results if required.

c Write down the degrees of freedom.

d Determine the results of a goodness of fit test at the 5% significance level to find out whether the data fits a binomial distribution. Remember to write down the null and alternative hypotheses.

The critical value for this test is 5.991.

e State the conclusion for the test and give a reason for your answer.

4 A multiple-choice test has five questions. Each question has four answers to choose from.

a Find the probability of getting any one question correct.

Five hundred students sit the test and the results are shown in the table.

Number correct, n	0	1	2	3	4	5
Frequency	38	66	177	132	51	36

b Using the binomial expansion $B(5, 0.25)$, find the expected probabilities of having 0, 1, 2, 3, 4 or 5 questions correct.

c Complete the table of expected frequencies.

Number correct, n	0	1	2	3	4	5
Expected frequency					7.32	

d State whether there are any expected values less than 5 and combine results if required.

e Write down the number of degrees of freedom.

f Determine the results of a goodness of fit test at the 5% significance level to find out whether the data fits a binomial distribution. Remember to write down the null and alternative hypotheses.

The critical value for this test is 9.488.

g State the conclusion for the test and give a reason for your answer.

Reflect What does the χ^2 goodness of fit test do?

Developing inquiry skills

	Area A	Area B
Small ($h \le 4.5$)	3	9
Medium ($4.5 < h \le 5.0$)	7	9
Large ($h > 5.0$)	14	6

From previous research it is known that this species of tree as a whole follows a normal distribution with a mean of 4.9 m and a standard deviation of 0.5 m.

Test the trees from each of the forest areas separately and see whether the observed values are consistent with both samples being taken from this distribution.

Combine the trees from both areas and carry out the test again.

What do your results suggest about the likelihood of the trees from area A being taller than those from area B?

8.4 The *t*-test

Maisy noticed that the sun shines more on one side of her garden than the other. She wanted to know whether this had any effect on the heights of the tulips on either side of the garden. She measured 20 tulips from each side of the garden. The data is shown in the table.

Length, in cm, of tulips on right side	Length, in cm, of tulips on left side
21	24
21	25
26	25
25	26
28	32
24	29
22	31
22	27
29	26
28	28
28	22
27	22
21	28
23	28
24	30
24	31
27	29
26	28
26	28
25	32

1 How can Maisy compare the two sets of data?

2 Discuss whether it would be fair to look at the mean values only. If not, give a reason.

3 Discuss whether you think that the spread of the data should also be taken into consideration. Give a reason for your answer.

4 Consider whether the two sets of data must be the same size or not.

The best way to compare these two sets of data is to use a *t*-test.

William Gosset was employed by Guinness to improve the taste and quality of their beer. In order to monitor the quality of the hops that were used in the brewing process, he invented the *t*-test. He published under the pen name "Student". Hence, it is sometimes referred to as Student's *t*-test.

The *t*-test is used for two data sets that are measuring the same thing (like the tulips above), and only applies to normally distributed data. There is a formula that is used to calculate it, but you won't need to use it as your GDC will do the work for you.

First of all, you need to set up your null and alternative hypotheses.

The null hypothesis is that the two means, $\bar{x}_1$ and $\bar{x}_2$, are equal:

$H_0: \bar{x}_1 = \bar{x}_2$

The alternative hypothesis is that the two means are not equal. For a **two-tailed test** this just means checking that $\bar{x}_1 \neq \bar{x}_2$. For a **one-tailed test**, it means checking either that $\bar{x}_1$ is less than $\bar{x}_2$ or that it is greater than $\bar{x}_2$.

For a two-tailed test, $H_1: \bar{x}_1 \neq \bar{x}_2$

For a one-tailed test, $H_1: \bar{x}_1 > \bar{x}_2$ or $\bar{x}_1 < \bar{x}_2$

For Maisy's tulips, the null and alternative hypotheses would be:

H_0: The mean of the tulips on the right side is the same as the mean of the tulips on the left side.

H_1: The mean of the tulips on the right side is not the same as the mean of the tulips on the left side.

Using your GDC you will be able to find the *t*-statistic and the *p*-value. To do the *t*-test with your GDC, put your data into two lists. Go to Statistics, Tests, 2-sample *t* Test, data (input), choose the correct alternative hypothesis, pooled (yes), enter. You will see the *t*-value and the *p*-value. However, in examinations you will only use the *p*-value.

Once you have found the p-value, you can compare it with the significance level just like you did in the χ^2 test.

If the p-value is greater than the significance level then you accept the null hypothesis; if it is smaller, then you do not accept the null hypothesis.

The t-test is mainly conducted at the 5% significance level, though, like the χ^2 test, it can also be conducted at the 1% or 10% significance levels.

The t-value is −3.07 and the p-value is 0.003 92.

$0.00392 < 0.05$, so the null hypothesis is rejected: there is a difference in the heights of the tulips on either side of the garden.

Example 7

Mr Arthur gives his two chemistry groups the same test. He wants to find out whether there is any difference between the achievement levels of the two groups.

The results are:

Group 1	54	62	67	43	85	69	73	81	47	92	55	59	68	72
Group 2	73	67	58	46	91	48	82	81	67	74	57	66		

a Write down the null and alternative hypotheses.

b Find the t-value and p-value for a t-test at the 5% significance level.

c Write down the conclusion to the test.

a $H_0: \bar{x}_1 = \bar{x}_2$ (there is no difference between the grades in Group 1 and the grades in Group 2)	Notice that the two groups do not need to be the same size.
$H_1: \bar{x}_1 \neq \bar{x}_2$ (there is a difference between the grades in Group 1 and the grades in Group 2)	This will be a two-tailed test as you want to know whether Group 1 is better or worse than Group 2.
b t-value = −0.235, p-value = 0.816	
c $0.816 > 0.05$, so you accept the null hypothesis: there is no significant difference between the two groups.	

Example 8

An oil company claims to have developed a fuel that will increase the distance travelled for every litre of fuel.

Ten scooters are filled with one litre of normal fuel and ten scooters are filled with one litre of the new fuel. The distances, in km, travelled on the one litre by each scooter are as follows:

Original fuel	36	38	44	42	45	39	48	51	38	43
New fuel	43	39	51	49	53	48	52	46	53	49

a Write down the null and alternative hypotheses.

b Find the t-value and p-value for a t-test at the 5% significance level.

c Write down the conclusion to the test.

a $H_0: \bar{x}_1 = \bar{x}_2$ (there is no difference between the distance travelled with the original and new fuels) $H_1: \bar{x}_1 < \bar{x}_2$ (the distance travelled with the original fuel is less than the distance travelled with the new fuel) **b** t-value $= -2.83$, p-value $= 0.00561$ **c** $0.00561 < 0.05$, so you do not accept the null hypothesis: the company's claim is correct.	Note that this is a one-tailed test as you are only considering that the distance travelled with the original fuel is less than the distance travelled with the new fuel.

Reflect How do we test for validity with statistical inferences?

TOK

What counts as understanding in mathematics?

Exercise 8G

1 Petra noticed that one of her apple trees grew in the shade and the others did not. She wanted to find out whether apples from the tree in the shade weighed less than those in the sun. She picked nine apples from each tree and weighed them in grams.

Tree in shade	75	82	93	77	85	78	91	83	92
Tree not in shade	74	81	95	79	95	82	93	88	90

a Write down the null and alternative hypotheses.

b State whether this a one-tailed test or a two-tailed test.

c Find the t-value and p-value for a t-test at the 10% significance level.

d Write down the conclusion to the test.

2 Fergus heard that babies born in the country weigh more than babies born in the town. He contacted two midwives, one who delivered babies in the country and one who delivered babies in the town, and asked them for the weights, in kg, of the babies that they had delivered during the previous week.

Country babies	2.8	3.2	2.7	3.5	3.0	2.9	4.1	3.9	
Town babies	3.1	3.5	2.8	3.7	4.2	2.6	3.2	2.9	3.8

a Write down the null and alternative hypotheses.

b State whether this a one-tailed test or a two-tailed test.

c Find the t-value and p-value for a t-test at the 10% significance level.

d Write down the conclusion to the test.

3 Jocasta picked some runner beans from two different plants and measured them, in cm. She wanted to find out whether there was any difference in the lengths of the beans from the two plants.

Length of beans on plant 1	19	23	21	25	24	18	25	18	24	16
Length of beans on plant 2	27	24	25	28	25	27	24	26	22	23

a Write down the null and alternative hypotheses.

b State whether this a one-tailed test or a two-tailed test.

c Find the t-value and p-value for a t-test at the 5% significance level.

d Write down the conclusion to the test.

4 The lifetimes of two different types of light bulb were tested to find out whether one was better than the other or not. The numbers of hours are listed in the table.

Bulb 1	1236	1350	1489	2052	1986	1875	2134	1985
Bulb 2	1567	1432	1267	2145	1879	1987	1679	1765

a Write down the null and alternative hypotheses.

b State whether this a one-tailed test or a two-tailed test.

c Find the t-value and p-value for a t-test at the 5% significance level.

d Write down the conclusion to the test.

5 The weights, in kg, of the 11-year-old boys and girls in Grade 6 were recorded to find out whether the boys weighed less than the girls.

Boys' weights	Girls' weights
33	35
32	39
35	43
36	45
41	39
32	44
38	38
34	32
36	
31	

a Write down the null and alternative hypotheses.

b State whether this a one-tailed test or a two-tailed test.

c Find the t-value and p-value for a t-test at the 5% significance level.

d Write down the conclusion to the test.

6 A pharmaceutical company claims to have invented a new remedy for weight loss. It claims that people using this remedy will lose more weight than people not using the remedy. A total of 20 people are weighed and tested. Ten people are given the new remedy and the other ten are given a placebo. After two months the people are weighed again and any weight loss, in kg, is noted in the table below.

New remedy	1.2	2.4	1.6	3.5	3.2	4.6	2.5	0.8	1.2	3.9
Placebo	0.6	0.0	1.0	1.3	2.1	0.7	1.9	2.4	0.3	1.0

a Write down the null and alternative hypotheses.

b State whether this a one-tailed test or a two-tailed test.

c Find the t-value and p-value for a t-test at the 1% significance level.

d Write down the conclusion to the test.

7 The lengths, in cm, of sweetcorn cobs in fields on either side of a main road are measured to find out whether there is any difference between them.

Field 1	17	18	15	21	22	24	19	23	25
Field 2	19	21	23	16	18	22	23	16	19

a Write down the null and alternative hypotheses.

b State whether this a one-tailed test or a two-tailed test.

c Find the t-value and p-value for a t-test at the 10% significance level.

d Write down the conclusion to the test.

Developing inquiry skills

It has already been confirmed that the trees in the opening problem are normally distributed, so you can perform a t-test on the data.

State your null and alternative hypotheses.

Write down whether this a one-tailed test or a two-tailed test.

Perform a t-test at the 5% significance level.

Write down the conclusion to the test.

Chapter summary

- The product moment correlation coefficient of the ranks of a set of data is called Spearman's rank correlation coefficient. The IB notation is r_s.
- A value of 1 means the set of data is strictly increasing and a value of −1 means it is strictly decreasing. A value of 0 means the data shows no monotonic behaviour.
- The advantages of Spearman's over the PMCC are:
  - It can be used on data that is not linear.
  - It can be used on data that has been ranked even if the original data is unknown or cannot be quantified.
  - It is not greatly affected by outliers.
- A χ^2 test for independence can be performed to find out whether two data sets are independent of each other or not. It can be performed at various significance levels. In the examination it will only be tested at the 1%, 5% or 10% significance level.
- The number of degrees of freedom is $v = (\text{rows} - 1)(\text{columns} - 1)$.
- Expected values must be greater than 5. If there are expected values less than 5 then you will need to combine rows or columns.
- The formula for the χ^2 test is $\chi^2 = \sum \frac{(f_o - f_e)^2}{f_e}$ where f_o are the observed values and f_e are the expected values.
- If the p-value is greater than the significance level (0.01, 0.05 or 0.10) then you accept the null hypothesis; if it is less than the significance level then you do not accept the null hypothesis.
- If the χ^2 test statistic is less than the critical value then you accept the null hypothesis; if it is greater than the critical value then you do not accept the null hypothesis.
- In a χ^2 goodness of fit test, $v = (n - 1)$.
- The t-test is used for two data sets that are measuring the same thing, and only applies to normally distributed data.

Developing inquiry skills

Thinking about the opening problem:

- Discuss how what you have learned in this chapter has helped you to answer the questions.
- Discuss how you decided whether the data was biased or not.
- Write down which statistical tests you were able to use from this chapter.
- State what claims you can make about the trees.
- Discuss what information you managed to find.

Continued on next page

- Discuss what assumptions you made.
- Comment on any other things that you wondered about.

Thinking about the inquiry questions from the beginning of this chapter:

- Discuss whether what you have learned in this chapter has helped you to think about an answer to these questions.
- Consider whether there are any that you are interested in and would like to explore further, perhaps for your internal assessment topic.

Chapter review

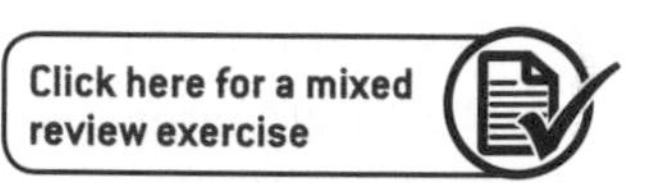

1 Prabu took a note of the heights of 12 of her classmates and timed how many seconds it took them to run the 100-metre dash. Her data is shown in the table.

Height, cm	Time, s
151	17.5
153	18
153	16.5
154	16
155	15.4
159	13.2
162	14
164	13.7
164	13.2
168	12.5
175	12
181	12

a Calculate Spearman's rank correlation coefficient (r_s) for this data.

b Interpret the value of r_s and comment on its validity.

2 Ibrahim belongs to a tennis club. He watched a match played by each of the top eight tennis players and recorded the number of aces that they hit. The results are:

Tennis rank	1	2	3	4	5	6	7	8
Number of aces	10	7	4	6	5	5	3	1

a Explain why it might not be appropriate to use the PMCC in this case.

b Calculate Spearman's rank correlation coefficient (r_s) for this data.

c Interpret the value of r_s and comment on its validity.

3 The colours of the eggs laid by three different types of hens were recorded.

Hen	Leghorn	Brahma	Sussex
White eggs	5	23	14
Brown eggs	25	7	16

Phoebe was interested to find out whether the colour of the eggs was independent of the type of hen. She decided to perform a χ^2 test at the 5% significance level on her data.

a Write down the null and alternative hypotheses.

b Show that the expected value of a Leghorn laying a white egg is 14.

c Write down the number of degrees of freedom.

d Find the χ^2 test statistic and the p-value.

e The critical value is 5.991. State the conclusion for this test.

4 Bert wanted to find out whether there was any relationship between gender and favourite colour of car. He collected data from 100 people and performed a χ^2 test at the 10% significance level.

Colour	Black	White	Red	Silver
Male	15	6	18	9
Female	16	14	7	15

a Write down the null and alternative hypotheses.

b Show that the expected number of males whose favourite colour of car is white is 9.6.

c Write down the number of degrees of freedom.

d Find the χ^2 test statistic and the p-value.

e The critical value is 6.251; state the conclusion for this test.

5 Each of the 140 members of a running club trains once a week in order to run a half-marathon. The number who train each day is shown in the table.

Day	Sun	Mon	Tue	Wed	Thu	Fri	Sat
Observed frequencies	21	15	17	18	16	21	32

a If the members are free to run on any day of the week, write down the table of expected values.

b Write down the number of degrees of freedom.

The critical value is 12.592.

c Determine the results of a goodness of fit test at the 5% significance level to find out whether the data fits a uniform distribution. Remember to write down the null and alternative hypotheses.

6 Benny tosses a fair die 90 times. The results are:

Number on die	1	2	3	4	5	6
Frequency	13	15	17	16	14	15

a Find the expected frequencies.

b Write down the number of degrees of freedom.

The critical value is 15.086.

c Determine the results of a goodness of fit test at the 1% significance level to find out whether Benny's data fits a uniform distribution. Remember to write down the null and alternative hypotheses.

7 Marilu tosses two unbiased coins 60 times. The number of tails that she gets is shown in the table.

Number of tails	0	1	2
Frequency	12	34	14

a Show that the expected frequency for getting 0 tails is 15.

b Find the table of expected frequencies.

c State whether there are any expected values less than 5 and combine results if required.

d Write down the number of degrees of freedom.

e Determine the results of a goodness of fit test at the 5% significance level to find out whether the data fits a binomial distribution. Remember to write down the null and alternative hypotheses.

The critical value for this test is 5.991.

f State the conclusion for the test and give a reason for your answer.

8 The heights of 14-year-old girls are normally distributed with a mean of 158 cm and standard deviation of 4 cm. Giorgio measures 500 14-year-old girls and his results are shown in the table.

Height, x cm	$x <$ 152	$152 \le x$ < 156	$156 \le x$ < 160	$160 \le x$ < 164	$x \ge$ 164
Frequency	12	133	201	109	45

a Find the table of expected frequencies.

b State whether there are any expected values less than 5 and combine results if required.

c Write down the degrees of freedom.

d Determine the results of a goodness of fit test at the 10% significance level to find out whether the data fits a normal distribution. Remember to write down the null and alternative hypotheses.

The critical value for this test is 7.779.

e State the conclusion for the test and give a reason for your answer.

9 Mrs Nelson gave her two Grade 12 classes the same history test. She wanted to find out whether one class was better than the other or not. The results of the test are:

Class 1	79	63	42	88	95	57	73	61	82	76	51	48
Class 2	65	78	85	49	59	91	68	74	82	56		

a Write down the null and alternative hypotheses.

b State whether this a one-tailed test or a two-tailed test.

c Find the t-value and p-value for a t-test at the 5% significance level.

d Write down the conclusion to the test.

Exam-style questions

10 P1: The head-teacher of Green High School wished to advertise that the IB results obtained from her students were better than those from the nearby Amber Academy. She examined the previous year's results and chose 12 students from the Academy and 12 students from her High School. She then tabulated the IB results for these students as follows.

Amber Academy	31	42	45	26	43	35	39	40	31	40	43	35
Green HS	40	30	36	41	39	41	37	42	36	44	40	41

a Write down the null and alternative hypotheses. (2 marks)

b State whether this a one-tailed test or a two-tailed test. (1 mark)

c Perform a *t*-test at the 5% significance level. (3 marks)

d Write down the conclusion to the test. (2 marks)

11 P1: A sample of ten 11-year-old students, and ten 16-year-old students were each given a questionnaire to determine how many books each of them had read in the previous year. The aim was to find out whether older students generally read fewer books than younger students. The results obtained were summarized as follows.

11 year olds	18	10	3	15	12	14	15	4	20	8
16 year olds	6	7	14	5	14	12	3	9	4	7

a Write down the null and alternative hypotheses. (2 marks)

b Perform a *t*-test at the 5% significance level. (3 marks)

c Write down the conclusion to the test. (2 marks)

12 P1: Two types of mobile phones were analysed to see whether there was a difference in their respective battery longevity. The phones were charged to full, then used normally until their batteries reduced to zero. The times taken (in seconds) for the batteries to discharge are shown in the following table.

Mobile 1	1250	1010	1500	1350	1310	1400	900	1550
Mobile 2	850	700	1130	1330	990	1440	880	1100

a Write down the null and alternative hypotheses. (2 marks)

b Perform a *t*-test at the 10% significance level. (3 marks)

c Write down the conclusion to the test. (2 marks)

13 P1: Eight people of varying ages undertook a small test to see whether there was any correlation between age and reaction times (given in seconds).

The following table shows each of their results.

Age	32	24	85	11	43	66	18	21
Reaction time	0.35	0.20	0.70	0.30	0.25	0.45	0.15	0.20

a Write down a table of ranks for this data. (3 marks)

b Calculate Spearman's rank correlation for this data. (2 marks)

c Briefly state what this result tells you. (2 marks)

d If the experiment were to be repeated, suggest two steps that could be taken to ensure the ensuing results were of greater validity. (2 marks)

14 P1: Neeve conducts a test to determine whether there is any correlation between a person's age and the number of hours they spend watching television per week.

Age	8	42	17	81	45	14	39	42	31	40	28	24
No. of hours	20	15	30	2	25	28	19	14	16	21	26	20

a Write down a table of ranks for this data. (3 marks)

b Calculate Spearman's rank correlation. (2 marks)

c Neeve concludes that your age affects how much TV per week you tend to watch. Using your calculations, comment on whether or not Neeve is correct. Suggest what conclusions you are able to make from your calculations. (3 marks)

15 P1: The following diagrams illustrate the relationships between six different sets of variables. In each case, match the diagram to the correct PMCC and Spearman's rank descriptions.

(6 marks)

Diagram	Description
A	1 PMCC = 0.8, Spearman = 1
B	2 PMCC = 1, Spearman = 1
C	3 PMCC = −1, Spearman = −1
D	4 PMCC = −0.8, Spearman = −1
E	5 PMCC = 0, Spearman = 0

C

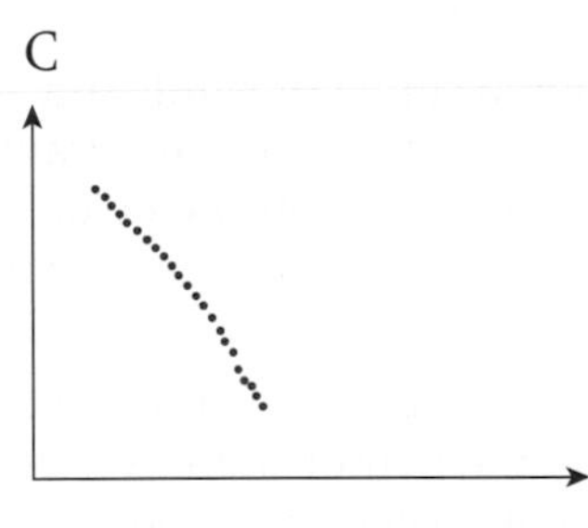

D

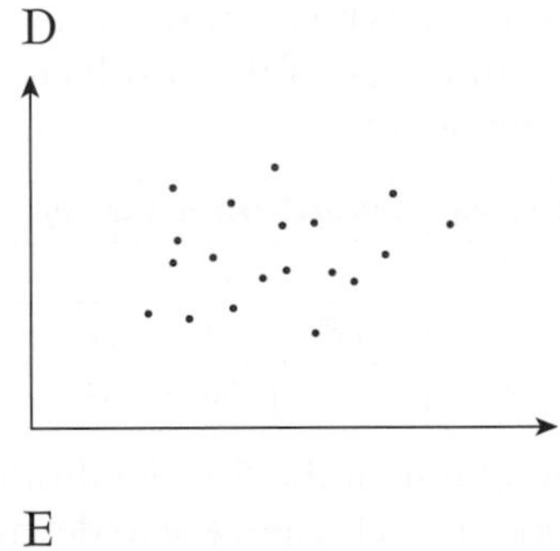

E

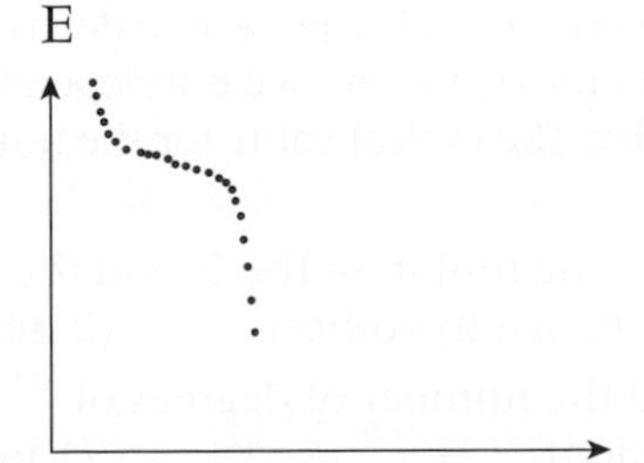

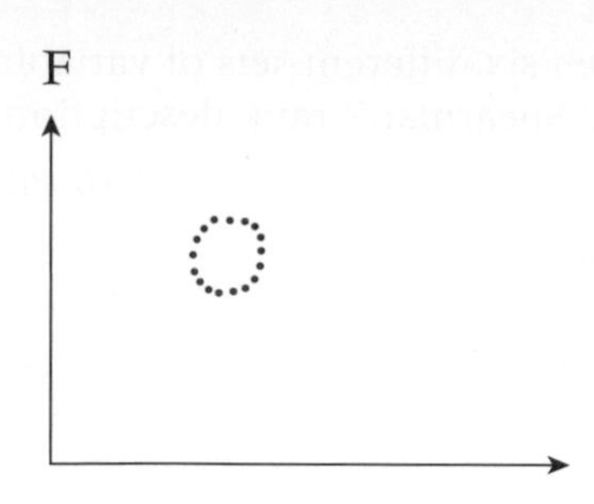

6 PMCC = 0.05, Spearman = 0.05

16 P1: A survey of 175 people of different ages was undertaken in which they were asked which type of burger they preferred to eat. A summary of the results is given in the table below.

	Chicken	Beef	Vegetarian	Totals
11–18 years	10	14	21	45
19–30 years	18	8	34	60
30–50 years	12	12	16	40
50+ years	14	8	8	30
Totals	54	42	79	175

Using the χ^2 test, at the 5% significance level, work through parts **a–d** to determine whether type of burger favoured is independent of age. The critical value for the test is 12.59.

a State the null hypothesis and the alternative hypothesis. (2 marks)

b Find the number of degrees of freedom. (1 mark)

c Calculate the χ^2 test statistic and the p-value for this data. (3 marks)

d State any conclusions obtained, justifying your answer. (2 marks)

17 P2: A survey was performed in a particular country in order to determine whether there was any evidence to support the claim that smoking is dependent on age. A summary of the results is given in the table below.

	Smoker	Non-smoker	Totals
16–25 years	31	59	90
26–60 years	19	99	118
Totals	50	158	208

a Copy and complete the table showing the expected frequencies.

	Smoker	Non-smoker	Totals
16–25 years			90
26–60 years			118
Totals	50	158	208

(3 marks)

b Using the χ^2 test with the assistance of technology, test at the 1% significance level, whether smoking is independent of age.

You should state your null and alternative hypotheses and draw conclusions based on the evidence obtained. The critical value for the test is 6.64. (7 marks)

18 P1: A survey of 266 people was conducted to determine whether there was any connection between an individual's gender and the type of movie they prefer. A summary of the results is given in the table below.

	Thriller	Romantic	Historical	Horror	Comedy	Totals
Male	68	23	11	36	36	174
Female	32	24	11	14	12	93
Totals	100	47	22	50	48	266

Using the χ^2 test, at the 5% significance level, work through parts **a–d** to determine whether movie preference is independent of gender. The critical value for the test is 9.49.

a State the null hypothesis and the alternative hypothesis. (2 marks)

b Find the number of degrees of freedom. (1 mark)

c Calculate the χ^2 test statistic and the p-value for this data. (3 marks)

d State any conclusions obtained, justifying your answer. (2 marks)

19 P1: Max tosses four coins a total of 80 times. The following table shows the distribution of the number of heads obtained in each toss of four coins.

Number of heads	0	1	2	3	4
Frequency	10	24	26	17	3

a Using the binomial distribution B(4, 0.5), find the expected probabilities of obtaining 0, 1, 2, 3 or 4 heads respectively. (2 marks)

b Perform a goodness of fit test at the 5% significance level to find out whether the data fits a binomial distribution. You should state the null and alternative hypotheses and justify any conclusions found. The critical value for this test is 9.49. (7 marks)

20 P2: 264 of the best mathematics students from Slovakia were invited to take part in a national mathematics competition. After their papers were marked, each student was awarded a mark out of 100. The following table shows the final results.

Mark obtained	Frequency
$0 \le x < 20$	1
$20 \le x < 40$	10
$40 \le x < 60$	114
$60 \le x < 80$	105
$80 \le x < 100$	34

In the previous year's competition, the data followed a normal distribution, with mean 62 and standard deviation 16.

a Copy and complete the following table of expected marks.

Mark obtained	Probability	Expected frequency
$0 \le x < 20$		
$20 \le x < 40$		
$40 \le x < 60$		
$60 \le x < 80$		
$80 \le x < 100$		

(3 marks)

b Use the χ^2 distribution at the 5% level of significance to test the hypothesis that the same distribution holds this year.

You should state the null and alternative hypotheses and re-write your frequency table if any values are found to be less than 5.

You should justify any conclusions found. (11 marks)

The correct critical value may be selected from the following table, where $\chi^2_{(5\%)}$ is the value such that $P(X > \chi^2_{(5\%)}) = 0.05$.

Degrees of freedom	$\chi^2_{(5\%)}$
1	3.84
2	5.99
3	7.82
4	9.49
5	11.07

Statistics and probability

Rank my maths!

Approaches to learning: Collaboration, Communication
Exploration criteria: Personal engagement (C), Use of mathematics (E)
IB topic: Spearman's rank

The task

In this task you will be designing an experiment that will compare the rankings given by different students in your class to determine how similar they are and what agreement there is

The experiment

Select one student in your group to be the experimenter. The other students in your group will be the subject of the experiment.

The experimenter is going to determine whether there is any similarity between the tastes of the other students in their group by asking them to rank a set of selections from the least to the most favourite.

Would you expect the rankings of the students in your group to be the same, similar or completely different?

What would help you to predict who might have similar tastes (where the strongest correlation would be)?

Under what circumstances might the rankings be similar and under what circumstances might they be different?

In your group, discuss:

What could you use the Spearman's rank method to test?

The experimenter should prepare their own set of selections for the other members of the group to compare in the area they have chosen to investigate.

They should make a prediction about how strong they think the rank correlation will be between the other students with regards to the experiment they are doing.

The experimenter will give their set of selections to the group.

The other students will rank the set of selections from favourite (1) to least favourite (n), where n is the number of selections.

What does the experimenter need to be aware of when providing instructions?

The experimenter should record the rankings in a table.

The students who are doing the ranking should not collaborate or communicate with each other.

Why is this important?

The results

Now find the Spearman rank correlation between the students.

Do the students display strong correlations?

Are the results what were expected?

Discuss as a group why the original hypothesis may have been accurate or inaccurate.

Write a conclusion for the experiment based on the results you have found.

You may wish to consider what Spearman's rank value would mathematically be considered to be high.

It is possible to test the significance of a relationship between the two rankings.

Extension

Investigate hypothesis testing for a Spearman's rank value.

9 Modelling relationships with functions: power functions

Concepts

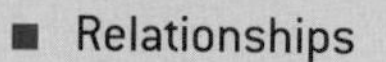
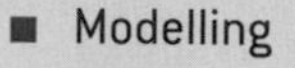

- Relationships
- Modelling

Microconcepts

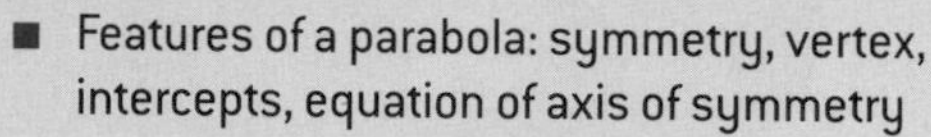

- Domain and range of a function
- Features of a parabola: symmetry, vertex, intercepts, equation of axis of symmetry
- Forms of a quadratic: general form and intercept form
- Messy data
- Cubic graphs and power functions
- Points of intersection
- Direct and inverse variation
- Optimization problems

If you are trying to use mathematics to model the path of a javelin, the shape of a bridge or the maximum volume of a container, for example, then you will need to study equations of curves. This chapter looks at ways of modelling real-life scenarios with curves and fitting equations to these curves, in order to predict the height of the curve (telling you, for example, the height of the javelin below) and the distance spanned by a curve (telling you the distance the javelin travels).

How can you predict where a javelin will land? How can you find out when its speed is fastest?

What is the maximum volume of a box made from a piece of card with squares cut from each corner?

How long does it take an object to fall, given that the distance varies directly with the square of the time taken?

How can you find the price of a car given that the price varies inversely with the age of the car?

Oliver is practising his basketball skills.

- What shape is the path of the ball? Sketch a path for the ball from Oliver's hands to the basketball hoop.
- Sketch a path for the ball from Oliver's hands to the hoop when he is standing
  - further away from the hoop
  - closer to the hoop.
- What do you notice about the shape of the ball's path when Oliver is standing in each position? What changes and what is the same?
- How can you model the path of a basketball from Oliver's hands to the hoop from any point on the court?
- How can the model help you to predict whether a ball will go into the hoop or not?
  - What information do you need to build this model?
  - What assumptions would you need to make in your model?
- How can you find the maximum height that the ball reaches? Will this height always be the same?

Developing inquiry skills

Write down any similar inquiry questions you might ask in order to model the path of something in a different sport; for example, determining where an archer's arrow will land, deciding whether a tennis ball will land within the baseline or considering whether a high-jumper will pass over the bar successfully.

What questions might you need to ask in these scenarios which differ from the scenario where Kazuki is playing basketball?

Think about the questions in this opening problem and answer any you can. As you work through the chapter, you will gain mathematical knowledge and skills that will help you to answer them all.

Before you start

You should know how to:	Skills check
1 Expand brackets: eg $(x+3)(x-2)=x^2+x-6$	**1** Expand **a** $(2x+3)(x-4)$ **b** $(7x-5)(2x+3)$
2 Factorize an expression: eg $3x^2-11x-4=(3x+1)(x-4)$	**2** Factorize **a** $2x^2+5x+2$ **b** $5x^2+13x-6$
3 Substitute coordinates into an equation: eg If $y=2x^2+4x+c$, find the value of c at the point $(1,3)$. Substitute $(1,3)$ into $y=2x^2+4x+c$ $3=2(1)^2+4(1)+c$ $3-2-4=c$ $-3=c$	**3 a** If $y=2x^2-3x+c$, find the value of c at the point $(2,-1)$. **b** If $y=5x^2+x+c$, find the value of c at the point $(1,-5)$.

Click here for help with this skills check 

9.1 Quadratic models

A dolphin jumps above the surface of the ocean. The path of the jump can be modelled by the equation

$f(x) = -0.09375x^2 + 1.875x - 3.375$

where:

- x represents the horizontal distance, in metres, that the dolphin has travelled from the point where it left the water
- $f(x)$ represents the vertical height, in metres, of the dolphin above the surface of the water.

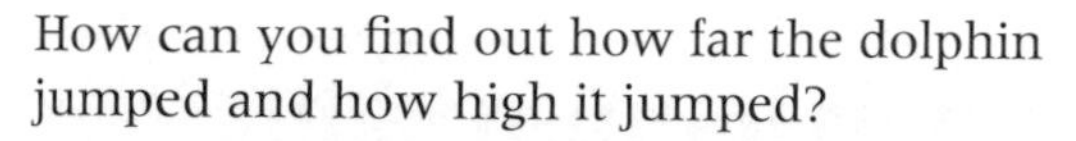

How can you find out how far the dolphin jumped and how high it jumped?

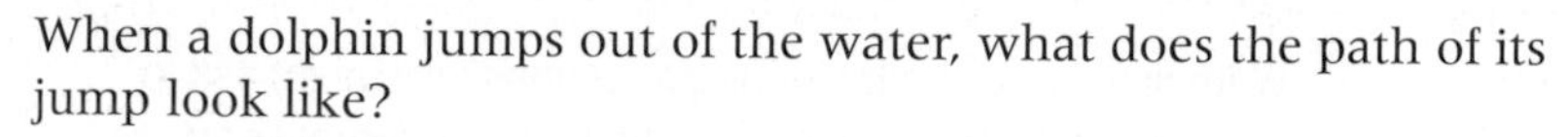

When a dolphin jumps out of the water, what does the path of its jump look like?

To investigate the dolphin's jump, you could use your GDC to plot a graph of the equation that is used to model the path of the jump. By finding the coordinates of certain points on the graph, you could tell how far and how high the dolphin jumped.

This section shows you how to do this.

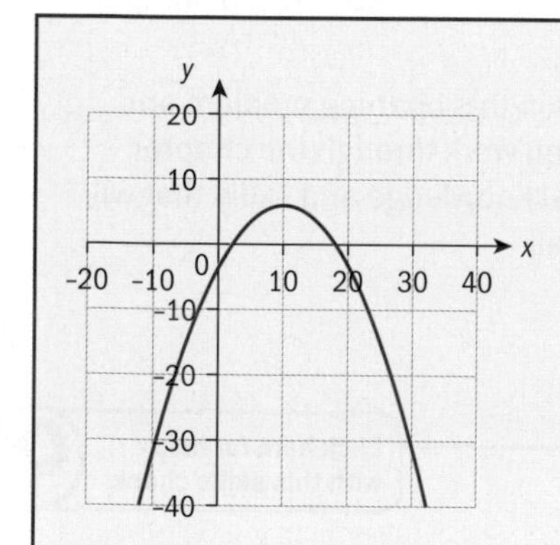

Quadratic functions are polynomial functions where the highest power of x is two.

For example, $f(x) = ax^2 + bx + c$; $a \neq 0$ and $a, b, c \in \mathbb{R}$ is a quadratic function.

The graph of a quadratic function is called a **parabola**.

International-mindedness

Frenchman Nicole Oresme was one of the first mathematicians to consider the concept of functions in the 14th century. The term "function" was introduced by the German mathematician Gottfried Leibniz in the 17th century, and the notation was coined by a Swiss mathematician, Leonard Euler, in the 18th century.

Reflect State what type of function this would be if $a = 0$.
How are quadratic graphs different from linear graphs?
How are they the same?

The maximum or minimum point on the graph of a quadratic function is called the **vertex**.

Reflect Describe how to find the domain of a function from its graph.
Describe how to find the range of a function from its graph.

The x-intercepts are the points where the graph $y = f(x)$ cuts the x-axis. They are also called the **zeros of the function**, because they are the x-values where $y = f(x) = 0$.

You could also find the x-intercepts by using the quadratic formula:

$$x = \frac{-b \pm \sqrt{b^2 - 4ac}}{2a}$$

The y-intercept occurs where $y = f(0)$.

International-mindedness

The Shulba Sutras in ancient India and the Bakhshali manuscript contained an algebraic formula for solving quadratic equations.

Transferring a graph from GDC to paper

You may be asked to sketch the graph of a function on paper. You need to be careful when copying the graph from your GDC onto the paper.

When you **"sketch"** a graph, you do not need to be as accurate as when you have to **"draw"** a graph on graph paper, but your sketch should:

- show the general shape of the graph accurately
- label the coordinates of any axes intercepts
- label the coordinates of any vertices.

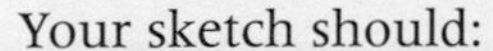

First you need to draw your axes and remember to label them and put at least one number on the axes for a scale. The x-values give the **domain** of the function and the y-values give the **range**.

Example 1

Plot the graph of $f(x) = -0.5x^2 + 7.5x - 18$ and then sketch this on paper.

Your sketch should:

- show the general shape of the graph accurately
- label the coordinates of any axes intercepts
- label the coordinates of any vertices.

Also state the domain and range of this function.

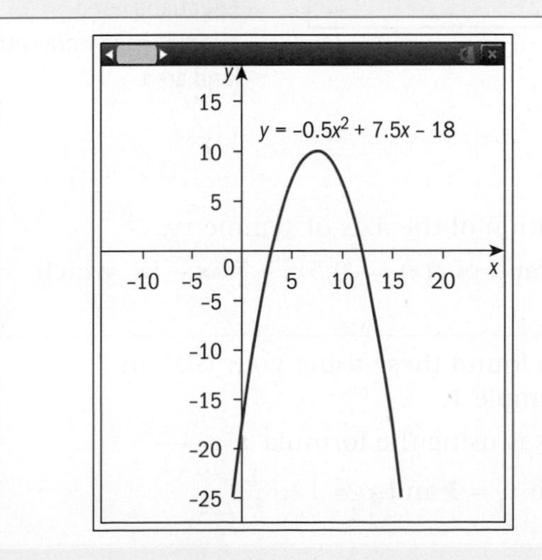

Graph the equation on your GDC.

Continued on next page

y-intercept is $(0, -18)$	Read the y-intercept from your GDC. You may need to adjust the viewing window in order to do this.
The zeros are at $x = 3$ and $x = 12$. So, $(3, 0)$ and $(12, 0)$ are also points on the graph. The vertex is $(7.5, 10.125)$.	Use CALC or Analyse Graph on your GDC to find the zeros of the function. Find the coordinates of the vertex using CALC or Analyse Graph.
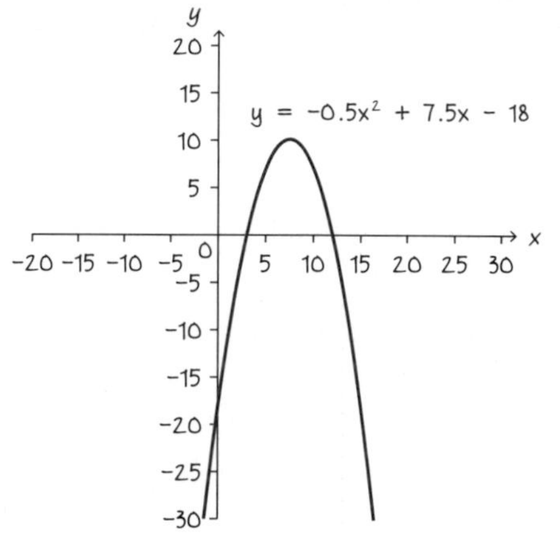	Draw and label your axes. You must draw the shape of the graph correctly and label the points where the graph cuts the axes, as well as the vertex. Since you are asked for a sketch, you do not need to put a full scale on the axes, but you should have at least one number on each axis.
Domain of f is $\{x \in \mathbb{R}\}$	You can evaluate the function f at any real value of x, so the domain is the set of all real numbers.
Range of f is $\{y \in \mathbb{R} \mid y \leq 10.125\}$	The vertex of the graph is a maximum and is at $(7.5, 10.125)$, so the maximum value of the function f is 10.125.

The line of symmetry of a quadratic graph is called the **axis of symmetry**.

If the x-coordinates of the x-intercepts are x_1 and x_2 then the equation of the axis of symmetry is

$$x = \frac{x_1 + x_2}{2}$$

International-mindedness

Over 2000 years ago, Babylonians and Egyptians used quadratics to work with land area.

Example 2

You can also use the formula above to find the equation of the axis of symmetry.

Find the equation of the axis of symmetry for the graph of $f(x) = -0.5x^2 + 7.5x - 18$, which you studied in Example 1.

The x-intercepts of the graph are $x = 3$ and $x = 12$. The equation of the axis of symmetry is $x = \frac{3+12}{2} = \frac{15}{2} = 7.5$.	You found these using your GDC in Example 1. This is using the formula $x = \frac{x_1 + x_2}{2}$ with $x_1 = 3$ and $x_2 = 12$.

Exercise 9A

1 Plot the graph of $y = f(x)$ for the following functions, and then sketch them on paper.

On each sketch, mark the coordinates of the vertex and the points where the graph intercepts the axes, and state the domain and range of each function.

a $f(x) = 2x^2 + 5x + 2$

b $f(x) = -x^2 + 6x + 7$

c $f(x) = 3x^2 - x - 4$

d $f(x) = -5x^2 - 4x + 12$

2 Consider the graph given by the equation $y = 0.4x^2 - 2x - 8$.

a Find the coordinates where the graph crosses the x-axis.

b Find the coordinates of the intercept with the y-axis.

c Find the equation of symmetry of the curve.

d Find the point of intersection of this curve with the curve given by the equation $y = -5x^3$.

Remember, we solved $f(x) = g(x)$ in Chapter 5.

Functions

At the beginning of this section you studied the equation of the path of a dolphin as it jumps out of the water. The path was modelled by the equation $f(x) = -0.09375x^2 + 1.875x - 3.375$. Use what you have learned to find out how high the dolphin jumped and how far it jumped.

TOK

We have seen the involvement of several nationalities in the development of quadratics in the chapter.

To what extent do you believe that mathematics is a product of human social collaboration?

Investigation 1

1 **a** By looking at the graph of $f(x) = x^2$ shown below, state the two values of x which both map to $f(x) = 4$.

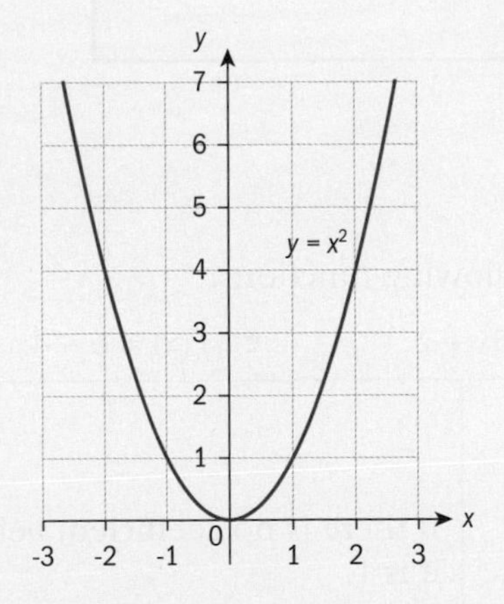

b Find the two values of x which both map to $f(x) = 16$.

Continued on next page

2 Use your GDC to plot the graphs of $y = f(x)$ for the functions $f(x) = -x^2 + 2x - 3$ and $f(x) = x^2 - 14x + 40$. For each function, find two values of x in the domain that both map to the same value of $f(x)$ in the range.

3 **Factual** What is the general shape of the quadratic function?

4 **Conceptual** How does the graph of a quadratic function show you that, for a given value y in the range, there will generally be more than one value of x for which $f(x) = y$?

Quadratic functions are called many-to-one functions.

5 **Factual** Why are quadratic functions called many-to-one functions?

6 **Conceptual** How are quadratic functions different from linear functions? Consider your answer to question **5**.

7 **Factual** How do you sketch the graph of an inverse function?

8 **Conceptual** Is it possible to find the inverse of a quadratic function?

9 **Conceptual** Do many-to-one functions have inverse functions?

Many-to-one functions do not have inverses.

Parameters of a quadratic function

For the function $f(x) = ax^2 + bx + c$, $a \neq 0$, a is the coefficient of x^2, b is the coefficient of x, and c is the constant.

a, b and c are called the parameters of the function.

Example 3

Write down the values of a, b and c in the following functions.

a $f(x) = 2x^2 + 3x - 4$ **b** $f(x) = x^2 - 5x + 2$ **c** $f(x) = 6 - 4x + 3x^2$

a $a = 2$, $b = 3$ and $c = -4$	
b $a = 1$, $b = -5$ and $c = 2$	If there is no coefficient before x^2 or x, then it is 1.
c $a = 3$, $b = -4$ and $c = 6$	Watch out for equations where the order of x^2, x or the constant have been rearranged.

Investigation 2

1 Plot the graphs of the following quadratic functions on your GDC and use this to help you complete the table below:

a $y=x^2+4x-5$ b $y=x^2+4x+3$ c $y=-x^2+x+2$
d $y=2x^2-8x-10$ e $y=-2x^2+3x+9$ f $y=4x^2-8x+3$

Equation	Value of a	Value of b	Value of c	y-intercept	Is there a maximum or minimum point?
a	1	4	−5	(0, −5)	minimum
b					
c					
d					
e					
f					

2 **Factual** How does the sign of a in a particular quadratic function determine whether a vertex is a maximum or minimum?

3 **Conceptual** What is the connection between the value of c and the y-intercept?

The parameters of a quadratic function determine whether the vertex is a maximum or minimum and the value of the y-intercept.

Exercise 9B

1 Given a general quadratic function of the form $f(x) = ax^2 + bx + c$, write down the values of a, b and c for each of the following functions.

a $f(x) = x^2 - 4x + 2$ b $f(x) = 3 - 2x + 2x^2$
c $f(x) = -2x^2 + x + 1$ d $f(x) = -x^2 + 2x - 3$
e $f(x) = -10 + x + 5x^2$

2 The graph of $y = ax^2 + bx + c$ passes through the points $(0, -9)$ and $(-1, 9)$. Given that the equation of symmetry of the graph is $x = 4$, find the values of a, b and c.

Developing inquiry skills

Look back at the opening problem for this chapter. Oliver was trying to throw a basketball through a hoop.

What type of function could you use to model the path of the basketball?

9.2 Problems involving quadratics

Example 4

A rectangular mirror has perimeter 260 cm.

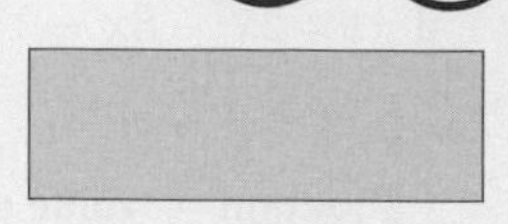

a The length of the mirror is x cm. Find the height of the mirror in terms of x.

b Find an equation for the area of the mirror, A cm^2, in terms of x.

c Plot a graph of your equation for the area of the mirror, showing area A on the y-axis and length x on the x-axis. Choose a suitable domain and range.

d Find the coordinates of the points where the graph intercepts the x-axis.

e State what these two values of x represent.

f Hence find the equation of the graph's line of symmetry.

g State what the equation of the line of symmetry tells you in this context.

a Let the height of the rectangle be y cm.

$2x + 2y = 260$

$x + y = 130$

$y = 130 - x$

Label the height y cm and form an equation in x and y for the perimeter of the rectangle.

b Area = length × height

$A = xy$

$A = x(130 - x)$

Substitute $y = 130 - x$ which you found in part **a**.

c

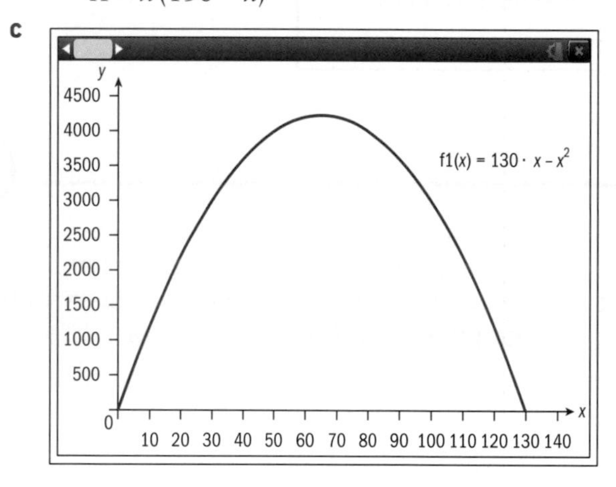

Use the ZOOM function on your GDC to locate the turning point. A reasonable domain would be from 0 to 150, and range from 0 to 4500.

d The x-intercepts are $(0, 0)$ and $(130, 0)$.

Use your GDC to find the zeros of the function.

e The x-coordinates 0 and 130 are the upper and lower limits between which the value of x must lie.

f The line of symmetry is $x = \frac{0+130}{2}$

$x = 65$

You find the line of symmetry by adding the x-intercepts together and halving the answer.

g This is the value of x which gives the largest area of the mirror.

The axis of symmetry passes through the vertex of the graph, which is a maximum in this case.

Exercise 9C

1 A rectangular picture frame has perimeter 70 cm.

   a The width of the frame is x cm. Find an expression, in terms of x, for the height of the frame.

   b Find an equation for the area of the frame, A cm^2, in terms of x.

   c Plot the graph which shows how A varies with x. Use a suitable domain and range. Hence sketch the graph on paper.

   d Find the x-intercepts of the graph in part **c**.

   e State what these two values of x represent.

   f Hence find the equation of the graph's line of symmetry, and state what this tells you in the context of the problem.

2 The first four terms of an arithmetic sequence are:

   6 10 14 18

   a Show that the sum to n terms can be written as $2n^2 + 4n$.

   b If the sum to n terms is 880, write a quadratic equation to represent this information. Rearrange to equal 0 and plot a graph of this equation on your GDC. Hence, sketch the graph on paper, showing the coordinates of the vertex and axes intercepts.

   c Find the positive x-intercept.

   d State what information the positive x-intercept tells you about the sequence.

3 A company produces and sells books.

   The weekly cost, in euros, for producing x books is €$(0.1x^2 + 400)$.

   The weekly income from selling x books is €$(-0.12x^2 + 30x)$.

   a Show that the weekly profit, $P(x)$, can be written as $P(x) = -0.22x^2 + 30x - 400$.

   b Plot a graph of this equation and hence sketch it on paper, showing the coordinates of the vertex and axes intercepts.

   c State what the x-intercepts represent in the context of the problem.

   d Find the equation of the axis of symmetry of the graph, and state what this tells you in the context of the problem.

4 The path of a football can be modelled by the quadratic equation

   $h(x) = -0.0125x^2 + 0.65x - 3.45$

   where $h(x)$ is the height of the football in metres, and x is the horizontal distance of the football in metres.

   a Plot a graph of this equation and hence sketch it on paper, showing the coordinates of the vertex and axes intercepts.

   b Find the x-intercepts and explain what these values represent.

   c Find the equation of the axis of symmetry, and state what this tells you in the context of the problem.

5 A ball is thrown vertically upwards.

   The path of the ball can be modelled by the equation $h(t) = 12t - 4t^2$ where $h(t)$ is the height of the ball after t seconds.

   a Plot a graph of this equation and hence sketch it on paper, showing the coordinates of the vertex and axes intercepts.

   b Find the t-intercepts and explain what these values represent.

   c Find the equation of the axis of symmetry, and state what this tells you in the context of the problem.

Investigation 3

1 In Example 4, we looked at how the area, A, of a rectangular mirror varied as the length, x, of the mirror varied.

   We found that the two variables were linked by the equation $A = x(130 - x)$.

Continued on next page

a **Factual** Is the graph of A against x symmetrical?

b **Conceptual** How could you use the coordinates of the points where the graph cuts the x-axis to find the equation of the axis of symmetry? Find the equation of the axis of symmetry in this case.

c **Factual** How you can use the equation of the axis of symmetry to find the y-coordinate of the maximum point? Find the y-coordinate of the maximum point in this case.

2 **Conceptual** How does the symmetry of a parabola allow us to solve real-life problems?

Suppose $(m, 0)$ and $(n, 0)$ are the x-intercepts of the quadratic graph $y = f(x)$.

The vertex of the graph has coordinates $\left(\frac{m+n}{2}, f\left(\frac{m+n}{2}\right)\right)$.

Example 5

The side view of a suspension bridge can be modelled by the equation $f(x) = 0.000\,972\,2x^2 - 1.167x$ where x is the horizontal distance in metres and $f(x)$ is the depth in metres.

a Plot the graph of the function.

b Find the coordinates of the points where the graph crosses the x-axis.

c Use these values to write down the equation of the axis of symmetry.

d Use part **c** to find the y-coordinate of the minimum.

e Describe, in this context, what the minimum point represents.

a

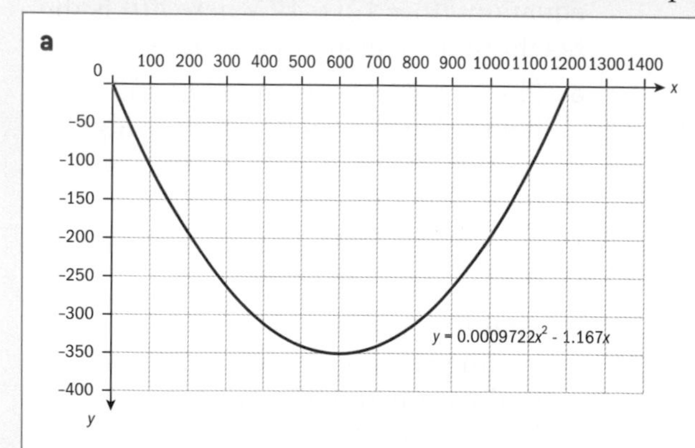

b $0.000\,972\,2x^2 - 1.167x = 0$

$x(0.000\,972\,2x - 1.167) = 0$

$x = 0 \text{ or } x = \frac{1.167}{0.000\,972\,2} = 1200$

Intercepts are $(0, 0)$ and $(1200, 0)$.

Set $f(x) = 0$ to find the x-intercepts.

c $x = \frac{0+1200}{2} = 600$	Given intercepts x_1 and x_2, the axis of symmetry is $x = \frac{x_1+x_2}{2}$.
d $y = f(600) = 0.0009722(600)^2 - 1.167(600)$ $y = -350$ **e** The lowest point of the suspension bridge.	This is $f\left(\frac{x_1+x_2}{2}\right)$.

Investigation 4

At a New Year celebration, Piotr sets off a firework rocket from a platform which is 1 m above the ground.

The path of the firework is described by a quadratic function with equation $f(t) = -0.2t^2 + 2t + 1$, where $t \geq 0$ represents the time, in seconds, since the firework took off, and $f(t)$ represents the height, in metres, of the firework above the ground.

Piotr wants to estimate the maximum height that his firework reaches and how long it will take before it lands on the ground.

1. **Factual** What are the two different methods that could be used to find the maximum height of the firework? Use either method to find the maximum height in this case.
2. Find the positive t-intercept and explain what this value represents in this context.
3. **Conceptual** Why it is useful to model real-life situations such as the path of a firework using a quadratic?
4. **Factual** Can a model give an accurate position of the firework 20 seconds after takeoff?

Example 6

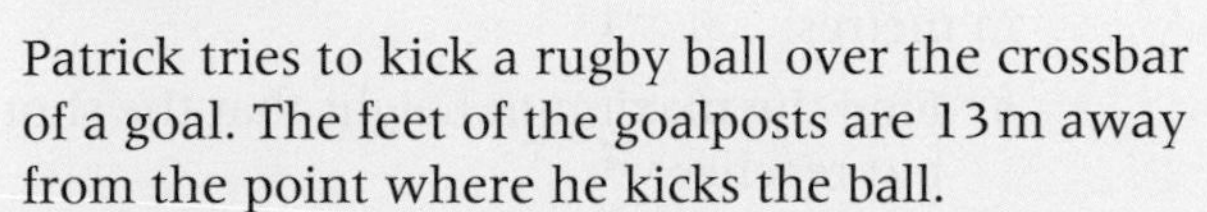

Patrick tries to kick a rugby ball over the crossbar of a goal. The feet of the goalposts are 13 m away from the point where he kicks the ball.

The vertical height of the ball (in metres) is approximately modelled by the function

$$f(x) = -0.1x^2 + 1.5x$$

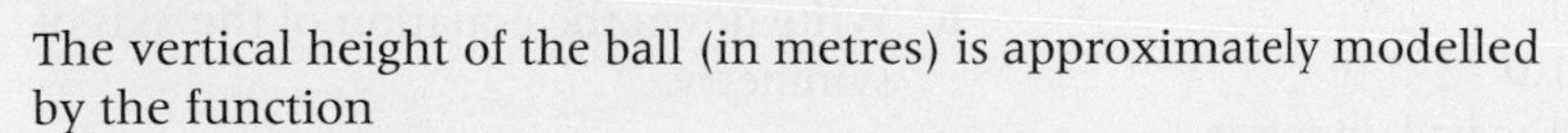

where x is the horizontal distance, in metres, of the ball from the point where it was kicked.

a Find the maximum height of the rugby ball.

TOK

We can successfully use mathematics to model real-world processes. Is this because we create mathematics to mirror the world or because the world is intrinsically mathematical?

Continued on next page

b Find the coordinates of the y-intercept. Explain what this tells you about the height from which Patrick kicks the ball.

c Given that the crossbar is 2.5 metres above the ground, determine whether the ball passes over the crossbar. Justify your answer.

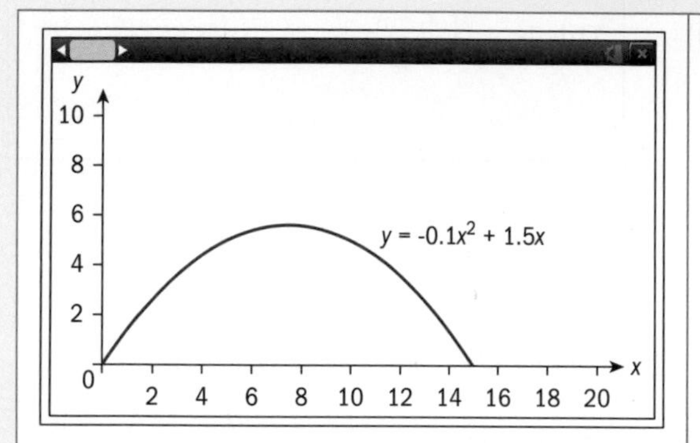	Plot $y = f(x)$ on your GDC.
a 5.63 metres (3 sf)	Use your GDC to find the maximum point.
b $(0, 0)$. Patrick kicks the ball from ground level.	Use your GDC to find the y-intercept. Since the initial vertical height of the ball is zero, the ball must initially be on the ground.
c Ball passes through the point $(13, 2.6)$, so it does pass over the crossbar.	Find the y-coordinate at the point where $x = 13$. If it is less than 2.5 m then the ball passes under the crossbar; if it is greater than 2.5 m then the ball passes over the crossbar.

Exercise 9D

1 For each of the following equations, find **i** the vertex and **ii** the equation of the axis of symmetry.

a $f(x) = x^2 - 4x + 2$ **b** $f(x) = 3 - 2x + 2x^2$

c $f(x) = -2x^2 + x + 1$ **d** $f(x) = -x^2 + 2x - 3$

e $f(x) = -10 + x + 5x^2$

2 Zander is playing a game of baseball. He hits the ball and the height of the ball is modelled by the formula

$y = -0.018x^2 + 0.54x + 1.0$

where y is the height of the ball, in metres, and $x > 0$ is the horizontal distance in metres.

a Find the maximum height that the ball reaches.

b Find the positive value for x when the graph crosses the x-axis and explain what this value represents.

c Find the y-intercept and explain what this represents.

3 Omar is on the school shot-put team. The path of the shot-put is modelled by a quadratic function with equation $y = 1.5 + 0.75x - 0.05x^2$, where y is the height of the shot-put in metres and $x > 0$ is the horizontal distance travelled in metres.

a Find the maximum height that the shot-put reaches.

b Write down the equation of the axis of symmetry.

c Find the positive value for x when the graph crosses the x-axis and explain what this value represents.

d Find the y-intercept and explain what this value represents.

4 Ziyue is the goalkeeper in his football team. He takes a free kick from the goal and the path of the ball is modelled by the function $f(x) = -0.06x^2 + 1.2x$ where $f(x)$ is the height of the ball, in metres, and x is the horizontal distance travelled by the ball, in metres.

   a Find the maximum height that the ball reaches.

   b Write down the equation of the axis of symmetry.

   c Find the x-intercepts and explain what these values represent.

5 A ramp in a skateboard park is modelled by a curve with equation $f(x) = 10.67 - 1.67x + 0.0417x^2$ where x is the horizontal distance in metres and $f(x)$ is the height above the ground in metres.

   a Find the maximum depth of the run.

   b Find the x-intercepts and explain what these values represent.

6 Jin throws a stone into the air. The height of the stone above the ground can be modelled by the equation $f(t) = 1 + 7.25t - 1.875t^2$ where t is the time, in seconds, that has passed since the stone was thrown, and $f(t)$ is the height of the stone, in metres.

   a Find the maximum height that the stone reaches.

   b Determine how long it takes for the stone to land on the ground.

7 A rectangular picture frame has perimeter 100 cm.

   a The width of the frame is x cm. Find an expression, in terms of x, for the height of the frame.

   b Find an expression for the area, A cm^2, in terms of x.

   c Plot the graph of A against x.

   d Find the x-intercepts.

   e Find the y-intercept.

   f Find the equation of the axis of symmetry.

   g Find the coordinates of the vertex.

   h Write down the maximum area of the picture frame, and the dimensions of the picture frame that give this maximum area.

8 A bullet is fired from the top of a cliff overlooking the sea. The path of the bullet may be modelled by the equation $y = -0.0147x^2 + 2x + 96$ $(x \geq 0)$ where x is the horizontal distance from the foot of the cliff and y is the vertical distance from the foot of the cliff.

   a State the height of the cliff.

   b Find the maximum height reached by the bullet.

   c Find the distance the bullet lies from the foot of the cliff, when it hits the water.

Different ways of writing a quadratic function

How can you use quadratic functions to decide whether or not the person lands in the safety net?

International-mindedness

How do you use the Babylonian method of multiplication?

Try 36×14.

Investigation 5

1 Expand the brackets of the following functions:
a $f(x)=2(x-3)(x-7)$ **b** $f(x)=(x-2)(x+1)$ **c** $f(x)=-2(x+2)(x+5)$.

2 **Factual** Are all the functions listed in question **1** quadratic functions? Justify your answer.

3 Plot the graphs of the three functions listed in question **1**, and then complete the table.

Function	x-intercepts	Coordinates of vertex	Equation of the axis of symmetry
$f(x)=2(x-3)(x-7)$			
$f(x)=(x-2)(x+1)$			
$f(x)=-2(x+2)(x+5)$			

4 **Factual** How would you find the x-intercepts for any quadratic function?

5 Use your results from question **4** to explain why it is easier to find the x-intercepts of a quadratic function when it is written in the form $f(x)=a(x-p)(x-q)$ (this form is called **intercept form**), rather than when it is in the form $f(x)=ax^2+bx+c$.

6 **Conceptual** Why is it useful to write quadratic functions in different forms?

7 **Factual** What is the equation of the axis of symmetry and the coordinates of the vertex of the quadratic function $f(x)=a(x-p)(x-q)$?

Example 7

a Plot the graph of $y = f(x)$ for $f(x) = 4(x - 5)(x + 3)$.

For parts **b**, **c** and **d**, you may only use your GDC to check your answer.

b Find the x-intercepts.

c Find the equation of the axis of symmetry.

d Find the coordinates of the vertex.

a

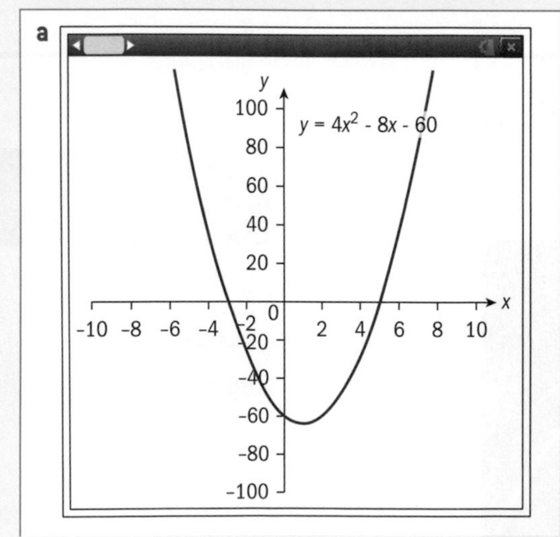

You may need to adjust the domain and range on your GDC in order to see the vertex and the x-intercepts clearly.

b $4(x-5)(x+3)=0$ $(x-5)=0$ or $(x+3)=0$ So, $x=5$ or $x=-3$	Set $f(x)=0$ and solve for x. You can verify your answers by looking at the x-intercepts on your GDC.
c $x=\frac{5+(-3)}{2}=1$	The axis of symmetry runs through the midpoint of the x-intercepts. $x=\frac{5+(-3)}{2}=1$
d $(1, -64)$	Substitute $x=1$ into the equation of the function.

Example 8

Sketch, on paper, the graph of $y=f(x)$ for the function $f(x)=-2x^2+8x+24$ showing clearly the x- and y-intercepts and the coordinates of the maximum or minimum point.

You should only use your GDC to verify your answer.

$f(x)=-2(x-6)(x+2)$	First, write the equation in intercept form.
Find the y-intercept when $x=0$ $y=-2(-6)(2)=24$	Find the y-intercept.
Find the x-intercepts when $f(x)=0$ $-2(x-6)(x+2)=0$ $x-6=0$ or $x+2=0$ $x=6$ or $x=-2$	Find the x-intercepts.
Midpoint of x-intercepts $=\frac{6+(-2)}{2}=2$	Find the x-coordinate of the vertex.
$f(2)=-2(2-6)(2+2)=32$	Find the y-coordinate of the vertex.
So $(2, 32)$ are the coordinates of the maximum value.	
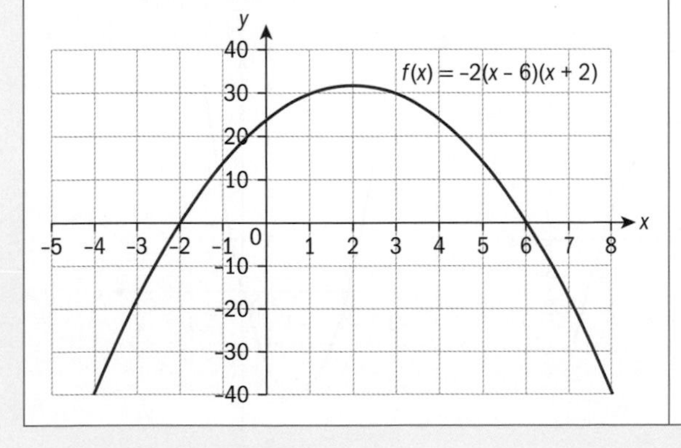	Plot these points and join them up with a smooth curve.

Exercise 9E

Given the equation $y = f(x)$ for each of the following functions:

a Find the intercept with the y-axis.

b Find the intercepts with the x-axis.

c Find the coordinates of the vertex.

d Find the equation of the axis of symmetry.

e Sketch the graph of the function on paper, showing clearly the x- and y-intercepts and the coordinates of the maximum or minimum point.

i $f(x) = (x-5)(x-9)$

ii $f(x) = 2(x+3)(x-1)$

iii $f(x) = -3(x-1)(x-3)$

iv $f(x) = -2(x+1)(x-2)$

v $f(x) = (2x+1)(x-1)$

HINT

Be careful with this function as one coefficient of x is not 1.

Two different ways that you can write a quadratic function are:

General form: $f(x) = ax^2 + bx + c$, $a \neq 0$

Intercept form: $f(x) = a(x-m)(x-n)$, $a \neq 0$

TOK

How would you choose which formula to use?

When is intuition helpful in mathematics? When is intuition harmful in mathematics?

Reflect What are the key features of a quadratic function?

How do the different ways of writing a quadratic function affect how you find the key features?

Finding the equation of a given curve

Investigation 6

Here are some graphs of quadratic functions.

a

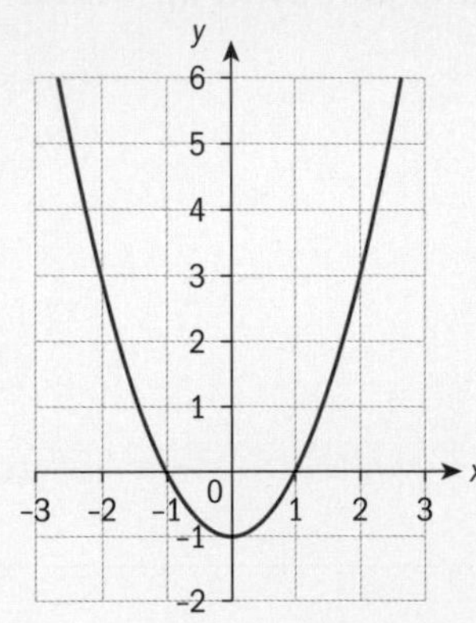

b

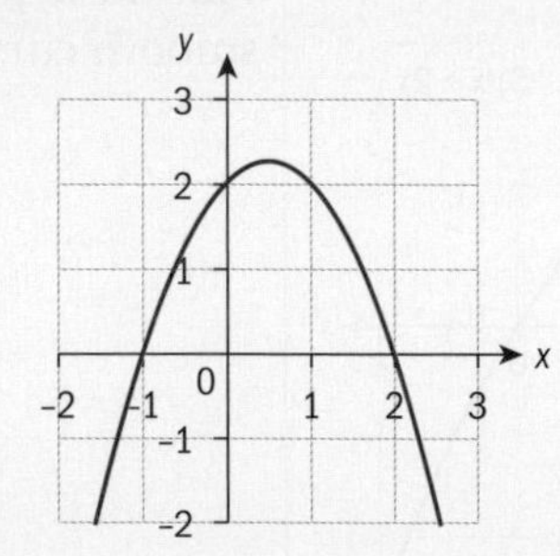

c

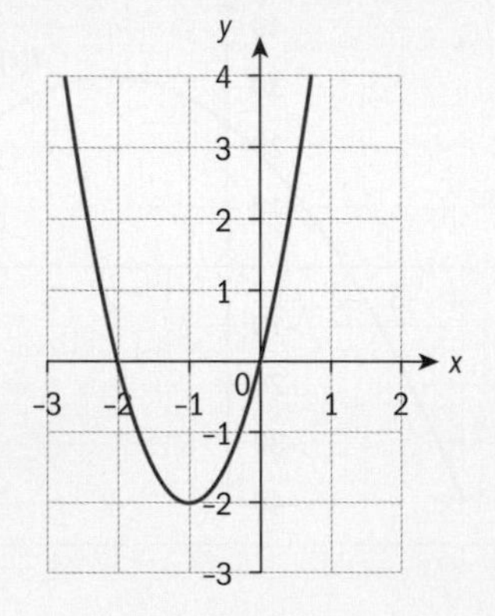

d

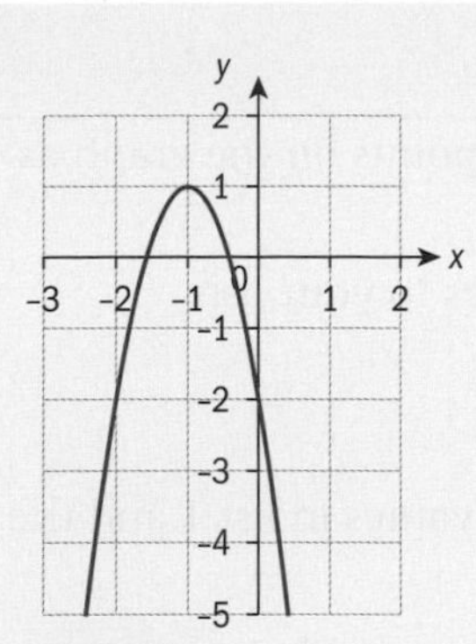

e

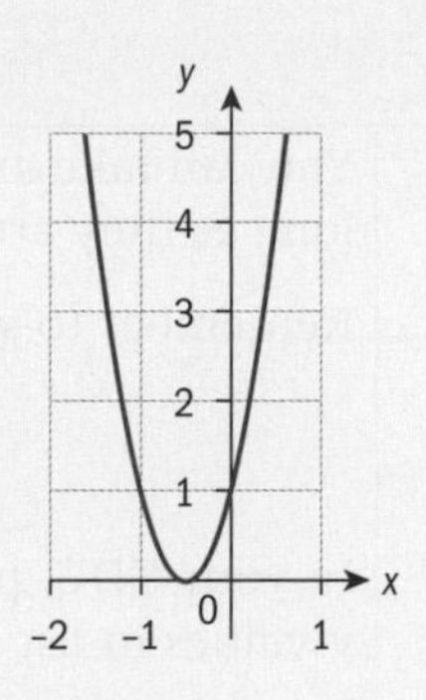

f

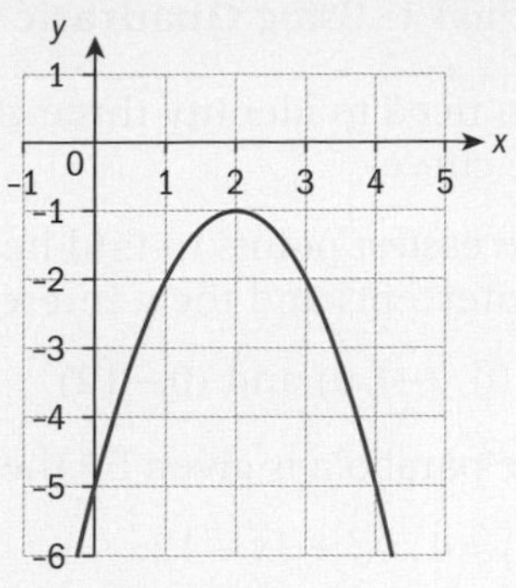

1 **Factual** By looking at the shape of the graph of $y = ax^2 + bx + c$, how can you tell whether a is positive or negative?

2 Write down coordinates of the vertex of each graph **a** to **f**.

3 Write down coordinates of the x- and y-intercepts for each graph **a** to **f**.

4 Follow these instructions.

- Open a spreadsheet in your GDC.
- For graph **a**, input the x-coordinate of the vertex into column A and the y-coordinate of the vertex into column B.
- Repeat, on a new line, for each of the axes intercept coordinates you found.
- Go to Menu – Statistics – Stat calculations – Quadratic regression.
- Check the "X List" takes the data from column A, and the "Y List" takes the data from column B.
- Select OK.
- Now you will see the values for the parameters a, b and c for the best fit quadratic equation through the three points you entered. Check the sign of a matches the shape of graph **a**.
- Write down the equation of graph **a**.

5 Repeat this for the other graphs **b** to **f**.

6 **Conceptual** How many points on a quadratic graph do you need in order to find the parameters? Investigate by inputting different numbers of points.

When you are given the graph of a quadratic function, you can find the function by entering the coordinates of three points on the graph into your GDC and then using quadratic regression.

Example 9

Find the equation of the following graph.

There are two methods you can use to do this on your GDC.

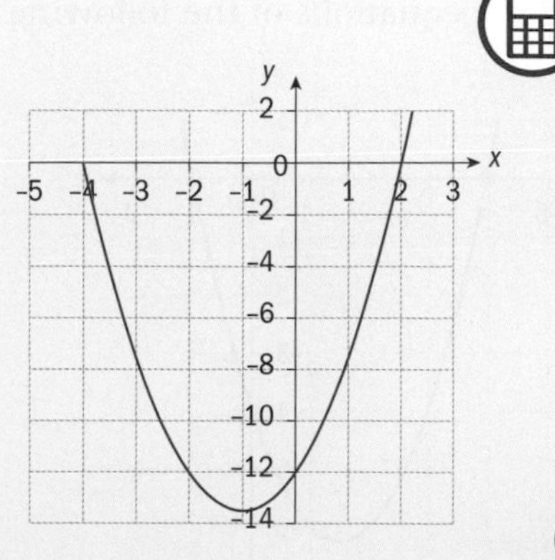

Continued on next page

Method 1: Using **Quadratic regression**

You need to identify three clear points on the curve.	You can take any three points on the graph as long as they are exact.
The easiest points to find here are the x-intercepts and the y-intercept:	Remember to give names to your lists.
$(2, 0)$, $(-4, 0)$ and $(0, -12)$	
The parabola is given by the equation $f(x) = 1.5x^2 + 3x - 12$.	In your GDC, put the x-values in list 1 and the y-values in list 2. Then go to Statistics – Stat calculations – Quadratic regression. Here you see the parameters for a, b and c.

Method 2: Using **Simultaneous equation solver**

Curve passes through $(2, 0)$, $(-4, 0)$ and $(0, -12)$.	First, find the three points on the curve as you saw in Method 1.
Substituting these points into the equation: $0 = a(2)^2 + 2b + c$ $\Rightarrow 4a + 2b + c = 0$	For each of the three points, substitute the x- and y-coordinates into the general equation of a parabola, $y = ax^2 + bx + c$.
$0 = a(-4)^2 - 4b + c$ $\Rightarrow 16a - 4b + c = 0$	This gives you three simultaneous equations in a, b and c.
$-12 = a(0)^2 + 0b + c$ $\Rightarrow -12 = c$	
$f(x) = 1.5x^2 + 3x - 12$	Solve these equations on your GDC to give the solution.

Exercise 9F

Find the equations of the following graphs.

1

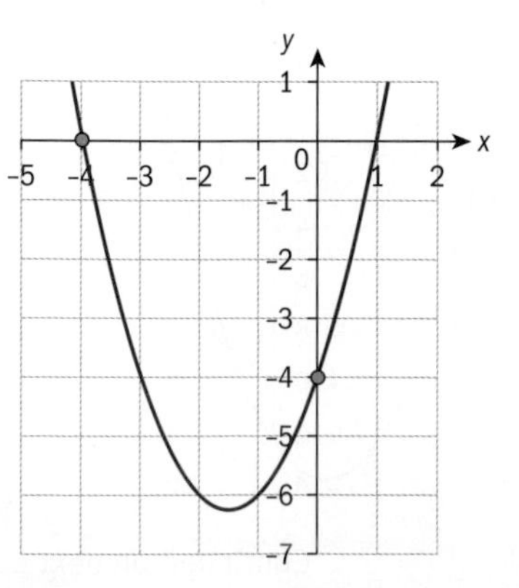

2

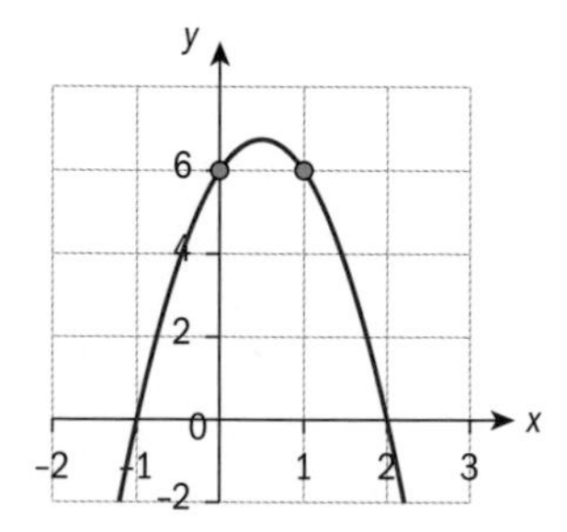

3

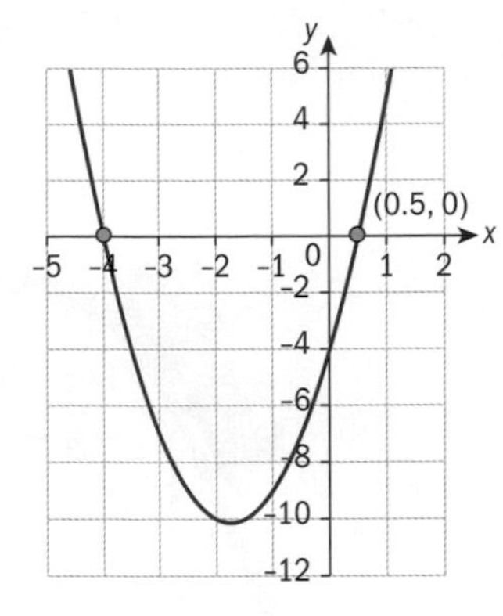

4

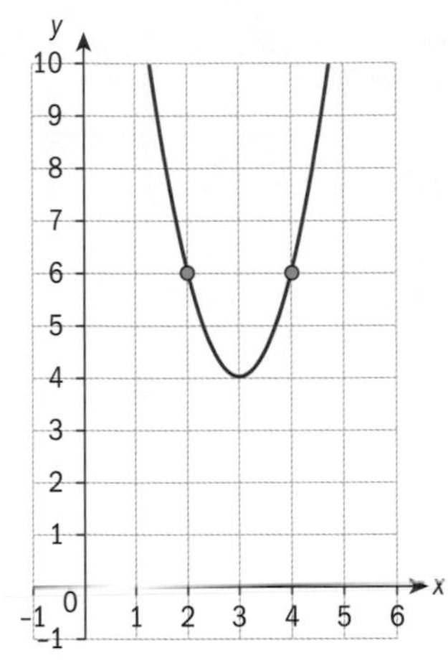

5

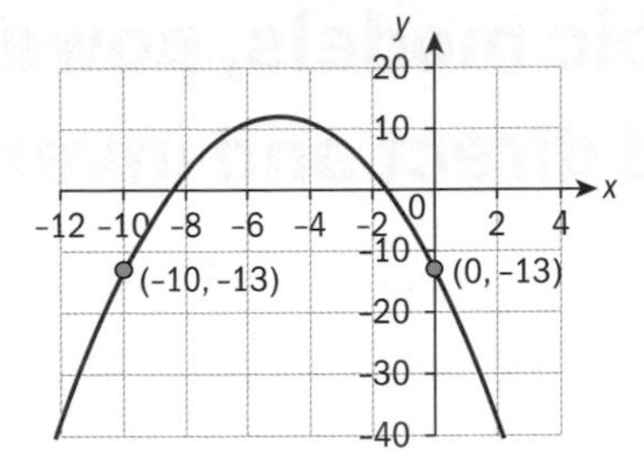

6

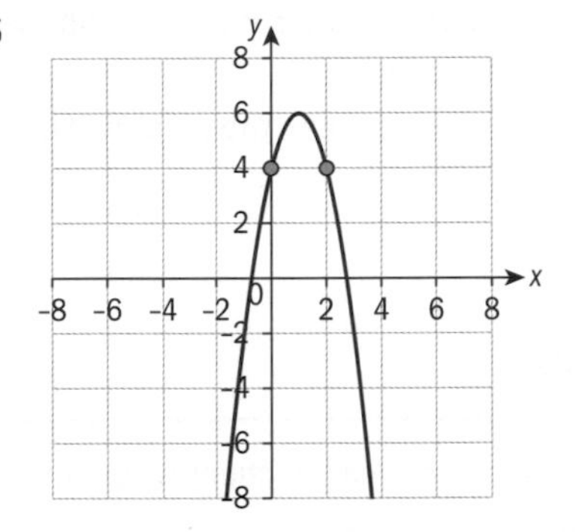

Developing inquiry skills

Look again at the opening problem where Oliver throws a basketball through a hoop.

Which three points are you able to find in the path of the basketball?

How could you use these points to model its path?

9.3 Cubic models, power functions and direct and inverse variation

Here is an illustration of part of a rollercoaster.

It is in the shape of the graph of a cubic function.

Reflect How are graphs of cubic functions different to graphs of quadratic functions?
Are there any similarities between the graphs of quadratic and cubic functions?

Cubic models

Cubic functions are polynomial functions where the highest power of x is three. For example, $f(x) = ax^3 + bx^2 + cx + d$; $a \neq 0$ and $a, b, c, d \in \mathbb{R}$ is a cubic function.

Reflect State what type of equation this would be if $a = 0$.

Investigation 7

On your GDC, plot the graphs of $y = f(x)$ for the following cubic functions.

a $f(x) = 2x^3$

b $f(x) = x^3 - 2x + 6$

c $f(x) = (x-5)^3$

d $f(x) = -3(x+1)^3$

e $f(x) = 2x^3 + x^2 + x + 1$

f $f(x) = 2x^3 - 3x^2 - 11x + 6$

g $f(x) = -x^3 - 4x^2 - x + 6$

Once you have plotted all of them, answer the following questions.

1. **Factual** State the number of turning points that different cubic equations have. Does every cubic equation have the same number of turning points?
2. **Conceptual** How do the number of turning points of a cubic function differ from the number of turning points of a quadratic function?
3. **Factual** Are the graphs of cubic functions symmetrical? If so, describe what type of symmetry they have.
4. **Factual** Do the graphs of cubic functions all have a maximum and minimum turning point?
5. **Factual** What are the number of roots that different cubic equations have? Does every cubic equation have the same number of roots?
6. **Factual** What are the parameters for a generic cubic function?
$$f(x) = ax^3 + bx^2 + cx + d; a \neq 0$$

7 By comparing each graph with its equation:
   a find which parameter gives the y-intercept
   b determine which parameter affects the width of the graph (try different values of this parameter to see if the graphs get wider or narrower)
   c state which parameter(s) you need to change, and how you need to change them, to reflect the graph in the y-axis.
   Hint: sketch $f(x) = 2x^3$ and $f(x) = -2x^3$, then $f(x) = x^3 + 3x^2 - x - 3$ and $f(x) = -x^3 - 3x^2 + x + 3$.

8 **Conceptual** From your discoveries in this investigation, which features of a graph do the parameters alter?

Not all cubic functions have a maximum and a minimum turning point. Some have neither. Also, cubic functions can have one, two or three roots.

Transferring the graph of a cubic function by hand from the GDC to paper

You need to be careful when copying the graph from your GDC onto paper.

First of all, you need to draw your axes and remember to label them and put the scale in. The x-values represent the **domain** and the y-values the **range**.

Make sure that any x- and y-intercepts are in the correct place on the graph.

Also, the maximum and minimum values need to be in the correct place.

You can also use the table of values on the GDC to plot the coordinates of some more points that lie on the curve if necessary.

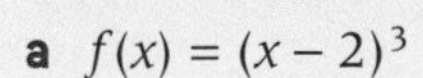

Example 10

Sketch the graphs of $y = f(x)$ for the following functions.

On your sketch, label the coordinates of points where the graphs intercept the axes, and any maximum or minimum points.

a $f(x) = (x - 2)^3$ **b** $f(x) = x^3 - 7x^2 + 4x - 12$

a

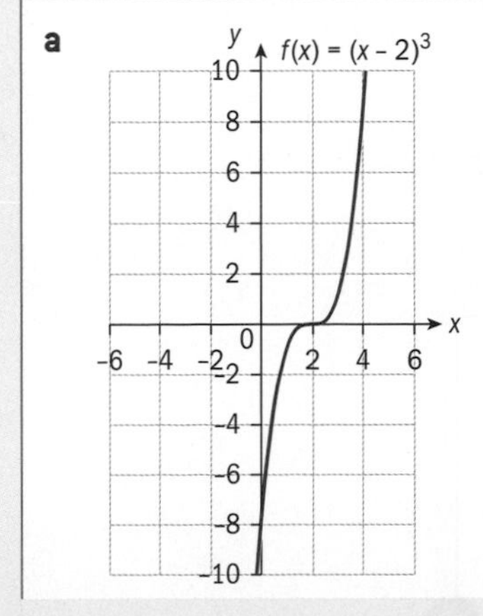

You can see from your GDC that this curve:

- does not have any maximum or minimum points
- cuts the y-axis at the point $(0, -8)$
- cuts the x-axis at $(2, 0)$.

Draw suitable axes and mark these points on.

You can take a few more points from the table of values to help you complete the graph.

eg $(-1, -27)$, $(1, -1)$, $(3, 1)$ and $(4, 8)$

Plot your points and draw a smooth line through the points.

Continued on next page

b

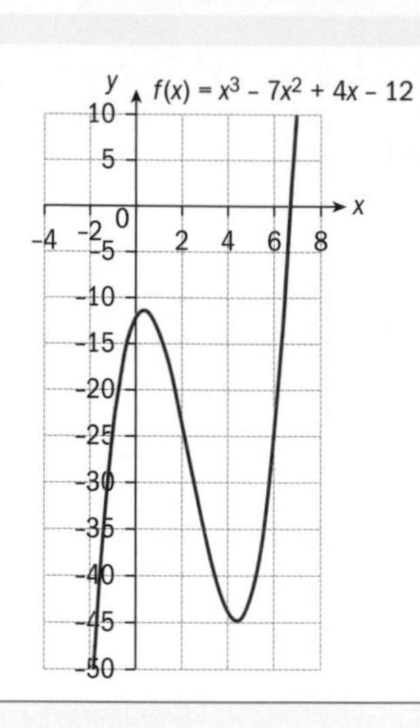

You can see from your GDC that this curve:

- has a maximum point at (0.306, −11.4)
- has a minimum point at (4.36, −44.7)
- cuts the y-axis at (0, −12)
- cuts the x-axis at (6.67, 0).

These four points are probably enough to help you sketch the curve, but you can look at other points from the table of values if you need to.

Draw suitable axes and mark these points on. Then sketch the curve.

Investigation 8

1. On your GDC, plot graphs of the following functions.
   a $y = 2x^3 + 1$ **b** $y = x^3 - 4x^2 - x + 4$
2. On each graph, perform the vertical line test and deduce whether or not the expressions in question **1** are functions—ie if all vertical lines cut the graph in one place only, then it is a function.
3. On each graph, perform the horizontal line test and deduce whether or not the expressions in question **1** are one-to-one—ie if all horizontal lines cut the graph in only one place then it is a one-to-one function.
4. **Factual** If a function is one-to-one then it has an inverse. Which of the graphs drawn have inverses?
5. Plot the line $y = x$ on the same axes.
6. **Conceptual** How is the graph and its inverse linked by the line $y = x$?
7. Can you plot a graph of the inverse function?

For a one-to-one cubic function $f(x)$, you can sketch the inverse function $f^{-1}(x)$ by reflecting the graph of $y = f(x)$ in the line $y = x$.

Example 11

Find the inverse of the following functions. Sketch each function with its inverse to confirm that they are symmetrical about the line $y = x$.

a $f(x) = x^3 - 2$ **b** $f(x) = 4(x+1)^3 + 5$

a $x \rightarrow \text{cubed} \rightarrow -2 = f(x)$

Performing the reverse operations:

$f^{-1}(x) = \text{cuberoot} \leftarrow +2 \leftarrow x$

So, $f^{-1}(x) = \sqrt[3]{(x+2)}$

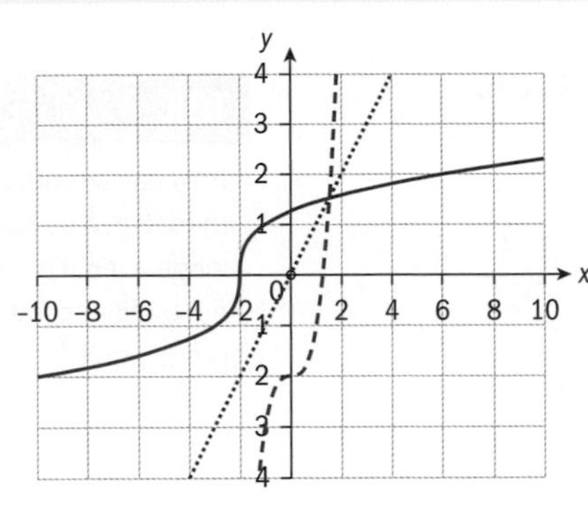

b $x \to +1 \to \text{cubed} \to \times 4 \to +5 = f(x)$

Performing the reverse operations:

$f^{-1}(x) = \leftarrow -1 \leftarrow \text{cuberoot} \leftarrow \div 4 \leftarrow -5 \leftarrow x$

So, $f^{-1}(x) = \sqrt[3]{\frac{(x-5)}{4}} - 1$

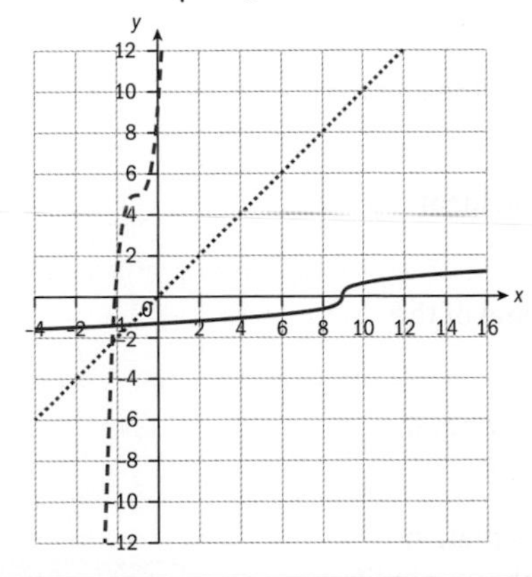

Exercise 9G

Sketch the graphs of $y = f(x)$ for the following functions.

Label the coordinates of points where the graphs intercept the axes, and any maximum or minimum points.

1 $f(x) = (x + 3)^3$

2 $f(x) = x^3 - 2x^2 - x + 3$

3 $f(x) = 2x^3 - 2x^2 - 12x$

4 $f(x) = 3(x + 2)^3 - 4$

5 $f(x) = 3x(x - 4)(x + 1)$

6 Plot each function and reflect it in the line $y = x$ to find the inverse function. Write down the equation of the inverse function in each case.

a $f(x) = x^3 + 3$ **b** $f(x) = 4x^3$

c $f(x) = 2x^3 + 1$

7 Sketch the graph of $f(x) = x^3 - 6x^2 + 3x + 10$.

a Find the coordinates of the x-intercepts and the y-intercept.

b Find the coordinates of the vertices.

c Write down the coordinates of the point of rotational symmetry.

d The graph is reflected in the y-axis. Write down the new equation.

Investigation 9

An open box is made from a piece of card measuring 12 cm by 10 cm, with squares of side x cm cut from each corner.

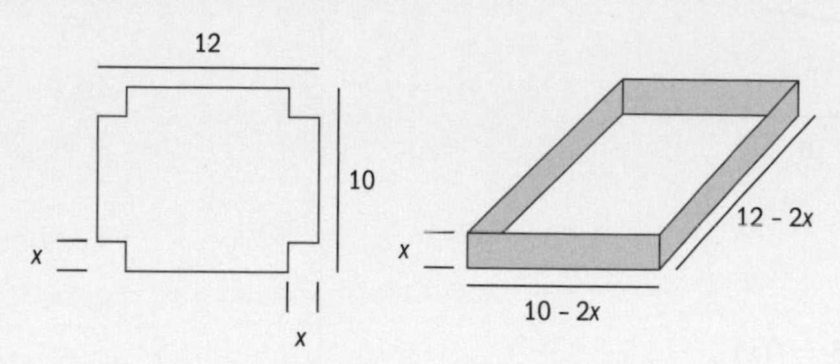

1 Explain why the width of the box is $(10-2x)$ cm, the length is $(12-2x)$ cm and the height is x cm.

2 Find the equation for the volume, V, of the open box in terms of x.

3 Plot a graph of V against x, using a suitable domain and range.

4 Find the x-intercepts.

5 **Factual** What are the upper and lower limits for the size of x in the context of this problem? Explain why.

6 **Factual** What are the coordinates of the local maximum and local minimum values of the graph?

7 **Factual** Which of these is not a possible value for the volume of the box? Justify your answer.

8 Given a certain value of x, could you use this model to predict what the volume of the box would be? What limitations would you have?

9 **Conceptual** Using your answer to question **8**, do you think that, in general, cubic models could be used to predict information about real-life situations?

TOK

How can a mathematical model give us knowledge even if it does not yield accurate predictions?

Example 12

A can of cat food has volume $V = 200\text{ cm}^3$.

The radius of the can is r cm and the height is h cm.

a Explain why $\pi r^2 h = 200$.

b Rearrange the equation from part **a** to make h the subject.

c Find an expression for the total surface area S of the can.

d Substitute your expression for h (from part **b**) into your expression for S (from part **c**) and hence show that $S = 2\pi r^2 + \dfrac{400}{r}$.

e Using suitable scales for your axes, plot the graph of this equation.

f Find the minimum surface area of the can and the value of r for which this occurs.

a A can of cat food can be modelled as a cylinder. The volume of a cylinder is given by $V = \pi r^2 h$. Since we are told that $V = 200$, it follows that $\pi r^2 h = 200$.	
b $h = \dfrac{200}{\pi r^2}$	Divide through by πr^2.
c $S = 2\pi r^2 + 2\pi rh$	$S = (2 \times \text{area of circle at end}) + (\text{curved surface area})$
d $S = 2\pi r^2 + 2\pi r\left(\dfrac{200}{\pi r^2}\right)$ $S = 2\pi r^2 + \dfrac{400}{r}$	Simplify the second term.
e 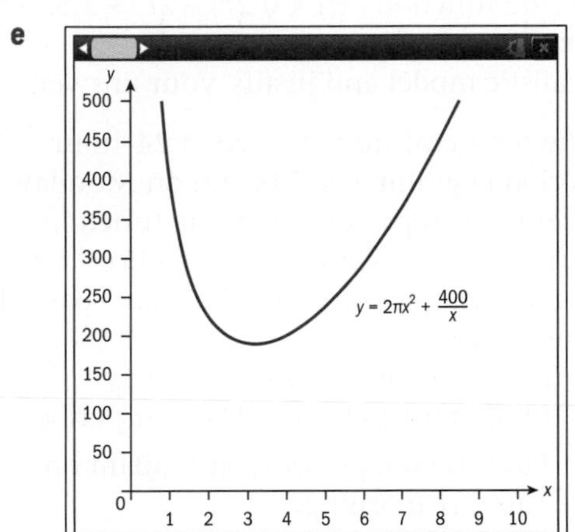	
f $189.3\,\text{cm}^2$ when $r = 3.17$	This is the minimum point on the graph.

You always need to keep in mind the following diagram when finding the best-fit curve for messy data. Sometimes it may be better to use a linear, cubic or other power function to find the best model for the data.

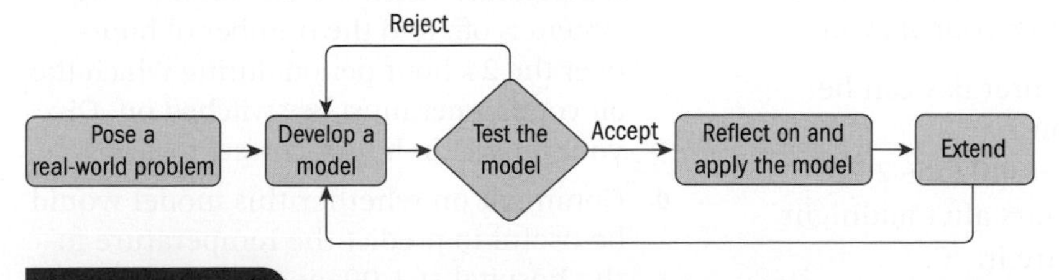

Exercise 9H

1 An open-top box is made from a piece of card measuring 8 cm by 6 cm with squares of side x cm cut out of the corners.

a Find an expression for the volume V of the open box in terms of x.

b Plot a graph of V against x.

c Find the x-intercepts.

d State what the x-intercepts tell you about the possible range of values for x.

e Find the local maximum and minimum values of V.

f Explain why, in real life, V cannot be equal to one of the values you found in part **e**.

2 A section of a toy rollercoaster is in the shape of a cubic curve with equation $f(x)=\frac{1}{6}x^3-2x^2+5\frac{5}{6}x+5$, where $f(x)$ is the height in metres.

- **a** Plot the graph of $y=f(x)$.
- **b** Find the local maximum and minimum values of the function.
- **c** Calculate the difference in height between the maximum and minimum points.

3 The number of fish, N, in a lake between 2005 and 2025 is modelled using the formula

$$N(x)=-0.025x^3+0.8x^2-6.75x+60$$

where x is the number of years after 2005.

- **a** Plot the graph of $N(x)$ for $0 \le x \le 20$.
- **b** Find the number of fish in the pond after 12 years.
- **c** Find the minimum number of fish in the lake.

4 A pandemic can be modelled by the equation $f(x)=(x-15)^3+4000$ where x is the number of weeks after the outbreak started and $f(x)$ is the number of cases reported.

- **a** Plot the graph of $f(x)$ between $x=0$ and $x=50$.
- **b** Does the graph have a local maximum or minimum value?
- **c** Find the number of cases when $x=20$ and when $x=40$.
- **d** State whether this is a suitable model for the scenario, and justify your answer.

5 The temperature on a winter day can be modelled by the equation $f(t)=-0.0066t^3+0.16t^2+0.025t-2.76$ where t is the time in hours after midnight and $f(t)$ is the temperature in °C.

- **a** Plot the graph of $f(t)$ between $x=0$ and $x=24$.
- **b** Find the local maximum and minimum temperatures.
- **c** Find the temperature at 10 am.

6 The number of bacteria, B, in a particular culture at time x minutes is conjectured to be given by the formula

$B(x)=5.17+1.25x-0.06x^2+0.00083x^3$

- **a** Write down the initial number of bacteria.
- **b** Find the local maximum and minimum number of bacteria.
- **c** State whether this is a suitable model or not and justify your answer.

7 The height, d cm, that a toy rollercoaster reaches after t seconds can be modelled by the function $d(t)=0.2t^3-2t^2+5.8t+5$. Explain whether or not this is a realistic model and justify your answer.

8 The temperature, x°C, over a 24-hour period beginning at 7.00 pm on Monday evening is represented by the function $x(t)=23.5-1.72t+0.2t^2-0.0056t^3$ where t is the number of hours that have passed since 7.00pm.

- **a** State the highest and lowest temperatures in this 24-hour period.
- **b** Find the temperature at 5.00am on Tuesday morning.

Regulations state that a hospital should be at a maximum temperature of 22°C.
If the temperature rises above this, the air conditioning system should be switched on.

- **c** Assuming that the temperature in the hospital is the same as the outside temperature when the air conditioning system is off, find the number of hours over the 24-hour period during which the air conditioner must be switched on. Give your answer in hours, correct to 1 dp.
- **d** Comment on whether this model would be useful to predict the temperature in the hospital at 1.00am on Wednesday morning. Justify your answer.

Finding points of intersection

Example 13

At the school sports day, Petra has to throw a shot-put. The path of the shot-put follows a parabolic curve given by the equation $f(t) = 1.75 + 0.75t - 0.0625t^2$, where t is the time (in seconds) that has elapsed since the shot-put left Petra's hand, and $f(t)$ is the height of the shot-put (in metres) above the ground.

a Find the height of the shot-put after 5 seconds.

b Find the times when the height of the shot is 3 metres.

a $f(5) = 1.75 + 0.75(5) - 0.0625(5)^2$

$= 1.75 + 3.75 - 1.5625$

$= 3.9375\text{ m}$

Substitute $t = 5$ into the equation for $f(t)$.

b

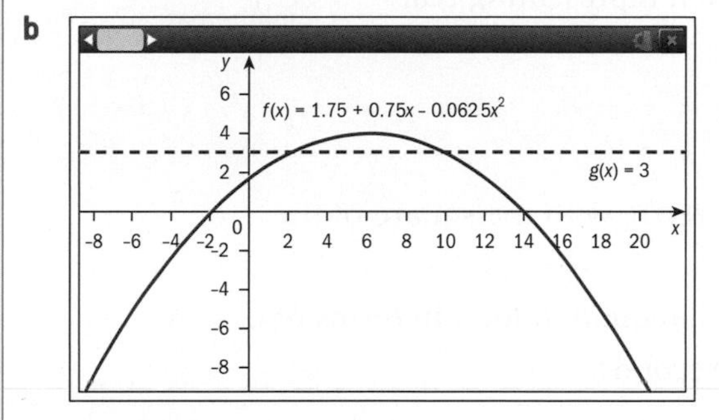

$x = 2$ and $x = 10$

The shot is 3 m above the ground when $t = 2$ s and $t = 10$ s.

The easiest way to find the times when the height of the shot is 3 metres is to use your GDC.

Graph the functions $f(x) = 1.75 + 0.75x - 0.0625x^2$ and $g(x) = 3$.

The values of x where $f(x) = g(x)$ give the times, t, when the height of the shot is 3 m.

Exercise 9I

1 Plot graphs of $y = f(x)$ and $y = g(x)$ for the following pairs of functions, and find the coordinates of the point where each pair intersect.

a $f(x) = x^5 - x^3 + x$ and $g(x) = 3 - x$

b $f(x) = 2x^{-2} + 4$ and $g(x) = 6$

c $f(x) = x^3 - 2x^2 + 1$ and $g(x) = 3x + 1$

2 The path of a ball is modelled by the equation $f(x) = -3.53x^2 + 13.7x + 1.75$ where $f(x)$ is the height of the ball in metres and x is the time in seconds.

Find the times when the height of the ball is 8 metres.

3 The path of an arrow can be modelled by the quadratic function $f(t) = -4.34t^2 + 25.77t + 1.8$ where t is the time in seconds and $f(t)$ is the height in metres.

Find the times when the height is 10 metres.

4 A company's profits, in US \$, are modelled by the equation $f(x) = -0.9x^2 + 52x - 360$, where x is the number of units sold each week.

Find the number of units sold to make a profit of US \$300 per week.

TOK

Does studying the graph of a function contain the same level of mathematical rigour as studying the function algebraically?

Direct and inverse variation

In Chapter 5 you learned about direct variation.

In this section you will be looking further at direct variation, and also at inverse variation.

The symbol for variation is $\propto$ and it stands for "$= k \times \ldots$" (that is, k multiplied by the variable which follows).

For direct variation, $y \propto x^n$ or $y = kx^n$

For inverse variation, $y \propto \frac{1}{x^n}$ or $y = \frac{k}{x^n}$

International-mindedness

Which do you think is superior: the Bourbaki group analytical approach or the Mandelbrot visual approach to mathematics?

In the following examples and exercises you will find out that variation provides an alternative to the use of functions when representing real-life situations.

Example 14

The distance, d metres, that a rock falls varies directly with the square of the time taken, t seconds.

a If the rock falls 6 metres in 2 seconds, write an equation for d in terms of t.

b Find the distance the rock has fallen after 5 seconds.

a $d \propto t^2$ So, $d = kt^2$ $6 = k(2)^2$ $k = 1.5$ The equation is $d = 1.5t^2$.	The first step is always to write the statement with the variation symbol and then write it as an equation involving "k". Notice that this gives us a quadratic equation for d in terms of t. Substitute the given values into the equation to find k.
b $d = 1.5(5)^2 = 1.5(25) = 37.5$ metres	Substitute $t = 5$ into your equation and find the distance d.

Exercise 9J

1 The price of a taxi fare, $\$p$ AUD, varies directly with the number of kilometres, n, travelled. The fare for travelling 12 kilometres is \$21 AUD. Write an equation connecting p and n, and hence find the fare for travelling 40 kilometres.

2 The distance, d metres, that a ball rolls down a slope varies directly with the square of the time, t seconds, it has been rolling for. In 2 seconds the ball rolls 9 metres.

a Write an equation connecting d and t.

b Find how far the ball rolls in 5 seconds.

c Find the time it takes for the ball to roll 26.01 metres.

3 The volume of a sphere varies directly with the cube of its radius. A certain sphere has radius 3 cm and volume 113.1 cm^3.

Find the volume of another sphere with radius 5 cm.

Investigation 10

1 A local authority pays their workers depending on the number of hours that they work each week. If the workers are paid €22 per hour, complete the following table:

Number of hours	20	25	30	35	40
Pay in €					

Plot a graph of this information on your GDC.

Describe how a worker's pay varies with the number of hours worked.

2 The local authority has decided to put artificial grass tiles on a football field.

If four people are available to lay the grass tiles, it takes them two hours to complete the work.

Fill in this table showing the number of people available and the number of hours it takes to complete the work.

Number of people	1	2	4	6	8	12
Number of hours			2			

Plot a graph of this information on your GDC.

3 **Factual** How do the number of hours to complete the work vary with the number of people available?

4 **Conceptual** For problems which involve direct and inverse variation, how does understanding the physical problem help you to choose the correct mathematical function to model the problem with?

Example 15

The number of hours N taken to build a wall varies inversely with the number of people x who are available to work on it.

a When three people are available the wall takes two hours to build. Find the time it takes to build the wall when four people are available to work on it.

b Given it takes three hours to build the wall, state how many people worked on it.

a $N \propto \frac{1}{x}$ and hence $N = \frac{k}{x}$	For inverse variation, the variable is written as $\frac{1}{x}$ or x^{-1} which is a power function.
$N = 2$ when $x = 3$ so $2 = \frac{k}{3}$ $k = 6$	Find the value of k using the given information.
$N = \frac{6}{x}$	Substitute the value of k you found into the equation for N.
$N = \frac{6}{4} = 1.5$	

Continued on next page

So, it takes 1.5 hours to build the wall when four people work on it.	Now find N when $x = 4$.
b $3 = \frac{6}{x}$ $3x = 6$ $x = \frac{6}{3} = 2$ So, two people were available to build the wall.	Substitute $N = 3$ into the equation $N = \frac{6}{x}$ which you found in part **a**.

Reflect Explain why direct variation and inverse variation are related to modelling real-life scenarios with mathematics.

Exercise 9K

1 In a lottery draw, the amount of money US $\$m$ awarded to each winner varies inversely with the number of winners w.

If there are 20 winners, each receives US $150.

a Write an equation connecting m and w.

b Determine how much each of 15 winners would receive in prize money.

2 The number of hours h that it takes to build a model varies inversely with the number of people p who work on it.

It takes two hours for six people to build the model.

a Write an equation connecting h and p.

b Find out how long it will take 10 people to build the model.

c The model needs to be ready for an exhibition which will take place in four hours' time. Calculate the number of people needed to build the model to ensure it is ready on time.

3 The variable y varies inversely with the square of x. $y = 3$ when $x = 4$.

a Write an equation connecting y and x.

b Find y when $x = 6$.

c Find x when $y = 12$.

4 The volume V of a gas varies inversely with the pressure p of the gas.

The pressure is 20 Pa when its volume is $180\,\text{m}^3$.

Find the volume when the pressure is 90 Pa.

5 A group of children attend a birthday party. The number of pieces of candy c that each child receives varies inversely with the number of children n.

You are given that when there are 16 children, each one receives 10 pieces of candy.

Find the number of pieces of candy each child receives when there are 20 children.

6 The intensity of a sound wave varies inversely with the square of the distance you are standing away from it.

a Given that you are initially standing 25 m away from the sound source, and then move so that you are 50 m away, find the factor by which the intensity of the sound has decreased.

b You then move from 50 m away, to only 40 m away. Determine how the sound intensity has changed on this occasion.

7 The variable y varies inversely with the square root of x.

When $x = 16$, $y = 3$.

a Find the value of y when $x = 4$.

b Find the value of x when $y = \frac{3}{2}$.

Developing inquiry skills

Now that you have learned different ways to model data from real-life situations, what type of function do you think would be the best way to model the path of Kazuki's basketball?

Can you find sufficient data points to model the path of the ball?

Can you use the official height of a basket to help find points?

Is it possible to find a general equation that will fit all possibilities?

Functions

9.4 Optimization

Investigation 11

A gardener has 24 metres of fencing to make a vegetable plot.

1. If the length of the plot is x metres and the width is y metres, show that $y = 12 - x$.
2. Find an expression for the area A of the vegetable plot in terms of x.
3. Plot a graph of A against x on your GDC.
4. **Factual** What is the maximum value of A?
5. **Factual** What dimensions of the plot give the maximum area?
6. **Factual** Why are the dimensions you found in question **5** described as the optimum solution?
7. **Conceptual** What does it mean to optimize a real-life situation? Is the optimal result always a maximum?

Different models can be used to optimize real-life problems.
When you are finding an optimal value, it can be a maximum value or a minimum value.

Example 16

Find the minimum surface area of a closed cylinder with volume 500 cm³.

$V = \pi r^2 h$ $500 = \pi r^2 h$ (1) $h = \dfrac{500}{\pi r^2}$	Rearrange the volume formula to make h the subject.
$A = 2\pi r^2 + 2\pi rh$ (2)	Surface area is the curved surface area plus the area of the top and bottom circles.

Continued on next page

$A = 2\pi r^2 + 2\pi r\left(\frac{500}{\pi r^2}\right)$ $A = 2\pi r^2 + \frac{1000}{r}$	Substitute (1) into (2) and simplify.
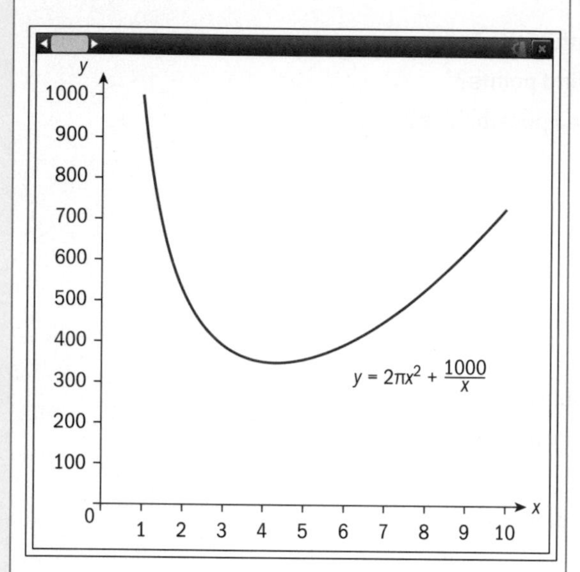	Use your GDC to graph this equation.
The minimum surface area is $348.7\,\text{cm}^3$ and this occurs when $r = 4.30$ cm.	Find the minimum value for r and A from your graph using the "minimum" function on your GDC.

Exercise 9L

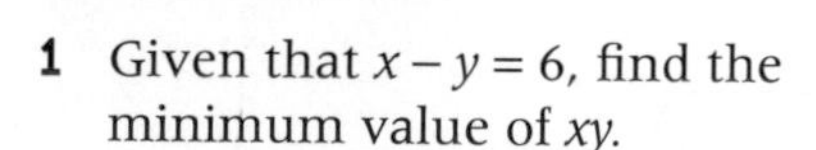

1 Given that $x - y = 6$, find the minimum value of xy.

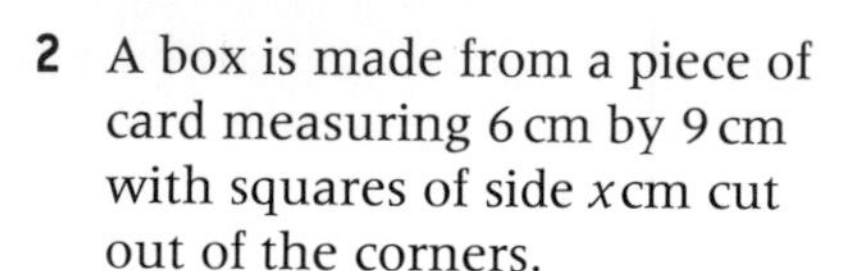

2 A box is made from a piece of card measuring 6 cm by 9 cm with squares of side x cm cut out of the corners.

a Show that the formula for the volume, $V\,\text{cm}^3$, is $V = 4x^3 - 30x^2 + 54x$.

b Plot a graph of the function and hence find the maximum volume.

3 Akshat owns a cake shop.

On a single day, the profit US P that Akshat makes is dependent on the number of cupcakes c that he bakes.

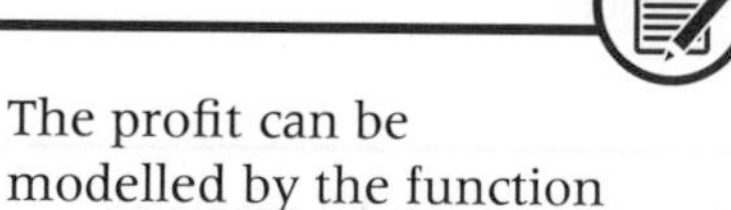

The profit can be modelled by the function $P(c) = -0.056c^2 + 5.6c - 20$.

Plot the graph of P against c. Find Akshat's maximum daily profit, and the number of cakes he needs to bake in order to make a maximum profit.

4 A company's profits, US f, are modelled by the equation $f(x) = -0.9x^2 + 52x - 360$, where x is the number of units sold each week.

Find the maximum profit.

TOK

How much do you agree with the following statement?

"Any technological influence is just a matter of speed and the quantity of data which can be processed."

5 The number of bacteria B, in millions, living in a particular culture after time x minutes is given by the formula

$$B(x) = 5.17 + 1.25x - 0.06x^2 + 0.00083x^3$$

Find the maximum and minimum number of bacteria in the culture.

6 The total surface area of a closed cylinder is $5000\,\text{cm}^2$. Find the dimensions of the cylinder which will maximize its volume, and state this maximum volume.

Developing your toolkit

Now do the Modelling and investigation activity on page 458.

Chapter summary

- Quadratic functions are polynomial functions where the highest power of x is two. For example, $f(x) = ax^2 + bx + c$; $a \neq 0$ and $a, b, c \in \mathbb{R}$, is a quadratic function.
- The maximum or minimum turning point on the graph of a quadratic function is called the **vertex**.
- The x-intercepts are also called the **zeros of the function**, because they are the x-values where $y = f(x) = 0$.
- You can find the coordinates of the x-intercepts by looking at where the graph of the function cuts the x-axis.

  You could also find the x-intercepts by using the quadratic formula:

$$x = \frac{-b \pm \sqrt{b^2 - 4ac}}{2a}$$

- The y-intercept occurs where $y = f(0)$.
- The line of symmetry of a quadratic graph is called the **axis of symmetry**.
- If the x-coordinates of the x-intercepts are x_1 and x_2 then the equation of the axis of symmetry is

$$x = \frac{x_1 + x_2}{2}$$

- Many-to-one functions do not have inverses.
- A quadratic function is a many-to-one function.
- For the function $f(x) = ax^2 + bx + c$; $a \neq 0$, a is the coefficient of x^2, b is the coefficient of x, and c is the constant.

  a, b and c are called the parameters of the function.

  The parameters of a quadratic function determine whether the vertex is a maximum or minimum, and the value of the y-intercept.
- Suppose $(m, 0)$ and $(n, 0)$ are the x-intercepts of the quadratic graph $y = f(x)$.

  The vertex of the graph has coordinates $\left(\frac{m+n}{2}, f\left(\frac{m+n}{2}\right)\right)$.
- Two different ways that you can write a quadratic equation are:

  General form: $f(x) = ax^2 + bx + c, a \neq 0$

  Intercept form: $f(x) = a(x - m)(x - n), a \neq 0$

Continued on next page

- When you are given the graph of a quadratic function, you can find the function by entering the coordinates of three points on the graph into your GDC and then using quadratic regression.
- Cubic functions are polynomial functions where the highest power of x is three.

  For example, $f(x) = ax^3 + bx^2 + cx + d$; $a \neq 0$ and $a, b, c, d \in \mathbb{R}$ is a cubic function.
- Not all cubic functions have a maximum and a minimum point. Some have neither. Also, cubic functions can have one, two or three roots.
- For a one-to-one cubic function $f(x)$, you can sketch the inverse function $f^{-1}(x)$ by reflecting the graph of $y = f(x)$ in the line $y = x$.
- A power function is a function of the form $f(x) = kx^n$ where $k, n \in \mathbb{R}$.

  The symbol for variation is $\propto$ and it stands for "$= k \times \ldots$" (that is, k multiplied by the variable which follows).

  For direct variation, $y \propto x^n$ or $y = kx^n$

  For inverse variation, $y \propto \frac{1}{x^n}$ or $y = \frac{k}{x^n}$
- Different models can be used to optimize real-life problems.
- When you are finding an optimal value, it can be a maximum value or a minimum value.

Developing inquiry skills

Oliver throws the basketball up and it passes through the hoop on its way down.

Describe how you could plot some points on the basketball's path of motion and use them to find a best-fit function to model the path.

Comment on whether it would be more suitable to model the path of Oliver's basketball using a quadratic or a cubic function. Show your working and justify your answer.

How many quadratic functions could be used to model different paths between Oliver and the hoop, given that the ball successfully passes through the hoop?

List the factors which might determine the particular path he chooses to throw the ball on.

Describe any limitations to modelling the path of the ball in this way.

Chapter review

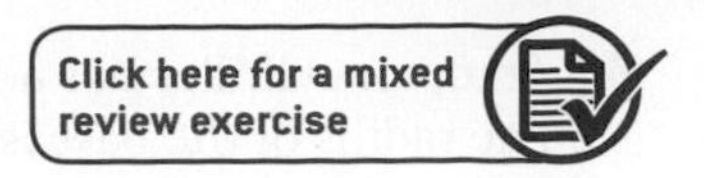

1 On paper, sketch the graphs of:

a $f(x) = x^2 - 2$

b $f(x) = x^2 - 5x + 4$.

On your sketches, label the coordinates of the vertex and any axes intercepts.

2 The perimeter of a picture is 400 cm.

a The length of the picture is x cm. Find the height in terms of x.

b Find an equation for the area, $A\,\text{cm}^2$, of the picture.

c On paper, sketch this graph using a suitable domain and range.

d Find the x-intercepts and explain what these represent.

3 For the equation $f(x) = x^2 + 6x - 7$ find:

a the values of a, b and c

b the coordinates of the y-intercept

c the coordinates of the x-intercepts

d the equation of the axis of symmetry

e the coordinates of the vertex.

4 Mimi throws a javelin. The path of the javelin is a parabola and is modelled by the equation $f(x) = -0.008\,16x^2 + 0.372x + 1.8$, where $f(x)$ is the height of the javelin in metres and x is the horizontal distance in metres.

a On paper, sketch the graph of the path that the javelin flies.

b Find the height of the javelin when $x = 20$ metres.

c Find the maximum height of the javelin.

d Find the point where the graph of the function crosses the x-axis and explain what this represents.

5 Anmol throws a stone in the air. The height of the stone, $h(t)$ metres, at time t seconds is modelled by the equation $h(t) = -2.262\,5x^2 + 8.575x + 1.9$.

a Find the y-intercept and explain what this represents.

b Find the maximum height of the stone.

c Find the rate of change in metres per second between $x = 1.8$ and $x = 1.9$ and decide whether the rate of change is increasing or decreasing.

d Find the time when the stone lands back on the ground.

6 For the following equations, find:

i the coordinates of the x-intercepts

ii the equation of the axis of symmetry

iii the coordinates of the vertex.

a $f(x) = 3(x - 2)(x - 4)$

b $f(x) = 4(x + 1)(x - 5)$

7 Find the equations of the following quadratic graphs.

a

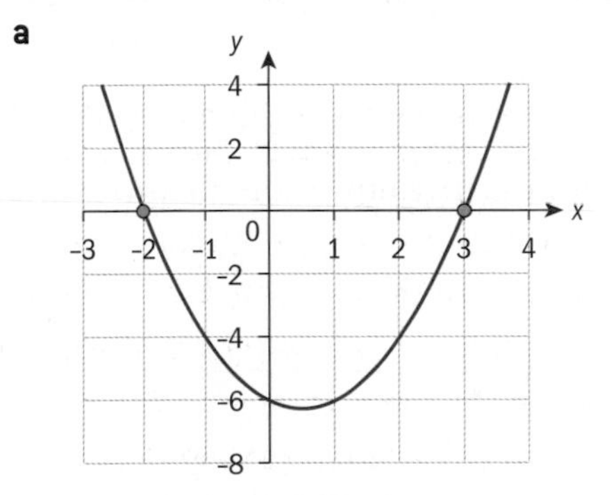

b

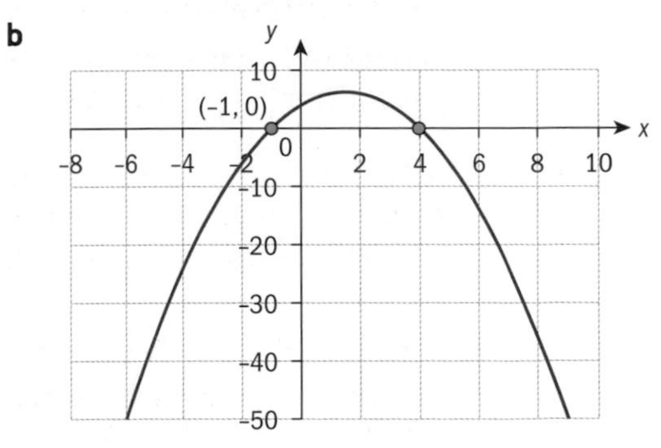

c

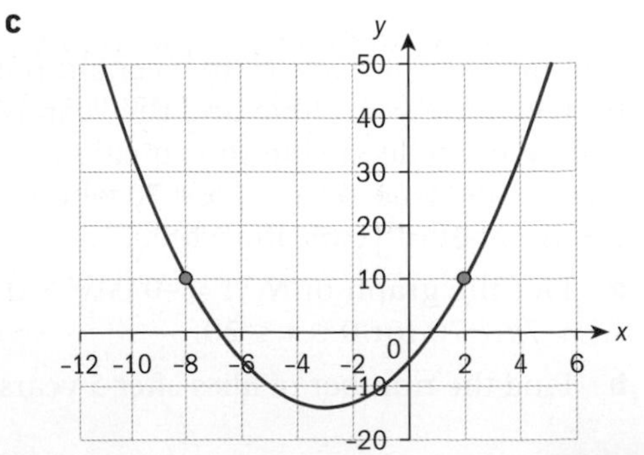

Functions

8 An open cylinder has a volume of 400 cm^3. The radius of the base is r cm and the height is h cm.

- **a** Explain why $\pi r^2 h = 400$.
- **b** Rearrange the equation in part **a** to make h the subject.
- **c** Write down an expression for the surface area, A, of the open cylinder.
- **d** Show that this can be written as

$$A = \pi r^2 + \frac{800}{r}$$

- **e** Plot the graph of $A = \pi r^2 + \frac{800}{r}$.
- **f** Find the minimum area and the value of r when this occurs.

9 Sketch the following cubic functions on paper, showing clearly the coordinates of any x- and y-intercepts.

- **a** $f(x) = 2x^3 - 1$
- **b** $f(x) = (x - 1)(x + 1)(x - 3)$

10 Find the equations of the following graphs.

a

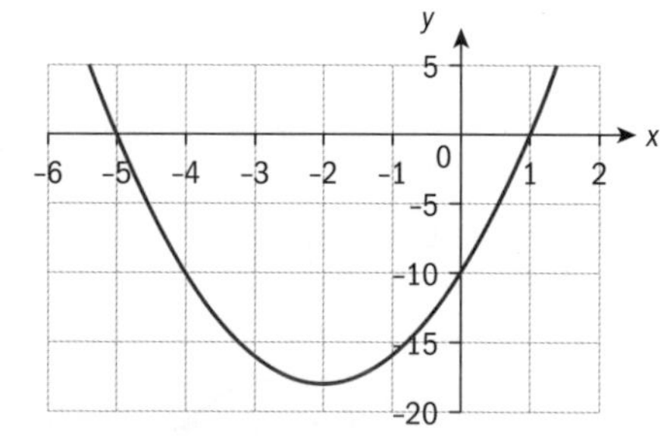

b

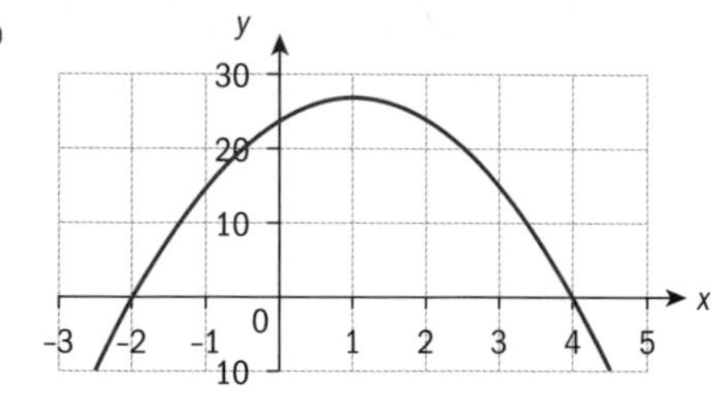

11 The number of lilies, N, in a certain pond between the year 2000 and the year 2020 can be modelled using the equation $N(x) = -0.04x^3 + 0.9x^2 - 7x + 70$ where x is the number of years after 2000.

- **a** Plot the graph of $N(x) = -0.04x^3 + 0.9x^2 - 7x + 70$ for $0 \le x \le 20$.
- **b** Find the number of lilies after 5 years.
- **c** Find the number of lilies after 12 years.
- **d** Find the maximum number of lilies and the year that this occurs.
- **e** Find the minimum number of lilies and the year that this occurs.
- **f** Find when there are 60 lilies in the pond.

12 The height of the handrail of a small bridge can be modelled by the equation $f(x) = 0.267x - 0.00889x^2$ where $f(x)$ is the height of the handrail above the walkway in metres, and x is the horizontal distance from the start of the bridge in metres.

The local authority wants to put a horizontal line of decorative lights at a height of 1 metre above the walkway. The two ends of the lights will be attached to the handrail.

- **a** Sketch the graph modelling the handrail, $f(x) = 0.267x - 0.00889x^2$ for $0 \le x \le 30$ and $0 \le y \le 5$.
- **b** Sketch the graph of $y = 1$ on the same screen.
- **c** Find the coordinates of the points where the lights are attached to the handrail.
- **d** Find the length of the lights needed.

13 The distance, d km, that a train travels varies directly with the time, t hours.

- **a** Given that the train travels 100 km in 1.25 hours, write an equation connecting d and t.
- **b** Find the distance the train has travelled after 2 hours.
- **c** Find how long it takes for the train to travel 300 km.

14 The number of slices of pizza, p, that a child receives varies inversely with the number of children, n, sharing it.

- **a** Given that two children receive six slices each, write an equation connecting p and n.
- **b** Find the number of slices that three children receive.
- **c** Find the number of children if each receives three slices.

Exam-style questions

15 P1: Boris plays a game of tennis. He serves from the base line to his opponent, Steffi.

The path of the ball may be modelled by the quadratic curve $H(x) = 2.103 + 0.1455x - 0.01932x^2$ for $0 \le x \le X$.

a Find the maximum height of the ball during its motion. (2 marks)

b The height of the net is 1.07 m. Boris is standing on the baseline of the court, at a horizontal distance of 11.98 m from the net. Show that the ball just passes over the net. (2 marks)

c Suggest a maximum value for X and explain why this would be a sensible value to take. (3 marks)

16 P1: The gravitational force (in Newtons), exerted on an object by the Earth, varies inversely with the square of the object's distance from the centre of the Earth.

The radius of the Earth may be estimated to be 6370 km, and an astronaut with a mass of 100 kg weighs approximately 980 N on the Earth's surface.

Find the approximate weight of the astronaut when he is 11 km above the Earth's surface. (4 marks)

17 P2: Mannie starts working as a police constable at the age of 20. He works for 40 years and retires at the age of 60. His annual salary, P (in thousands of UK pounds), is given by the formula $P(t) = 0.0045(t - 30)^3 + 0.1215(t - 30)^2 + 0.3585t + 23$, where t is the number of years Mannie has been working for the police force.

a Find Mannie's starting salary. (2 marks)

b Find Mannie's maximum salary over the course of his working life (2 marks)

c Find the times (to the nearest whole number of years) when Mannie's salary was UK£35 000. (3 marks)

d Over what proportion of his working life was Mannie's salary increasing? (3 marks)

Functions

18 P1: The following table illustrates the mean daily temperature for the city of Paphos in Cyprus over the course of 12 months.

Month	1	2	3	4	5	6	7	8	9	10	11	12
Temperature (°C)	13	13	14	17	20	23	25	26	24	22	18	15

a Find the best fit quadratic equation through these points. (3 marks)

b Using your equation, estimate for how long the average temperature lies above 16°C. (2 marks)

c By comparing your quadratic equation with the given data, suggest two reasons why a quadratic model may be inaccurate for analysing this particular weather pattern. (2 marks)

19 P1: The population P of ferrets in a ferret sanctuary after t months is given by $P(t) = 21 + 2.91t - 0.087t^2 + 0.0007t^3$.

a Sketch the graph of P against t for the first 80 months. (2 marks)

b Find the maximum ferret population during the first two years. (1 mark)

c Find the time(s) when the ferret population is under 40. (4 marks)

20 P1: The length (l) of a violin string varies inversely with the frequency (f) of its vibrations.

A violin string 13 cm long vibrates at a frequency of 400 Hz.

Find the frequency of a 10 cm violin string. (4 marks)

Click here for further exam practice

Hanging around!

Approaches to learning: Thinking skills: Create, Generating, Planning, Producing
Exploration criteria: Presentation (A), Personal engagement (C), Reflection (D)
IB topic: Quadratic modelling, Using technology

Investigate

Hang a piece of rope or chain by its two ends. It must be free hanging under its own weight. It does not matter how long it is or how far apart the ends are.

What shape curve does the hanging chain resemble?

How could you test this?

Import the curve into a graphing package

A graphing package can fit an equation of a curve to a photograph.

Take a photograph of your hanging rope or chain.

What do you need to consider when taking this photo graph?

Import the image into a graphing package.

Carefully follow the instructions for the graphing package you are using.

The image should appear in the graphing screen.

Fit an equation to three points on the curve

Select three points that lie on the curve.

Does it matter which three points you select?

Would two points be enough?

In your graphing package, enter your three points as x- and y-coordinates.

Now use the graphing package to find the best fit quadratic model to your three chosen points.

Carefully follow the instructions for the graphing package you are using.

Test the fit of your curve

Did you find a curve which fits the shape of your image exactly?

What reasons are there that may mean that you did not get a perfect fit?

The shape that a free-hanging chain or rope makes is actually a **catenary** and not a parabola at all. This is why you did not get a perfect fit.

Research the difference between the shape of a catenary and a parabola.

International-mindedness

The word "catenary" comes from the Latin word for "chain".

Extension

Explore one or more of the following—are they quadratic?

The cross section of a football field

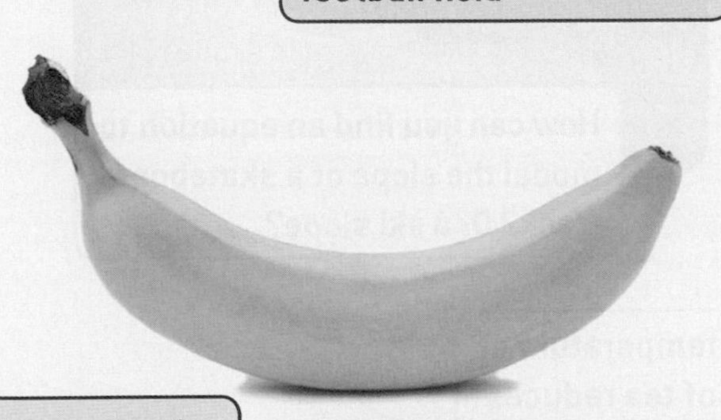

The curve of a banana

The path of a football when kicked in the air—here you would need to be able to use available software to trace the path of the ball that is moving

A well-known landmark—perhaps the Sydney Harbour Bridge or the arches at the bottom of the Eiffel Tower.

Other objects that look like a parabola—for example, the arch of a rainbow, water coming from a fountain, the arc of a satellite dish.

10 Modelling rates of change: exponential and logarithmic functions

The reduction in temperature of a hot drink, the slope of a skateboard track, compound interest, depreciation of the value of a car, even the half-life of radioactive decay all give rise to curves. This chapter looks at ways of modelling these curves with equations in order to predict various outcomes.

Concepts

- Change
- Modelling

Microconcepts

- Exponential functions
- Horizontal asymptotes
- Geometrical sequences and series
- Common ratio
- Compound interest
- Annual depreciation
- Annuity and amortization
- Growth and decay
- Logarithms

How long will it take you to become a millionaire if you receive US$1 the first month, US$2 the second month, US$4 the third month, US$8 the fourth month and so on?

How can you find an equation to model the slope of a skateboard track? Or a ski slope?

How can you work out which rate gives you the most interest on your savings?

If the temperature of a cup of tea reduces at a given rate, will it ever reach 0°C?

How can you work out the monthly payments on a loan?

Since the Olympic Games began, the height that sportsmen and women have been able to jump in the pole vault has increased. Can you predict the smallest height that no athlete will ever be able to jump over?

What information will you need?

What assumptions will you have to make?

How will you be able to construct a model?

Will this be different for men and women? If so, why?

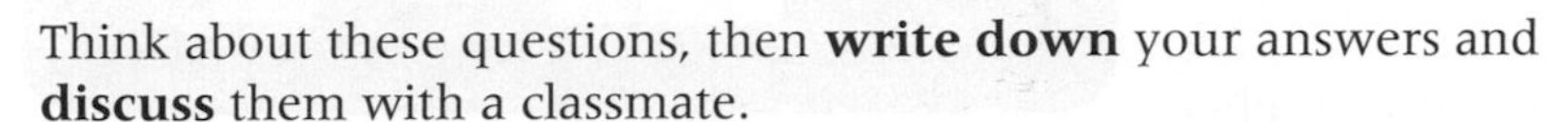

Think about these questions, then **write down** your answers and **discuss** them with a classmate.

What further information do you think that you will need? What other things do you notice? What do you wonder?

Developing inquiry skills

Think about the questions you would need to ask to model other sorts of activity and achievements, like the shortest time for a marathon or the longest long jump.

Before you start

Click here for help with this skills check

You should know how to:	Skills check
1 Laws of exponents eg $2^3 \times 2^4 = 2^7$	**1** Find the values of: **a** $4^2 \times 4^9$ **b** $\frac{5^8}{5^2}$.
2 Percentages eg 6% of 24 $= \frac{6}{100} \times 24 = 1.44$	**2** Find the values of: **a** 3% of 24 **b** 28% of 150.
3 Sigma notation $\sum_{n=1}^{5} n^2 = 1^2 + 2^2 + 3^2 + 4^2 + 5^2$ $= 1 + 4 + 9 + 16 + 25$ $= 55$	**3** Find the values of: **a** $\sum_{i=1}^{4} 2^i$ **b** $\sum_{k=1}^{6} (k+3)$.

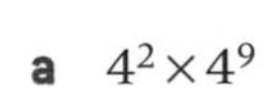

10.1 Geometric sequences and series

The game of chess was invented in India by a man named Sissa ibn Dahir. The king, Shihram, was so pleased with the game that he offered Sissa any reward that he wanted. Sissa said that he would take this reward: the king should put one grain of wheat on the first square of a chessboard, two grains of wheat on the second square, four grains on the third square, eight grains on the fourth square, and so on, doubling the number of grains of wheat with each square.

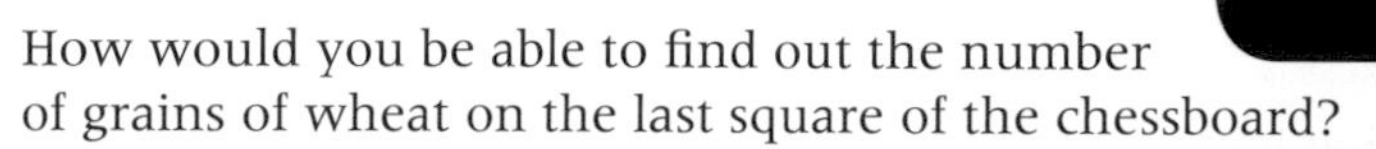

How would you be able to find out the number of grains of wheat on the last square of the chessboard?

How would you be able to find the total number of grains of wheat on the chessboard?

This is one example of a geometric sequence.

A tennis competition involves 256 players. Each game has two players and there is only one winner of each game. In the first round all 256 players will take part, in the second round only the winners from the first round will take part and so on. How can you find out the total number of rounds that take place before there is a winner? How can you find out the total number of games that are played in the competition?

This is also an example of a geometric sequence.

> A sequence of numbers in which each term can be found by multiplying the preceding term by a **common ratio** is called a geometric sequence.
> A geometric series is the sum of the terms of a geometric sequence.

International-mindedness

The Elements, Euclid's book from 300BC, contained geometric sequences and series.

Investigation 1

For each of the following sequences, find the next term and describe the rule for finding the next term.

a 2, 4, 6, 8, ... **b** 1, 3, 9, 27, ... **c** 1, 1, 2, 3, 5, ...

d −3, −1.5, 0, 1.5, ... **e** 1, 4, 9, 16, ... **f** 5, 15, 45, 135, ...

Conceptual How can you tell whether a sequence is geometric or not? Which of these sequences are geometric?

Investigation 2

1 Complete the following geometric sequences:

a 2, 4, 8, 16, __, __, __, __ **b** −1, 3, −9, 27, __, __, __, __

c __, __, −6, −3, $-\frac{3}{2}$, $-\frac{3}{4}$, __, __ **d** __, __, __, __, 1, 5, 25, 125

2 Complete this table for these sequences.

	$\frac{\text{term 2}}{\text{term 1}}$	$\frac{\text{term 3}}{\text{term 2}}$	$\frac{\text{term 4}}{\text{term 3}}$
a	$\frac{4}{2}=2$	$\frac{8}{4}=2$	$\frac{16}{8}=2$
b			
c			
d			

3 What do you notice about the ratio of consecutive terms for each sequence?

You have found the common ratio of each geometric sequence.

Each term of a geometric sequence can be written in terms of its first term and common ratio.

Sequence **a** can be written in terms of its first term, 2, and common ratio, 2:

2, 2(2), $2(2)^2$, $2(2)^3$, $2(2)^4$, $2(2)^5$, ...

4 Find an expression for the nth term of sequence **a**.

Repeat for the other three sequences.

5 **Conceptual** Can you always write a geometric sequence in terms of its first term and common ratio?

The first term of a geometric sequence is called u_1, the common ratio is called r, and the nth term is called u_n.

$$r = \frac{u_2}{u_1} = \frac{u_3}{u_2} = \frac{u_4}{u_3} = \ldots$$

6 Complete the following:

$u_1 = u_1$

$u_2 = u_1 \times r = u_1 r$

$u_3 = u_2 \times r = u_1 r \times r = u_1 r^2$

$u_4 = u_3 \times r = u_1 r^2 \times r = u_1 r^3$

$u_5 =$

$u_6 =$

$\vdots$

$u_n =$

7 **Factual** What is the formula for the nth term of a geometric sequence?

8 **Conceptual** How can you tell if a sequence is geometric?

TOK

Why is proof important in mathematics?

Number and algebra

Functions

To check whether a sequence is geometric, find the ratio of pairs of consecutive terms. If this ratio is constant, it is the common ratio and the sequence is geometric.

The nth term in a geometric sequence is given by the formula $u_n = u_1 r^{n-1}$.

Example 1

1 For each of these geometric sequences, write down:

  i the first term, u_1 ii the common ratio, r iii u_{10}.

  a 2, 6, 18, 54, … b −3, 6, −12, 24, … c 16, 8, 4, 2, …

2 Carolien is starting a new job. She earns €48 000 in her first year, and her salary increases by 5% each year. Show that Carolien's annual salary follows a geometric sequence, and state the common ratio. Calculate how much Carolien will earn in her fifth year at work.

3 The third term of a positive geometric sequence is 63 and the fifth term is 567. Find the common ratio and the first term.

1 a i $u_1 = 2$	
ii $r = \frac{6}{2} = 3$	
iii $u_{10} = 2(3)^9 = 39\,366$	
b i $u_1 = -3$	
ii $\frac{6}{-3} = -2$	
iii $u_{10} = -3(-2)^9 = 1536$	
c i $u_1 = 16$	
ii $r = \frac{8}{16} = 0.5$	
iii $u_{10} = 16(0.5)^9 = 0.031\,25$	
2 $u_1 = 48\,000$	
$u_2 = 48\,000 + 5\% \times 48\,000$	
$= 48\,000(1 + 5\%)$	
$= 48\,000(1 + 0.05)$	
$= 48\,000 \times 1.05$	
So, $r = 1.05$	
Therefore,	
$u_5 = u_1 r^4 = 48\,000(1.05)^4 = €58\,344.30$	
3 $u_3 = 63$	Remember that $u_3 = u_1 r^2$.
So, $u_1 r^2 = 63$	
$u_5 = 567$	$u_5 = u_1 r^4$
So, $u_1 r^4 = 567$	

Then you can write: $\frac{u_5}{u_3} = \frac{u_1r^4}{u_1r^2} = \frac{567}{63} = 9$ $r^2 = 9$ $r = \pm 3$ You are told that all the terms are positive, so $r = 3$ Substituting back into the first equation: $u_1 = \frac{63}{9} = 7$	Here you can cancel the u_1 and r^2 to get r^2 on the left-hand side.

Exercise 10A

1 The first three terms of a geometric sequence are 3, 6, 12.

a Write down the common ratio.

b Calculate the value of the 15th term.

2 A marrow plant is 0.5 m long. Every week it grows by 20%.

a Find the value of the common ratio.

b Calculate how long the plant is after 12 weeks.

c Comment on the predicted length of the marrow after one year.

3 The first four terms of a geometric sequence are 27, x, 3, 1.

a Find the value of x.

b Write down the common ratio.

c Calculate the value of the eighth term.

4 There is a flu epidemic in Cozytown. On the first day, 2 people have the flu. On the second day, 10 people have the flu. On the third day, 50 people have the flu.

a Show that the number of people with the flu forms a geometric sequence.

b Hence, assuming the conjecture is true, calculate how many people have the flu after one week (seven days).

c Hence calculate how many people will have the flu after one year. Comment on your answer.

5 The second term of a geometric sequence is −32 and the fourth term is −2. Given that all the terms are negative, find the common ratio and the first term.

Example 2

Mr Farmer buys machinery for US$25 000. The machinery depreciates at a rate of 8% per annum.

a Find the value of the machinery after five years.

b Find how many years it is before Mr Farmer's machinery is worth half its original value.

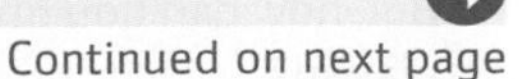

Continued on next page

a If the machinery depreciates at 8% each year, then after one year it will be worth only $(100 - 8)\% = 92\% = 0.92$ of its original value.

So, after five years it will be worth US\$25 000 $\times 0.92^5 =$ US\$16 477.04.

b Half of the original value is US\$12 500

So, $25\,000 \times 0.92^n = 12\,500$

Using your Solver, $n = 8.31$

So, Mr Farmer's machinery is worth half its original value after 8.31 years.

Exercise 10B

1 At the end of 2016 the population of a city was 200 000. At the end of 2018 the population was 264 500.

- **a** Assuming that these end of year figures follow a geometric sequence, find the population at the end of 2017.
- **b** Calculate the population at the end of 2020.
- **c** Comment on whether this increase will continue. Give a reason for your answer.

2 One kilogram of tomatoes costs US\$2.20 at the end of 2015. Prices rise at 2.65% per year. Find the cost of a kilogram of tomatoes at the end of 2019.

3 Petra buys a camper van for €45 000. Each year the camper van decreases in value by 5%. Find the value of the camper van at the end of six years.

4 Beau spends €15 000 buying computer materials for his office. Each year the material depreciates by 12%.

- **a** Find the value of the material after three years.
- **b** Find how many years it takes for the materials to be worth €5 000.

Developing inquiry skills

Return to the chess problem on p. 463. You should now be able to find out how many grains of wheat are on the last square.

But, how can you find the total number of grains of wheat on all the squares?

You should also be able to find the number of rounds of tennis played until the final.

But, how can you find the total number of games played?

Investigation 3

The following are geometric sequences:

a 1, 5, 25, 125, 625 **b** 6, 12, 24, 48, 96 **c** 2, −6, 18, −54, 162

Complete the following table for each sequence:

	u_1	r	r^2	r^2-1	$r-1$	Sum of first 2 terms
a						
b						
c						
	u_1	r	r^3	r^3-1	$r-1$	Sum of first 3 terms
a						
b						
c						
	u_1	r	r^4	r^4-1	$r-1$	Sum of first 4 terms
a						
b						
c						
	u_1	r	r^5	r^5-1	$r-1$	Sum of first 5 terms
a						
b						
c						

See if you can find a connection between:

- the sum of the first 2 terms, r^2-1, $r-1$ and u_1
- the sum of the first 3 terms, r^3-1, $r-1$ and u_1
- the sum of the first 4 terms, r^4-1, $r-1$ and u_1
- the sum of the first 5 terms, r^5-1, $r-1$ and u_1.

Using the connections you discovered, suggest a formula for the sum of the first n terms of a geometric series in terms of u_1, r and n. Are there any values that r cannot take?

The sum to n terms of a geometric series is written as S_n. The formula for S_n is

$$S_n = \sum_{i=1}^{n} u_i = \frac{u_1(r^n-1)}{r-1} = \frac{u_1(1-r^n)}{1-r}, r \neq 1$$

You can use either form.

To see how we might get that formula, consider the series 1, 2, 4, 8, 16, 32.

Here, $u_1 = 1$ and $r = 2$. If we take the sum of this series we have

$S_6 = 1 + 2 + 4 + 8 + 16 + 32$ (1)

Multiply this by 2:

$2S_6 = 2 + 4 + 8 + 16 + 32 + 64$ (2)

Now do (1) – (2). You'll see that a lot of the terms cancel out:

$S_6 - 2S_6 = 1 - 64$ (3)

Solving this for S_6, you get $S_6 = 63$.

Check this against the formula:

$$S_6 = \frac{1(1-2^6)}{1-2} = \frac{(1-64)}{-1} = 63$$

If you look at (3) and compare it with the formula you will see that on the left-hand side we have $(1-2)S_6$, which is $(1-r)S_6$, and on the right-hand side we have $1-2^6$, which is $1-r^6$.

So, $S_6 = \dfrac{u_1(1-r^6)}{(1-r)}$.

Example 3

1 Find the sum to eight terms of the geometric series 7, 28, 112, 448, …

2 When Kenzo starts Bright Academy, the school fees are JYN 2 500 000 (Japanese yen). The fees increase by 2% each year. Kenzo attends the school for a total of six years.

a Write down the common ratio.

b Calculate the school fees in year six.

c Calculate the total fees paid for the six years that Kenzo attends Bright Academy.

3 The second term of a geometric sequence is 6 and the fourth term is 24. All the terms are positive.

a Find the first term and the common ratio.

b Find the sum of the first 10 terms.

1 $u_1 = 7$ $r = \frac{28}{7} = 4$ So, $S_8 = \frac{7(4^8-1)}{4-1} = 152\,915$	
2 **a** $r = 1.02$	$1 + 2\% = 1 + 0.02 = 1.02$
b $u_6 = u_1 r^5 = 2\,500\,000 \times 1.02^5$ $= \text{JYN } 2\,760\,202$	Remember, $u_n = u_1 r^{n-1}$
c $S_6 = \frac{u_1(r^6-1)}{(r-1)} = \frac{2\,500\,000(1.02^6-1)}{(1.02-1)}$ $= \text{JYN } 15\,770302$	

3 a $u_2 = u_1 r = 6$ $u_4 = u_1 r^3 = 24$ $\frac{u_4}{u_2} = \frac{u_1 r^3}{u_1 r} = \frac{24}{6} = 4$ $r^2 = 4$ $r = 2$ $u_1 r = 6$ $u_1(2) = 6$ $u_1 = 3$ **b** $S_{10} = \frac{3(2^{10} - 1)}{(2 - 1)} = 3069$	You do not need $r = -2$ because all the terms are positive.

TOK

How do mathematicians reconcile the fact that some conclusions conflict with intuition?

Exercise 10C

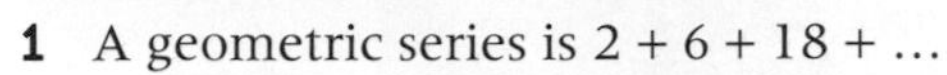

1 A geometric series is $2 + 6 + 18 + \ldots$

a Find the common ratio.

b Find the sum of the first 10 terms.

2 Aziza receives an allowance of EGP 2500 (Egyptian pounds) each month. Every month this amount is increased by 2%.

a Find how much Aziza receives in month six.

b Calculate the total amount that Aziza receives in one year.

3 The population of a small island is 12 000. The population is expected to grow by 1.2% each year.

a Find the population at the end of seven years.

At the beginning of each year, every inhabitant of the island receives a gift from the government.

b Find the total number of gifts that the government has given to the inhabitants by the beginning of the 8th year.

4 A geometric sequence with all positive terms has third term 8 and fifth term 128.

a Find the common ratio and the first term.

b Find the eighth term.

c Find the sum of the first eight terms.

5 When Chen starts at a private school, the school fees are CNY 270 000 (Chinese yuan) a year. Every year the fees increase by 4%.

a Calculate the school fees for the second and third years.

b Find the total cost of the school fees, given that Chen stays at the school for six years.

6 Kyle joins a sports club. The fee for the first year is €75.

a Given that the fee is increased by 2% every year, find the cost for the second year.

b Given that Kyle is a member of the sports club for five years, calculate how much he pays in total over that period.

7 Find the sum to nine terms of this geometric series:

$24 + 36 + 54 + 81 + \ldots$

8 Alexa wins a prize in a lottery. She receives CAD 8000 (Canadian dollars) the first month, CAD 6000 the second month, CAD 4500 the third month and so on for a total of six months.

a Calculate how much she receives in month six.

b Calculate the total prize money for the six months.

9 Find the sum of this geometric series: $1 + 3 + 9 + \ldots + 19683$.

10 Giovanna works on a farm and feeds the chickens each week. At the beginning the farm has 500 chickens. The number of chickens increases each week by 1%.

a Find the number of chickens after 15 weeks.

b Find the total number of feeds in 15 weeks.

Developing inquiry skills

Now you can find the total number of grains of wheat from the chess-board problem.

You can also find the total number of games of tennis that are played.

10.2 Compound interest, annuities, amortization

On her 10th birthday, Yun Lu inherits some money from her grandmother. She can either have US$10 each month now to invest in a bank which offers 2% interest compounded annually until she is 21, or she can have US$1400 on her 21st birthday.

How can you work out which option Yun Lu should choose?

When you invest money in a bank or other financial institution, you will be given interest at regular intervals such as annually, half-yearly, quarterly or monthly.

If the interest paid is **simple interest**, then the interest remains the same for each year that you have your money in the bank.

If the interest paid is **compound interest**, then the interest is added to the original amount and the new value is used to calculate the interest for the next period.

Investigation 4

For each of the following amounts deposited in a bank, calculate how much interest would be paid at the end of three years.

Amount	Simple interest at 5%	Compound interest at 5% compounded annually
US$100	Year 1: 5% of 100 = US$5 Year 2: also US$5 Year 3: also US$5 So, total is US$15	Year 1: 5% of 100 = US$5 Year 2: 5% of 105 = US$5.25 Year 3: 5% of 110.25 = US$5.5125 So, total is US$15.76

US\$500	Year 1: Year 2: Year 3: Total =	
US\$2000	Year 1: Year 2: Year 3: Total =	

Factual Which pays out more interest: simple or compound? Explain why.

Conceptual What is the connection between compound interest and geometric sequences?

If you invest US\$100 in a bank that offers 5% interest compounded annually, then:

- At the end of year 1 you have $100 + 5\% = 100(1 + 0.05) = 100\left(1+\frac{5}{100}\right)$
- At the end of year 2 you have $100\left(1+\frac{5}{100}\right) + 5\%$ of $100\left(1+\frac{5}{100}\right) = 100\left(1+\frac{5}{100}\right)^2$
- At the end of year 3 you have $100\left(1+\frac{5}{100}\right)^2 + 5\%$ of $100\left(1+\frac{5}{100}\right)^2 = 100\left(1+\frac{5}{100}\right)^3$

And so on. At the end of year n you would have $100\left(1+\frac{5}{100}\right)^n$.

In general, if you invest a **present** value of PV, the rate is r% compounded annually, and the number of years is n, then the **future** value (FV) is found by the following formula:

$$FV = PV\left(1+\frac{r}{100}\right)^n$$

Example 4

Peter invests AUD 2000 (Australian dollars) in a bank that offers simple interest at a rate of 3% per annum.

Paul invests AUD 2000 in another bank that offers 2.9% interest compounded annually.

a Calculate how much money they each have in the bank after 10 years.

b Determine after how many years they will each have at least AUD 5000 in their accounts.

a Peter: 3% of 2000 = AUD 60 60 × 10 = 600	

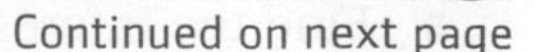

Continued on next page

So at the end of 10 years, Peter has AUD 600 + his original AUD 2000 = AUD 2600

Paul:

Using the formula, Paul has

$$FV = 2000\left(1+\frac{2.9}{100}\right)^{10}$$

$FV =$ AUD 2661.85

b Peter:

3% of 2000 = AUD 60

So, Peter earns AUD 60 interest each year.

In order to have AUD 5000, Peter will have to make 5000 − 2000 = AUD 3000 interest.

So, it will take $\frac{3000}{60} = 50$ years before he has AUD 5000 in the bank.

Paul:

Using the formula:

$$5000 = 2000\left(1+\frac{2.9}{100}\right)^{n}$$

So $n = 32.05$

It will take Paul just over 32 years to have AUD 5000 in the bank.

Exercise 10D

1 Merel and Misty each have €5000 to invest. Merel invests her money in a bank that pays 4.5% simple interest each year. Misty invests her money in a bank that pays 4.4% interest compounded annually.

Calculate who has more money at the end of 15 years.

2 Visay invests LAK 500 000 000 (Lao Kip) in a bank that pays 3.2% interest compounded annually.

a Calculate how much Visay has in the bank after eight years.

b Determine how many years it will take for his money to double.

3 Silvia invests UK£4500 in a bank that pays r% interest compounded annually. After five years, she has UK£5066.55 in the bank.

a Find the interest rate.

b Calculate how many years it will take for Silvia to have UK£8000 in the bank.

4 Sal wants to buy a scooter that costs US$1500. He deposits US$1000 in a bank that pays 7.5% interest compounded annually.

Calculate how long it will take before he can buy the scooter.

5 Huub invests €5000 in a bank that pays 1.2% interest compounded annually.

a Find how much money he has in the bank after five years.

He then removes the money from this bank and puts it all into another bank that pays a higher interest rate compounded annually.

After a further five years, Huub now has €5675.33 in the bank.

b Find the interest rate for the second bank.

Investigation 5

A bank pays interest at 6 % per annum compounded every half-year.

1 How many times a year does it pay interest?

2 What percentage interest does it pay each time?

3 Complete the following table:

6% interest per annum	Number of times interest is paid	Percentage each time
Compounded yearly	1	6%
Compounded half-yearly	2	3%
Compounded quarterly		
Compounded monthly		
r% interest per annum		
Compounded yearly	1	r%
Compounded half-yearly		
Compounded quarterly		
Compounded monthly		

4 Find the formula for the (future value) FV for a (present value) PV invested for n years at r% interest compounded half-yearly.

5 Find the formula for the FV for a PV invested for n years at r% interest compounded quarterly.

6 Find the formula for the FV for a PV invested for n years at r% interest compounded monthly.

7 **Conceptual** In terms of time periods and the principal how is the compound interest calculated?

The **compounding period** is the time period between interest payments. For example, for interest compounded quarterly, the compounding period is three months.

If PV is the present value, FV the future value, r the interest rate, n the number of years and k the compounding frequency, or number of times interest is paid in a year (ie $k = 1$ for yearly, $k = 2$ for half-yearly, $k = 4$ for quarterly and $k = 12$ for monthly), then the general formula for finding the future value is

$$FV = PV\left(1 + \frac{r}{k \times 100}\right)^{k \times n}$$

You can also use the Finance app on your GDC.

Example 5

1 Rafael invests BRL 5000 (Brazilian real) in a bank offering 2.5% interest compounded annually.

a Calculate the amount of money he has after five years.

After the five years, Rafael withdraws all his money and puts it in another bank that offers 2.5% interest per annum compounded monthly.

b Calculate the amount of money that he has in the bank after three more years.

2 Alexis invests RUB 80 000 (Russian ruble) in a bank that offers interest at 3% per annum compounded quarterly.

a Calculate how much money Alexis has in the bank after six years.

b Calculate how long it takes for his original amount of money to double.

1 a $FV = 5000\left(1+\frac{2.5}{1\times 100}\right)^{1\times 5}$

$= \text{BRL } 5657.04$

$PV = 5000,\ r = 2.5,\ k = 1,\ n = 5$

b $FV = 5657.04\left(1+\frac{2.5}{12\times 100}\right)^{12\times 3}$

$= \text{BRL } 6097.16$

$PV = 5657.04,\ r = 2.5,\ k = 12,\ n = 3$

Or, using the Finance app on your GDC:

a N = 5

I% = 2.5

PV = −5000

PMT = 0

FV =

PpY = 1

CpY = 1

Move the cursor back to FV and press enter to get the answer.

PV is usually negative because you have given it to the bank.

PMT is periodic money transfers (also called annuity payment) and you do not need it here.

PpY is periods in the year; when you are dealing with years, this is always 1.

CpY is compounding periods: 1, 2, 4 or 12 depending on how often interest is paid.

b N = 3

I% = 2.5

PV = −5657.04

PMT = 0

FV =

PpY = 1

CpY = 12

Move the cursor to FV and press enter.

2 a $FV = 80\,000\left(1+\dfrac{3}{4\times100}\right)^{4\times6}$ $= \text{RUB } 95\,713.08$	N = 6 I% = 3 PV = −80 000 PMT = 0 FV = PpY = 1 CpY = 4 Move the cursor back to FV and press enter.
b $160\,000 = 80\,000\left(1+\dfrac{3}{4\times100}\right)^{4\times n}$ Using the Finance app: $n = 23.19$ So it would take 23 years for his money to double.	Double the original amount is RUB 160 000 N = I% = 3 PV = −80 000 PMT = 0 FV = 160 000 PpY = 1 CpY = 4 Move the cursor back to N and press enter.

Exercise 10E

1 Ambiga invests MYR 8000 (Malaysian ringgits) in a bank offering interest at a rate of 4.6% per annum compounded monthly.

a Calculate the amount of money Ambiga has in the bank after seven years.

b Calculate how long it will take for her money to double.

2 A bank is offering a rate of 3.4% per annum compounded quarterly. Mrs Safe invests €3500 in this bank.

a Calculate the amount of money she has after six years.

Mr Secure invests €x in this bank. After six years, the amount in his bank is €4000.

b Calculate the value of x, correct to the nearest euro.

c Calculate the number of years it would take for Mr Secure's money to double.

3 Rik invests SGD 40 000 (Singaporean dollars) in an account that pays 5% interest per year, compounded half-yearly.

a Calculate how much he has in the bank after four years.

The bank then changes the interest rate to 4.9% per annum, compounded monthly.

b Calculate how much Rik has in the bank after another four years.

4 Mr Chen invests CNY 20 000 (Chinese yuan) in a bank that offers interest at a rate of 3.8% per annum compounded quarterly. Mrs Chang also invests CNY 20 000 in a bank that offers interest at a rate of 3.9% per annum compounded yearly.

Calculate who has earned more interest after five years.

5 Peter invests UK£400 in a bank that offers interest at a rate of 4% per annum compounded monthly.

a Calculate how much money he has in the bank after 10 years.

b Calculate how long it takes for his money to double.

6 Yvie invests US$1200 in a bank that offers interest at a rate of r% per annum compounded monthly. Her money doubles in 10 years. Find the value for r.

7 Colin, Ryan and Kyle each have €1500 to invest. Colin invests his money in a bank that offers 2.6% interest compounded quarterly. Ryan invests his money in a bank that offers 2.55% interest compounded monthly. Kyle invests his money in a bank that offers 2.75% interest compounded annually.

a Calculate who has the most money in their account after six years.

b Find how long it will take before Ryan has €2500.

c Find how long it will take for Kyle to double his money.

Inflation measures the rate that prices for goods increase over time and, as a result, how much less your money can buy.

An **inflation adjustment** is the change in the price of an article that is a direct result of inflation.

Example 6

The inflation rate of a country was calculated as 4.48% per annum.

a Given that the same rate continues for the next five years, work out the percentage increase due to inflation at the end of the five years.

b A computer game costs US$35 today. Calculate what you would expect it to cost next year due to an inflation adjustment.

a Over five years, US$1 will become $\left(1+\frac{4.48}{100}\right)^5 = 1.245$, so the percentage increase is 24.5%.	
b The computer game will cost $35 \times 1.0448 = 36.568$, or US$36.57.	

What happens if you want to make payments at **regular intervals**?

An **annuity** is a fixed sum payable at specified intervals, usually annually, over a period, such as the recipient's life, in return for a premium paid either in instalments or in a single payment.

Investigation 6

Abraham wants to save up so that he can retire early. He decides to save US$5000 every year in an annuity that pays 4% interest compounded annually. If he wants to retire in 20 years' time, determine how you can work out how much he will have in the annuity at the end of the 20 years.

Complete the following spreadsheet:

Year	Add 5000	+ 4% interest
1	5000	5200
2	10 200	10 608
3	15 608	16 232.32
4		
5		
6		
7		
8		
9		
10		
11		
12		
13		
14		
15		
16		
17		
18		
19		
20		

How much more is this than 20 × 5000?
Discuss the benefits of saving money regularly.

TOK

"Debt certainly isn't always a bad thing. A mortgage can help you afford a home. Student loans can be a necessity in getting a good job. Both are investments worth making, and both come with fairly low interest rates"—Jean Chatzky

Do all societies view investment and interest in the same way?

What is your stance?

The formula for working out an annuity is

$$FV = A\frac{(1+r)^n - 1}{r}$$

where FV is the future value, A is the amount invested each year, r is the interest rate and n is the number of years.

You can rearrange this formula if you want to calculate any of the other unknowns.

Fortunately, your GDC works all this out for you!

You can use the PMT button on the finance icon of your GDC.

Example 7

Anmol decides to save for a yacht. He would like to have TRY 1 000 000 (Turkish lira) at the end of 10 years. He saves every year in an annuity that pays 4% interest. Calculate how much he has to save each year.

N = 10 I% = 4 PV = 0 PMT = FV = 1 000 000 PpY = 1 CpY = 1 Move the cursor to PMT and press Enter (or ALPHA ENTER) This gives a PMT of TRY 83 290.94. So, he will have to save TRY 83 290.94 every year for the next 10 years.	$P = \frac{FV(r)}{(1+r)^n - 1}$

The formula for working out monthly payments is $P = \frac{rPV}{1-(1+r)^{-n}}$ where P is the payment, r is the rate, PV is the present value and n is the number of months.

Fortunately, you do not need to remember this formula because you can use your calculator in the exams.

Example 8

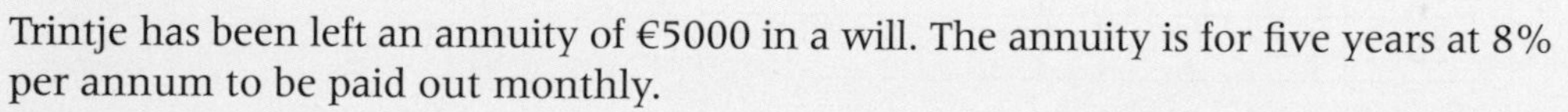

Trintje has been left an annuity of €5000 in a will. The annuity is for five years at 8% per annum to be paid out monthly.

Find the monthly payments.

N = 60 I% = 8/12 PV = 5000 PMT = FV = 0 PpY = 1 CpY = 1 Move the cursor to PMT and press Enter (or ALPHA ENTER) This gives PMT = €101.38. So, Trintje receives €101.38 each month for five years.	Because the money is paid each month, you have to multiply the number of years by 12. The interest rate per month is 8/12 = 0.666...%

These are examples of annuities.

Reflect Discuss the pros and cons of an annuity.

Exercise 10F

1 Sarah-Jane starts saving for her pension. She puts UK£1500 into an annuity each year. Given that the annuity pays 3.5% interest, calculate how much she will have saved after 30 years. Is this a good investment?

2 Pedro puts MXN 2500 (Mexican pesos) into an annuity that pays 2.8% interest per annum every month for five years. Calculate how much he has saved at the end of the five years.

3 Stijn has been left an annuity of €25 000 in a will. The annuity is for 10 years at 6% per annum to be paid out annually. Work out the amount that Stijn receives each year.

4 Fiona has won an annuity of US$6000. It is for five years at 4.8% per annum to be paid out monthly. Find the monthly payments. Contrast the benefits of the full US$6000 in one payment or the annuity.

5 Mikey wants to receive an annuity of UK£500 a month for five years. The monthly interest rate is 0.8%. Determine the present value of the annuity.

6 You have US$20 000 and want to get a monthly income for 10 years. If the monthly interest rate is 0.9%, find how much you would receive each month. Comment on why you think that you would do this.

There are also similar examples but in the following cases the value of the item decreases each term. This is called **amortization**.

Investigation 7

Pim borrows US$1000 from a bank that charges 4% interest compounded annually. He wants to pay the loan back in six months in monthly instalments. The bank informs Pim that he must pay US$168.62 back each month.

1 Work out how much he has to pay in total.

If the interest is 4% compounded annually, then that would be 2% for half a year.

2 Work out 2% of US$1000.

3 Explain why the amount that Pim pays back in total is less than US$1000 + 2% of US$1000.

Pim sets up a spreadsheet to monitor his payments.

He pays US$168.62 each month, but the bank also charges interest each month on the amount owed.

4 If the interest rate is 4% per annum, work out what it is per month.

5 Calculate the interest for the first month.

6 Describe how you worked out how much is paid off the loan each month.

Continued on next page

7 The spreadsheet shows the first two lines of Pim's payments. Fill in the next few lines.

Amount owed	Payment	Interest	Payment – interest	Remaining loan
1000	168.62	3.33	165.29	834.71
834.71	168.62	2.78	165.84	668.87

After six months, the remaining loan should be US$0.

The formula to find the payments is

$$A = PV\frac{r(1+r)^n}{(1+r)^n - 1}$$

where A is the amount, PV is the present value, r is the rate and n is the number of periods.

Fortunately, you do not need to remember this formula because you can use your calculator in the exams.

You can use your GDC to work out payments, and so on.

You can also use any of the online websites such as www.amortization-calc.com or www.bankrate.com to calculate payments, and so on. You can adjust the length of time for the loan or the interest rate to see what effect this has.

Example 9

Tejas takes out a loan of €35 000 for a car. The loan is for 10 years at 1% interest per month. Find how much he has to repay each month.

N = 120

I% = 1

PV = 35 000

PMT =

FV = 0

PpY = 1

CpY = 1

Using your GDC, this gives PMT = €502.15.

So, Tejas has to pay €502.15 each month for 10 years.

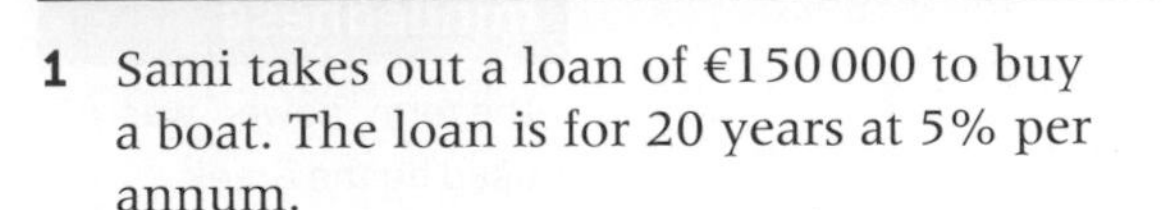

Exercise 10G

1 Sami takes out a loan of €150 000 to buy a boat. The loan is for 20 years at 5% per annum.

a Find how much he must pay each month.

b Another loan was for 10 years at 5% per annum. Determine how much Sami would pay each month.

c For each of these calculations find out the difference in the monthly payments and the amount of interest paid.

2 Mr and Mrs Jones take out a mortgage of UK£350 000 for a house. The mortgage is for 30 years at 2.3% per annum. Find their monthly repayments and discuss whether this is a reasonable amount.

3 Zak takes out a loan of €2000 at 4%. He repays the loan in five end-of-year instalments. Find his yearly payment.

4 A car costs US$28 000. Benji takes out a loan at 10% per annum for five years. Calculate the amount that Benji must pay each month and comment on your answer.

Developing inquiry skills

You should now be able to work out the answer to the original problem at the beginning of this section. What do you think that Yun Lu should do?

TOK

"A government's ability to raise and lower short-term interest rates is its primary control over the economy"—Alex Berenson

How can knowledge of mathematics result in individuals being exploited or protected from extortion?

10.3 Exponential models

The temperature, T, of a cup of tea is modelled by the function $T(x) = 21 + 55(1.9)^{-x}$, where T is measured in °C and x in minutes.

How long will the tea stay warm?

What information do you need to know?

Will the temperature of the tea ever reach 0°C?

Can you find out the temperature at different times?

Investigation 8

1 Using technology, sketch the following functions:

a $f(x)=2^x$ b $f(x)=1.5^x$ c $f(x)=0.5^x$

d $f(x)=0.3^x$ e $f(x)=e^x$

In part **e**, e is a special number that we will discuss shortly.

2 Sort the functions into two groups, increasing and decreasing.

3 For the function $f(x)=e^x$, what value of x gives $f(x)=e$? Use this to find the value of e from your graph.

These are all examples of **exponential** functions.

International-mindedness

The term "power" was used by the Greek mathematician Euclid for the square of a line.

An exponential function has a number, called the base, raised to a power.
The independent or input variable is the power.
The general equation is of the form $f(x)=ka^x+c, \quad a>0,\ k\neq 0$

Reflect How can you tell from its equation whether an exponential function is increasing or decreasing?

Which of the graphs you drew could represent a cooling temperature? Which could represent the growth of bacteria?

Which type of exponential equation represents growth, and which decay?

e is an irrational number with a value of 2.718 28…

Investigation 9

Fill in the following table by using technology to find the gradients of the lines that are tangents to the given curves at the given points. Recall that the gradient of the tangent line gives the rate of change of a function.

Curve	$y=2^x$		$y=e^x$		$y=3^x$	
	y	Rate	y	Rate	y	Rate
$x=1$	2	1.386				
$x=2$	4					
$x=3$						
$x=4$						

How does the value of the base in an exponential function affect the rate of change?

What is special about the number e in this context?

The functions we have been graphing do not cross the x-axis. As the value of x decreases they get closer and closer to the axis but never cross it. So, the x-axis is called an **asymptote**. Its equation is $y = 0$.

Investigation 10

The temperature of a cup of coffee is modelled by the equation $f(x) = 65(1.9)^{-x} + 20$, where $f(x)$ is the temperature in °C and x is the time in minutes.

1. Using technology, sketch the graph of $f(x) = 65(1.9)^{-x} + 20$ for $0 \le x \le 12$.
2. Describe how you can find the initial temperature of the coffee.
3. Add the line $x = 4$ to your graph. What does the intersection of these two lines represent?
4. Add the line $f(x) = 50$ to your graph. What does the intersection of these two lines represent?
5. **Conceptual** What does the intersection of two graphs represent?
6. **Conceptual** What is the equation of the horizontal asymptote? Explain what this represents.

Now remember two things from Section 10.2:

- When something changes by a constant percentage, we can model it as a geometric sequence.
- The nth term of a geometric sequence is $u_1 r^{n-1}$. Can you see how this relates to the general equation of an exponential function?

Exponential functions can be used to model real-life situations where there is a constant percentage change.

Investigation 11

1. Sketch the following functions using technology and complete the table:

   a $f(x) = 3^x$ **b** $f(x) = 0.5^x$

   c $f(x) = 2^x + 3$ **d** $f(x) = 0.4^x - 2$

   e $f(x) = 2(3^x) + 1$ **f** $f(x) = 4(2^x) - 5$

   g $f(x) = 3(0.5)^x - 2$

	Value of k	Value of a	Value of c	y-intercept	Equation of asymptote	Increasing or decreasing
a	1	3	0	(0, 1)	$y=0$	increasing
b						
c						
d						
e						
f	4	2	−5	(0, −1)	$y=-5$	increasing
g						

Continued on next page

2 What is the connection between the value of c and the equation of the horizontal asymptote?

3 What is the connection between the values for k and c and the coordinates of the y-intercept?

4 What is the connection between the value for a and whether the function is increasing or decreasing?

5 Describe how you can decide which variable is independent and which is dependent.

6 **Conceptual** What does the horizontal asymptote represent? How can you decide if a function has a horizontal asymptote?

7 What are the parameters of an exponential function?

8 Now sketch the function $f(x) = 3^{-x}$ and compare it with your sketch for part **a**.

9 Explain the difference between growth and decay.

The function $f(x) = ka^x + c$ has the horizontal asymptote $y = c$ and crosses the y-axis at the point $(0, k + c)$.

International-mindedness

The word "exponent" was first used in the 16th century by German monk and mathematician Michael Stifel. Samuel Jeake introduced the term indices in 1696.

Example 10

For each of the following functions, find:

i the equation of the horizontal asymptote

ii the coordinates of the point where the curve cuts the y-axis

iii state whether the function is increasing or decreasing.

a $f(x) = 3^x + 2$

b $f(x) = 5 \times 0.2^x - 3$

a i The horizontal asymptote has equation $y = 2$.	2 is the value for c
ii The curve cuts the y-axis at the point $(0, 3)$.	$k = 1$ and $c = 2$, so, $1 + 2 = 3$
iii The function is increasing because 3 is greater than 1.	
b i The horizontal asymptote has equation $y = -3$.	-3 is the value for c
ii The curve cuts the y-axis at the point $(0, 2)$.	$k = 5$ and $c = -3$, so $5 - 3 = 2$
iii The function is decreasing because 0.2 lies between 0 and 1.	

Exercise 10H

For the following equations, find:

a the y-intercept

b the equation of the horizontal asymptote

c state whether the function shows growth (increasing) or decay (decreasing).

1 $f(x) = 4^x + 1$ **2** $f(x) = 0.2^x - 3$

3 $f(x) = e^{2x}$ **4** $f(x) = 3^{0.1x} + 2$

5 $f(x) = 3(2)^x - 5$ **6** $f(x) = 4(0.3)^{2x} + 3$

7 $f(x) = 5(2)^{0.5x} - 1$ **8** $f(x) = 2(0.4)^x - 1$

Example 11

Bailey invests US\$2000 in a bank that pays interest at a rate of 4.5% per annum compounded monthly.

a Sketch a graph to model how much Bailey has in the bank after x years.

b Deduce whether an exponential model is a reasonable choice to model how much she has in the bank.

c Calculate how much money Bailey has in the bank after four years.

d Calculate how long it will take for her to have US\$3000 in the bank.

a

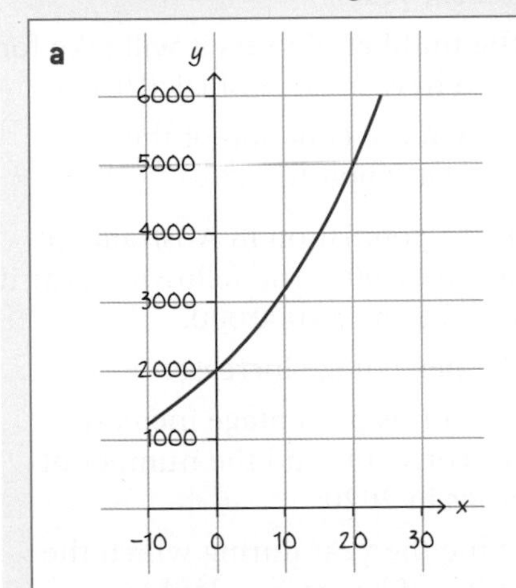

If US\$$y$ represents how much Bailey has in the bank, and x represents the time in years, then the formula to use is:

$$y = 2000\left(1 + \frac{4.5}{1200}\right)^{12 \times x} \text{ or}$$

$$y = 2000(1.003\,75)^{12x}$$

b Yes, because the independent variable is the exponent.

c US\$2393.63

Put $x = 4$ into the equation or use your GDC.

d 9.03 years, so after 10 years she will have US\$3000 in the bank.

Sketch the line $y = 3000$ and find the point of intersection.

Example 12

Saul buys a tractor for US\$51 000. The price of the tractor reduces by 8% each year.

a Find the price of the tractor after four years.

b Find the number of years until the tractor is worth half its original value.

Continued on next page

a If the price reduces by 8% each year then the multiplying factor will be 100% − 8% = 92%, which can be written as 0.92.

So, after four years, it will be worth $51\,000 \times 0.92^4 =$ US$36 536.04

b Half of 51 000 = 25 500, so $51\,000 \times 0.92^n = 25\,500$, which gives $n = 8.31$ years.

Exercise 10I

1 The increase in the length of an ivy plant, in metres, is modelled by the equation $f(x) = 2(1.1)^x$, where x is the time in weeks.

a Find the initial length of the ivy plant.

b Find the length after 10 weeks.

c Find how long it takes for the plant to reach a length of 10 metres.

2 A radioactive material has been spilled on farmland and the quantity remaining can be modelled by the equation $f(x) = 130(0.85)^x$, where x is the time in years and $f(x)$ is the mass in grams.

a Find the initial mass of the material.

b Find the mass after four years.

c Write down the equation of the horizontal asymptote to the graph.

d Calculate how many years it takes before only half the material remains.

3 The temperature, T°C, of a cup of hot chocolate can be modelled by the equation $T(x) = 70(1.45)^{-x} + 21$, where x is the time in minutes.

a Sketch the graph of $T(x)$ for $0 \le x \le 12$.

b Find the initial temperature of the hot chocolate.

c Find the temperature after five minutes.

d Find the number of minutes it takes for the hot chocolate to reach 50°C.

e Write down the equation of the horizontal asymptote of the graph.

f Write down the temperature of the room.

4 The value of a car depreciates each year. When new, the car is worth UK£36 000. The following year it is worth UK£34 200.

a Find the percentage decrease in value.

b Given that this percentage decrease remains constant, find the value of the car after six years.

c Find the number of years it will take for the car to have a value of UK£20 000.

d Write down the equation of the horizontal asymptote.

5 The kangaroo population in Australia, in 2017, was 50 000 000. The following year it was predicted to be 55 000 000.

a Find the percentage increase.

b Given that this percentage increase remains constant, find the number of kangaroos in 2020.

c Determine the year during which the population of kangaroos doubles.

6 The length, in cm, of a plant is measured each week and the results are shown in the table.

Week	1	3	5	7	9	11	13	15	17
Length	5	7	11	15	23	31	45	64	92

a Plot these points on your GDC.

b Determine whether this is a linear growth or an exponential growth.

c Find a suitable equation to represent this data.

d Hence estimate the length after 10 weeks and state whether this is a reasonable estimate.

7 The time, in minutes, that it takes Terry to solve a crossword puzzle in the newspaper each day is shown in the table.

Day	1	2	3	4	5	6	7
Time	45	40	36	31	27	22	17

a Plot these points on your GDC.

b Determine whether this is a linear decay or an exponential decay.

c Find a suitable equation to represent this data.

d Use your equation to estimate how long it will take Terry to complete the crossword puzzle on day 10.

TOK

The phrase "exponential growth" is used popularly to describe a number of phenomena.

Do you think that using mathematical language can distort understanding?

10.4 Exponential equations and logarithms

Investigation 12

Recall from Chapter 5 that if $f(a) = b$, then $f^{-1}(b) = a$.

Suppose $f(x) = 10^x$. Complete this mapping table:

a	$f(a) = 10^a$
0	
1	
2	
3	

1 Explain using the mapping, the equation or a graph of $f(x)$ why its inverse function exists.

2 Complete the mapping for the inverse function:

b	$f^{-1}(b)$
1	
10	
100	
1000	

We call the inverse of $f(x) = 10^x$ the **logarithmic function**, $g(x) = \log_{10} x$ or $\log x$.

Here, 10 is called the **base** of the logarithm. Any positive number can be used as a base for logarithms, but the only ones you need to know about are 10 and e.

Continued on next page

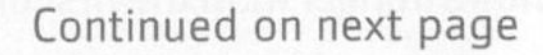

3 Use your GDC to complete this table:

Exponent form	Logarithm form
$10^2 = 100$	$\log_{10} 100 = 2$
$10^3 =$	$\log_{10} 1000 =$
$10^4 =$	$\log_{10} 10\,000 =$
$10^5 =$	$\log_{10} 100\,000 =$
$e^2 =$	$\log_e 7.389 =$
$e^3 =$	$\log_e 20.086 =$
$e^4 =$	$\log_e 54.598 =$
$e^5 =$	$\log_e 148.413 =$

A disease spreads according to the model $y = 10^t$, where t is the time in days.

4 Find the number of people who are infected initially.

5 Calculate how many people are infected after five days.

6 Find out how long it takes for 3 000 000 people to be infected.

7 **Conceptual** How are logarithms related to exponential functions?

8 **Conceptual** What kind of problems can you solve using logarithms?

Logarithms represent the inverse functions of exponential functions.
If $10^x = b$ then $\log_{10} b = x$, and if $e^x = b$ then $\log_e b = \ln b = x$.

On your GDC, $\log_{10}$ is called "LOG" and $\log_e$ is called "LN", which stands for "natural logarithm".

Example 13

a Find x if $10^x = 12$. **b** Find x if $\ln x = 3$.

a $10^x = 12$ can be written as $\log_{10} 12 = x$
Using your GDC, $\log_{10} 12 = 1.079$
So $x = 1.079$

b $\ln x = 3$ is the same as $\log_e x = 3$
So $e^3 = x$
$x = 20.1$

Investigation 13

You have all heard about earthquakes being measured on the "Richter scale", but do you really know what this means?

The Richter scale is not a linear scale but a logarithmic scale. So, if one earthquake has a value of 5.2 and another has a value of 6.4, then the second one is $10^{1.2} = 15.85$ or about 16 times stronger than the first one $(1.2 = 6.4 - 5.2)$.

International-mindedness

Logarithms do not have units but many measurements use a log scale such as for earthquakes, the pH scale and human hearing.

So, using a logarithmic scale, an increase (or decrease) of 1 means an actual increase (or decrease) of the thing being measured of 10. A change of 2 on the scale would mean an actual change of 100.

Complete the following table to find out how many times stronger each subsequent earthquake is than an initial earthquake measuring 5.2 on the Richter scale.

First earthquake measurement	Second earthquake measurement	Number of times stronger
5.2	5.8	
5.2	6.9	
5.2	7.5	
5.2	8.1	
5.2	8.7	

Explain why logarithms can be useful in measuring some things.

Discuss with your partner and share.

Conceptual What kind of situations do logarithms model? Give examples.

Exercise 10J

1 The length of a pumpkin plant, in metres, can be modelled by the equation $f(x) = e^x$, where x is the time in weeks after the pumpkin is planted.

  a Find the initial length of the pumpkin plant.

  b Find the number of weeks it takes for the plant to reach 10 metres.

2 The spread of a virus can be modelled by the equation $f(x) = 10^x$, where x is the time in weeks.

  a Find how many people have the virus after two weeks.

  b Find the number of weeks it takes for 500 people to have the virus.

  c Find the number of weeks it takes for 2000 people to have the virus.

3 **a** Find the value of $2 + \log 3$.

  b Find the value of x when $\log(x) = 1$.

  c Find the value of x when $\log(x) = 3$.

4 **a** Find the value of $0.5 + \ln 3$.

  b Find x if $\ln(x) = 4.2$.

  c Find x if $3^x = 5$.

5 A funny joke is spreading through the office. The speed that it is spreading can be modelled by the equation $f(x) = 4 + e^x$, where x is the time in minutes and $x \geq 0$.

  a Find out many employees heard the joke initially.

  b Find how many minutes it took before 150 employees heard the joke.

Developing your toolkit

Now do the Modelling and investigation activity on page 496.

Chapter summary

- A sequence of numbers in which each term can be found by multiplying the preceding term by a **common ratio** is called a geometric sequence.
- A geometric series is the sum of the terms of a geometric sequence.
- To check whether a sequence is geometric, find the ratio of pairs of consecutive terms. If this ratio is constant, it is the common ratio and the sequence is geometric.
- The nth term in a geometric sequence is given by the formula $u_n = u_1 r^{n-1}$.
- The sum to n terms of a geometric series is written as S_n. The formula for S_n is

$$S_n = \sum_{i=1}^{n} u_i = \frac{u_1(r^n - 1)}{r-1} = \frac{u_1(1-r^n)}{1-r}, r \neq 1$$

  You can use either form.
- In general, if you invest a **present** value of PV, the rate is r% compounded annually, and the number of years is n, then the **future** value (FV) is found by the following formula:

$$FV = PV\left(1+\frac{r}{100}\right)^n$$

- The **compounding period** is the time period between interest payments. For example, for interest compounded quarterly, the compounding period is three months.
- If PV is the present value, FV the future value, r the interest rate, n the number of years and k the compounding frequency, or number of times interest is paid in a year (ie $k=1$ for yearly, $k=2$ for half-yearly, $k=4$ for quarterly and $k=12$ for monthly), then the general formula for finding the future value is

$$FV = PV\left(1+\frac{r}{k \times 100}\right)^{k \times n}$$

- **Inflation** measures the rate that prices for goods increase over time and, as a result, how much less your money can buy.
- An **inflation adjustment** is the change in the price of an article that is a direct result of inflation.
- An **annuity** is a fixed sum payable at specified intervals, usually annually, over a period, such as the recipient's life, in return for a premium paid either in instalments or in a single payment.
- The formula for working out an **annuity** is

$$FV = A \times \left(\frac{(1+r)^n - 1)}{r}\right)$$

  where FV is the future value, A is the amount invested each year, r is the interest rate and n is the number of years.
- The formula for working out monthly payments for annuities is

$$P = \frac{rPV}{1-(1+r)^{-n}}$$

  where P is the payment, r is the rate, PV is the present value and n the number of years.
- The formula to find the payments for amortization is

$$A = PV\frac{r(1+r)^n}{(1+r)^n - 1}$$

  where A is the amount, PV is the present value, r is the rate and n is the number of periods.

- An exponential function has a number, called the base, raised to a power; the independent or input variable is the power.
- The general equation is of the form $f(x) = ka^x + c,\ \ a > 0,\ \ k \neq 0$. This has the horizontal asymptote $y = c$ and crosses the y-axis at the point $(0, k + c)$.
- e is an irrational number with a value of 2.718 28...
- Exponential functions can be used to model real-life situations where there is a constant percentage change.
- Logarithms represent the inverse functions of exponential functions.
- If $10^x = b$ then $\log_{10} b = x$, and if $e^x = b$ then $\log_e b = \ln b = x$.

Developing inquiry skills

At the beginning of this chapter you were asked "Can you predict the smallest height that no athlete will ever be able to jump over?"

Has what you have learned in this chapter helped you to answer the question?

What information did you manage to find?

What assumptions did you make?

How will you be able to construct a model?

Will this be different for men and women? If so, explain why?

What other things did you wonder about?

You were also asked these questions at the beginning of the chapter:

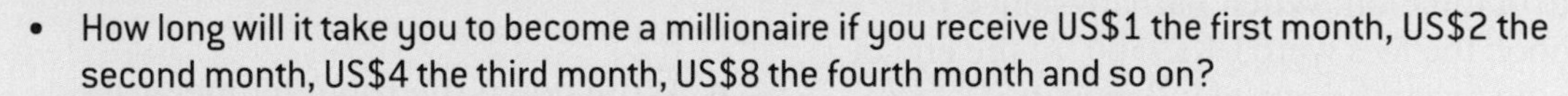

- How long will it take you to become a millionaire if you receive US$1 the first month, US$2 the second month, US$4 the third month, US$8 the fourth month and so on?
- How can you find an equation to model the slope of a skateboard track? Or a ski slope?
- How can you work out which rate gives you the most interest on your savings?
- How can you work out the monthly payments on a loan?
- If the temperature of a cup of tea reduces at a given rate, will it ever reach 0°C?

Has what you have learned in this chapter helped you to think about an answer to most of these questions?

Are there any that you are interested in and would like to explore further, perhaps for your IA topic?

Chapter review

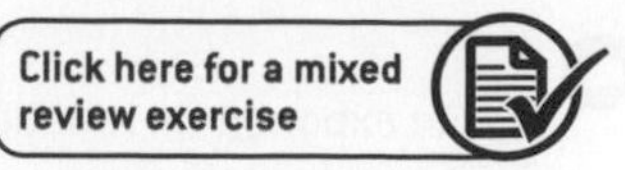

1 The first four terms of a geometric sequence are 2, 5, 12.5 and 31.25.

a Find the common ratio.

b Find the eighth term.

c Find the sum of the first eight terms.

2 Saul bought a new bicycle for US$350. Every year the value of the bicycle decreased by 12%. Find the value of the bicycle at the end of five years.

3 Molly bought a flat. At the beginning of the third year, the value of Molly's flat had increased to €363 000. At the beginning of the fifth year, the value had increased to €439 230.

a Calculate the rate of increase.

b Calculate how many euros Molly originally paid for the flat.

c Find how much the flat is worth at the beginning of the ninth year.

4 A geometric sequence has terms 2, 6, 18, ..., 118 098.

a Calculate the number of terms in this sequence.

b Find the sum of the corresponding series.

5 The school fees increase each year by the rate of inflation. When Barnaby joined the school, the fees were UK£9500. At the end of the first year the rate of inflation was 1.16%.

a Find the cost of the school fees for the second year.

The following year the rate of inflation was 1.14%.

b Find the cost of the school fees for the third year.

6 Wei invested SGD 3000 (Singapore dollars) in a bank that paid 2.35% interest per year compounded monthly.

a Find how much Wei had in the bank after six years.

b Find the number of years before he had SGD 5000 in the bank.

7 Marek invested PLN 4500 (Polish złoty) in a bank that paid r% interest per annum compounded quarterly. After six years he had PLN 5179.27 in the bank. Find the interest rate.

8 Silvia is left an annuity of US$4000 in her uncle's will. The annuity is for five years at 4% per annum and is to be paid out monthly. Find the monthly payments.

9 Nathalie borrows €6500 for a motorcycle. The loan is for five years at 2.5% interest per annum. Find how much Nathalie's monthly payments are.

10 The population of rabbits can be modelled by the function $f(x) = 24\,000 \times 1.12^x$, where x is the time in years.

a Find the number of rabbits after five years.

b Calculate how long it will take for the population of rabbits to double.

11 The temperature, T°C, of a cup of soup can be modelled by the equation $T(x) = 21 + 74 \times (1.2)^{-x}$, where x is the time in minutes.

a Find the initial temperature.

b Find the temperature after 10 minutes.

c Find how many minutes it takes for the soup to reach 40°C.

d Write down the room temperature.

12 An exponential equation is $f(x) = 4 \times 2.3^x + 16$.

a Write down the equation of the horizontal asymptote.

b Write down the coordinates of the point where the curve cuts the y-axis.

13 The spread of a disease can be modelled by the equation $y = 4 + e^x$, where x is the time in days.

a Find the number of people with the disease after seven days.

b Find the number of days it takes for 25 000 people to be affected.

14 The height, in metres, of a runner bean plant increases each week according to the formula $h(t) = 0.25 + \log 2t$, where t is the time in weeks.

a Find the height of the plant after four weeks.

b Find the number of weeks it takes for the plant to reach a height of 2 m.

Exam-style questions

15 P1: k^2, $6 + k$, $12 - k$ are consecutive values of a sequence $\{a_n, n \geq 0\}$.

Find the value of k for which $\{a_n\}$ is a non-constant arithmetic sequence, stating its general term. (6 marks)

16 P1: A company is considering producing a new product. The new production requires the installation of new machines and equipment. For this purpose, the company wants to borrow money by selling bonds, each priced at US $100 000, lasting for a 10-year period. The interest which the company pays to the bondholders is 5% per year.

Calculate the amount of interest to be paid per bond over the 10-year period in each of the following cases. Give your answers exactly, or correct to two decimal places where necessary.

a Simple interest is charged. (2 marks)

b Interest is compounded annually. (2 marks)

c Interest is compounded monthly. (2 marks)

17 P2: In this question, give all answers to two decimal places.

Carlos decides to take out a loan of 20 000 Peruvian Soles (SOL) to buy a car. His bank offers two options to finance the loan.

Option A: A five-year loan with an annual interest rate of 12.8% compounded *quarterly*. No deposit required.

Option B: A five-year loan with an annual interest rate of r% compounded *monthly*. Terms of the loan require a 10% deposit and monthly repayments of SOL 400.

a If Carlos chooses option A, find:

i the repayment he makes each quarter

ii the total amount he pays back

iii the total interest he pays on the loan. (5 marks)

b If Carlos chooses option B, find:

i the total amount Carlos pays to the bank

ii the annual interest rate, r. (5 marks)

c State which option Carlos should choose. Justify your reasoning. (2 marks)

18 P1: Consider the graph of the exponential function f defined by $f(x) = 16 - 4 \times 2^{-x}$.

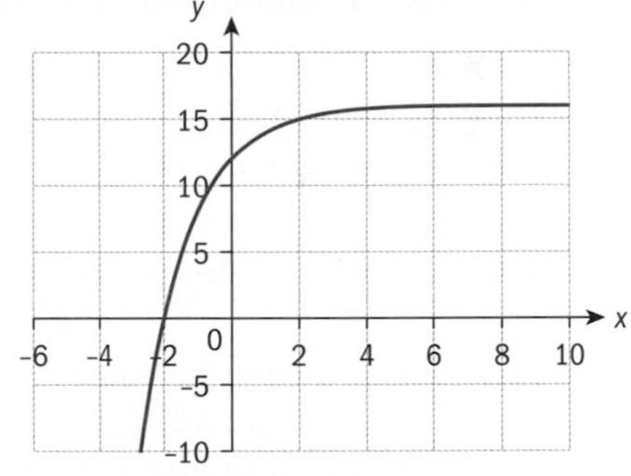

a State the domain of f. (1 mark)

b State the zero of f. (1 mark)

c State the y-intercept of the graph of f. (1 mark)

d State the range of f. (1 mark)

e State the interval where f is positive. (1 mark)

f State the equation of the horizontal asymptote of f. (1 mark)

19 P1: In a controlled experiment, the temperature T°C of a liquid, t hours after the start of the experiment, is $T = 25 + e^{0.4t}$, $0 \leq t \leq 12$.

a Sketch the graph of the temperature $T = T(t)$ for $0 \leq t \leq 12$. (2 marks)

b State the temperature halfway through the experiment, to the nearest 0.1°C. (1 mark)

c Find the time at which the temperature of the liquid reaches 100°C. Give your answer in hours and minutes, to the nearest minute. (3 marks)

20 P2: A controlled experiment took place in a laboratory. Biologists estimated the number of microbes N present in a culture t days after the start of their experiment. The first results obtained are shown in the top two rows of the table below.

t	3	5	10	15	20
N	115	178	480	900	1600
$\log N$					

a Complete the last row of the table, giving values to 2 decimal places. (2 marks)

b Draw a graph of $\log N$ against t, using appropriate scales. (4 marks)

c Hence, state the type of model that best fits the data displayed in part **b**. (1 mark)

d Dr Wise claim that the number of microbes can be modelled by $N(t) = A \times B^t$.

Explain why Dr Wise is correct. (4 marks)

e Hence, determine the values of the parameters A and B of Dr Wise's model, showing all your calculations. (5 marks)

21 P1: In a research laboratory, biologists studied the growth of a culture of bacteria. From the data collected hourly, they concluded that the culture increases in number according to the formula

$$N(t) = 35 \times 1.85^t$$

where N is the number of bacteria present and t is the number of hours since the experiment began.

Use the model to calculate:

a the number of bacteria present at the start of the experiment (1 mark)

b the number of bacteria present after four hours, giving your answer correct to the nearest whole number of bacteria (2 marks)

c the time it would take for the number of bacteria to exceed 1000. (2 marks)

Due to lack of nutrients, the culture cannot exceed 1000 bacteria.

d State the domain of validity of the model. (1 mark)

22 P1: Mr Williams is about to retire, and he needs to decide how his pension should be paid. His pension provider offers him three payment options:

Option 1: US \$10 000 each year for 10 years.

Option 2: US \$2500 in the first year, and then the amount increases by US \$2000 in each subsequent year for a for a total of 10 years.

Option 3: US \$100 in the first year, and then the amount doubles in each subsequent year for a for a total of 10 years.

Determine which option has the greatest total value. Justify your answer by showing all appropriate calculations. (6 marks)

23 P2: A large city is concerned about pollution and decides to look at the number of people using taxis.

At the end of the year 2010, there were 50 thousand taxis in the city. At the end of the nth year after 2010, the number of thousand taxis is given by $50 \times 1.1^{n-1}$.

a i Find the number of taxis in the city at the end of 2015. (2 marks)

ii Assuming that this model remains valid, find the year in which the number of taxis is double the number of taxis at the end of 2010. (2 marks)

At the end of 2010 there were 2 million people in the city who used taxis. n years later, the number of million people P_n who use taxis is given by $P_n = 2 \times 1.01^{n-1}$.

b Assuming that this model remains valid, find the value of P at the end of 2020, giving your answer to the nearest thousand. (3 marks)

Let R be the ratio of the number of people using taxis to the number of taxis. At the end of 2010, R was 43.6 : 1.

c Find the ratio R at the end of

i 2015 (2 marks)

ii 2020. (2 marks)

d Comment on the values found in part **c**. (1 mark)

24 P2: The diagram shows three graphs.

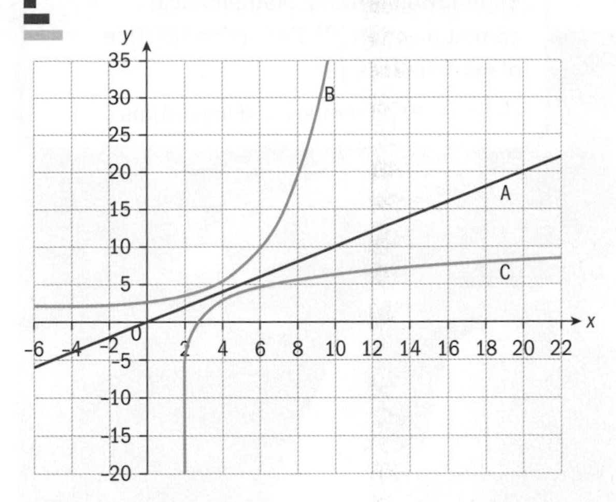

A is part of the line $y = x$.

B is part of the graph of $f(x) = \left(\frac{3}{2}\right)^{x-1} + 2$.

C is the reflection of graph of $f(x) = \left(\frac{3}{2}\right)^{x-1} + 2$ in the line $y = x$.

a State:

i the y-intercept of $y = f(x)$ (1 mark)

ii the coordinates of the point where C cuts the x-axis. (1 mark)

b Determine an equation for C in the form $y = g(x)$. (4 marks)

c Write down:

i the equation of the horizontal asymptote to the graph of $y = f(x)$ (1 mark)

ii the equation of the vertical asymptote to the graph of $y = g(x)$. (1 mark)

d Hence state the domain of g. (1 mark)

e Solve the equation $f(x) + g(x) = 0$. (2 marks)

25 P1: Katharina and Carolina go to a swimming pool. They both swim the first length of the pool in 2 minutes 6 seconds.

The time that Katharina takes to swim each subsequent length is 5 seconds more than she took to swim the previous length.

The time that Carolina takes to swim each subsequent length is 1.04 times that which she took to swim the previous length.

a **i** State the time Katharina takes to swim the third length.

ii Show that Carolina takes 2 minutes and 16 seconds (to the nearest second) to swim the third length. (5 marks)

Katharina and Carolina swim a total of 10 lengths of the pool. They start at exactly at the same time.

b Show that Katharina completes the 10 lengths before Carolina. (4 marks)

c Hence state the value of the time difference, correct to the nearest second, between their total times. (1 mark)

26 P2: Mr Garcia is offered a 10-year contract with a starting annual salary of 20 million Colombian pesos (COP) and yearly increases of 5% on the previous annual salary.

a Justify that Mr Garcia's annual salaries over the 10-year period form a geometric sequence, and state its common ratio. (2 marks)

b Determine Mr Garcia's annual salary during the third year of his contract. (2 marks)

c Find the total amount Mr Garcia will earn during his 10 years at the company, correct to the nearest thousand pesos. (3 marks)

After considering the company offer, Mr Garcia asks for a different payment plan. He wants to start with the same annual salary, but have fixed annual pay rises of x thousand pesos (COP). The company financial manager agrees to this new payment scheme if the total amount for the 10 years remains the same as it was the original payment offer.

d Determine the value of x, correct to the nearest peso, that satisfies the financial manager's conditions. (5 marks)

e Hence, determine the number of years for which the second plan gives Mr Garcia a higher income than the first. (4 marks)

A passing fad?

Approaches to learning: Communication, Research
Exploration criteria: Mathematical communication (B), Reflection (D), Use of mathematics (E)
IB topic: Exponentials and logarithms

Fortnite was released by Epic Games in July 2017 and quickly grew in popularity.

Here is some data for the total number of registered players worldwide from August 2017 to June 2018:

Date	August 2017	November 2017	December 2017	January 2018	June 2018
Months since launch, t	1	4	5	6	11
Number of registered players, P (millions)	1	20	30	45	125

The data is taken from the press releases of the developers, Epic Games.
Is this data reliable?
Are there any potential problems with the data that has been collected?
What other data might be useful?

Plotting this data on your GDC or other graphing software gives this graph:

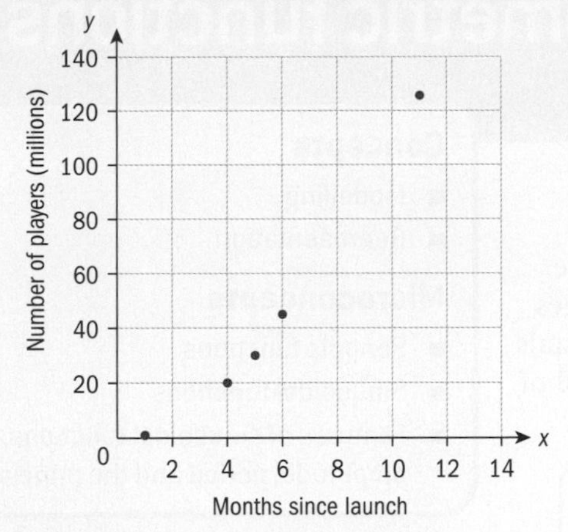

Describe the shape of the graph.

What could be some possible explanations for this growth?

Model the data

Why might it be useful to find a model that links the number of players of Fortnite against the number of months since the launch of the game?

Who might this be useful for?

Let t be the number of months since July 2017.

Let P be the number of players in millions.

Assume that the data is modelled by an exponential function of the form $P = a \cdot b^t$ where a and b are constants to be found.

Use the techniques from the chapter or previous tasks to help you.

Consider different models **by hand** and **using technology**.

How do these models differ? Why?

Which one is preferable? Why?

What alterations could be made to the model?

Use the model to predict the number of users for the current month.

Do you think this is likely to be a reliable prediction?

How reliable do you think the models are at predicting how many Fortnite players there are now? Justify your answer.

Research the number of players there are in this current month who play Fortnite.

Compare this figure with your prediction based on your model above. How big is the error?

What does this tell you about the reliability of your previous model?

Plot a new graph with the updated data and try to fit another function to this data.

Will a modified exponential model be a good fit?

If not, what other function would be a better model that could be used to predict the number of users now?

Extension

Think of another example of data that you think may currently display a similar exponential trend (or exponential decay).

Can you collect reliable and relevant data for your example?

Find data and present it in a table and a graph.

Develop a model or models for the data (ensure that your notation is consistent and your variables are defined). You could use technology or calculations by hand.

For how long do you think your model will be useful for making predictions?

Explain.

Modelling periodic phenomena: trigonometric functions

Concepts

- Modelling
- Representation

Microconcepts

- Periodic functions
- Sinusoidal functions
- Features of sinusoidal functions: amplitude, period and the principal axis

Many natural, and human-made, phenomena repeat themselves regularly over time. Tides rise and fall, the sun rises and sets, the moon changes its appearance, a clock's pendulum swings, crystals vibrate when an electrical current is applied. All of these things happen in a predictable fashion.

How can you model features of weather systems that change with time and geographical location?

Riga average temperature in degrees Celsius
2014 and 2015

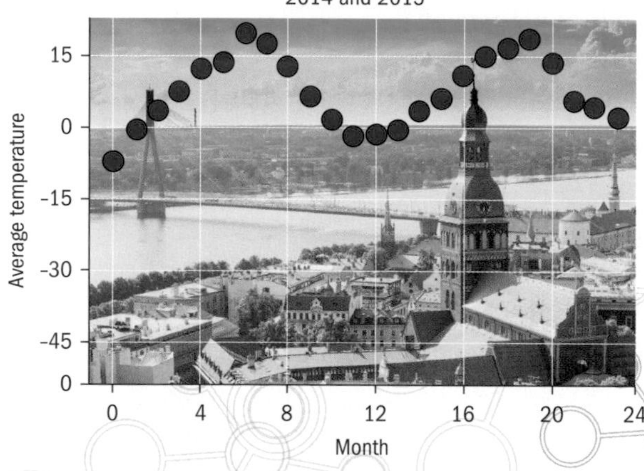

How can a biologist model the changes of populations in a predator–prey system?

Given the times of high and low tides on a given day, how could you decide when there would be enough water in a harbour to enter with a boat?

If there is a full moon tonight, how can we predict when the next new moon will occur?

The graph on the previous page shows the average monthly temperature for Riga, Latvia, compiled in 2014 and 2015. The graph on the right shows the number of babies born on each day of June 1978 in the USA.

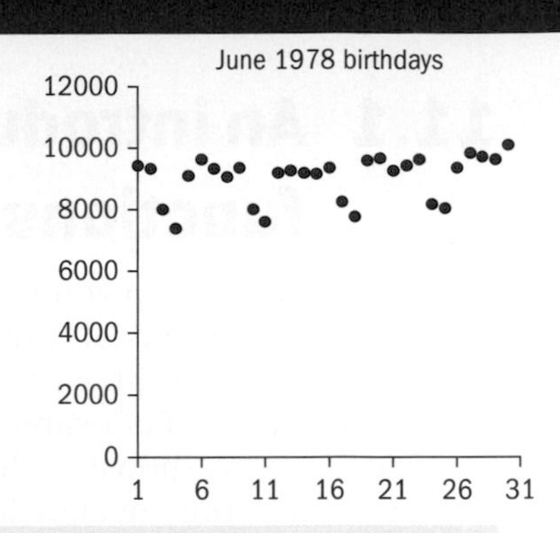

- What patterns can you see in these graphs?
- To what extent is your perception of these patterns dependent on the viewing window that has been chosen for you?
- How often do any repeating patterns repeat?
- What reasons could there be for any repeating patterns and the shape of the functions?
- If you developed models for these data sets, what would you expect to happen outside the domain of your functions?

Developing inquiry skills

Which other climate phenomena could have a repeating pattern? What questions could you ask for your country? What data would you need?

Think about the questions in this opening problem and answer any you can. As you work through the chapter, you will gain mathematical knowledge and skills that will help you to answer them all.

Before you start

You should know how to:

1 Calculate the length of the shortest side in triangle ABC.

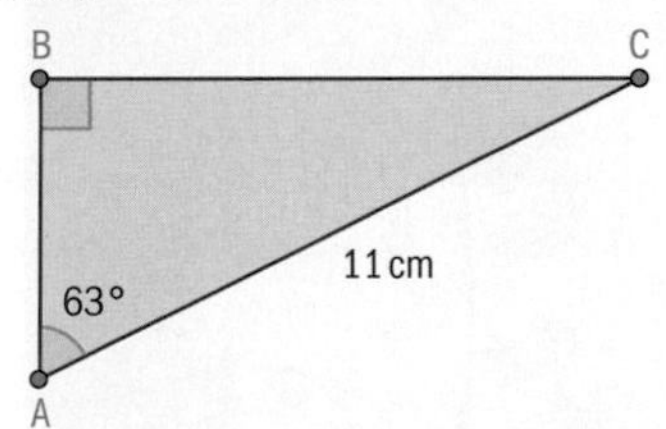

Since $\hat{C} = 27°$, AB will be the shortest side.
$AB = 11\sin 27° = 4.99$ cm to 3sf

2 Use technology to solve the equation $x^3 - x = 10x^2 - 200$ where $x > 5$.

a Graph each side of the equation as a function.

b Find the coordinates of the point(s) required.

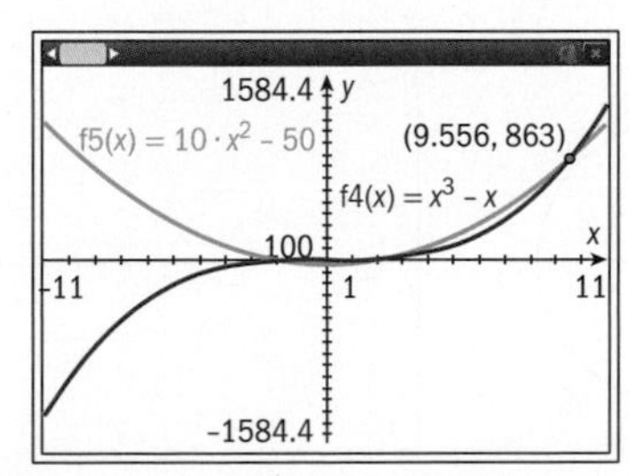

c Write the solution correct to three significant figures: $x = 9.56$.

Skills check

Click here for help with this skills check

1 Calculate all the angles and sides in the triangle PQR where $\hat{R} = 90°$, PQ = 13 m and QR = 5 m.

2 Karolina stands 73 m from the base of an apartment block and measures the angle of elevation to her apartment as 43°. What is the height of Karolina's apartment?

3 Use technology to solve the equations:

a $2^x - 17 = -10x - x^2$ where $x \leq 0$

b $19 + \dfrac{3}{t-11} = t^4$ where $t > 0$.

11.1 An introduction to periodic functions

Periodic functions help you to model a set of data in the special case where there is a repeating pattern of **identical** y-values. For example, the periodic functions shown to the right come from the study of electricity. The values of the pink function quantify power (watts) and the values of the green quantify current (amps). The independent variable is time in milliseconds.

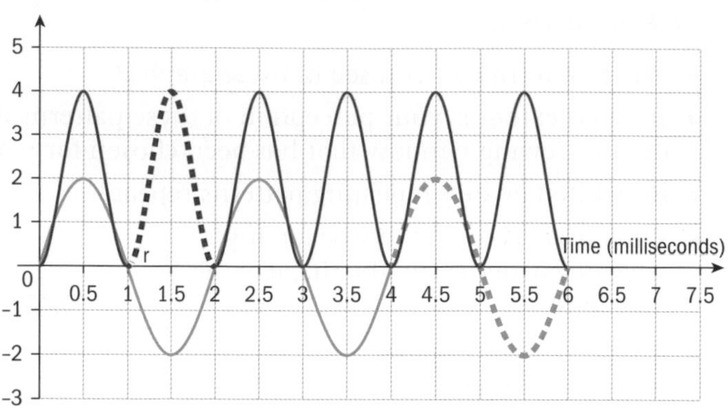

Notice that the values of the functions repeat regularly. The entire graph is a series of congruent shapes, which form a continuous curve. The **period** of the function is the length of the **shortest** interval for which the function values when repeated give the **entire** graph. Hence the period of the power function is 1 millisecond and of the current 2 milliseconds, as shown in the diagram by the dashed lines. This means that the relationship between time and power repeats itself each millisecond, and the relationship between time and current repeats itself every 2 milliseconds.

TOK

Is there always a trade-off between accuracy and simplicity?

In the following investigation you will explore a physical structure that gives rise to a periodic function.

For example, a new Ferris wheel attraction is planned for Alphapark to carry passengers in pods. Safety laws require that the operators of the wheel must know the exact position of each pod: the extent to which it is above or below the x-axis and the extent to which it is to the right or left of the y-axis.

You will model this context physically, and then with mathematics.

Investigation 1

Part A: Construction of a spaghetti curve

You will need: spaghetti (or drinking straws), protractor, string, ball of colourful wool, scissors, glue and poster paper. Work in groups of two or three.

1 At one end of the paper, construct a circle on a set of axes with a radius of one spaghetti length.

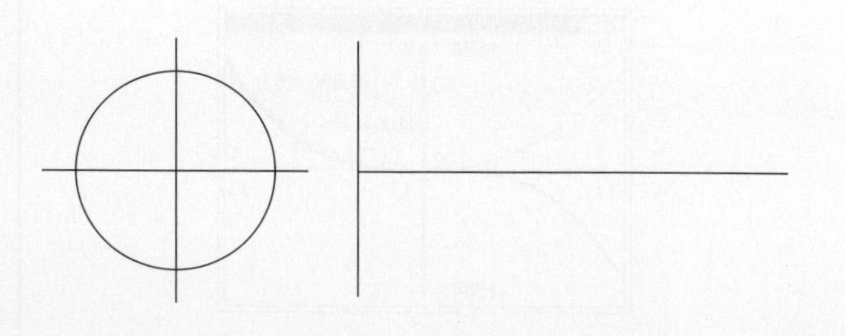

2 Use a protractor to measure and mark every 30° around the circle.

3 To the right of the circle draw a graph with an x-axis that is about 6.5 spaghetti lengths long and a y-axis 2 spaghetti lengths tall.

4 Place the string around the circle with the end at 0°. Mark the 30° marks onto the string. Then put the string on the x-axis of the right-hand set of axes and transfer the marks labelling every 30°.

5 Place a piece of spaghetti from the origin to the 30° mark on the circle. Take another piece of spaghetti and measure the vertical distance from the pod (P) to the x-axis. Cut the piece of spaghetti where it crosses the x-axis and glue this piece at 30° on the x-axis of the graph as shown. (Diagram not to scale.)

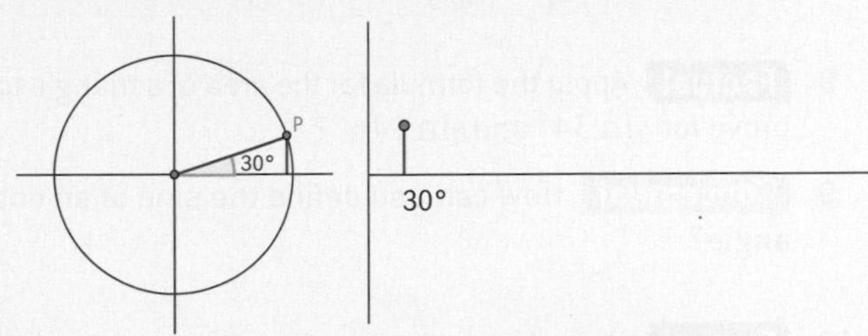

6 Now repeat for 30°, 60° and so on, until you have gone completely around the circle to 360°. If the pod is above the x-axis of the circle, then place the spaghetti above the x-axis of the graph; otherwise place it below the x-axis.

7 Take the wool and glue it to the graph on your poster from zero degrees, along the top of the spaghetti pieces to form a smooth curve.

Part B: Reflection on the properties of your spaghetti curve

1 **Factual** Your spaghetti curve is a periodic function. Write down its period.

2 **Factual** What do the y-values of your spaghetti curve represent in the context of the Ferris wheel?

3 **Factual** What are the maximum and minimum values of your curve and at what values of x do they occur? At what values of x is the value of your curve zero?

4 Graph the function $f(x) = \sin x$ on your GDC. Compare and contrast your spaghetti curve with the curve shown on your GDC.

Represent the angle shown in the **right-angled triangle** formed by the pod, the origin and the intersection with the x-axis as θ and let the radius of the circle be 1.

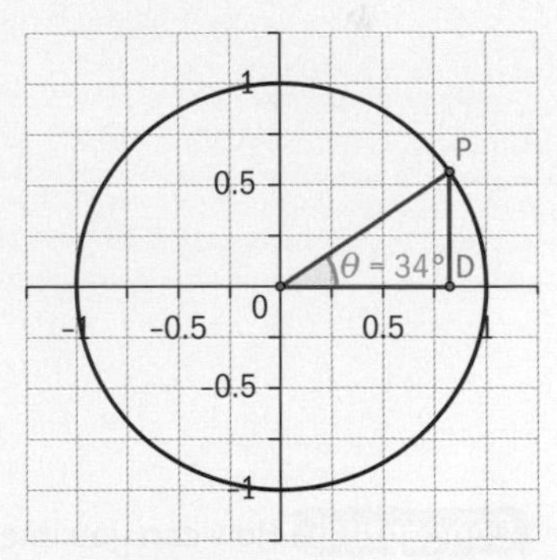

5 **Factual** Find the height of the pod above the x-axis in terms of θ.

6 **Factual** For what values of θ is your answer to **5** valid?

Now imagine the pod moving over the y-axis so that θ is greater than 90° but less than 180°.

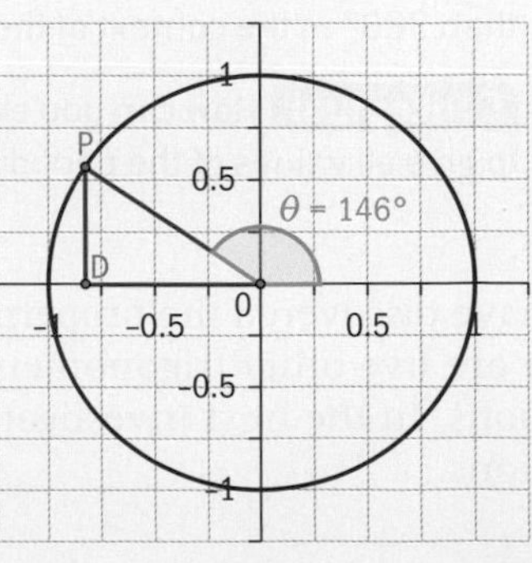

7 **Factual** Compare and contrast the position of the pod when $\theta = 146°$ and when $\theta = 34°$.

What does this suggest about $\sin 34°$ and $\sin 146°$?

Continued on next page

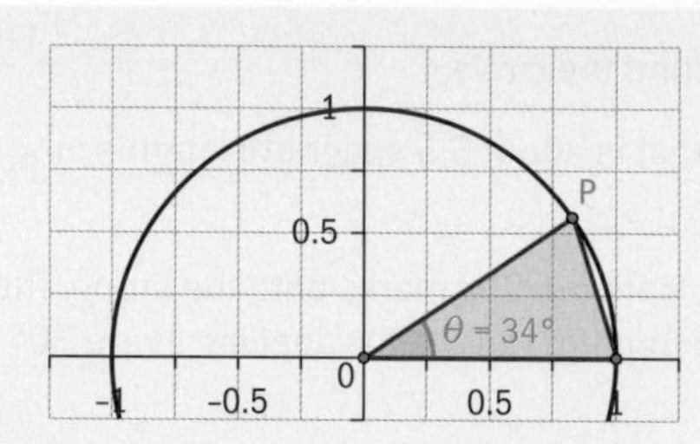

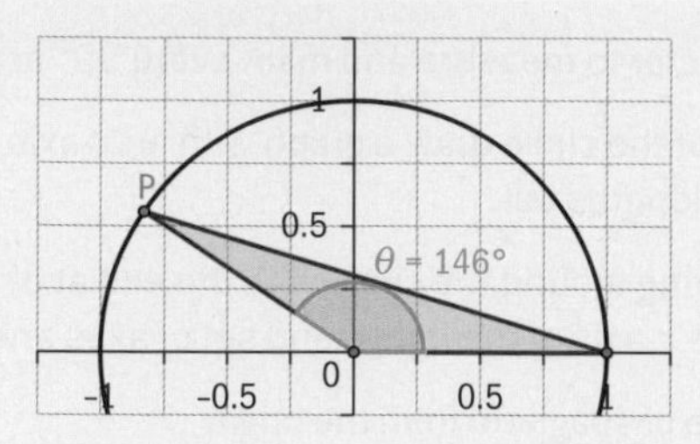

8 **Factual** Apply the formula for the area of a triangle to the two triangles shown here. What does this prove for $\sin 34°$ and $\sin 146°$?

9 **Conceptual** How can you define the sine of an obtuse angle in terms of the sine of an acute angle?

10 **Factual** When the pod is at position P, use geometry and trigonometry to find how far it is below the x-axis.

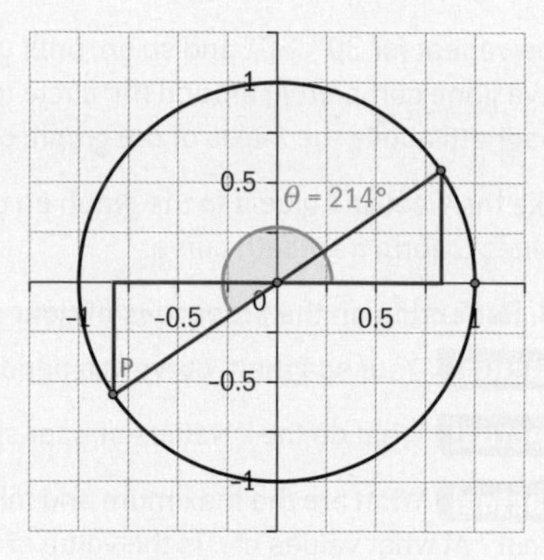

11 **Factual** When the pod is at position P, use geometry and trigonometry to find how far it is below the x-axis.

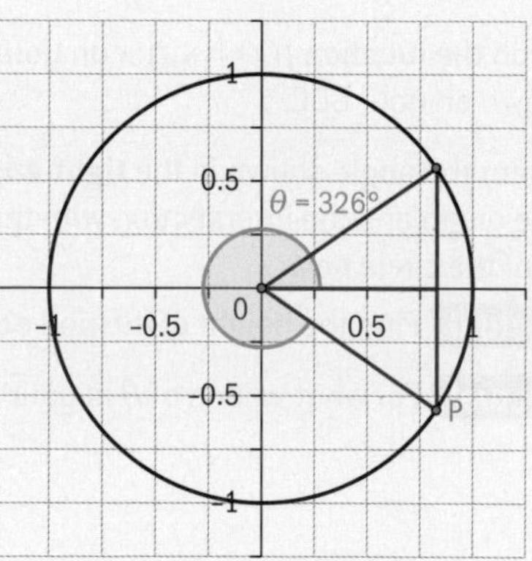

12 **Conceptual** How can you interpret the negative values of $\sin \theta$ when θ is greater than 180° but less than 360° in the context of the Ferris wheel?

13 **Conceptual** How can you extend the concept of the trigonometric ratio $\sin \theta$ in a right-angled triangle to give all values of the periodic function $f(x) = \sin x$?

You have discovered the properties of the periodic function $f(x) = \sin x$. There are five other trigonometric ratios that are also periodic functions. In the next investigation you will explore one of these functions.

Investigation 2

Part A: Construction of a new spaghetti curve

You will make a new periodic graph. Follow the same steps as in Investigation 1 part A, except this time measure the **horizontal** distance from the pod (P) to the y-axis. If the pod is to the right of the y-axis, assign a positive sign, if to the left of the y-axis, assign a negative sign.

Part B: Reflection on the properties of your new spaghetti curve

1 **Factual** Compare and contrast your function with that of $f(x) = \sin x$.

2 **Conceptual** Why should your new periodic function be periodic?

Now use your GDC to graph $g(x) = \cos x$ and $h(x) = \tan x$. Compare and contrast the graphs of these functions and your new spaghetti curve.

3 **Factual** Which of $g(x) = \cos x$ and $h(x) = \tan x$ model your new spaghetti curve?

4 **Factual** What are the maximum and minimum values of your curve and at what values of x do they occur? At what values of x is the value zero?

5 **Conceptual** Why is your new spaghetti curve modelled so well by one of these functions but not the other?

6 **Conceptual** How can you define the cosine of an obtuse angle in terms of the cosine of an acute angle?

Definition: $f(x) = \sin x$ and $g(x) = \cos x$ are examples of **sinusoidal functions**.

You can use technology to solve simple trigonometric equations in a given domain and hence solve problems.

Example 1

Given that the radius of the Ferris wheel planned for Alphapark is 70 m and that θ represents the angle measured anticlockwise from the x-axis, find all the angles for which the pod P is:

a a distance of 40 m vertically above the x-axis

b a distance of 20 m horizontally from the y-axis.

a 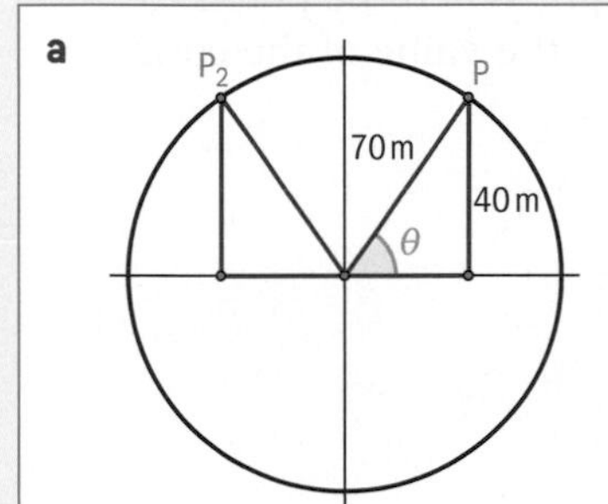	Sketching a labelled diagram is a good problem-solving strategy.
$\sin\theta = \dfrac{40}{70}$ can be graphed on the domain $0° \le \theta \le 360°$:	The diagram shows that there are two possible positions of the pod that are 40 m above the x-axis. The problem involves the vertical distance, hence sin is the right trigonometrical function to apply. This fact is consistent with the given quantities in the right-angled triangle sketched: the hypotenuse and the side opposite to the angle.

Continued on next page

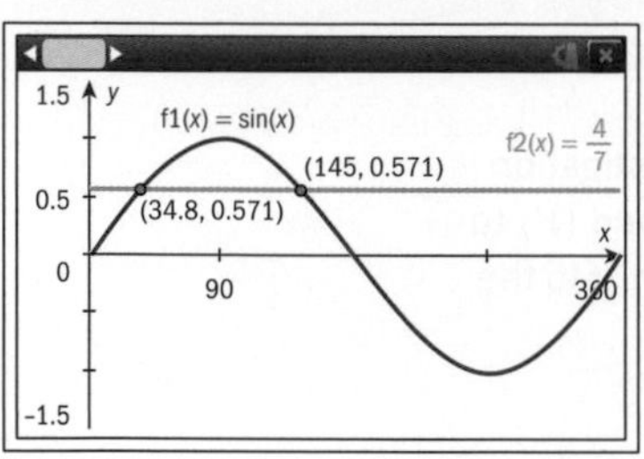

Make sure the viewing window shows the domain and the range of the function so that you can identify all solutions.

It is essential that you ensure your GDC is in degree mode at all times for both calculations **and** for graphing.

Find the intersections of the sine curve with the line $y = 4/7$ in order to find the solutions.

There are two values θ for which the pod is 40 m above the x-axis: 34.8° and 145° correct to three significant figures.

Reflect that this solution is consistent with the diagram sketched at the start. The GDC rounded the angles to 3 sf, however, if the angles were given to 1 dp, the values would be 34.8° and 145.2°, which add to 180° as required by the diagram.

b

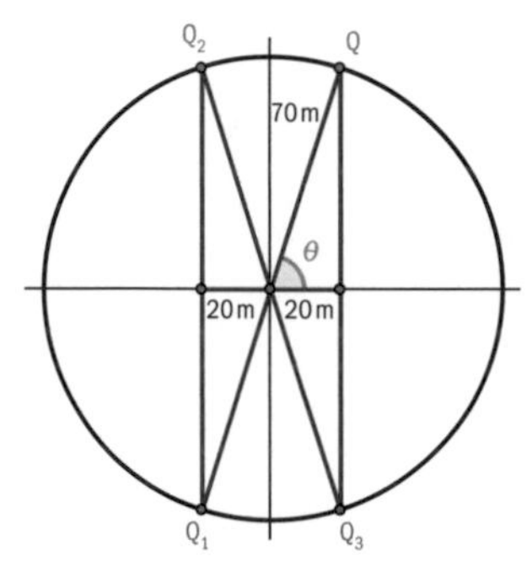

A sketch shows there are four solutions to be found.

The problem involves the horizontal distance, hence cos is the right trigonometrical function to apply.

$\cos\theta = \dfrac{20}{70}$ can be graphed on the domain $0° \le \theta \le 360°$:

This fact is consistent with the given quantities in the right-angled triangle sketched: the hypotenuse and the side adjacent to the angle.

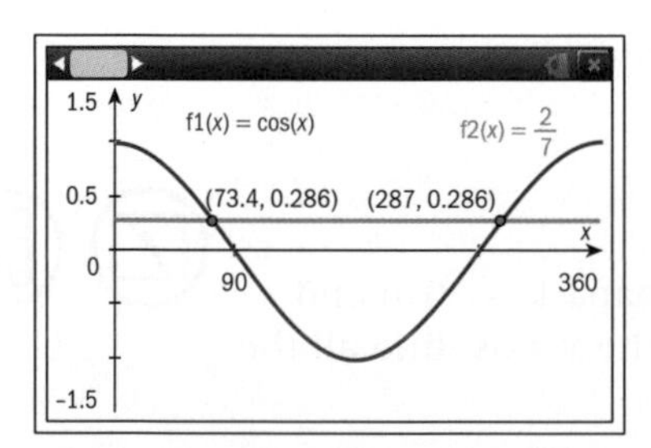

There are two values of θ for which the pod is 20 m to the right of the y-axis: and 73.4° and 287° correct to three significant figures.

Reflecting on the context of the problem helps you understand that more solutions can be found from the graph. When the pod is to the left of the y-axis, the value of the cosine function is negative.

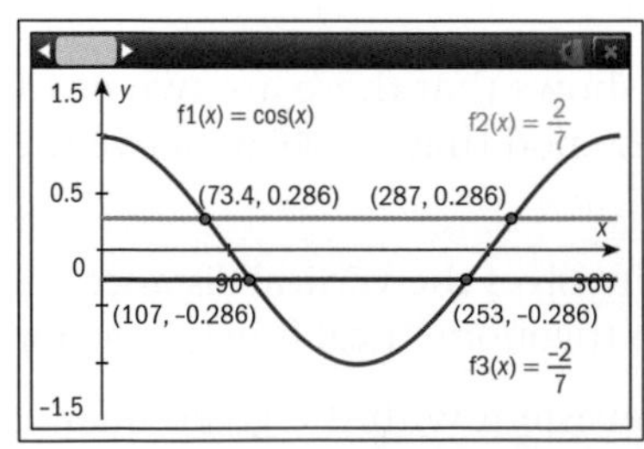

Altogether, there are four values of θ for which the pod is 20 m from the y-axis: 73.4°, 107°, 253° and 287° correct to three significant figures.

Only now is the solution consistent with the context of the problem.

Exercise 11A

1 Given that the radius of the Ferris wheel planned for Alphapark is 70 m and that θ represents the angle measured anti-clockwise from the x-axis, find all the angles for which the pod P is:

a a distance of 30 m from the x-axis

b a distance of 20 m to the right of the y-axis.

2 **a** Solve the following equations.

i $\sin\theta_1 = \cos\theta_1$ for $0° \le \theta_1 \le 360°$

ii $\sin\theta_2 = 0.8$ for $0° \le \theta_2 \le 360°$

iii $\cos\theta_3 = -0.1$ for $0° \le \theta_3 \le 360°$

b Hence find the coordinates of the pod on the Alphapark Ferris wheel for each value of θ_1, θ_2 and θ_3 found in part **a**.

3 Jakub models the average monthly temperature T degrees Celsius in Warsaw with the function $T = -11\cos(30t) + 7.5$ on the domain $0 \le t \le 11$. t is time measured in months, with $T(0)$ representing the average temperature in January.

a Predict the average temperature in May.

b Predict when the average temperature would be zero.

4 Zuzanna models the depth D metres of sea water in a harbour h hours after midnight with the function $D = 1.8\sin(30h) + 12.3$.

a Predict the depth of water in the harbour at 5.30am.

b Predict when the depth of water in the harbour will be 10.9 m.

Developing inquiry skills

Return to the opening problem. Do the functions appear to be periodic? If so, state their period.

Identify the type of function you could use to model this data.

11.2 An infinity of sinusoidal functions

In the previous section, we dealt with one Ferris wheel, and used a radius of 70 m to represent it. How would the graph change if the Ferris wheel were twice as tall?

When you alter the parameters k, r and c of the function $f(x) = ke^{rx} + c$, an infinity of different exponential models are created. Similarly, when you alter the parameters a, b and d of the sinusoidal functions $f(x) = a\sin(bx) + d$ and $g(x) = a\cos(bx) + d$ you create an infinite number of sinusoidal models.

In this section you will explore how to model this and other more general sinusoidal phenomena.

TOK

Solving an equation has given you an answer in mathematics, but how can an equation have an infinite number of solutions?

Definition: In a sinusoidal function with range $\min \le y \le \max$, the **principal axis** is the line $y = \frac{\min + \max}{2}$, the horizontal line located exactly in the middle of the range of the function, and the **amplitude** is the vertical distance from this line to any maximum or minimum points of the function.

For example,

$f(x) = \sin x$ and $g(x) = \cos x$ both have:

- range $-1 \le y \le 1$
- domain $x \in \mathbb{R}$
- period 360°
- principal axis $y = 0$
- amplitude 1.

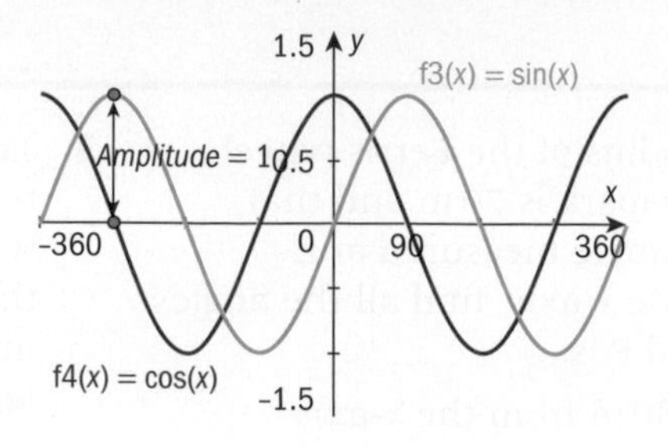

The function $h(x) = 3\sin(2x) - 2$ has:

- range $-5 \le y \le 1$
- domain $x \in \mathbb{R}$
- period 180°
- principal axis $y = \frac{-5+1}{2} = -2$
- amplitude 3.

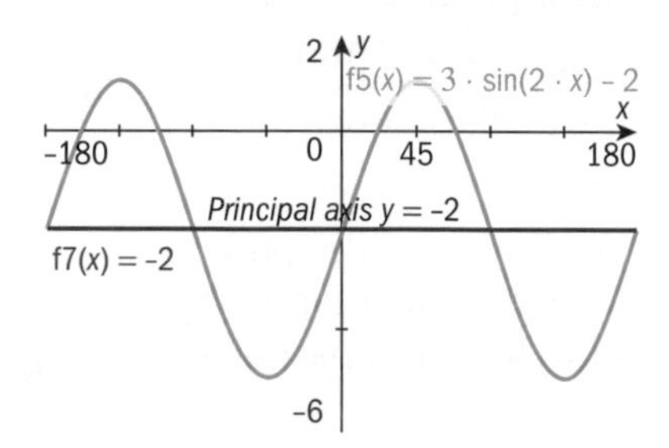

Investigation 3

1 Use technology to graph each set of functions and fill in the table:

	$y=2\sin(x)$	$y=2\sin(3x)$	$y=2\sin(3x)+1$
Amplitude			
Period			
Principal axis			
	$y=0.8\cos(x)$	$y=0.8\cos\left(\frac{x}{2}\right)$	$y=0.8\cos\left(\frac{x}{2}\right)-3$
Amplitude			
Period			
Principal axis			

Explore: Use dynamic geometry software or your GDC to create images like these and widen your exploration to include positive and negative values of a, b and d.

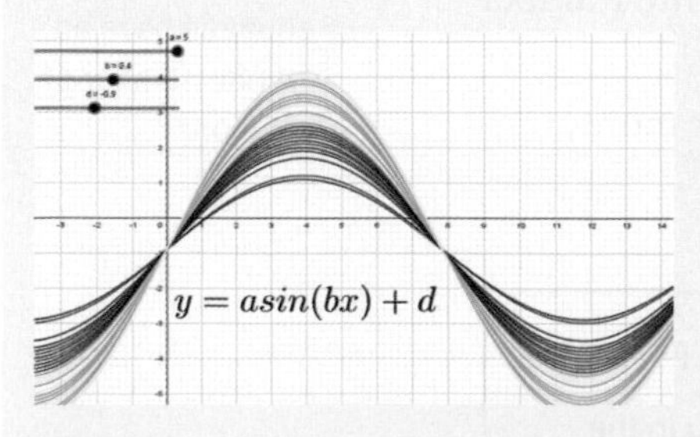

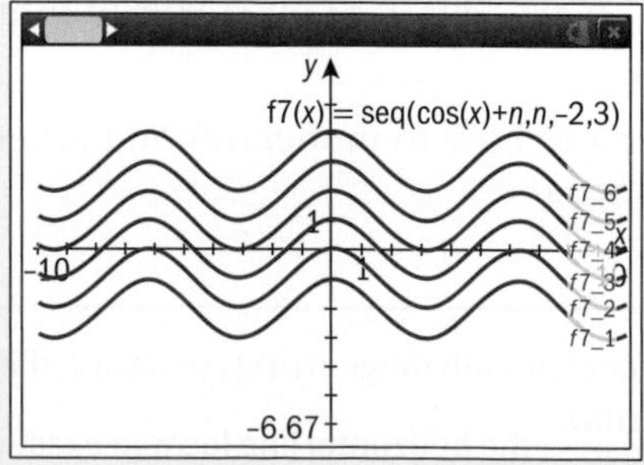

TOK

Why are there 360 degrees in a complete turn?

Write down more functions of the form $f(x) = a\sin(bx) + d$ and $g(x) = a\cos(bx) + d$. Try to predict the amplitude, period and principal axis before you graph with technology.

2 **Factual** Find the amplitude, period and principal axis of $y = 3.7\sin(2x) - 2$. Predict without technology and confirm with technology.

3 **Factual** Find the amplitude, period and principal axis of $y = -1.6\cos(0.25x) + 5$. Predict without technology and confirm with technology.

4 **Factual** Identify the correct statements about the sinusoidal functions

$f(x) = a\sin(bx) + d$ and $g(x) = a\cos(bx) + d$ with range $\min \le y \le \max$.

a i The amplitude is $|a|$, ii The amplitude is a.

b i $a = \dfrac{\max - \min}{2}$ ii $|a| = \dfrac{\max - \min}{2}$

c The period is equal to: i b ii $\dfrac{360°}{|b|}$ iii $\dfrac{360°}{b}$

d The principal axis has equation: i $y = |d|$ ii $y = d$

e i $d = \dfrac{\max + \min}{2}$ ii $|d| = \dfrac{\max + \min}{2}$

5 **Factual** What effect does changing the sign of a have on the orientation of the graph?

6 **Conceptual** How does changing the parameters of a sinusoidal function change its graph?

Example 2

The graph of a function of the form $y = -20\cos(3x)$ is shown on Clara's GDC.

a Write down the amplitude of the function and justify your answer.

b Write down the period of the function and justify your answer.

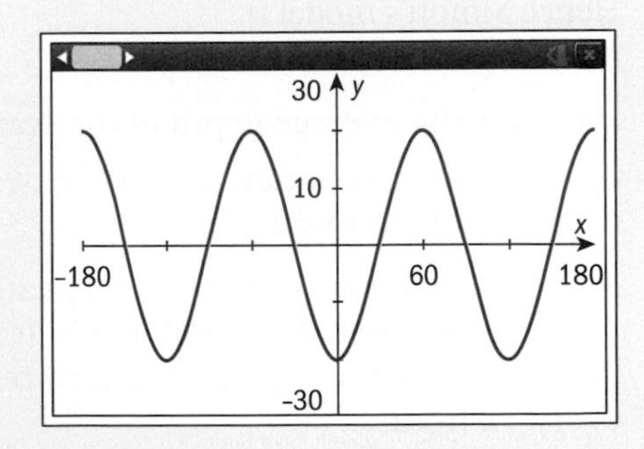

a Since the vertical distance from the principal axis to any turning point is 20 the amplitude is 20.	The graph is a reflection of the standard cosine graph in the x-axis; the amplitude is the same as that of $y = 20\cos(3x)$.
b It can be seen that the period is 120° from the graph. This is confirmed by the formula for the period $= \dfrac{360°}{3} = 120°$.	120° is the length of the shortest interval in which the values of the function when repeated give the entire graph.

Functions

Example 3

Simon explores tidal patterns by measuring the depth of water at the end of a pier. He finds that the data can be modelled by the function $f(x) = a\sin(bx) + d$ where x is the number of hours after 12 midnight. Simon's laptop crashes, but he finds a paper with some of his data:

Time	Coordinates	Description
3.00am	(3, 11.2)	Maximum
9.00am	(9, 5)	Minimum

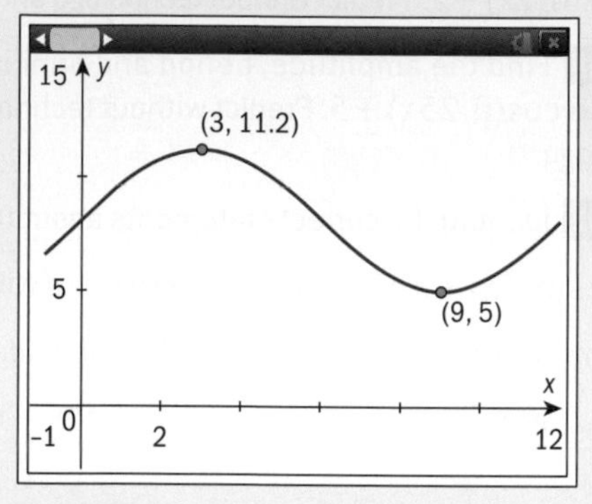

a Determine the values of a, b and d in Simon's model.

b Interpret the values of a, b and d in context.

a The principal axis is $y = \frac{11.2+5}{2} = 8.1$. Hence d is 8.1. The amplitude is $\frac{11.2-5}{2} = 3.1$. Hence a is 3.1. One half of a cycle is $9 - 3 = 6$ hours long. Hence the period is 12 hours.	Apply the formulae for the principal axis and the amplitude.
$b = \frac{360}{12} = 30$	Apply the formula for the period.
Hence Simon's model is $f(x) = 3.1\sin(30x) + 8.1$.	This can be verified with technology—it gives the exact graph shown above.
b $d = 8.1$ is the average depth of the water. $a = 3.1$ is the maximum amount by which the depth differs from 8.1. $b = 30$ means that the function repeats 30 times in a domain of 360 hours. It is more useful to state that the tidal pattern repeats every 12 hours.	Reflect back to the original context.

TOK

"Beauty is the first test; there is no permanent place in the world for ugly mathematics,"—Godfrey Harold Hardy (1877–1947), a Professor of Geometry at the University of Oxford has been called the most important British pure mathematician of the first half of the 20th century.

What do we mean by elegance in mathematical proof?

Exercise 11B

1 For each function, predict the amplitude, the period, the equation of the principal axis and the range.

a $f(x) = 3\sin(4x) + 1$

b $f(x) = 0.5\sin(0.5x) - 3$

c $f(x) = 7.1\cos(3x) + 1$

d $f(x) = -5\cos\left(\dfrac{x}{2}\right) + 8.1$

2 For the following graphs, determine the numerical values of the unspecified constants:

a

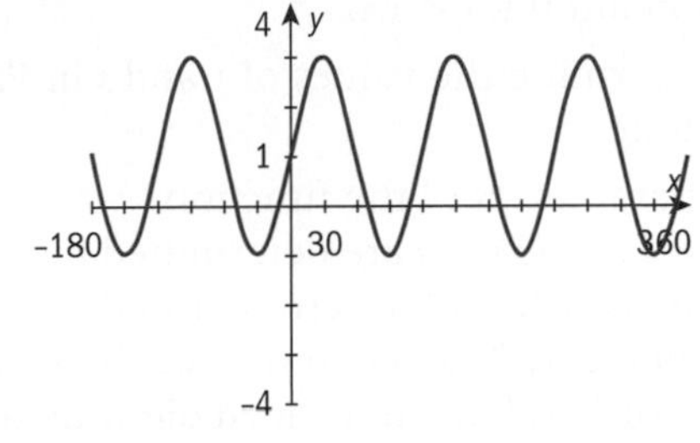

$y = a\sin(3x) + 1$

b

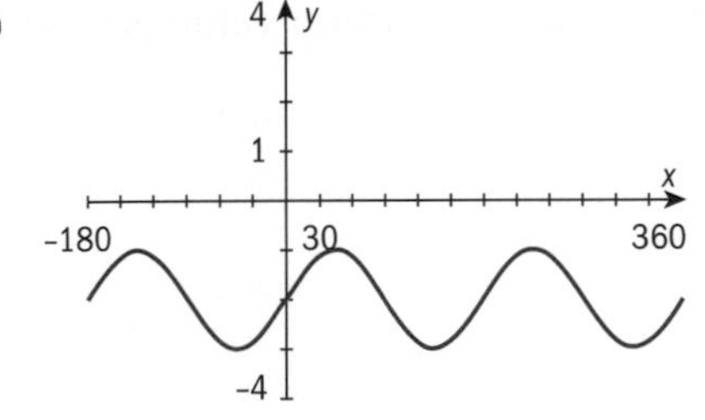

$y = a\sin(bx) - 2$

c

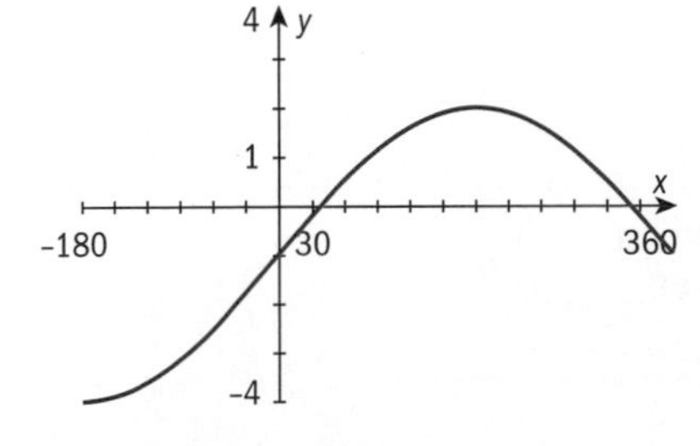

$y = a\sin(bx) + d$

3 For the following graphs, determine the numerical values of the unspecified constants:

a

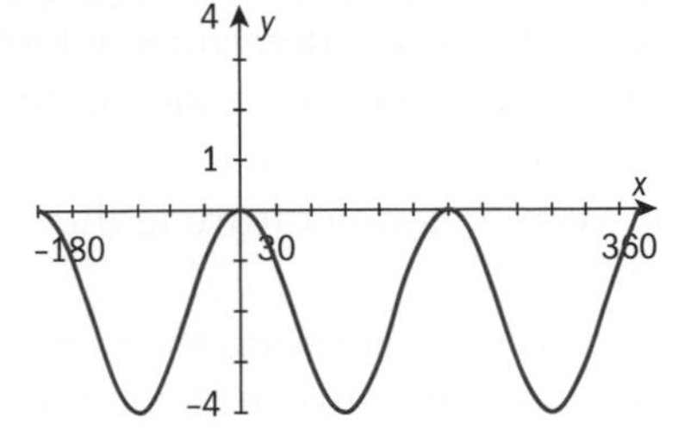

$y = 2\cos(bx) - 2$

b

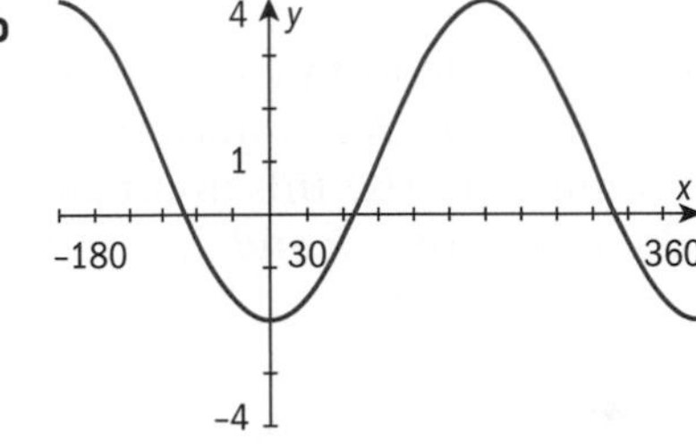

$y = a\cos(x) + d$

c

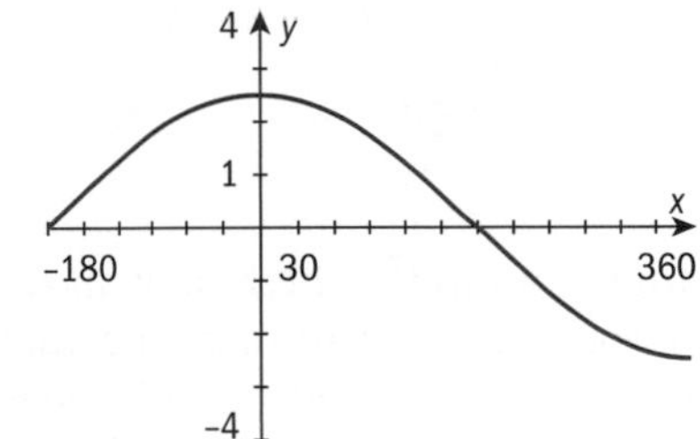

$y = a\cos(bx) + d$

4 a A function in the form $y = t + s\cos(rx)$ is shown. The point Q is a maximum and P is a minimum. Use this information to find the values of t, s and r.

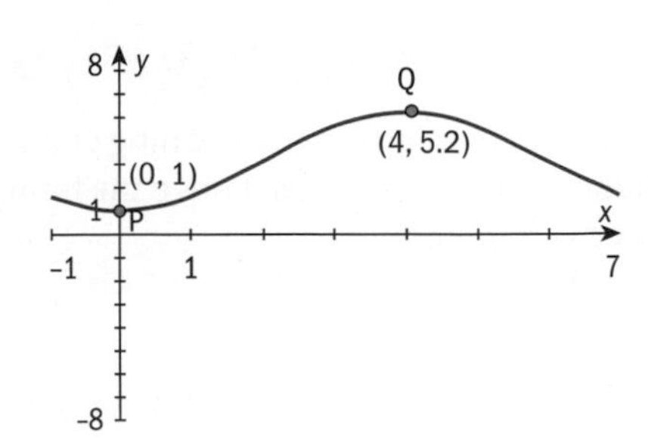

b This function models the temperature in a refrigerator unit. Given that y is measured in degrees Celsius and x is time measured in hours, determine which of t, s and r are related to the following quantities:

i the average temperature in the unit

ii the difference between the minimum temperature in the unit and the average temperature

iii the time taken for the unit to complete an entire cycle of operation.

5 a A function in the form $y = a\sin(bx) + d$ is shown. The point A is a maximum and B is a minimum. Use this information to find the values of a, b and d.

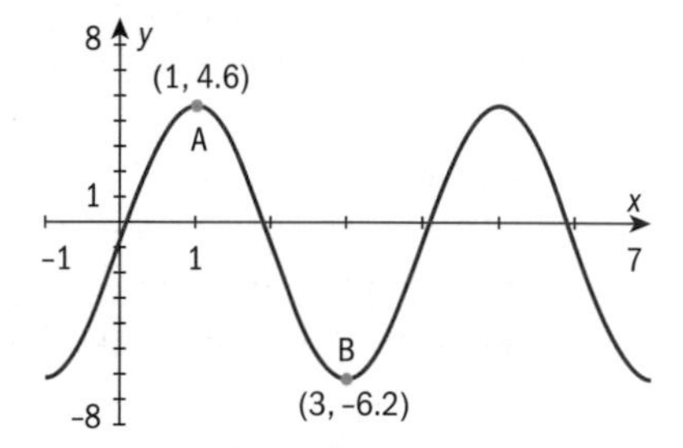

b This function models the height above or below sea level of a part of a renewable energy generator. Given that y is measured in metres and x is time measured in minutes, interpret each of the values of a, b and d.

6 Pam explores ocean level patterns influenced by currents by measuring how far the water level varies above and below a point on an oil rig. She finds that the data can be modelled by the function $f(x) = r\cos(sx)$ where x is the number of hours after 12 midnight. Due to safety concerns, Pam can only observe the first two minimum points and she defines the principal axis as $y = 0$.

Time	Coordinates	Description
2.30am	(2.5, −10)	Minimum
7.30am	(7.5, −10)	Minimum

a Sketch a graph of Pam's model for the domain $0 \le x < 15$.

b Determine the values of r and s in Pam's model.

7 The parameters of the function $g(x) = a\cos(bx) + d$ are determined by throwing dice. A fair tetrahedral die numbered 1, 2, 3, 4 determines the value of a and of b. A fair three-sided die numbered 1, 2, 3 determines the value of d.

Find the probability that the resulting function has a maximum value greater than 5.

Developing inquiry skills

If one of the functions in the opening problem can be modelled by a sinusoidal function, does it have the form $y = a\sin(bx) + d$ or $y = a\cos(bx) + d$?

11.3 A world of sinusoidal models

In this section you will apply your knowledge and understanding of periodic functions to explore whether a certain context can be modelled by a sinusoidal function. As an example, consider the following:

TOK

Sine curves model musical notes and the ratios of octaves. Does this mean that music is mathematical?

1 **Situation:** Maria is planning an ecosystems trip to Verkhoyansk, a Russian town near the Arctic Circle, in order to study the permafrost. She knows that Verkhoyansk has a very cold climate and that the temperatures vary greatly.

2 **Problem:** To aid her planning, Maria wants to predict:

  a At what times in the year does the temperature change fastest? This has implications for the clothing she needs to pack.

  b At what dates will the temperature be between −20°C and −5°C? This has implications for the design of her experiments.

3 **Data:** Maria gathered this data from the internet:

Month	Average temperature, °C
1	−47
2	−43
3	−30
4	−13
5	2.4
6	12.9
7	15.5
8	11.1
9	2.5
10	−14.5
11	−35.9
12	−42

4 **Develop a model**

Maria graphed her data using technology and noticed the data resembles a sinusoidal function with a period of 12 months. She then applied the knowledge you learned in the previous section to calculate the parameters:

$$|a| = \frac{\max - \min}{2} = \frac{15.5 - -47}{2} = 31.25$$

$$b = \frac{360^\circ}{12} = 30$$

$$d = \frac{\max + \min}{2} = \frac{15.5 + -47}{2} = -15.75$$

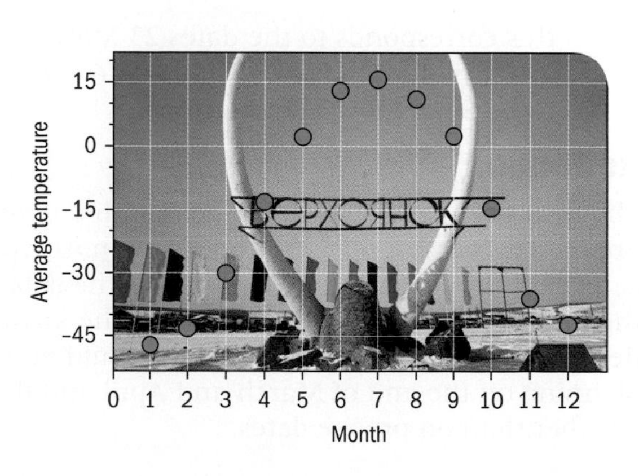

5 Test the model using technology

Maria notices that the data follows the orientation of a cosine graph reflected in the x-axis so she proposes that $y = -31.25\cos(30x) - 15.75$ would be an appropriate model and she tests it with technology.

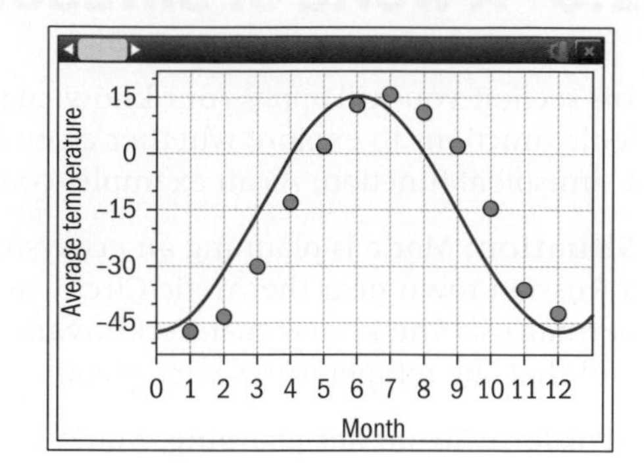

Maria reflects on the fact that the model seems to be the correct shape but not in the correct position. She then remembers that the maximum value of $\cos x$ occurs at $x = 0$; however, her data set starts at $x = 1$.

So Maria represents January with $x = 0$, February with $x = 1$ and so on in order to make the function easier to apply.

Finally, Maria is satisfied with her model.

6 Apply the model

a From the graph, Maria sees that between the x-coordinates of 2 and 4 the gradient of the graph increases by approximately 15°C per month and between 8 and 10 there is a decrease of approximately 15°C per month. So, during the periods March–May and September–November she can expect considerable changes in temperature and needs to pack accordingly.

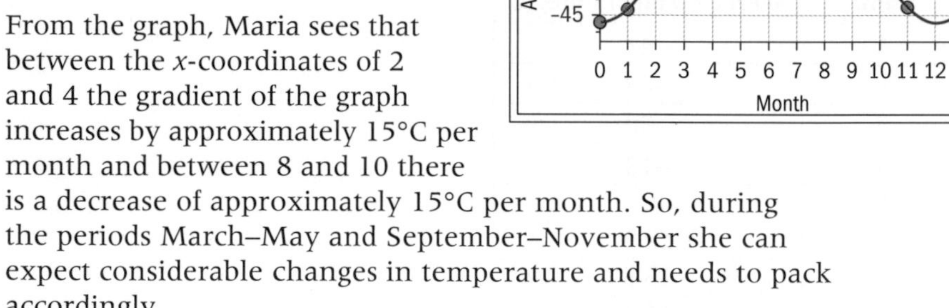

b Allowing for the x-axis starting at 0, Maria can estimate that between 3.74 and 4.67 months the temperatures will be between −20°C and −5°C.

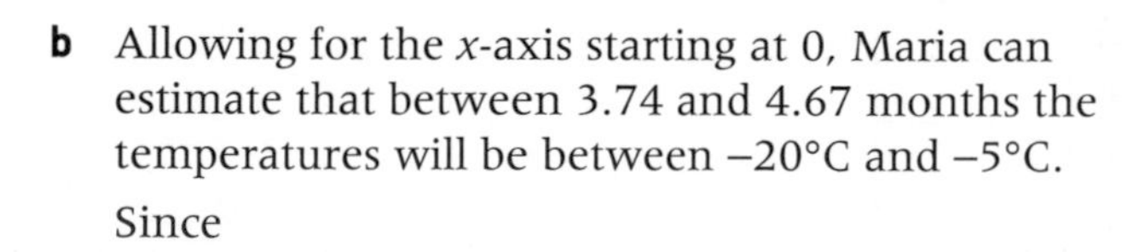

Since

$0.74 \times 31 \approx 23$ and $0.67 \times 30 \approx 20$

this corresponds to the dates 23 March and 20 April. Similarly, 9.33 and 10.26 correspond to 10 September and 8 October.

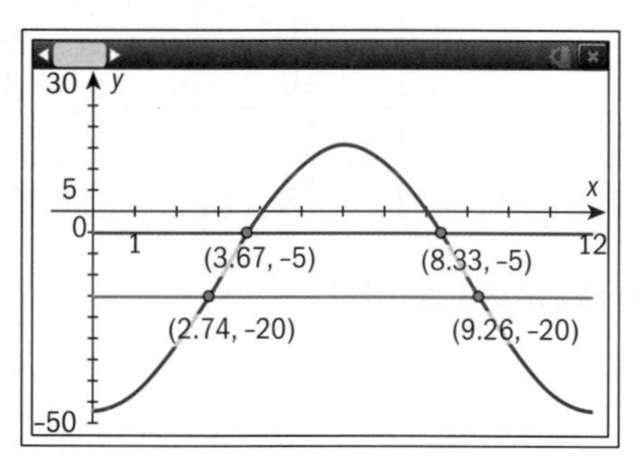

Reflection

The data used represents average monthly temperature. Therefore, the predictions cannot be considered any more precise than an average is, even if the graph of the function fits the data well. As a consequence, the dates found are only guidelines. The advantage of this model lies in simplicity not precision. Maria could more reasonably focus her planning on the end of March and April and the start of September and October than on precise dates.

Investigation 4

Aim: Collaborate to create a world map annotated with temperature graphs across different latitudes and longitudes in order to look for patterns, practise the modelling process and consider temperature data in a global perspective.

Three longitudinal lines for 80°W, 20°E and 110°E have been added to this map of the world in order to guide the sample of temperature data from large vertical landmasses. Pick one or more lines.

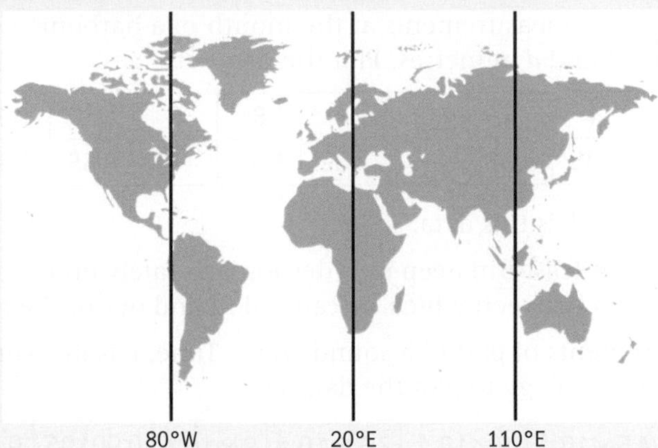

In a group, choose cities close to your line with a wide range of latitudes. For example, Ushuaia is the most southerly city in the world and Quito is very close to the equator. Choose one city each and follow the steps **3**, **4** and **5** of Maria's permafrost example.

Then **apply the model** to predict **a** the temperature in your chosen city on 31 October this year; **b** the temperature in your chosen city on your 40th birthday.

1 **Factual** Which of **a** and **b** is a more valid answer and why?

Collaborate: Collect all your graphs together in order to look for patterns in temperature data models in relation to latitude. Use technology to create a presentation of a sequence of temperature graphs to present to your peers.

2 **Factual** In which longitudes does the sinusoidal model fit best? Are there any cities in which the sinusoidal model is less useful? What factors influence the data apart from latitude?

3 **Factual** How does latitude affect the amplitude, period and principal axis of the models?

4 **Factual** When using technology to present a sequence of functions, is it always best to use the window chosen by default by the device? Explain your answer.

5 **Conceptual** When using technology, how does the choice of viewing window affect your ability to communicate patterns effectively?

6 Which natural phenomena can you think of that could be modelled by sinusoidal functions?

TOK

Trigonometry was developed by successive civilizations and cultures.

To what extent is mathematics a product of human social interaction?

Developing your toolkit

Now do the Modelling and investigation activity on page 520.

Exercise 11C

1 **a** Plot this data set.

x	10	11	12	13	14	15	16	17	18	19	20
y	13.4	11.2	8.3	5.2	4.1	5.2	8.5	11.8	12.7	11.5	8.3

b Find a model for the data using a function of the form $y = a\sin(bx) + d$.

2 **a** This data set represents water depth measurements at the mouth of a harbour. T is measured in hours from midnight and d in metres. Plot the data set.

T	0	1	2	3	4	5	6	7	8	9	10	11	12
d	10	9.68	8.55	7	5.45	4.32	4	4.32	5.45	7	8.55	9.68	10.1

b Show that a sinusoidal function models this data.

c Sam's boat needs the water to be at least 6 m deep in order for it to safely enter the harbour. Predict the range of hours between which he can sail in and out of the harbour.

3 **a** This data set represents measurements of part of a sound wave. Time, t, is measured in seconds and D in decibels. Use technology to plot the data set.

t	0	1.5×10^{-4}	3×10^{-4}	4.5×10^{-4}	6×10^{-4}	7.5×10^{-4}	9×10^{-4}	0.00105	0.0012	0.00135
D	0	0.4029	0.7375	0.9471	0.9961	0.8763	0.6079	0.2365	−0.175	−0.5569

b Find a model for the data using a function of the form $y = a\sin(bx) + d$.

c Hence predict the decibels of the sound after 0.00011 seconds.

d Hence predict the decibels of the sound after 0.002 seconds.

e Comment on which of your answers to **c** and **d** are more reliable, giving a reason for your answer.

4 Pam is researching old steam engines. The following diagram shows a representation of a mechanism she is exploring.

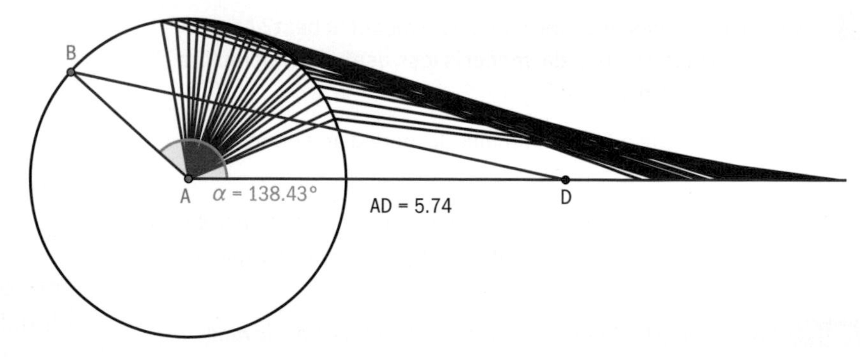

AB is fixed at 2.4 m and BD is fixed at 7.7 m. D can move only horizontally back and forth as B moves anticlockwise around the circle. AD is the distance from the point D to the centre of the circle. α is the angle $D\hat{A}B$.

The following data is collected.

α	1.06	12.2	18.2	26.6	45.9	72.1	100	149	174	196	224	252	277	295
AD	10.1	10.0	9.94	9.77	9.17	8.10	6.91	5.54	5.31	5.37	5.79	6.61	7.62	8.38

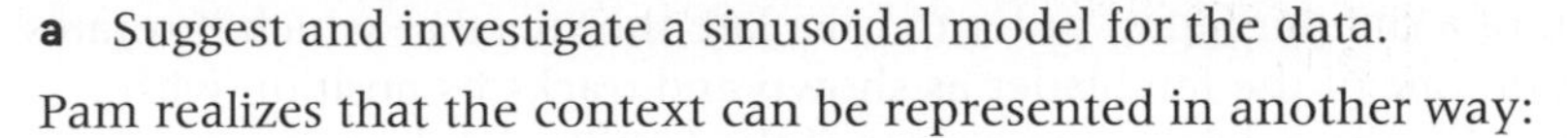

a Suggest and investigate a sinusoidal model for the data.

Pam realizes that the context can be represented in another way:

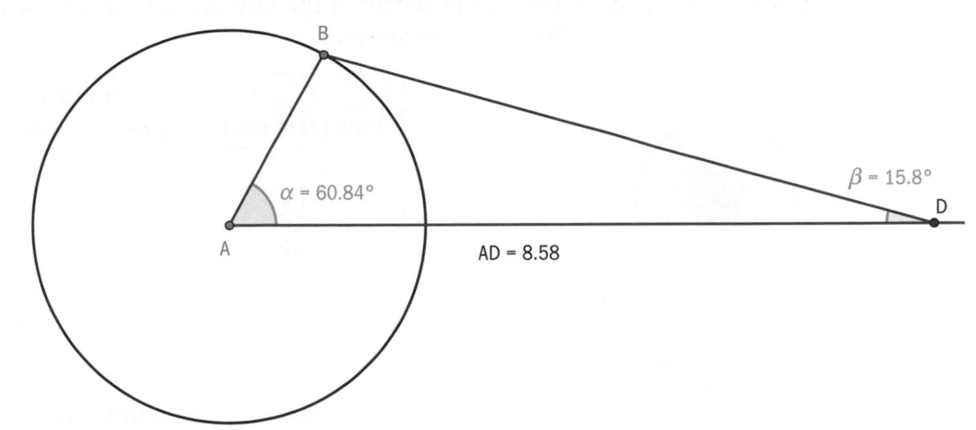

b Calculate AD in terms of the angles α and β and the lengths of AB and BD, using the cosine rule.

c Hence or otherwise comment critically on the fit of your model.

---

5 The times of sunrise and of sunset in Reykjavik, Iceland, on the 1st of each month are represented in this table using hours and minutes.

Month (t)	Rise	Set
0	11.20	15.42
1	10.09	17.13
2	8.37	18.43
3	6.47	20.16
4	5.01	21.49
5	3.24	23.27
6	3.05	23.57
7	4.34	22.34
8	6.09	20.47
9	7.35	19.00
10	9.09	17.13
11	10.44	15.49

a Calculate the times in decimal notation. Explain why this is necessary.

b Find sinusoidal models for sunrise $r(t)$ and for sunset $s(t)$.

c Draw the graph of $d(t) = s(t) - r(t)$, hence estimate:

i the number of days in the year with at least 18 hours of daylight

ii the fraction of a year with at least 12 hours of daylight but no more than 18.

6 Karim collects sunrise data for Cairo on the 15th of each month from www.sunrisesunset.com and represents them in this table as decimals.

Month (t)	Rise
0.5	6.87
1.5	6.58
2.5	6.08
3.5	5.48
4.5	5.03
5.5	4.88
6.5	5.07
7.5	5.37
8.5	5.65
9.5	5.95
10.5	6.33
11.5	6.72

a Find a sinusoidal model for the sunrise $r(t)$.

b Karim wishes to take time-lapse photographs at the Pyramid of Cheops on the Giza Plateau on the outskirts of Cairo. He wants to arrive there at least one hour before sunrise to set up his equipment and to capture a range of images as the sun rises. However, he does not want to arrive any earlier than 5.00am. Predict from your model between which dates he can arrive.

Functions

7 Zaida analyses the motion of a circular fan in order to calculate the speed at which it rotates. She places a coloured dot on one of the fan blades as shown and tracks its position with video tracking software, placing the origin at the centre of the fan. Zaida's laptop runs out of memory; however, she manages to collect the data shown:

Time (t) in secs	y-coordinate of green dot (cm)
0	0
0.01	15
0.02	25.98076211
0.03	30
0.04	25.98076211
0.05	15
0.06	3.675×10^{-15}

a Find a model of the y-coordinate of the green dot as a function of time.

b Hence calculate the speed of the fan.

Chapter summary

- Periodic functions help you to model a set of data in the special case where there is a repeating pattern of **identical** y-values.
- The **period** of the function is the length of the **shortest** interval for which the function values when repeated give the entire graph.

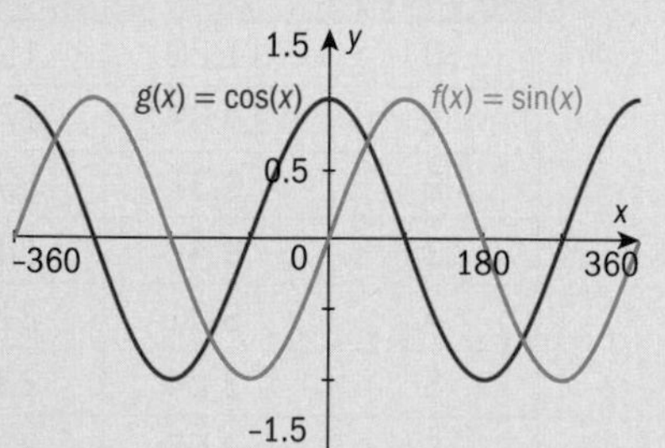

- Two examples of periodic functions are $f(x) = \sin x$ and $g(x) = \cos x$. They both have range $-1 \le y \le 1$ and are examples of **sinusoidal functions.** Both have period 360°.
- In a sinusoidal function with range $\min \le y \le \max$, the **principal axis** is the line $y = \dfrac{\min + \max}{2}$ and the **amplitude** is the vertical distance from this line to any turning point of the function, as shown in this example.

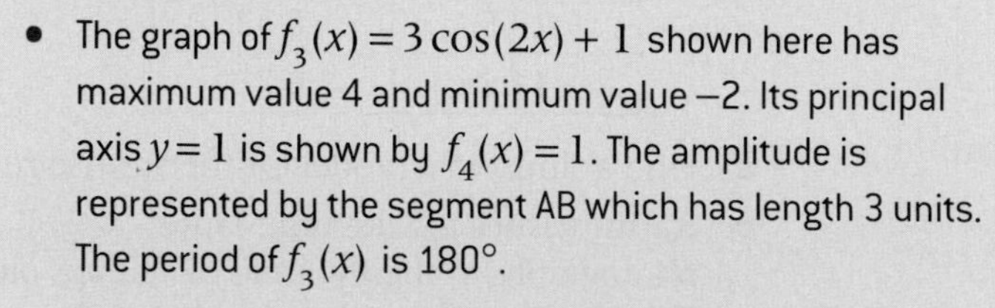

- The graph of $f_3(x) = 3\cos(2x) + 1$ shown here has maximum value 4 and minimum value −2. Its principal axis $y = 1$ is shown by $f_4(x) = 1$. The amplitude is represented by the segment AB which has length 3 units. The period of $f_3(x)$ is 180°.

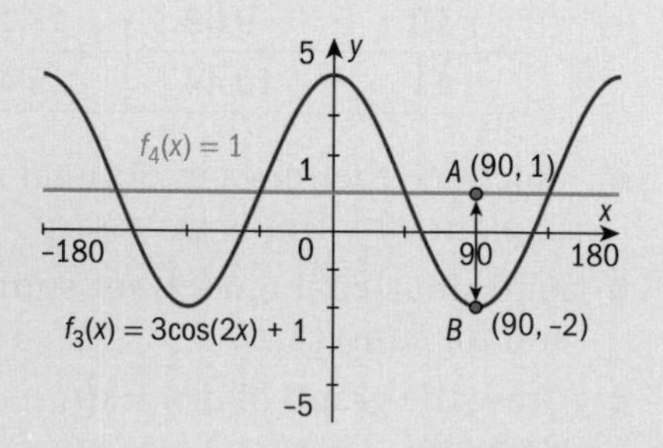

- The parameters a, b and d in the functions $f(x) = a\sin(bx) + d$ and $g(x) = a\cos(bx) + d$ enable you to **predict** the amplitude, period and principal axis respectively.

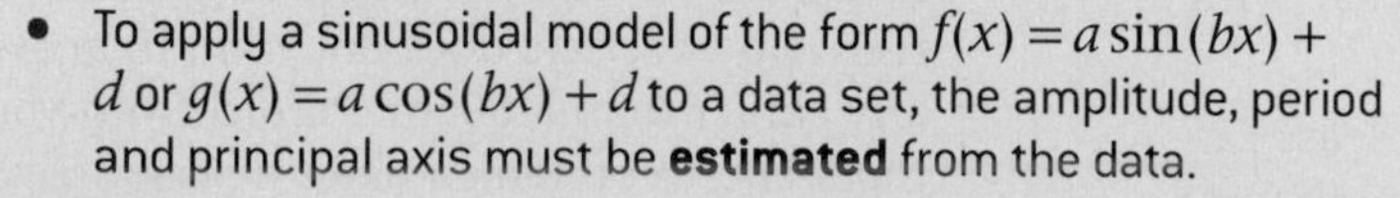

- To apply a sinusoidal model of the form $f(x) = a\sin(bx) + d$ or $g(x) = a\cos(bx) + d$ to a data set, the amplitude, period and principal axis must be **estimated** from the data.

Developing inquiry skills

Return to the opening problem. You were given a graph showing the average monthly temperature for Riga compiled over a period of two years and a graph showing the number of babies born on each day of June 1978 in the USA.

Now that you have learned about sinusoidal models, you can reflect on the fact that a sinusoidal model fits only one of these sets of data well.

Which one?

Why is that?

What kind of phenomena can be modelled by periodic functions?

What kind of phenomena can be modelled by sinusoidal models?

Chapter review

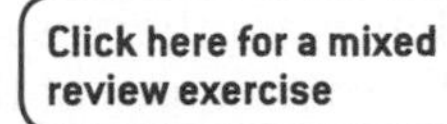

1 Solve the equation $-0.4\cos t + 1 = 1.3$ where $301° \leq t \leq 700°$.

2 For the following graphs, find the equation of the function.

a

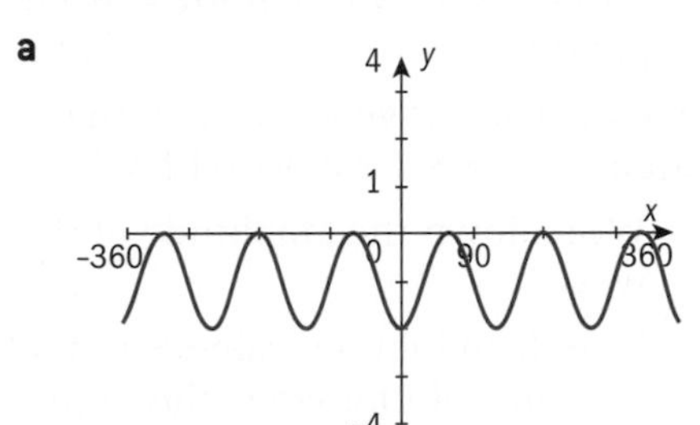

b

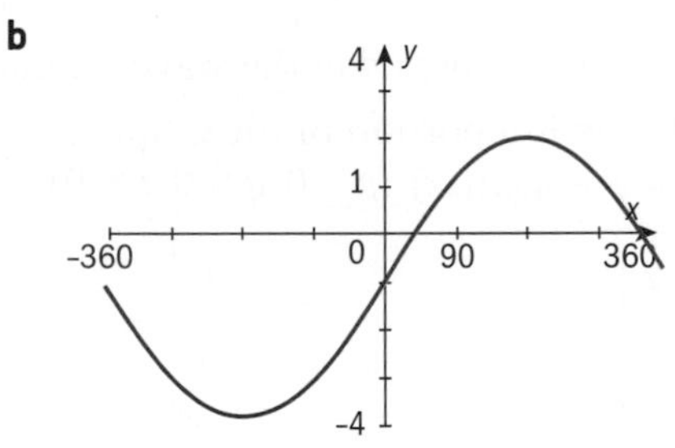

c

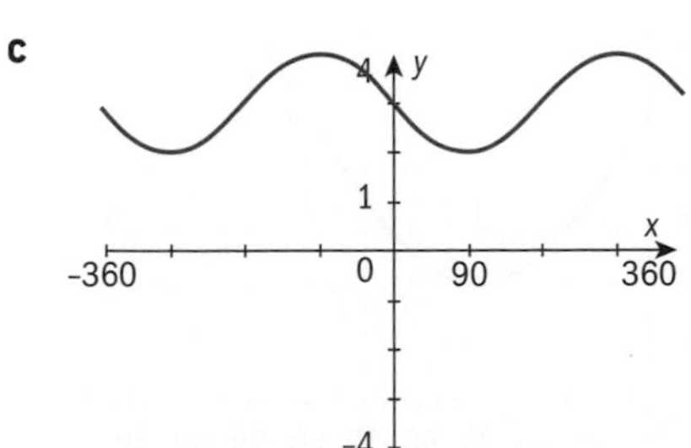

3 Given $f(x) = 3x$ and $g(x) = -2\cos x$, consider the functions

$a(x) = f(x) \times g(x)$, $b(x) = f(x) + g(x)$, $c(x) = f(g(x))$ and $d(x) = g(f(x))$.

Find the period, amplitude and equation of the principal axis for any of these functions that are periodic.

4 The graph of a sinusoidal function $f(x)$ on the domain $[-360°, 360°]$ is shown below.

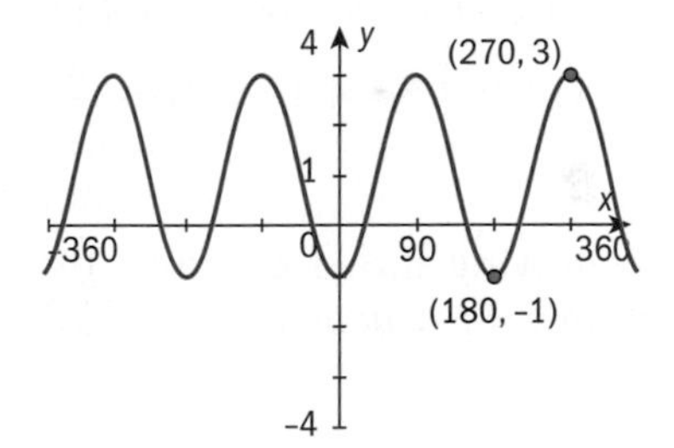

There is a maximum at $(270, 3)$ and a minimum at $(180, -1)$.

a Find the equation of the function.

b The equation $f(x) = \beta$ has exactly five solutions. Find the value of β.

5 A water wave passes a sensor in a physics laboratory. As the wave passes, the depth of the water in metres (d) after time t seconds is modelled by the function $d = 18.3 - 0.47\cos(20.9t)$.

a What are the maximum and minimum values of d?

b Find the first time after 19 seconds at which the depth is 18 m.

6 Teodora examines old photos of the Neue Elbbrücke bridge in Hamburg. She finds out that the three sections shown are each 102 m long and she estimates that the vertical distance from the surface of the road to the top of the bridge is 16 m and that the road is 6 m above the river. She uses this information to model the bridge with functions as shown below:

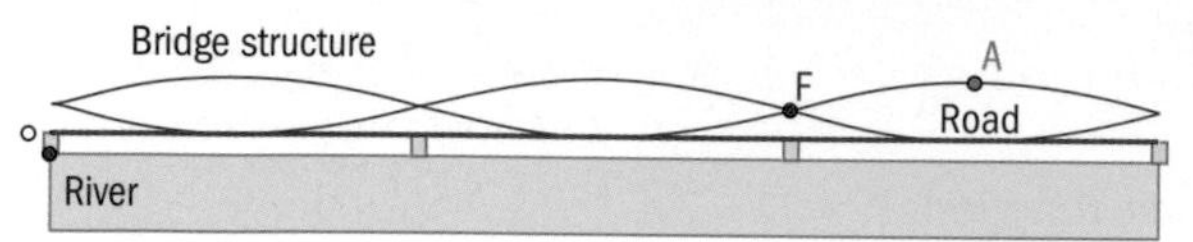

a Find a sinusoidal model of the bridge and draw a diagram like Teodora's.

b Hence or otherwise find an estimate for the coordinates of the points shown.

Exam-style questions

7 P1: Consider the curve given by the equation $y = 4.1 + 2.3\cos\left(\frac{x}{2}\right)$.

a Find the minimum and maximum values for y. (2 marks)

b Find the first two positive values of x for which $2.5 = 4.1 + 2.3\cos\left(\frac{x}{2}\right)$. (3 marks)

8 P1: The following shows a portion of the graph of $y = p + q\cos(rx)$.

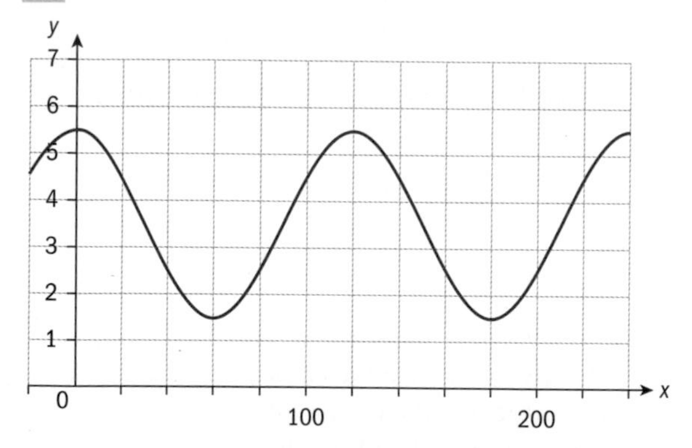

Determine the values of the constants p, q and r. (6 marks)

9 P1: A sinusoidal function is given by the equation $y = 20\sin(4x)$ for $0° \le x \le 180°$.

a Sketch the graph of $y = f(x)$. (3 marks)

b State the equation of the principal axis of the curve. (1 mark)

c Using technology, solve the inequality $f(x) < -5$. (4 marks)

10 P1: The following diagram shows a sinusoidal curve of the form $y = p\sin(qx + r)$.

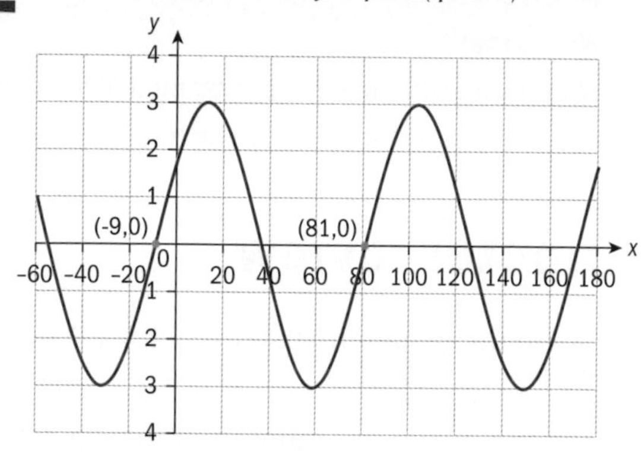

Determine the values of the constants p, q and r. (8 marks)

11 P1: A sinusoidal wave is given by the equation $y = 5.7 + 0.3\cos(12.5x)$.

a Write down the amplitude of the wave. (1 mark)

b Find the minimum value of y and first the value of x for when this occurs. (3 marks)

c Find the period of the wave. (2 marks)

12 P1: Below is a portion of the graph of $y = p - q\sin(rx)$ $(p > 0, q > 0, r > 0)$.

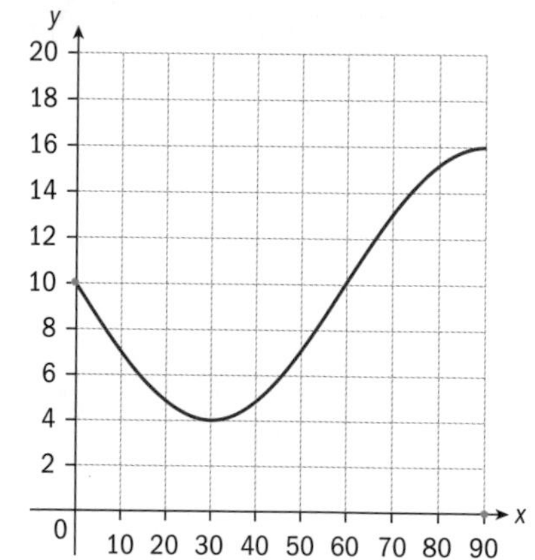

a On the same axes, sketch the graph of $y = p + \frac{q}{2}\sin(rx)$ (2 marks)

b Determine the values of the constants p, q and r (7 marks)

13 P1: A sinusoidal curve $y = a + b\sin(cx)$ $(a > 0,$ $b > 0,\ c > 0)$ has two concurrent maximum points at coordinates $(60°, 5)$ and $(300°, 5)$.

Given that it also has an amplitude of 2 units, determine the values of a and b, and determine the least possible value of c. (8 marks)

14 P2: Over the course of a single December day in Limassol, Cyprus, the highest temperature was found to be 22°C. The lowest temperature was 12°C, which occurred at 03.00am

If t is the number of hours since midnight, the temperature T may be modelled by the equation $T = A\sin(B(t - C)) + D$.

a Find the values of A, B, C and D. (9 marks)

b Hence, by using technology, find for how many hours the daily temperature was above 20°C. (4 marks)

The sound of mathematics!

Modelling and investigation activity

Approaches to learning: Research, Critical thinking, Using technology
Exploration criteria: Mathematical communication (B), Personal engagement (C), Use of mathematics (E)
IB topic: Trigonometric functions

Brainstorm

In small groups, brainstorm some ideas that link music and mathematics.

Construct a **mindmap** from your discussion with the topic "MUSIC" in the centre.

Share your mindmaps with the whole class and discuss.

This task concentrates on the relationship between **sound waves** and **trigonometric functions.**

Research

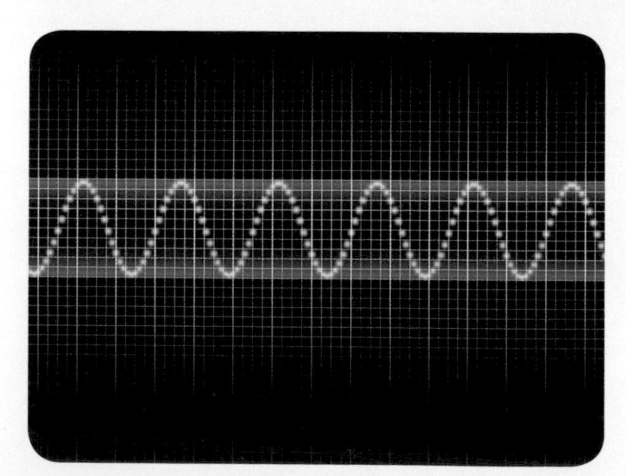

Research how sound waves and trigonometric functions are linked.

Think about:

How do vibrations cause sound waves?

How do sound waves travel?

What curve can be used to model a sound wave?

The fundamental properties of a basic sound wave are its **frequency** and its **amplitude.**

What is the frequency of a sound?

What units is it measured in?

What is the amplitude of a sound?

Use what you studied in this chapter to answer these questions:

If you have a sine wave with the basic form $y = a \sin(bt)$, where t is measured in seconds, how do you determine its period, frequency and its amplitude?

What do the values of a and b represent? What does y represent in this function?

With this information, determine the equivalent sine wave for a sound of 440 Hz.

Did you know?

This is in fact the sine wave equation for the note A.

Technology

There are a large number of useful programs that you can use to consider sound waves.

Using these programs it is possible to record or generate a sound and view a graphical representation of the sound wave with respect to time.

If you have a music department in your school and/or access to such a program, you could try this.

Design an investigation

Using the available technology and the information provided here, what could you investigate and explore further?

What experiment could you design regarding sounds?

Disucss your ideas with your group.

What exactly would be the aim of each investigation, exploration or experiment that your group thought of?

Select one of the ideas in your group and plan further.

What will you need to think about as you conduct this exploration?

How will you ensure that your results are reliable?

How will you know that you have completed the exploration and answered the aim?

Extension

Trigonometric functions also occur in many other areas that aid your understanding of the physical world.

Some examples are:

- temperature modelling
- tidal measurements
- the motion of springs and pendulums
- the electromagnetic spectrum.

They can be thought of in similar terms of waves with different frequencies, periods, amplitudes and phase shifts too.

Think about the task that you have completed here and consider how you could collect other data that can be modelled using a trigonometric function. You could use one of the examples here or research your own idea.

12 Analyzing rates of change: differential calculus

Calculus is a mathematical tool for studying change. This chapter will focus on differentiation, which looks at how quickly one variable is changing **with respect to** another.

Change is all around us. Examples in physics include change of position, velocity, density, current and power; in chemistry we have rates of reaction and in biology the growth of bacteria; in economics rates include marginal cost and profit; in sociology we might want to measure the speed at which a rumour spread or periodic changes in fashion. All of these can be applications of differentiation. But how do we go about measuring these changes?

Concepts

- Change
- Relationships

Microconcepts

- Limit
- Rate of change
- Derivative
- Increasing and decreasing functions
- Related rate of change
- Tangent line
- Normal line
- Local maximum and minimum

How fast can an athlete throw a javelin?

How much should a new car cost?

How can a trader determine whether profit margins are increasing or decreasing?

How fast can a signal travel along a fibre-optic cable?

A firm conducts a survey in order to find the optimum price for a new product. The results are shown in the table:

Price (US$$x$)	Percentage who would buy at this price (d)
5	68
10	55
15	43
20	33
25	22
30	14
35	8
40	4

The firm decides that a **demand** equation is needed to model these results. Two models are suggested: a power model of the form $d = \frac{a}{x^b}$ and an exponential model of the form $d = a(c)^x$.

- Plot the data on your GDC.
- Explain why these two models might be suitable.
- Find best fit equations for each of the models.
- The business would like to maximize its profit. What other information would it need?
- If this information were available, how might you work out the maximum value of a profit function?
- The marketing team have a model that links extra demand with the amount of money spent on advertising. How can this be incorporated into your model?

Developing inquiry skills

What other information might you need to work out the profit?

What could the company do in order to persuade more people to buy the product?

How might this affect the model?

Think about the questions in this opening problem and answer any you can. As you work through the chapter, you will gain mathematical knowledge and skills that will help you to answer them all.

Before you start

Click here for help with this skills check

You should know how to:

1 Expand brackets

For example, expand and simplify

$(2x - 3)(x + 4)$

$= 2x^2 + 8x - 3x - 12$

$= 2x^2 + 5x - 12$

2 Find the equation of a line

eg Find the equation of the line with gradient 5 passing through the point (2, 1).

$1 = 5(2) + c$

$c = -9 \quad y = 5x - 9$

3 Find the area and volume of solids

eg Find the curved surface area and volume of a cylinder with radius 3 cm and height 5 cm.

Curved area $= 2\pi \times 3 \times 5 = 94.25\ \text{cm}^2$

Volume $= \pi \times 3^2 \times 5 = 141.4\ \text{cm}^3$

4 Laws of indices

eg

a $\dfrac{x^5}{x^3} = x^{5-3} = x^2$ **b** $\dfrac{1}{x^2} = x^{-2}$

5 Find maximum and minimum values of a function using your GDC

eg Find the coordinates of the maximum value of $y = x^3 - 8x^2 + 15x$

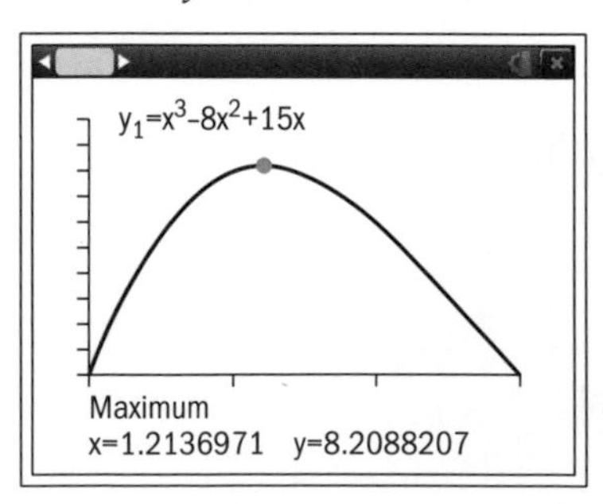

(1.21, 8.21)

Skills check

1 Expand the following:

a $(x - 5)(3x + 2)$

b $x(2x - 1)(3x + 5)$

2 Find the equation of the line perpendicular to the line $y = 2x - 3$ and passing through the point (3, 4).

3 Find the total surface area and the volume of a cone with radius 4 cm and vertical height 6 cm. Give your answers in terms of π.

4 Simplify the following:

a $\dfrac{x^2}{x^7}$ **b** $\dfrac{1}{x}$

5 Find the minimum value of $y = x^4 - 5x^3 - 4x^2 + 20,\ 0 \le x \le 2$

12.1 Limits and derivatives

The graph below shows how the profits of a company increase with the number of widgets it sells.

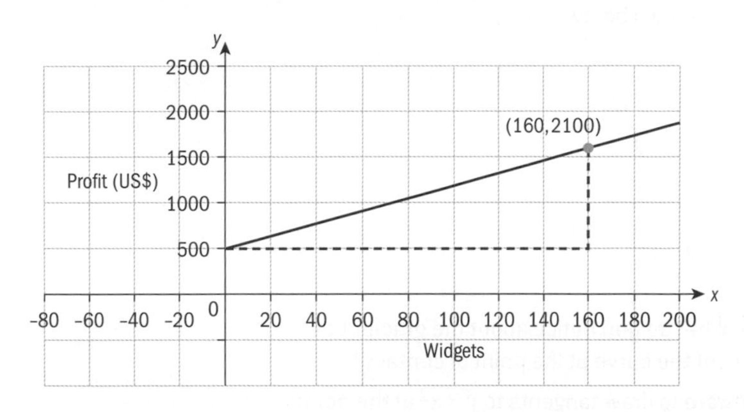

From previous work on linear functions you will know that the rate at which the profit increases for each new widget made (profit per widget) is $\frac{2100-500}{160-0}=10$ US$ per widget.

This is equivalent to the **gradient** of the curve. If the curve had a steeper gradient, the profit per widget would be higher; if the gradient had a less steep gradient, the profit per widget would be lower.

Hence the **rate of change** of one variable with respect to another is equivalent to the **gradient** of the line.

This is easy to calculate when the functions are **linear**. The following investigation will consider how the gradient of a curve at a point might be calculated.

TOK

What value does the knowledge of limits have?

Investigation 1

Consider the curve $y=x^2$ shown below.

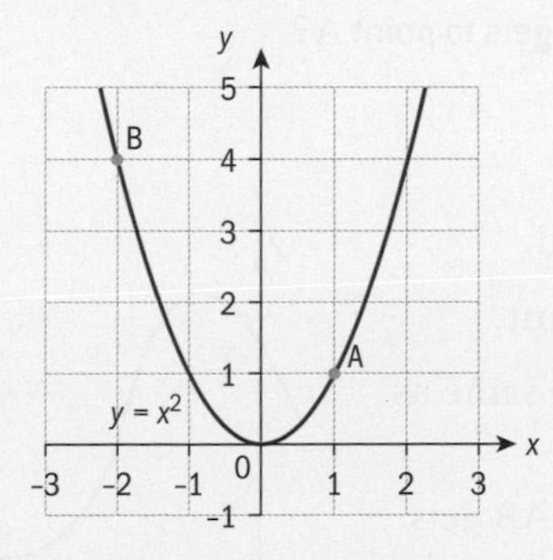

1 a Give the range of x for which the gradient of the curve is
i negative **ii** positive **iii** equal to zero.

b At which point is the gradient of the curve greatest, A or B?

Continued on next page

Calculus

The **tangent** to a curve at a point is the straight line that just touches, but does not cross, the curve at that point.

2 a On your GDC or on other software plot the curve $y = x^2$ and draw the tangent at the point $(1, 1)$, as shown below.

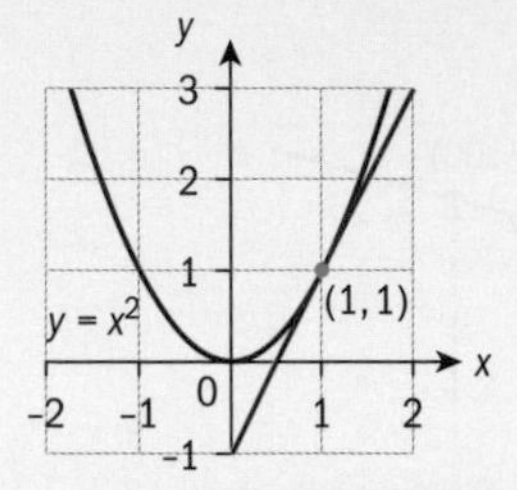

b Zoom in to the point $(1, 1)$. What do you notice about the gradient of the tangent and the gradient of the curve at the point of contact?

c Use your GDC or online software to draw tangents to $y = x^2$ at the points listed below and, in each case, write down the gradient of the tangents.

x	−3	−2	−1	0	1	2	3
Gradient of tangent at x					2		

d What do you think the gradient of the tangent at $x = 5$ is?

e What do you think the gradient of the tangent at $x = -8$ is?

f Can you find a general expression for the gradient of the tangent to $y = x^2$ at the point on the curve with coordinates (x, x^2)?

g To justify your expression in part **f**, find the slope of the chord from point A $(1, 1)$ to a point B $(1.1, 1.21)$.

h Find the slope of the chord from point A $(1, 1)$ to a point B $(1.01, 1.0201)$.

i Find the slope of the chord from point A $(1, 1)$ to a point B $(1.001, 1.002001)$.

j What happens to the slope the closer point B gets to point A?

k Does this value fit your expression in part **f**?

The line joining points A and B is called a **chord**.

The line through points A and B is called a **secant**.

The **average rate of change** from A to B is the same as the **gradient** of the line AB.

As B moves closer to A, the gradient of the line AB gets closer to the **gradient of the tangent** at A.

The gradient of a secant identifies the average rate of change while the gradient of a tangent identifies the instantaneous rate of change.

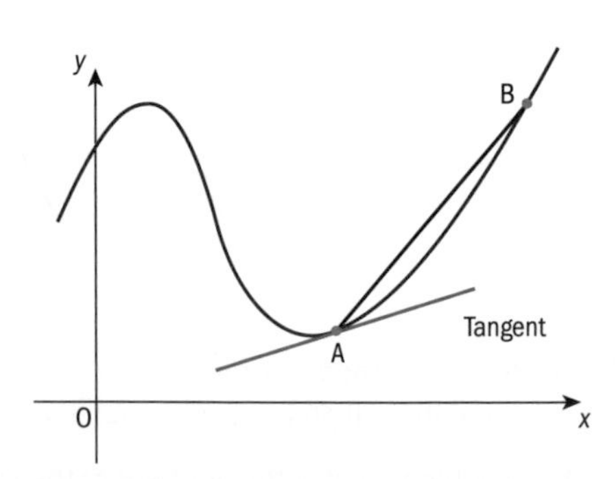

The **gradient** of a tangent is the **limit** of a secant.

The **average rate of change** between two points is the same as the gradient of the secant between the two points.

The **instantaneous rate of change** at a point is the gradient of the tangent at that point.

The gradient function is called the **derivative** and it is denoted by $f'(x)$.
The derivative of $f(x)$ with respect to x is defined as $f'(x)$.

An alternative notation for $f'(x)$ is $\frac{dy}{dx}$ or, if different variables are used, $\frac{ds}{dt}$ or $\frac{dV}{dt}$ etc.

International-mindedness

The $\frac{dy}{dx}$ notation was used by German mathematician, Gottfried Leibniz in the 17th century

Investigation 2

1 Write down the gradients of the following functions:

$y = 3, y = -5, y = 8, y = -10$

- a What is the gradient of $y = c$?
- b What is the derivative of $y = c$?

2 Write down the gradients of the following functions:

$y = 3x, y = -7x, y = 4x, y = x$

- a What is the gradient of $y = mx$?
- b What is the derivative of $y = mx$?

3 a i Using technology, find the gradients of the tangents to $f(x) = 2x^2$ at $x = -2, -1, 0, 1, 2$.

ii Find a general expression for the derivative of $f(x) = 2x^2$ at any point x.

b i Using technology, find the gradients of the tangents to $f(x) = 3x^2$ at $x = -2, -1, 0, 1, 2$.

ii Find a general expression for the derivative of $f(x) = 3x^2$ at any point x.

c State an expression for the derivative of $f(x) = ax^2$.

HINT
There are many interactive demos that you can use to find these answers.

4 a i Using technology, find the gradients of the tangents to $f(x) = x^3$ at $x = -2, -1, 0, 1, 2$.

ii Find a general expression for the derivative of $f(x) = x^3$ at any point x.

b i Using technology, find the gradients of the tangents to $f(x) = 2x^3$ at $x = -2, -1, 0, 1, 2$.

ii Find a general expression for the derivative of $f(x) = 2x^3$ at any point x.

c State an expression for the derivative of $f(x) = ax^3$.

5 a State the derivative for $y = x^4$.

b Generalize the derivatives for $y = x^n$.

c Generalize the derivatives for $y = ax^n$.

6 **Conceptual** What does the derivative of a function at any point represent?

The derivative of $f(x) = ax^n$ is $f'(x) = anx^{n-1}, n \in \mathbb{Z}$.
To find the derivative of a polynomial, you have to find the derivative of each term separately.

For example, if $f(x) = ax^n + bx^{n-1} + cx + d$, then
$f'(x) = anx^{n-1} + b(n-1)x^{n-2} + c$.

Remember that the derivative of $y = mx$ is m and the derivative of $y = c$ is 0.

HINT

Note the last two also follow the same rule as the first if you write mx as mx^1 and c as cx^0.

Example 1

Find the derivatives of the following functions and find the gradient of the tangent at the point where $x = 2$:

1 $y = 3x^4$ **2** $y = 5x^3$ **3** $y = 2x^2 + 3x - 5$

4 $y = x^3 + 2x + 1$ **5** $y = \frac{2}{x} + x, x \neq 0$ **6** $y = \frac{3}{x^2}, x \neq 0$

For parts **3** and **5**:

a Sketch the function and its derivative on the same axes and say whether the function is increasing or decreasing at the point where $x = 2$.

b Write down the range of values of x for which the function is increasing.

1 $\frac{dy}{dx} = 12x^3$

$3 \times 4x^{4-1}$

When $x = 2$,

$\frac{dy}{dx} = 12(2)^3 = 12(8) = 96$

To find the gradient of the tangent at $x = 2$, substitute 2 for x in the derivative. Your GDC will also give you the numerical value of the derivative at any point.

2 $\frac{dy}{dx} = 15x^2$

$5 \times 3x^{3-1}$

When $x = 2$, $\frac{dy}{dx} = 15(2)^2 = 60$

3 $\frac{dy}{dx} = 4x + 3$

$2 \times 2x^{2-1} + 3 \times 1x^{1-1} - 0$

When $x = 2$, $\frac{dy}{dx} = 11$

The gradient of $y = mx$ is m. So, the derivative of $y = 3x$ is 3.

The gradient of $y = c$ is 0. So, the derivative of $y = -5$ is 0.

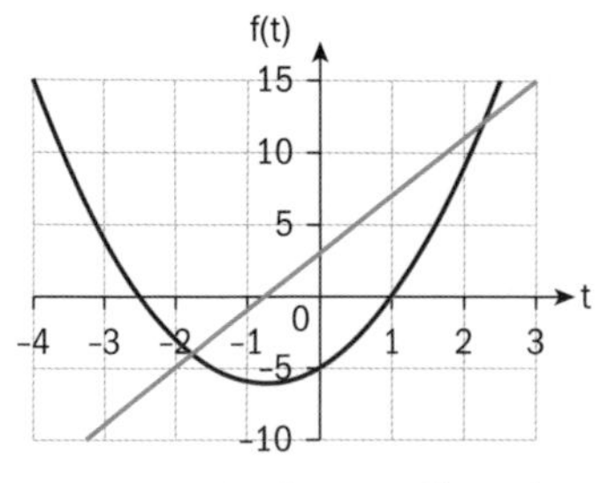

a At $x = 2$, the gradient is positive and so is increasing.

b $x > -0.75$

This can be found by either finding the minimum point of $f(x) = 2x^2 + 3x - 5$ or by finding where $f'(x) > 0$.

4 $\frac{dy}{dx} = 3x^2 + 2$

When $x = 2$, $\frac{dy}{dx} = 14$

5 $\frac{2}{x} + x = 2x^{-1} + x$, $\frac{dy}{dx} = -2x^{-2} + 1 = \frac{-2}{x^2} + 1$

When $x = 2$, $\frac{dy}{dx} = 0.5$

a At $x = 2$, the gradient is positive and so is increasing.

In order to use the rule for differentiation you need to first write the expression using negative indices.

$\frac{2}{x} = 2x^{-1}$, so $\frac{dy}{dx} = 2 \times -1x^{-1-1}$

and the derivative of x is 1.

b $x > 1.41$, $x < -1.41$

Again the range in which the curve is increasing can be found by finding the maximum and minimum points using a GDC or by finding where $f'(x) > 0$.

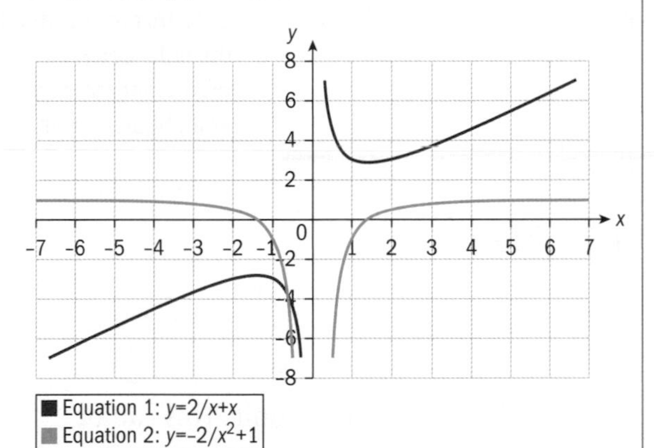

6 $\frac{dy}{dx} = -6x^{-3} = \frac{-6}{x^3}$

When $x = 2$, $\frac{dy}{dx} = -0.75$

$\frac{3}{x^2} = 3x^{-2}$, so $\frac{dy}{dx} = 3 \times -2x^{-2-1}$

Reflect What is the relationship between the gradient of a curve and the sign of its derivative?

A function is increasing if $f'(x) > 0$, is stationary if $f'(x) = 0$ and is decreasing if $f'(x) < 0$.

Usually the prime notation, $f'(x)$, is used when dealing with functions and the fractional notation, $\frac{dy}{dx}$, is used when dealing with curves.

There are no set rules, however; y' and $\frac{df}{dx}$ are both correct, though less frequently seen.

International-mindedness

French mathematician Joseph Lagrange, a close friend of Leonhard Euler, introduced the $f'(x)$ notation in the 18th century.

The following exercise contains a mixture of both notations. Remember that they both indicate that the function gives the gradient, or rate of change, of the variable.

Exercise 12A

i Find the derivative of each of the following functions with respect to x.

ii Find the gradient of the tangent at the point where $x = 1$ and state whether the function is increasing or decreasing at this point.

iii Write down the range of values of x for which the function is increasing.

a $y = 6$ **b** $y = 4x$

c $f(x) = 3x^2$ **d** $f(x) = 5x^2 - 3x$

e $f(x) = 3x^4 + 7x - 3$

f $f(x) = 5x^4 - 3x^2 + 2x - 6$

g $y = 2x^2 - \frac{3}{x}$

h $y = \frac{6}{x^3} + 4x - 3$

i $y = (2x - 1)(3x + 4)$

j $f(x) = 2x(x^3 - 4x - 5)$

k $y = \frac{7}{x^3} + 8x^4 - 6x^2 + 2$

International-mindedness

The seemingly abstract concept of calculus allows us to create mathematical models that permit human feats, such as getting a man on the Moon. What does this tell us about the links between mathematical models and physical reality?

Using differentiation in everyday life problems

Example 2

1 In economics the marginal cost is the cost of producing one more unit. This can be approximated by the gradient of a cost curve.

A company produces motorcycle helmets and the daily cost function can be modelled as $C(x) = 600 + 7x - 0.0001x^3$ for $0 \le x \le 150$ where x is the number of motorcycle helmets produced and C the cost in US dollars.

a Write down the daily cost to the company if no helmets are produced.

b Find an expression for the marginal cost, $C'(x)$, of producing the helmets.

c Find the marginal cost of producing **i** 20 helmets and **ii** 80 helmets.

d State appropriate units for the marginal cost.

2 A hammock can be modelled by the function $f(x) = 2x^2 + 3x - 4$. A mosquito, travelling in a straight line of gradient 11, touches the hammock at point A.

Find the coordinates of A.

1 a US$ 600

b $C'(x) = 7 - 0.0003x^2$

c i $C'(20) = 6.88$

ii $C'(80) = 5.08$

d US$ per helmet

2 $f'(x) = 4x + 3$ $4x + 3 = 11$ $4x = 8$ $x = 2$ $f(x) = 2(2)^2 + 3(2) - 4 = 10$ So, the coordinates of A are (2, 10).	First, find the derivative. You know that the derivative is the same as the gradient of the tangent. Therefore, equate the derivative to 11 and solve for x. Now substitute 2 for x into the original equation.

Exercise 12B

1 The area, A, of a circle of radius r is given by the formula $A = \pi r^2$.

a Find $\frac{dA}{dr}$.

b Find the rate of change of the area with respect to the radius when $r = 2$. Give your answers in terms of π.

2 The profit, US\$$P$, made from selling cupcakes, c, is modelled by the function $P = -0.056c^2 + 5.6c - 20$.

a Find $\frac{dP}{dc}$.

b Find the rate of change of the profit with respect to the number of cupcakes when $c = 20$ and $c = 60$.

c Comment on your answers for part **b**.

3 The distance of a bungee jumper below his starting point can be modelled by the function $f(t) = 80t^2 - 160t$, $0 \le t \le 2$, where t is the time in seconds.

a Find $f'(t)$.

b State the quantity represented by $f'(t)$.

c Find $f'(0.5)$ and $f'(1.5)$ and comment on the values obtained.

d Find $f(2)$ and comment on the validity of the model.

4 Points A and B lie on the curve $f(x) = x^3 + x^2 + 2x$ and the gradient of the curve at A and B is 3.

Find the coordinates of points A and B.

5 The outline of a building can be modelled by the function $h = 2x - 0.1x^2$ where h is the height of the building and x the horizontal distance from the start of the building.

An observer stands at the point A. The angle of elevation from his position to the highest point he can see on the building is 45°.

Calculate the height above the ground of the highest point he can see on the building.

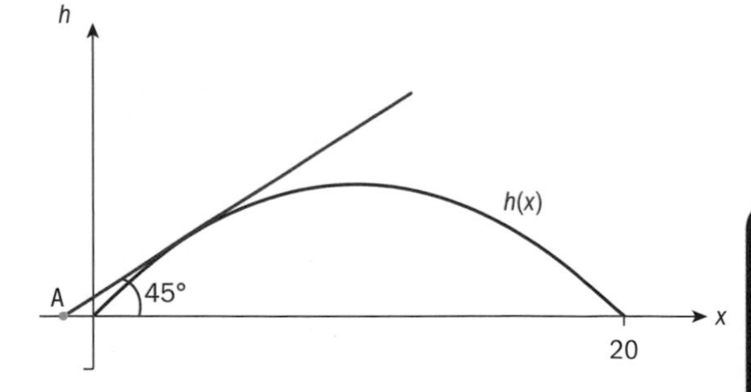

12.2 Equations of tangent and normal

The equation for the tangent to a curve at a given point can be easily found using the gradient to the curve and the coordinates of the point.

Example 3

The side of a hill can be modelled by the function $f(x) = -0.02x^2 + 3.4x$ for certain values of x.

a Sketch the graph of $f(x) = -0.02x^2 + 3.4x$.

There is a monastery at the top of the hill which could be an interesting tourist attraction. So, it is decided to build a funicular railway to reach to almost the top of the hill. The funicular can be modelled by the tangent to the curve where $x = 50$.

b Find the derivative of $f(x) = -0.02x^2 + 3.4x$.

c Find the gradient of the tangent at the point where $x = 50$.

d Find $f(x)$ when $x = 50$.

e Hence, find the equation of the tangent at $x = 50$ and sketch this on the graph.

a

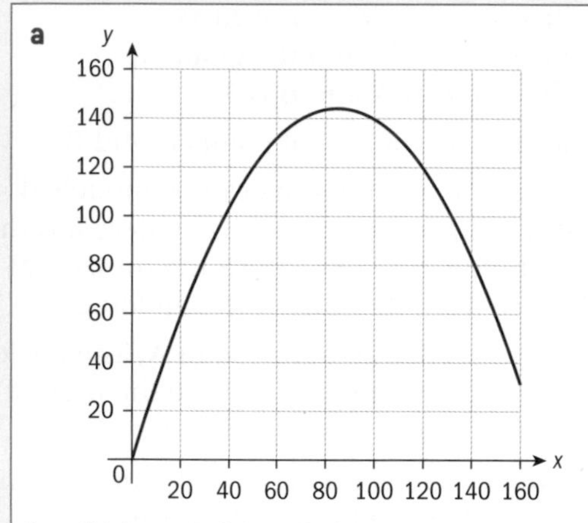

b $f'(x) = -0.04x + 3.4$

c $f'(50) = -0.04(50) + 3.4 = 1.4$

So, 1.4 is the gradient of the tangent at $x = 50$.

Substitute 50 for x into the derivative. You can also use your GDC to find this value.

d $f(50) = -0.02(50)^2 + 3.4(50) = 120$

e Using $y = mx + c$, $120 = 1.4(50) + c$

So, $c = 50$.

Therefore, the equation of the tangent is

$y = 1.4x + 50$

Substitute 50 for x into the original function to find $f(50)$. Your GDC will also give you the equation of the tangent.

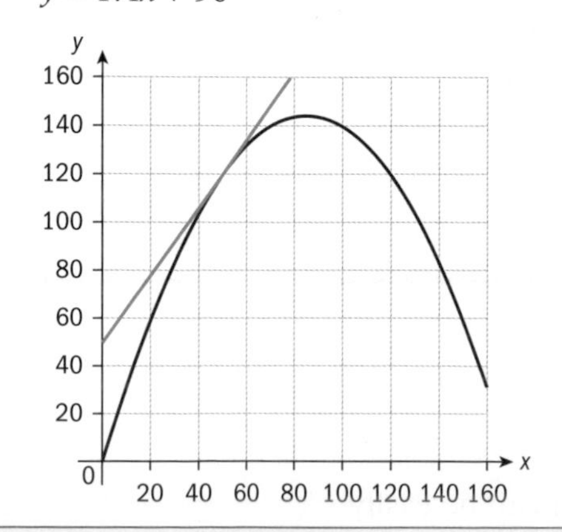

TOK

Who do you think should be considered the discoverer of calculus?

Investigation 3

Part of a pond is in the shape of a parabola which can be modelled by the function

$f(x) = -\frac{8}{90}x^2 + \frac{8}{3}x.$

The local authority decides to build a path on either side of the pond. The two paths will meet at the drinking fountain that is already in place at a point with coordinates (15, 29).

It is decided that the most economical way to achieve this is to build paths that just touch the side of the pond at 5 m and 25 m from the edge, as shown in the diagram. These paths would then be tangents to the curve.

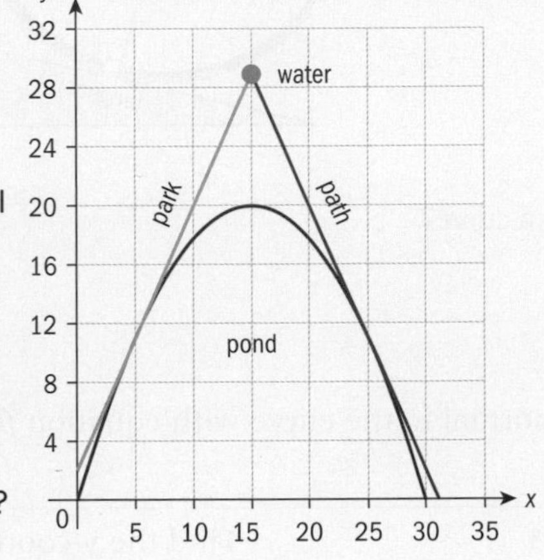

1 How can the local authority work out the equations for these paths?

2 How can they be certain that the paths will meet at the drinking fountain?

3 Using technology, sketch the graph of $f(x) = -\frac{8}{90}x^2 + \frac{8}{3}x$.

4 Find the derivative of $f(x) = -\frac{8}{90}x^2 + \frac{8}{3}x$.

5 What does this represent?

6 Find the gradient of the tangent to the curve when $x = 5$.

7 Find the value of $f(x)$ when $x = 5$.

8 Hence, find the equation of the tangent when $x = 5$.

9 Use technology to draw this line on your graph.

10 Repeat this process for $x = 25$.

Now that you have the equations of the two tangents, how can you find out where they meet?

Do they meet at the location of the drinking fountain?

11 **Conceptual** What is the relation between the gradient of a tangent at a given point and the derivative of a function at the same point?

The derivative of a function at a point on its graph represents the gradient of the tangent at that point.

The **normal** line is perpendicular to the tangent line.

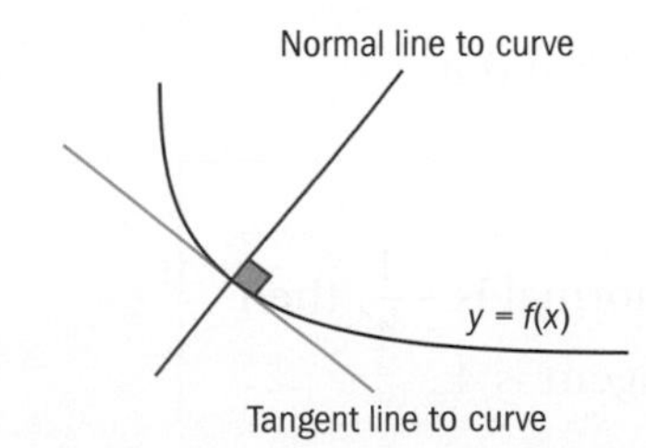

For example, on a bicycle the spokes are at right angles to the tyre at the point of contact. Hence the direction of the road will be a tangent to the wheel at the point of contact and the spoke at that point will be a normal to the point of contact.

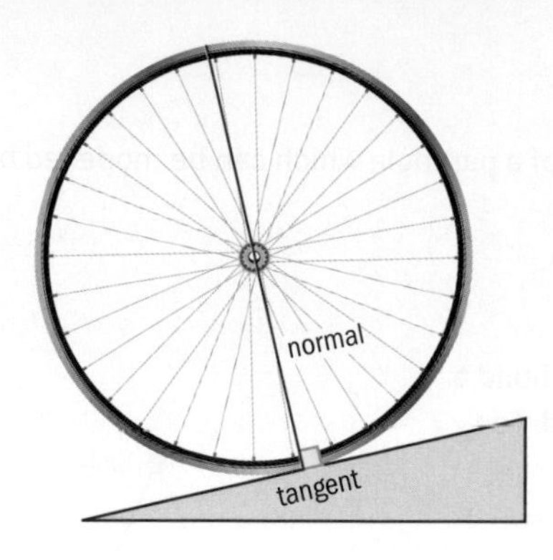

HINT

If two lines are perpendicular then the product of their gradients is -1.

If the gradient of the tangent is $\frac{a}{b}$, then the gradient of the normal is $-\frac{b}{a}$.

Reflect What is a normal to a curve?

Example 4

Find the equation of the normal to the curve with equation $f(x) = 2x^3 + 3x - 2$ at the point where $x = 1$.

$f(1) = 2(1)^3 + 3(1) - 2 = 3$	Find the y-coordinate of the point.
The point is $(1, 3)$.	
$f'(x) = 6x^2 + 3$	Next find $f'(x)$.
$f'(1) = 6(1)^2 + 3 = 9$	Find $f'(1)$ to work out the gradient of the tangent line.
The gradient of the tangent line is 9.	
Gradient of the normal line $= -\frac{1}{9}$	Gradient of normal line $= \frac{-1}{\text{tangent gradient}}$
$3 = -\frac{1}{9}(1) + c$	Substitute the point $(1, 3)$ into the equation of the normal line.
$c = 3\frac{1}{9}$	
The equation of the normal is $y = -\frac{1}{9}x + 3\frac{1}{9}$	

Example 5

The gradient of the normal to the curve with equation $f(x) = kx^3 - 2x + 1$ at the point $(1, b)$ is $-\frac{1}{4}$. Find the values of k and b.

$f'(x) = 3kx^2 - 2$	
If the gradient of the normal is $-\frac{1}{4}$, then the gradient of the tangent is 4.	

So, $f'(1) = 3k(1)^2 - 2 = 4$ $3k = 6$ $k = 2$	The derivative is the gradient of the tangent at any point. Here the point is $x = 1$. So, you put $x = 1$ into the derivative and equate it to 4, which is the gradient of the tangent at that point.
$f(1) = 2(1)^3 - 2(1) + 1 = 1$ So, $b = 1$	To find b you need to substitute 1 for x into the original equation.

Exercise 12C

1 Find the equation of the tangent to the curve with equation $f(x) = 2x^2 - 4$ at the point where $x = 3$.

2 Brian makes a seesaw for his children from part of a log and a plank of wood. The shape of the log can be modelled by the equation $f(x) = -x^2 + 2x$ for certain values of x.

The plank of wood is a tangent to the log at the point where $x = 1$.

Find the equation of the plank.

3 A cyclist is cycling up a hill. The path of the bicycle on the hill can be modelled by part of the function $f(x) = -2x^2$.

a Find the gradient of the wheel at the point where $x = 1$.

b Find the gradient of the spoke that is perpendicular to the wheel when it touches the hill at the point where $x = 1$.

4 Find the equation of the normal to the curve with equation $f(x) = 3x^2 - 4x + 5$ at the point $(1, 4)$.

5 Find the equation of the tangent and the normal to the curve with equation $y = x^4 - 6x + 3$ at the point where $x = 2$.

6 The edge of a lake can be modelled by part of the function $f(x) = x^2$.

A fountain is to be placed in the lake at the point where the equations of the normals at $x = 2$ and $x = -2$ meet.

Find the equations of the normals and the point where the fountain will be placed.

7 The perimeter of a park can be modelled by part of the function $f(x) = -0.112x^2 + 5.6x$.

a Sketch the graph of this function.

The local authority wants to place a memorial statue inside the park. They decide to place it where the two normal lines to the curve at $x = 15$ and $x = 35$ meet.

b Find the equations of the two normal lines.

c Find the point of intersection of the two normal lines.

d State whether or not this a suitable place for the memorial. Explain your answer.

8 The gradient of the tangent to the curve with equation $f(x) = ax^2 + 3x - 1$ at the point $(2, b)$ is 7. Find the values of a and b.

9 The gradient of the tangent to the function with equation $f(x) = x^2 + kx + 3$ at the point $(1, b)$ is 3. Find the values of k and b.

10 The gradient of the normal to the function with equation $y = ax^2 + bx + 1$ at the point $(1, -2)$ is 1. Find the values of a and b.

HINT

There are many websites that will give the derivative of a function as an equation and indeed some calculators do this as well. These are **not** allowed in exams.

However, in examinations, it will be expected that you use your GDC to find numerical values for the derivatives at given points and also to draw the graph of the derivative. This widens the range of functions that you might need to find the gradient for. Unless told otherwise, always use your GDC if it makes answering the question simpler.

Example 6

Consider $y = \frac{x+2}{x-1}$, $x \neq 1$.

Find the gradient of the curve at the points where $x = 2$ and $x = 3$.

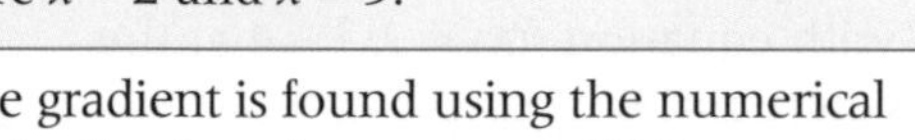

-3 and -0.75	The gradient is found using the numerical derivative function on your GDC.

Exercise 12D

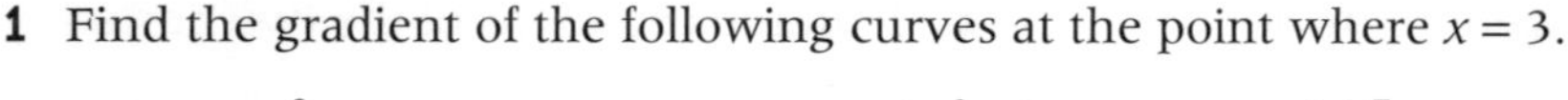

1 Find the gradient of the following curves at the point where $x = 3$.

a $y = \frac{x^2}{x-1}$ **b** $y = x\ln x$ **c** $f(x) = \frac{2x^2-1}{x+3}$ **d** $s = \frac{t+5}{t^3+1}$ **e** $y = xe^{2x}$ **f** $g(x) = (x^2-3)(x-1)^5$

12.3 Maximum and minimum points and optimization

Investigation 4

The number of bacteria, B, in thousands, in a culture at time, t minutes, is given by the formula:

$B(t) = 0.00314x^3 - 0.1926x^2 + 2.848x + 5$

a Sketch the graph of $B(t) = 0.00314x^3 - 0.1926x^2 + 2.848x + 5$ for $0 \leq t \leq 40$.

b Find $B'(t)$.

c Find the values of t when there are the most and the least bacteria.

d Find the gradient of the function at the two points found in part **b**.

e Calculate the gradient of the function at a point just before and just after the values of t found in part **b**.

f Use your results to explain the difference between the local maximum and the local minimum.

g **Conceptual** How do stationary points help solve real-life problems?

International-mindedness

Maria Agnesi, an 18th century, Italian mathematician, published a text on calculus and also studied curves of the form $y = \frac{a^2}{x^2} + a^2$.

TOK

How can you justify the rise in tax for plastic containers eg plastic bags, plastic bottles etc using optimization?

You can also use technology and solve $f'(x) = 0$ to find the value of stationary points.

For a function $f(x)$, if $f'(c) = 0$, then the point $(c, f(c))$ is called a **stationary point** as the instantaneous rate of change is 0 at that point.

If the signs of the gradients are different on both sides of the stationary point then it means it is either a maximum or minimum point.

In many cases, we are interested in when a function reaches its maximum or its minimum as in real-life cases these points can be important. For example, if it is a function of profit versus the number of products produced, we would like to know where the profit is a maximum.

A stationary point can be a local **maximum** point or a local **minimum** point. At a local maximum or minimum point $f'(x) = 0$.

HINT

The local maximum or local minimum points may not be the maximum or minimum values of the graph.

If the sign of the derivative goes from negative to positive, there is a minimum point.

If the sign of the derivative goes from positive to negative, there is a maximum point.

Maximum Minimum

Example 7

Consider the curve $s = 3t^3 + 2t^2 - 4t + 2$.

a Find $\frac{ds}{dt}$.

b On the same axes, sketch $s = 3t^3 + 2t^2 - 4t + 2$ and its derivative.

c Solve the equation $\frac{ds}{dt} = 0$.

d State the feature of $s = 3t^3 + 2t^2 - 4t + 2$ indicated by these points.

e If the domain of the function is restricted to $-2 \le t \le 2$, find the actual maximum and minimum values of the function.

a $\frac{ds}{dt} = 9t^2 + 4t - 4$

Using the usual rules for differentiating but with the new notation.

b

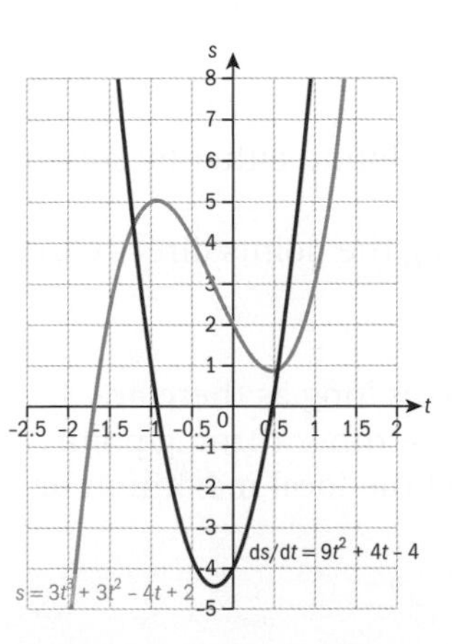

You can either plot the curve found in part **a** or use the derivative function on the GDC to do so.

Use your GDC to find the roots of $9t^2 + 4t - 4 = 0$.

Continued on next page

c $t = -0.925, 0.481$	This observation provides an alternative method for solving part **c**: using the GDC to find the maximum and minimum points on the original curve.
d Each of these values of t the curve gives a stationary point, one is a local maximum point and the other a local minimum point. **e** 26 and -6	The maximum and minimum values (sometimes called the **global** maximum and minimum) are at the end points of the curve.

Example 8

Consider the curve $y = x + \frac{1}{x}, x \neq 0$.

a Find an expression for $\frac{dy}{dx}$.

b Sketch the curve.

c Find the coordinates of any local maximum or minimum points.

d Find the gradient of two points, close to but on either side of the minimum point.

e Explain why this shows that the point found is a local minimum.

a $\frac{dy}{dx} = 1 - \frac{1}{x^2}$ **b** 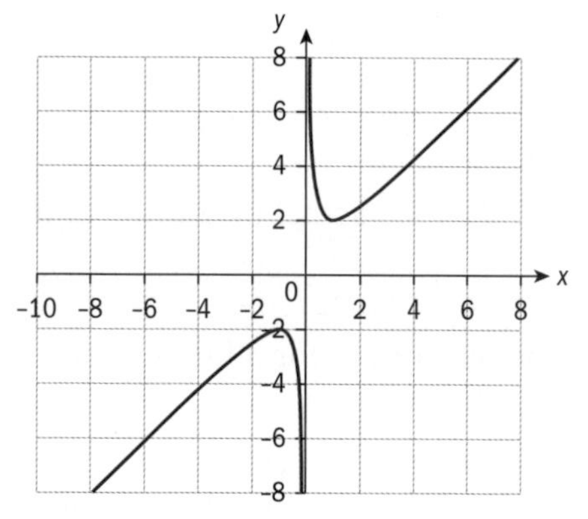	
c Local maximum point is $(-1, -2)$. Local minimum point is $(1, 2)$.	These points can be found either by solving $\frac{dy}{dx} = 0$ or by finding the points directly from the GDC.
d $x = 0.5, \frac{dy}{dx} = 1 - 4 = -3$ $x = 1.5, \frac{dy}{dx} \approx 1 - 0.444 = 0.556$ **e** Gradient changes from negative to positive.	Any points will do as long as there are no discontinuities or other turning points between them and the local maximum or minimum.

Exercise 12E

1 Jules throws a stone into the air. The height of the stone can be modelled by the equation $f(t) = 1 + 7.25t - 1.875t^2$ where t is the time, in seconds, that has passed since the stone was thrown, and $f(t)$ is the height of the stone, in metres.

- **a** Find $f'(t)$.
- **b** Find the stationary point.
- **c** Show that this is a maximum point.

2 A company's profits, in thousands of dollars, can be modelled by the function:

$P(x) = 0.08x^3 - 1.9x^2 + 12.5x$, where x is the number of units sold (in millions) each week.

- **a** Sketch a graph of $P(x)$ for $0 \leq x \leq 15$.
- **b** Calculate the average rate of change between $x = 2$ and $x = 3$, including the units. Explain the meaning of the value you find.
- **c** Calculate the instantaneous rate of change at $x = 3$, $x = 8$ and $x = 13$. Explain the meaning of these values.
- **d** State the values for which the instantaneous rate of change is positive. State the values of x for which the instantaneous rate of change is negative. Explain the meaning of each of these results.
- **e** Write down the values of x for which the instantaneous rate of change is zero. Justify your answer.
- **f** Find the maximum and minimum values and show that these are maximum and minimum values.

3 The path of a shot-put can be modelled by the function $f(t)=1.75 + 0.75t - 0.0625t^2$, where t is the time (in seconds) and $f(t)$ is the height of the shot-put (in metres) above the ground.

Find the maximum height of the shot-put and the time when this occurred.

4 The profit $\$P$ that Yarona makes each day is dependent on the number of brownies, n, that she bakes.

Her daily profit can be modelled by the function $P(n) = -0.056n^2 + 5.6n - 20$.

Find the maximum profit that Yarona makes in a day and the number of brownies she needs to sell to make this profit.

5 A business buys engine parts from a factory. The business is currently deciding which of three purchasing strategies to use for the next stage of its development.
Its researchers produce models for each of the strategies. P is the expected profit (in €10 000) and n is the number of parts it buys (in 1000s). The largest number of parts the factory can sell to the business is 5000 and there is no minimum.

- **a** For each model find the maximum profit and the number of parts the business needs to buy to make this profit.
- **b** State which strategy it should adopt.

  - **i** $P = 0.5n + 1.5 + \dfrac{4}{n+1}$
  - **ii** $P = \dfrac{n^3}{3} - \dfrac{5n^2}{2} + 6n - 4$
  - **iii** $P = \dfrac{n^3}{24} - \dfrac{5n^2}{8} + 3n$

6 John is a keen cyclist and is planning a route in the Alps. The profile of the route he would like to take can be modelled by the function:

$y = -0.081x^4 + 0.89x^3 - 2.87x^2 + 3x$, $0 \leq x \leq 6$, where y (× 100 m) is the height of the point on the route that is a horizontal distance x (× 10 km) from his starting point.

Plot the graph and find the maximum height John will climb to on this route.

International-mindedness

The Greeks' mistrust of zero meant that Archimedes' work did not lead to calculus.

Optimization

Many practical problems involve finding maximum or minimum values. For example, we may want to maximize an area or minimize cost. Such problems are called optimization problems.

Example 9

A can of dog food contains 500 cm³ of food. The manufacturer wants to receive the maximum profits by making sure that the surface area of the can has optimal dimensions. Let the radius of the can be r cm and the height, h cm.

Find the dimensions of the can that will have the minimum surface area.

$V = \pi r^2 h = 500$ So, $h = \frac{500}{\pi r^2}$ Surface area, $S = 2\pi rh + 2\pi r^2$	In order to make maximum profits, the surface area of the can needs to be as small as possible. You are given the volume of the can and you use this to find an expression for the height in terms of the radius.
$S = 2\pi r\left(\frac{500}{\pi r^2}\right) + 2\pi r^2 = \frac{1000}{r} + 2\pi r^2$ $= 1000r^{-1} + 2\pi r^2$	This expression can be substituted into the formula for the surface area. Then you have an equation in terms of r only.
$\frac{dS}{dr} = -1000r^{-2} + 4\pi r$ $\frac{dS}{dr} = 0$ at maximum or minimum point So, $-1000r^{-2} + 4\pi r = 0$ Using your solver, $r = 4.3$ Check $\frac{dS}{dr}$ at $r = 4$ and $r = 5$ At $r = 4$, $\frac{dS}{dr} = -12.2$ and at $r = 5$, $\frac{dS}{dr} = 22.8$ The derivative goes from negative to positive and so it is a minimum turning point. Therefore, the best dimensions for the can are radius = 4.3 cm and height = 8.6 cm.	You can then graph the function to find the local minimum point or you can find the derivative and let $\frac{dS}{dr} = 0$ and solve for r.

Investigation 5

1 A semicircular water trough has surface area 10 m^2 as shown.

Explain why troughs can be designed that have the same surface area but different volumes.

2 Find the formula for the outside surface area in terms of r and h.

3 As you increase r, describe what must happen to h to keep the surface area at 10 m^2. Does the volume of the trough increase as r increases? Describe what happens to the volume.

4 Rearrange your formula from part **2** to make h the subject.

5 Find the formula for the volume of the trough. Why can you not find the maximum volume of the trough yet?

6 Substitute your expression for h from **4** to show that $V = 5r - \frac{1}{2}\pi r^3$.

7 Draw a sketch or use your GDC to plot the graph of $f(r) = 5r - \frac{1}{2}\pi r^3$.

8 How do you know that the trough has a maximum volume? How does the maximum value relate to $f(r) = 5r - \frac{1}{2}\pi r^3$?

9 Find $f'(r)$ and solve $f'(r) = 0$. What does your solution represent?

10 Find the maximum volume of the trough.

11 Explain how you can check that it is indeed the maximum value.

12 **Conceptual** How does finding the maximum or minimum values help in solving real-life problems?

HINT

Optimization is covered in Chapter 9, but this does not include differentiation.

TOK

Does the fact that Leibniz and Newton came across the calculus at similar times support the argument of Platonists over Constructivists?

If a function to be optimized has only one variable then either a GDC or differentiation can be used directly to find the maximum or minimum point. If a function to be optimized has more than one variable then a constraint must also be given. This constraint can then be written as an equation and substituted into the function to eliminate one of the variables.

Exercise 12F

1 An open top cylinder has a volume of 400 cm^3. The radius of the base is r cm and the height is h cm.

a Show that $\pi r^2 h = 400$.

b Write down an expression for the surface area, A, of the open cylinder.

c Show that this can be written as $A = \pi r^2 + \frac{800}{r}$.

d Sketch the graph of

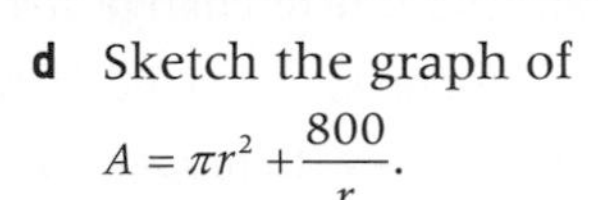

$A = \pi r^2 + \frac{800}{r}$.

e Find the minimum area and the value of r when this occurs.

f Verify that it is a minimum point.

2 The total surface area of a closed cylinder is 5000 cm^2.

Find the dimensions of the cylinder that maximize its volume and state this maximum volume.

Verify that it is a maximum point.

3 A vegetable garden is in the shape of a rectangle. The garden is surrounded by 100 m of fencing.

a If the width of the garden is x metres, find an expression for the length.

b Show that the area of the garden is $A = x(50 - x)$ m^2.

c Find $\frac{dA}{dx}$.

d Find the maximum area and the dimensions of the garden at this point.

4 A cone has radius, r cm, and height $(18 - r)$ cm.

Find the maximum volume of the cone and the values for the radius and height that give this volume.

5 A rectangular piece of card measures 20 cm by 24 cm. Equal squares of side x cm are cut out of each corner.

The rest of the card is folded up to make a box.

a Find the volume of the box in terms of x.

b Find the value of x for which the volume is a maximum and find the maximum volume.

6 A closed cylinder has a volume of 300 cm^3.

a If the radius of the base is r cm and the height is h cm, find an expression for the volume.

b Find an expression for the total surface area in terms of r.

c Find the minimum surface area and the values for r and h that give this area.

7 The profit, $\$P$, that a firm makes per day can be modelled by the function $P(n) = -0.092n^2 + 33.3n - 313$, where n represents the number of goods sold per day.

Find the maximum profit and the value of n when this occurs.

8 A company's weekly profit, in dollars, in relation to the number of units sold each week, x, can be modelled by the function $f(x) = -0.9x^2 + 52x - 360$.

Find the maximum profit and the value of x when this occurs.

Developing your toolkit

Now do the Modelling and investigation activity on page 546.

Chapter summary

- The gradient of a tangent is the **limit** of a secant.
- The **average rate of change** between two points is the same as the gradient of the secant between the two points.
- The **instantaneous rate of change** at a point is the gradient of the tangent at that point.
- The gradient function is called the **derivative** and it is denoted by $f'(x)$.
- The **derivative** of $f(x)$ **with respect to** x is written as $f'(x)$.
- An alternative notation for $f'(x)$ is $\frac{dy}{dx}$ or, if different variables are used, $\frac{ds}{dt}$ or $\frac{dV}{dt}$ etc.
- The derivative of $f(x) = ax^n$ is $f'(x) = anx^{n-1}$, $n \in \mathbb{Z}$.

- To find the derivative of a polynomial, you have to find the derivative of each term separately.
  - For example, if $f(x) = ax^n + bx^{n-1} + cx + d$, then $f'(x) = anx^{n-1} + b(n-1)\ x^{n-2} + c$.
  - Remember that the derivative of $y = mx$ is m and the derivative of $y = c$ is 0.
  - Note these last two also follow the same rule as the first if you write mx as mx^1 and c as cx^0.
- A function is increasing if $f'(x) > 0$, is stationary if $f'(x) = 0$ and is decreasing if $f'(x) < 0$.
- Usually the prime notation, $f'(x)$, is used when the question is about functions and the fractional notation, $\frac{\mathrm{d}y}{\mathrm{d}x}$, is used for curves.
- There are no set rules however. $y' = 2x - 1$ and $\frac{\mathrm{d}f}{\mathrm{d}x}$ are both correct, though less frequently seen.
- The derivative of a function at any point represents the gradient of the tangent at that point.
- The equation of the **normal** is perpendicular to the equation of the tangent.
- If two lines are perpendicular then the product of their gradients is -1.
- If the gradient of a tangent is $\frac{a}{b}$, then the gradient of the normal is $-\frac{b}{a}$.
- For function $f(x)$, if $f'(c) = 0$, then the point $(c, f(c))$ is called a **stationary point** as the instantaneous rate of change is 0 at that point.
- If the signs of the gradients are different on both sides of the stationary point then it means it is either a maximum or minimum point.
- In many cases, we are interested in when a function reaches its maximum or its minimum as in real-life cases these points can be important. For example, if it is a function of profit versus the number of products produced we would like to know where the profit is a maximum.
- A stationary point can be a local **maximum** point or a local **minimum** point.
- At a local maximum or minimum point $f'(x) = 0$.
  - The local maximum or local minimum points may not be the maximum or minimum values of the graph.
  - If the sign of the derivative goes from negative to positive, there is a minimum point.
  - If the sign of the derivative goes from positive to negative, there is a maximum point.

Maximum Minimum

- If a function to be optimized has only one variable then either a GDC or differentiation can be used directly to find the maximum or minimum point.
- If a function to be optimized has more than one variable then a **constraint** must also be given. This constraint can then be written as an equation and substituted into the function to eliminate one of the variables.

Developing inquiry skills

Return to the opening problem. Has what you have learned in this chapter helped you to answer the questions?

What information did you manage to find?

What assumptions did you make?

How will you be able to construct a model?

What other things did you wonder about?

Chapter review

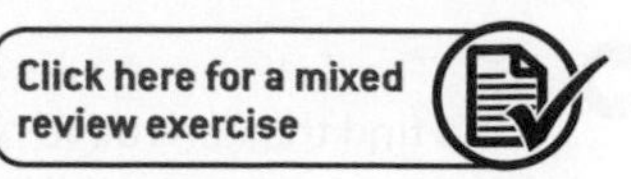

1 Find the derivative of each of the following functions with respect to x and find the value of the gradient of the tangent to the curve at $x = 1$.

a $f(x) = 6$

b $y = 3x + 5$

c $g(x) = 2x^2 - 4x - 1$

d $y = 6x^3 - 3x^2 + x - 7$

e $f(x) = \frac{2}{x} + 3x - 9$

f $f(x) = \frac{6}{x^3} + 2x^2 + 8$

2 Find the equation of the tangent and normal to the curve $f(t) = 0.5t^2 - 3t + 2$ at the point where $t = 4$.

3 The gradient of the tangent to the curve $f(x) = x^2 - 5x - 4$ at point A is 1.

Find the coordinates of point A.

4 The gradient of the normal to the curve $f(x) = 3x^2 + 4x - 3$ at point B is $\frac{1}{2}$.

Find the coordinates of point B.

5 Jacek is practising on his skateboard. His journey along the track can be modelled by a curve with equation $f(t) = 10.667 - 1.667t + 0.0417t^2$, where t is the time in seconds and $f(t)$ is the distance in metres.

a Find $f'(t)$.

b Find $f'(12)$ and $f'(32)$ and comment on these values.

c Find the time when Jacek is at the minimum point on the track and explain how you know that it is a minimum.

6 A can of cat food has volume $V = 300$ cm^3.

The radius of the can is r cm and the height is h cm.

a Show that $\pi r^2 h = 300$.

b Find an expression for the total surface area S of the can.

c Substitute an expression for h (from part **a**) into your expression for S (from part **b**) and hence show that $S = 2\pi r^2 + \frac{600}{r}$.

d Find $\frac{dS}{dr}$.

e Find the minimum value of S and the values of r and h when this occurs and show that it is a minimum value.

Exam-style questions

7 **P1:** A ball is thrown from the top of a cliff. The path of the ball may be modelled by the equation $y = -0.05x^2 + 1.5x + 82$ $(x \geq 0)$, where x is the horizontal distance from the foot of the cliff and y is the height above the foot of the cliff.

a Find an expression for $\frac{dy}{dx}$. (2 marks)

b Hence find the maximum height reached by the ball. (3 marks)

c Justify that the height you found in part **b** is a maximum point. (3 marks)

8 **P2:** A rectangular piece of paper, measuring 40 cm by 30 cm, has a small square of side length x cm cut from each corner. The flaps are then folded up to form an open box in the shape of a cuboid.

a Show that the volume V of the cuboid may be expressed as $V = 1200x - 140x^2 + 4x^3$. (3 marks)

b Find an expression for $\frac{dV}{dx}$. (2 marks)

c Hence show that the cuboid will have a maximum volume when $x^2 - \frac{70}{3}x + 100 = 0$. (2 marks)

d Using technology, find the maximum possible volume of the cuboid. (4 marks)

9 P1: The following table shows the temperature (T) in a town, measured hours (h) after midnight.

Hours after midnight	Temperature T (°C)
0	8
5	11
10	12.5
15	12
20	9.5
24	2

It is suggested that T and h are connected by the relationship $T = ah^2 + bh + c$, where a, b and c are constants.

a Determine the values of a, b and c. (4 marks)

b Find an expression for $\frac{dT}{dh}$. (2 marks)

c Use your expression for $\frac{dT}{dh}$ to determine at what time during the day the temperature is at its highest. (2 marks)

d Referring to your answer to part **c**, suggest one reason why a quadratic model may not be accurate in this instance. (1 mark)

10 P1: Using your GDC, find the stationary points on the curve $y = 2x^4 + 0.2x^3 - 8x^2 + 5x + 3$, giving each coordinate to 3 sf. (6 marks)

11 P1: A curve is given by the equation

$$y = \frac{2x^3}{3} - \frac{7x^2}{2} + 2x + 5.$$

a Determine the coordinates on the curve where the gradient is −3. You must show all your working, and give your answers as exact fractions. (6 marks)

b Find the range of values of x for which the curve is decreasing. (2 marks)

12 P1: Find the equation of the normal to the curve $y = 2 - \frac{x^4}{2}$ at the point where $x = 1$. (7 marks)

13 P1: The gradient of the tangent to the function with equation $y = ax^2 + bx + 3$ at the point $(2, -1)$ is 8. Find the value of a and the value of b. (7 marks)

14 P2: Two numbers x and y are related by the equation $x + y = 20$.

a Find the maximum possible value of xy. (5 marks)

b Find the minimum possible value of $x^2 + y^2$. (5 marks)

c Justify that your answer to part **b** is a minimum value. (3 marks)

River crossing

Approaches to learning: Thinking skills: Evaluating, Critiquing, Applying
Exploration criteria: Personal engagement (C), Reflection (D), Use of mathematics (E)
IB topic: Differentiation, Optimization

The problem

You are standing at the edge of a slow-moving river which is one kilometre wide. You want to return to your campground on the opposite side of the river. You can swim at 3 km/h and run at 8 km/h. You must first swim across the river to any point on the opposite bank. From there you must run to the campground, which is 2 km from the point directly across the river from where you start your swim.

What route will take the least amount of time?

Visualize the problem

Here is a diagram of this situation:

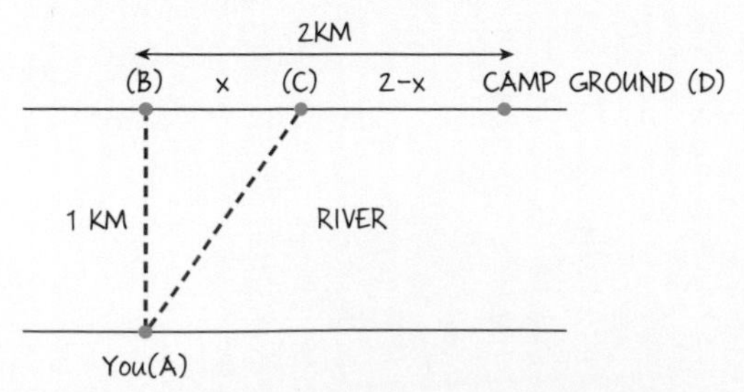

Discuss what each label in the diagram represents.

Solve the problem

What is the length of AC in terms of x?

Using the formula for time taken for travel at a constant rate of speed from this chapter, write down an expression in terms of x for:

1 the time taken to swim from A to C

2 the time taken to run from C to D.

Hence write down an expression for the total time taken, T, to travel from A to D in terms of x.

You want to minimize this expression (find the minimum time taken).

Find $\frac{dT}{dx}$.

Now solve $\frac{dT}{dx} = 0$ to determine the value of x that minimizes the time taken.

How do you know this is a valid value?

For this value of x, find the minimum time possible and describe the route.

Assumptions made in the problem

The problem is perhaps more accurately stated as:

You are standing at the edge of a river. You want to return to your campground which you can see further down the river on the other side. You must first swim across the river to any point on the opposite bank. From there you must run to the campground.

What route will take the least amount of time?

Look back at the original problem.

What additional assumptions have been made in the original question?

What information in the question are you unlikely to know when you are standing at the edge of the river?

What additional information would you need to know to determine the shortest time possible?

The original problem is a simplified version of a real-life situation. Criticize the original problem, and the information given, as much as possible.

Extension

In an exploration it is important to reflect critically on any assumptions made and the subsequent significance and limitations of the results.

Consider the open-box problem in Q5 of Exercise 12F on page 542 and Example 9 involving the "volume of a cylindrical can" on page 540 and/or some of the questions in Exercise 12F on pages 541–2.

If you were writing an exploration using these problems as the basis or inspiration of that exploration, then:

- What assumptions have been made in the question?
- What information in the question are you unlikely to know in real life?
- How could you find this missing information?
- What additional information would you need to know?

Criticize the questions as much as possible!

13 Approximating irregular spaces: integration

Concepts

- Space
- Approximation

Microconcepts

- Lower limit
- Upper limit
- Definite integrals
- Area under a curve
- Trapezoidal rule
- Numerical integration
- Antiderivatives
- Indefinite integral
- Constant of integration

What do the graphs of functions that have the same derivative have in common? How do they differ?

This chapter explores integration, the reverse of differentiation. The area of an island and the amount of glass needed for the windows of a building can be represented by integrals. Integrals give you a way to estimate the values of areas that cannot be found using existing area formulae.

How can you estimate the area covered by oil spills out at sea?

How can you find the amount of glass in this building?

How can you estimate the area affected by a hurricane?

San Cristóbal is the easternmost island of the Galapagos. Here is a map of the island.

It is claimed that the total area of the island is 558 km². How can you test this value?

Use a rectangle to estimate the area of the island.

How did you use the map scale?

Does your result underestimate or overestimate the claimed area? Why?

How could you improve your estimate?

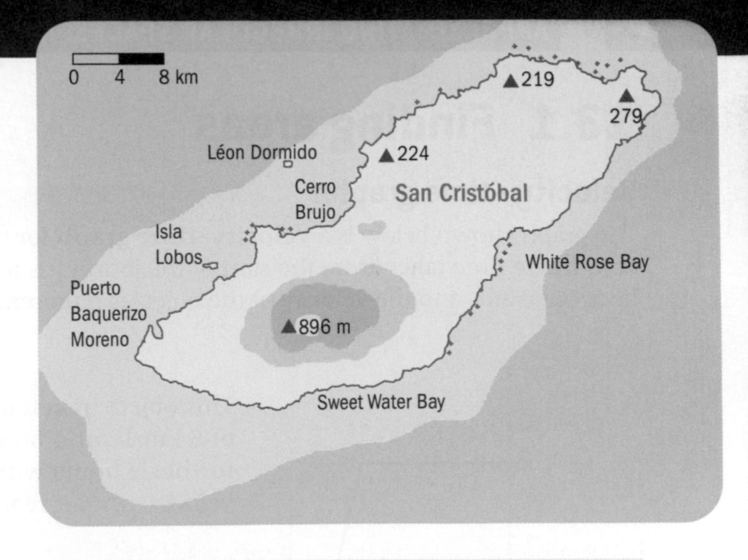

Developing inquiry skills

Write down any similar inquiry questions you might ask to model the area of something different; for example, the area of a national part, city of lake in your country.

Think about the questions in this opening problem and answer any you can. As you work through the chapter, you will gain mathematical knowledge and skills that will help you to answer them all.

Before you start

You should know how to:

1 Use geometric formulae to find area

For example, the area of a trapezoid:

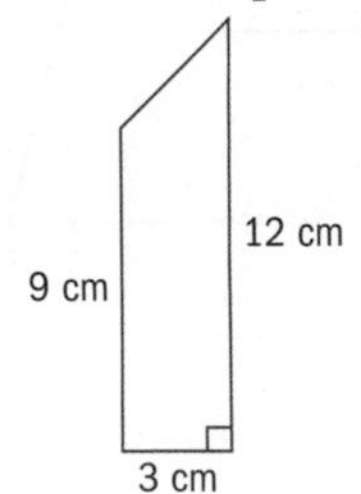

$$A = \frac{1}{2}(b_1 + b_2)h$$

$$= \frac{1}{2}(9 + 12) \times 3$$

$$= 31.5\text{ cm}^2$$

2 Find the derivative of functions of the form $f(x) = ax^n + bx^{n-1} + \ldots$ where all the exponents of x are integers

For example, the derivative function of

$f(x) = 3x^2 + \frac{x}{2} - 0.5$ is $f'(x) = 6x + \frac{1}{2}$

Skills check

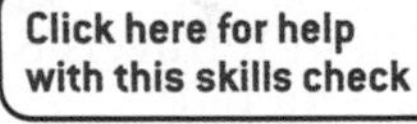

1 Find the areas.

a

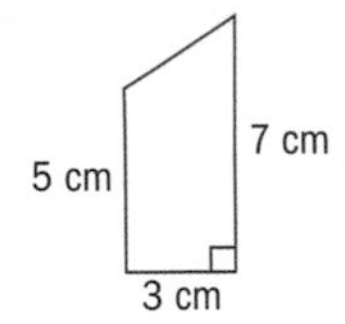

b

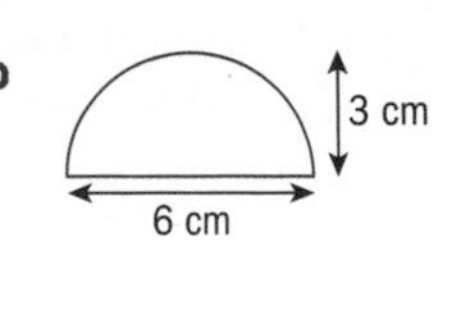

2 Find the derivative of each of these functions.

a $f(x) = x^3 + \frac{5x}{2} - 3$

b $g(x) = 4x^2 - x$

13.1 Finding areas

Velocity–time graphs

The graph shown below is a **velocity–time graph** for the journey of an object. The time taken from the start of the journey is represented on the horizontal axis and the velocity of the object is represented on the vertical axis.

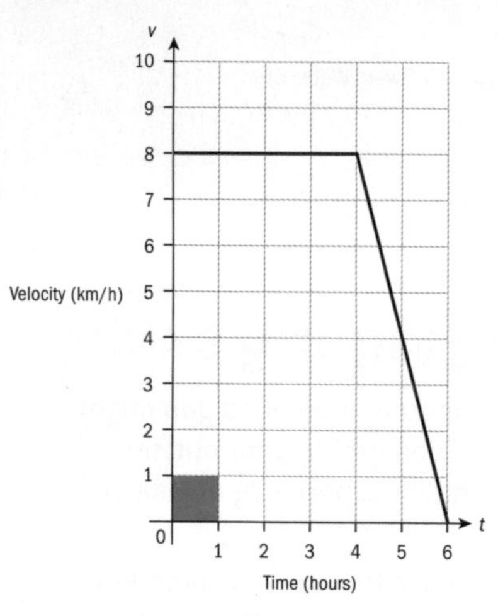

This object travels at a constant velocity of 8 km/h for 4 hours and then it travels a further 2 hours with a steadily decreasing velocity until it reaches 0 km/h.

What distance did this object travel during the entire journey?

The graph is a piecewise linear function. To answer the question each part will be considered separately.

First 4 hours

The velocity is 8 km/h.

In 1 hour it travels 8 km therefore in 4 hours it travels $8 \times 4 = 32$ km.

What is the relationship between 32 and **the area under the graph** in this first part?

The green square shown in the graph above represents 1 km. Why?

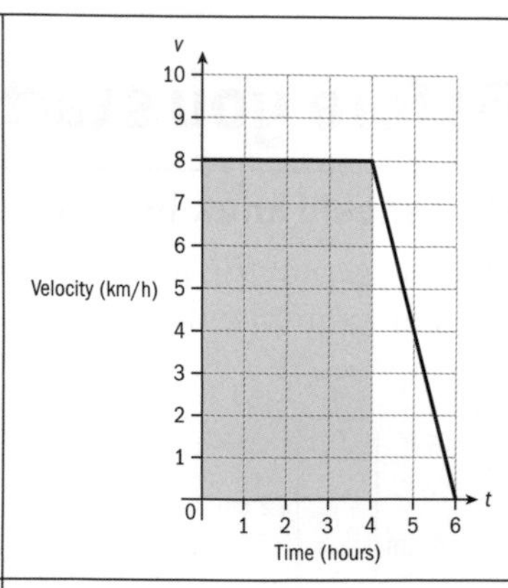

Last 2 hours

How many green squares will fit under the second part of the piecewise function?

Find the area under the graph.

The shape is triangular. Applying the formula for the area of a triangle:

$$\frac{8\,\text{km/h} \times 2\,\text{h}}{2} = 8\,\text{km}$$

Total distance travelled = 32 km + 8 km = 40 km

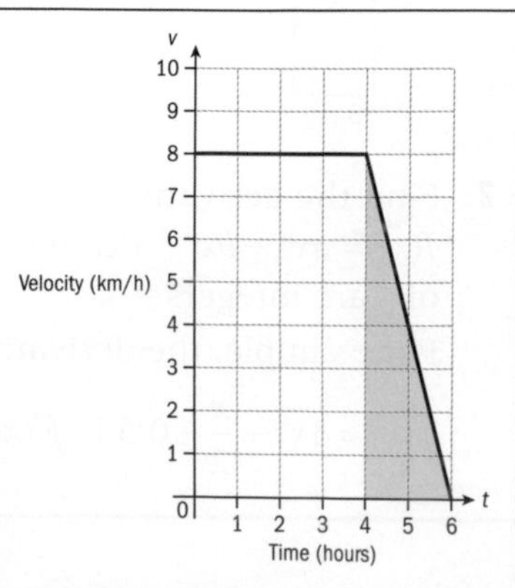

Reflect How does the area under a velocity–time graph relate to the distance travelled?

Now, if the velocity–time graph were curved, how would you calculate the area?

In this chapter you will study different methods to find or approximate areas between the graph of a function and the x-axis.

Example 1

For this velocity–time graph, find the distance travelled.

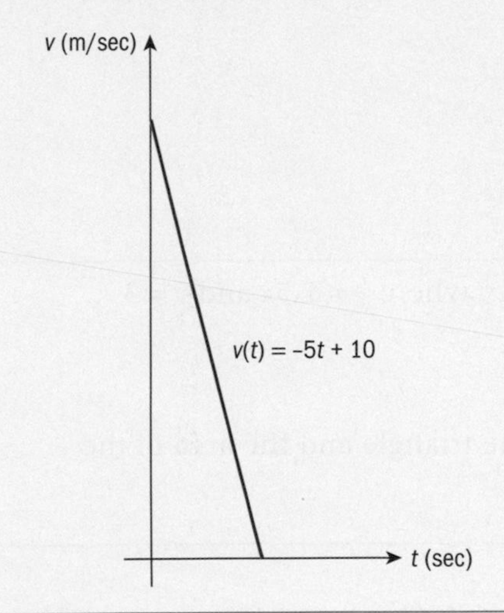

	The distance travelled is equal to the area under the velocity–time graph.
When $v = 0$: $-5t + 10 = 0$ then $t = 2$	To calculate the width of the triangle, find the point where the graph cuts the t-axis.
When $t = 0$: $v(0) = 10$	To calculate the height, find the point where the graph cuts the v-axis. At this point $t = 0$. 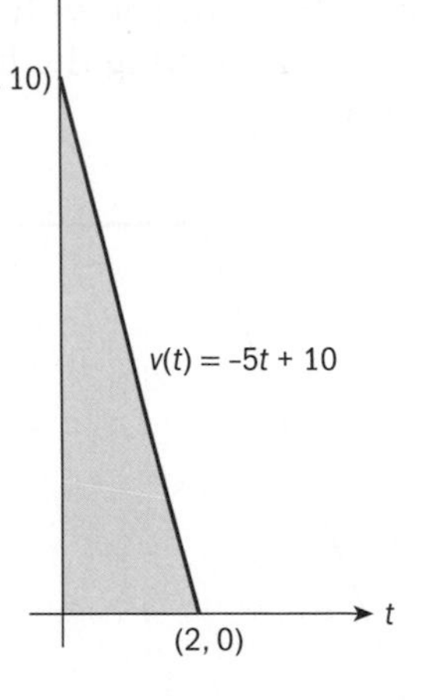
Area $= \frac{1}{2} \times 10 \times 2 = 10$ Distance travelled is 10 m.	Substitute into the area formula using $b = 2$ and $h = 10$.

Example 2

Find the area under the graph.

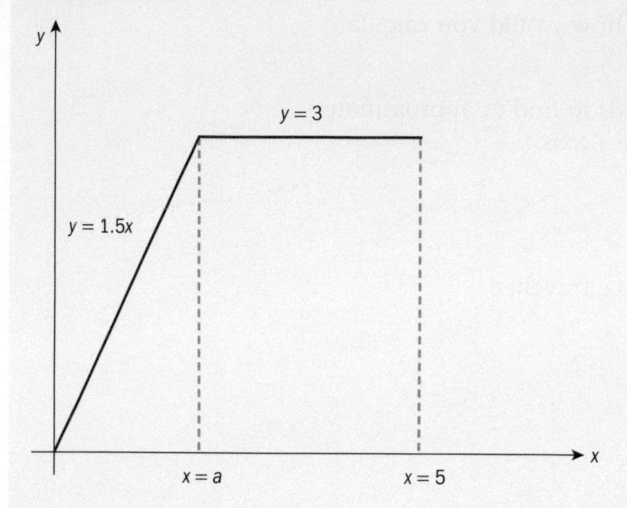

$1.5a = 3 \Rightarrow a = 2$ Area of triangle $= \dfrac{2 \times 3}{2} = 3$	Find the value of a, where $y = 1.5x$ and $y = 3$ intersect.
Area of rectangle $= 3 \times 3 = 9$ Area under graph $= 3 + 9 = 12$	Add the area of the triangle and the area of the rectangle.

Exercise 13A

1 The graph below shows how the velocity of a car changes during the first 60 seconds of a journey.

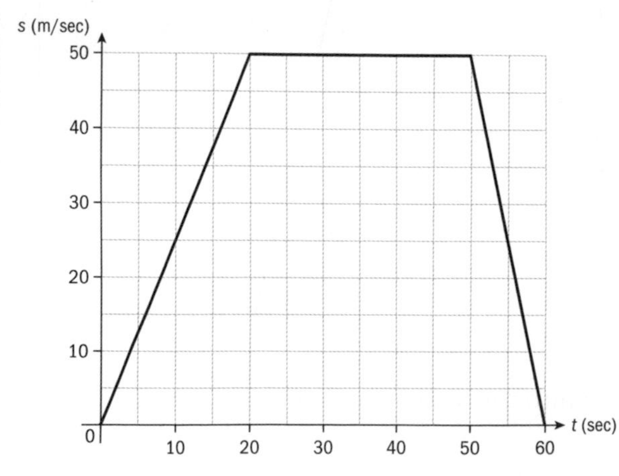

Find the distance travelled during the 60 seconds.

2 For each velocity–time graph, find the distance travelled.

a

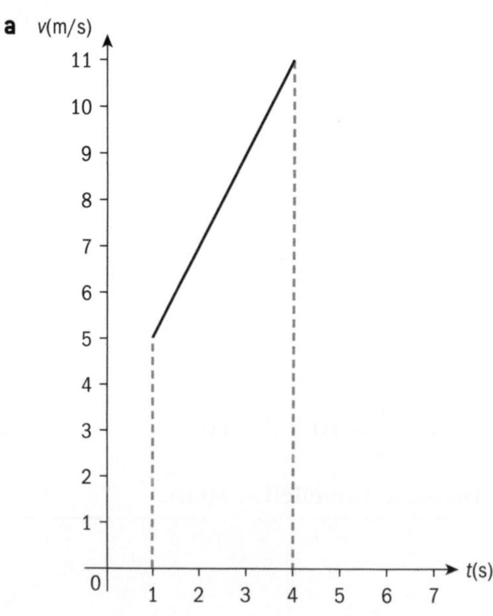

b

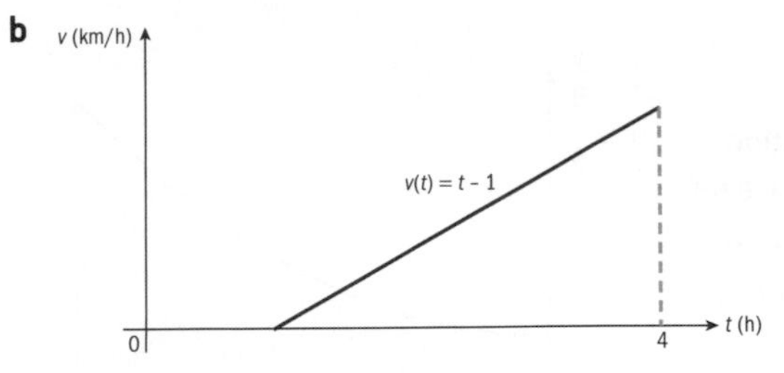

c

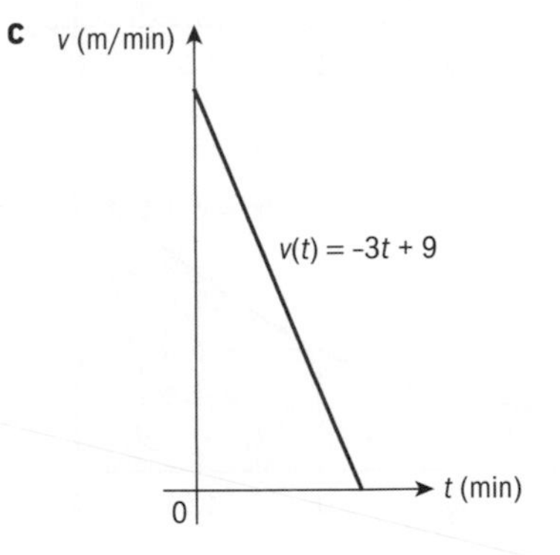

3 For each graph, find the shaded area.

a

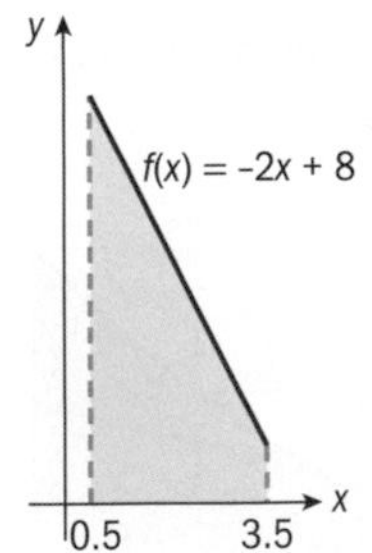

b

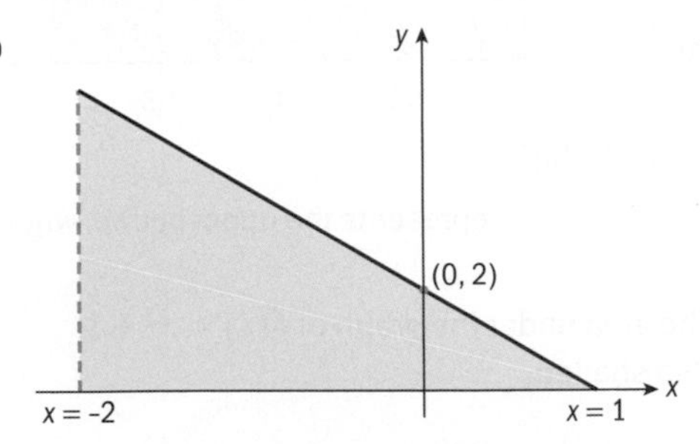

c

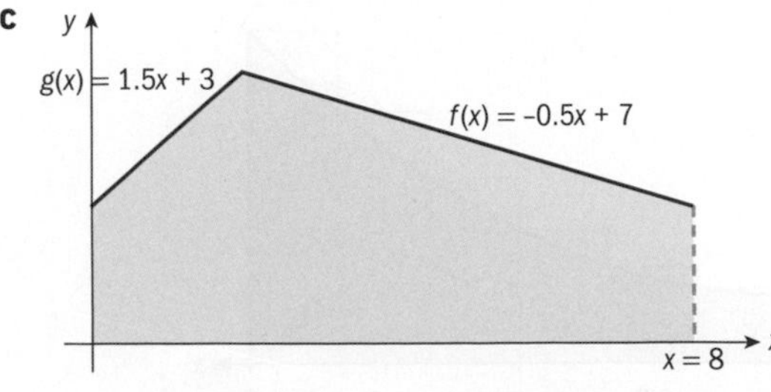

d

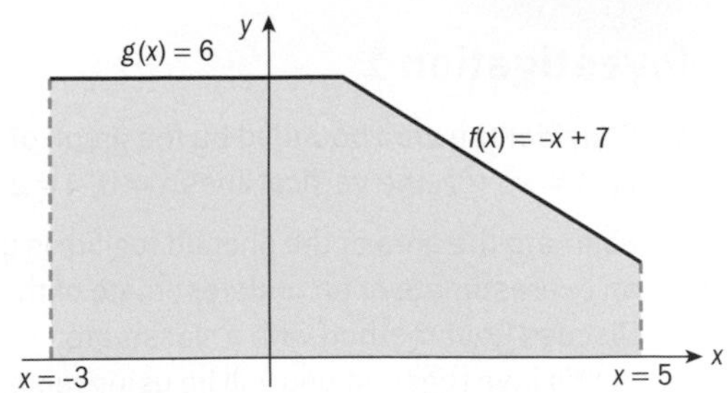

4 **a** Sketch the graph of the function $f(x) = 0.5x + 3$.

b Shade the region enclosed between the graph of f, the x-axis and the vertical lines $x = 1$ and $x = 6$.

c Find the area of the shaded region.

5 **a** Sketch the graph of the function $f(x) = -2x + 6$.

b Shade the region enclosed between the graph of f, the vertical line $x = 0$ and the x-axis.

c Find the area of the shaded region.

6 **a** Sketch the graph of the piecewise linear function

$$f(x) = \begin{cases} x, & 0 \le x \le 5 \\ 5, & 5 < x \le 9 \end{cases}$$

b Shade the region under the graph of $f(x)$ and above the x-axis.

c Find the area of the shaded region.

Calculus

Investigation 1

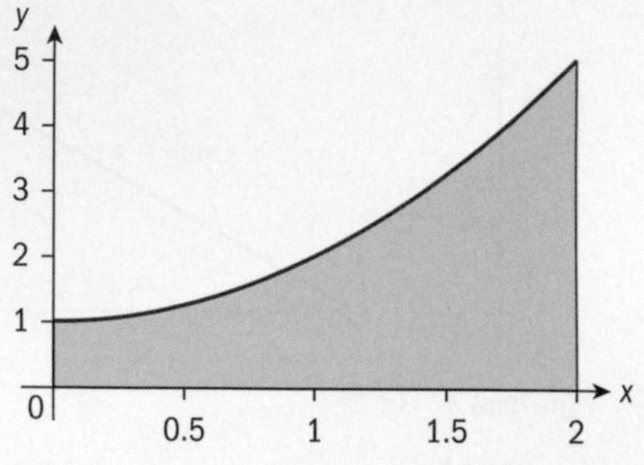

1 Consider the area bounded by the graph of the function $f(x) = x^2 + 1$, the vertical lines $x = 0, x = 2$ and the x-axis.

Estimate the area of the shaded region. Is your estimate an overestimate or an underestimate of the actual area? Discuss your method with a classmate.

2 In this investigation you will be using rectangles or vertical strips to estimate this area. At the end of the investigation you can check how close your estimate was to the actual area.

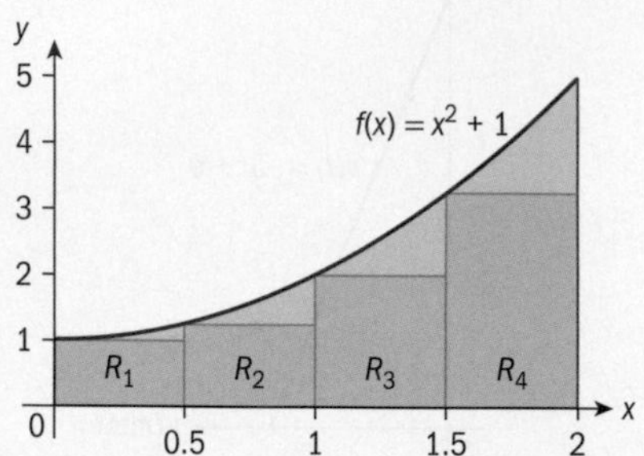

- The graph shows four rectangles with the same width. The area under the graph of the function $f(x) = x^2 + 1$ between the vertical lines $x = 0, x = 2$ is also shown.
- What is the width of the rectangles? How do you calculate it?
- What is the relationship between the height of each rectangle and the graph of the function?
- Find the heights of each of these rectangles.
- Find the area of each of these rectangles and then find **the sum** of the areas of these rectangles.
- Is this an underestimate or an overestimate of the actual area? Why? The sum of these areas will be a **lower bound** of the area of the shaded region. This will give an **underestimate** of the area.

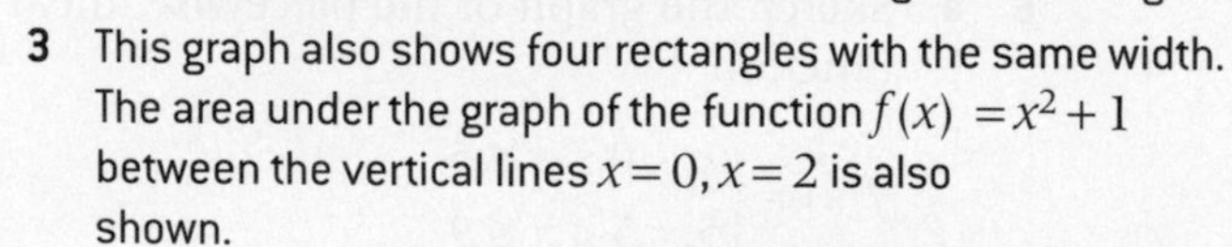

3 This graph also shows four rectangles with the same width. The area under the graph of the function $f(x) = x^2 + 1$ between the vertical lines $x = 0, x = 2$ is also shown.

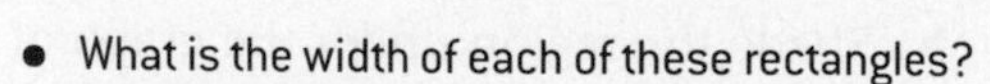

- What is the width of each of these rectangles?
- Find the heights of each of these rectangles.
- Find the area of each of these rectangles and then find **the sum** of the areas of these rectangles.
- Is this an underestimate or an overestimate of the actual area? Why?

- If L_S represents the lower bound, A represents the actual area and U_S represents the upper bound, write an inequality relating L_S, U_S and A.

4 In each of the following graphs there are six rectangles. The area under the graph of $f(x) = x^2 + 1$ between the vertical lines $x = 0, x = 2$ and the x-axis is also shaded.

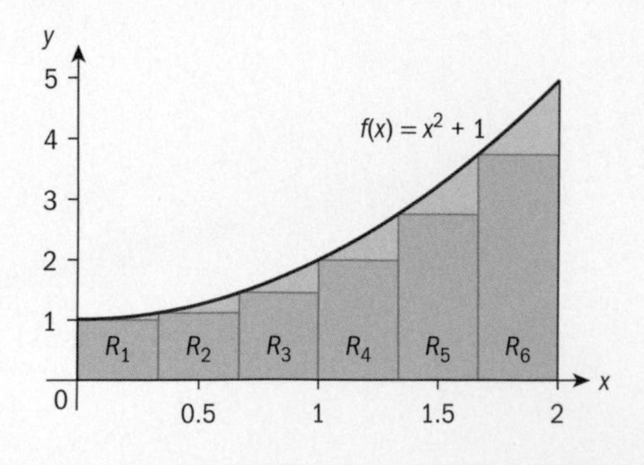

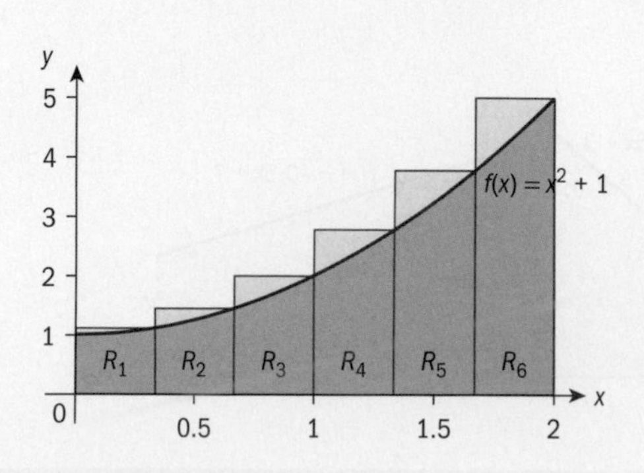

- Approximate the area under the curve by considering a lower bound and an upper bound. Why do you think that more rectangles are being used?
- Complete the following tables to organize the information. You can create a table with your GDC to calculate the heights of the rectangles. How would you calculate the widths? Remember that they are all equal.
- Lower bound with six rectangles:

Rectangle	Width	Height	Area
R_1			
R_2			
R_3			
R_4			
R_5			
R_6			
			$L_S=$

- Upper bound with six rectangles:

Rectangle	Width	Height	Area
R_1			
R_2			
R_3			
R_4			
R_5			
R_6			
			$U_S=$

- What do you notice? Are the lower and upper bounds closer to each other than when you had four rectangles? How do you think these estimates can be improved?
- Write a new inequality relating L_S, U_S and A.

5 Look at the following graphs. The number of rectangles, n, has been increased in each case. L_S and U_S are also given.

$n=20,\ L_S=4.47,\ U_S=4.87$

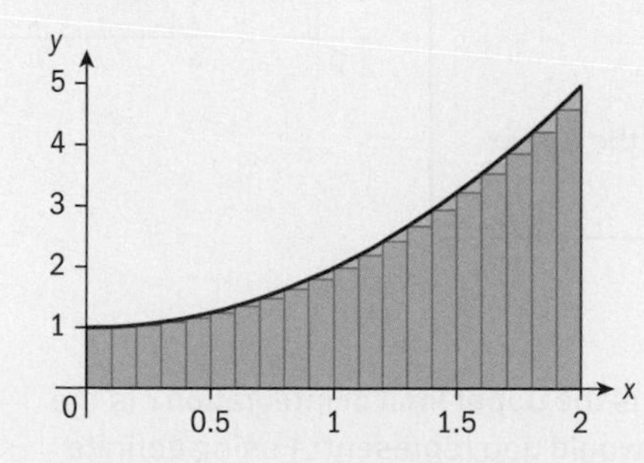

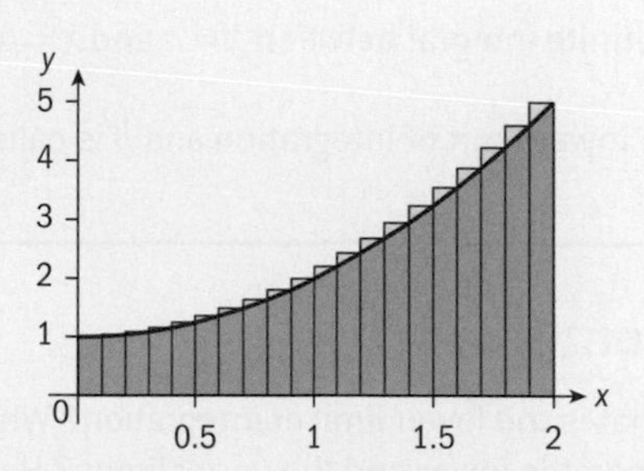

Continued on next page

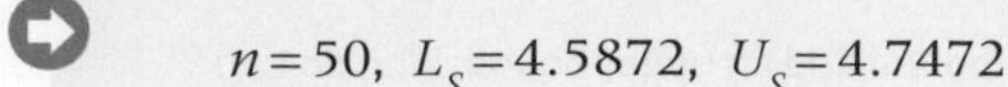
$n=50,\ L_S=4.5872,\ U_S=4.7472$

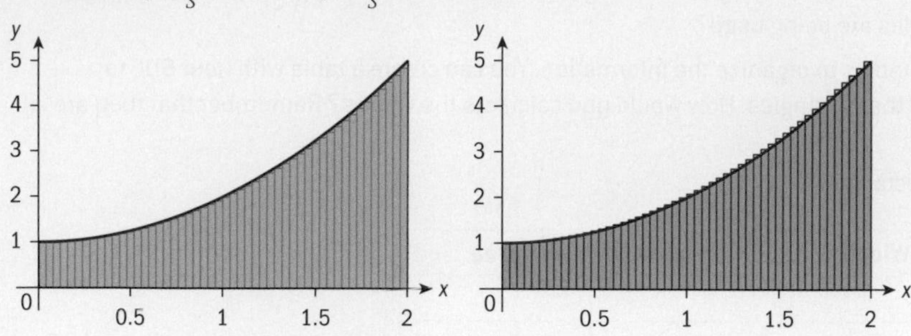

$n=100,\ L_S=4.6268,\ U_S=4.7068$

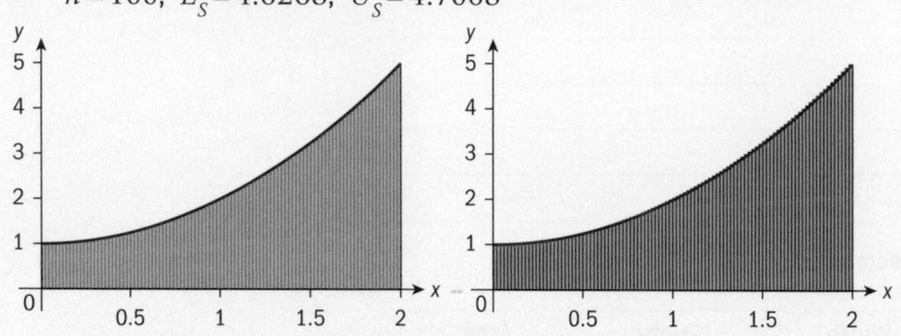

- What can you say about the values of L_S as n increases?
- What can you say about the values of U_S as n increases?
- Here are some more values for n, L_S and U_S.

n	L_S	U_S
500	4.65867	4.67467
1000	4.66267	4.67087
10000	4.66627	4.66707

- What happens when n tends to infinity (gets larger and larger)? What can you say about the value of A?

International-mindedness

A Riemann sum, named after 19th century German mathematician Bernhard Riemann, approximates the area of a region, obtained by adding up the areas of multiple simplified slices of the region.

TOK

We are trying to find a method to evaluate the area under a curve.

"The main reason knowledge is produced is to solve problems."

To what extent do you agree with this statement?

When f is a non-negative function for $a \le x \le b$, $\int_a^b f(x)\,dx$ gives the area under the curve from $x=a$ to $x=b$.

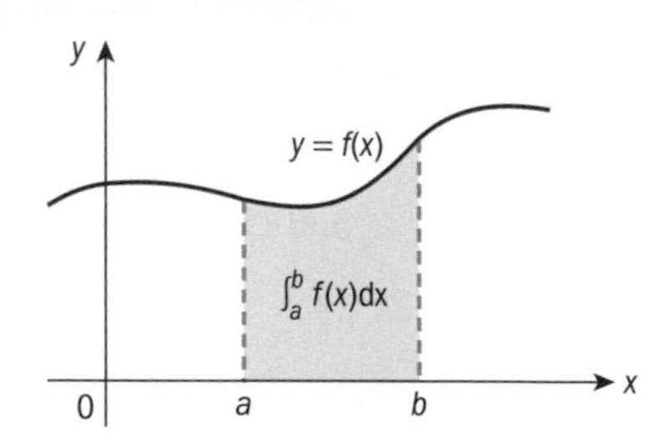

$\int_a^b f(x)\,dx$ is read as "the definite integral between $x=a$ and $x=b$."

The number a is called the **lower limit** of integration and b is called the **upper limit** of integration.

Investigation 1 (continued)

6 In the investigation, what is the lower limit of integration? What is the upper limit of integration? Is the function positive between the lower and the upper limits? How would you represent A using definite integral notation?

7 **Conceptual** How do areas under curves within a given interval relate to the definite integral and to lower and upper rectangle sums on the same interval?

Reflect What is a definite integral?

Example 3

a Write down a definite integral that gives the area of the shaded region.

b Find the value of the definite integral.

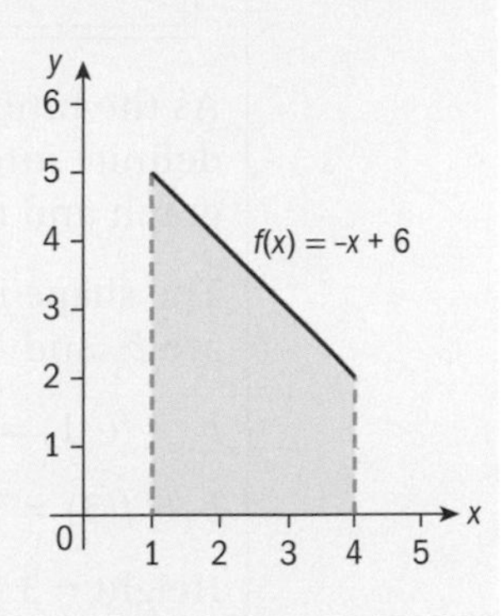

a $\int_1^4 (-x+6)\,dx$	The lower limit is $x = 1$. The upper limit is $x = 4$. The function is $f(x) = -x + 6$.
b $\int_1^4 (-x+6)\,dx = \frac{1}{2} \times (2+5) \times 3 = 10.5$	The shape is trapezoidal. Bases are: $b_1 = f(1) = -1 + 6 = 5$ $b_2 = f(4) = -4 + 6 = 2$ Height = 3 Substitute into the area of a trapezoid formula.

Example 4

For the definite integral $\int_{-1}^{3} (x+4)\,dx$:

a State clearly the function being integrated, the lower limit of integration and the upper limit of integration.

b Sketch the graph of the function. Shade the region whose area the definite integral represents.

c Find the value of the definite integral by using an area formula.

a The function is $f(x) = x + 4$. The lower limit is $x = -1$ and the upper limit is $x = 3$.	Given that this is a linear function, you should expect the graph to be a line segment in the interval $-1 \leq x \leq 3$.

Continued on next page

Calculus

	You can use technology to see the graph. 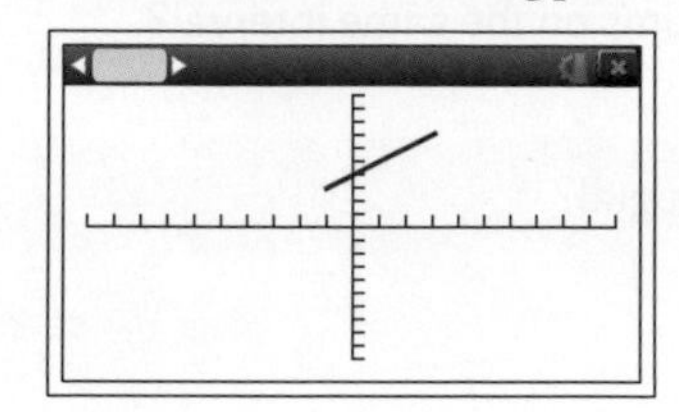
b 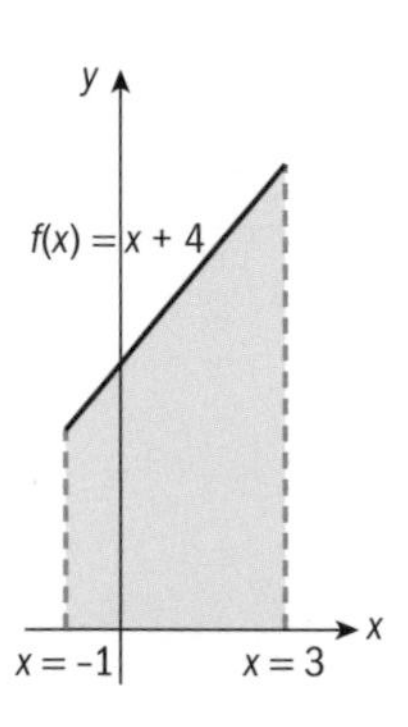	As the function is positive in the given interval, the definite integral represents the area between the graph and the x-axis. The shape is trapezoidal. The bases of the trapezoid are b_1 and b_2. $b_1 = f(-1) = -1 + 4 = 3$ $b_2 = f(3) = 3 + 4 = 7$ Height $= 3 - (-1) = 4$
b $\int_{-1}^{3}(x+4)\,dx = \frac{1}{2}\times(3+7)\times 4 = 20$	Substitute into the area of a trapezoid formula.

Exercise 13B

1 a Write down a definite integral that gives the area of each of the following regions.

i

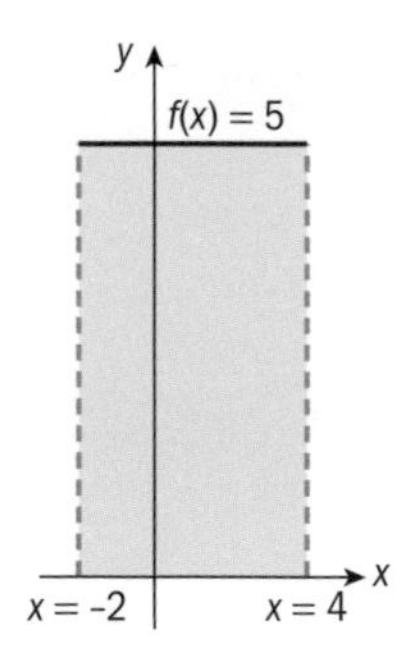

ii

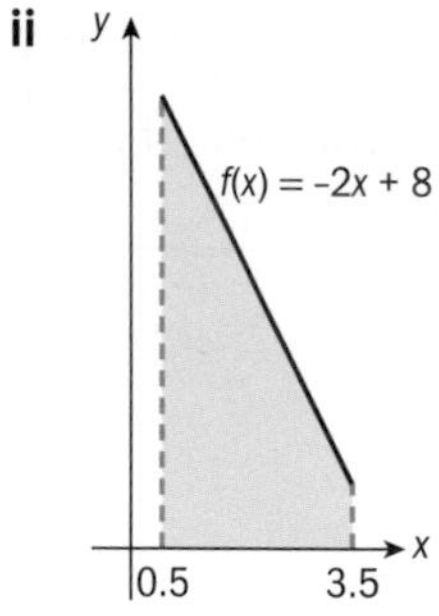

iii

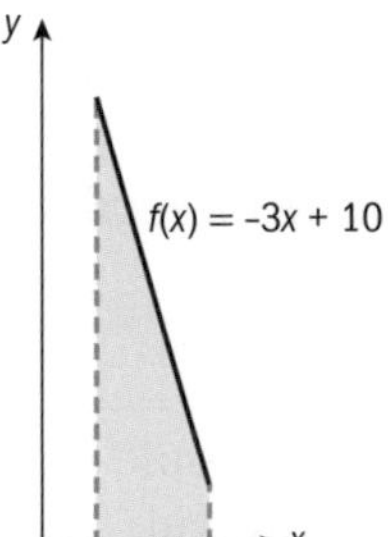

iv

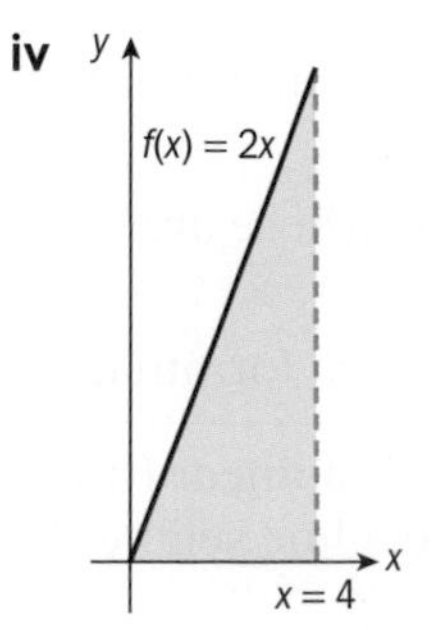

b Calculate the definite integrals from part **a** using existing area formulae.

2 Find the value of the following definite integrals by using existing area formulae. In each case sketch the function in an appropriate interval and shade the region whose area the definite integral represents.

a $\int_{2}^{6}(x+1)\,dx$

b $\int_{0}^{4}(-2x+8)\,dx$

c $\int_{-2}^{0}(-0.5x+4)\,dx$

Example 5

Calculate the definite integral $\int_0^2 (x^2+1)\,dx$.

Compare your answer with the value found for A in Investigation 1.

TOK

Is imagination more important than knowledge?

$\int_0^2 (x^2+1)\,dx = \frac{14}{3}$ (4.67 to 3 sf).

This value is the same as the one found for A, the area under the graph of the function $f(x) = x^2 + 1$ between the vertical lines $x = 0$ and $x = 2$, in Investigation 1.

Method 1

Draw the graph of the function with your GDC.

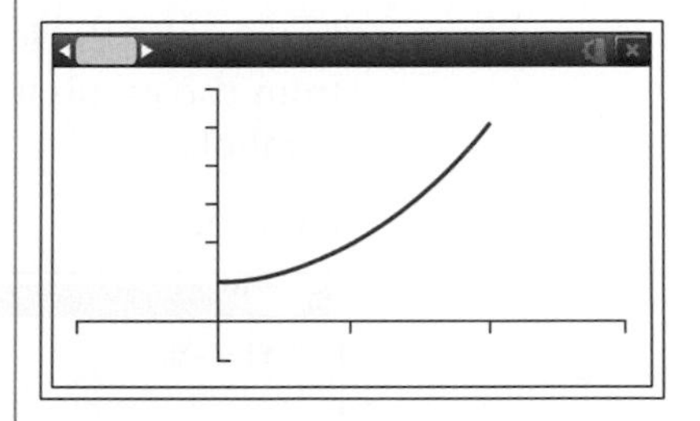

Find the area.

Then enter the lower and upper bounds.

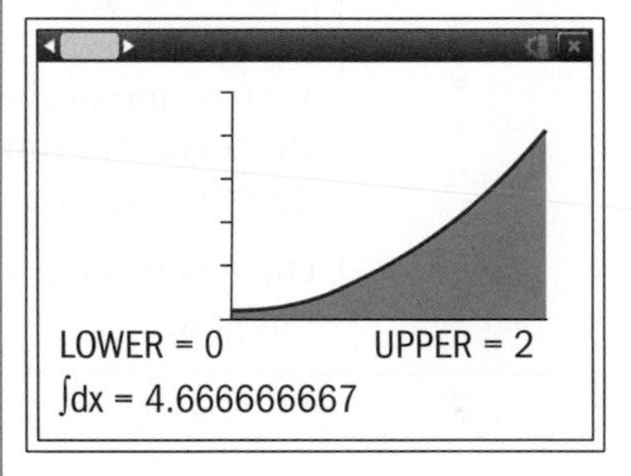

The definite integral is equal to 4.67.

Method 2

Evaluate the definite integral.

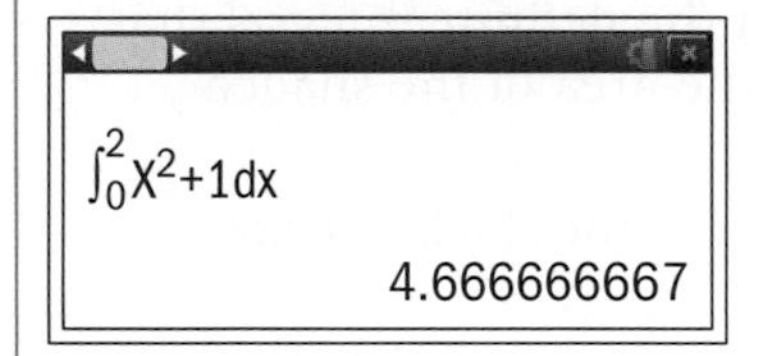

Calculus

Example 6

Consider the region A enclosed between the curve $y = -x(x-3)$ and the x-axis.

a Write down the definite integral that represents the area of A.

b Find the area of A.

Continued on next page

a $\int_0^3 -x(x-3)\,dx$	You first have to identify the region using your GDC. 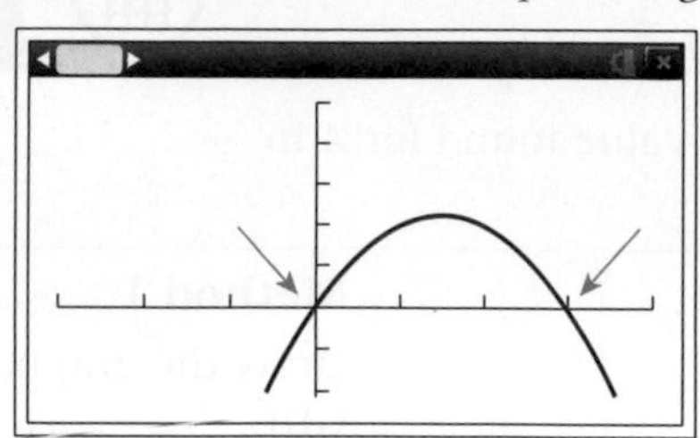 In this case the lower and upper bounds are not given but from the graph it can be seen that these are the **roots** of the parabola. Find the roots. 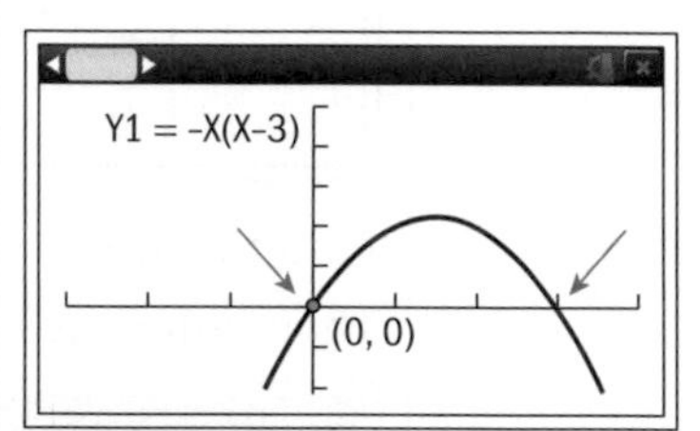 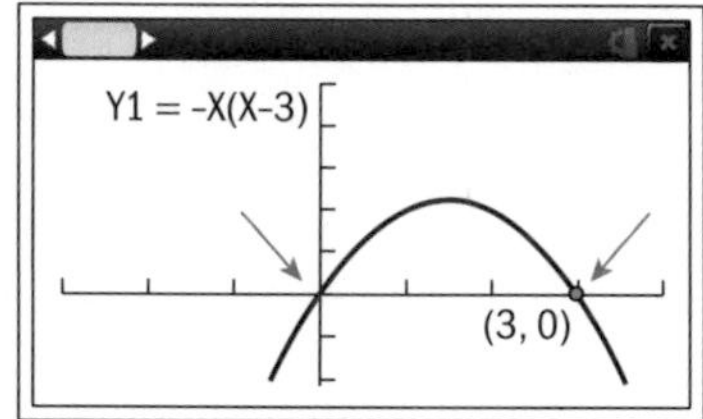 $x = 0$ is one of the roots, the lower bound of the definite integral. $x = 3$ is the other root, the upper bound of the definite integral.
b $A = 4.5$	Once you have identified the region, write down the definite integral.

Exercise 13C

1 For each of the following diagrams:

i Write down the definite integral that represents the area of the shaded region.

ii Find the area of the shaded region.

a

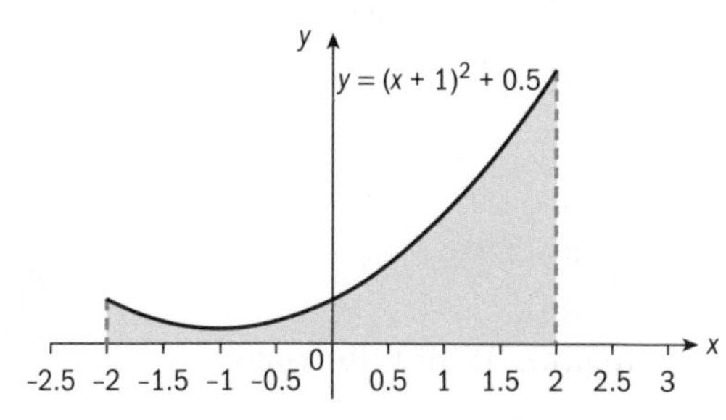

b

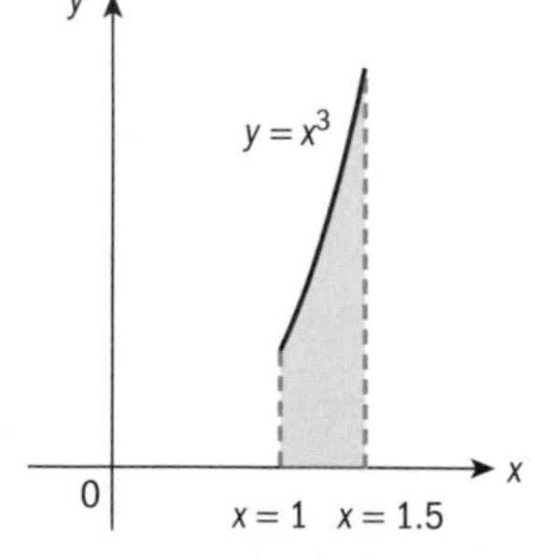

c

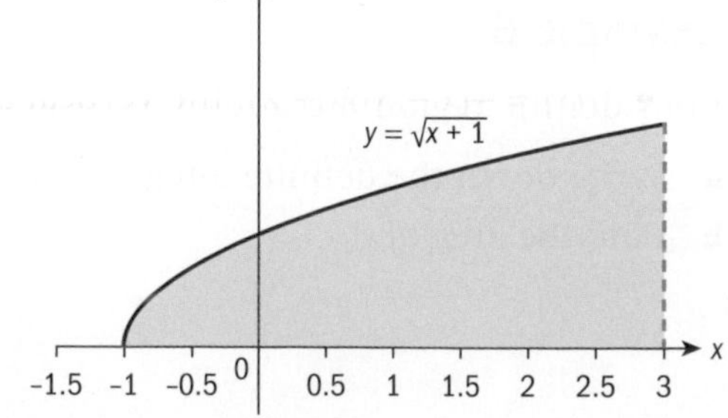

d

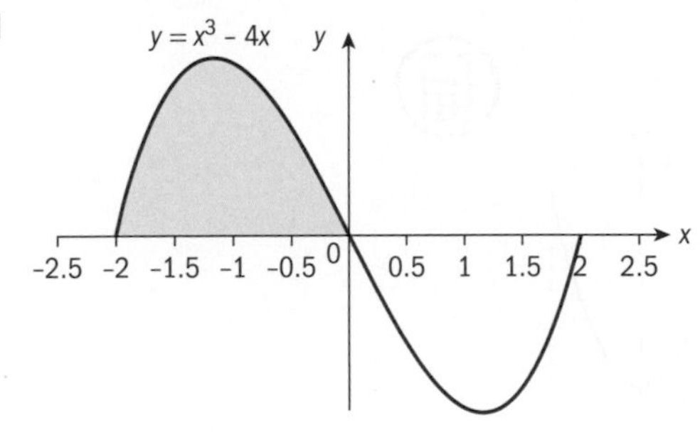

e

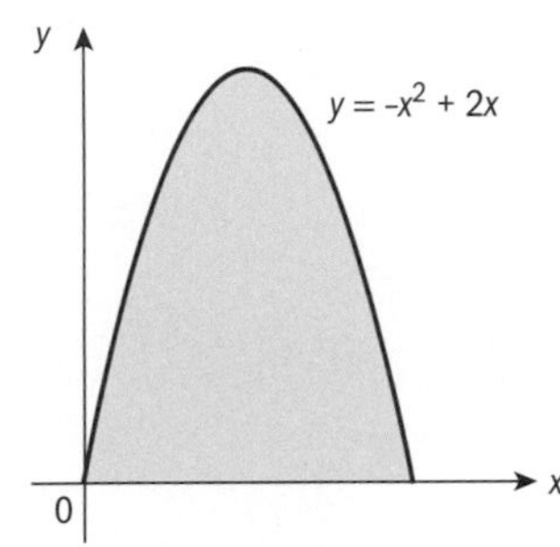

f

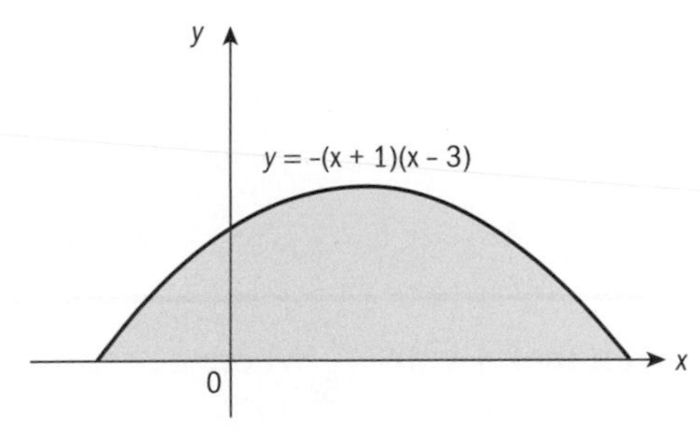

2 In each of the following:

i Write down the definite integral that represents the enclosed area.

ii Find the area.

a $y = x^2$, the x-axis and the vertical lines $x = 2$ and $x = 4$

b $y = 2x$, the x-axis and the vertical lines $x = -1$ and $x = 1$

c $y = \dfrac{1}{1+x^2}$ and the x-axis in the interval $-1 \le x \le 1$

d $y = \dfrac{1}{x}$ and the x-axis in the interval $0.5 \le x \le 3$

e $f(x) = -(x-3)(x+2)$, the vertical axis and the vertical line $x = 1$

f $f(x) = -(x-3)(x+2)$, the vertical axis and the horizontal axis

g $f(x) = -(x-3)(x+2)$ and the horizontal axis

h $f(x) = -x^2 + 2x + 15$ and the vertical lines $x = -2$ and $x = 4.5$

i $f(x) = -x^2 + 2x + 15$ and the line $y = 0$

j $f(x) = 3 - e^x$, the vertical line $x = -1$ and the x-axis

k $y = (x+2)^3 + 5$ and the coordinate axes.

3 Consider the curve $y = -x^2 + 4$.

a Find the zeros of this curve.

b Find the point where this curve cuts the y-axis.

Below is shown the graph of a piecewise function f made up by a horizontal line segment and part of the parabola $y = -x^2 + 4$. The area under the graph of f and above the x-axis has been shaded.

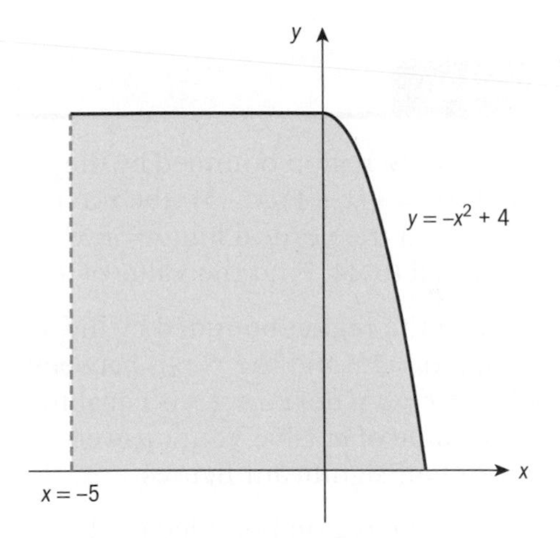

c Find the area under the graph of f in the interval $-5 \le x \le 0$.

d **i** Write down an expression for the area under the graph of f and above the x-axis for $x > 0$.

ii Find the area.

e Find the area of the whole shaded region.

Example 7

The area of the region bounded by the graph of $f(x) = x^2 + 3$, the x-axis and the vertical lines $x = -1$ and $x = a$ with $a > -1$ is equal to 12. Find the value of a.

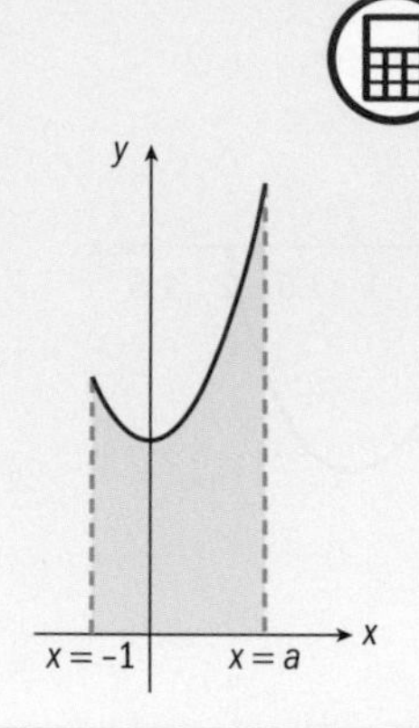

$a = 2$	The unknown, a, is the upper bound of this area. First write down the definite integral $\int_{-1}^{a}(x^2+3)\,dx = 12.$ Use technology to find the value of a.

Exercise 13D

1 The area of the region bounded by the graph of $f(x) = -(x + 1)(x - 5)$, the x-axis, the y-axis and the vertical line $x = a$ where $a > 0$ is equal to 24. Find the value of a.

2 The area of the region bounded by the graph of $f(x) = 2^{-x}$ and the x-axis between $x = -3$ and $x = a$ where $a > -3$ is equal to 9. Find the value of a. Give your answer correct to four significant figures.

3 The area of the region bounded by the graph of $f(x) = x + \frac{1}{x}$ and the x-axis between $x = a$ and $x = 3$ where $0 < a < 3$ is equal to 6.

- **a** Describe this region using a partially shaded diagram.
- **b** Find the value of a. Give your answer correct to four significant figures.

4 Given that $\int_{-2}^{a} x^2\,dx = \frac{7}{3}$ where $a > -2$:

- **a** Find the value of a.
- **b** Describe the region whose area is defined by the definite integral on a partially shaded diagram on a set of axes.

5 Given that $\int_{-1}^{b}(1+x^3)\,dx = 2$ where $b > -1$:

- **a** Find the value of b.
- **b** Describe the region whose area is defined by the definite integral on a partially shaded diagram on a set of axes.

6 Given that $\int_{-1}^{t}\sqrt{x+1}\,dx = \frac{16}{3}$ where $t > -1$:

- **a** Find the value of t.
- **b** Describe the region whose area is defined by the definite integral on a partially shaded diagram on a set of axes.

Numerical integration

You will now study a new **numerical method** to estimate areas between a curve and the x-axis in a given interval. Numerical integration is used, among other techniques, when we do not have a mathematical function to describe the area. Instead we have a set of points.

However, throughout the investigation we will use a function to illustrate this new method.

Investigation 2

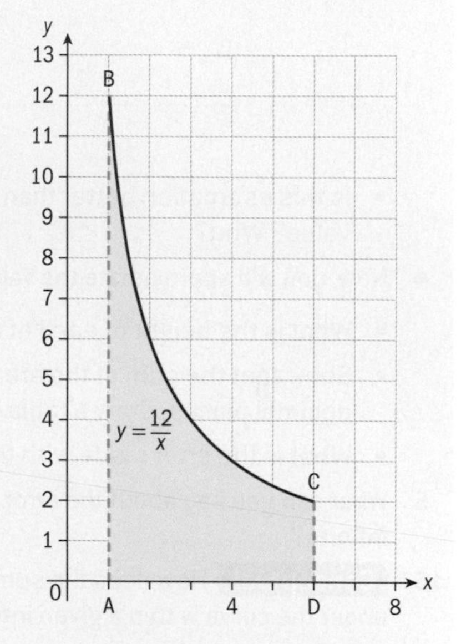

1 Consider the curve $y = \dfrac{12}{x}$, where $1 \le x \le 6$.

The area of the region enclosed between the graph of $f(x) = \dfrac{12}{x}$ and the x-axis in the interval $1 \le x \le 6$ will be called S.

Write down an expression for S and find its value. Give your answer correct to two decimal places.

2 Consider the trapezoid ABCD.

- What is the relationship between the base AB of the trapezoid ABCD and the graph of the function $f(x) = \dfrac{12}{x}$?
- What is the relationship between the base DC of the trapezoid ABCD and the graph of the function $f(x) = \dfrac{12}{x}$?
- Find the area of the trapezoid ABCD. Is this value an overestimate or an underestimate of S? What is the **error** in this approximation?

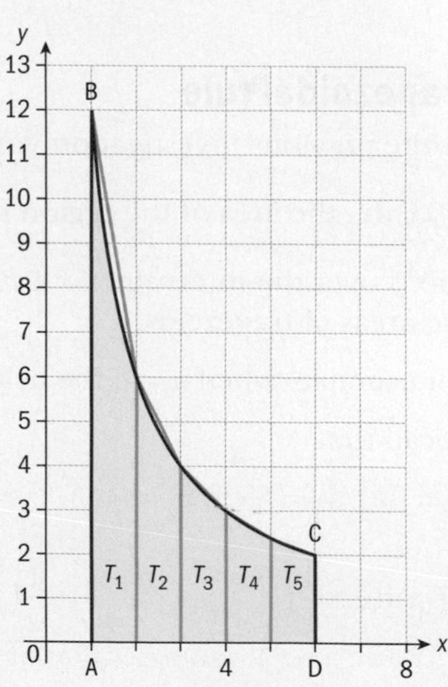

3 To find a better approximation to the value of S you can subdivide the shaded area into strips with **equal width**. The area of every strip can be approximated with the area of a trapezoid and then all these areas added.

The graph shows the shaded area subdivided into five strips. First, you will find the area of trapezoids T_1, T_2, T_3, T_4 and T_5.

You will approximate the value of S by adding up the areas of these five trapezoids.

- The height of each of these trapezoids (or the width of the strips) is equal to 1 unit. How is it found?
- Let the parallel sides of the trapezoids be y_0, y_1, y_2, y_3, y_4 and y_5 where y_0 is the length of AB and y_5 is the length of DC. How can you calculate the lengths of y_0, y_1, y_2, y_3, y_4 and y_5?
- The table on the next page will help you organize the calculations. Use a table on your GDC to complete it.

Continued on next page

Trapezoid	Base 1 (b_1)	Base2 (b_2)	h	$A=\frac{1}{2}(b_1+b_2)h$
T_1	$y_0=f(1)=\frac{12}{1}=12$	$y_1=$	1	
T_2	$y_1=$	$y_2=$	1	
T_3	$y_2=$	$y_3=$	1	
T_4	$y_3=$	$y_4=$	1	
T_5	$y_4=$	$y_5=f(6)=\frac{12}{6}=2$	1	
		Sum of the areas of the five trapezoids		

- Is this estimation better than the estimate for S found in **1**? Why? How could you improve this value? Why?

4 Now, you will approximate the value of S by adding up the area of $n=8$ trapezoids.

- What is the height of each of the trapezoids now? How did you find this value?
- Show that the sum of the area of the eight trapezoids is now equal to 21.87, correct to two decimal places. Draw a table similar to the one from part **3** to organize the calculations.
- What is the error made with this approximation?

5 What can you say about the error made in the approximation when the number of trapezoids, n, tends to infinity?

6 **Conceptual** How does the sum of the areas of trapezoids defined by a curve approximate the area under the curve within a given interval?

Trapezoidal rule

In the previous investigation you saw that the definite integral $\int_a^b f(x)\,\mathrm{d}x$, the area of the region bounded by the curve $y=f(x)$ and the x-axis over the interval $a \le x \le b$, can be approximated by the sum of the areas of trapezoids.

For example, when $n=4$, the height of each of the trapezoids is $h=\frac{b-a}{4}$.

Recall that

$y_0=f(a)$, $y_1=f(x_1)$, $y_2=f(x_2)$, $y_3=f(x_3)$ and $y_4=f(b)$ so that

$$\int_a^b f(x)\,\mathrm{d}x \cong \underbrace{\frac{1}{2}(y_0+y_1)\times\frac{b-a}{4}}_{\text{Area of } T_1}+\underbrace{\frac{1}{2}(y_1+y_2)\times\frac{b-a}{4}}_{\text{Area of } T_2}+\underbrace{\frac{1}{2}(y_2+y_3)\times\frac{b-a}{4}}_{\text{Area of } T_3}+\underbrace{\frac{1}{2}(y_3+y_4)\times\frac{b-a}{4}}_{\text{Area of } T_4}$$

The right-hand side can be simplified because there is a common factor $\frac{1}{2}\times\frac{b-a}{4}$:

$$\int_a^b f(x)\mathrm{d}x \cong \frac{1}{2}\times\frac{b-a}{4}\left[(y_0+y_1)+(y_1+y_2)+(y_2+y_3)+(y_3+y_4)\right]$$

The sum inside the square brackets can be simplified to:

$$\int_a^b f(x)\mathrm{d}x \cong \frac{1}{2}\times\frac{b-a}{4}\{y_0+2(y_1+y_2+y_3)+y_4\}$$

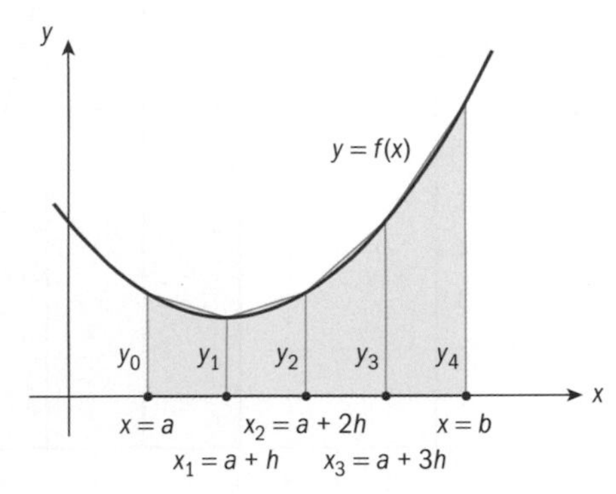

More generally,

The trapezoid rule is $\int_a^b f(x)\mathrm{d}x \cong \frac{1}{2}\times\frac{b-a}{n}\times\{y_0+y_n+2(y_1+y_2+\ldots+y_{n-1})\}$

where the interval $a \le x \le b$ is divided into n intervals of equal width.

What is the value of x_i when $i = 0$?

What is the value of x_i when $i = n$?

What is the meaning of $\frac{b-a}{n}$ in this formula?

Reflect What geometric methods can be used to approximate integrals?

Example 8

Estimate the area under a curve over the interval $4 \le x \le 12$, with x- and y-values given in the following table.

x	4	6	8	10	12
y	5	13	10	3	4

Area of trapezoid 1 $= \frac{1}{2}(5+13)\times 2 = 18$	In this example, there is no formula of the form $y = f(x)$. You need to find the area of each trapezoid and then sum these areas.
Area of trapezoid 2 $= \frac{1}{2}(13+10)\times 2 = 23$	The table suggest that there are four trapezoids.

Continued on next page

Calculus

Area of trapezoid 3 $= \frac{1}{2}(10+3)\times 2 = 13$

Area of trapezoid 4 $= \frac{1}{2}(3+4)\times 2 = 7$

Area under the curve = 18 + 23 + 13 + 7 = 61

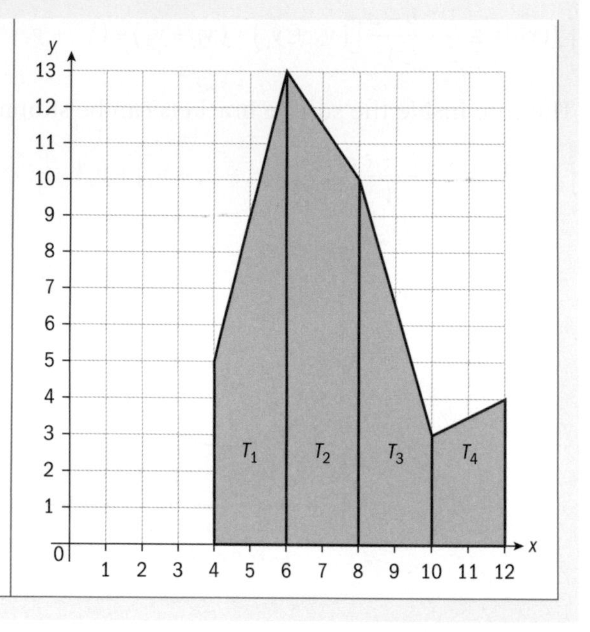

Example 9

Estimate the area under the graph of $f(x) = e^x$ over the interval $0 \le x \le 1$ using five trapezoids.

x	0	0.2	0.4	0.6	0.8	1
y	1	$e^{0.2}$	$e^{0.4}$	$e^{0.6}$	$e^{0.8}$	e

Draw a table of values.

$$\int_a^b e^x \, dx \cong \frac{1}{2}\times 0.2\times\left(1+2(e^{0.2}+e^{0.4}+e^{0.6}+e^{0.8})+e\right)$$

$$\cong 1.7240...$$

$$\cong 1.72$$

Substitute into the trapezoid rule using $a = 0,\ b = 1,\ n = 5$, and with heights of trapezoids $= \frac{1-0}{5} = 0.2$.

Example 10

The cross-section of a river is shown here. If the water is flowing at 0.8 m/s use the trapezoidal rule, with seven trapezoids, to find an approximation for the volume of water passing this point in one minute. All lengths are in metres.

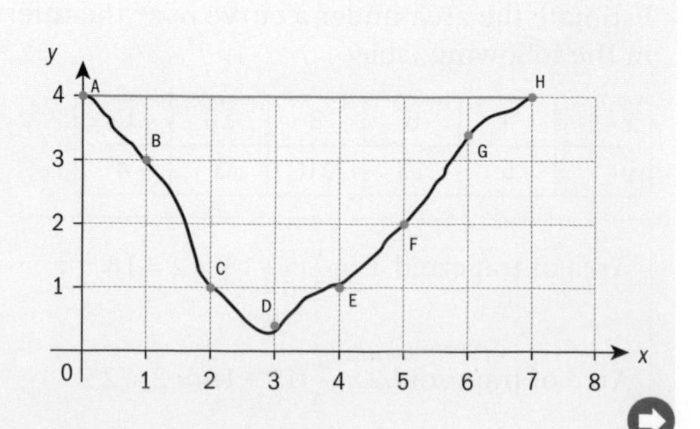

A	B	C	D	E	F	G	H
(0,4)	(1, 3)	(2, 1)	(3, 0.4)	(4, 1)	(5, 2)	(6, 3.4)	(7, 4)

Use the trapezoidal rule to find the area under the curve:	The lengths of the parallel lines are given by the y-coordinates in the table.
$A \approx \frac{1}{2} \times 1(4+4+2(3+1+0.4+1+2+3.4))$	The trapezoidal rule is applied to these values.
$= \frac{1}{2} \times 29.6 = 14.8 \text{ m}^2$	
Cross-sectional area of river is:	
$28 - 14.8 = 13.2 \text{ m}^2$	
Volume of water per minute:	Volume of water is the amount that passes per second multiplied by 60.
$= 13.2 \times 0.8 \times 60 \approx 634 \text{ m}^3$	

Exercise 13E

1 Estimate the area under a curve over the interval $1 \le x \le 9$, with the x- and y-values given in the following table.

x	1	3	5	7	9
y	5	7	6	10	4

2 Estimate the area under a curve over the interval $0 \le x \le 6$, with the x- and y-values given in the following table.

x	0	1.5	3	4.5	6
y	1	4	2	5.5	0

3 Estimate the area under a curve over the interval $-1 \le x \le 4$, with the x- and y-values given in the following table.

x	−1	0.25	1.5	2.75	4
y	5	7	3.5	6	8

4 Estimate the area under the graph of $y = f(x)$ using the data points given in the diagram.

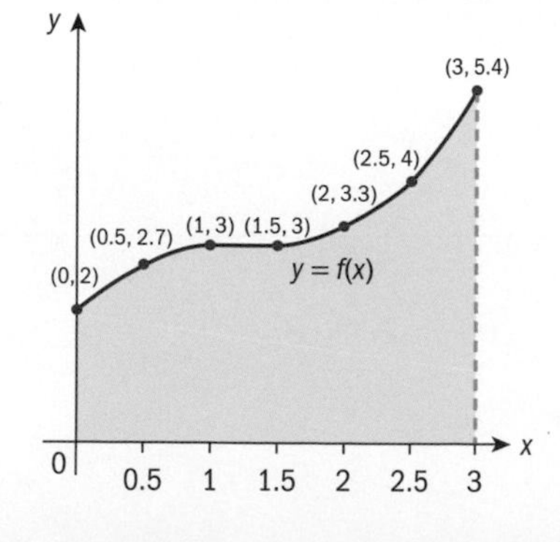

5 Estimate the area under the graph of $y = f(x)$ using the data points given in the diagram.

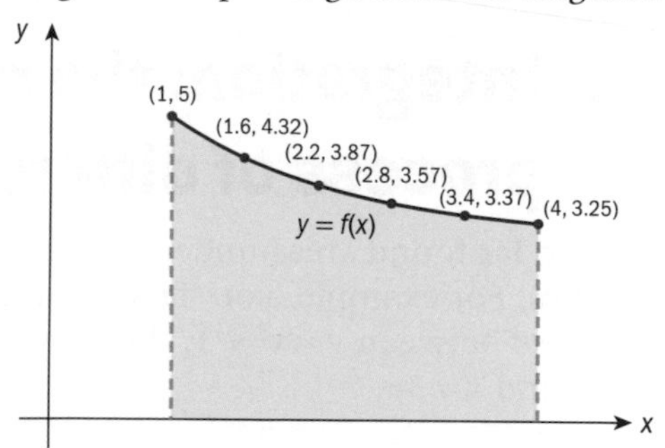

6 Estimate the area between each curve and the x-axis over the given interval, using the specified number of trapezoids. Give your answers to four significant figures.

a $f(x) = \sqrt{x}$, interval $0 \le x \le 4$ with $n = 5$

b $f(x) = 2^x$, interval $-1 \le x \le 4$ with $n = 4$

c $f(x) = \frac{10}{x} + 1$, interval $2 \le x \le 5$ with $n = 6$

d $y = -0.5x(x - 5)(x + 1)$, interval $0 \le x \le 5$ with $n = 5$

7 Consider the region enclosed by the curve $y = -2(x - 3)(x - 6)$ and the x-axis.

a Sketch the curve and shade the region.

b **i** Write down a definite integral that represents the area of this region.

ii Find the area of this region.

c Estimate the area of this region using six trapezoids.

d Find the percentage error made with the estimation made in part **c**.

8 Consider the region enclosed by the graph of the function $f(x) = 1 + e^x$, the x-axis and the vertical lines $x = 0$ and $x = 2$.

a Sketch the function f and shade the region.

b **i** Write down a definite integral that represents the area of this region.

ii Find the area of this region. Give your answer correct to four significant figures.

c Estimate the area of this region using five trapezoids. Give your answer correct to four significant figures.

d Find the percentage error made with the estimation found in part **c**.

Developing inquiry skills

In the opening scenario for this chapter you looked at how to estimate the area of an island.

How could you improve your estimation of the area of the island using what you have studied in this section?

How close is your estimate to the claimed area? Why is your answer an estimate?

TOK

Galileo said that the universe is a grand book written in the language of mathematics.

Where does mathematics come from? Does it start in our brains or is it part of the universe?

13.2 Integration: the reverse process of differentiation

You have so far found areas under curves between two given values. For example, you have seen how to find the area enclosed between $y = x^2 + 1$, the x-axis and the vertical lines $x = 1$ and $x = 3$.

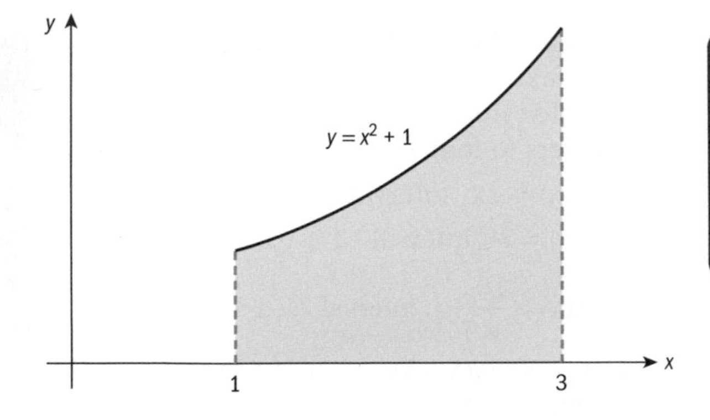

HINT

The area of the shaded region is $\int_1^3 (x^2 + 1)\,dx = \frac{32}{3}$.

You are now going to find expressions for areas when one of the limits is fixed and the other is variable.

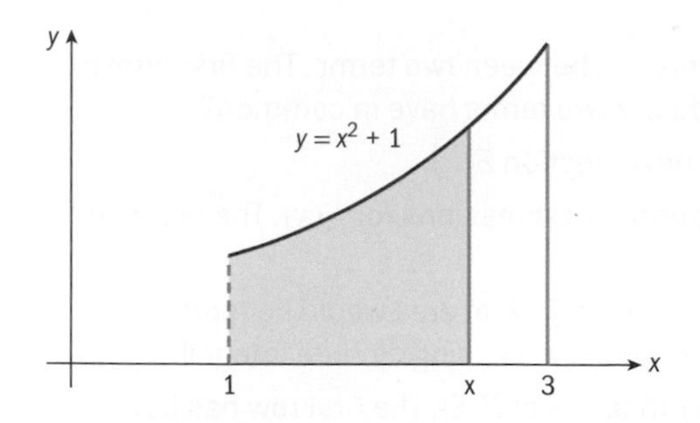

HINT

As the value of x increases from 1 to 3, the area of the shaded region also increases.

There is a function, the **area function,** which maps every value of x to the area of the shaded region.

What is the formula of the area function?

And what would the area function be in a different region?

Consider a positive and continuous function $y = f(t)$ over the interval $a \le t \le b$. The area enclosed between the graph of $y = f(t)$, the t-axis and the vertical lines $t = a$ and $t = x$ where $a \le x \le b$ is defined by

$A(x) = \int_a^x f(t)\,dt$.

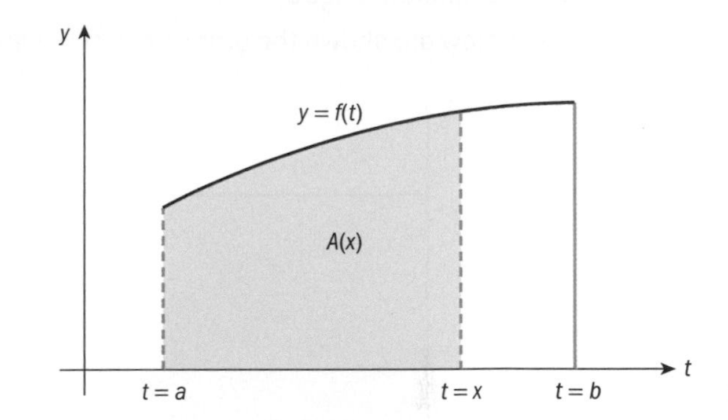

In this investigation you will find the relationship between the two major branches of calculus: integration and differentiation.

Investigation 3

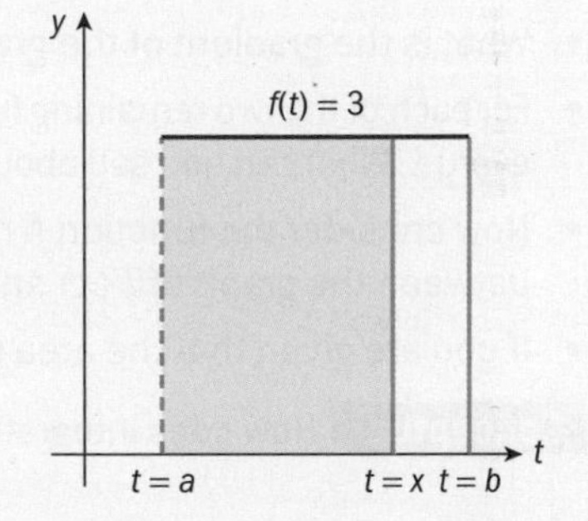

1 Consider the area under the graph of $f(t) = 3$ between $t = a$ and $t = x$, where $a < x$.

Show that the area function is $A(x) = 3x - 3a$.

2 Consider the area under the graph of $f(t) = t$ between $t = a$ and $t = x$, where $a < x$. Show that the area function can be written as $A(x) = \frac{x^2}{2} - \frac{a^2}{2}$. Draw the graph of the function f and shade the area enclosed between this graph and the t-axis over the interval $a \le t \le x$.

3 Consider the area under the graph of $f(t) = 2t$ between $t = a$ and $t = x$, where $a < x$. Show that the area function can be written as $A(t) = x^2 - a^2$. Draw the graph of the function f and shade the area enclosed between this graph and the t-axis over the interval $a \le t \le x$.

4 The results from **1** to **3** are summarized in the first two columns from the table shown below. The last two columns will be completed later.

$f(t)$	$A(x)$	$F(x)$	$A(x)$
$f(t) = 3$	$A(x) = 3x - 3a$	$F(x) = 3x$	$F(x) - F(a)$
$f(t) = t$	$A(x) = \frac{x^2}{2} - \frac{a^2}{2}$		
$f(t) = 2t$	$A(x) = x^2 - a^2$		

Continued on next page

- The expressions for $A(x)$ have been written as the difference between two terms. The first term is a function of x and the second term is constant. What do the two terms have in common?

Let the first term in each of the expressions for $A(x)$ be a new function $F(x)$.

- Complete the third column of the table with the corresponding expressions for $F(x)$. The first row has been completed for you.
- For which value of a are the expressions for $F(x)$ and $A(x)$ equal? What area would be represented by $F(x)$ when a takes this value? How would you represent this area using definite integrals?
- Complete the fourth column of the table by writing $A(x)$ in terms of $F(x)$. The first row has been completed for you.
- Below are shown the graphs of $f(t) = 3$ and $F(x) = 3x$.

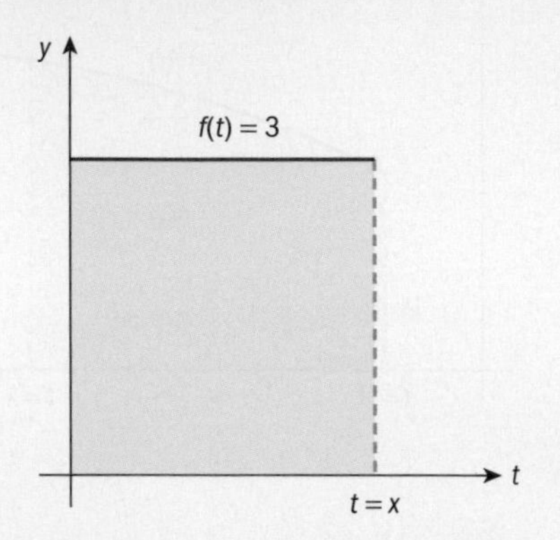

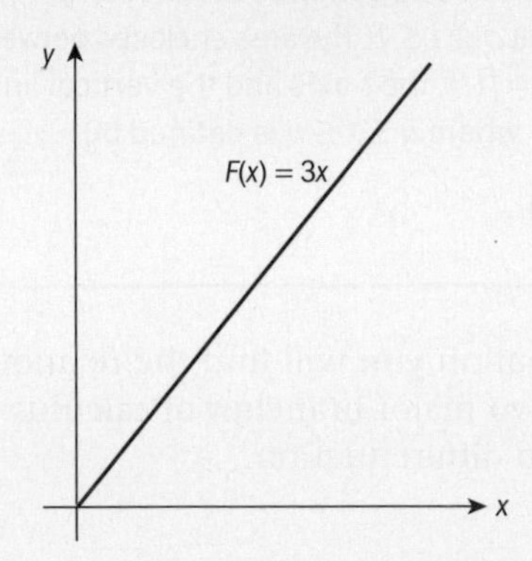

- What is the gradient of the graph of F at every x? How does this relate to the graph of f?
- For each of the two remaining functions f, find an expression for the gradient of the function F at every x. What can you say about the relationship between the gradient function of F and the function f?
- Now consider the function $f(t) = 3t + 1$. What would be $F(x)$ in this case? What is the relationship between the graph of $F(x)$ and the graph of f?
- If you are given that the area function is $F(x) = x^3$, how would you find the formula of f?

5 **Conceptual** How does integration relate to differentiation?

If $f(t) \geq 0$ is a continuous function, the area enclosed between the graph of f and the x-axis over the interval $a \leq t \leq x$ can be found with the definite integral $\int_a^x f(t)\,\mathrm{d}t$.

Also $\int_a^x f(t)\,\mathrm{d}t = F(x) - F(a)$ where $F'(x) = f(x)$.

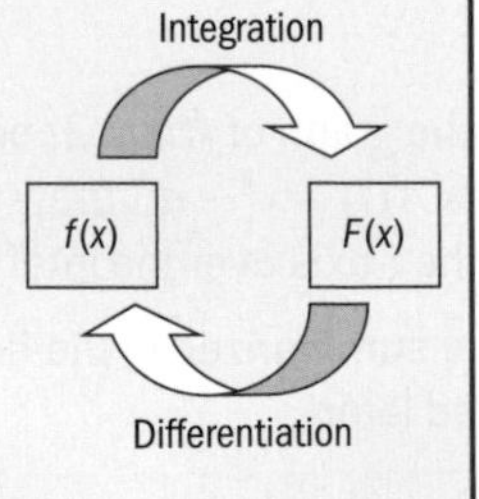

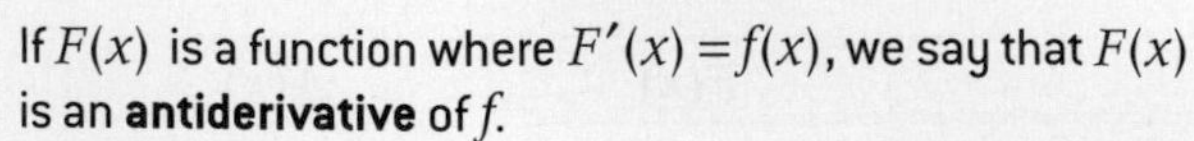

If $F(x)$ is a function where $F'(x) = f(x)$, we say that $F(x)$ is an **antiderivative** of f.

The process of finding an antiderivative is called **antidifferentiation.**

International-mindedness

The **fundamental theorem of calculus** shows the relationship between the derivative and the integral and was developed in the 17th century by Gottfried Leibniz and Isaac Newton.

For example,

$F(x) = 3x$ is an antiderivative of $f(x) = 3$ because $F'(x) = 3$.

$F(x) = \dfrac{x^2}{2} - 3$ is an antiderivative of $f(x) = x$ because $F'(x) = x$.

$F(x) = x^2 + 1$ is an antiderivative of $f(x) = 2x$ because $F'(x) = 2x$.

Can you think of an antiderivative for $f(x) = 3x^2$?

Reflect What is an antiderivative of a function?

How is the antiderivative related to the definite integral?

TOK

How do "believing that" and "believing in" differ?

How does belief differ from knowledge?

Example 11

a Find an antiderivative of $y = 4 + 2x$.

b Find a function $y = g(x)$ when the gradient function is $\frac{dy}{dx} = 3x$.

a $F(x) = 4x + x^2$	An antiderivative of $y = 4 + 2x$ is a function F whose derivative is $4 + 2x$. The derivative of $4x$ is 4. The derivative of x^2 is $2x$. The derivative of $4x + x^2$ is $4 + 2x$.
b $g(x) = 1.5x^2$	If the gradient function of $y = g(x)$ is $3x$ then $g'(x) = 3x$. Function g is a function whose derivative is $3x$. The derivative of $1.5x^2$ is $3x$.

Example 12

Consider the definite integral $\int_0^x (3-2t)\,dt$ where $0 \le x \le 1.5$.

a Draw on a diagram the function $f(t) = 3 - 2t$ and shade the area represented by $\int_0^x (3-2t)\,dt$.

b By using antiderivatives find an expression for $\int_0^x (3-2t)\,dt$ in terms of x.

a

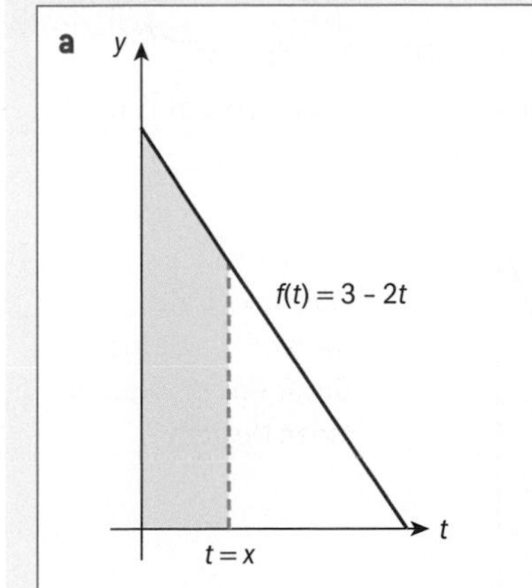

You can use your GDC to draw the graph.

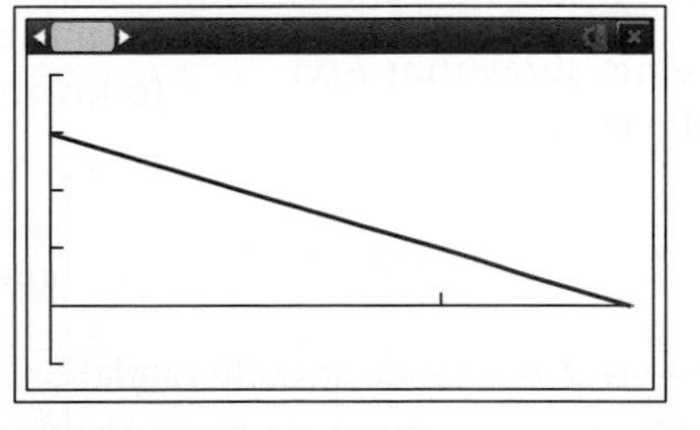

HINT

Remember that $0 \le x \le 1.5$.

Shade the area under the graph of f between $t = 0$ and $t = x$.

Continued on next page

b $\int_0^x (3-2t)\,dt = 3x - x^2$	Find an antiderivative of $f(t) = 3 - 2t$, a function whose derivative is $f(t) = 3 - 2t$. The derivative of $3t$ is 3. The derivative of t^2 is $2t$. The derivative of $3t - t^2$ is $3 - 2t$. $F(x) = 3x - x^2$ $F(0) = 0$ $F(x) - F(0) = 3x - x^2$

HINT

Use the formula for the area of a trapezoid to check that this answer is correct.

Exercise 13F

1 a For each of the following, show that $F(x)$ is an antiderivative of $f(x)$.

i $f(x) = 1$ and $F(x) = x$

ii $f(x) = 1$ and $F(x) = x + 3$

iii $f(x) = 1$ and $F(x) = x - 6$

b Find two further antiderivatives for $f(x) = 1$.

c Write down the general form of an antiderivative of $f(x) = 1$.

2 a For each of the following, show that $F(x)$ is an antiderivative of $f(x)$.

i $f(x) = 2x$ and $F(x) = x^2$

ii $f(x) = 2x$ and $F(x) = x^2 + 1$

iii $f(x) = 2x$ and $F(x) = x^2 - 4$

b Find two further antiderivatives for $f(x) = 2x$.

c Write down the general form of an antiderivative of $f(x) = 2x$.

3 a For each of the following, show that $F(x)$ is an antiderivative of $f(x)$.

i $f(x) = x$ and $F(x) = \frac{x^2}{2}$

ii $f(x) = x$ and $F(x) = \frac{x^2}{2} + 2.5$

iii $f(x) = x$ and $F(x) = \frac{x^2}{2} - 12$

b Write down the general form of an antiderivative of $f(x) = x$.

4 a For each of the following, show that $F(x)$ is an antiderivative of $f(x)$.

i $f(x) = 2x + 1$ and $F(x) = x^2 + x$

ii $f(x) = 2x + 1$ and $F(x) = x^2 + x - 3.2$

iii $f(x) = 2x + 1$ and $F(x) = x^2 + x + 4$

b Write down the general form of an antiderivative of $f(x) = 2x + 1$.

5 Find a function $y = g(x)$ when the gradient function is $\frac{dy}{dx} = 3$.

6 Find a function $y = g(x)$ when the gradient function is $\frac{dy}{dx} = -2$.

7 Find a function $y = g(x)$ when the gradient function is $\frac{dy}{dx} = \frac{x}{2}$.

8 Find an antiderivative of $f(x) = x^2$.

9 Find the area function, $A(x)$, in each of the following situations.

a

y
f(t) = 2.5t
t
t = x

b

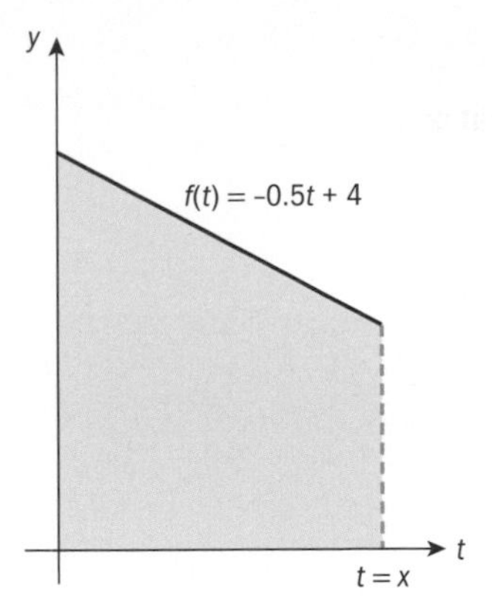

d

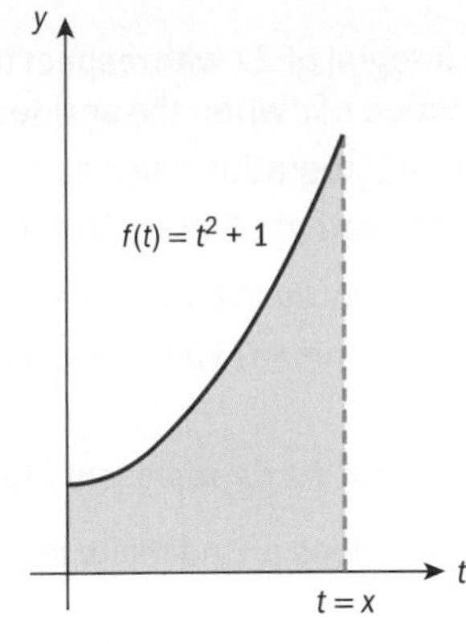

c

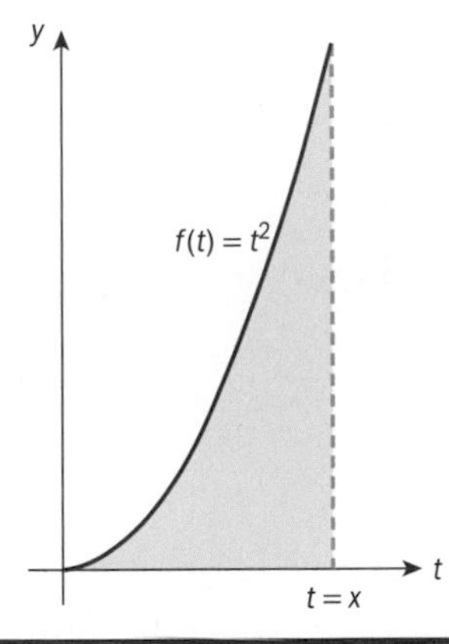

10 Consider the graph of the function $f(t) = 4$ over the interval $0 \le t \le x$. Find $\int_0^x 4\,dt$.

11 Consider the graph of the function $f(t) = 2t + 1$ over the interval $0 \le t \le x$. Find $\int_0^x (2t+1)\,dt$.

12 Consider the graph of the function $f(t) = 4t$ over the interval $2 \le t \le x$. Find $\int_2^x 4t\,dt$.

Investigation 4

1 On the same set of axes, sketch the graphs of

$y = x^2$ $\quad y = x^2 + 1$ $\quad y = x^2 - 2$ $\quad y = x^2 + 3$

over the interval $-4 \le x \le 4$.

- How can you describe their relative position?
- Find the gradient of each of these curves at $x = 1$. What do you notice?
- Now find the gradient of each of these curves at $x = -2$. What can you say about these answers?
- What is the gradient of each of these curves at any x?
- Find another curve for which the gradient at x is the same as the gradient of any of the curves from **1**.
- How would you write the formula of any curve with the gradient the same as the gradient of the curves from **1**?
- All these curves make up a family of functions that are antiderivatives of $2x$. Why is this?

The notation used to indicate this family of functions is $\int 2x\,dx = x^2 + c$, where c is a constant.

Continued on next page

This is read as "the integral of $2x$ with respect to x is $x^2 + c$".

- What is the value of c when the antiderivative is $y = x^2 + 1$? What is the constant of integration when the antiderivative is $y = x^2$?
- What is the derivative of $x^2 + c$ with respect to x?

2 What is the family of functions that are antiderivatives of $f(x) = 3$? What do their graphs have in common? Write this family of functions using integral notation.

3 What does the integral $\int x\,dx$ represent? Calculate it.

4 **Conceptual** How does an indefinite integral define a family of antiderivatives?

If $F'(x) = f(x)$ then $\int f(x)\,dx = F(x) + c$ where $c \in \mathbb{R}$.

The expression $\int f(x)\,dx$ is called an **indefinite integral** and $\int f(x)\,dx$ is read as "the integral of f with respect to x".

Note: c is called the **constant of integration**.

Reflect What does an indefinite integral represent?

How is an antiderivative related to an indefinite integral?

Example 13

a Find the family of antiderivatives of $f(x) = 5$.

b Find $\int 4x\,dx$.

a $F(x) = 5x + c$ where $c \in \mathbb{R}$	These are the functions F for which $F'(x) = 5$.
b $2x^2 + c$ where $c \in \mathbb{R}$	This represents the family of antiderivatives of $4x$, the functions whose derivatives are $4x$. This is $2x^2 + c$.

Exercise 13G

1 Find the family of antiderivatives of 2.

2 Find the family of antiderivatives of $x + 1$.

3 Find all the functions whose derivatives are equal to -1.

4 Find all the functions whose derivatives are equal to $2x$.

5 Calculate the following indefinite integrals.

a $\int 1\,dx$ b $\int 6\,dx$ c $\int \frac{1}{2}\,dx$

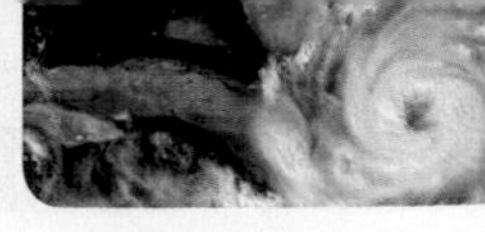

6 Calculate the following indefinite integrals.

a $\int x\,dx$ **b** $\int 2x\,dx$

c $\int 5x\,dx$ **d** $\int \frac{1}{2}x\,dx$

e $\int ax\,dx$ (a is a non-zero constant)

7 Calculate the following indefinite integrals.

a $\int x^2\,dx$ **b** $\int 3x^2\,dx$ **c** $\int 4x^2\,dx$

d $\int \frac{1}{2}x^2\,dx$ **e** $\int \frac{x^2}{3}\,dx$ **f** $\int \frac{3}{2}x^2\,dx$

g $\int ax^2\,dx$ (a is a non-zero constant)

8 Calculate the following indefinite integrals.

a $\int x^3\,dx$ **b** $\int 4x^3\,dx$ **c** $\int 2x^3\,dx$

d $\int \frac{4}{5}x^3\,dx$ **e** $\int \frac{x^3}{3}\,dx$ **f** $\int -x^3\,dx$

g $\int ax^3\,dx$ (a is a non-zero constant)

In questions **5** to **8** you may have noticed that all the functions were of the form ax^n.

$\int ax^n\,dx = \frac{ax^{n+1}}{n+1} + c$ where a and n are constants, $a \neq 0$ and $n \neq -1$. n is an integer.

This is an integration rule and is called the **power rule.**

Reflect Why is the condition $n \neq -1$ given in the power rule?

Example 14

a Find $\int \frac{x^5}{4}\,dx$. **b** Find $\int (\frac{x^2}{2} + 3x - 1)\,dx$. **c** Find $\int \frac{4}{x^2}\,dx$.

a $\int \frac{x^5}{4}\,dx = \frac{x^6}{24} + c$	Apply the power rule with $a = \frac{1}{4}$ and $n = 5$: **HINT** Do not forget the constant of integration. $\int \frac{x^5}{4}\,dx = \int \frac{1}{4}x^5\,dx = \frac{1}{4} \times \frac{x^{5+1}}{5+1} + c = \frac{1}{4} \times \frac{x^6}{6} + c$
b $\int (\frac{x^2}{2} + 3x - 1)\,dx = \frac{x^3}{6} + \frac{3x^2}{2} - x + c$	As the derivative of a sum is the sum of the derivatives, calculate an antiderivative for each term by applying the power rule and then add the three terms up. $\int \frac{x^2}{2}\,dx = \frac{1}{2} \times \frac{x^{2+1}}{2+1} + c_1 = \frac{1}{6}x^3 + c_1$

Continued on next page

Calculus

	$\int 3x\,dx = 3 \times \frac{x^{1+1}}{1+1} + c_2 = \frac{3x^2}{2} + c_2$
	$\int -1\,dx = -x + c_3$
c $\int \frac{4}{x^2}dx = -\frac{4}{x} + c$	Apply the power rule $a = 4;\ n = -2$ $\int \frac{4}{x^2}dx = \int 4x^{-2}\,dx = 4 \times \frac{x^{-2+1}}{-2+1} + c$

HINT

c_1, c_2, c_3 are three constants that added up give another constant c.

Reflect What is the integral of a sum of multiples of powers of x?

How can you tell that the indefinite integrals are correct?

Exercise 13H

1 Find the following indefinite integrals.

a $\int 10\,dx$ **b** $\int 0.6x^2 dx$

c $\int x^5 dx$ **d** $\int (7-2x)\,dx$

e $\int (1+2x)\,dx$ **f** $\int (5+x-\frac{1}{3}x^2)\,dx$

g $\int (-x+\frac{3x^2}{4}+0.5)\,dx$ **h** $\int (1-x+\frac{x^3}{2})\,dx$

i $\int (x^2-\frac{1}{2}x+4)\,dx$ **j** $\int \frac{1}{2}x^{-2}\,dx$

k $\int \frac{2}{x^4}dx$ **l** $\int (4x+\frac{5}{x^3})\,dx$

2 For $f(x) = x^2 - \frac{x}{3} + 4$, find:

a $f'(x)$ **b** $\int f(x)\,dx$

3 Find $\int (t-3t^2)\,dt$.

4 Find $\int (4t^3 - 3t + 1)\,dt$.

5 Find **all** the functions F for which the gradient equals $3 + x - \frac{x^2}{4}$.

6 Find **all** the functions F for which the gradient is $y = -x + 0.5x^2$.

7 **a** Expand $(x+1)(x-2)$.

b Hence, find all the functions $y = f(x)$ for which $\frac{dy}{dx} = (x+1)(x-2)$.

8 Find the indefinite integral of the function $g(x) = x^3 - \frac{2}{x^2} + 1$.

Example 15

a The curve $y = f(x)$ passes through the point $(1, 3)$. The gradient of the curve is given by $f'(x) = 2 + \frac{x}{3}$. Find the equation of the curve.

b If $\frac{dy}{dx} = 2x - 4x^2$ and $y = -1$ when $x = 3$, find y in terms of x.

a $f(x)=\int(2+\frac{x}{3})\mathrm{d}x = 2x+\frac{1}{3}\times\frac{x^2}{2}+c$ $f(x)=2x+\frac{x^2}{6}+c$	Apply the power rule to find an antiderivative of $f'(x)$.
$f(1)=2\times1+\frac{1^2}{6}+c$ $3=2+\frac{1}{6}+c$ $c=\frac{5}{6}$ $f(x)=2x+\frac{x^2}{6}+\frac{5}{6}$	If the curve passes through the point $(1, 3)$ then $f(1)=3$.
b $y=\int(2x-4x^2)\mathrm{d}x = x^2-\frac{4x^3}{3}+c$	Integrate $\frac{\mathrm{d}y}{\mathrm{d}x}$ to find y in terms of x.
$-1=3^2-\frac{4\times3^3}{3}+c$ $-1=9-36+c$ $c=26$ $y=x^2-\frac{4x^3}{3}+26$	Use the fact that $y=-1$ when $x=3$ to find the value of the constant c.

Example 16

Find the cost function, $C(x)$, when the marginal cost is $M(x) = 1 + 2x$ and the fixed cost is US$40.

$C(x)=x+x^2+40$	The cost function is an antiderivative of the marginal cost function. To find $C(x)$, integrate $M(x)$ $\int(1+2x)\,dx = x+x^2+c$ The fixed cost is used to find the constant of integration. $C(0) = 40$ then $0 + 0^2 + c = 40$ and $c = 40$

Calculus

Exercise 13I

1 The derivative of the function f is given by $f'(x) = 3x + 4x^2$. The point $(-1, 0)$ lies on the graph of f. Find an expression for f.

2 It is given that $\frac{dy}{dx} = x + \frac{x^2}{5} + 2$ and that $y = 3$ when $x = 4$. Find an expression for y in terms of x.

3 It is given that $f'(x) = 3 - x$ and $f(2) = 1$. Find $f(x)$.

4 It is given that $f'(x) = 3x^2 - \frac{x^4}{3}$ and $f(0) = 2$. Find $f(x)$.

5 It is given that $\frac{dy}{dx} = 0.2x + 3x^2 + 1$ and that $y = -1$ when $x = 1$. Find the value of y when $x = 0.5$.

6 The derivative of the function f is given by $f'(x) = -x + 3$. The point $(-2, 0)$ lies on the graph of f.

  a Find an expression for f.

  b Find $f(2)$.

7 A company's marginal cost function is $M(x) = x - \frac{2}{x^2} + 1$ and the fixed cost is US$145. Find the cost function.

8 A company's marginal cost function is $M(x) = 3 - 4x + x^2$ and the fixed cost for the company is US$1000. Find the cost function.

9 A refrigerator factory's marginal revenue function is $M(t) = t^2 - 80$, where t is the number of refrigerators produced by the factory and the revenue is given in US$.

The factory earns US$ 567,000 in revenue from selling 120 refrigerators.

  a Find the revenue function, $R(t)$.

  b Find the factory's total revenue from producing 150 refrigerators.

International-mindedness

Ibn Al Haytham, born in modern-day Iraq in the 10th century, was the first mathematician to calculate the integral of a function in order to find the volume of a paraboloid.

Developing your toolkit

Now do the Modelling and investigation activity on page 584.

TOK

Why do we study mathematics?

What's the point?

Can we do without it?

Chapter summary

- When f is a non-negative function for $a \le x \le b$, $\int_a^b f(x)\,dx$ gives the area under the curve from $x = a$ to $x = b$.

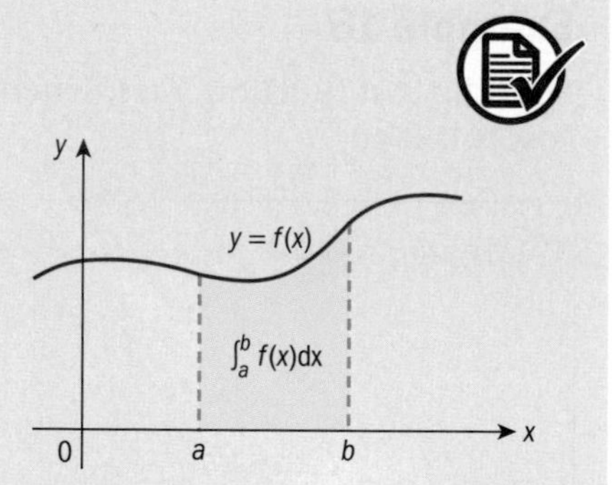

- If $v(t)$ is a velocity–time function and $v(t) \ge 0$ over the interval of time $a \le t \le b$ then

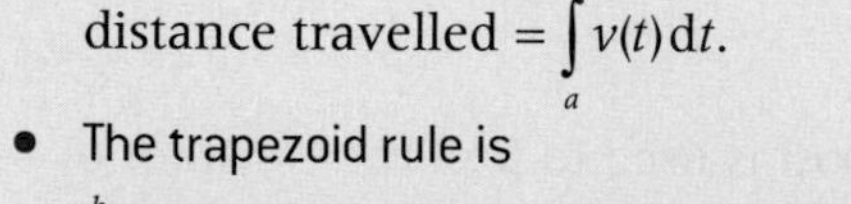

distance travelled $= \int_a^b v(t)\,dt$.

- The trapezoid rule is

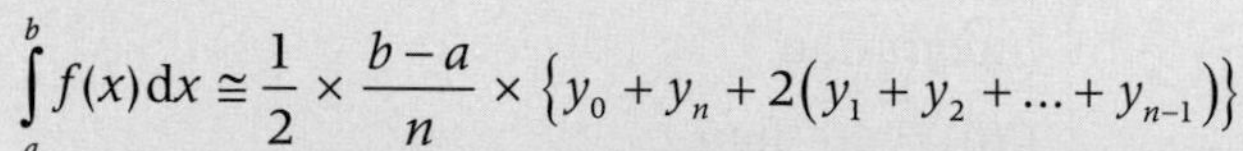

$$\int_a^b f(x)\,dx \cong \frac{1}{2} \times \frac{b-a}{n} \times \{y_0 + y_n + 2(y_1 + y_2 + \ldots + y_{n-1})\}$$

where the interval $a \le x \le b$ is divided into n intervals of equal width.

- Consider a positive and continuous function $y = f(t)$ over the interval $a \le t \le b$. The area enclosed between the graph of $y = f(t)$, the t-axis and the vertical lines $t = a$ and $t = x$ where $a \le x \le b$ is defined by $A(x) = \int_a^x f(t)\,dt$.

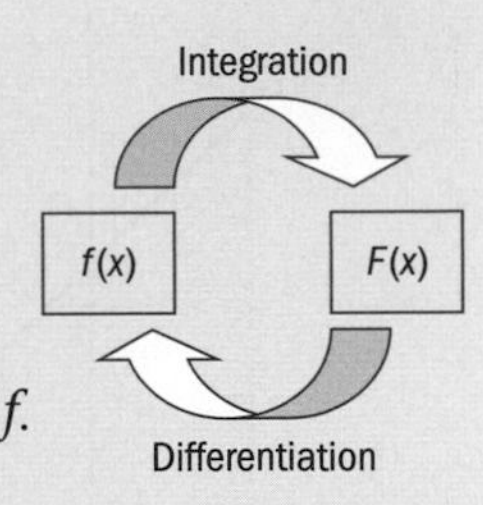

- If $f(t) \geq 0$ is a continuous function, the area enclosed between the graph of f and the x-axis over the interval $a \leq t \leq x$ can be found with the definite integral $\int_a^x f(t)\,dt$. Also $\int_a^x f(t)\,dt = F(x) - F(a)$ where $F'(x) = f(x)$.

  If $F(x)$ is a function where $F'(x) = f(x)$, we say that $F(x)$ is an **antiderivative** of f. The process of finding an antiderivative is called **antidifferentiation.**

- If $F'(x) = f(x)$ then $\int f(x)\,dx = F(x) + c$ where $c \in \mathbb{R}$.

  The expression $\int f(x)\,dx$ is called an **indefinite integral** and $\int f(x)\,dx$ is read as "the integral of f with respect to x".

  HINT

  c is called the **constant of integration**.

- $\int ax^n\,dx = \frac{ax^{n+1}}{n+1} + c$ where a and n are constants, $a \neq 0$ and $n \neq -1$.

  This is an integration rule and is called the **power rule.**

Developing inquiry skills

Write down any further inquiry questions you could ask and investigate how you could find the areas of irregular shapes and curved shapes.

Chapter review

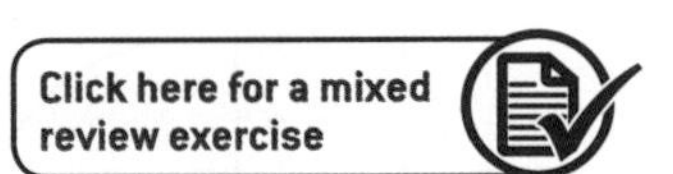

1 For each of the following shaded regions:

 i Write down a definite integral that represents the area of the region.

 ii Hence or otherwise, find the area of these shaded regions.

a

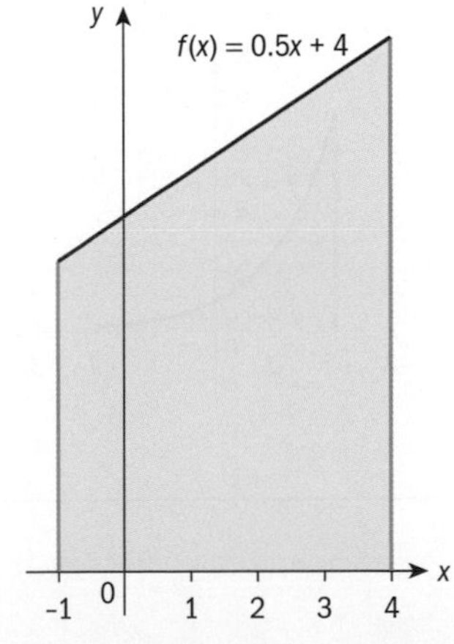

b

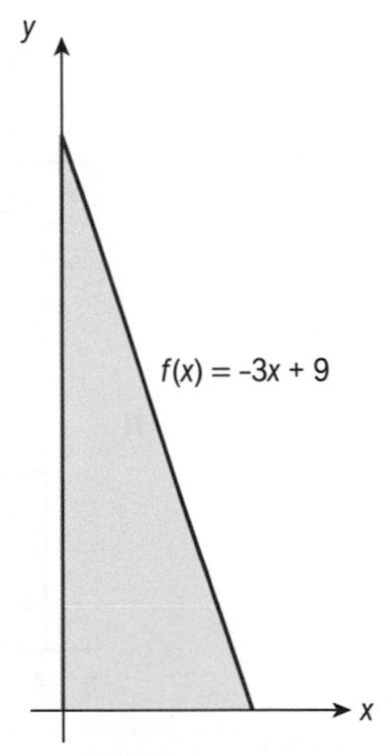

2 For each of the following regions:

 i Write down a definite integral that represents the area of the region.

 ii Hence or otherwise, find the area of these shaded regions.

a

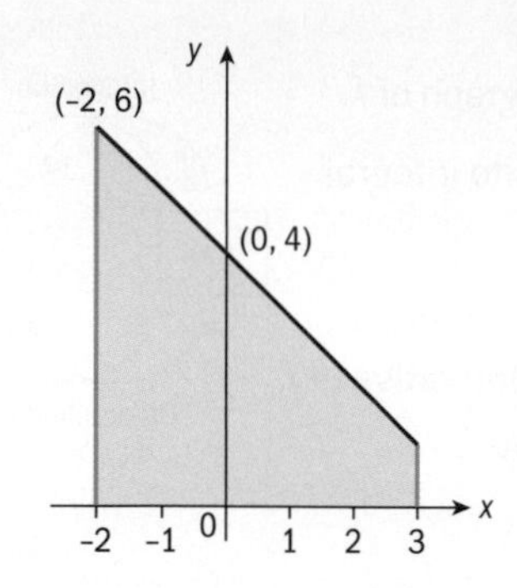

b

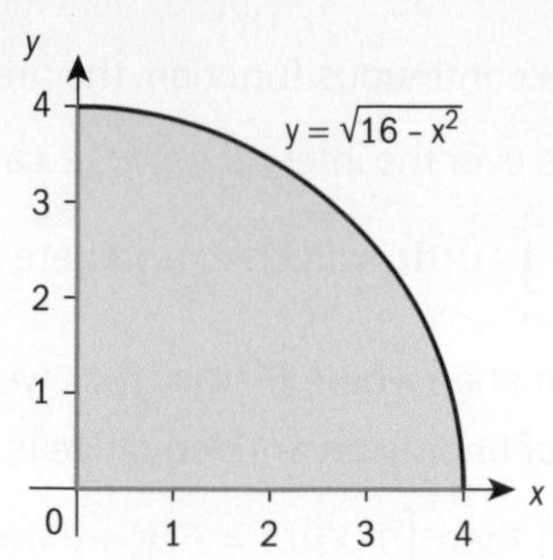

3 For each of the following regions:

i Write down a definite integral that represents the area of the region.

ii Find the area of the region.

a

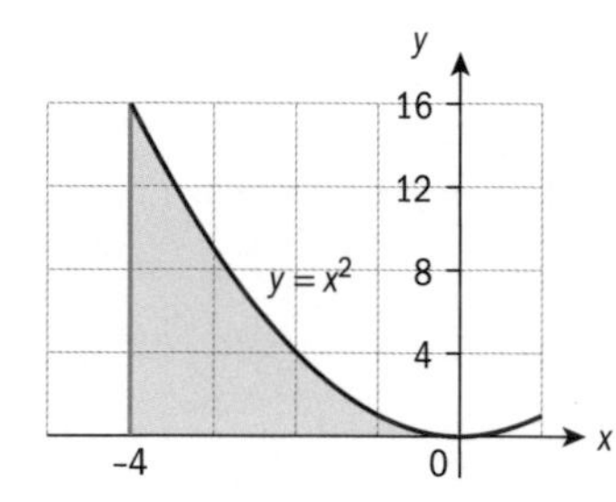

b

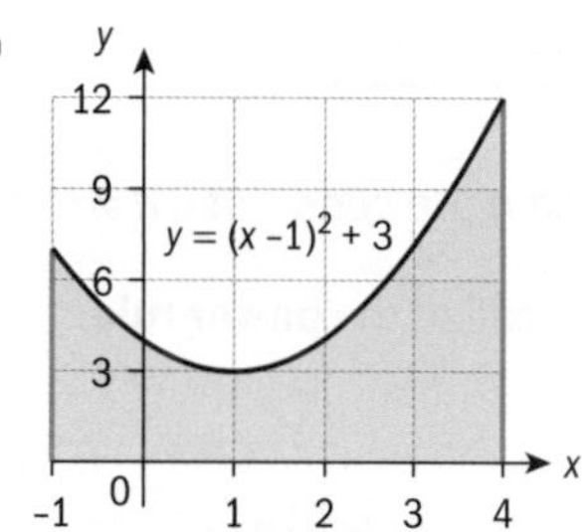

c

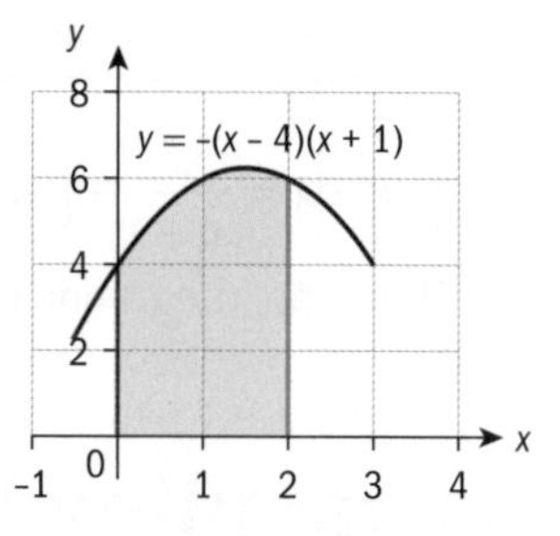

d

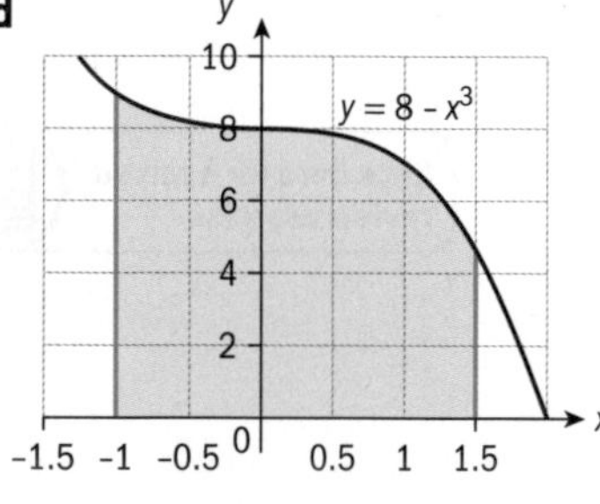

e

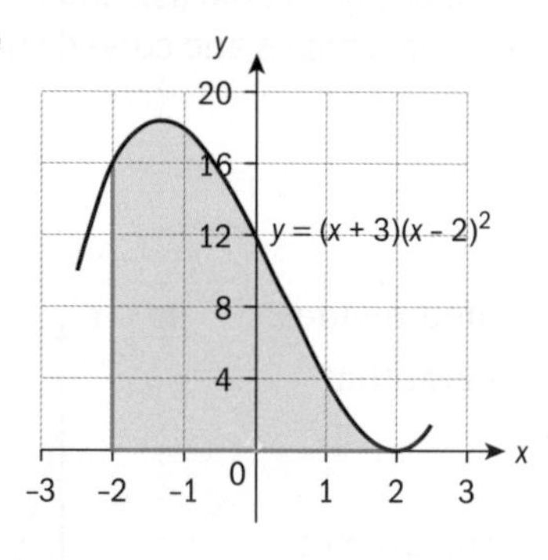

f

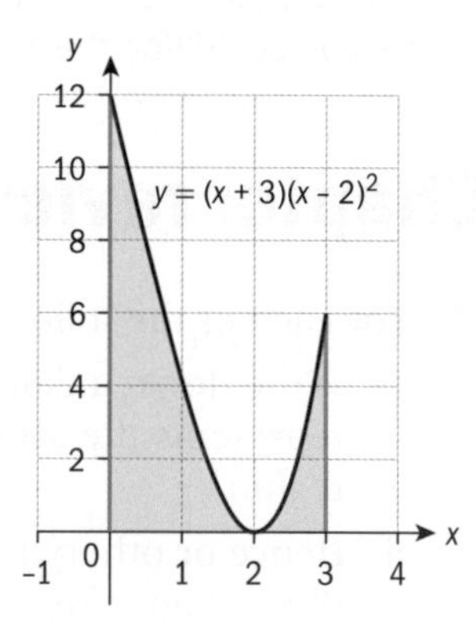

g

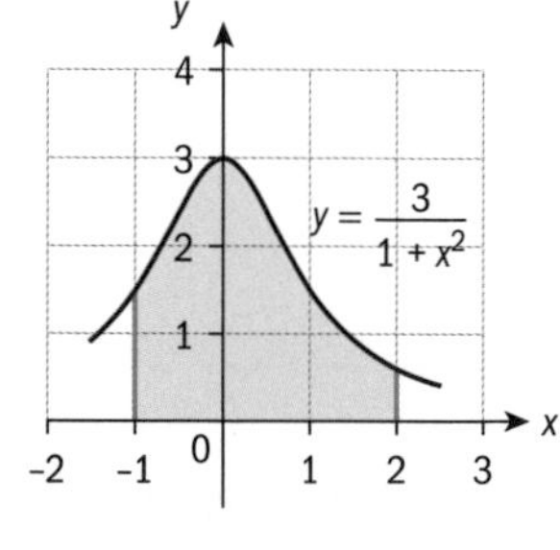

h

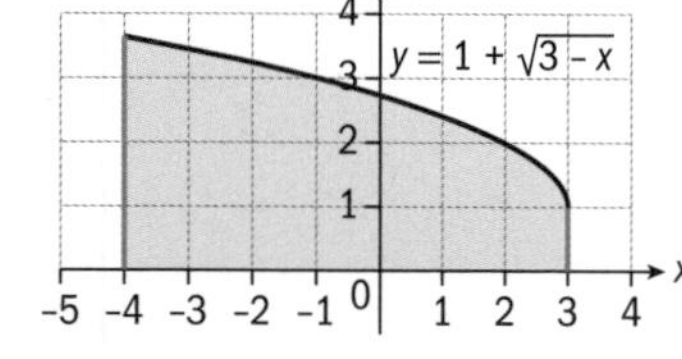

i

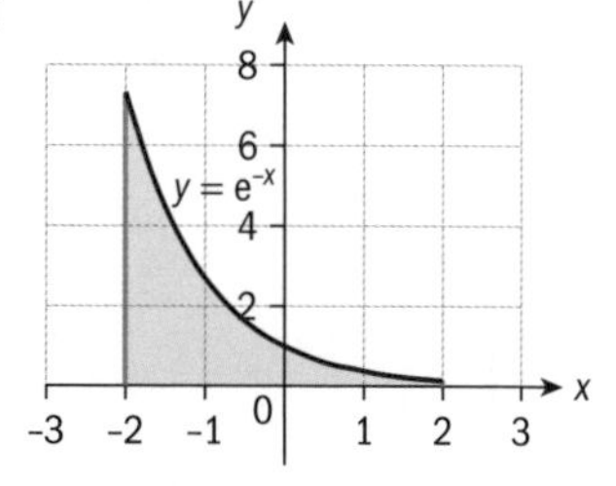

4 For each of the following definite integrals:

i On a diagram shade the region that they represent.

ii Find their value.

a $\int_0^4 \sqrt{x}\,dx$ **b** $\int_{-2}^{2} x^2 dx$

c $\int_{2.5}^{3} -(x-2)(x-4)\,dx$

d $\int_{2}^{4} -(x-2)(x-4)\,dx$

e $\int_2^5 \frac{10}{x+1}\,dx$ **f** $\int_{-1}^{1} (3^x+2)\,dx$

g $\int_{-2}^{3} (x^2-2x+3)\,dx$

5 Consider the region A enclosed between the graph of $y=-(x+1)(x-4)$ and the x-axis.

a Write down a definite integral that represents the area of A.

b Find the value of this area.

6 Consider the curve $y = x(x-4)^2$. Let A be the region enclosed between this curve and the x-axis.

a Write down the x-intercepts of this curve.

b Write down a definite integral that represents the area of A.

c Find the value of this area.

7 Consider the curve $y = x^3$. Let A be the region enclosed between this curve, the x-axis and the vertical line $x = 2$.

a Write down the x-intercept of this curve.

b Sketch the curve and clearly label A.

c Write down a definite integral that represents the area of A.

d Find the value of this area.

8 Consider the graph of the function $f(x) = (x+2)^2 + 1$.
The region bounded by the graph of f, the x-axis, the y-axis and the vertical line $x = b$ with $b > 0$ has an area equal to 42.

a Sketch the region.

b Find the value of b.

9 The table below shows the coordinates (x, y) of five points that lie on a curve $y = f(x)$.

x	1	3.25	5.5	7.75	10
$y=f(x)$	3	9	7	12	5

Estimate the area under the curve over the interval $1 \le x \le 10$.

10 Estimate the area under the graph of $f(x) = \sqrt{x-2}$ over the interval $2 \le x \le 4$ using five trapezoids. Give your answer correct to four significant figures.

11 Consider the graph of the function $f(t) = -t^2 + 2t$ where $f(t) \ge 0$.

a Draw the graph of the function f and shade the area enclosed between this graph and the t-axis over the interval $0 \le t \le x$.

b Find an expression for the area under the graph of f over the interval $0 \le t \le x$.

12 Consider the graph of the function $f(t) = 4t$ over the interval $3 \le t \le x$. Find $\int_3^x 4t\,dt$.

13 Calculate $\int (2+x)\,dx$.

14 Find $\int \left(1 + x - \frac{x^3}{4}\right)dx$.

15 Line L passes through the points $(-0.25, 0)$ and $(1, 10)$. Consider the region bounded by the graph of line L for $-0.25 \le x \le 1$, the curve $y = (x-4)^2 + 1$ for $1 \le x \le 6$ and the x-axis. The region is shown below.

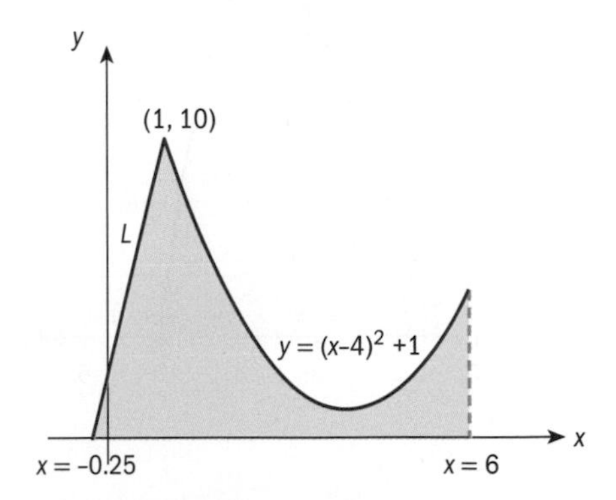

a Find the area under the graph of L for $-0.25 \le x \le 1$.

b Write down an expression for the area under the curve $y = (x-4)^2 + 1$ for $1 \le x \le 6$.

c Hence, find the area of the shaded region.

16 The diagram shows the graph of three linear functions, g, f and h.

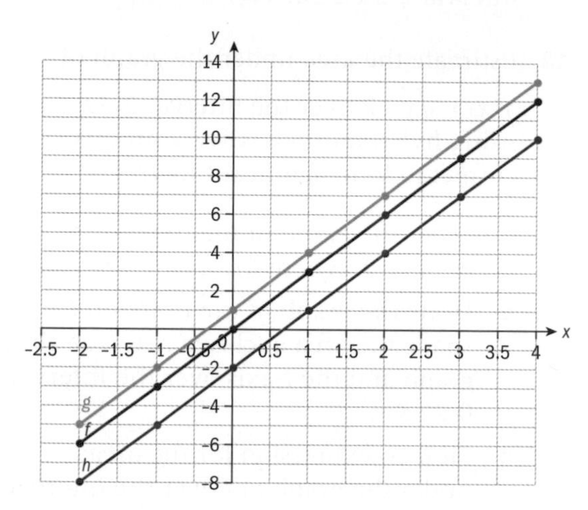

a Find the equation of each of these functions.

The three functions are antiderivatives of $y = t(x)$.

b Find the equation of $y = t(x)$.

c Find $\int t(x)\mathrm{d}x$.

17 Estimate the area under the graph of $y = f(x)$ in the interval $1.5 \le x \le 5$ using the data points given in the diagram.

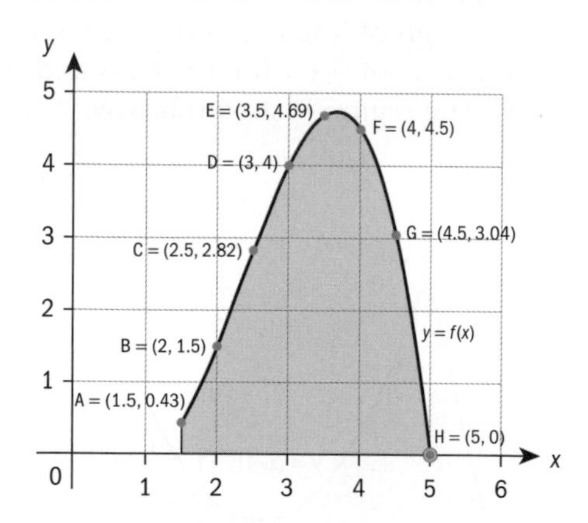

18 a Estimate the area A under the graph of $f(x) = e^{-x^2}$ over the interval $0 \le x \le 1$ using five trapezoids. Give your answer correct to four significant figures.

b i Write down a definite integral that represents A.

ii Hence, find the actual area. Give your answer correct to four significant figures.

c Find the percentage error made with the estimation found in part **a**.

Exam-style questions

19 P1: Find $\int \left(5 - 12x^2 + 4x^3\right) \mathrm{d}x$, simplifying your answer as far as possible. (4 marks)

20 P1: The derivative of the function f is given by $f'(x) = \frac{3}{2}x^2 + x + 3$ and the curve $y = f(x)$ passes through the point $\left(-1, \frac{13}{2}\right)$.

Find an expression for f. (6 marks)

21 P2: a Find the coordinates of the points of intersection of the graphs of $y = 6x - x^2$ and $y = 10 - x$. (4 marks)

b On the same axes, sketch the graphs of $y = 6x - x^2$ and $y = 10 - x$. (2 marks)

c Find the exact value for the area bounded by the two curves. (7 marks)

22 P1: A particle P is travelling in a straight line. After t seconds, the particle has velocity $v = 0.6t^2 + 4t + 1$ m/s for $t \ge 0$.

a Find an expression for the displacement of the particle from the origin after t seconds. (2 marks)

b Hence, find the distance travelled by the particle during the third second of motion. (3 marks)

23 P1: Consider the curve $y = -\frac{x^2}{10}(x-10)$.

The area A is defined as the region bounded by the curve and the x-axis.

a Sketch the curve, clearly showing the area defined as A. (3 marks)

b Write down a definite integral that represents A. (1 mark)

c Find the exact value of A. (5 marks)

24 P2: a By using technology, find the coordinates of the points of intersection of the graphs of $y = x^3$ and $y = \sqrt[3]{x}$. (4 marks)

b On the same axes, sketch the graphs of $y = x^3$ and $y = \sqrt[3]{x}$. (2 marks)

c Find an exact value for the total area bounded by the two curves. (6 marks)

25 P1: The diagram below shows an area bounded by the x-axis, the line $x = 6$, the line $y = 2x - 2$ and the curve $y = \frac{36}{x^2}$.

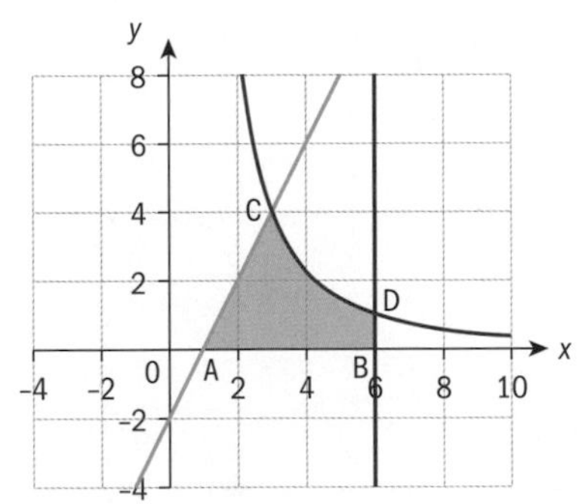

a Using technology or otherwise, find the coordinates of points A, B, C and D. (4 marks)

b Show that the shaded area is exactly 10 units^2. You must show all of your working. (6 marks)

26 P2: Consider the area enclosed by the curve $y = 5 - \frac{x^3}{25}$ and the positive x- and y-axis.

a Sketch the curve, shading the area described above. (3 marks)

b Using the trapezium rule with five strips, determine an approximation for the shaded area. (5 marks)

c Explain why your answer to part **b** will be an underestimate. (2 marks)

d Using integration, determine the exact value of the shaded area. (4 marks)

e Find the percentage error of your approximation, compared with the exact value. (2 marks)

Click here for further exam practice

In the footsteps of Archimedes

Approaches to learning: Research, Critical thinking
Exploration criteria: Mathematical communication (B), Personal engagement (C), Use of mathematics (E)
IB topic: Integration, Proof, Coordinate geometry

The area of a parabolic segment

A parabolic segment is a region bounded by a parabola and a line.

Consider this shaded region which is the area bounded by the line $y = x + 6$ and the curve $y = x^2$:

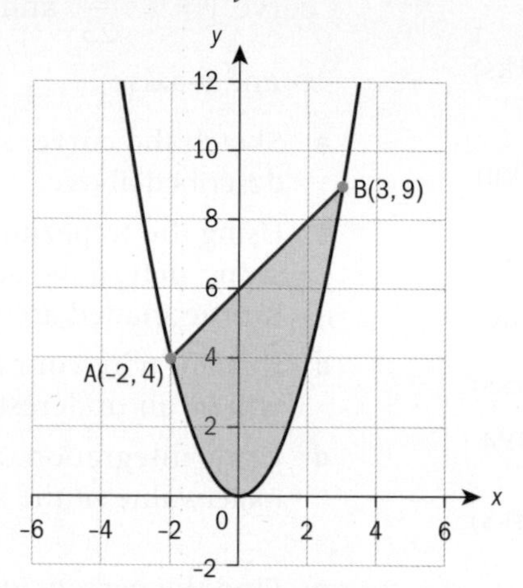

From this chapter you know that you can calculate the shaded area using integration.

On the diagram points A (−2, 4) and B (3, 9) are marked.

Point C is such that the x-value of C is halfway between the x-values of points A and B.

What are the coordinates of point C on the curve?

Triangle ABC is constructed as shown:

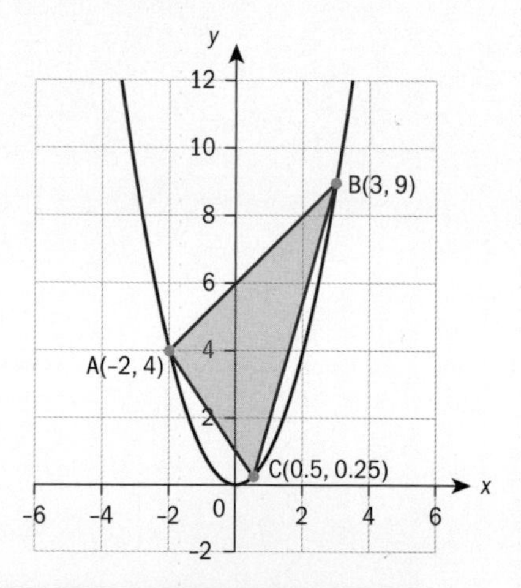

Archimedes showed that the area of the parabolic segment is $\frac{4}{3}$ of the area of triangle ABC.

Calculate the area of the triangle shown.

What methods are available to calculate the area of the triangle?

Use integration to calculate the area between the two curves.

Hence verify that Archimedes' result is correct for this parabolic segment.

You can show that this result is true for any parabola and for any starting points A and B on the parabola.

Consider another triangle by choosing point D on the parabola such that its x-value is halfway between the x-values of A and C, similar to before.

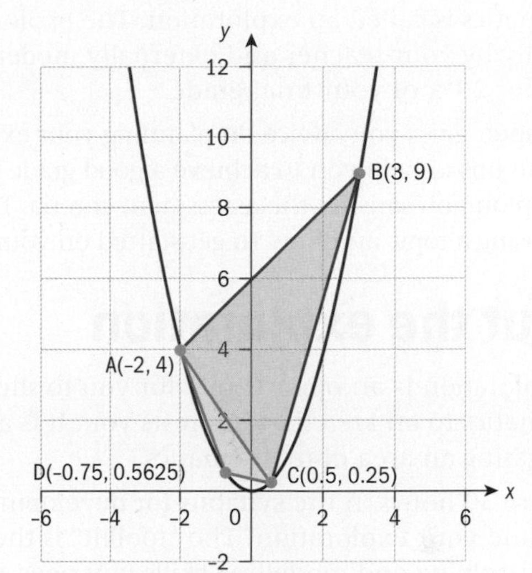

What are the coordinates of point D?

Calculate the area of triangle ACD.

Similarly, for line BC, find E such that its x-value is half-way between C and B.

What are the coordinates of point E?

Hence calculate the area of triangle BCE.

Calculate the ratio between the areas of the new triangles and original triangle ABC.

What do you notice?

You can see already that if you add the areas of triangles ABC, ACD and BCE, you have a reasonable approximation for the area of the parabolic segment.

You can improve this approximation by continuing the process and forming four more triangles from sides AD, CD, CE and BE.

If you add the areas of these seven triangles, you have an even better approximation.

How could the approximation be improved?

Generalize the problem

Let the area of the first triangle be X.

What is the total area of the next two, four and eight triangles in terms of X?

If you continued adding the areas of an *infinite* number of such triangles, you would have the *exact* area for the parabolic segment.

By summing the areas of all the triangles, you can show that they form a geometric series.

What is the common ratio?

What is the first term?

What is the sum of the series?

What has this shown?

Extension

This task demonstrates the part of the historical development of the topic of limits which has led to the development of the concept of calculus.

Look at another area of mathematics that you have studied on this course so far.

- What is the history of this particular area of mathematics?
- How does it fit into the development of the whole of mathematics?
- How significant is it?
- Who are the main contributors to this branch of mathematics?

14 Exploration

All IB Diploma subjects have an internal assessment (IA). The IA in mathematics is called an exploration. The exploration will be assessed internally by your teacher and externally moderated by the IB and counts for 20% of your final grade.

This chapter gives you advice on planning your exploration, as well as hints and tips to help you to achieve a good grade by making sure that your exploration satisfies the assessment criteria. There are also suggestions on choosing a topic and how to get started on your exploration.

About the exploration

The exploration is an opportunity for you to show that you can apply mathematics to an area that interests you. It is a piece of written work investigating an area of mathematics.

There are 30 hours in the syllabus for developing your mathematical toolkit and your exploration. The "toolkit" is the inquiry, investigative, problem-solving and modelling skills you need to write a good exploration. You can build these skills throughout this book—in particular, in the Investigations, Developing inquiry skills and Modelling and investigation activities in each chapter.

You should expect to spend around 10–15 hours of class time on your exploration and up to 10 hours of your own time.

During **class time** you will:

- go through the assessment criteria with your teacher
- brainstorm to come up with suitable topics or titles
- look at previous explorations and the grading
- meet with your teacher to discuss your choice of topic and your progress.

During **your own time** you will:

- research the topic you have chosen, to make sure that it is appropriate for an exploration (if not, you will have to conduct further research to help you select a suitable topic)
- collect and organize your information or data and decide which mathematical processes to apply
- write your exploration
- submit a draft exploration to your teacher (your teacher will set a deadline for this)
- present your draft exploration to some of your peers, for their feedback
- submit the final exploration (your teacher will set a deadline for this). If you do not submit an exploration then you receive a grade of "N" and will not receive your IB Diploma.

How the exploration is marked

After you have submitted the final version of your exploration your teacher will mark it. This is "internal assessment" (in school). Your teacher submits these marks to the IB, from which a random sample of explorations is selected automatically. Your teacher uploads these sample explorations to be marked by an external moderator. This external moderation of internal assessment ensures that all teachers in all schools are marking students' work to the same standards.

To begin with, the external moderator will mark three of your school's explorations. If the moderator's mark is within 2 marks of your teacher's mark, then all your teacher's marks stay the same.

If the moderator's mark is more than 2 marks higher or lower than your teacher's mark, the external moderator will mark the remaining explorations in the sample. This may increase the mark if the teacher marked too harshly or decrease the mark if the teacher marked too leniently. The moderator sends a report to the school to explain the reason for any change in the marks.

IA tip

When coloured graphs are uploaded in black and white, they are very difficult to follow, as information such as colour-coded keys is lost. Make sure your exploration is uploaded in colour if it contains colour diagrams.

Internal assessment criteria

Your exploration will be assessed by your teacher, against the criteria given below. The IB external moderator will use the same assessment criteria.

The final mark for each exploration is the sum of the scores for each criterion. The maximum possible final mark is 20. This is 20% of your final mark for Mathematics: applications and interpretation Standard level.

The criteria cover five areas, A to E:

Criterion A	Presentation
Criterion B	Mathematical communication
Criterion C	Personal engagement
Criterion D	Reflection
Criterion E	Use of mathematics

IA tip

Make sure that **all** the pages are uploaded. It is almost impossible to mark an exploration with pages missing.

IA tip

Make sure you understand these criteria. Check your exploration against the criteria frequently as you write it.

Criterion A: Presentation

This criterion assesses the organization, coherence and conciseness of your exploration.

Achievement level	Descriptor
0	The exploration does not reach the standard described by the descriptors below.
1	The exploration has some coherence or some organization.
2	The exploration has some coherence and shows some organization.
3	The exploration is coherent and well organized.
4	The exploration is coherent, well organized, concise.

To get a good mark for Criterion A: Presentation

- A **well organized** exploration has:
  - an **introduction** in which you discuss the context of the exploration
  - a statement of the **aim** of the exploration, which should be clearly identifiable
  - a **conclusion**.
- A **coherent** exploration:
  - is logically developed and easy to follow
  - should "read well" and express ideas clearly
  - includes any graphs, tables and diagrams where they are needed—not attached as appendices to the document.
- A **concise** exploration:
  - focuses on the aim and avoids irrelevancies
  - achieves the aim you stated at the beginning
  - explains all stages in the exploration clearly and concisely.
- References must be cited where appropriate. Failure to do so could be considered academic malpractice.

IA tip

For more on citing references, academic honesty and malpractice, see pages 591–593.

Criterion B: Mathematical communication

This criterion assesses how you:

- use appropriate mathematical language (notation, symbols, terminology)
- define key terms, where required
- use multiple forms of mathematical representation, such as formulae, diagrams, tables, charts, graphs and models, where appropriate.

IA tip

Only include forms of representation that are relevant to the topic. For example, do not draw a bar chart and pie chart for the same data.

If you include a mathematical process or diagram without using or commenting on it, then it is irrelevant.

Achievement level	Descriptor
0	The exploration does not reach the standard described by the descriptors below.
1	There is some relevant mathematical communication which is partially appropriate.
2	The exploration contains some relevant, appropriate mathematical communication.
3	The mathematical communication is relevant, appropriate and is mostly consistent.
4	The mathematical communication is relevant, appropriate and consistent throughout.

To get a good mark for Criterion B: Mathematical communication

- Use appropriate mathematical language and representation when communicating mathematical ideas, reasoning and findings.

- Choose and use appropriate mathematical and ICT tools such as graphic display calculators, screenshots, mathematical software, spreadsheets, databases, drawing and word-processing software, as appropriate, to enhance mathematical communication.
- Define key terms that you use.
- Express results to an appropriate degree of accuracy.
- Label scales and axes clearly in graphs.
- Set out proofs clearly and logically.
- Define variables.
- Do not use calculator or computer notation.

IA tip

Use technology to enhance the development of the exploration—for example, by reducing laborious and repetitive calculations.

Criterion C: Personal engagement

This criterion assesses how you engage with the exploration and make it your own.

Achievement level	Descriptor
0	The exploration does not reach the standard described by the descriptors below.
1	There is evidence of some personal engagement.
2	There is evidence of significant personal engagement.
3	There is evidence of outstanding personal engagement.

IA tip

Students often copy their GDC display, which makes it unlikely they will reach the higher levels in this criterion. You need to express results in proper mathematical notation. For example,

use 2^x and not $2^\wedge x$

use × not *

use 0.028 and not 2.8E-2.

To get a good mark for Criterion C: Personal engagement

- Choose a topic for your exploration that you are interested in, as this makes it easier to display personal engagement.
- Find a topic that interests you and ask yourself "What if ...?"
- Demonstrate personal engagement by using some of these skills and practices from the mathematician's toolkit:
  - creating mathematical models for real-world situations
  - designing and implementing surveys
  - running experiments to collect data
  - running simulations
  - thinking and working independently
  - thinking creatively
  - addressing your personal interests
  - presenting mathematical ideas in your own way
  - asking questions, making conjectures and investigating mathematical ideas
  - considering historical and global perspectives
  - exploring unfamiliar mathematics.

IA tip

Just showing personal interest in a topic is not enough to gain the top marks in this criterion. You need to write in your own voice and demonstrate your own experience with the mathematics in the topic.

Criterion D: Reflection

This criterion assesses how you review, analyse and evaluate your exploration.

Achievement level	Descriptor
0	The exploration does not reach the standard described by the descriptors below.
1	There is evidence of limited reflection.
2	There is evidence of meaningful reflection.
3	There is substantial evidence of critical reflection.

To get a good mark for Criterion D: Reflection

- Include reflection in the conclusion to the exploration, but also throughout the exploration. Ask yourself "What next?"
- Show reflection in your exploration by:
  - discussing the implications of your results
  - considering the significance of your findings and results
  - stating possible limitations and/or extensions to your results
  - making links to different fields and/or areas of mathematics
  - considering the limitations of the methods you have used
  - explaining why you chose this method rather than another.

IA tip

Discussing your results without analysing them is not meaningful or critical reflection. You need to do more than just describe your results. Do they lead to further exploration?

Criterion E: Use of mathematics

This criterion assesses how you use mathematics in your exploration.

Achievement level	Descriptor
0	The exploration does not reach the standard described by the descriptors below.
1	Some relevant mathematics is used.
2	Some relevant mathematics is used. Limited understanding is demonstrated.
3	Relevant mathematics commensurate with the level of the course is used. Limited understanding is demonstrated.
4	Relevant mathematics commensurate with the level of the course is used. The mathematics explored is partially correct. Some knowledge and understanding are demonstrated.
5	Relevant mathematics commensurate with the level of the course is used. The mathematics explored is mostly correct. Good knowledge and understanding are demonstrated.
6	Relevant mathematics commensurate with the level of the course is used. The mathematics explored is correct. Thorough knowledge and understanding are demonstrated.

To get a good mark for Criterion E: Use of mathematics

- Produce work that is commensurate with the level of the course you are studying. The mathematics you explore should either be part of the syllabus, at a similar level or beyond.
- If the level of mathematics is not commensurate with the level of the course you can only get a maximum of 2 marks for this criterion.
- Only use mathematics relevant to the topic of your exploration. Do not just do mathematics for the sake of it.
- Demonstrate that you fully understand the mathematics used in your exploration.
  - Justify **why** you are using a particular mathematical technique (do not just use it).
  - Generalize and justify conclusions.
  - Apply mathematics in different contexts where appropriate.
  - Apply problem-solving techniques where appropriate.
  - Recognize and explain patterns where appropriate.

IA tip

Make sure the mathematics in your exploration is not only based on the prior learning for the syllabus. When you are deciding on a topic, consider what mathematics will be involved and whether it is commensurate with the level of the course.

Academic honesty

This is very important in all your work. Your school will have an Academic honesty policy which you should be given to discuss in class, to make sure that you understand what malpractice is and the consequences of committing malpractice.

According to the IB Learner Profile for Integrity:

"We act with integrity and honesty, with a strong sense of fairness and justice, and with respect for the dignity and rights of people everywhere. We take responsibility for our actions and their consequences."

Academic honesty means:

- that your work is authentic
- that your work is your own intellectual property
- that you conduct yourself properly during examinations
- that any work taken from another source is properly cited.

IA tip

Reference any photographs you use in your exploration, including to decorate the front page.

Authentic work:

- is work based on your own original ideas
- can draw on the work of others, but this must be fully acknowledged in footnotes and bibliography
- must use your own language and expression
- must acknowledge all sources fully and appropriately in a bibliography.

Malpractice

The IB Organization defines malpractice as "behaviour that results in, or may result in, the candidate or any other candidate gaining an unfair advantage in one or more assessment components."

Malpractice includes:

- plagiarism—copying from others' work, published or otherwise, whether intentional or not, without the proper acknowledgment
- collusion—working together with at least one other person in order to gain an undue advantage (this includes having someone else write your exploration)
- duplication of work—presenting the same work for different assessment components
- any other behaviour that gains an unfair advantage such as taking unauthorized materials into an examination room, stealing examination materials, disruptive behaviour during examinations, falsifying CAS records or impersonation.

IA tip

Plagiarism detection software identifies text copied from online sources. The probability that a 16-word phrase match is "just a coincidence" is $\frac{1}{10^{12}}$.[1]

Collaboration and collusion

It is important to understand the distinction between collaboration (which is allowed) and collusion (which is not).

IA tip

Discussing individual exploration proposals with your peers or in class before submission is collaboration.

Individually collecting data and then pooling it to create a large data set is collaboration. If you use this data for your own calculations and write your own exploration, that is collaboration. If you write the exploration as a group, that is collusion.

Collaboration

In several subjects, including mathematics, you will be expected to participate in group work. It is important in everyday life that you are able to work well in a group situation. Working in a group entails talking to others and sharing ideas. Every member of the group is expected to participate equally and it is expected that all members of the group will benefit from this collaboration. However, the end result must be your own work, even if it is based on the same data as the rest of your group.

Collusion

This is when two or more people work together to intentionally deceive others. Collusion is a type of plagiarism. This could be working with someone else and presenting the work as your own or allowing a friend to copy your work.

References and acknowledging sources

The IB does not tell you which style of referencing you should use—this is left to your school.

IA tip

Be consistent and use the same style of referencing throughout your exploration.

The main reasons for citing references are:

- to acknowledge the work of others
- to allow your teacher and moderator to check your sources.

[1]Words & Ideas. The Turnitin Blog. Top 15 misconceptions about Turnitin. Misconception 11: matched text is likely to be completely coincidental or common knowledge (posted by Katie P., 9 March 2010).

To refer to someone else's work:

- include a brief reference to the source in the main body of your exploration—either as part of the exploration or as a footnote on the appropriate page
- include a full reference in your bibliography.

The bibliography should include a list with full details of **all** the sources that you have used.

Choosing a topic

You need to choose a topic that interests you, as then you will enjoy working on the exploration and you will be able to demonstrate personal engagement by using your own voice and demonstrating your own experience.

Discuss the topic you choose with your teacher and your peers before you put too much time and effort into developing the exploration. Remember that the work does not need to go beyond the level of the course which you are taking, but you can choose a topic that is outside the syllabus and is at a commensurate level. You should avoid choosing topics that are too ambitious, or below the level of your course.

These questions may help you to find a topic for your exploration:

- What areas of the syllabus are you enjoying most?
- What areas of the syllabus are you performing best in?
- Would you prefer to work on purely analytical work or on modelling problems?
- Have you discovered, through reading or talking to peers on other mathematics courses, areas of mathematics that might be interesting to look into?
- What mathematics is important for the career that you eventually hope to follow?
- What are your special interests or hobbies? Where can mathematics be applied in this area?

One way of choosing a topic is to start with a general area of interest and create a mind map. This can lead to some interesting ideas on applications of mathematics to explore. The mind map on pages 594–5 shows how the broad topic "Transport" can lead to suggestions for explorations into such diverse topics as baby carriage design, depletion of fossil fuels and queuing theory.

On page 596 there is an incomplete mind map for you to continue, either on your own or by working with other mathematics students.

IA tip

Cite references to others' work even if you paraphrase or rewrite the original text.

You do not need to cite references to formulae taken from mathematics textbooks.

IA tip

You must include a brief reference in the exploration as well as in the bibliography. It is not sufficient just to include a reference in the bibliography.

IA tip

Your exploration should contain a substantial amount of mathematics at the level of your course, and should not just be descriptive. Although the history of mathematics can be very interesting it is not a good exploration topic.

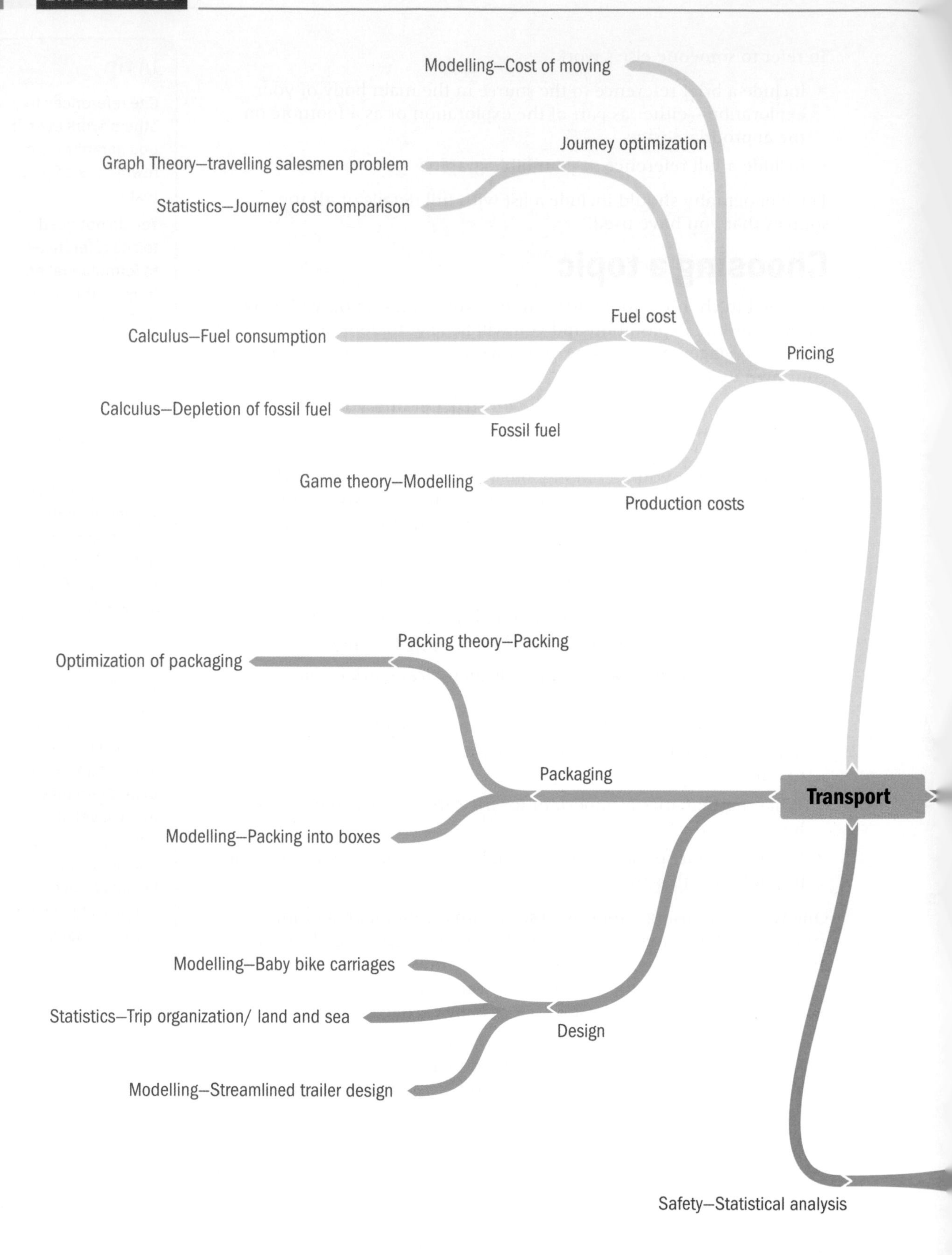
Modelling–Cost of moving
Journey optimization
Graph Theory–travelling salesmen problem
Statistics–Journey cost comparison
Fuel cost
Calculus–Fuel consumption
Pricing
Calculus–Depletion of fossil fuel
Fossil fuel
Game theory–Modelling
Production costs
Packing theory–Packing
Optimization of packaging
Packaging
Transport
Modelling–Packing into boxes
Modelling–Baby bike carriages
Statistics–Trip organization/ land and sea
Design
Modelling–Streamlined trailer design
Safety–Statistical analysis

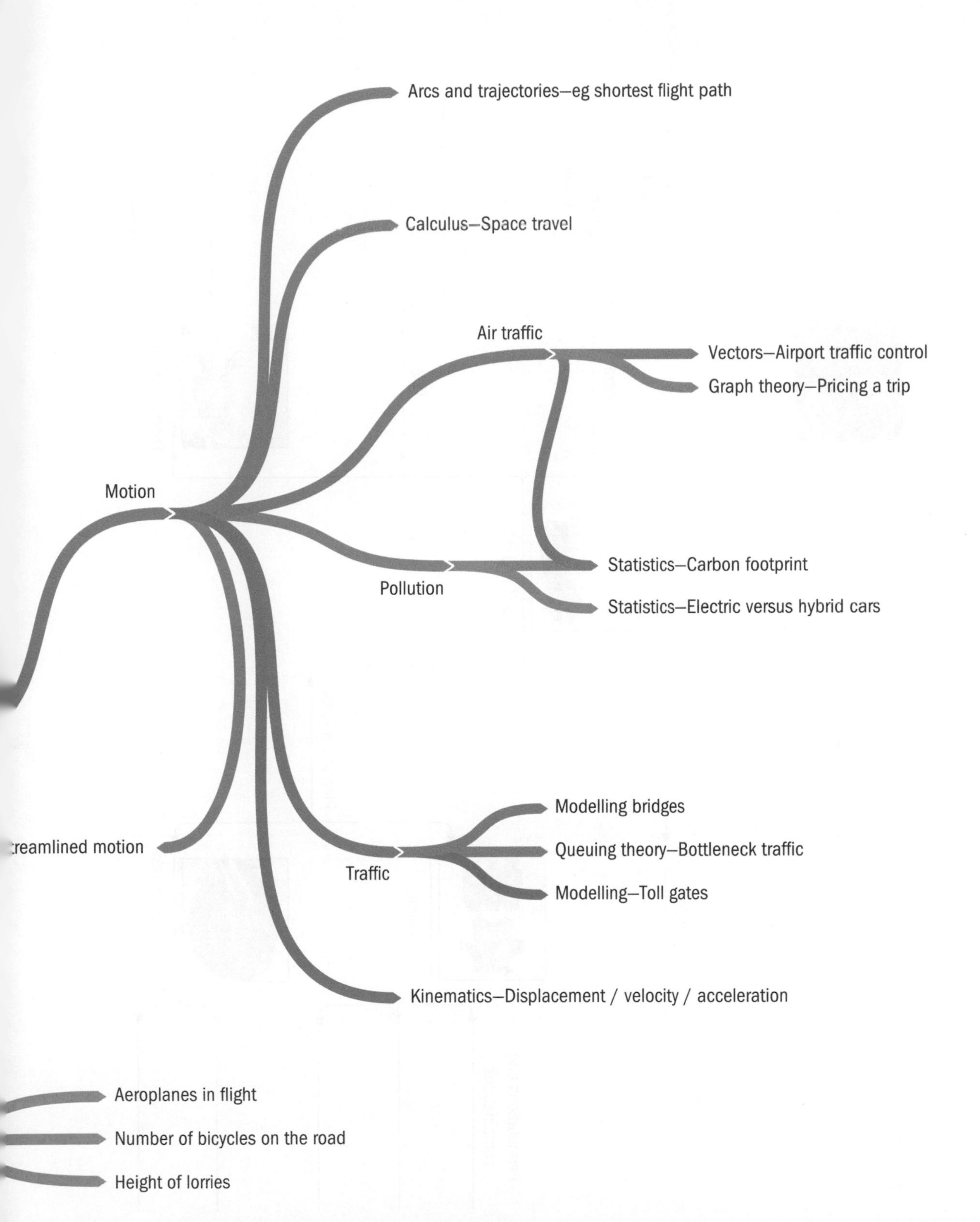
Arcs and trajectories—eg shortest flight path
Calculus—Space travel
Air traffic
Vectors—Airport traffic control
Graph theory—Pricing a trip
Motion
Statistics—Carbon footprint
Pollution
Statistics—Electric versus hybrid cars
Modelling bridges
treamlined motion
Queuing theory—Bottleneck traffic
Traffic
Modelling—Toll gates
Kinematics—Displacement / velocity / acceleration
Aeroplanes in flight
Number of bicycles on the road
Height of lorries

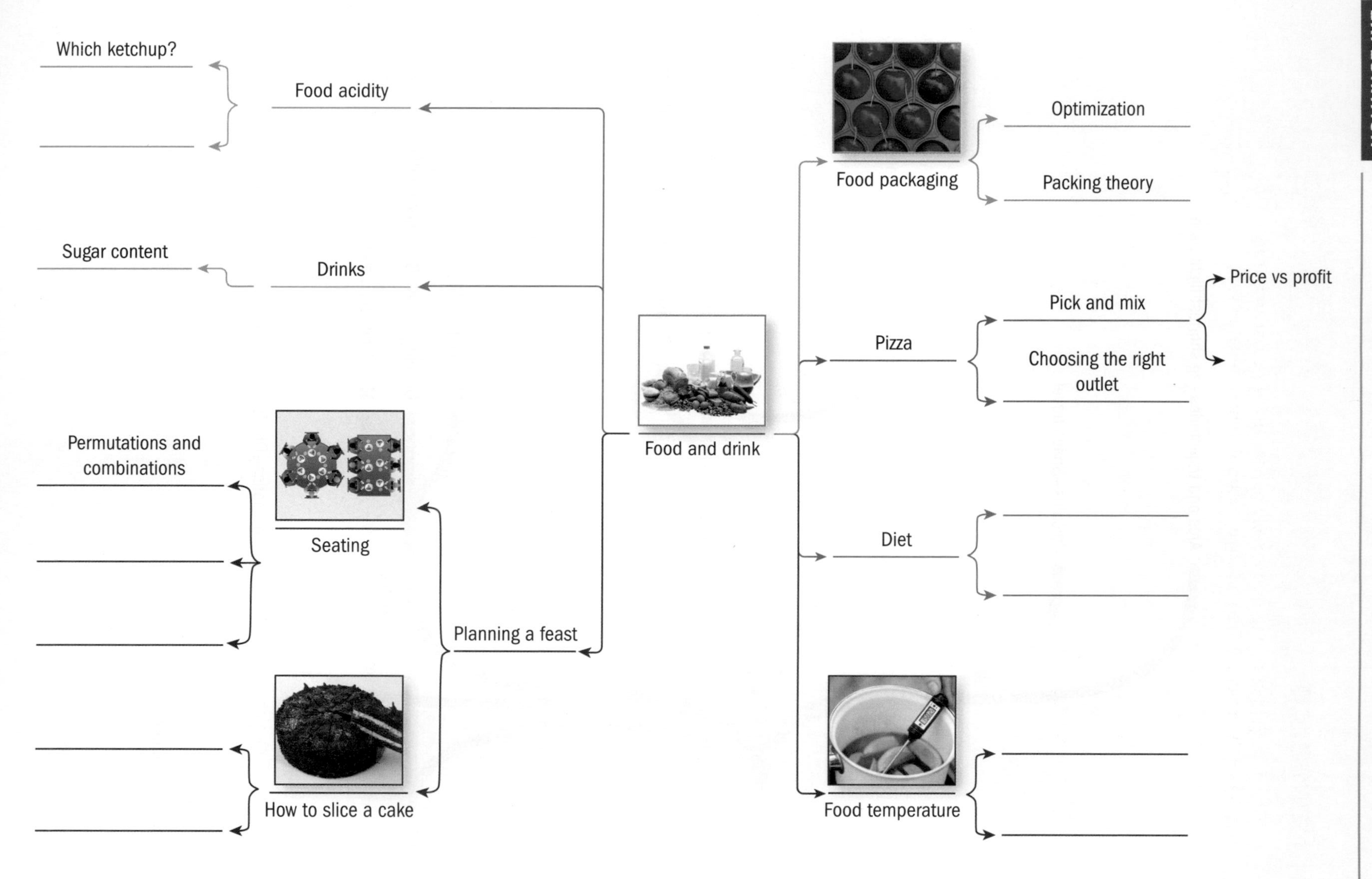

Food and drink
Food acidity
Which ketchup?
Drinks
Sugar content
Planning a feast
Seating
Permutations and combinations
How to slice a cake
Food packaging
Optimization
Packing theory
Pizza
Pick and mix
Choosing the right outlet
Price vs profit
Diet
Food temperature

Research

Once you have chosen a topic, you will need to do some research. The purpose of this research is to help you determine how suitable your topic is.

- Do not rely on the internet for all your research—you should also make use of books and academic publications.
- Plan your time wisely—make sure that you are organized.
- Do not put it off—start your research in good time.
- For internet research: refine your topic so you know exactly what information you are looking for, and use multiple-word searches. It is very easy to spend hours on the internet without finding any relevant information.
- Make sure that you keep a record of all the websites you use—this saves so much time afterwards. You will need to cite them as sources, and to include them in your bibliography.
- Make sure that the sources are reliable—who wrote the article? Are they qualified? Is the information accurate? Check the information against another source.
- Research in your own language if you find this easier.

> **IA tip**
>
> Try to avoid writing a research report in which you merely explain a well-known result that can easily be found online or in textbooks. Such explorations have little scope for meaningful and critical reflection and it may be difficult to demonstrate personal engagement.

These questions will help you to decide whether the topic you have chosen is suitable.

- What areas of mathematics are contained in the topic?
- Which of these areas are contained in the syllabus that you are following?
- Which of these areas are not in the syllabus that you are following but are contained in the other IB mathematics course?
- Which of these areas are in none of the IB mathematics courses? How accessible is this mathematics to you?
- Would you be able to understand the mathematics and write an exploration in such a way that a peer is able to understand it all?
- How can you demonstrate personal engagement in your topic?
- Will you manage to complete an exploration on this topic and meet all the top criterion descriptors within the recommended length of 12 to 20 pages (double spaced and font size 12-point)?

> **IA tip**
>
> Learning new mathematics is not enough to reach the top levels in Criterion C: Personal engagement.

Writing an outline

Once you think you have a workable topic, write a brief outline including:

- why you chose this topic
- how your topic relates to mathematics

- the mathematical areas in your topic, eg algebra, geometry, calculus, etc
- the key mathematical concepts covered in your topic, eg modelling data, areas of irregular shapes, analysing data, etc
- the mathematical skills you will use in the exploration, eg integration by parts, working with complex numbers, using polar coordinates, etc
- any mathematics outside the syllabus that you need to learn
- technology you could use to develop your exploration
- new key terms that you will need to define or explain
- how you are going to demonstrate personal engagement
- a list of any resources you have used or will use, in the development of your exploration. If this list includes websites you should include the URL and the date when this was accessed.

Share this outline with your teacher and with your peers. They may ask questions that lead you to improve your outline.

This template may help you write the outline for the exploration when presenting a formal proposal to your teacher.

IA tip

Popular topics such as the Monty Hall problem, the Birthday paradox, and so on are not likely to score well on all the criteria.

Mathematics exploration outline

Topic:
Exploration title:
Exploration aim:
Exploration outline:
Resources used:
Personal engagement:

IA tip

As you write your exploration, remember to refer to the criteria on pages 505–591.

Writing your exploration

Now you should be ready to start writing your exploration in detail.

You could ask one of your classmates to read the exploration and give you feedback before you submit the draft to your teacher. If your exploration is related to another discipline, eg economics, it would be better if the peer reading your exploration is someone who does not study economics.

IA tip

Remember that your peers should be able to read and understand your work.

Mathematical exploration checklist

Work through this checklist to confirm that you have done everything that you can to make your exploration successful.

- ☐ Does your exploration have a title?
- ☐ Have you ensured your exploration does not include any identifying features—for example, your name, candidate number, school name?
- ☐ Does your exploration start with an introduction?
- ☐ Have you clearly stated your aim?
- ☐ Does your exploration answer the stated aim?
- ☐ Have you used double line spacing and 12-point font?
- ☐ Is your exploration 12–20 pages long?
- ☐ Have you cut out anything that is irrelevant to your exploration?
- ☐ Have you checked that you have not repeated lots of calculations?
- ☐ Have you checked that tables only contain relevant information and are not too long?
- ☐ Is your exploration easy for a peer to read and understand?
- ☐ Is your exploration logically organized?
- ☐ Are all your graphs, tables and diagrams correctly labelled and titled?
- ☐ Are all graphs, tables and diagrams placed appropriately and not all attached at the end?
- ☐ Have you used appropriate mathematical language and representation (not computer notation, eg *, ^, etc)?
- ☐ Have you used notation consistently through your exploration?
- ☐ Have you defined key terms (mathematical and subject specific) where necessary?
- ☐ Have you used appropriate technology?
- ☐ Have you used an appropriate degree of accuracy for your topic or exploration?
- ☐ Have you shown interest in the topic?
- ☐ Have you used original analysis for your exploration (eg simulation, modelling, surveys, experiments)?
- ☐ Have you expressed the mathematical ideas in your exploration in your own way (not just copy-and-pasted someone else's)?
- ☐ Does your exploration have a conclusion that refers back to the introduction and the aim?
- ☐ Do you discuss the implications and significance of your results?
- ☐ Do you state any possible limitations and/or extensions?
- ☐ Do you critically reflect on the processes you have used?
- ☐ Have you explored mathematics that is commensurate with the level of the course?
- ☐ Have you checked that your results are correct?
- ☐ Have you clearly demonstrated understanding of why you have used the mathematical processes you have used?
- ☐ Have you acknowledged direct quotes appropriately?
- ☐ Have you cited all references in a bibliography?
- ☐ Do you have an appendix if one is needed?

Paper 1

Time allowed: 1 hour 30 minutes
Answer all the questions.
All numerical answers must be given exactly or correct to three significant figures, unless otherwise stated in the question.

Answers should be supported by working and/or explanations. Where an answer is incorrect, some marks may be awarded for a correct method, provided this is shown clearly.

You need a graphic display calculator for this paper.

1 Diego receives COP 1 000 000 from his grandfather. He invests some of his money into a saving account which pays annual interest at a rate of 6%, which is compounded monthly.

a Find the least amount of money that Diego must initially deposit into the account if he is to have COP1 500 000 in this account after 10 years. Give your answer to two decimal places. [2 marks]

With the remainder of the money that Diego does not invest, he buys a car. The value of the car decreases at an average of 10% every year.

b Find the value of the car after 10 years. [3 marks]

2 Alessandra starts a three-week fitness programme that requires her to walk at a fast pace over a distance which increases each day.

The first day, a Monday, she walks a distance of 200 m. On Friday, she walks a distance of 360 m. The distance she walks increases by a constant amount d m every day.

a Write down the value of d. [1 mark]

b Calculate the distance Alessandra will walk on the second Sunday after the start of her programme. [2 marks]

c Find the total distance Alessandra will have walked at the end of the three-week fitness programme. [3 marks]

3 Consider the relation $C = C(A)$ between the circumference of a circle C and its area A defined for $A \geq 0$.

a Justify that:

i $C = C(A)$ is a function.

ii C has an inverse function C^{-1}. [2 marks]

b Sketch the graphs of $C(A)$ and $C^{-1}(A)$ on the same axes. [2 marks]

c Hence, find the area of a circle with circumference 25. [2 marks]

4 A cup of hot chocolate is left on a counter for several hours. Initially its temperature was 86°C. After 30 minutes the temperature had already dropped to 28°C.

The temperature, T°C, of the hot chocolate is modelled by the function $T(t) = 22 + a2^{bt}$, where t is the number of hours that have elapsed since the hot chocolate was first left to stand.

a Find the value of:

i a **ii** b [4 marks]

b Write down the equation of the horizontal asymptote of the graph of T. [1 mark]

c State the meaning of the asymptote found in part **b**. [1 mark]

5 Kathy is collecting information for a statistics project. She asks a group of students who have pet dogs about the number of times that they usually walk the dog per day.

The data collected is shown in the following table.

Number of walks	1	2	3	4	5
Number of students	4	8	10	5	1

a State, with a reason, whether "number of walks" is a discrete or continuous variable. [2 marks]

b For the students that Kathy surveyed, find:

i the modal number of dog walks per day [1 mark]

ii the mean number of dog walks per day. [2 marks]

6 Jasmine plays a computer game. In the game, she collects tokens of different colours. Each colour token gives the player a different number of points.

Jasmine records the relative frequencies of obtaining tokens of each colour. She uses this to estimate the probability of obtaining a token of a certain colour.

Colour	Red	Blue	Green	Yellow	Pink	Orange
Points	2	3	5	1	3	4
Estimate of probability	$\frac{1}{9}$	$\frac{1}{6}$	$\frac{1}{12}$	$\frac{2}{9}$	$\frac{1}{6}$	p

a Find the value of p. [2 marks]

b Find the expected score if Jasmine plays the game once. [2 marks]

7 Jeanny has six coloured pencils and eight coloured pens in her colouring box. Every afternoon from Monday through to Friday, Jeanny returns from kindergarten and picks a pen or pencil at random from this box. She then begins to draw with the object she picked.

a State the probability that, on any one day, Jeanny picks a pen. [1 mark]

b Find the probability that, in one week (Monday through to Friday), Jeanny picks a pen on exactly two days. [3 marks]

8 A farmer fed two groups of cows according to different feeding programmes.

The first group was fed according to the "Intensive" programme, and the second group was fed according to the "Extensive" programme.

After two years, the size of the fat layer on the belly of each cow was measured. The data, in centimetres, is given below.

Intensive programme: 20.5, 20, 19, 24, 24.5, 25, 26

Extensive programme: 19, 19, 21, 20, 20.5, 18.5

a On the same axis, construct two box-and-whisker diagrams to represent the results of each set of cows. [3 marks]

b Describe what your box-and-whisker diagrams tell you about the size and spread of the fat layers on the cows from each of the feeding programmes. [2 marks]

9 The loudness of a sound, L decibels (dB), is a function of the sound intensity, S watts per square metre (W/m^2).

Loudness is given by the formula $L = 10 \cdot \log_{10}(S \times 10^{12})$.

Humans can hear sounds between 0 and 140 decibels.

a Determine whether a human can hear a sound with intensity of 0.000 001 W/m^2 [3 marks]

In some countries, workers must wear hearing protection when the loudness of sounds reaches 80 decibels or more.

b In these countries, determine the maximum intensity a sound can be before a worker must wear hearing protection. [2 marks]

10 As part of a study into colour preference, Alicia surveyed 100 students at her school. She recorded each student's gender, and their favourite colour from blue, red, brown or black. Alicia's results are shown in the table below.

	Colour			
	Blue	**Red**	**Brown**	**Black**
Female	16	15	10	13
Male	23	12	9	2

Alicia conducted a χ^2 test for independence at a 5% level of significance.

a State the null hypothesis. [1 mark]

b Write down the associated p-value. [1 mark]

c State, giving a reason, whether the null hypothesis should be accepted. [2 marks]

11 Consider the function $f(x) = x(x-2)(x-4)$.

a Sketch the graph of f and shade the two regions bounded by the graph of f and the x-axis. Label the two regions A and B. [3 marks]

b Find:

i the area of region A

ii the area of region B

iii the **total** shaded area. [4 marks]

12 Lisa measures the arm spans (the length from finger tips on one hand to finger tips on the other hand, when arms are held horizontally outstretched) of students in two mathematics classes.

She is interested to see whether the mean arm span, μ_1, of class A is the same as the mean arm span, μ_2, of class B. Lisa's data is shown in the table below.

Class A	161	178	194	204	173	162	177	196	194	160		
Class B	165	146	159	160	170	173	199	140	178	200	155	194

In this question, you will use a *t*-test to compare the means of the two groups at the 10% level of significance.

You may assume the data is normally distributed and the standard deviations are equal between the two groups.

a **i** State the null hypothesis.

ii State the alternative hypothesis. [2 marks]

b Find the associated *p*-value for this test. [2 marks]

c State, giving a reason, whether Lisa should accept the null hypothesis. [2 marks]

13 Points A(2, 9), B(2, 3), C(8, 4), D(8, 7) and E(4, 6) represent wells in the Savannah National Park. These wells are shown in the coordinate axes below.

Horizontal scale: 1 unit represents 1 km.

Vertical scale: 1 unit represents 1 km.

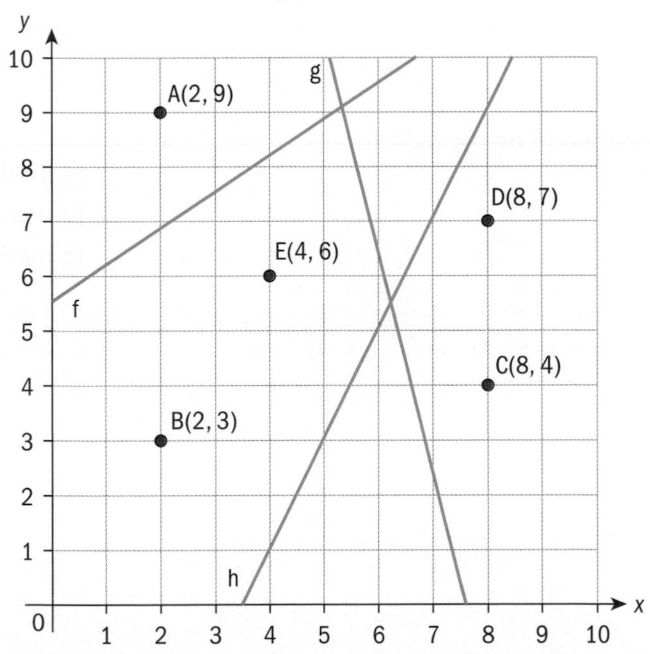

a Calculate the distance between the wells A and E. [2 marks]

Pablo, the Park Ranger draws three lines, f, g and h, around point E and obtains an incomplete Voronoi diagram around the point E.

b Find the equation of the line which would complete the Voronoi cell around point E. Give your answer in the form $ax + by + d = 0$ where $a, b, d \in \mathbb{Z}$. [5 marks]

c In the context of the question, explain the significance of the cell around point E. [1 mark]

14 A lighthouse is situated at point D. Light is emitted from point C, 65 m above sea level.

The beam of light from the lighthouse traces a semicircular arc which, on the surface of the water, is 50 m wide.

The edges of the beam pass through the two sets of points A and B. The distance from any point A to C is 100 m.

Not to scale

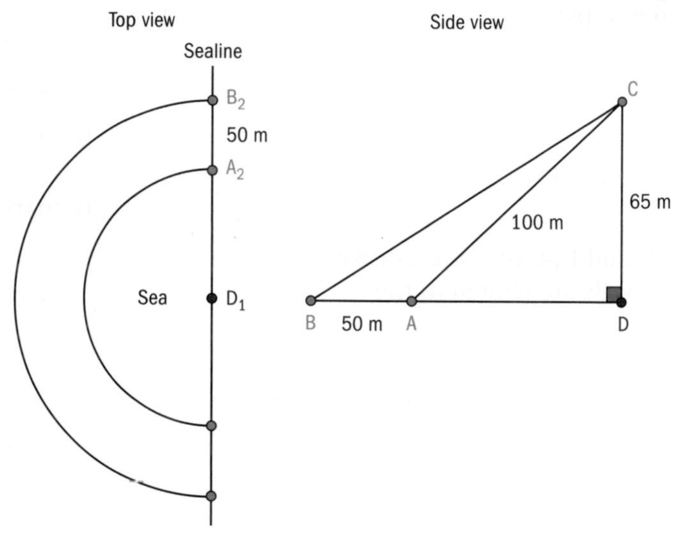

a Find the area of the surface of the sea swept by the light beam, correct to the nearest square metre. [4 marks]

b Find $C\hat{A}D$. [2 marks]

c Hence find the length BC. [3 marks]

Paper 2

Time allowed: 1 hour 30 minutes

Answer all the questions.

All numerical answers must be given exactly or correct to three significant figures, unless otherwise stated in the question.

Answers should be supported by working and/or explanations. Where an answer is incorrect, some marks may be awarded for a correct method, provided this is shown clearly.

You need a graphic display calculator for this paper.

1 A surveyor is marking out an ornamental garden in the shape of a triangle. She places a marker in the ground at point A and then travels 80 m due North to a point B, where she places a second marker. She then travels 70 m on a bearing of 125° and places a third marker at point C. Finally she walks in a straight line back to the first marker at A.

a Sketch this information on a diagram. [2 marks]

b Calculate the length of the side AC. [3 marks]

c Calculate the area of the triangular garden. [2 marks]

d Find the bearing that the surveyor was walking on when she travelled from C to A. [5 marks]

2 The time, T, that competitors take to complete a 10 000 m race is normally distributed with mean $\mu = 50$ minutes and standard deviation $\sigma = 5$ minutes.

a Sketch a diagram to represent this information with the numbers 50 and 5 indicated on it. [3 marks]

b Find the probability that a random competitor takes between 45 and 55 minutes. [2 marks]

c Find the probability that a random competitor takes less than 40 minutes to complete the race. [2 marks]

d The fastest 75% of competitors receive a medal. Find the time that the race has to be completed under, in order for a competitor to receive a medal. [2 marks]

e Given that a competitor received a medal, find the probability that they finished the race in less than 40 minutes. [3 marks]

f If 10 000 competitors ran the race, estimate (to the nearest minute) how many competitors completed the race in less than 33 minutes. [2 marks]

3 On a particular day, the depth of water (d m) at a harbour entrance can be modelled by the equation $d = 11 + 5\sin 30t°$. Here, t is the time (in hours) that have elapsed since midnight.

a State the depth of water at midnight. [1 mark]

b State the period of this function. [2 marks]

c **i** State the maximum value of the depth and the times during the day when this will occur.

ii State the minimum value of the depth and the times during the day when this will occur. [4 marks]

d Sketch the graph of d as a function of t, for $0 \le t \le 24$. [3 marks]

e A large boat can only come into the harbour when the depth of water is greater than 14 m. Find the morning time interval, in hours and minutes, during which the boat can enter. [3 marks]

4 **a** When Mich takes a free throw in a basketball game, the probability that he scores is always constant at $\frac{2}{3}$.

Mich takes nine free throws. Let X be the number of times he scores.

i State the distribution that X satisfies, and state the parameters.

ii Write down the mean of this distribution.

iii Calculate the probability that Mich scores exactly seven times.

iv Calculate the probability that he scores four or more times. [8 marks]

b When Ken takes a free throw at basketball the probability that he scores is always constant at $\frac{1}{3}$. Find the minimum number of free throws that Ken would have to take to be at least 99% certain of scoring at least once. [5 marks]

5 Eleven students, $A - K$, revised for and then took a maths exam. Let h represent the number of hours that each student spent revising, and let s represent the score (out of 100) that they gained. The data for each student is shown in the following table.

	A	B	C	D	E	F	G	H	I	J	K
h	10	9.5	9	8.5	8	7.5	7	6.5	6	5	0
s	100	91	93	90	80	85	79	70	69	60	65

a Identify any outliers in the values of h. Justify your answers. [4 marks]

b Calculate the Pearson's product moment correlation coefficient, r, for this bivariate data. [2 marks]

c Write down the equation of the line of best fit of s on h. [2 marks]

d Hence estimate the score, to the nearest integer, of a twelfth student who spent 5.5 hours revising. [2 marks]

e Rank the students from 1 to 11 according to the number of hours they spent revising, h. Rank 1 should represent the most hours spent revising.

In a similar way, rank the students according to the test score, s, they obtained. Rank 1 should represent the highest test score.

Copy and complete the table below to show the rankings for each student.

	A	B	C	D	E	F	G	H	I	J	K
Hour rank	1	2	3								
Score rank	1	3	2								

[2 marks]

f Calculate the Spearman's rank correlation coefficient, r_s, for this data. [2 marks]

g Explain why $r_s > r$. [1 mark]

6 Alison rides a rollercoaster. The height of the rollercoaster over the first part of the ride can be modelled by the equation

$h(x) = \frac{x^3}{3} - \frac{5x^2}{2} + 4x + 5$ for $0 \le x \le 7$, where x m is the horizontal distance from the start, and h m is the height.

a Find $h'(x)$. [2 marks]

b Solve $h'(x) = 0$. [2 marks]

c Hence, or otherwise, find the coordinates of the stationary points of $h(x)$, and determine the nature of each. [4 marks]

d Sketch the graph of $h(x)$ against x. [2 marks]

e If Alison is more than 10 m above the ground she suffers from acrophobia. Find the values of x for which this happens. [1 mark]

f If the gradient is smaller than -1, Alison will scream. Find the values of x for which this happens. [2 marks]

Answers

Chapter 1

Skills check

1 33.3 cm

2 133 cm^2

3 a 79.1 cm^2 b 41.0 cm^2
c 104 cm^2

Exercise 1A

1 a 9.48 m b 5.32 g
c 1.81 cm d 7.00 in
e 4580 km f 13 200 kg

2 a 9.48 m b 5.32 g
c 1.81 cm d 7.00 in

3 a 3 b 4 c 3
d 4 e 1, 2, 3, 4, 5, 6 or 7

4 39.7

Exercise 1B

1 a 8890 b 3.75
c 30 000 d 0.0064
e 7.9 (to 1 d.p.)

2

	Number	Round to the nearest ten	Round to the nearest hundred	Round to the nearest thousand
a	2815	2820	2800	3000
b	75 391	75 390	75 400	75 000
c	316 479	316 480	316 500	316 000
d	932	930	900	1000
e	8253	8250	8300	8000

3 a 502 EUR b 1000 USD
c 10 BGN d 1400 JPY

4 207 cm (to 3 s.f.)

5 2.5 cm (to 2 s.f.)

6 110 cm^3 (to 2 s.f.)

Exercise 1C

1 a Lower bound = 23.5 mm, upper bound = 24.5 mm, range is 23.5 mm ≤ x < 24.5 mm

b Lower bound = 3.15 m, upper bound = 3.25 m, range is 3.15 m ≤ x < 3.25 m

c Lower bound = 1.745 kg, upper bound = 1.755 m, range is 1.745 kg ≤ x < 1.755 kg

2 Lower bound = 6.32875 cm^2, upper bound = 6.69375 cm^2

3 181.25 million km per day

4 Maximum = 0.35, minimum = 0.30

Exercise 1D

1 0.0507% (to 3 s.f.)

2 0.0904% (to 3 s.f.)

3 16.5% (to 3 s.f.)
b 1.436 km%

4 a actual temperature = 10 °C, approximate temperature = 9 °C
b 10%

Exercise 1E

1 a 2^6 b 5^3
c 6^2 d 20^2
e 8^3 f 3^8
g 3^{-15} h 2^0
i 15^4 j 5^3

2 a x^{12} b x^7

Exercise 1F

1 a 1.61×10^{-3} g cm^{-3}
b 1.97×10^{-10} m
c 9.46×10^{12} km
d 1.675×10^{-24} g

2 a 1.2×10^{-1}
b 5.04×10^7
c 4.005×10^{-5}
d 1×10^{-3}

3 0.24 mm (2.4×10^{-1} mm)

4 a 4.6×10^9 years b 5000

5 2.505×10^{25} molecules

6 a 6.515×10^{25} kg
b 11 times larger than earth (to the nearest digit)

7 a 5 : 1 b 1.5×10^9

c

Country	Percentage Change (relative to 1985 value
India	44.7% (increase)
United States	25% (increase)
China	18.2% (increase)
Bulgaria	19.1% (decrease)

d The data does not support the hypothesis.

Exercise 1G

1 a 17.8 m b 7.5 mm
c 6.6 mm d 10.3 mm

2 a Right-angled
b Right-angled
c Right-angled
d Not right-angled

3 a 1.41 units
b 1.73 units
c 2 units d 4 units

4 15.0 cm^2

Exercise 1H

1 a $y = 7.89$ cm, $z = 4.34$ cm
b $x = 15.1$, $z = 18.2$
c $x = 9.31$, $y = 23.2$

2 a 7.36 m
b 4.25 m c 6.96 m

3 a Angle $\angle ACB = 58°$

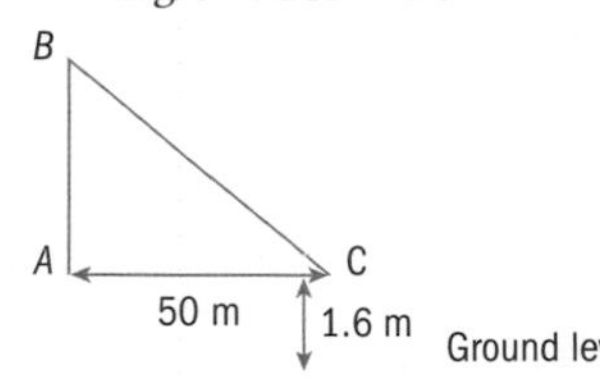

b 81.6 m

Exercise 1I

1 a

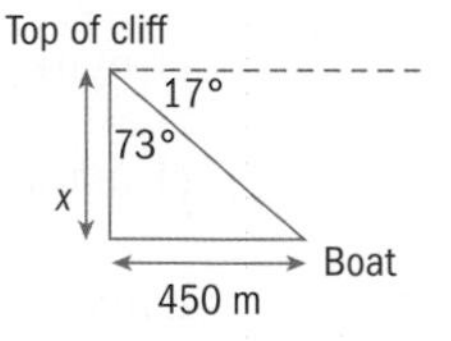

b 138 m

2 1.02 m

3 1300 m

4 52.6°

5 a

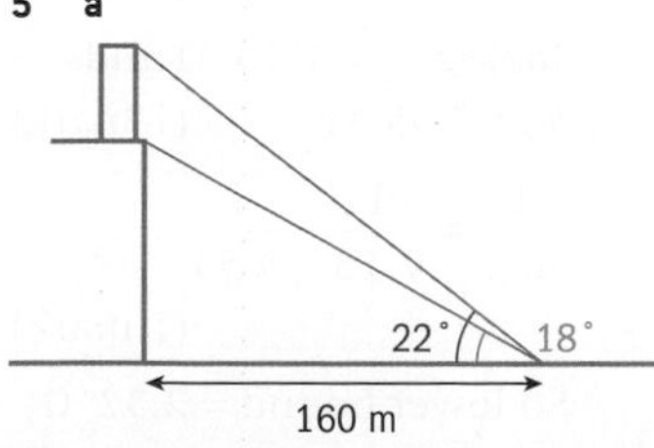

b 12.7 m

Exercise 1J

1 a 6.11 cm b 3.14 cm
c 23.82 cm

2 32.7 cm

3 a 209 m b 13.4 min
c 24 min

4 a 227 cm b 536 cm

Chapter review

1 Answers will vary

2 a 9.23 b 0.511 c 2.12

3 a 4×10^{-5} m b 1×10^{12}

4 Lower bound = 5.18×10^6 g, upper bound = 5.22×10^6 g

5 Areas of the semicircles are $\frac{1.5^2\pi}{2}, \frac{2^2\pi}{2}, \frac{2.5^2\pi}{2}$; this rule holds for semi-circles on the sides of right-angled triangles.

6 a 1677 cm
b 4.59%

7 400 ft

8 4.44 m

9 3.60 m

10 a

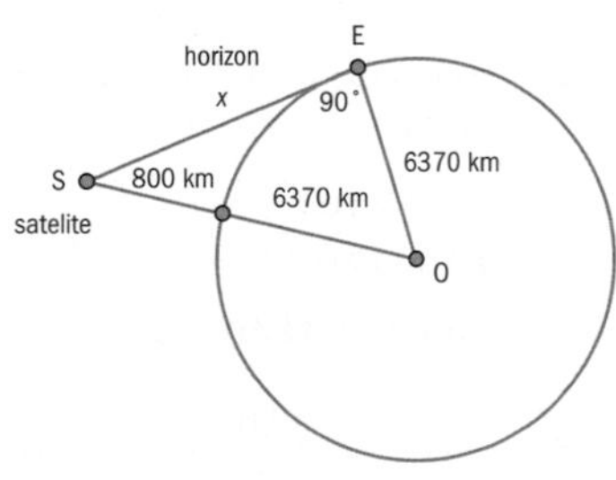

b 3291 km

11 a 26.11 m
b 15.64 m

12 237 m

13 a

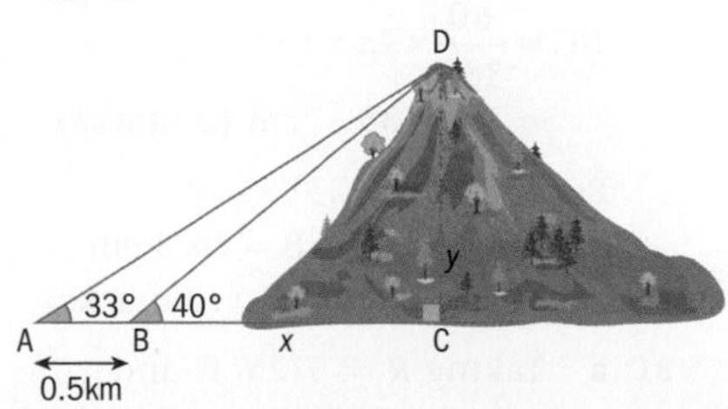

b 1.436 km

14 81.1 cm

Exam-style questions

15 a $AY^2 = 5^2 + 6^2$ (1 mark)
$\Rightarrow AY = 7.81$ cm (1 mark)

b $\tan X\hat{A}Y = \frac{6}{7.8102...}$ (1 mark)

$X\hat{A}Y = 37.5°$ (1 mark)

16 a $V = \pi \times 1.8^2 \times 14.5$ (1 mark)
$= 148$ cm^3 (1 mark)

b $SA = (2\pi \times 1.8 \times 14.5) + (2\pi \times 1.8^2)$ (1 mark)
$= 184$ cm^3 (1 mark)

17 Arc $AB = \frac{40}{360} \times 2 \times \pi \times 7$
$(= 4.8869...)$ (2 marks)
Perimeter = 4.8869... + 7 + 7 (1 mark)
= 18.9 cm (1 mark)

18 a $\frac{10x^3 \times 3x^4}{2x^{-6}} = \frac{30x^7}{2x^{-6}}$ (2 marks)

$= 15x^{13}$ (1 mark)

b $\frac{x^2 \times 4x^{-3}}{x^{-1}} = \frac{4x^{-1}}{x^{-1}} = 4$ (3 marks)

c $\sqrt{3x^3 \times 12x^0 \times 4x^5} = \sqrt{144x^8}$ (2 marks)
$= 12x^4$ (1 mark)

d $\frac{(x^{-2})^5}{(x^3)^{-4}} = \frac{x^{-10}}{x^{-12}}$ (2 marks)
$= x^2$ (1 mark)

19 $23.5 - 1.6 = 21.9$ (1 mark)
$\tan\theta = \frac{21.9}{100}$ (2 marks)
$\theta = 12.4°$ (1 mark)

20 $\left|\frac{\pi - \frac{355}{113}}{\pi}\right| \times 100$ (2 marks)
$= 8.49 \times 10^{-6}\%$ (1 mark)

21 Let x be the horizontal distance from Sharon to both the yacht and the kittiwake.

$\tan 19° = \frac{90}{x}$ (2 marks)

Yacht / kittiwake is at a horizontal distance of $\frac{90}{\tan 19°} = 262.3789\ldots$ m from Sharon (1 mark)

Vertical height of kittiwake from cliff is $261.3789\ldots \tan 15° = 70.0362\ldots$ m (2 marks)

Required distance is $90 + 70.0362\ldots = 160$ m (1 mark)

22 $7.1 \sin 70° = 6.6718\ldots$ m (2 marks)

$7.1 \sin 80° = 6.9921\ldots$ m (2 marks)

So the minimum height is 6.67 m and the maximum height is 6.99 m

23 a $\text{time} = \frac{1.496 \times 10^8 \times 10^3}{3 \times 10^8}$
$= 498.6666\ldots$ seconds (2 marks)

$\frac{498.6666\ldots}{60} = 8.31$ minutes (1 mark)

b $\text{time} = \frac{4.014 \times 10^{13} \times 10^3}{3 \times 10^8}$
$= 1.338 \times 10^8$ seconds (2 marks)

$\frac{1.338 \times 10^8}{60 \times 60 \times 24 \times 365} = 4.24$ years (1 mark)

c $3 \times 10^8 \times (2.5 \times 10^6 \times 365 \times 24 \times 60 \times 60) = 2.3652 \times 10^{22}$ m (2 marks)

2.3652×10^{22} m $= 2.37 \times 10^{19}$ km (1 mark)

24 a $1 \times (3 \times 10^8)^2 = 9 \times 10^{16}$ J (2 marks)

b $\frac{9 \times 10^{16}}{60} = 1.5 \times 10^{15}$ seconds (2 marks)

$\frac{1.5 \times 10^{15}}{60 \times 60 \times 24 \times 365} = 4.76 \times 10^7$ years (1 mark)

25 a $R_{MIN} = \frac{5.5}{0.25} = 22\ \Omega$ (2 marks)

$R_{MAX} = \frac{6.5}{0.15} = 43.3\ \Omega$ (1 mark)

b $\frac{43.3 - 30}{30} \times 100 = 44.3\%$ (2 marks)

26 Valid attempt to find a horizontal distance. (2 marks)

cg $\tan 31° = \frac{120}{x} \Rightarrow x = \frac{120}{\tan 31°}$ (1 mark)

Other horizontal distance is $\frac{120}{\tan 39°}$ (1 mark)

Distance required is $\frac{120}{\tan 31°} - \frac{120}{\tan 39°}$ (1 mark)
$= 51.5$ m (1 mark)

27 a $V_{MIN} = \frac{4\pi \times 3.45^3}{3} = 172\text{ cm}^3$ (2 marks)

$V_{MAX} = \frac{4\pi \times 3.55^3}{3} = 187\text{ cm}^3$ (2 marks)

b Re-arranging $V = \frac{4\pi r^3}{3}$

gives $r = \sqrt[3]{\frac{3V}{4\pi}}$ (2 marks)

$r_{MIN} = \sqrt[3]{\frac{3 \times 495}{4\pi}} = 4.91$ cm (2 marks)

$r_{MAX} = \sqrt[3]{\frac{3 \times 505}{4\pi}} = 4.94$ cm (2 marks)

28 a $\pi \times (1.3 \times 10^{-6})^2 \times 4.5 \times 10^{-6} = 2.39 \times 10^{-17}\text{ m}^3$ (2 marks)

b Percentage error $= \frac{2.3891\ldots \times 10^{-17} - 2 \times 10^{-17}}{2.3891\ldots \times 10^{-17}} \times 100 = 16.3\%$ (2 marks)

29 a $\frac{(1.675 \times 10^{-27}) + (1.673 \times 10^{-27}) + (9.109 \times 10^{-31})}{3} = 1.12 \times 10^{-27}$ kg (2 marks)

b $2((1.675 \times 10^{-27}) + (1.673 \times 10^{-27}) + (9.109 \times 10^{-31}))$
$= 6.70 \times 10^{-27}$ kg (2 marks)

c $\frac{1.675 \times 10^{-27}}{9.109 \times 10^{-31}} = 1838.84$ (1 mark)

Ratio is 1:1840 (1 mark)

d $\frac{1 \times 10^{-30} - 9.109 \times 10^{-31}}{9.109 \times 10^{-31}} \times 100 = 9.78\%$ (2 marks)

30 Arc length AD $= \frac{60}{360} \times 2\pi \times 5 = 5.2359\ldots$ cm (2 marks)

Arc length BC $= \frac{60}{360} \times 2\pi \times 13 = 13.6135\ldots$ cm (2 marks)

Perimeter $= 5.2359\ldots + 13.6135\ldots + 8 + 8 = 34.8$ cm (2 marks)

31 a Taking $R_1 = 7.25\ \Omega$ and $R_2 = 3.65\ \Omega$. (1 mark)

$\frac{1}{R_{TOT}} = \frac{1}{7.25} + \frac{1}{3.65}$ (1 mark)

So upper bound $= 2.43\ \Omega$ (1 mark)

Taking $R_1 = 7.15\ \Omega$ and $R_2 = 3.55\ \Omega$. (1 mark)

$\frac{1}{R_{TOT}} = \frac{1}{7.15} + \frac{1}{3.55}$ (1 mark)

So lower bound $= 2.37\ \Omega$ (1 mark)

b $\dfrac{2.4277... - 2.40}{2.40} \times 100 =$
1.1563...% (2 marks)

$\left|\dfrac{2.3721... - 2.40}{2.40}\right| \times 100 =$
1.1584...% (1 mark)

So range of percentage errors is anything from 0 to 1.16% (1 mark)

Chapter 2

Skills check

1 **a** 15.8 km **b** 4.90 km

2 **a** 34.7° **b** 45.6°

3 37.5 cm^2

4 20°

Exercise 2A

1 **a** 48.8° **b** 66.6° **c** 41.7°

2 AB = 14.8 cm, BC = 11.0 cm

3 14.6 km

Exercise 2B

1 **a** $\angle A = 125°$, $\angle B = 26.2°$, $\angle C = 28.4°$

b $\angle A = 37.7°$, $\angle B = 51.0°$, $\angle C = 91.3°$

c $\angle C = 72°$, $a = 9.77$ m, $b = 23.0$ m.

2 **a** 41.0°, 87.0°, 106 mm

b 66°, 17.2 cm, 17.2 cm

c 40°, 60°, 80°, 22.0 m, 16.3 m

3 374 m

4 12.1 km

5 $\hat{A} = 97.1°$, $\hat{B} = 32.0°$, $\hat{C} = 50.9°$

6 Use Velina's suggestion as we have an angle side pair. AC = 42.7 m.

Exercise 2C

1 **a** 87.5 cm^2

b 33.1 cm^2

c 67.0 cm^2

2 **a** 47.2 cm^2

b 68.6 m^2

c 62.4 cm^2

3 **a** 176 m^2

b 0.585 kg

4 54.8 m^2

5 210 cm^2

6 965 m^2

Exercise 2D

1 **a** 22.0 cm^2

b 3.85 cm^2

c 117 cm^2

2 **a** 99.0 cm^2

b 173° (or 172.8°)

c 7.59 cm

3 **a** 18.3 cm^2

b 41.1 cm^2

4 1.13 cm^2

5 1.22 m^2

Exercise 2E

1

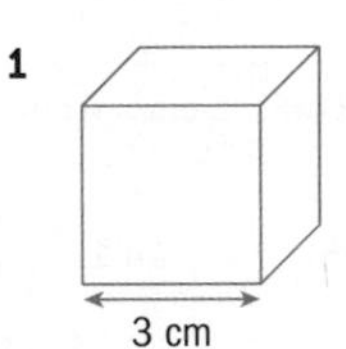

2

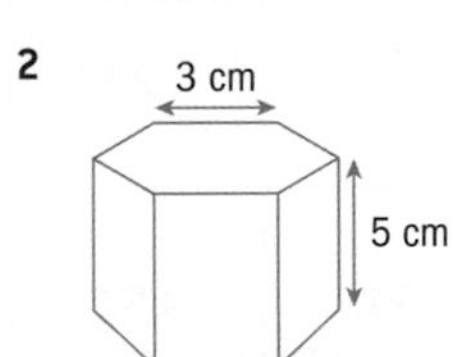

3 **a**

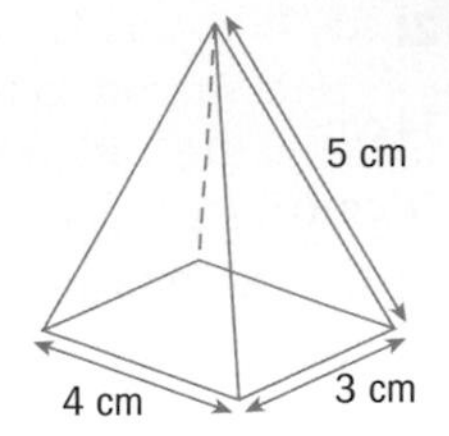

b

5 cm
4 cm
3 cm

Exercise 2F

1 **a** 13.5 m **b** 51.1°

c 26.4°

2 **a** 77.8° **b** 12.1 cm

c 12.1 cm

3 **a** 68.2° **b** 5.39 cm

c 37.5°

4 **a** 5.64 cm

b 12.4 cm

c 62.9°

Exercise 2G

1 \$126

2 133 g (or 132.912 g)

3 34

4 6.17 cm

5 8.77 cm

Exercise 2H

1 **a** 650 cm^3

b 33.5 cm^3

c 33.7 cm^3

d 62.4 cm^3

e 227 cm^3

2 400 litres, 900 litres

3 47.9 m

Exercise 2I

1 a $154\,cm^2$

b $176\,cm^2$

2 a $124\,m^2$ b $755\,m^2$

c $111\,cm^2$

3 $151\,cm^2$

4 a $396\,m^3$

b 33.3 L

Chapter review

1 $2830\,m^3$

2 60.3 km

3 a 89 km b 6.64°

4 $A = 9.82\,cm^2$, $l = 3.93\,cm$

5 $2.79\,cm^2$

6 a 10.5 cm

b 20.9 cm

c 31.4 cm

d BH is double AG, CI is three times AG. DJ = 41.9 cm, EK = 52.4 cm

e $1885\,cm^3$

7 a i

Side S	1 cm	2 cm	4 cm
Surface area: $6S^2$	6	24	96
Volume: S^3	1	8	64
Surface area to volume ratio, SA:V: $\frac{6}{S}$	6:1	3:1	1.5:1

ii

Diameter D	1 cm	2 cm	4 cm
Surface area: $4\pi\left(\frac{D}{2}\right)^2$	3.14	12.6	50.3
Volume: $\frac{4}{3}\pi\left(\frac{D}{2}\right)^3$	0.523	4.19	33.5
Surface area to volume ratio, SA:V: $\frac{6}{D}$: 1	6:1	3:1	1.5:1

iii

Diameter $D \times$ length L	1 cm × 1 cm	1 cm × 2 cm	1 cm × 4 cm
Surface area: $2\pi\left(\frac{D}{2}\right)^2 + \pi DL$	4.71	7.85	14.1
Volume: $\pi\left(\frac{D}{2}\right)^2 L$	0.78	1.57	3.14
Surface area to volume ratio, $SA:V = \frac{2D+4L}{DL}:1$	6:1	5:1	4.5:1

iv

Base side (b) × length (L)	1 cm × 1 cm	1 cm × 2 cm	1 cm × 4 cm
Surface area: $2b^2 + 4bL$	6	10	18
Volume: b^2L	1	2	4
Surface area to volume ratio, $SA:V$: $\frac{2b+4L}{bL}:1$	6:1	5:1	4.5:1

b Smaller cells have better SA : V ratio no matter the shape. However, it's better to have the shape of a cylinder or a prism rather than cube or sphere, and to grow in length rather than in width.

8 a $379\,cm^3$

b $451\,cm^2$

9 $2336\,m^3$

10 $19.6\,cm^2$

11 $82.6\,cm^2$

12 $V = 213\,cm^3$, (or $211.5\,cm^3$)
$A = 210\,cm^2$

Exam-style questions

13 a Attempt to use cosine rule: (1 mark)

$$\cos A = \frac{195^2 + 170^2 - 210^2}{2 \times 195 \times 170}$$ (1 mark)

$= 0.3442\ldots$

$A = 69.9°$ (1 mark)

b Area $= \frac{1}{2} \times 170 \times 195 \times \sin 69.8628\ldots$ (2 marks)

$= 15\,600\ m^2$ (1 mark)

14 $AC^2 = 560^2 + 1200^2 - 2 \times 560 \times 1200 \times \cos 110°$ (2 marks)

AC = 1487.7079... m (1 mark)

Distance required is
560 + 1200 − 1487.7079...
= 272 m (2 marks)

15 Area horse can reach

$= \frac{70}{360} \times \pi \times 85^2$

(= 4413.5013...)(2 marks)

Area of triangle

$= \frac{1}{2} \times 85 \times 300 \times \sin 70°$

(= 11981.0809...) (2 marks)

Area horse cannot reach

$= \frac{1}{2} \times 85 \times 300 \times \sin 70° - \frac{70}{360} \times \pi \times 85^2$ (1 mark)

$= 7570$ m². (1 mark)

16 Use of either $\frac{\theta}{360} \times \pi \times 16^2$ or $\frac{\theta}{360} \times \pi \times 8^2$ (1 mark)

Area ABCD $= \frac{\theta}{360} \times \pi \times 16^2 - \frac{\theta}{360} \times \pi \times 8^2 = 200$ (2 marks)

$\frac{\pi\theta}{360}(16^2 - 8^2) = 200$

$\pi\theta = 375$

$\theta = \frac{375}{\pi} = 119°$ (2 marks)

17 a $BD^2 = 9^2 + 16^2 - 2 \times 9 \times 16 \times \cos 33°$ (2 marks)

($\Rightarrow$BD = 9.7705... cm)

$AD = \sqrt{9.7705...^2 - 7^2}$ (1 mark)

= 6.8163... cm (1 mark)

So perimeter = 7 + 9 + 16 + 6.8163... = 38.8 cm (1 mark)

b Total area

$= \left(\frac{1}{2} \times 6.8163... \times 7\right) + \left(\frac{1}{2} \times 9 \times 16 \times \sin 33°\right)$ (2 marks)

= 63.1 cm² (1 mark)

18 a $\frac{\sin A}{5.7} = \frac{\sin 67°}{6.9}$ (2 marks)

$\Rightarrow A = 49.5009...°$ (1 mark)

b $B = 180 - 67 - 49.5009 = 63.4990...°$ (1 mark)

Area $= \frac{1}{2} \times 6.9 \times 5.7 \times \sin 63.4990...$ (1 mark)

= 17.6 cm² (1 mark)

19 $A\hat{B}D = 40°$ (1 mark)

$\frac{BD}{\sin 115°} = \frac{5.6}{\sin 40°}$ (2 marks)

$\Rightarrow$BD = 7.8958... cm (1 mark)

$\frac{AB}{\sin 25°} = \frac{5.6}{\sin 40°}$ (1 mark)

$\Rightarrow$AB = 3.6818... cm (1 mark)

Total length required = 2 × (3.6818... + 5.6) + 7.8958... = 26.5 cm (1 mark)

20 a Volume $= \frac{2}{3}\pi r^3 + \frac{1}{3}\pi r^2 h$ (2 marks)

$= \frac{2}{3}\pi(4^3) + \frac{1}{3}\pi(4^2) \times 9$

$= \frac{128\pi}{3} + \frac{144\pi}{3}$ (1 mark)

$= \frac{272\pi}{3}$ (1 mark)

b Surface area hemisphere = $2\pi r^2 = 2\pi \times 4^2 = 32\pi$ (2 marks)

Surface area cone

$= \pi r l = \pi \times 4 \times \sqrt{9^2 + 4^2}$

$= 4\pi\sqrt{97}$ (2 marks)

Total surface area

$= 32\pi + 4\pi\sqrt{97} = 224$ cm² (1 mark)

21 a

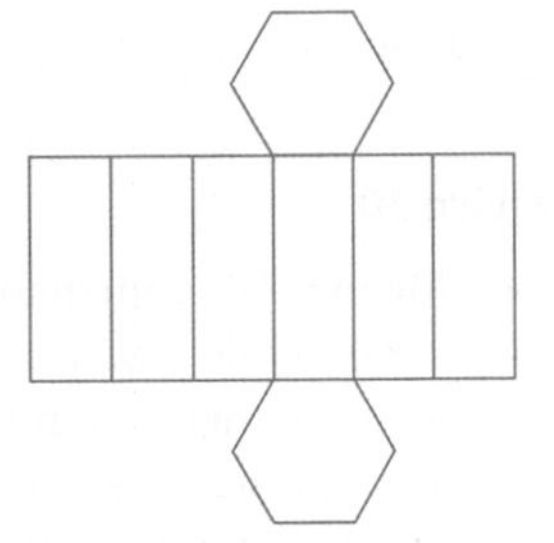

Note that other nets are also possible. (2 marks)

b Area of one hexagonal face

$= 6\left(\frac{1}{2} \times 4 \times 4 \times \sin 60°\right)$

$= 24\sqrt{3}$ (2 marks)

Therefore volume

$= 24\sqrt{3} \times 12 = 288\sqrt{3}$

= 499 cm³ (1 mark)

c Total surface area

$= 2 \times 24\sqrt{3} + 6 \times 12 \times 4$

$= 48\sqrt{3} + 288 = 371$ cm² (2 marks)

22 a Let X be the position of the bird.

Angle $A\hat{B}X = 180 - 42 = 138°$

$\Rightarrow$ Angle $A\hat{X}B = 180 - 138 - 27 = 15°$ (1 mark)

$\frac{AX}{\sin 138°} = \frac{115}{\sin 15°}$ (2 marks)

$\Rightarrow$AX = 297.3120...

AX = 297 m (1 mark)

b 297.3120...sin 27° (1 mark)

= 135 m (1 mark)

23 a If X is the foot of the vertical line from E, then letting AX = x, by Pythagoras it follows that $x^2 + x^2 = 16^2$. (1 mark)

$x^2 = 128$

$x = 8\sqrt{2}$ (1 mark)

Let α be the angle $E\hat{A}X$.

$\cos\alpha = \frac{8\sqrt{2}}{20}$ (1 mark)

$\alpha = 55.6°$ (1 mark)

b Let Y be the mid-point of BC.

Then $EY = \sqrt{20^2 - 8^2} = \sqrt{336}$ (2 marks)

$\cos\beta = \frac{8}{\sqrt{336}}$ (1 mark)

$\beta = 64.1°$ (1 mark)

c From part a), $A\hat{E}X = 180 - 90 - 55.5500... = 34.4499...°$ (1 mark)

Therefore $A\hat{E}C = 2 \times 34.4499... = 68.9$ (1 mark)

d Let A be the surface area of the base. Then

$V = \dfrac{Ah}{3}$

$= \dfrac{16^2 \times 20\sin 55.5500...°}{3}$

$= 1410\text{ cm}^3.$ (2 marks)

e $A = 4\left(\dfrac{1}{2} \times 16 \times \sqrt{20^2 - 8^2}\right)$

$+16^2$ (2 marks)

$= 843\text{ cm}^2$ (1 mark)

Chapter 3

Skills check

1 a ☺ = 1 child

0	☺☺☺☺
1	☺☺☺☺☺
2	☺☺☺☺☺☺☺☺
3	☺☺☺☺☺☺
4	☺☺

b

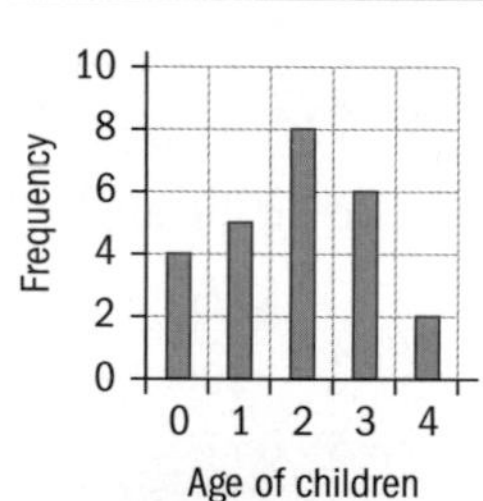

c

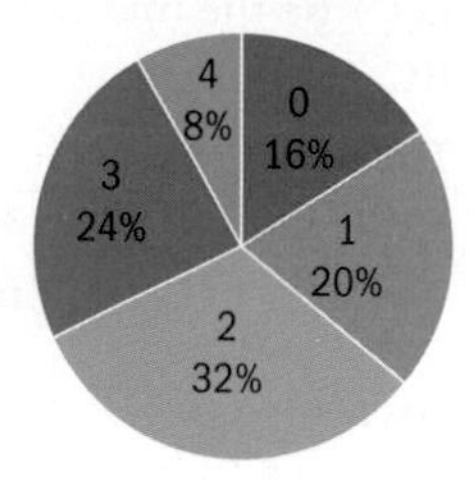

2 Mean = 7, median = 8, mode = 9, range = 6

3 Major gridlines correspond to 1 cm

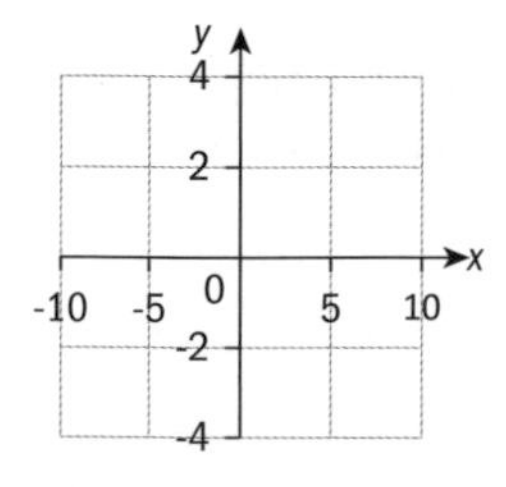

Exercise 3A

1 a Discrete **b** Continuous

c Discrete **d** Discrete

e Discrete **f** Continuous

g Continuous **h** Discrete

i Continuous

g Continuous

2

Number of sweets	Frequency
21	4
22	6
23	5
24	5
25	4
26	1

3

Height, in metres, h	Frequency
$2 \le h < 3$	6
$3 \le h < 4$	6
$4 \le h < 5$	5
$5 \le h < 6$	3

4

Weight, in kilograms, w	Frequency
$0 \le w < 10$	11
$10 \le w < 20$	8
$20 \le w < 30$	3
$30 \le w < 40$	3

Exercise 3B

1 a Mean = 10.1, median = 9.0, mode = 8.6; mean/median are most appropriate measures in this case, because the data is continuous.

b Mean = 8.64, median = 8.5, mode does not exist; both mean and median are appropriate measures in this case.

c Mean = 32.62, median = 30, mode = 15; both mean and median are appropriate measures but mode is too small.

2 a i Modal class: $150 \le n < 180$

ii Mean ≈ 112

iii Median ≈ 105
Data is *skewed*, median is most appropriate measure as this is least affected by the skewed values.

b i Modal class: $50 \le s < 55$

ii Mean ≈ 54.4

iii Median ≈ 52.5
Data set is well centred with all three measures agreeing well, so mean is best measure of the central tendency as this minimises the error for the guess of the next value.

c i Modal class: $7 \le t < 8$

ii Mean ≈ 5.86

iii Median ≈ 5.5
Data has no clear tendency. Median is best measure to use because the mean is affected by the high modal class.

Exercise 3C

1 a Mean = 6.1, median = 5.2, mode = 7.5, possible outliers = 17.8 and 25.0

b Mean = 3.5, median = 3.6, mode = 2.5, possible outlier = 6.1

c Mean = 65, median = 62, mode = 62, possible outlier = 22

Exercise 3D

1 a Discrete

b Mean = 13.9, mode = 15, median = 14, all are appropriate to use

c 7.3, indicates data points quite spread out

d Range = 32, IQR = 9

2 a Modal class: $30 \le c < 40$

b Mean ≈ 34, median ≈ 35

c 10.1, indicates data points very spread out

d Variance ≈ 102.3, range ≈ 40, IQR ≈ 10. All estimates because they use the mid value of the class intervals.

3 New mean = US$ 3600, new standard deviation = US$ 250. When every data point is shifted by an equal amount, the mean shifts by the same amount while standard deviation does not change as the data is spread out by the same amount.

4 a Mean = 8.9, median = 10, mode = 12. Since mode is quite a bit higher than the other two measures, the mean and the median are the most appropriate to use.

b Standard deviation = 4.10, showing data points have medium spread in relation to the mean.

c Range = 14, IQR = 6. Both relatively big, implying data is reasonably spread out.

5 New mean = 60, new standard deviation = 6

6 a Modal class = $180 \le x < 190$

b Mean ≈180.2, standard deviation ≈11.0, data are strongly spread out.

7 a Mrs Ginger: mean = 84, standard deviation = 16; Mr Ginger: mean = 80, standard deviation = 20; Miss Ginger: mean = 76, standard deviation = 24

b

	Mrs Ginger	Mr Ginger	Miss Ginger
Matty	44	30	16
Zoe	70	62.5	55
Ans	92	90	88

8 a Basketball players: mean = 200.8, standard deviation = 10.5; males: mean = 172.3, standard deviation = 10.3

b Basketball players much taller on average, however both samples have nearly the same standard deviation, so the spread of the data is nearly the same.

9 a Males: mean = $2546.30, standard deviation = $729.78; females: mean = $2114.58, standard deviation = $635.25

b Salaries of males have higher mean and greater spread than salaries of females.

Exercise 3E

1 a 155.4 cm

b For example, use stratified sampling: 160 students in total, so a sample of 50 students should be chosen; take 9, 9, 8, 8, 8, 8 students from grades 7, 8, 9, 10, 11, 12 respectively. The final answer will differ depending on the random numbers. This method will be unbiased.

2 a Mean = 25.92, which is well below 60 so should not lose much revenue.

b Use a random number generator to obtain 35 numbers between 1 and 100 and average the ages represented by those numbers. Answers will differ.

c Depends on the starting point. Starting with the first number, mean = 26.21.

d Usually the systematic sample mean will be closer to the population mean than the random sample mean. However, that depends heavily on the set of the random numbers used.

3 a 3.58

b Students' own answers

c Estimated mean should be close to actual mean but answers can differ.

Exercise 3F

1 Theo's data:

Number	Frequency
1	8
2	8
3	6
4	7
5	5
6	6

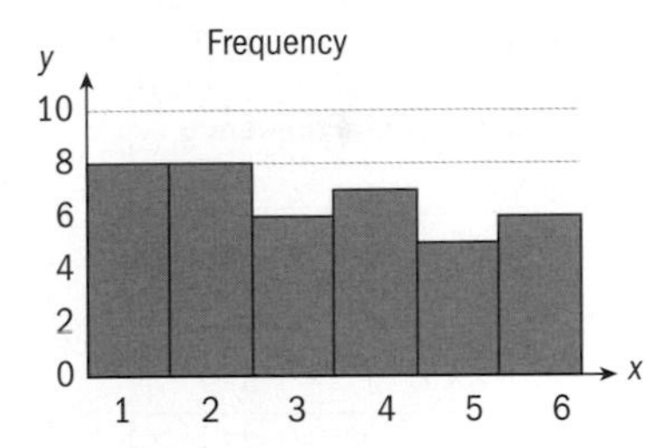

Millie's data:

Number	Frequency
1	7
2	7
3	8
4	5
5	7
6	6

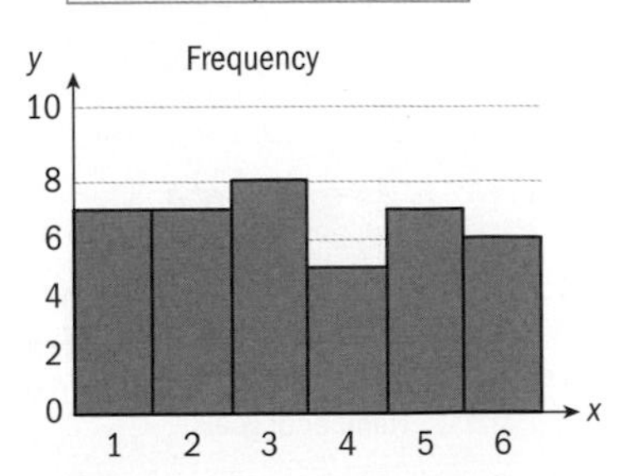

The distribution of the numbers are similar and the frequencies for each number thrown are very similar.

2 Female data:

Number of goals	Frequency
0	4
1	5
2	5
3	5
4	3
5	2
6	1

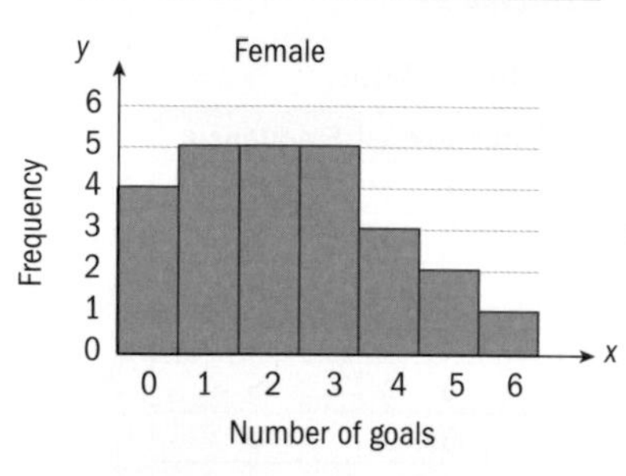

Male data:

Number of goals	Frequency
0	3
1	4
2	5
3	3
4	3
5	2
6	3
7	1
8	1

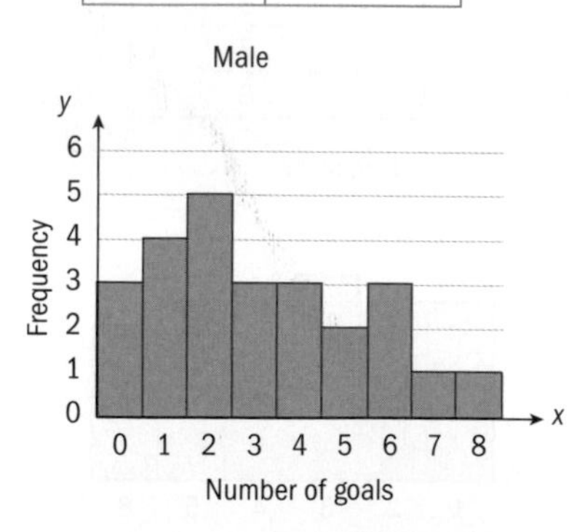

The range of the number of goals is bigger for males. The female data is more uniform than male data and more females scored no goals.

3 Frequency table and histogram: **a, b**

Height, in metres, h	Frequency
$140 < h \leq 145$	2
$145 < h \leq 150$	17
$150 < h \leq 155$	9
$155 < h \leq 160$	2
$160 < h \leq 165$	2

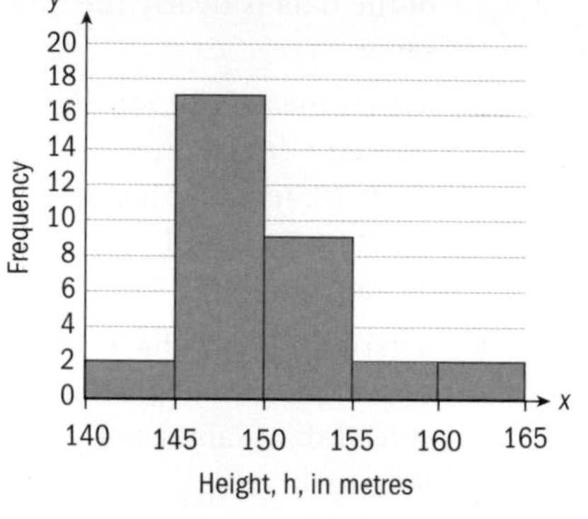

c

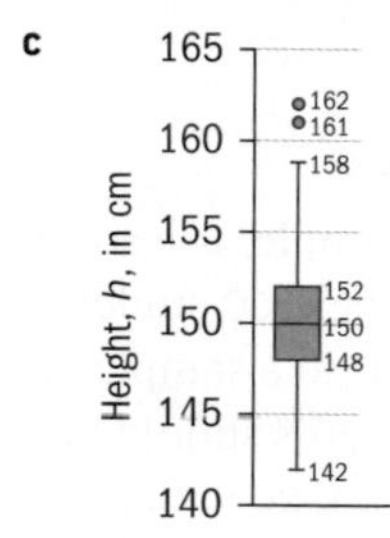

d From the box and whisker diagram, the data is symmetrical.

4 a

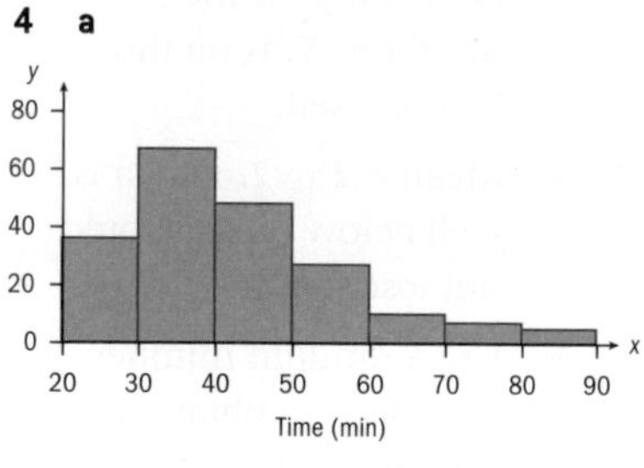

b Mean = 42, median = 35, LQ = 35, UQ = 45, range = 70, outliers: data points in $60 \leq x \leq 90$

c

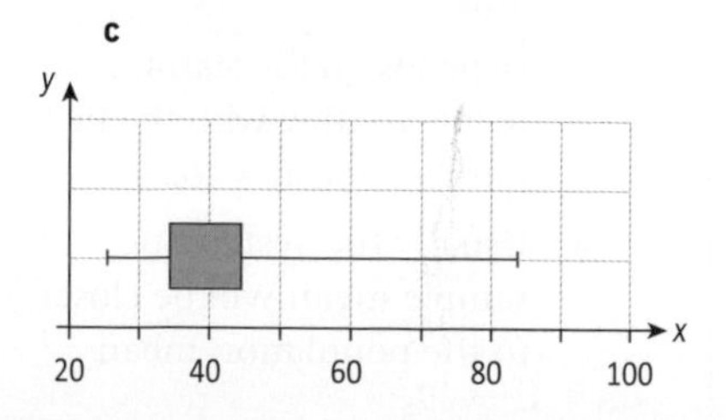

d Marcus did worse than 75% of participants, so he may not be satisfied.

5 a Boys: mean = 6, median = 6, LQ = 5, UQ = 7, range = 7

Girls: mean = 6, median = 6, LQ = 5, UQ = 7, range = 8 There are no outliers.

b Plots are almost identical except that boys have higher minimum value.

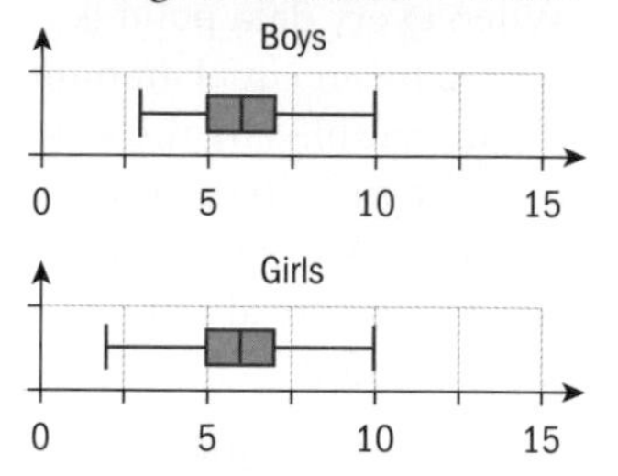

c Both data sets are symmetrical

6 a Mean = 17, median = 16, LQ = 12, UQ = 21, range = 37, outliers = 35, 43

b

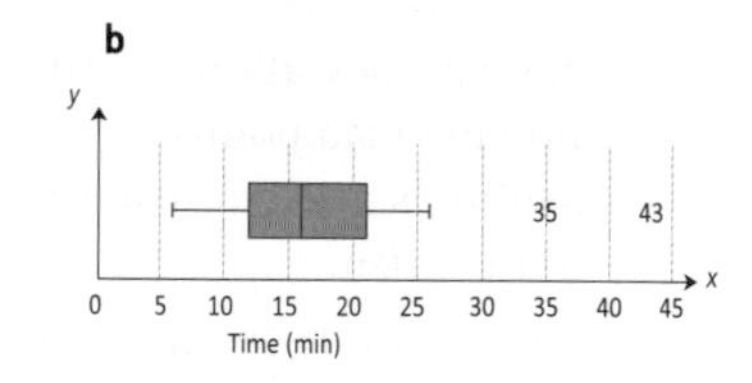

c 17

7 a Median = 3 **b** IQR = 2

c Data is almost symmetrical

8 a Boys: 55, girls: 65

b Boys: 30, girls: 30

c 25% **d** 50%

e 15 **f** 45

g Neither of the data sets is perfectly symmetrical.

9 a 120 kg **b** 70

c 10 **d** 50% **e** 20

f Distribution of weight is skew with respect to average weight.

g Mostly males

Exercise 3G

1 a

Upper boundary	Cumulative frequency
2	5
4	16
6	39
8	70
10	89
12	97
14	100

b

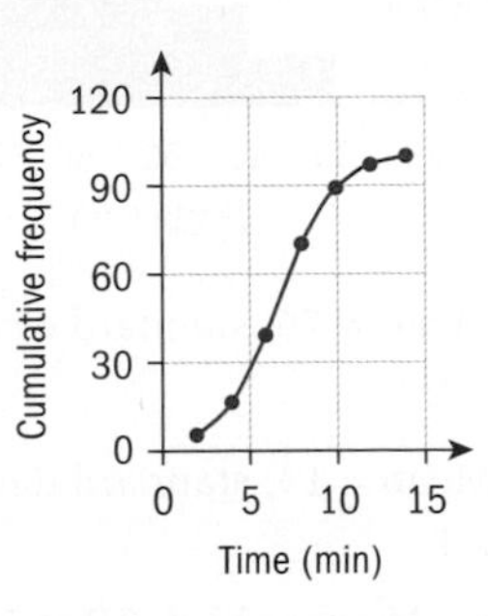

c Median ≈ 6.75 min, IQR ≈ 3.5 min

d 3 min **e** 5 people

2 a

Upper boundary	Cumulative frequency
4	5
8	37
12	78
16	106
20	128
24	140
28	147
32	150

b

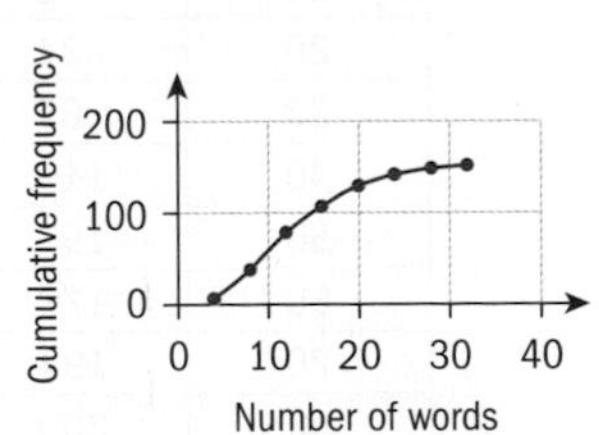

c Median ≈ 12, IQR ≈ 10

d No outliers **e** 22

f

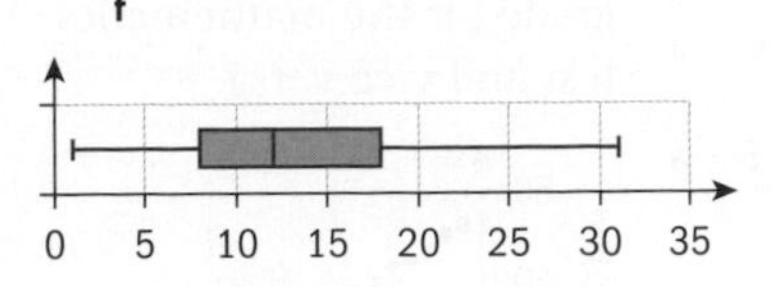

g Adult books, because mean 15 falls inside IQR while the mean 8 does not.

3 a

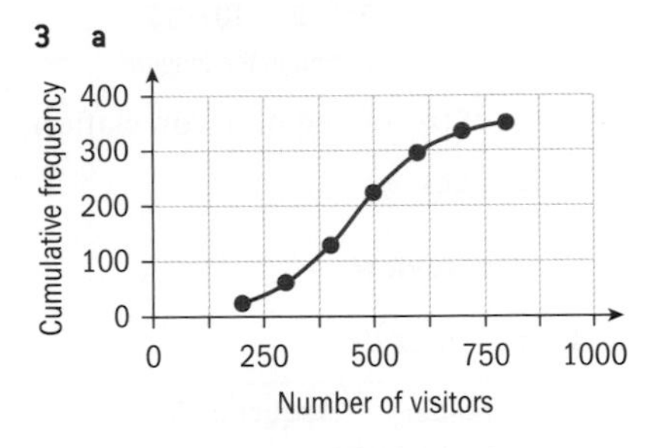

b Median ≈ 450, IQR ≈ 205

c No outliers.

d

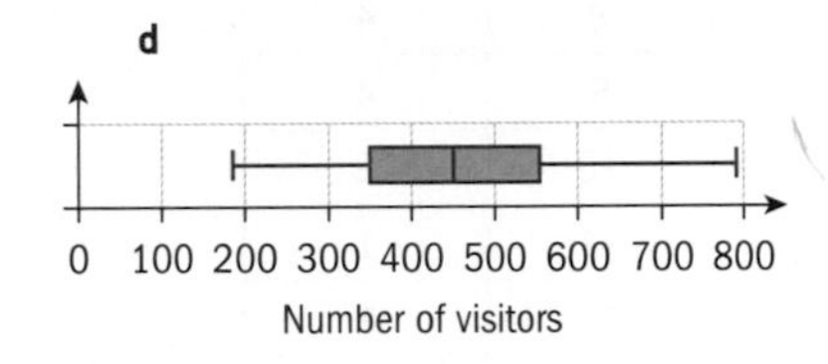

e 90 days

4 a Both cumulative frequency and box-and-whiskers graphs are useful for analysing this situation.

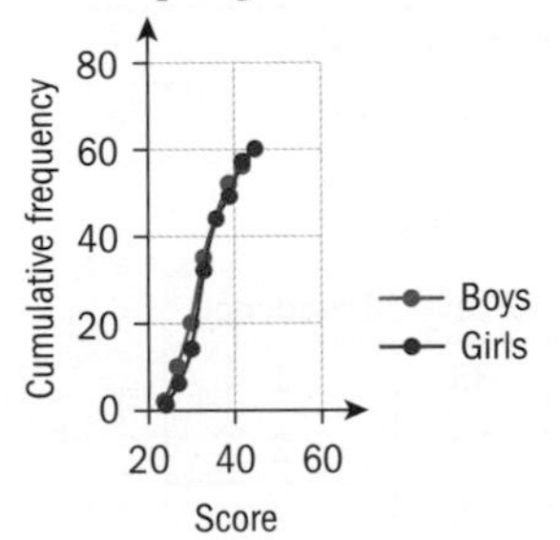

Girls

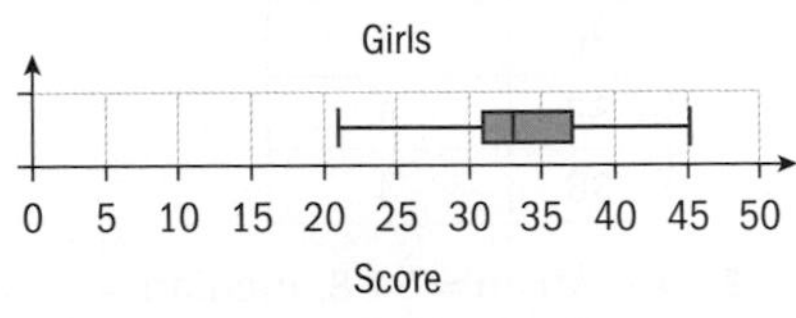

Boys

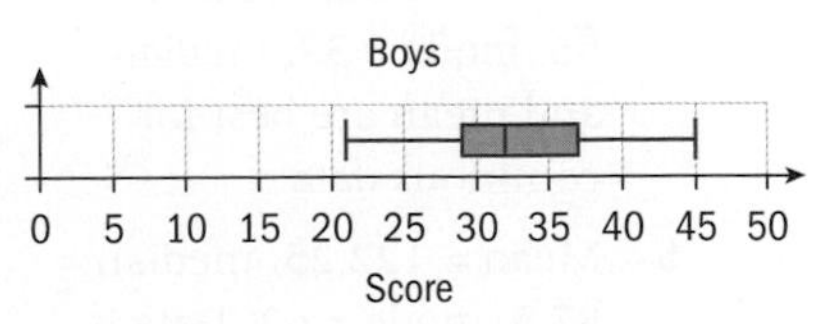

b Both graphs suggest fairly similar results. Cumulative frequency graph shows that more boys than girls acquired the lower range scores. Box-and-whiskers graphs show that both data sets are not perfectly symmetrical. IQR of the boys' scores is wider than the girls' scores. The median score of the girls is just above the median score of the boys.

c Martin did better than 25% of the boys. However, more than 75% of girls did better than Mary.

5 a Median ≈ 95, IQR ≈ 60, 90th percentile ≈ 150

b No outliers

6 a Median ≈ 300, LQ ≈ 225, UQ ≈ 380

b

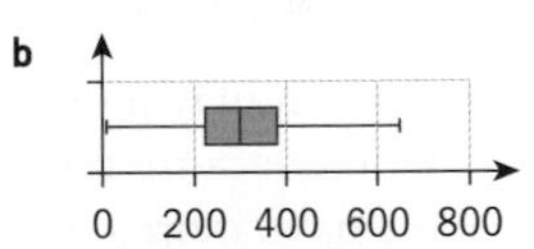

c

Length, l, in cm	Cumulative frequency
$50 \le l < 100$	5
$100 \le l < 150$	5
$150 \le l < 200$	10
$200 \le l < 250$	12
$250 \le l < 300$	18
$300 \le l < 350$	15
$350 \le l < 400$	13
$400 \le l < 450$	8
$450 \le l < 500$	4
$500 \le l < 550$	4
$550 \le l < 600$	3
$600 \le l < 650$	2
$650 \le l < 700$	1

d Mean = 312, standard deviation = 132

Exercise 3H

1 a Negative, moderate
b Positive, strong
c Negative, strong
d Positive, weak
e No positive or negative correlation
f No positive or negative correlation (but could split into intervals of strong positive and strong negative correlations)

2 a

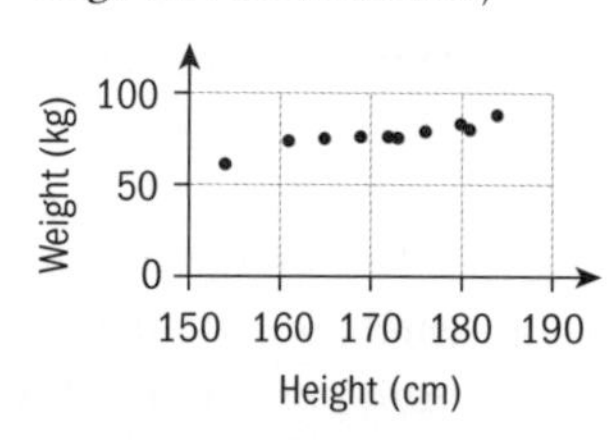

b Strong positive correlation; hence, the taller the football player, the heavier he or she is expected to be
c Correlation may indicate causation because taller people have more body mass.

3 a

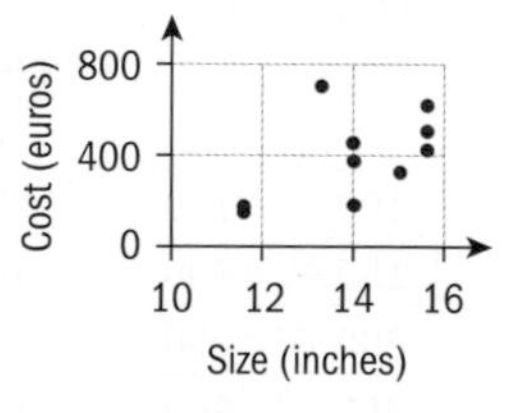

b Weak positive correlation
c Since the correlation is weak, the size of the screen has little influence on the cost of the laptop.

4 a

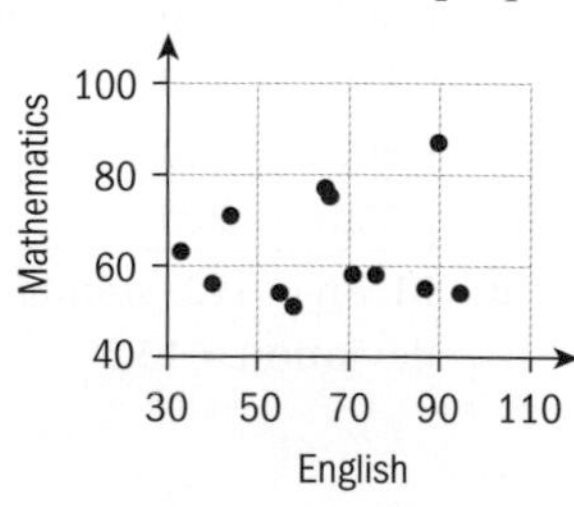

b No visible positive or negative correlation
c The grade for the English test does not influence the grade for the mathematics test and vice versa.

5 a

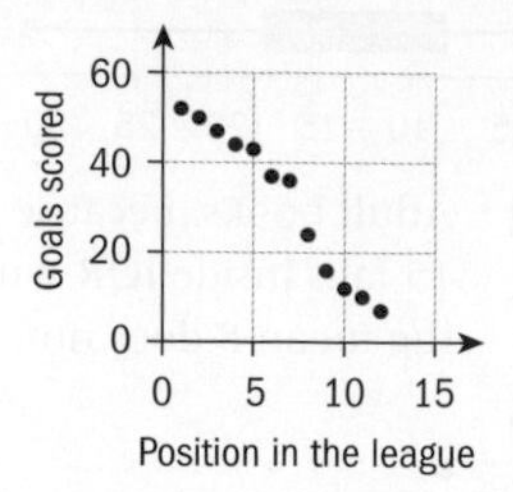

b Strong negative correlation
c Yes

Chapter review

1 a Discrete

Number of apples	Frequency
7	4
8	4
9	4
10	3
11	1
12	1

b Continuous

Length, l (cm)	Frequency
$7 \leq l < 9$	5
$9 \leq l < 11$	2
$11 \leq l < 13$	2
$13 \leq l < 15$	4

c Discrete

Size	Frequency
33	1
34	2
35	4
36	3
37	4
38	3

2 a Mean = 54.8, median = 56, mode = 32, median and mean are best for continous data
b Mean = 122.25, median = 87.5, mode = 62. Data is very skewed so best to use median.
c Mean = 6.42, median = 6, mode = 6. All measures agree so can use any of them.

3 a $50 \leq l < 60$
b Median ≈ 55, mean ≈ 54.42, SD ≈11.28
c

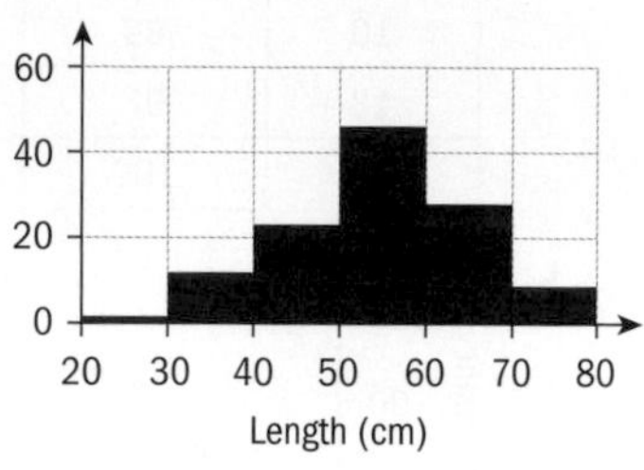

4 Mean = 76, standard deviation = 14

5 Mean = 13, standard deviation = 1

6 a Mean = 32.8, SD = 7.51, indicates data points are reasonably spread out
b Range = 25, IQR = 12
c Min = 20, LQ = 27, median = 34, UQ = 39, max = 45, no outliers
d

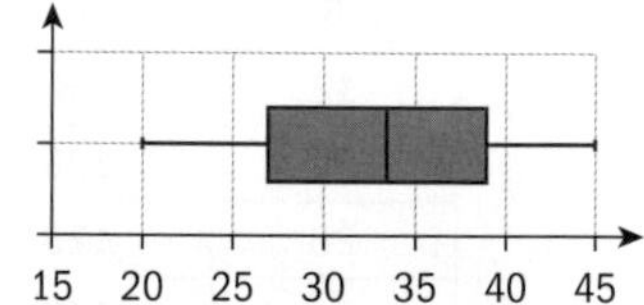

7 a

Upper boundary	Cumulative frequency
10	8
20	24
30	65
40	119
50	155
60	177
70	194
80	200

b

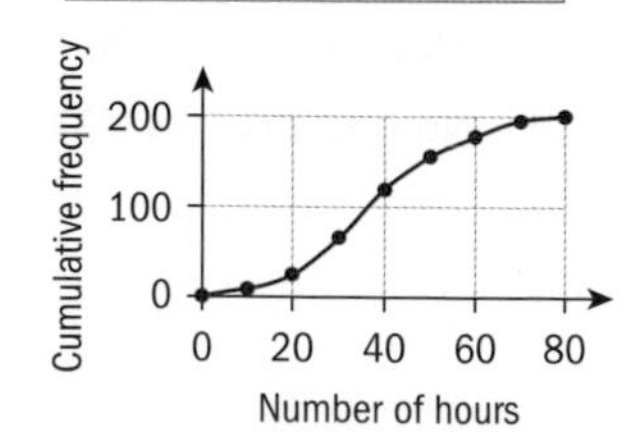

c Median ≈ 36, IQR ≈ 21

d

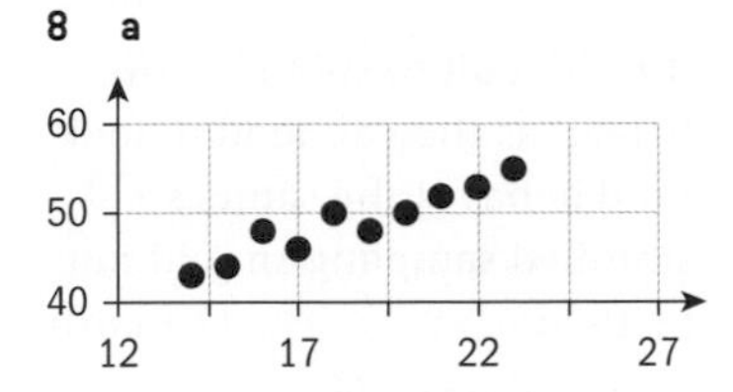

8 a

60
50
40
12 17 22 27

b Strong and positive correlation

c It is unlikely that weather is the cause of the increase/decrease in the number of eggs.

Exam-style questions

9 a $40 \le x < 60$ (1 mark)

b Use of mid-point (1 mark)

Use of GDC to find estimate of mean gives $\bar{x}$ = 48.3% (2 marks)

c Use of GDC to find estimate of standard deviation gives $\sigma = 22.6$ (2 marks)

d The second class had slightly lower marks on average. (1 mark)

Their standard deviation was also lower, meaning their marks were more consistent (or less spread out) than the first class. (1 mark)

10 a $\frac{398}{25} = 15.92\text{g}$ (2 marks)

b 13th mouse has weight of 16 g (1 mark)

So median is 16 g (1 mark)

c $Q_1 = 14$ (1 mark)

$Q_3 = 19$ (1 mark)

Interquartile range is

$Q_3 - Q_1 = 19 - 14$

$= 5$ (2 marks)

d $Q_3 + 1.5 \times \text{IQ range} =$

$19 + 1.5 \times 5$ (1 mark)

$= 26.5\text{g}$ (1 mark)

11 a Using GDC, mean = 23.58°C (2 marks)

b Using GDC, SD = 3.78°C (2 marks)

c Using GDC, mean = 22.83°C (2 marks)

d Using GDC, SD = 5.52°C (2 marks)

e Tenerife has a higher mean temperature (23.58°C), so on average, (1 mark) temperatures could be said to be higher in Tenerife. (1 mark)

The standard deviation of temperatures in Tenerife (3.38°C) is lower (1 mark)

than the standard deviation of temperatures in Malta. Therefore, the temperatures in Tenerife can also be said to be more consistent. (1 mark)

12 a $5.25 + 3 = 8.25$ years (2 marks)

b 1.2 years (1 mark)

As the ferrets age, their spread of ages will not change (1 mark)

13 Total population is 65.12 million. (1 mark)

Number required from England: $\frac{54.8}{65.12} \times 5000 = 4208$ (2 marks)

Number required from Wales: $\frac{3.10}{65.12} \times 5000 = 238$ (1 mark)

Number required from Scotland: $\frac{5.37}{65.12} \times 5000 = 412$ (1 mark)

Number required from Northern Ireland:

$\frac{1.85}{65.12} \times 5000 = 142$ (1 mark)

14 a $45 - 15 = 30$ (2 marks)

b $37 - 22 = 15$ (2 marks)

c 75% (1 mark)

d Interquartile range is 15

Outlier is anything above $Q_3 + 1.5 \times \text{IQ range} = 37 + 1.5 \times 15$ (1 mark)

= 59.5 minutes (1 mark)

Yes, 60 minutes would count as an outlier. (1 mark)

15 a 69 minutes (2 marks)

b 50 minutes (2 marks)

c 83 minutes (2 marks)

d 170 minutes (2 marks)

e Interquartile range is

$83 - 50 = 33$ (1 mark)

$Q_3 + 1.5 \times \text{IQ range} = 83 + 1.5 \times 33$ (1 mark)

$= 132.5$

$Q_1 - 1.5 \times \text{IQ range} = 50 - 1.5 \times 33$ (1 mark)

$= 0.5$

Therefore 180 is an outlier (1 mark)

f

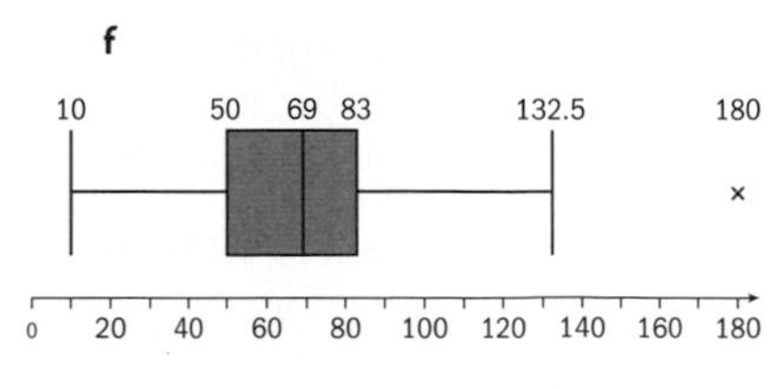

(3 marks)

16 a $45 \le x < 60$ (1 mark)

b $45 \le x < 60$ (2 marks)

c

Time, (t minutes)	c.f.
$0 \le x < 15$	4
$15 \le x < 30$	9
$30 \le x < 45$	21
$45 \le x < 60$	45
$60 \le x < 75$	63
$75 \le x < 90$	70
$90 \le x < 100$	75

(3 marks)

d

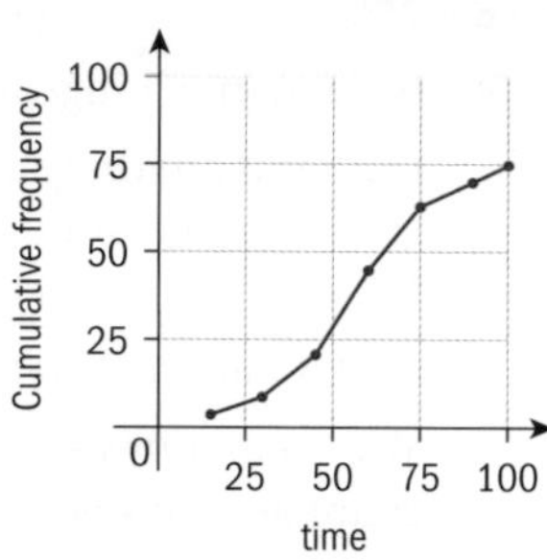

(2 marks)

e From graph, 37.5 on y-axis gives median of 55.

(2 marks)

Interquartile range is approximately $70 - 40 = 30$.

(2 marks)

17 a Using GDC, mean = \$43 600 (2 marks)

a Median = $\frac{1}{2}(11+1)$th value = 6th value, which is \$25 000

(2 marks)

c $\text{LQ} = \frac{1}{4}(11+1) = 3\text{rd value}$, which is \$25 000

$\text{UQ} = \frac{3}{4}(11+1)\text{th} = 9\text{th}$ value, which is \$80 000

(1 mark)

$\text{IQR} = 80\,000 - 25\,000 = 55\,000$ (1 mark)

d Analyst 8 would use the mean average (1 mark)

in order to suggest that their salary of \$25 000 was significantly below the 'average' of \$43 600.

(1 mark)

e The managing director would use the median (or mode) (1 mark)

and say that Analyst 8 was already earning the 'average' salary of \$25 000. (1 mark)

f **Either:** In this case, the median would be fairest

(1 mark)

as there is no one earning a salary close to the mean value of \$43 600, and the majority of workers earn the median salary.

(1 mark)

Or: The mean would be fairest (1 mark)

because it takes into account the significantly higher salaries of the Line Managers and Director.

(1 mark)

18 a In a random sample, each member of the population has an equal probability of being chosen. (1 mark)

A stratified sample is where the population is divided into strata, based on shared characteristics. (1 mark)

Random samples within each strata are chosen, where the sample size is in the same proportion to the size of the strata within the population.

(1 mark)

A systematic sample is where members are listed and then chosen from a random starting point with a fixed periodic interval. (1 mark)

The interval is the population size divided by the sample size. (1 mark)

b i Population size is unknowable, so specific periodic choice of interval is difficult to determine. (2 marks)

ii Random sampling.

(1 mark)

It is difficult to split rats into 'strata' as they all tend to look (and behave) the same, so stratified sampling should not be used. (1 mark)

19 a

Height, (x cm)	Frequency
$20 \le x < 25$	3
$25 \le x < 30$	3
$30 \le x < 35$	4
$35 \le x < 40$	7
$40 \le x < 45$	4
$45 \le x < 50$	2
$50 \le x < 55$	1

b Using GDC, mean height = 35.8 cm (2 marks)

c Using GDC, variance = 68.9 cm^2 (2 marks)

d Using GDC, standard deviation = 7.99 cm

(2 marks)

e On average, the neighbour's garden's flowers had a slightly lower height compared to Eve's. (1 mark)

The neighbour's flowers also had a smaller standard deviation, indicating they were grown to a more consistent length.

(2 marks)

20 a $\frac{611}{1200} \times 718\,824 = 366\,001$

(2 marks)

b Geographical location of residents; leisure / work travellers; age (anything appropriate or connected to travel). (2 marks)

Chapter 4

Skills check

1 **a** $x = \frac{7}{2}$ **b** $x = \frac{5}{3}$

2

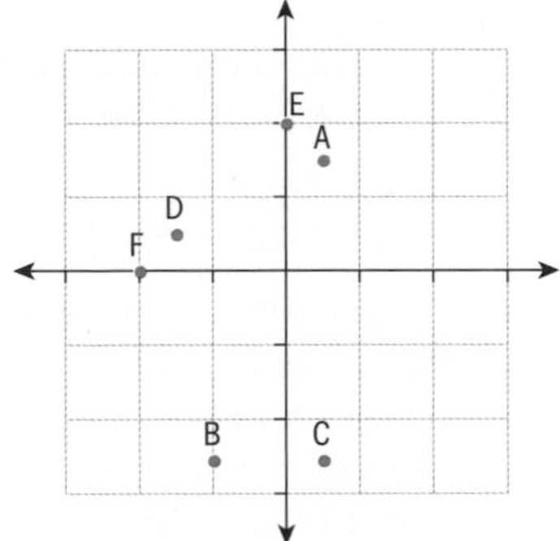

3 **a** 8 **b** $-\frac{1}{2}$

4 8 cm

Exercise 4A

1 **a** (1.71, 1.92)

b (3, 6.75, −7.6)

2 $\left(0, -\frac{3}{2}\right)$

3 (0, 2)

4 **a** B(6, 7, 0)

b A(6, 7, 4)

c $M\left(3, \frac{7}{2}, 2\right)$

Exercise 4B

1 **a** 9.71

b 8.83

2 3

3 Calculate the lengths (and leave as roots for now):

$AB = \sqrt{(-4-6)^2 + (6-2)^2} = \sqrt{116}.$

$BC = \sqrt{(6+8)^2 + (2+4)^2} = \sqrt{232},$

$AC = \sqrt{(-4+8)^2 + (6+4)^2} = \sqrt{116}.$

Therefore, as $AB^2 + AC^2 = BC^2$, this triangle is a right-angled triangle.

4 21.1

5 $x = 7, y = 4.5$

6 **a** 8.54 km

b The radar will be able to detect one, but not both, of the aircraft.

Exercise 4C

1 **a** A(x, −8), where x is any number

b A(−3, y), where y is any number

2 **a** $\frac{-1}{3}$ **b** $\frac{5}{13}$ **c** −1

3 Stair 1 gradient = 0.48, stair 2 gradient = 0.5, stair 3 gradient = 0.833

a Stair 3 has the greatest gradient

b Stair 1 has the least gradient.

4 **a**

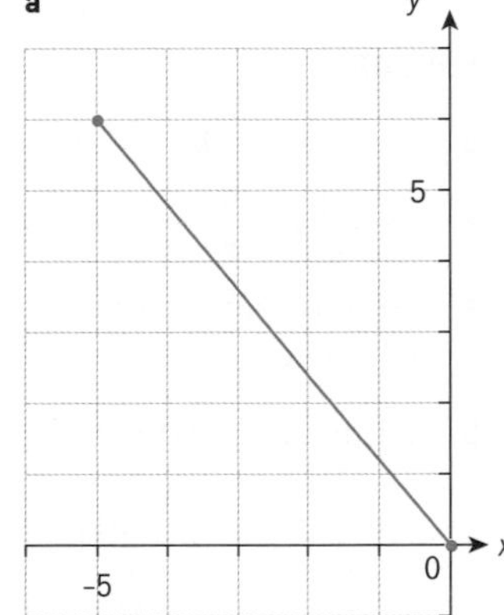

b

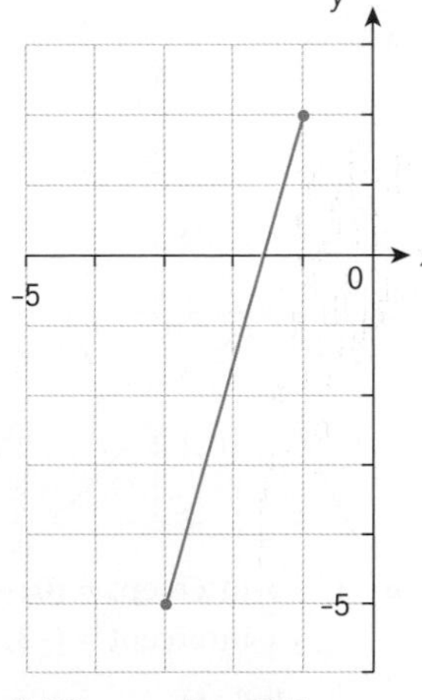

5 **a** 0.75 **b** −7.5

6 The slope with gradient −3 is steeper, as the skier changes height more (3 units vs $\frac{1}{3}$ units) over a single unit distance travelled horizontally, compared to the slope with gradient $\frac{-1}{3}$.

7 150 m

Exercise 4D

1 0.0833, the ramp does not conform to safety regulations.

2 **a**

Number of days worked	City guide payment (USD)	Flower shop assistant payment (USD)
1	160	60
5	320	300
10	520	600

b

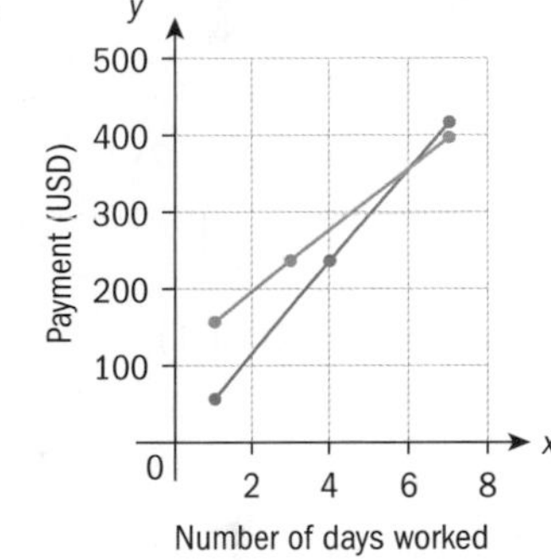

c Students should decide depending on the number of days they plan to work. If they plan to work fewer days than 6 then they should take the city guide job. If they plan to work longer than 6 days, the flower shop will pay more and they should take that job.

3 **a** 3.5 m **b** 0.457 (3 s.f.)

c The roof does not satisfy the requirements, as the gradient is greater than the maximum specified by the regulations (0.457 > 0.17).

4 a 25%

b The road with a sign indicating 15% is the steeper road.

c 0.75 km

5 a i black **ii** red

iii green **iv** blue

b

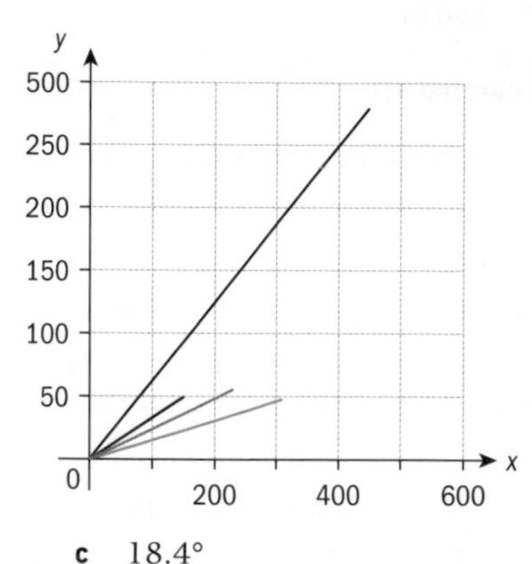

c 18.4°

Exercise 4E

1 a $y - 4 = 3(x - 1)$

b $y + 4 = -5(x - 7)$

c $y - 3 = -\frac{x+1}{2}$

d $y - 4 = 2(x + 1)$

2 a

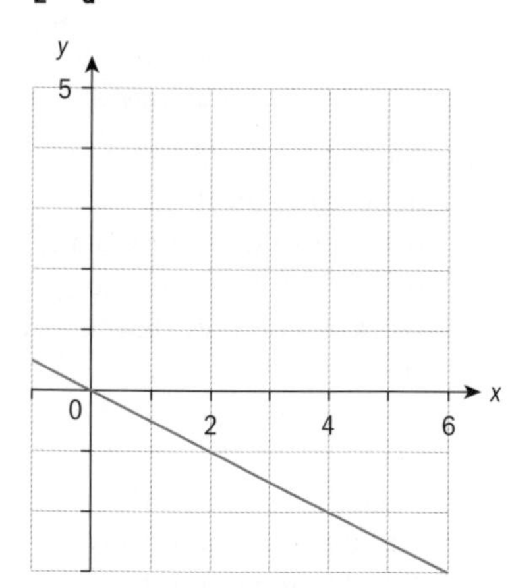

b

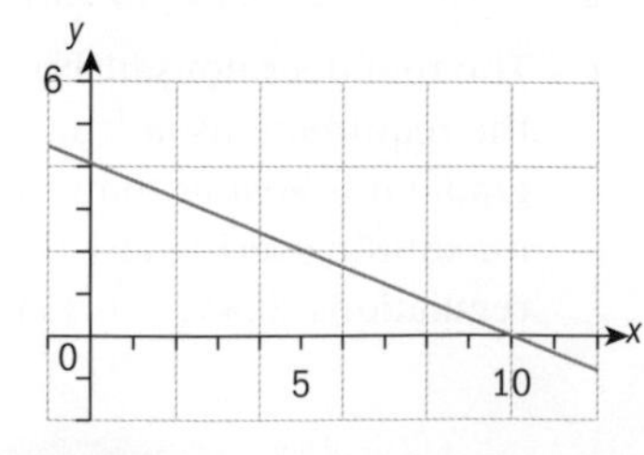

c

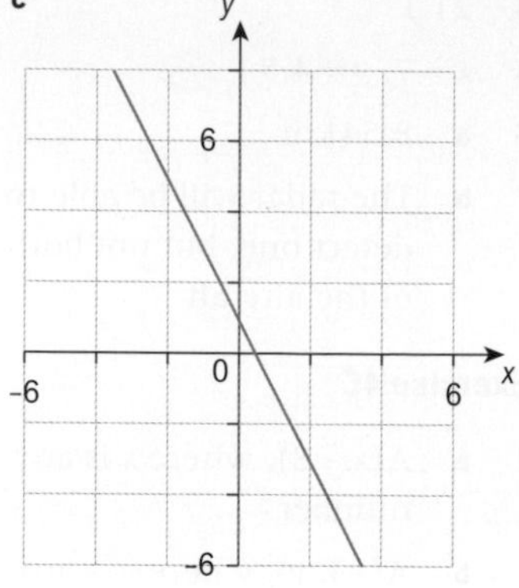

3 a $y = 4x + 13$ **b** $y = \frac{2x}{7} - 4$

c $y = -\frac{1}{5}x + \frac{2}{5}$ **d** $y = \frac{2}{3}x - \frac{4}{3}$

4 a A does lie on the line $y = -x - 9$

b A does not lie on the line $5x + 4y = 9$

5 a 7 **b** 10

Exercise 4F

1 a $y = 3x - 2$ **b** $y = -4x + 1$

c $y = \frac{-2}{3}x - \frac{5}{3}$

2 a x-intercept = (0, 0), y-intercept = (0, 0)

b y-intercept = $(0, 3)$, x-intercept = $\left(\frac{15}{2}, 0\right)$

c y-intercept = (0, 1), x-intercept = (−5, 0)

3 a

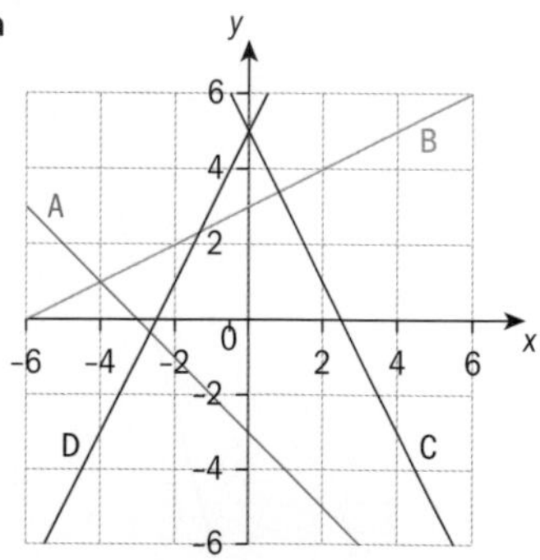

b i y-intercept = (0, −3), x-intercept = (−3, 0)

ii (−4, 1)

c Both C and D have y-intercept (0, 5)

4 a(i) (ii)

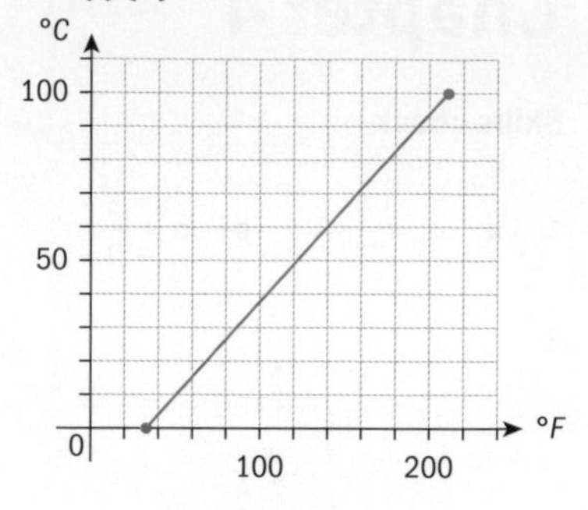

b $C = \frac{5}{9}F - \frac{160}{9}$ **c** $\frac{5}{9}$

d The gradient is the amount that the temperature changes, measured in °C when the temperature measured in °F changes by 1.

e $-\frac{160}{9}$

f The temperature in °C at 0 °F

g 28.3 °C **h** 14 °F

i Adding $\frac{160}{9}$ to both sides of the expression in part **b** gives $C + \frac{160}{9} = \frac{5}{9}F$.

Then multiplying both sides by $\frac{9}{5}$ gives the expression $F = \frac{9}{5}C + 32$.

5 L_1: $y = \frac{1}{2}x$, L_2: $x = -1$ L_3: $y = 5$, L_4: $y = -x + 4$

6 a $A = 30 + 0.02x$

b 0.02, represents the amount Maria earns (in USD) per dollar spent in the restaurant.

c 30, represents the amount Maria would earn in a day if no food was sold.

d

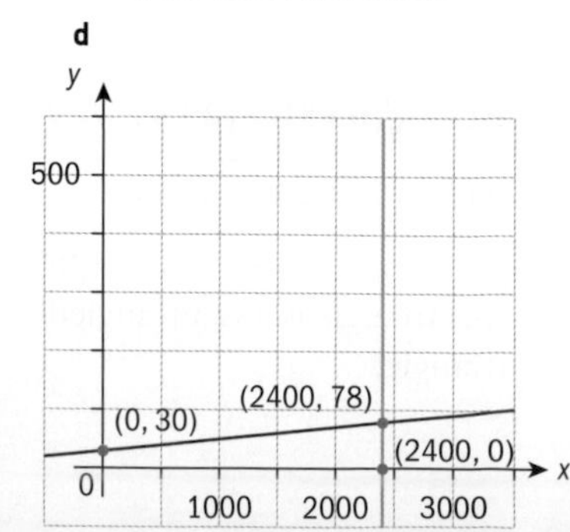

e \$78

Exercise 4G

1 **a** -1

b parallel to the y-axis

c $\frac{1}{3}$ **d** $\frac{-2}{5}$

2 **a** $\frac{1}{3}$ **b** -8

c undefined gradient

d $\frac{-3}{2}$

3 **a** Perpendicular

b Parallel **c** Neither

4 Intersect at $(-5, -4)$

5 -4

6 $\frac{-7}{2} = \frac{y+0.2}{x+1}$

(or $y = \frac{-7}{2}x - \frac{37}{10}$)

7 Gradients:

$M_{AB} = \frac{-1-0}{-3-2} = \frac{1}{5}, M_{BC} = \frac{3-0}{5-2} = 1,$

$M_{CD} = \frac{2-3}{0-5} = \frac{1}{5}, M_{DA} = \frac{2+1}{0+3} = 1.$

Therefore lines between AB and CD are parallel and those between BC and DA are parallel. so the quadrilateral is a parallelogram.

Exercise 4H

1 **a** Bernard is at a point with $x = 50, y = -75$. Since $\frac{1}{2}x - 100 = 25 - 100 = -75 = y$, Bernard is on the line with equation $y = \frac{1}{2}x - 100$.

b (340, 70)

2 **a** $y - 4 = \frac{-1}{2}(x - 3)$ or

$y = \frac{-1}{2}x + \frac{11}{2}$

b $y = x - 1$

3 5.83

4 (6.36, 0.337)

5 (3, 0)

Exercise 4I

1 **a**

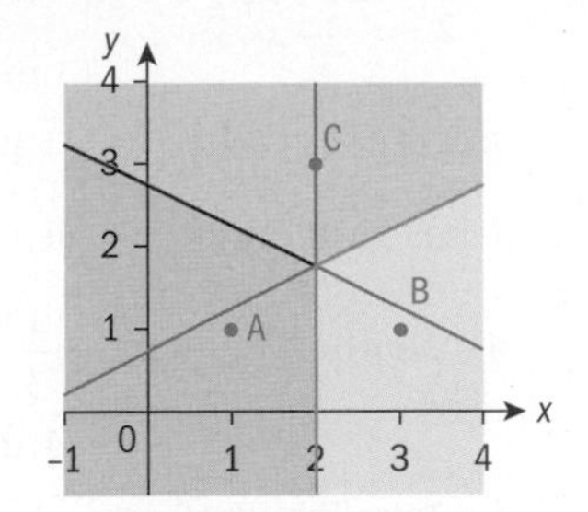

b

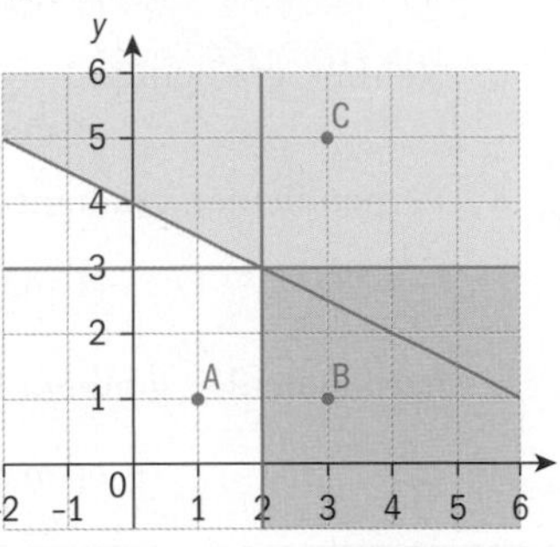

2 **a** and **c**

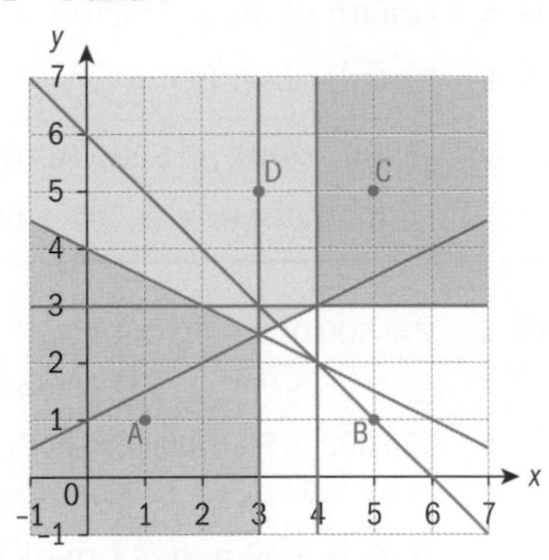

b bookstore A

d bookstore D

Exercise 4J

1 **a** The intersection of the perpendicular bisectors will be a vertex in the Voronoi diagram. It will be equally distant from A, B, and C and the center of a circle passing through the three points. Hence the position of the solution to 'the toxic waste problem'.

b $y = -5x + 21$ and $y = 2.5$

c (3.7, 2.5)

d 2.75 km

2 **a** **i** $y = x - 10$

ii $y = -2x + 140$

b

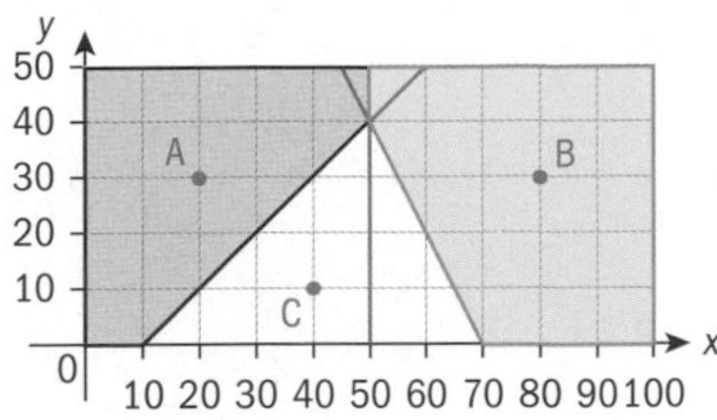

c **i** 1200 m^2 **ii** 1700 m^2

d **i** (50,40) **ii** 31.6 m

Chapter review

1 For example:

a $-2x + 2y = 4$, $y = 2 + x$, $-3x + 3y - 6 = 0$

b $y + 2x - 19 = 0$, $2y - 6 = -4(x - 8)$, $y = -2x + 19$

c $2x + 4y + \frac{1}{3} = 0$, $12x + 24y = -2$, $y = \frac{1}{2}x - \frac{1}{12}$

2 **a** $\left(\frac{34}{9}, -\frac{80}{7}\right)$

b (5, 0)

c $\left(-\frac{16}{7}, -\frac{10}{7}\right)$

3 Square

4 **a** $P_1(5, -2)$, $P_2\left(\frac{21}{11}, \frac{12}{11}\right)$, $P_3(-3, -3)$

b 4.37, 8.06, 6.39

c 18.8

5 **a** 0.75 km

b Maria walks at $\frac{5}{1.6} = 3.125$ km per hour; so Maria walks faster.

c Petya: $y = \frac{3}{4}x$,

Maria: $y = 3.125x - 9.5$

6 a Incorrect; correct equation is $y = -\frac{1}{5}x - 3$

b Incorrect; correct equation is $y = 5$

c Correct

7 a $y = -\frac{1}{4}x$ **b** $y = \frac{x}{6} - \frac{1}{2}$

8 a $y = 3x + 11$ **b** $y = \frac{1}{3}x + \frac{7}{6}$

9 a i $y = \frac{5}{8}x - \frac{35}{2}$

ii $m_{AB}(140, 70)$

iii $y = -\frac{8}{5}x + 294$

b i $y = \frac{1}{9}x + \frac{40}{3}$

ii $m_{AC}(150, 30)$

iii $y = -9x + 1380$

c i $y = -4x + 1000$

ii $m_{AC}(230, 80)$

iii $y = \frac{1}{4}x + \frac{45}{2}$

d $\left(\frac{5430}{37}, \frac{2190}{37}\right)$

e A wind turbine at point T would meet the regulations.

f $17\,700\,\text{m}^2$

10 a

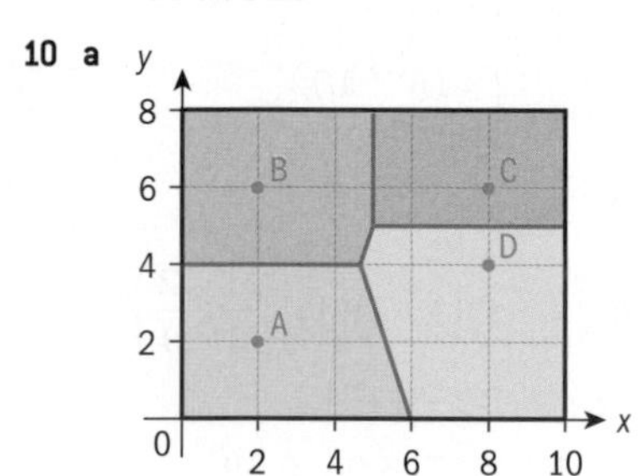

b $x = \frac{14}{3}$

c i $21\frac{1}{3}$ **ii** $19\frac{5}{6}$

iii 15 **iv** $23\frac{5}{6}$;

Yes, B can support

Exam-style questions

11 Let $B = (x, y, z)$ (1 mark)

$\frac{x+4}{2} = 7 \Rightarrow x = 10,$

$\frac{y-6}{2} = 3 \Rightarrow y = 12,$

$\frac{z+10}{2} = -5 \Rightarrow z = -20$

(3 marks)

$B = (10, 12, -20)$ (1 mark)

12 L has gradient of 3 (1 mark)

a Neither, gradient is $\frac{1}{3}$ (1 mark)

b Parallel, gradient is 3 (1 mark)

c Neither, gradient is 2 (1 mark)

d Perpendicular, gradient is $\frac{-1}{3}$ (1 mark)

e Perpendicular, gradient is $\frac{-1}{3}$ (1 mark)

13 a $M = \left(\frac{-3+5}{2}, \frac{8+3}{2}\right)$

$= \left(1, \frac{11}{2}\right)$ (2 marks)

b $\text{gradient} = \frac{3-8}{5-(-3)} = -\frac{5}{8}$ (2 marks)

c i $\frac{8}{5}$ (1 mark)

ii $y = \frac{8}{5}x + c$

through $\left(1, \frac{11}{2}\right)$

$\Rightarrow \frac{11}{2} = \frac{8}{5} + c \Rightarrow c = \frac{39}{10}$ (1 mark)

equation is $y = \frac{8}{5}x + \frac{39}{10}$ (1 mark)

14 a Length of lift $BP = \sqrt{500^2 + 400^2 + 300^2} = 707\,\text{m}$ (2 marks)

b Length of lift PQ =

$\sqrt{(900-500)^2 + (600-400)^2 + (700-300)^2} = 600\,\text{m}$ (2 marks)

So total distance is 707 + 600 = 1307 m (1 mark)

15 a $V = 500\,000 - 50\,000t$ (1 mark)

b $125\,000 = 500\,000 - 50\,000t$

$\Rightarrow t = 7.5$ (2 marks)

Time is 3.30 p.m. (1 mark)

c $500\,000 - 50\,000t$

$= 800\,000 - 100\,000t$

$\Rightarrow t = 6$ (2 marks)

Time is 2.00 p.m. (1 mark)

$V(6) = 500\,000 - 50\,000 \times 6 = 200\,000$ litres (1 mark)

16 a L_1 is $y = \frac{a}{3}x - 9$ which has gradient $\frac{a}{3}$ (1 mark)

L_1 and L_2 perpendicular, so

$\frac{a}{3} \times \frac{2}{3} = -1 \Rightarrow a = -\frac{9}{2}$ (2 marks)

b Intersection is (–3.23, 1.85) (2 marks)

17 a $2 \times 6 \times 5 + 2 \times 6 \times 3 + 2 \times 3 \times 5 = 126$ (2 marks)

b $\sqrt{3^2 + 5^2 + 6^2} = \sqrt{70}$

$= 8.37\,(3sf)$ (2 marks)

c i M = (1.5,2.5,3) (1 mark)

ii Triangle AMB is isosceles. Let Q be the midpoint of AB (1 mark)

QB = 2.5 $\quad BM = \frac{\sqrt{70}}{2}$ (2 marks)

$\sin QMB = \frac{2.5}{\frac{1}{2}\sqrt{70}}$

$\Rightarrow QMB = 36.69...$ (2 marks)

AMB = 73.4° (3sf) (1 mark)

18 a $(1200 - a)r = 765$

$(500 - a)r = 315$ (2 marks)

$\frac{1200-a}{500-a} = \frac{765}{315} \Rightarrow 378\,000$

$-315a = 382\,500 - 765a$

$\Rightarrow 450a = 4500$ (1 mark)

$a = 10$ (1 mark)

$r = \frac{765}{1190} = \frac{9}{14} = 0.643(3sf)$

(2 marks)

b $(1200 + 500 - 10) \times \frac{9}{14}$

$= 1086.43$ Euros

(2 marks)

19 a Minimum distance is the perpendicular distance.

(1 mark)

Gradient of road is 1 so line ST has gradient −1.

(1 mark)

ST is the line $y = -x + c$, and passes through (80,140)

$140 = -80 + c \Rightarrow c = 220$

(2 marks)

Intersection of $y = -x + 220$ and $y = x - 80$ is S = (150, 70). (2 marks)

b

$\text{ST} = \sqrt{(150-80)^2 + (70-140)^2}$

$= 99.0$ km

(2 marks)

20 a

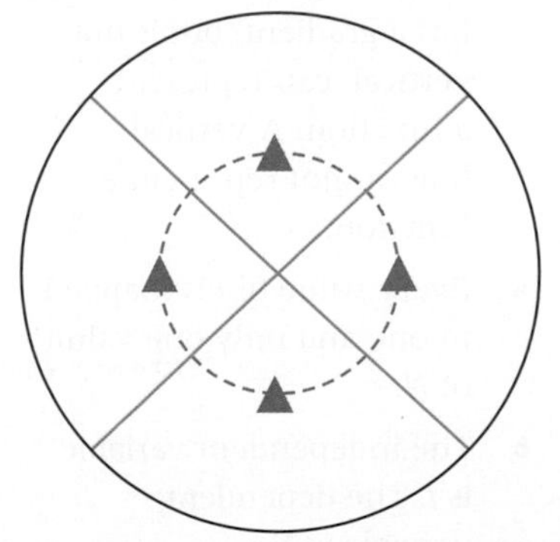

(2 marks)

b $\frac{\pi \times 10^2}{4} = 78.5(3\text{sf})\text{m}^2$

(2 marks)

c

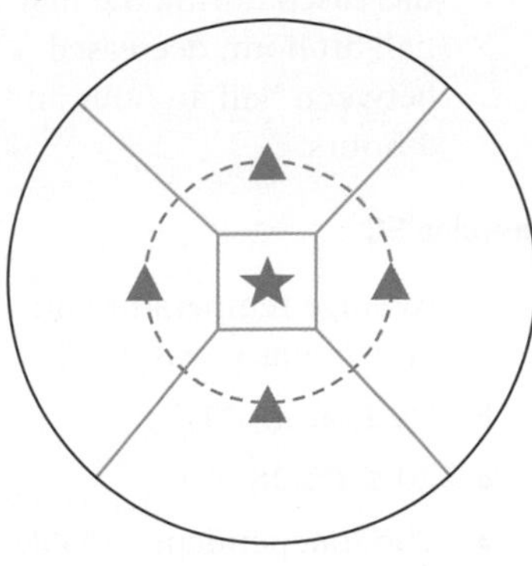

(4 marks)

d Square of side 5 has area of 25 m² (2 marks)

e $\frac{\pi \times 10^2}{4} - \frac{25}{4} = 72.3\text{m}^2$

(2 marks)

f 4 (1 mark)

Chapter 5

Skills check

1 a $x = \frac{10}{3}$ **b** $x \le 2$

2

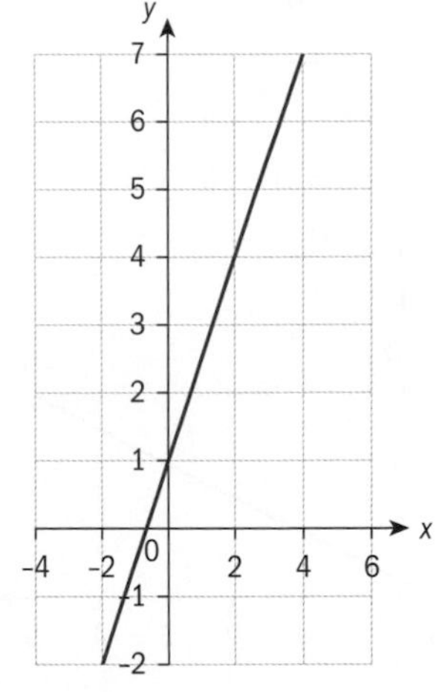

3 $y = -2x + 4$

Exercise 5A

1 a One-to-many

b One-to-one

c Many-to-one

d One-to-many

e Many-to-many

2 a R_1 is a function as every independent variable is mapped to one and only one dependent variable.

b R_2 is not a function as the input value 1 is mapped to two distinct output values (6 and 4).

c R_3 is a function as every input value is mapped to one and only one output value.

3 a $B = \{-1, 0, 0.5, 1, 2\}$

b $B = \{0, 2, 3, 4, 6\}$

c $B = \{1, 1.25, 2, 5\}$

4 a $x \Rightarrow \frac{1}{x}$

b R is a function as every input value is mapped to one and only one output value.

c $\frac{1}{2}$ **d** $\frac{1}{3}$

5 a $y = -1$

b If the output is −64 then the input is −4.

c $B = \{-1, 0, 1, 8, 27, 64\}$

d Function, since every input value is mapped to one and only one output

Exercise 5B

1 a {2, 3, 4, 5}

b $\{y: 2 \le y \le 5\}$ **c** $\mathbb{R}$

2 a i $y = 5$ **ii** $y = -3$

b $x = \frac{1}{2}$

c

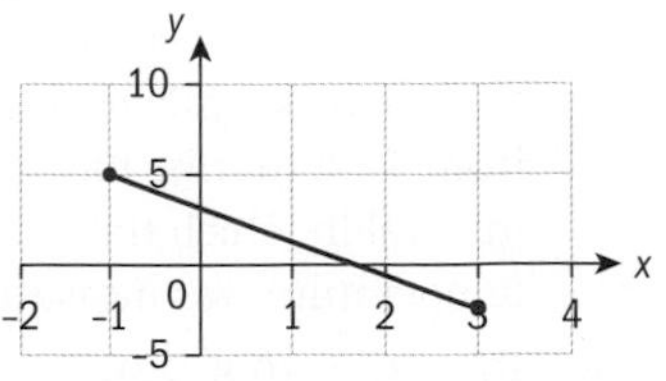

d $\{y: -3 \le y \le 5\}$

3 b and **d** are both functions as every input value is mapped to one and only once output. **a**, **c** and **e** all have at least one input which is mapped to at least two distinct output elements so do not correspond to functions.

4 **a** Domain = {−5, −4, −3, −2, −1, 0, 2, 4}, range = {−2, 0, 2, 4, 6, 8}

b Domain = $\{x: -8 \le x \le 6\}$, range is $\{y: -4 \le y \le 3\}$

c Domain = $\{x: -7 < x \le 9\}$, range is $\{y: 0 \le y < 4\}$

d Domain = $\{x: -7 < x < 7\}$, range is $\{y: -4 \le y < 3\}$

e Domain = $\{x: -2 \le x \le 1\}$, range is $\{y: -2 \le y \le 2\}$

5 **a** **i** True **ii** False **iii** True

b **i** True **ii** False **iii** True

c **i** True **ii** True **iii** False

d **i** True **ii** True **iii** False

e **i** False **ii** True **iii** True

6 **a** {−2, −1.5, −1, −0.5, 0, 0.5}

b {−6, −4, −2, 0, 2, 4}

c {5, 4, 3, 2, 1, 0}

7 **a** Yes, it is a function: at every moment there is only one value for V.

b Domain = $\{t: 0 \le t \le 2\}$, and represents the interval of time in which the temperature was measured.

Range = $\{V: 0 \le V \le 4\}$, and represents the variation of the person's temperature from 37 °C during the interval in which the temperature was measured.

c (0.2, 3.6), (0.8, 2.9), (1.5, 2), (2, 0)

d 40 °C **e** 2 hours

f Increased during the first half an hour, decreased between half an hour and 2 hours.

Exercise 5C

1 **a** Average temperature on 2nd January was 25 °C

b {1, 2, 3, …, 31}

c $20 \le T \le 28$

2 **a** The independent variable is x

b **i** 2 **ii** 12

c $f(2.5) = 10 - 4(2.5)$
$= 10 - 10 = 0$

d 4

3 **a** 80

b **i** (0, 100) **ii** (−50, 0) **iii** (−25, 50)

c

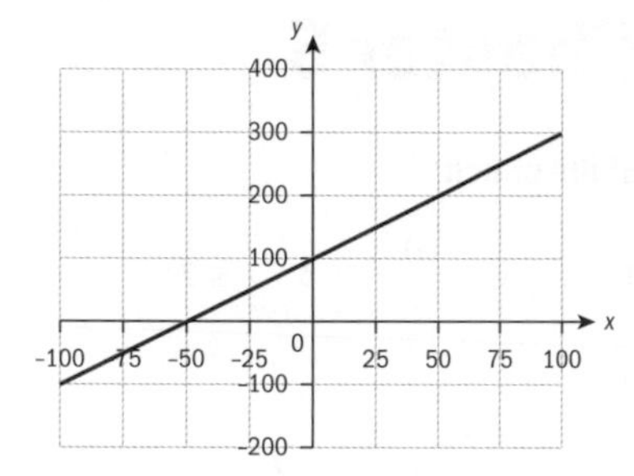

4 **a** **i** 5 **ii** −3

b **i** $x = 0$ **ii** $x = 3$

c $\{x > 3\}$

5 **a** **i**

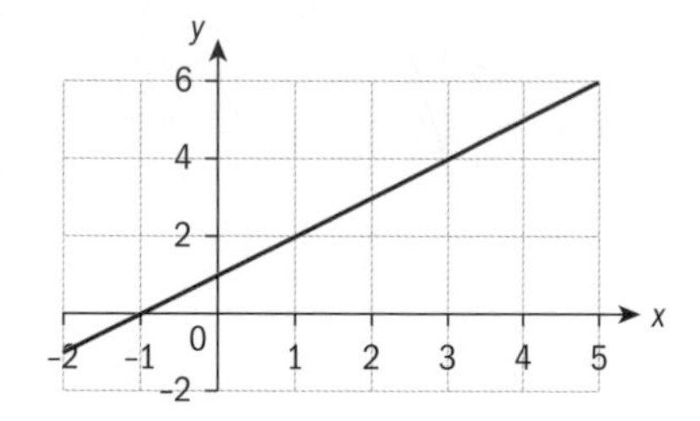

ii

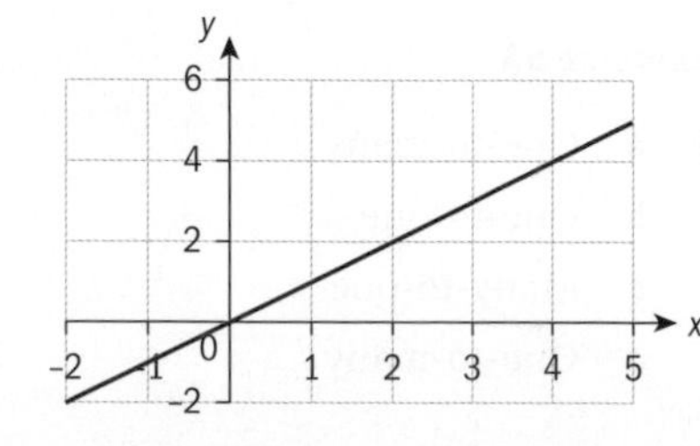

iii

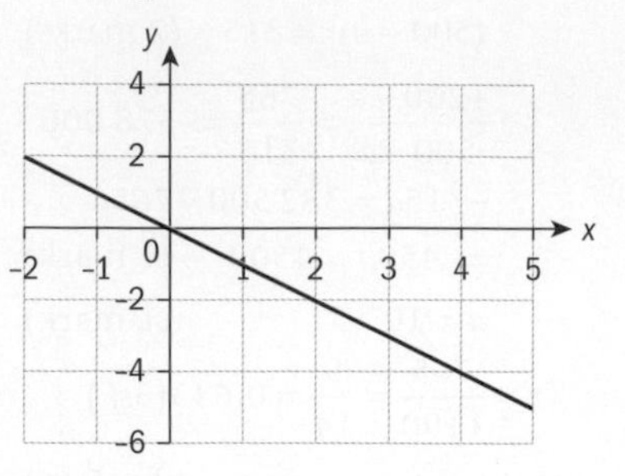

iv

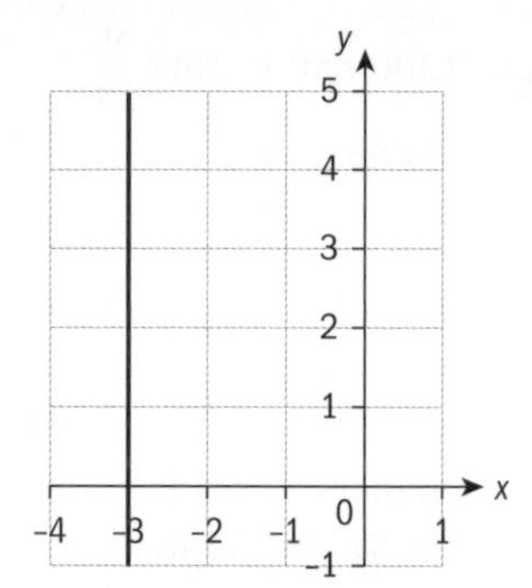

v

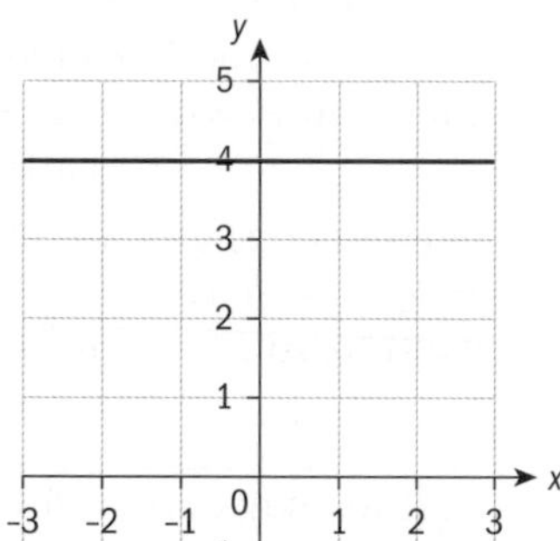

b Graphs **i**, **ii**, **iii** and **v** are functions

c Any straight line that has a gradient, but is not vertical, can represent a function. A vertical line cannot represent a function.

6 **a** Every value of t is mapped to one and only one value of N.

b The independent variable is t. The dependent variable is N.

c 4 **d** 8

e 12 **f** $16 \le t \le 20$

g $4 \le N \le 128$

7 **a** \$300 **b** \$510

c 225 km

8 **a** $v(0) = 50\,\mathrm{m\,s^{-1}}$. This is the initial speed.

b $v(2) = 34\,\mathrm{m\,s^{-1}}$

c $t = 4.3\,\mathrm{s}$

d The body comes to rest at 6.25 seconds.

Exercise 5D

1 **a** Linear **b** Linear

c Non-linear **d** Linear

2 **a** **i** Independent = time (hours); dependent = price (dollars)

ii Linear, rate of change = \$12.50/hour

b **i** Independent = time (years), dependent = population of fish.

ii Non-linear reaction, rate of change not constant

c **i** Independent = purchase price (Euros), dependent = VAT (Euros)

ii Linear, rate of change is 0.22 cents/Euro

d **i** Independent = temperature (°C), dependent = number of passes sold.

ii Linear, rate of change is −8 passes per degree.

3 **a** **i** Increasing

ii Increasing

iii Decreasing

iv Neither

b **i**

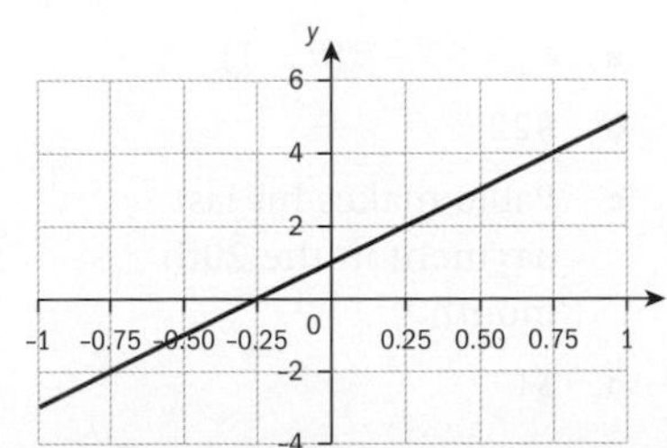

ii

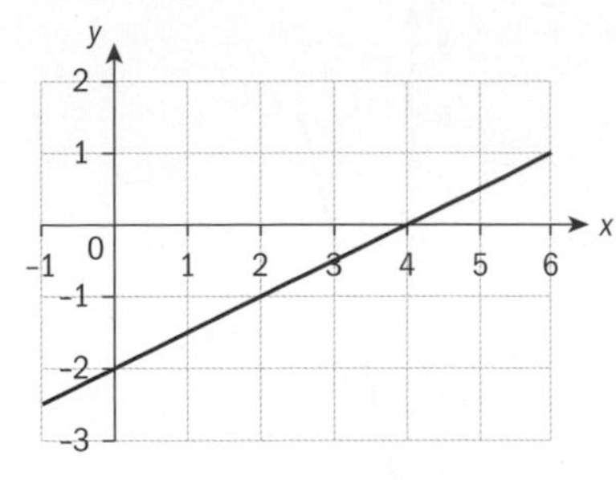

iii

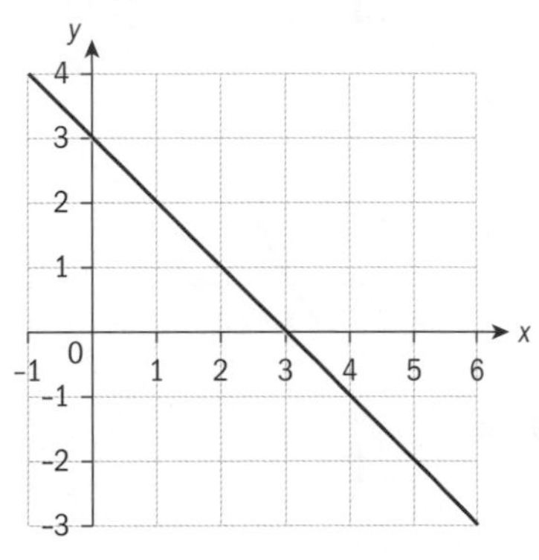

iv

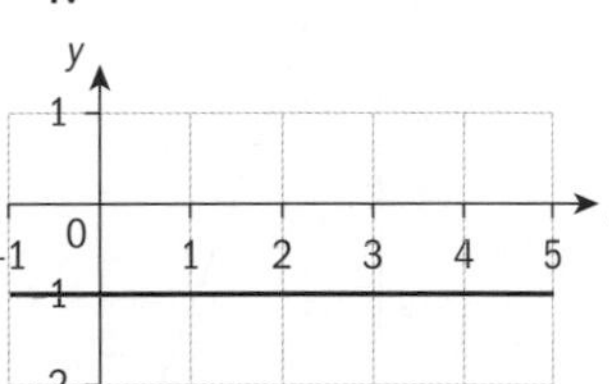

4 Non-linear

5 **a** 8 **b** 0.75 **c** 4.25

6 **a** $N(t) = 50t + 950$

b 950

7 **a** $f(x) = -x + 6$ **b** 9

8 **a** $120\,\mathrm{m\,s^{-1}}$

b 8 seconds

c $-15\,\mathrm{m\,s^{-2}}$

d $v(t) = -15t + 120$

Exercise 5E

1 **a** e.g. (9,12) **b** $\frac{4}{3}$

c $u(x) = \frac{4}{3}x$

d **i** £1 = \$1.33 **ii** £75

2 **a** 1.6

b $k(m) = 1.6m$

Exercise 5F

1 **a** $(-1, -2)$

b $\left(-\frac{1}{3}, \frac{7}{3}\right)$

2 **a**

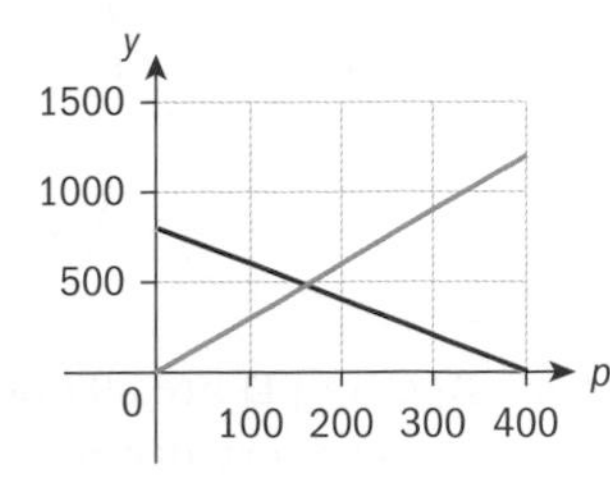

b (160, 480)

c Demand is twice the supply.

d When the price is \$250, supply is 2.5 times demand

3 **a** **i** \$500 **ii** \$380

b Option A: $A(x) = 5x$, option B: $B(x) = 2.3x + 150$

c $\frac{500}{9}$ km

4 **a** 2.4

b $(-1.25, 0)$

Exercise 5G

1 **a** All three functions are one-to-one.

b The inverse to function **b** is $k = \frac{p}{2.2}$, the inverse to function **c** is $R = \frac{100}{85}P$, The inverse to function **d** is $W = 0.9V$.

2 **a** No inverse, not one-to-one

b Inverse, one-to-one

c Inverse, one-to-one

d No inverse, not one-to-one

3 **a** f^{-1} exists as f is one to one.

b **i** 1 **ii** 5

iii 3 **iv** −3

c

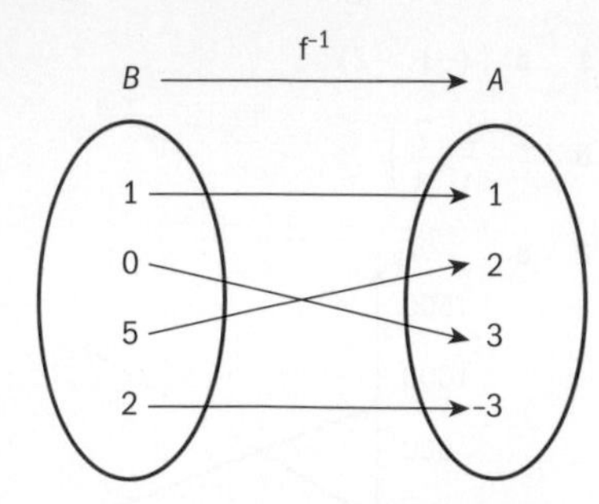

4 a **B, C** and **D** are all one-to-one as every distinct value of x is mapped to a distinct value of y. **A** is not one-to-one as every value of x is mapped to the same value of y.

b A linear function with a non-zero gradient will have an inverse, while any linear function with a gradient of zero will not have an inverse.

5 a $A = \{x: -2 \leq x \leq 1\}$

b $B = \{y: -1 \leq y \leq 2\}$

c f is a one-to-one function as it has a positive gradient.

d i 2 ii 0

iii 0 iv −2

6 a True b False

Exercise 5H

1 $\frac{4}{\pi}$. This is the radius of a circle with a circumference of 8.

a

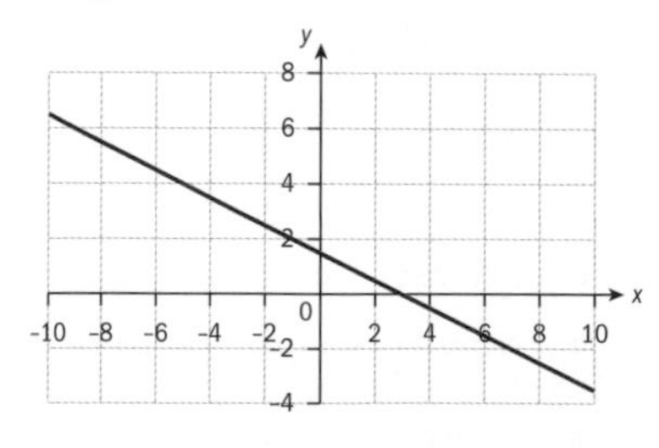

b

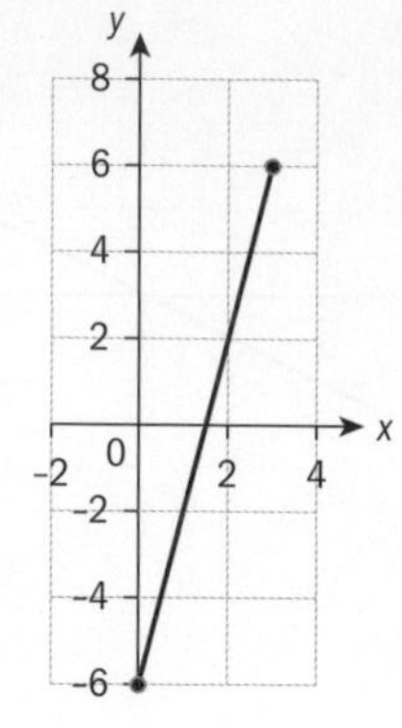

c

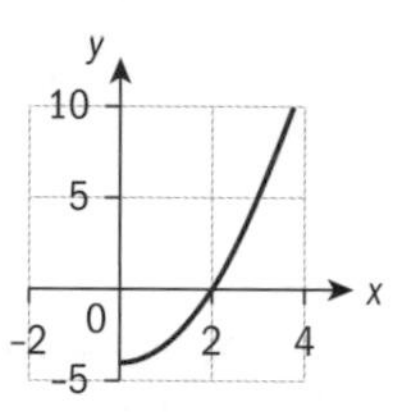

3 a

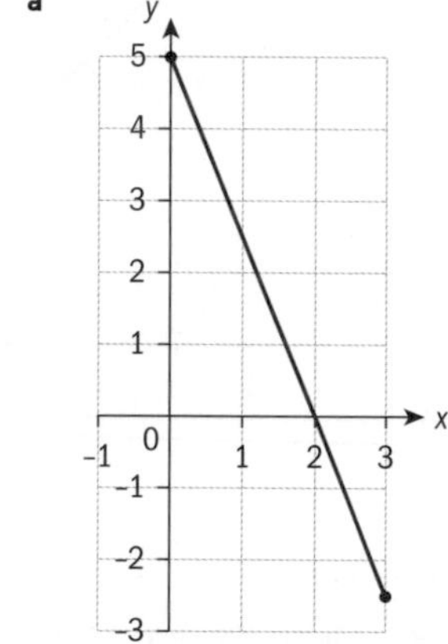

b $b = -2.5$

c

Function	Domain	Range
f	$0 \leq x \leq 3$	$-2.5 \leq y \leq 5$
f^{-1}	$-2.5 \leq x \leq 5$	$0 \leq y \leq 3$

d

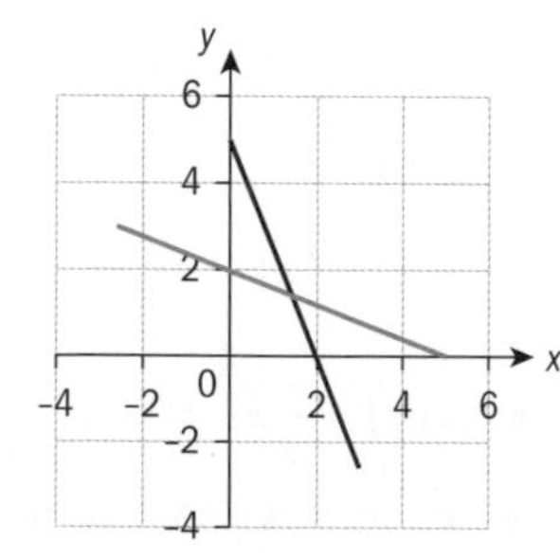

e $\left(\frac{10}{7}, \frac{10}{7}\right)$

4 a Domain is $\{x: -1 \leq x \leq 2\}$, range is $\{y: y \geq 0\}$

b 1 c Decreasing

d $x = 2$

Exercise 5I

1 a 23, 28, 33 b 13, 21, 34

c 17, 23, 30 d 16, 32, 64

e −1, 1, −1 f $\frac{5}{6}, \frac{6}{7}, \frac{7}{8}$

g 0, −25, −50

2 a 2, 3, 4 b 4, 7, 10

c 2, 4, 8 d 3.5, 3, 2.5

3 a $u_n = n^2$ b $u_n = n^3$

c $u_n = n$ d $u_n = \frac{2}{n}$

4 a 23 b 52 c 28

d 400 is not a term in the sequence.

Exercise 5J

1 a 1.5 b 11

2 a $b_1 = 3, b_2 = 10, b_3 = 21$

b Difference between consecutive terms is not the same, so it is not arithmetic

3 a $u_n = 5 + 4(n - 1)$

b Not a term as 111 is not a multiple of 4.

4 a $12 = u_1 + 2d$. $40 = u_1 + 9d$

b $u_1 = 4, d = 4$

c $u_{100} = 4 + 99 \times 4 = 400$

5 a Second year: 95, third year: 105

b 175

c The company will have 285 employees after 20 years.

6 a $u_n = 55 - 3(n - 1)$

b \$22

c Pablo makes his last payment in the 20th month.

d \$1

7 a $a_n = 2.6 + 1.22(n - 1)$

b 35.54 m **c** 2065

8 a −25 **b** 775 **c** 50

9 a The difference between successive terms is constant.

b 2 **c** −2

d $u_n = -2 + 2(n - 1)$

e (20, 36) lies on this graph.

10 a Pattern 5 requires 16 sticks and pattern 6 requires 19 sticks.

b 61 **c** Pattern 42

11 31

Exercise 5K

1 323

2 −450

3 3720

4 a 42 **b** 10 794

5 a $S_n = \frac{n}{2}(2 + 7(n - 1))$

b 40

6 a $S_n = \frac{n}{2}(8 - 3(n - 1))$

b −95 **c** 15

7 a 7.5 km **b** 97.5 km

8 a 4.5 **b** $a = 9.5$. $b = 18.5$

c 252.5 **d** 15

9 a $2k + 1 + d = -k + 10$.
$-k + 10 + d = k - 1$
$3k - 9 = -2k + 11$
$5k = 20$ so $k = 4$

b 9, 6, 3 **c** −3

d −390

10 a 10, 13, 16 **b** $x - 2$

c 18

Exercise 5L

1 a \$1350 **b** \$1275

c \$1554.58

2 \$17 142.86 **3** 4.6%

4 16 years **5** False

Exercise 5M

1 a

x	9	5	2	0.5	0.1
y	1	5	8	9.5	9.9

b $y(x) = 10 - x$

c i $\{x: 0 < x < 10\}$

ii $\{y: 0 < y < 10\}$

d

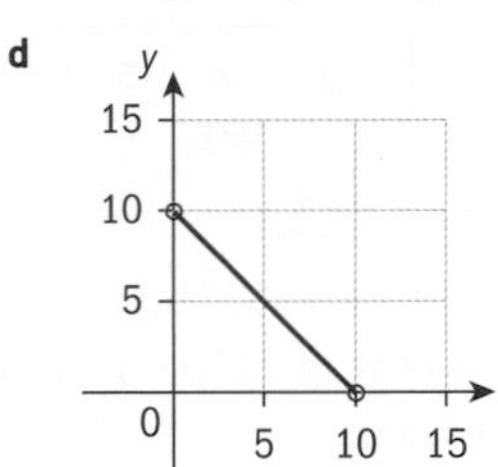

e Rate of change = −1: for every centimetre base increases, height decreases by the same amount

2 a

Width (x m)	5	8	20	34
Length (y m)	60	54	30	2

b

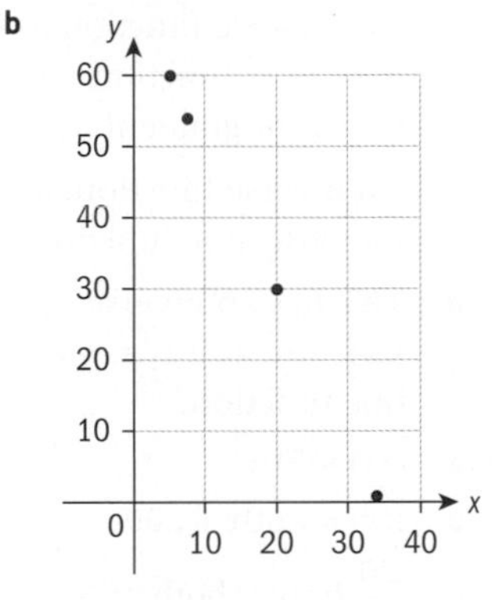

c Linear, $y = 70 - 2x$

d It is not reasonable as a width of 36 m would require more than 72 m fence.

e Domain = $\{x: 0 < x < 35\}$, range = $\{y: 0 < y < 70\}$

3 a

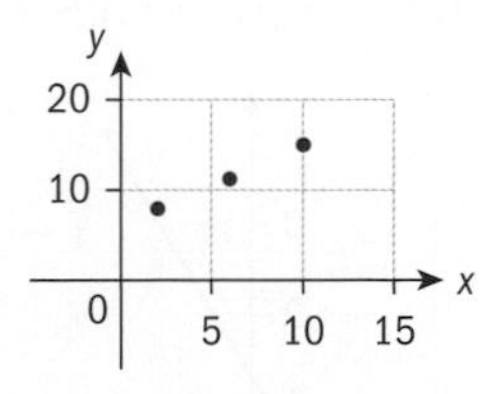

b Linear model because the three points seem to lie on a straight line

c $S = 0.888x + 6.14$

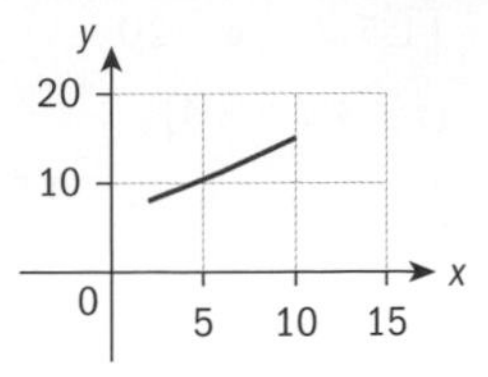

d Model fits data well

e i \$12.4 million

ii \$19.5 million

Chapter review

1 a $R = \{(-3, -1), (-2, -2), (-1, 1), (0, 2), (1, 2), (2, 0)\}$

b Every value of x is mapped to one and only one value of y.

c $\{-3, -2, -1, 0, 1, 2\}$

d $\{-2, -1, 0, 1, 2\}$

2 a Not a function: fails vertical line test

b Function: every value of x is mapped to one and only one value of y

c Function: every value of x is mapped to one and only one value of y

d Not a function: fails vertical line test

3 a i One-to-one

ii $\{x: -5 \leq x \leq 5\}$

iii $\{y: -5.5 \leq y \leq 9.5\}$

b i Many-to-one

ii $\{x: x \geq -4\}$

iii $\{y: -8 \leq y \leq 8\}$

c i Many-to-one

ii $\{x: -3 \leq x \leq 4\}$

iii $\{y: -2 \leq y \leq 3\}$

d i Many-to-one

ii $\{x: -2 \leq x \leq 1\}$

iii $\{y: -2 \leq y \leq 2\}$

4 **a** $b = 13, m = -0.5$

b 11.5 **c** 20

5 **a** $\{x: -3 \le x \le 4\}$

b 11 **c** $x = 0.5$

d

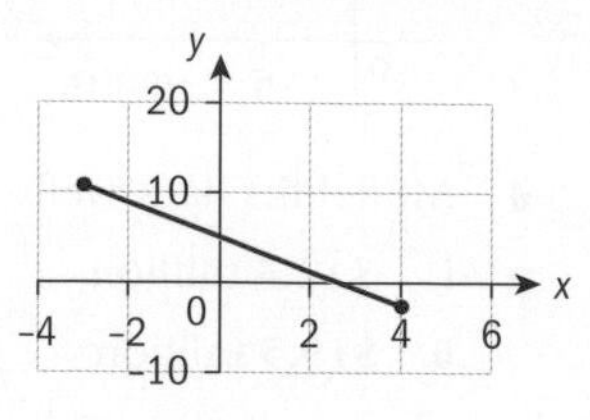

e $\{y: -3 \le y \le 11\}$

6 $\{5, 4, 0, 2.5, 0\}$

7 **a** $\mathbb{R}$

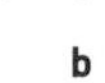

b

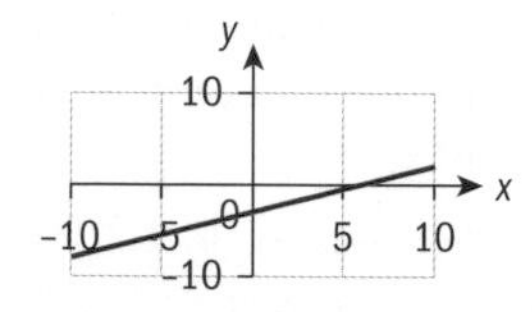

c $\mathbb{R}$

d **i** A is on the graph.

ii B is not on the graph.

8 **a** **i** -1 **ii** 1 **iii** 1

b **i** 4 and -4 **ii** 0

iii -7

c -8 **d** $\{x: -5 < x < 5\}$

9 **a** $T(x) = 150 - 2.5x$

b $T(6) = 135$. This is the amount on the card after the card has been used 6 times.

c 18 times

d **i** $\{0, 1, 2, 3, \ldots, 60\}$

ii $\{150, 147.5, 145, \ldots 0\}$

10 **a** $d(t) = -40t + 480$

b 480 m **c** 12 minutes

d

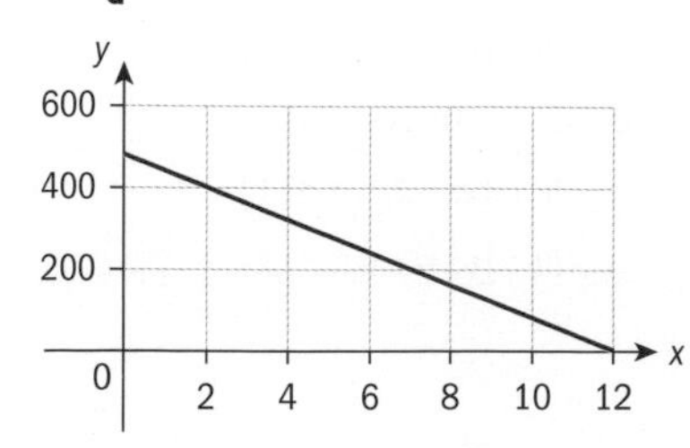

11 **a** $f(x) = 0.25x - 0.5$

b 0.75

12 **a** AUD 121.10

b

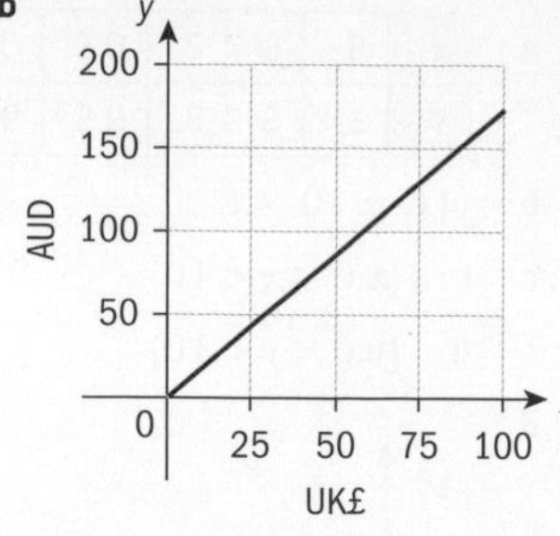

c $a(p) = 1.73p$

d $a(100) = 173$. This is the amount of AUD equivalent to £100.

e $a^{-1}(50) = 28.90\ldots$ This is the amount of UKP equivalent to AUD 50.

13 $\left(\frac{2}{3}, \frac{13}{3}\right)$

14 **a** No inverse function as a many-to-one function.

b Has inverse function as it is a linear function with a non-zero gradient.

c No inverse function; a many-to-one function.

d This has an inverse function as it is a one-to-one function.

15 **a** $c(x) = 70x$

b $t(x) = -60x + 200$

c $\frac{20}{13}$ hours, both $\frac{1400}{13}$ km from Sun City.

16 **a** **i** 8 **ii** $\frac{8}{3}$

b **i** -1 **ii** -1

c Domain is $\{x: -7 \le x \le 8\}$, range is $\{y: -1 \le y \le 4\}$

d

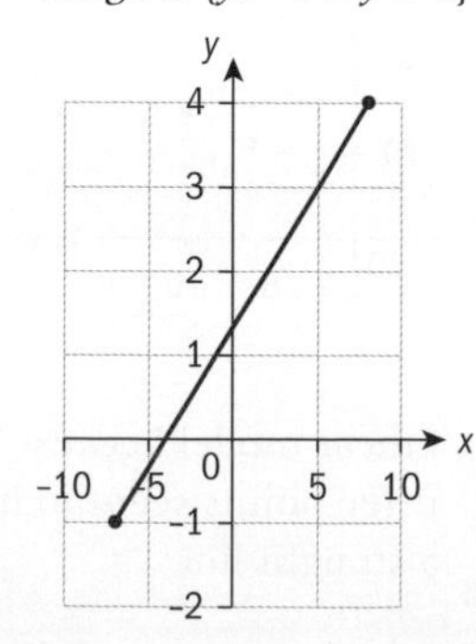

17 **a** $u_2 = 7.5, u_3 = 10$

b $u_n = 5 + 2.5(n - 1)$

c $n = 150$

18 **a** 45 **b** 1296

19 **a** $S_n = \frac{n}{2}(2 \times 105 + (n - 1) \times (-5))$

b $n = 19$ or 24

20 **a** 15

b $251 = 3 + 2(n - 1)$

$248 = 2(n - 1)$

$n - 1 = 124$. $n = 125$

c $S_{100} = 10\,200$

21 **a** The difference between successive terms is $a_{n+1} - a_n = (3 - 10n + 10) - (3 - 10n) = 10$. This is constant, so the sequence is arithmetic.

b 101

22 **a** £324 **b** 6

23 **a**

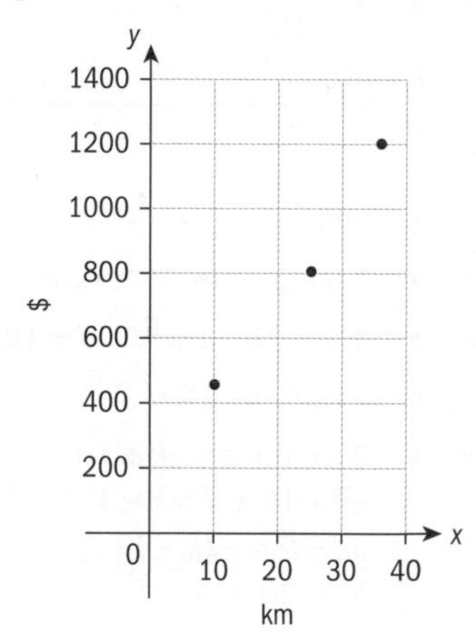

b Linear model

c $P = 28.6x + 144$

The model fits the data well.

d **i** \$915

ii P\$1290

e 27 km is more appropriate as within the given data range, while 40 km is less appropriate as outside the data range.

Exam-style questions

24 a $-24 \le f(x) \le 26$ (2 marks)

b $f(x) = \{-4, -2, 0, 2, 4, 6\}$ (2 marks)

c $0 \le f(x) \le 100$ (2 marks)

d $125 \le f(x) \le 250$ (2 marks)

25 a

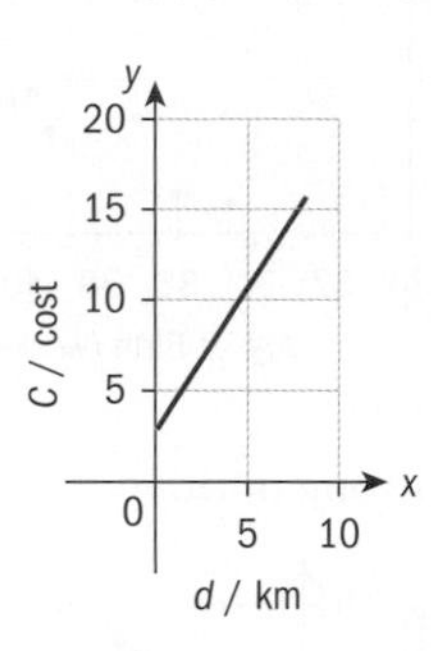

(3 marks)

b the fixed charge for a single journey (1 mark)

c the extra cost per km of travel (1 mark)

d $2.8 \le C \le 15.6$ (2 marks)

e Either by reading from graph, or by solving $10 = 2.8 + 1.6d$, gives $d = 4.5$ km (2 marks)

26 a $u_1 = 23 + 7 = 30$ (1 mark)

b $u_{50} = 23 + 7 \times 50 = 373$ (2 marks)

c Solving $23 + 7n = 1007$ gives $n = 140.6$ (2 marks)

n is not an integer, therefore 1007 is not a term in this sequence. (1 mark)

27 a $C = 430 + 14.5P$ (2 marks)

b $C = 430 + 14.5 \times 25$ (1 mark)

$= \$792.50$ (1 mark)

c $1000 = 430 + 14.5P$ (1 mark)

$P = \dfrac{1000 - 430}{14.5} = 39.3$ (1 mark)

She can therefore invite a maximum of 39 people (1 mark)

d $C = 430 + 14.5 \times 16 = \$662$ (1 mark)

$\dfrac{662}{16} = 41.375$ (1 mark)

Denise will therefore need to charge a minimum of \$41.38 per person (1 mark)

28 a Solving $3x - 10 = 5$ gives $x = 5$ (2 marks)

Solving $3x - 10 = 50$ gives $x = 20$ (1 mark)

Domain is $5 < x < 20$ (1 mark)

b Range is $5 < f^{-1}(x) < 20$ (2 marks)

29 a $P = 1200T - 850$ (2 marks)

b $5000 = 1200T - 850$ (1 mark)

$T = 4.875$ (1 mark)

So a total of 5 months (1 mark)

30 a

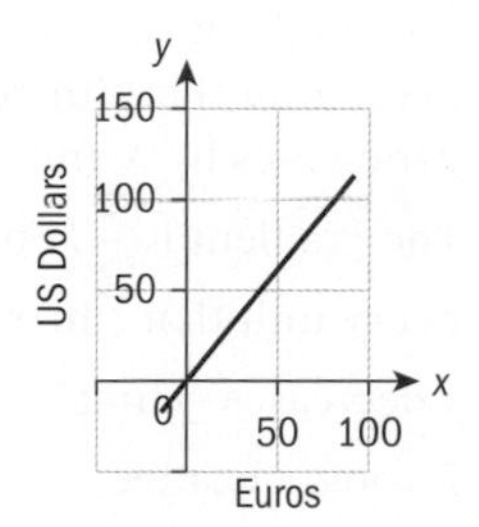

(3 marks)

b By reading correctly from the graph, approximately €60 (2 marks)

c By reading correctly from the graph, approximately \$37 (2 marks)

31 a NOT a function, since eg the value of $x = 5$ is related to more than one co-ordinate on the y-axis. (2 marks)

b This is a function. Each value of x is related to only one value for y. (2 marks)

c This is a function. Each value of x is related to only one value for y. (2 marks)

d This is a function. Each value of x is related to only one value for y. (2 marks)

32 Value of Ben's car is $V = 25\,600 - 1150t$ (2 marks)

Value of Charlotte's car is $V = 18\,000 - 480t$ (2 marks)

Either: Plot respective graphs and read off intersection point, or: solve the equation $25\,600 - 1150t = 18\,000 - 480t$ (1 mark)

$t = 11.3$ years (1 mark)

At this time, both cars are worth £12 555 (1 mark)

33 a $f(x) = 128\left(\dfrac{3}{2}\right) - 15 = 177$ (2 marks)

b $f(-3) = 128(-3) - 15$
$= -399$ (2 marks)

$f(15) = 128(15) - 15$
$= 1905$ (1 mark)

Range is $-399 < f(x) < 1905$ (1 mark)

c Solving $128a - 15 = 1162.6$ (1 mark)

$a = 9.2$ (1 mark)

34 a Using GDC, £1200 (2 marks)

b Using GDC (1 mark)

$a = 1200$ (1 mark)

$b = 75$ (1 mark)

c $V = 1200 + 75 \times 50$
$= £4950$ (2 marks)

d This is probably not a realistic model, (1 mark)

as it indicates the value of the painting will increase indefinitely. (1 mark)

35 a Domain is $-3 \le x \le 3$ (1 mark)

Range is $-1 \le f(x) \le 1$ (1 mark)

b Domain is $-1.5 \le x \le 5$ (1 mark)

Range is $-5 \le f(x) \le 4$ (1 mark)

c Domain is $0 \le x \le 24$ (1 mark)

Range is $0 \le f(x) \le 12$ (1 mark)

d Domain is $-3 \le x \le 3$ (1 mark)

Range is $0 \le f(x) \le 9$ (1 mark)

36 a $115 = 20a + b$;
$137.5 = 25a + b$ (1 mark)

$a = 4.5$ (1 mark)

$b = 25$ (1 mark)

b $4.5 \times 23 + 25 = 128.5$ mg (2 marks)

c The relationship may no longer be linear outside the range of children's weights. (1 mark)

Using the relationship involves extrapolation to adults' weights, which is not mathematically sound. (1 mark)

37 a Substituting $C = 0$ gives 32°F (1 mark)

b Solving $C = \frac{9C}{5} + 32$ (2 marks)

$-\frac{4C}{5} = 32$

$C = -40$ (1 mark)

c Attempting to make C the subject of $F = \frac{9C}{5} + 32$ (1 mark)

$F - 32 = \frac{9C}{5}$ (1 mark)

$C = \frac{5(F-32)}{9}$ (1 mark)

d When $F = 46.4$,
$C = \frac{5(46.4-32)}{9} = 8$ (2 marks)

When $F = 84.2$, $C = 29$ (1 mark)

The range is therefore 8°C to 29°C (1 mark)

38 a Solving $30 + 12.5d = 70 + 8.35d$ (1 mark)

$d = 9.64$ (1 mark)

So Abel's holiday lasts a minimum of 10 days (1 mark)

b Using $C = 70 + 8.35d$ (1 mark)

$70 + 8.35 \times 14 = £186.90$ (1 mark)

$70 + 8.35 \times 21 = £245.35$ (1 mark)

So $£186.90 \le C \le £245.35$ (1 mark)

Chapter 6

Skills check

1 a The gradient is 3. For every unit that x increases, y increases by 3 units.

b The gradient is $\frac{1}{2}$. For every unit that x increases, y decreases $\frac{1}{2}$ unit.

2 a Strong, negative

b No correlation

c Weak, positive

Exercise 6A

1 a Positive, nonlinear

b Positive, linear

c Positive, linear

d Positive, linear

e No correlation

f Positive, non-linear

2 a Zero

b Positive perfect linear

c Strong negative linear

d Weak negative linear

e Weak positive linear

f Perfect negative linear

g Moderate negative linear

Exercise 6B

1 a No **b** Yes **c** No

2 a

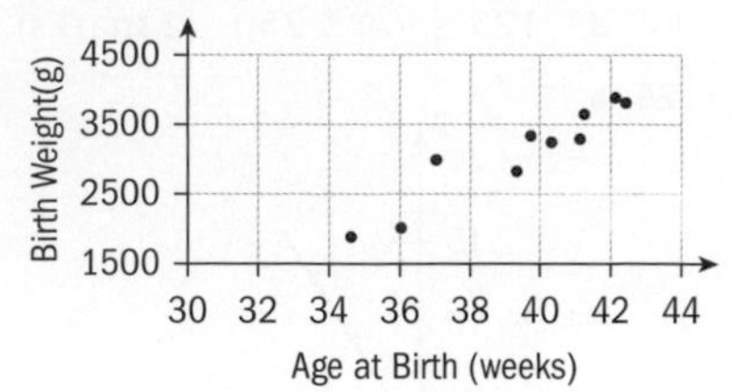

b 0.945

c Strong positive

3 a

b 0.987 **c** Strong positive

4 a

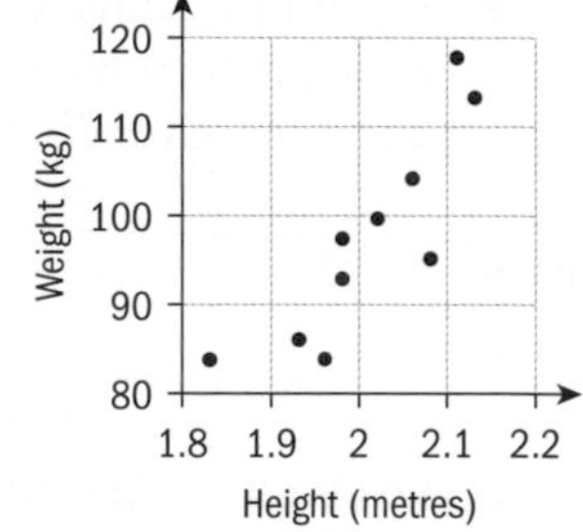

b 0.863 **c** Strong positive

Exercise 6C

1 a 0.9 **b** 1 **c** −0.6

2 a −0.9 **b** 0 **c** 0.7

3 a −1 **b** −0.3 **c** 0.5

4 a 0.883 **b** Strong positive

c The more hours spent studying, the better the grade achieved.

5 a 0.987 **b** Strong positive

Exercise 6D

1 a

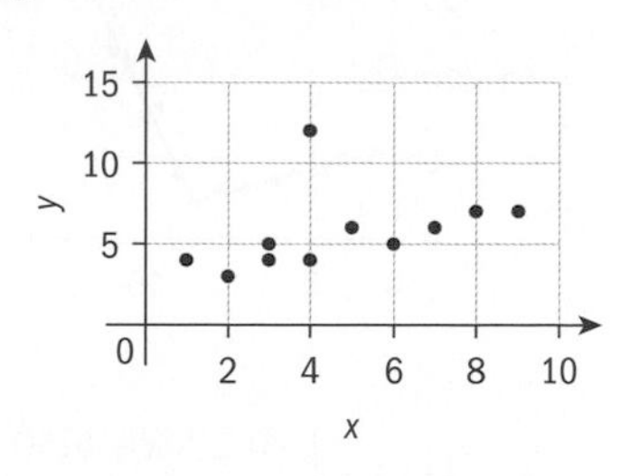

b 0.389

c Outlier (4,12) does not follow the trend.

d 0.891

2 a

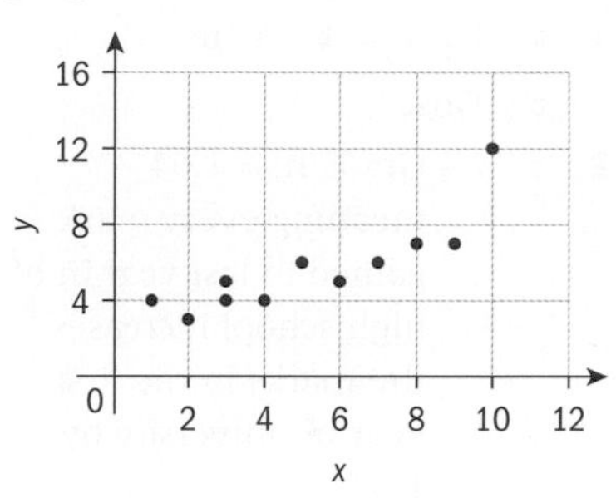

b 0.849

c Outlier (10, 12) follows trend

d 0.891

3 a

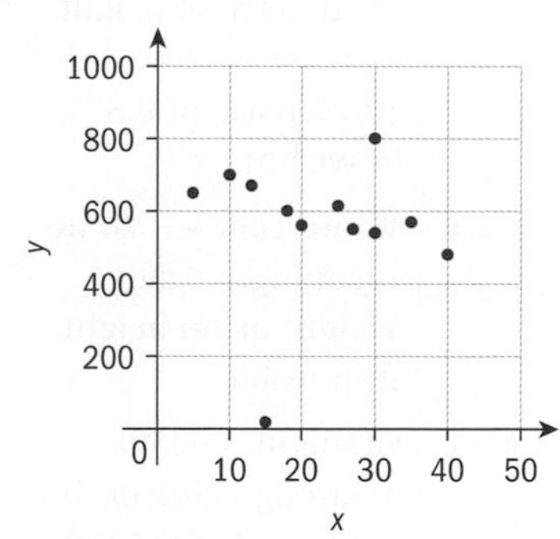

b −0.001

c Two outliers: (15, 20) and (30, 800)

d −0.865

Exercise 6E

1 a **i** and **iv**

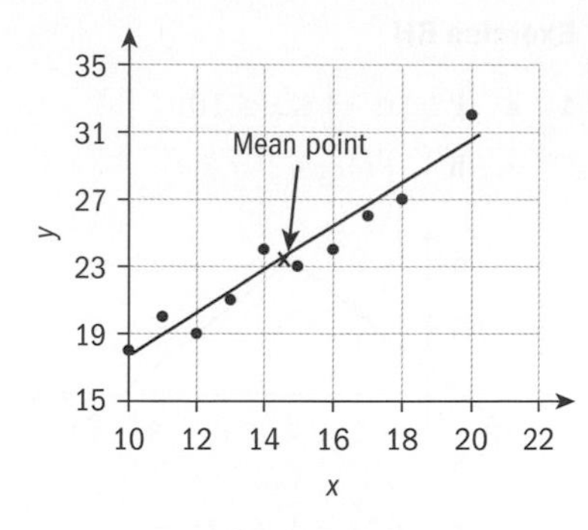

ii Positive strong correlation

iii x: 14.6, y: 23.4

b **i** and **iv**

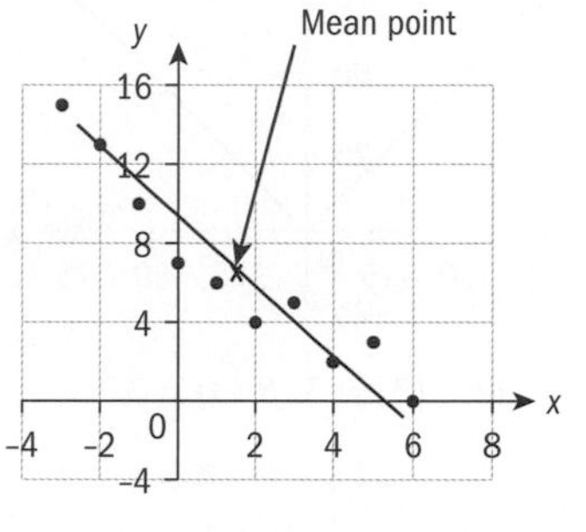

ii Strong negative correlation

iii x: 1.5, y: 6.5

2 a, e, f

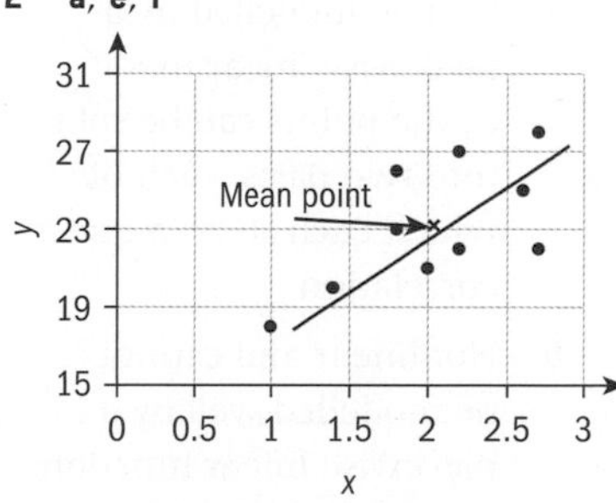

b Moderate positive correlation

c 2.04 million square feet

d 23.2 million people

3 a, e, f

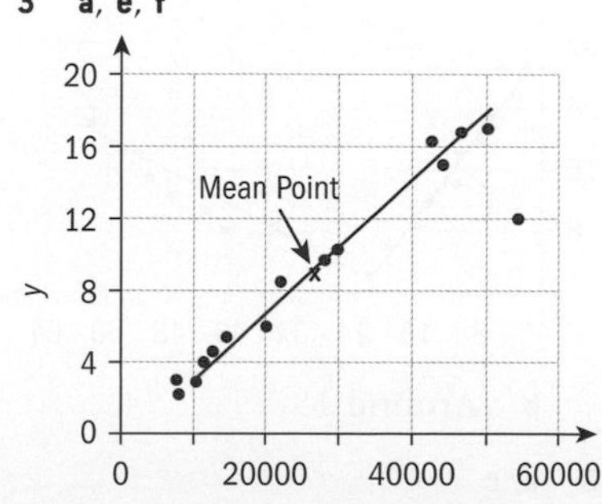

b Strong positive linear

c 26 845 **d** 8.9

Exercise 6F

1 a 0.768. Strong positive

b $y = 0.161x + 9.84$

c 117.5 min **d** 28.74 Euros

e

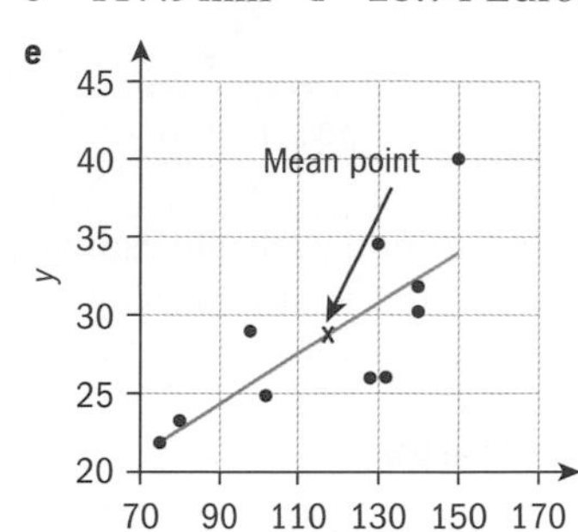

2 a $r = 0.824$, Strong positive

b $y = 0.0706x + 1.19$

c 13.8 **d** 2.16 min

e

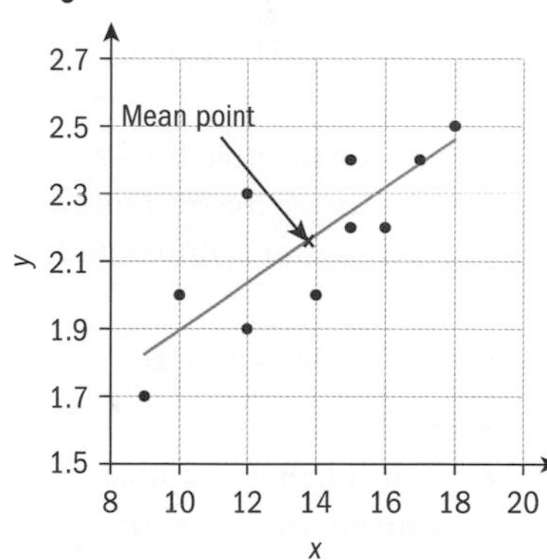

3 a $r = 0.930$, strong positive linear

b $y = 93.7x + 92.8$

c 1.853 m **d** 266.4 kg

e

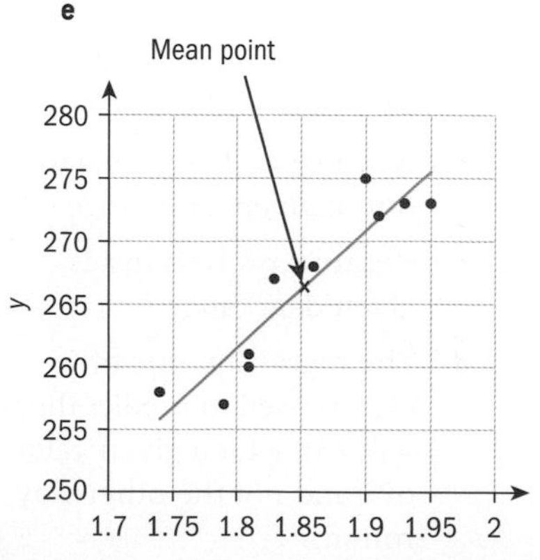

4 a -0.971, strong negative

b $y = -0.168x + 6.18$

c (15, 3.66)

d

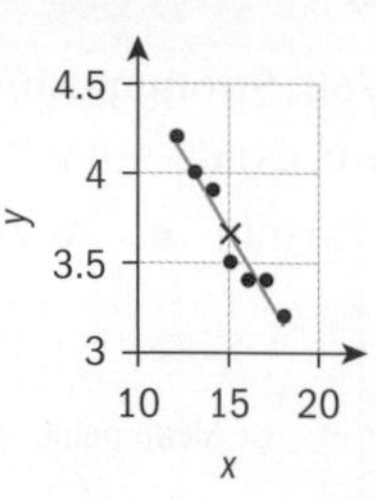

e 4; not a reliable estimate as 11 is outside range of given data

Exercise 6G

1 a

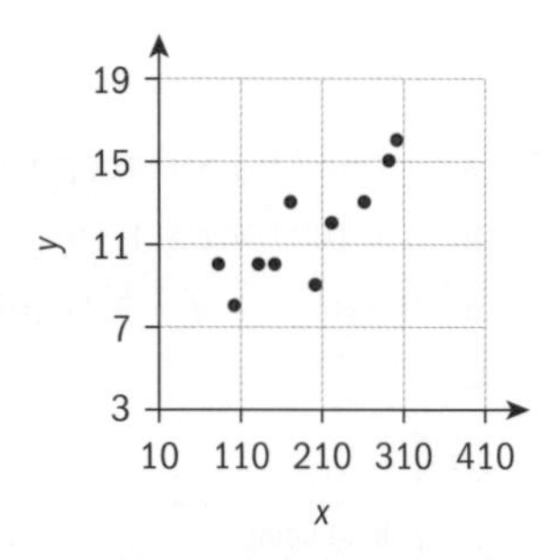

b Strong positive linear; can use regression line as correlation is linear and strong

c $y = 0.0286x + 6.16$

d Around 14

e Not reliable as 400 is outside the given data range.

2 a $-0,772$

b Strong negative

c $y = -1.30x + 17.3$

d 9.5 min

3 a Because correlation is moderate and linear

b Because $x = 0$ is quite far out of given date range

c Because $x = 10$ is inside given data range

d The regression line of y on x is used to predict the value of y for a given value of x and not the other way round.

Exercise 6H

1 a i $\{x: -3 \leq x \leq 10\}$

ii

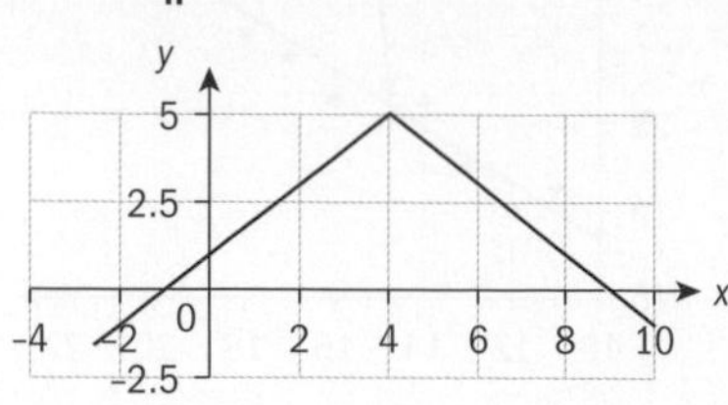

iii $f(3) = 4, f(-3) = -2$

b i $\mathbb{R}$

ii

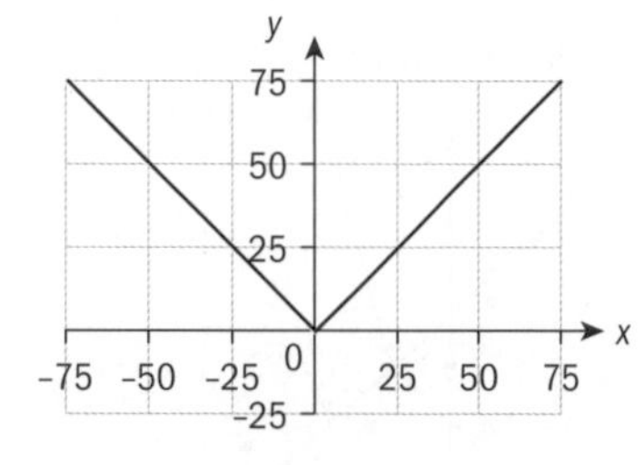

iii $f(3) = 3, f(-3) = 3$

2 a $f(x) = \begin{cases} x+5, -4 \leq x < 0 \\ 5, 0 \leq x \leq 3 \end{cases}$

b $f(x) = \begin{cases} 3-x, -2 \leq x < 2 \\ x-1, 2 \leq x \leq 4 \end{cases}$

3 a Can be modelled by a piecewise linear model as the points can be split into two parts, each of which each show a linear correlation.

b Not linear and cannot be modelled well by a piecewise linear function

c Can be modelled by a single linear model, does not need a piecewise linear model

4 a

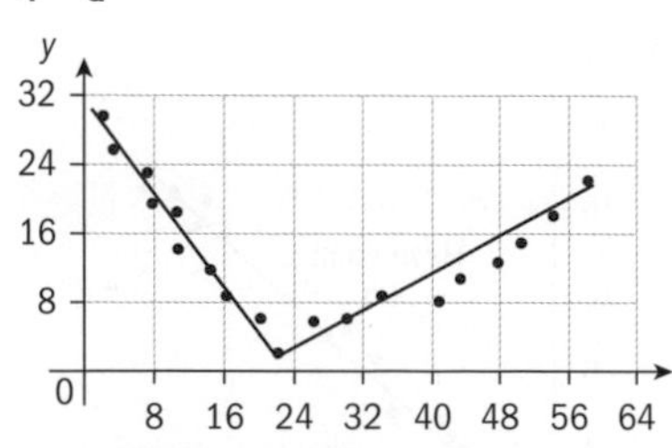

b Around 15

5 a, c

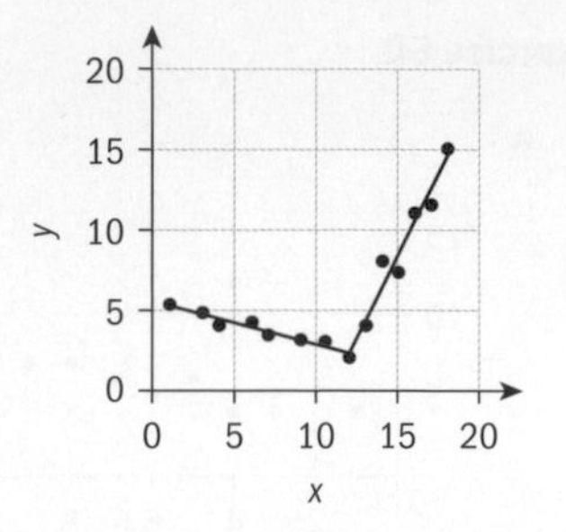

b $f(x) = \begin{cases} -0.270x + 5.50 & 0 \leq x \leq 12 \\ 2.04x - 22.1 & 12 < x \leq 20 \end{cases}$

d i $f(8) = 3.34$

ii $f(15.5) = 9.52$

Exercise 6I

1 a False b True

c False

2 a i Gradient = 1.04, meaning every mark gained in last year in of high school increases the marks in the first year of university by 1.04

ii y-intercept = -2.50; no meaning, not possible to get negative marks

b i Gradient = 0.87; meaning each centimetre of height increase corresponds to an increase of 0.87 kg in weight

ii y-intercept = -70; no meaning, negative weight or no height impossible

c i Gradient = -250; meaning value of the car goes down \$250 each year

ii y-intercept = 9000; meaning when bought the car is worth \$9000

3 a 0.955 b Strong positive

c i 0.0871 g: for every gram weight goes up, length goes up 0.0871 g

ii 18.1; not relevant as outside the data range

Chapter review

1 **a** I **b** V **c** III **d** II

2 **a** 0 **b** 0.86 **c** −1
d 1 **e** −0.99 **f** −0.9

3 **a, b**

c 62 cm

d Reliable as 6.5 months is inside the data range

4 **a** −0.790

b Negative strong

c $y = -0.0652x + 5.11$

d Not reliable, as 40 is outside the data range

5 **a** **i** 130 **ii** 90

b

c $\{y: 50 \le y \le 150\}$

6 **a** −0.899

b Strong negative

c **i** $m = -6824.11$

ii For every position lower in the league the average attendance goes down by 6824

d 74 321 **e** 40 200

f 15 is outside given data range.

7 **a** $r = 0.984$

b Strong positive linear

c Because correlation between the two variables is strong and linear

d $y = 0.849x + 6.28$, e.g. $25 < x < 85$

e 53%

f Leave out, it is an outlier

Exam-style questions

8 **i** Perfect positive (1 mark)

ii Strong negative (1 mark)

iii Weak positive (1 mark)

iv Weak negative (1 mark)

v Zero (1 mark)

9 **a**

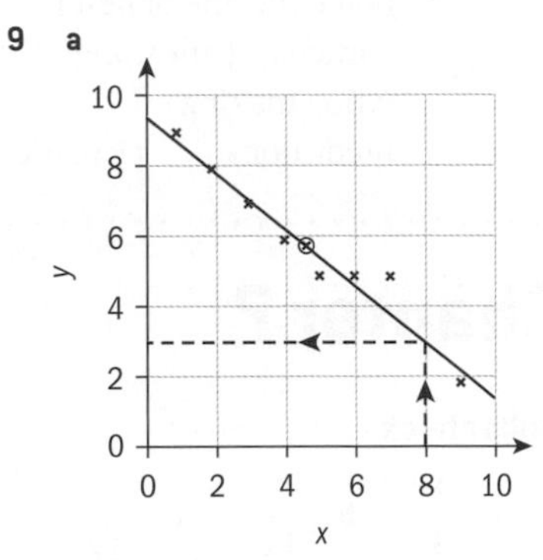

A1 scales A3 points (A2 6 points, A1 3 points)

b strong, negative (2 marks)

c **i** $\bar{x} = 4.625$ (2 marks)

ii $\bar{x} = 5.875$ (2 marks)

iii See above. (1 mark)

d See above (line passes through $(\bar{x}, \bar{y})$ gains 1 mark; correct line gains (2 marks)

e 3.2 (see above for lines drawn on) (2 marks)

10 **a** $100 = 70m + c$

$140 = 100m + c$

$40 = 30m \quad m = \frac{4}{3} \quad c = \frac{20}{3}$

(3 marks)

b Positive (1 mark)

c Line goes through $(\bar{x}, \bar{y})$ (1 mark)

$\bar{y} = \frac{4}{3} \times 90 + \frac{20}{3} = \frac{380}{3}$

(2 marks)

d Estimate is

$\frac{4}{3} \times 60 + \frac{20}{3} = \frac{260}{3}$

(2 marks)

11 **a** 40°C (1 mark)

b 70°C (1 mark)

c 100°C (1 mark)

d **i**

(2 marks)

ii $T \ge 80$

$40 + 2t = 80 = t = 20$

(1 mark)

$130 - t = 80 \Rightarrow t = 50$

(1 mark)

Interval is $20 \le t \le 50$.

(2 marks)

12 **a**

x	13	14	15	16	16	17	18	18	19	19
y	2	0	3	1	4	1	1	2	1	2

(3 correct: 1 mark; 5 correct: 2 marks; all correct: 3 marks)

b $r = -0.0695$ (3 s.f.) (2 marks)

c Very weak (negative) correlation so line of best fit is almost meaningless. (1 mark)

It would be extrapolation to use this data to predict for a 25-year-old. (1 mark)

13 i Gradient $m = \frac{0.6}{3} = 0.2$ (2 marks)

ii $l = 0.6$ (1 mark)

iii $k = 3$ (1 mark)

iv $a = 5$ (1 mark)

v $b = 0.6$ (1 mark)

vi Gradient $p = \frac{0.9 - 0.6}{8 - 5} = 0.1$ (2 marks)

vii $0.6 = 0.1 \times 5 + q \Rightarrow q = 0.1$ (2 marks)

viii $r = 8$ (1 mark)

14 a $r = -0.358$ (3s.f.) (2 marks)

b $Q_1 = 12.3$ $Q_3 = 12.7$ (2 marks)

IQR = 0.4 $12.3 - 1.5 \times 0.4 = 11.7$ (1 mark)

So 10.8 is an outlier (1 mark)

c $r = -0.860$ (3s.f.) (2 marks)

d Changed from weak negative to strong negative (1 mark)

e $t = -0.0627y + 138.63$ (2 marks)

f $-0.0627 \times 2010 + 138.63 = 12.6$ seconds (2 marks)

15 a $0.51 \times 120 + 7.5 = 68.7$ (2 marks)

b The line of best fit goes through $(\bar{x}, \bar{y})$ (1 mark)

$\bar{y} = 0.51 \times 100 + 7.5 = 58.5$ (1 mark)

c Strong, positive (2 marks)

d x on y (1 mark)

16 a $r = 0.979$ (3s.f.) (2 marks)

b Strong, positive (2 marks)

c i $y = 1.23x - 21.3$ (2 marks)

ii $x = 0.776y + 20.8$ (2 marks)

d $1.23 \times 105 - 21.3 = 108$ (1 mark)

e $0.776 \times 95 + 20.8 = 95$ (1 mark)

f It is extrapolation (1 mark)

17 a i 0.849 (3s.f.) (2 marks)

ii strong, positive (2 marks)

iii $y = 0.937x + 0.242$ (2 marks)

b i 0.267 (3s.f.) (2 marks)

ii weak, positive (2 marks)

iii The Pearson product moment correlation coefficient is too small to make the line of best fit particularly meaningful when making predictions. (1 mark)

Chapter 7

Skills check

1 a $\frac{1}{3}$ **b** $\frac{7}{12}$ **c** $\frac{5}{12}$

2 a $\frac{57}{116}$ **b** $\frac{3}{29}$ **c** $\frac{49}{58}$

3 2.5

Exercise 7A

1 $\frac{1}{3}$

2 a $\frac{17}{20}$ **b** $\frac{7}{10}$ **c** 1 **d** $\frac{7}{10}$ **e** $\frac{2}{5}$ **f** $\frac{1}{5}$ **g** 0

3 a $\frac{6}{11}$ **b** $\frac{3}{11}$ **c** $\frac{5}{11}$ **d** $\frac{2}{11}$ **e** 0 **f** $\frac{4}{11}$ **g** $\frac{1}{11}$

4 a $\frac{1}{10000}$ **b** $\frac{99}{10000}$

b $\frac{1}{10}$ **c** $\frac{9987}{10000}$

Exercise 7B

1 a $\frac{73}{239}$ **b** $\frac{37}{136}$ **c** 544

2 a $\frac{3}{8}$ **b** 60

3 a 77 **b** 64 **c** 26

4 108 **5** 340 **6** 36

7 a 9.74

b Assuming the same proportions of cars are observed in July as in June

8 2.20 **9** 23.5

10 Compare your answer with that of a classmate

Exercise 7C

1 a 5 **b** $\frac{41}{127}$ **c** 4094

2 a $\frac{9}{20}$ **b** $\frac{3}{4}$ **c** 27

3 a

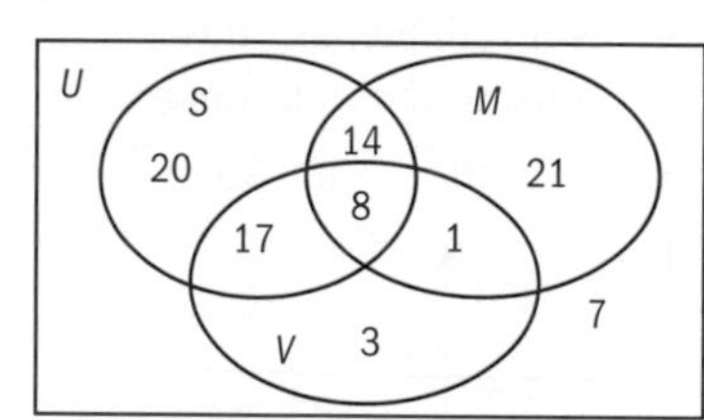

b $\frac{44}{91}$ **c** $\frac{32}{91}$

4 a

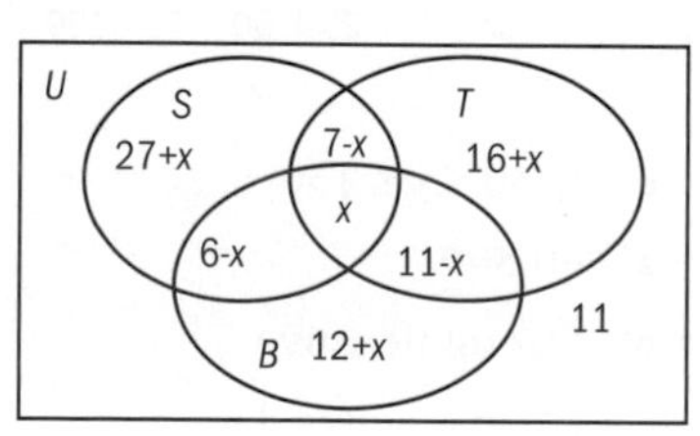

b 4 **c** $\frac{8}{47}$

Exercise 7D

1 a $\frac{5}{12}$ **b** $\frac{1}{3}$

2 a $\frac{4}{9}$ **b** 709

3 a $\frac{1}{4}$ **b** $\frac{3}{16}$ **c** $\frac{1}{8}$

d $\frac{1}{4}$ **e** $\frac{13}{36}$ **f** $\frac{5}{9}$

g For example, M and N are even

4 a

		Chromosome inherited from mother	
		X	X
Chromosome inherited from father	X	XX	XX
	Y	XY	YX

b There are 4 outcomes in total, 2 of which result in XX pairs. Hence, $p = \frac{2}{4}$ $= 0.5$.

Exercise 7E

1 $\frac{5}{7}$

2 $\frac{7}{9}$

3 $\frac{3}{8}$

4 15.4

5 $T = 5$

6 $P(A) = \frac{2}{3}$, $P(B) = \frac{1}{3}$, $P(C) = \frac{2}{3}$, $P(D) = \frac{1}{3}$

Exercise 7F

1 a $P(A) = \frac{3}{10}$, $P(B) = \frac{2}{5}$,

$P(A \cap B) = \frac{1}{10}$ and

$P(A \cup B) = \frac{3}{5}$

b $P(A) + P(B) - P(A \cap B)$

$= \frac{3}{10} + \frac{4}{10} - \frac{1}{10} = \frac{6}{10} = \frac{3}{5}$

$= P(A \cup B)$

c Events A and B are not mutually exclusive because $P(A \cap B)$ is not 0.

2 a $P(C) = \frac{7}{12}$, $P(D) = \frac{5}{12}$,

$P(C \cap D) = 0$ and

$P(C \cup D) = 1$

b $P(C) + P(D) =$

$\frac{7}{12} + \frac{5}{12} = \frac{12}{12} = 1 = P(C \cup D)$

c Events C and D are mutually exclusive because $P(C \cap D) = 0$.

3 a

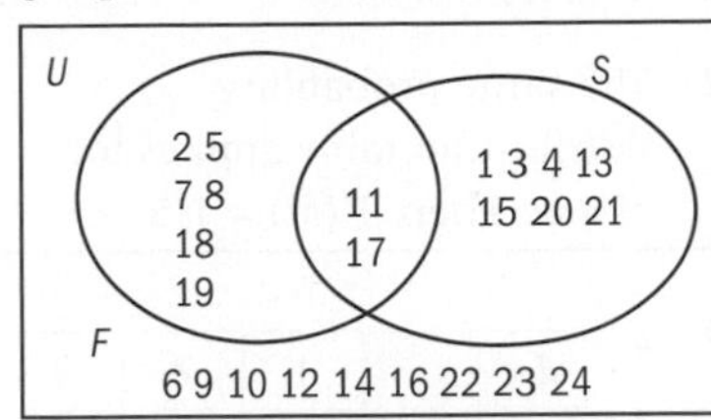

b Events **not** mutually exclusive because intersection of sets S and F is not empty

c $\frac{5}{8}$

4 a

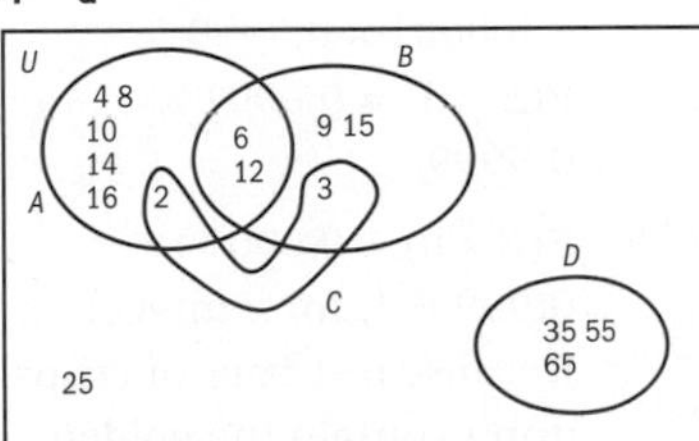

b $A\&D$, $B\&D$ and $C\&D$

Exercise 7G

1 a Independent

b Neither

c Neither

d Independent

e Mutually exclusive

f Neither

g Independent

2 $P(A \cup V) = 0.6373$; and represents the probability of either the event A, or B, or both A and B happening.

3 a $P(S) \times P(M) = P(S \cap M)$ $= 0.275$. Since they are equal to $P(S)$, the events are independent.

b $P(S) = 0.61$ and $P(S|M) = \frac{15}{24} = 0.63$. As they are not equal, events not independent.

4 a

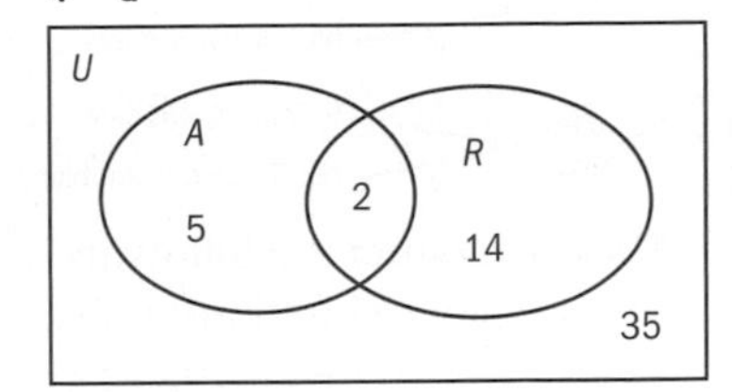

b 0.964

5 a i $\frac{2}{11}$ **ii** $\frac{2}{11}$

iii $\frac{4}{121}$

b i $\frac{2}{11}$ **ii** $\frac{1}{5}$ **iii** $\frac{2}{55}$

6 a 0.52 **b** 0.43

c $x = 7$, $y = 11$, $z = 6$

d A and B, and A and C are pairs of dependent events.

Exercise 7H

1 a

b 0.41

2 0.66

3 $1 - \left(\frac{5}{6}\right)^4$. Although drawing a tree diagram with 1296 branches might take a while, the other representation is an incorrect probability calculation.

4 a 0.915 **b** 0.31

c 170 **d** 60%

5 0.491

6 a 0.4

b

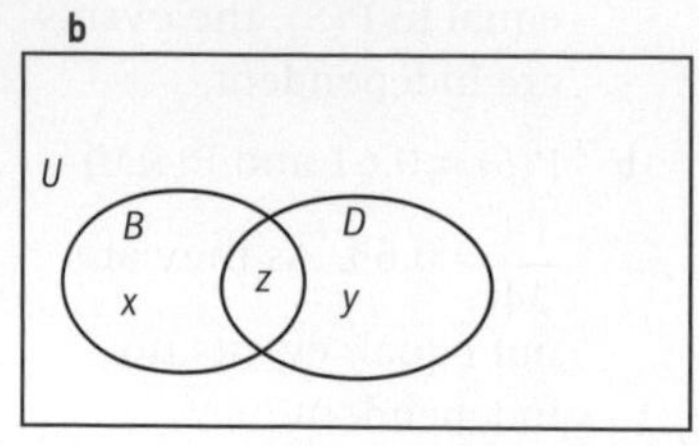

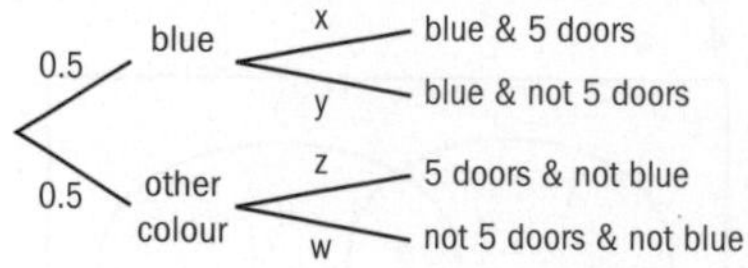

Then, construct the following simultaneous equations: $0.5(x + z) = 0.3$, $0.5(1 + z) = 0.6$, $z = 0.2$, $x = 0.4$, $y = 0.6$, $w = 0.8$, so the probability that the car hasn't got 5 doors and is not blue is $0.5w = 0.4$ as before.

7 0.872

Exercise 7I

1 Table **a** does not represent a discrete probability distribution because the probabilities of all possibilities don't add up to 1 and Table **b** does not represent a discrete probability distribution because $P(B = 2) = -0.2$ is negative. Table **c**, however, could represent a discrete probability distribution because all values $0 \le P(C = c) \le 1$, and the probabilities add up to 1.

2 Since $\frac{1+2+3+4+5+6}{21} = 1$,

$f(t) = \frac{t-4}{21}$ defines a discrete probability distribution on a given domain.

t	5	6	7	8	9	10
$f(t)$	$\frac{1}{21}$	$\frac{2}{21}$	$\frac{3}{21}$	$\frac{4}{21}$	$\frac{5}{21}$	$\frac{6}{21}$

3 6

4 **a** {MMM, MMF, MFM, FMM, MFF, FMF, FFM, FFF}

b

F	0	1	2	3
$P(F=f)$	$\frac{1}{8}$	$\frac{3}{8}$	$\frac{3}{8}$	$\frac{1}{8}$

5 **a** 0.11 **a** 0.14 **b** 0.59

c 0.41 **d** 0.38

6 **a** Probabilities don't add up to 1.

b 0.13

7

t	1	2	3	4	5
$P(T=t)$	0.2	0.15	0.1	0.44	0.11

Exercise 7J

1 $k = 0.1$, $E(B) = 1$

2 The same probability distribution table applies for $M = m$. Then, $E(M) = 1.5$.

Expected number of male and female births in a set of triplets is expected to be equal, and $E(M) + E(F) = 3$.

3 $\frac{3}{5}$

4 $\frac{16}{17}$

5 Expected prize E = US\$ 3.25 ≠ US\$ 5, so the game is not fair.

6 **a** 0.0001, 0.0005, 0.001

b –US\$ 8.80

c US\$ 1.20

7 **a**

d	1	2	3	4	6	8	9	12	16
$P(D=d)$	$\frac{1}{16}$	$\frac{2}{16}$	$\frac{2}{16}$	$\frac{3}{16}$	$\frac{2}{16}$	$\frac{2}{16}$	$\frac{1}{16}$	$\frac{2}{16}$	$\frac{1}{16}$

b $\frac{2}{5}$ **c** US\$ 7.50

8 **a** $P(B = 1) = 0.0001$, $P(B = 2) = 0.0001 \times 0.9999$, $P(B = 3) = 0.0001 \times 0.9999^2$.

b $P(B = n) = 0.0001 \times 0.9999^{n-1}$, for n integer. Because $n-1$ bags of crisps don't contain the golden ticket and the nth bag does, $f(b) = P(B = b) = 0.0001(0.9999)^{b-1}$

c $f(b)$ is defined for b positive integers, $b \ge 1$

d 0.0009996

Exercise 7K

1 **a** $X \sim B(7, \frac{1}{2})$

b The 'success' probability is not constant because the die is not replaced.

c $X \sim B(4, \frac{1}{2})$

d $X \sim B(4, 0.3)$

e No 'success' probability

2 **a** 0.313 **b** 0.891 **c** 0.641

3 **a** 0.130 **b** 0.860

c 0.0000154

4 **a** 0.00361 **b** 0.0000130

5 **a** 0.999 **b** 0.995

6 **a** 0.101 **b** 0.863

c 0.964

7 Model this as $X \sim B(5, 0.964)$, $P(X = 4) = 0.15545$ while $Y \sim B(5, 0.5)$, $P(Y = 4) = 0.15625$

Exercise 7L

1 **a** 0.0535 **b** 0.991

c 0.809 **d** 0.558

e 0.979

f Dependent, because $P(X \le 4) \ne P(X \le 4 | X \ge 2)$

g 1.74 **h** 1.24

2 **a** 0.773 **b** 3.5

c 1.75

3 **a** **i** 0.224 **ii** 0.157

b $P(A) = 0.617$, $P(B) = 0.000160$, $P(A|B) = 0$

c Events A and B are **not** independent because $P(A|B) \neq P(A)$.

d Events A and B are mutually exclusive because they cannot occur together.

4 a 0.00163 b 0.678

5 a $E(R) = 5 \times 0.964 = 4.82$, $E(B) = 2.50$. On average, R scores higher than B.

b Variance of $R = 0.174$, variance of $B = 1.25$. The data for R is less spread than for B.

6 a $A \sim B(25, 0.2)$

b $P(A \leq 5) = 0.617$

c $P(A \geq 7) = 0.220$

d $P(A \leq 3) = 0.234$

e $E(A) = np = 25 \times 0.2 = 5$; on average, Alex can expect to get 5 answers right by randomly guessing.

f 0.383 g 0 points h 0.212

7 a $T \sim B(538, 0.91)$, assuming that whether one passenger turns up on time is independent of any other passenger.

b 9.21×10^{-23} means it is close to impossible for everyone to turn up on time.

c 0.000 672 means it is highly unlikely for 510 or more passengers to turn up on time.

d 551 e 591

f $P(T = 538) = 0.0573$ and $P(T > 538) = 0.468$, so it is quite likely that more people than there are seats would show up and not very likely that exactly as many people as there are seats would show up.

8 0.5

Exercise 7M

1 a

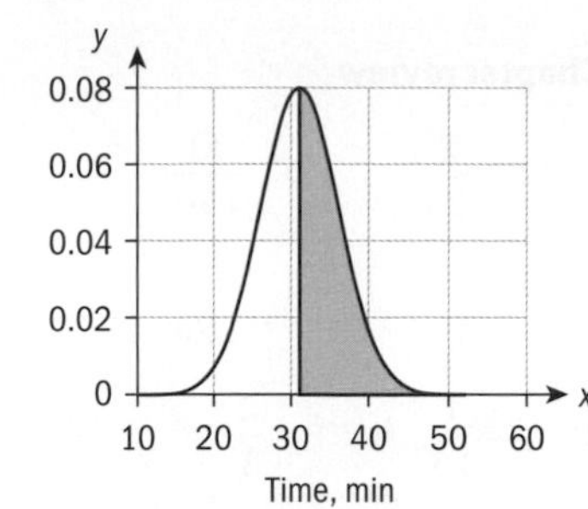

b

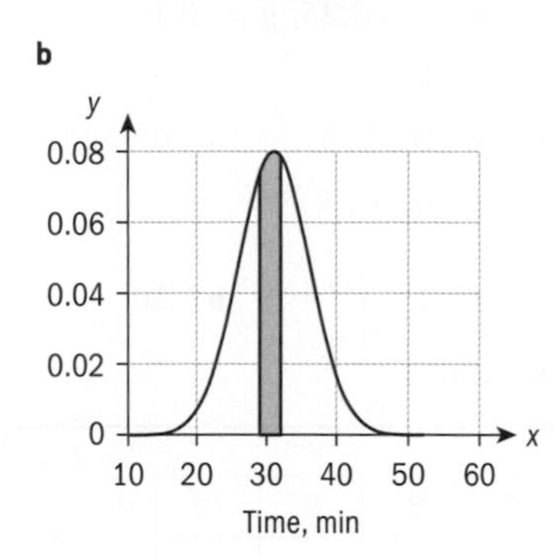

c

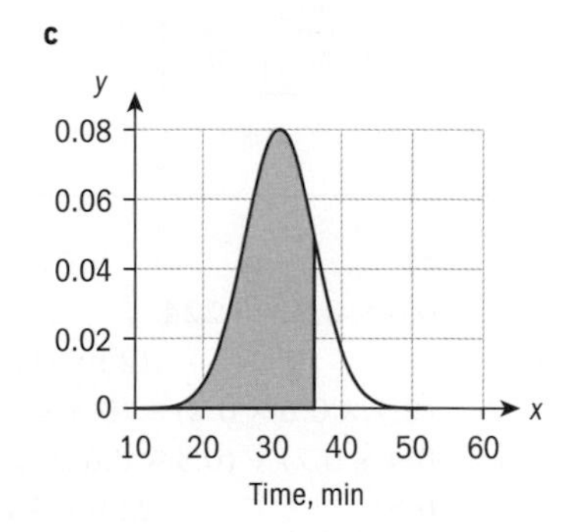

2 a

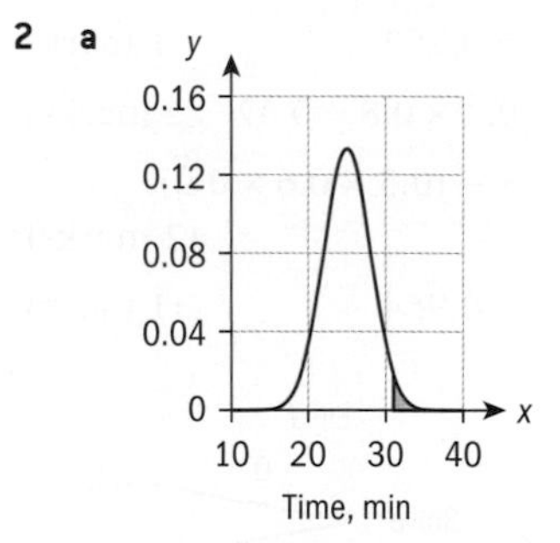

b

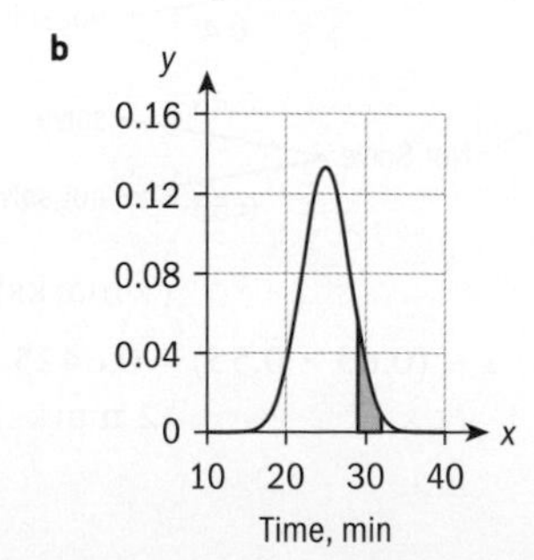

c

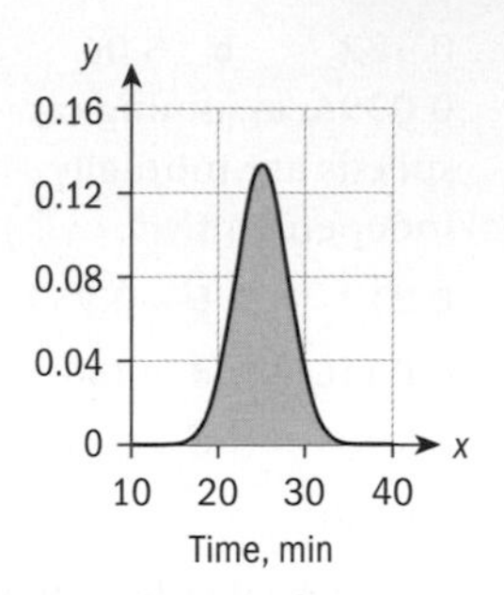

3 a

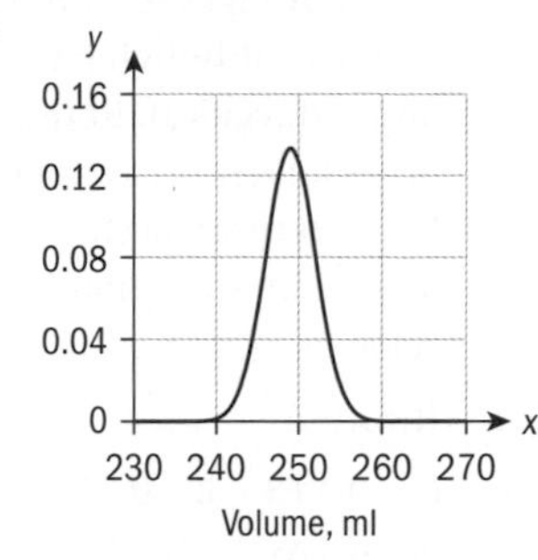

b 15.9%

c $P(S < 246) = 0.1587 \approx 16\%$

d 74

4 a

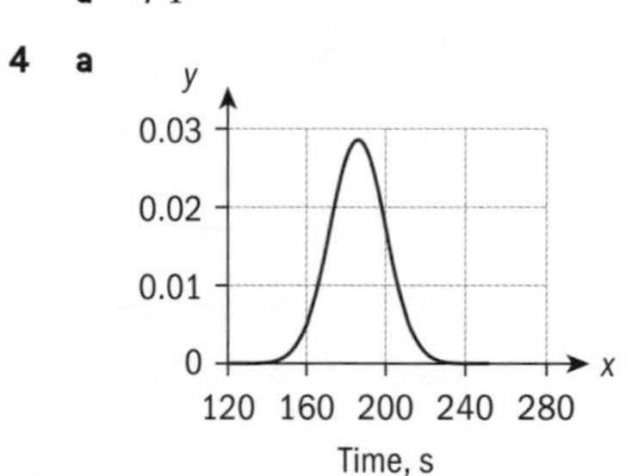

b 2.28%

c $P(T \geq 214) = 1 - P(T < 214) = 0.02275 \approx 2.28\%$

d 28

5 a 0.5 b 0.16

c 0.0225 d 0.815

e 0.0015 f 0.844

6 a 0.483 b 0.184

c 0.0829 d 0.525

e 0.887

7 a A2, B4, C5, D1, E3

b C c p and q

Exercise 7N

1 990 g

2 $t = 384$ g

3 a t and u b 405 g

4 a 0.382 b 3.06

c 0.0396, assuming car speeds are mutually independent

5 a 0.212 b 0.959

c 0.0446 d 446

6 a 0.170 b 21.6 km

c 13 800

7 a Route A takes less time on average, although has a larger deviation. Route B takes longer on average, but has a very small standard deviation, so is more reliable.

b Route A

c i 0.113 ii 0.759

iii 0.101

8 a $Q_3 = 73.4$

b Less than 10 years

Chapter review

1 a $\frac{2}{25}$ b $\frac{51}{150}$

c $\frac{1}{5}$ d $\frac{11}{150}$

2 $\beta = \frac{1}{15}$, $E(K) = \frac{2}{3}$

3 a $p = 0.25, q = -0.1$

b 0.5

4 a 0.8115 b 0.5525

c 0.9025

5 a $1 - \left(\frac{5}{6}\right)^n$ b 30

6 0.6

7 a

t	4	5	6	7	8	9	10
P($T=t$)	$\frac{1}{36}$	$\frac{4}{36}$	$\frac{8}{36}$	$\frac{10}{36}$	$\frac{8}{36}$	$\frac{4}{36}$	$\frac{1}{36}$

b 18

8 a 0.325 b 14.8 kg

c 81.0 kg d 0.759

e 325

9 a 0.629

b $p = 0.636$ so Judith is more likely to win.

Exam-style questions

10 a $\left(\frac{3}{16} \times \frac{13}{15}\right) + \left(\frac{13}{16} \times \frac{3}{15}\right)$ (2 marks)

$\frac{13}{40}$ (1 mark)

b $1 - \left(\frac{13}{16} \times \frac{12}{15}\right)$ (2 marks)

$\frac{7}{20}$ (1 mark)

11 a $0.7 \times 0.4 \times 0.8 = 0.224$ (2 marks)

b $(0.7 \times 0.6 \times 0.2) + (0.3 \times 0.4 \times 0.2) + (0.3 \times 0.6 \times 0.8)$ (2 marks)

$= 0.252$ (1 mark)

c $0.4 \times 0.8 = 0.32$ (2 marks)

d $1 - (0.3 \times 0.6 \times 0.2)$ (2 marks)

$= 0.964$ (1 mark)

12 a

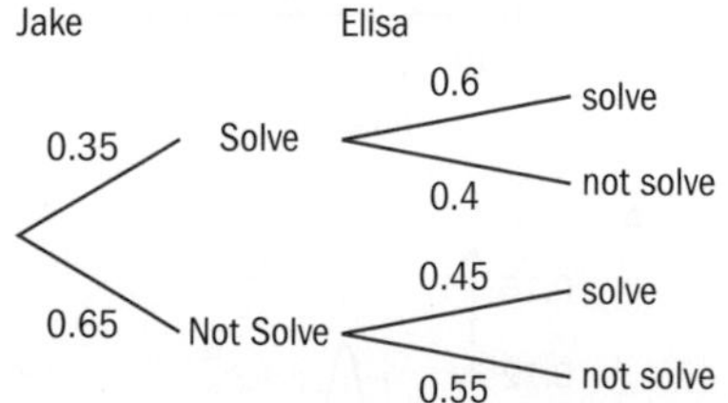

(3 marks)

b $1 - (0.65 \times 0.55) = 0.6425$ (2 marks)

c $\frac{\text{P(Jake and Elisa solve)}}{\text{P(Elisa solve)}}$

$= \frac{0.35 \times 0.6}{(0.35 \times 0.6) + (0.65 \times 0.45)}$ (3 marks)

$= 0.418$ (1 mark)

13

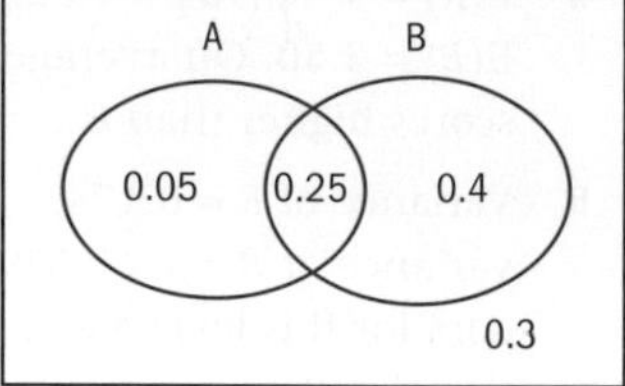

a 0.4 (attempt to use their Venn diagram: 1 mark; correct answer: 2 marks)

b 0.6 (2 marks)

c 0.75 (2 marks)

14 a $32 + 25 - 48$ (1 mark)

$= 9$ (1 mark)

b $\frac{32-9}{48}$ (2 marks)

$= \frac{23}{48}$ (1 mark)

c $P(E|U) = \frac{9}{25}$ and

$P(E) = \frac{32}{48} = \frac{2}{3}$. (2 marks)

$P(E|U) \neq P(E)$, so not independent. (1 mark)

15 a $\frac{43}{50}$ (2 marks)

b $\frac{7}{25}$ (2 marks)

c $\frac{5}{11}$ (2 marks)

d 0 (2 marks)

e $\frac{10}{17}$ (2 marks)

16 a $P(A \cap B) = P(A)P(B)$ (1 mark)

$= 0.3 \times 0.5 = 0.15$ (1 mark)

b $P(A \cup B) = P(A) + P(B) - P(A \cap B)$ (1 mark)

$= 0.3 + 0.5 - 0.15 = 0.65$ (1 mark)

c $P(B' \cap A) = P(A) - P(A \cap B)$ (1 mark)

$= 0.3 - 0.15 = 0.15$ (1 mark)

d $P(B|A') = \frac{P(B \cap A')}{P(A')}$ (1 mark)

$= \frac{P(B) - P(A \cap B)}{P(A')}$

$= \frac{0.5 - 0.15}{1 - 0.3}$ (1 mark)

$= 0.5$ (1 mark)

17 a $1 - P(\text{no scoring}) =$

$1 - 0.72^4$ (2 marks)

$= 0.731$ (1 mark)

b $4 \times 0.28 \times 0.72^3$ (2 marks)

$= 0.418$ (1 mark)

c $1 - P(\text{no goals}) - P(\text{exactly one goal})$ (1 mark)

$= 1 - 0.72^4 - 0.418$ (1 mark)

$= 0.313$ (1 mark)

18 a Let X be the discrete random variable 'number of boys'.

So $X \sim B(10, 0.512)$

$P(X = 6) = \binom{10}{6} 0.512^6 (1 - 0.512)^4$ (2 marks)

$= 0.215$ (1 mark)

b $P(X = 0) = \binom{10}{0} 0.512^0 (1 - 0.512)^{10}$ (1 mark)

$= 0.000766$ (1 mark)

c Find $P(X \le 4)$ using GDC (1 mark)

$= 0.348$ (1 mark)

19 a $\frac{k}{2} + k + k^2 + 2k^2 + \frac{k}{2} = 1$ (1 mark)

$3k^2 + 2k - 1 = 0$ (1 mark)

$(3k - 1)(k + 1) = 0$ (1 mark)

$\Rightarrow k = \frac{1}{3}$ (1 mark)

b $E(X) = 0 \times \frac{k}{2} + 0.5 \times k + 1 \times k^2 + 1.5 \times 2k^2 + 2 \times \frac{k}{2}$ (1 mark)

$E(X) = \frac{k}{2} + k^2 + 3k^2 + k$

$= 4k^2 + \frac{3k}{2}$ (1 mark)

$= 4\left(\frac{1}{3}\right)^2 + \frac{3}{2}\left(\frac{1}{3}\right)$ (1 mark)

$= \frac{4}{9} + \frac{1}{2}$

$= \frac{17}{18}$ (1 mark)

c $P(X \ge 1.25) = 2k^2 + \frac{k}{2}$ (1 mark)

$= 2\left(\frac{1}{3}\right)^2 + \frac{1}{6}$ (1 mark)

$= \frac{2}{9} + \frac{1}{6}$

$= \frac{7}{18}$ (1 mark)

20 a Let X be the discrete random variable 'time taken for Blossom to walk to her cafe'. So $X \sim N(35, 3.4^2)$

Using GDC, $P(X > 37)$ (1 mark)

$= 0.278$ (1 mark)

b $P(X < 36.5) - P(X < 34)$ (1 mark)

$= 0.670 - 0.384$ (1 mark)

$= 0.286$ (1 mark)

c $P(X < 30) = 0.0707$ (1 mark)

$0.0707 \times 25 = 1.77$ (1 mark)

So approximately two occasions. (1 mark)

21 a Let X be the discrete random variable 'mass of a can of baked beans'.

Then $X \sim N(415, 12^2)$

$P(X > m) = 0.65 \Rightarrow P(X \le m) = 1 - 0.65 = 0.35$ (1 mark)

Using inverse normal distribution on GDC $\Rightarrow m = 410.4$ (2 marks)

b You require $P(X > 422.5 | X > 420)$. (1 mark)

$P(X > 422.5 \mid X > 420) = \frac{P(X > 422.5)}{P(X > 420)}$ (1 mark)

$= \frac{1 - P(X \le 422.5)}{1 - P(X \le 420)}$

$= \frac{0.266}{0.338}$ (1 mark)

$= 0.787$ (1 mark)

c Using GDC

$P(X < 413.5) = 0.450$ (2 marks)

Let Y be the random variable 'Number of cans of beans having a mass less than 413.5 g'.

In Ashok's experiment, Y is Binomially distributed across 144 trials with probability of 'success' (ie mass less than 413.5 g) being 0.450.

So $Y \sim B(144, 0.450)$ (1 mark)

$P(Y \ge 75) = 0.0524$ (1 mark)

Chapter 8

Before you start

1 $P(6) = \frac{1}{6}$, $P(H) = \frac{1}{2}$, $P(6 \cap H) = \frac{1}{12}$, $P(6) \times P(H)$

2 0.479

3 0.027, 0.189, 0.441, 0.343

Skills check

1 $P(S) = \frac{2}{6} = \frac{1}{3}$ while $P(S|E) = \frac{1}{3}$. $P(S) = P(S|E)$ so the events are independent.

2 0.631

3

x	0	1	2	3	4
$P(X=x)$	0.0016	0.0256	0.1536	0.4096	0.4096

Exercise 8A

1 **a** 1 **b** 1 **c** −1 **d** 0

2 $r_s = 0.2$ so there is only weak positive correlation between the taste and value for money.

3 **a** $r_s = -0.964$ **b** $r_s = 0.893$

4 **a** PMCC used for linear relationships and scatter plot shows relationship is not linear

Positive correlation of medium strength, strong but non-linear positive correlation between the scores.

b $r_s = -0.942$

c Strong negative correlation. Since force does not change significantly at high values of velocity, value of r_s could be significantly affected by small changes in data.

c $r_s = 0.883$, indicating strong positive correlation between the scores; a more realistic result, given the scatter plot

d Spearman's rank correlation because the data points are not linear

5 **a** $r = 0.670$

b

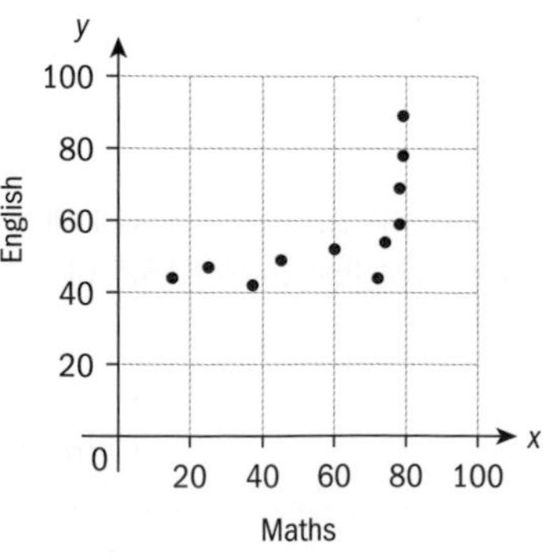

6 **a** Because ranks given instead of quantitative data

b $r_s = 0.886$, so strong positive correlation between price and taste of coffee

7 **a** **i** $r = 0.874$

ii $r = 0.0776$

b **i** $r_s = 0.304$ **ii** $r_s = 0.0418$

c Although both measures are affected by the outlier, Spearman's rank correlation coefficient is affected less.

Exercise 8B

1 **a** Probability that a person chosen at random likes black cars best is $\frac{22}{80}$. Probability that a person chosen at random is male is $\frac{38}{80}$. If the two events are independent, the expected number of males who prefer black cars is $80 \times \frac{22}{80} \times \frac{38}{80} = 10.45$.

b Probability that a person chosen at random likes white cars best is $\frac{18}{80}$. Probability that a person chosen at random is male is $\frac{42}{80}$. If the two events are independent, the expected number of males who prefer black cars is $80 \times \frac{18}{80} \times \frac{42}{80} = 9.45$.

c 4.69

2 **a** Probability that a person chosen at random buys small coffee is $\frac{34}{110}$. Probability that a person chosen at random is male is $\frac{54}{110}$. If the two events are independent, the expected number of males who prefer black cars is $110 \times \frac{34}{110} \times \frac{54}{110} = 16.7$.

b Probability that a person chosen at random buys large coffee is $\frac{46}{110}$. Probability that a person chosen at random is female is $\frac{56}{110}$. If the two events are independent, the expected number of males who prefer black cars is $110 \times \frac{46}{110} \times \frac{56}{110}$ $= 23.4$.

c 5.21

3 a Probability that a pet chosen at random eats carrots is $\frac{70}{130}$. Probability that a pet chosen at random is a rabbit is $\frac{50}{130}$. If the two events are independent, the expected number of rabbits who eat carrots is $130 \times \frac{70}{130} \times \frac{50}{130}$ $= 26.9$.

b Probability that a pet chosen at random eats lettuce is $\frac{60}{130}$. Probability that a pet chosen at random is a hamster is $\frac{46}{130}$. If the two events are independent, the expected number of rabbits who eat carrots is

$130 \times \frac{60}{130} \times \frac{46}{130} = 21.2$.

c 8.05

4 a Probability that a person chosen at random comes by bicycle is $\frac{43}{120}$. Probability that a person chosen at random is a male is $\frac{60}{120}$. If the two events are independent, the expected number of males who come by bicycle is

$120 \times \frac{43}{120} \times \frac{60}{120} = 21.5$.

b Probability that a person chosen at random comes by car is $\frac{33}{120}$. Probability that a person chosen at random is a female is $\frac{60}{120}$. If the two events are independent, the expected number of females who come by car is

$120 \times \frac{33}{120} \times \frac{60}{120} = 16.5$.

c 6.90

Exercise 8C

1 a

Sport	Cycling	Basketball	Football	Totals
Males	7	10	6	23
Females	9	8	10	27
Totals	16	18	16	50

b $H_{0:}$ favourite sport is independent of gender. H_1: favourite sport is not independent of gender.

c 2

d $\chi^2 = 1.16$ and $p = 0.560$

e Expected values:

Sport	Cycling	Basketball	Football
Males	7.36	8.28	7.36
Females	8.64	9.72	8.64

f $\chi^2 = 1.16 < 4.605$ so H_0 is accepted.

g $p = 0.560 > 0.10$ so H_0 is accepted and p-value supports the conclusion.

2 a $H_{0:}$ favourite bread is independent of gender. H_1: favourite bread is not independent of gender.

b E(Female liking white bread) = P(Female) × P(likes white bread) × total = $\frac{41}{80} \times \frac{31}{80} \times 80 =$ $15.8875 \approx 15.9$

c 3

d $\chi^2 = 2.12$ and $p = 0.548$

e $\chi^2 < \chi^2_c$ so H_0 is accepted.

3 a $H_{0:}$ favourite genre of film is independent of age. H_1: favourite genre of film is not independent of age.

b E(20 – 50 year-olds prefer horror films) = P(20 – 50 year-olds) × P(prefers horror films) × total = $\frac{130}{300} \times \frac{77}{300} \times 300 = 33.367 \approx 33.4$

c v = (rows – 1)(columns – 1) = (3 – 1) × (3 – 1) = 4

d $\chi^2 = 45.2$ and $p = 3.6 \times 10^{-9}$

e $p < 0.10$ (for 10% test) so H_0 is rejected.

4 a $H_{0:}$ favourite flavour of dog food is independent of breed. H_1: favourite flavour of dog food is not independent of breed.

b

Flavour	Beef	Chicken	Lamb
Boxer	11.7	7.00	9.33
Labrador	15.8	9.50	12.7
Poodle	14.6	8.75	11.7
Collie	7.92	4.75	6.33

c

Flavour	Beef	Chicken	Lamb	Totals
Boxer	14	6	8	28
Labrador	17	11	10	38
Poodle/Collie	19	13	22	54
Totals	50	30	40	120

d $\chi^2 = 3.14$ and $p = 0.535$

e $p = 0.535 > 0.05$ so H_0 is accepted.

5 **a** H_0: favourite flavour of chocolate is independent of gender. H_1: favourite flavour of chocolate is not independent of gender.

b 2

c $\chi^2 = 9.52$ and $p = 0.00856$

d $\chi^2 = 9.52 > 9.210$ so H_0 is rejected.

6 **a** H_0: GPA is independent of number of hours spent on social media. H_1: GPA is not independent of number of hours spent on social media.

b E(0 – 9 hours and high GPA) = P(0 – 9 hours) × P(high GPA) × total = $\frac{85}{270} \times \frac{99}{270} \times 270 = 31.167$

≈ 31.2

c 4

d $\chi^2 = 78.5$ and $p = 3.6 \times 10^{-16}$

e $\chi^2 = 78.5 > 7.779$ so H_0 is rejected.

7 H_0: the number of people walking their dog is independent of the time of the day. H_1: the number of people walking their dog is not independent of the time of the day.

$v = 4$, $\chi^2 = 5.30$ and $p = 0.258$. Since $0.258 > 0.05$ and $5.30 < 9.488$, H_0 is accepted.

8 **a** H_0: the number of bottles of water sold is independent of temperature. H_1: the number of bottles of water sold is not independent of temperature.

b 4

c $\chi^2 = 3.30$ and $p = 0.509$

d Since $0.509 > 0.01$ and $3.30 < 13.277$, H_0 is accepted.

9 **a** H_0: annual salary is independent of the type of degree. H_1: annual salary is not independent of the type of degree.

b 4

c $\chi^2 = 24.4$ and $p = 6.53 \times 10^{-5}$

d Since $6.53 \times 10^{-5} < 0.05$ and $24.4 > 9.488$, H_0 is rejected.

Exercise 8D

1 **a**

Colour	Frequency
Yellow	120
Orange	120
Red	120
Purple	120
Green	120

b 4

c H_0: The data satisfies a uniform distribution. H_1: The data does not satisfy a uniform distribution.
$\chi^2 = 10.45$ and $p = 0.0335$

d Since $0.0335 < 0.05$ and $10.45 > 9.488$, H_0 is rejected.

2 **a**

Month	Jan	Feb	Mar	Apr	May	Jun	Jul	Aug	Sep	Oct	Nov	Dec
Freq	5	5	5	5	5	5	5	5	5	5	5	5

b 11

c H_0: The data satisfies a uniform distribution. H_1: The data does not satisfy a uniform distribution. $\chi^2 = 6$ and $p = 0.873$

d Since $0.873 > 0.10$ and $6 < 17.275$, H_0 is accepted.

3 **a** E(number of calls) = $\frac{840}{7} = 120$ **b** 6

c H_0: The data satisfies a uniform distribution. H_1: The data does not satisfy a uniform distribution. $\chi^2 = 86.1$ and $p = 1.97 \times 10^{-16}$

d Since $1.97 \times 10^{-16} < 0.05$ and $86.1 > 12.592$, H_0 is rejected.

4 **a**

Last digit	0	1	2	3	4	5	6	7	8	9
Frequency	49	49	49	49	49	49	49	49	49	49

b 9

c H_0: The data satisfies a uniform distribution. H_1: The data does not satisfy a uniform distribution. $\chi^2 = 9.06$ and $p = 0.432$

d Since $0.432 > 0.10$ and $9.06 < 14.684$, H_0 is accepted.

Exercise 8E

1 **a** H_0: The lengths are normally distributed with mean of 19 cm and standard deviation of 3 cm. H_1: The lengths are not normally distributed with mean of 19 cm and standard deviation of 3 cm.

b 0.00939 **c** 2.35 fish

d

Length of fish, x cm	Probability	Expected frequency
$9 \le x < 12$	0.009386	2.35
$12 \le x < 15$	0.081396	20.35
$15 \le x < 18$	0.278230	69.56
$18 \le x < 21$	0.378066	94.52
$21 \le x < 24$	0.204702	51.18
$24 \le x < 27$	0.043960	10.99
$27 \le x < 30$	0.003708	0.927

d Combine the rows with expected frequencies less than five with the rows next to them, ie the top row with the second row and the last row with the second to last row.

e

Length of fish, x cm	Frequency	Expected frequency
$9 \le x < 15$	27	22.70
$15 \le x < 18$	71	69.56
$18 \le x < 21$	88	94.52
$21 \le x < 24$	52	51.18
$24 \le x < 30$	12	11.92

g 4 h $\chi^2 = 1.31$ and $p = 0.860$

i Since $0.860 > 0.05$ and $1.31 < 9.488$, H_0 is accepted.

2 a Table with the expected frequencies:

Weight, w kg	$w < 45$	$45 \le w < 50$	$50 \le w < 55$	$55 \le w < 60$	$w \ge 60$
Expected frequency	1.96	48.54	117.77	30.96	0.77

b

Weight, w kg	$w < 50$	$50 \le w < 55$	$55 \le w$
Observed frequency	56	82	62
Expected frequency	50.5	117.77	31.73

c 2

d H_0: The weights are normally distributed with mean of 52 kg and standard deviation of 3 kg. H_1: The weights are not normally distributed with mean of 52 kg and standard deviation of 3 kg. $\chi^2 = 40.3$ and $p = 1.74 \times 10^{-9}$

e Since $1.74 \times 10^{-9} < 0.05$ and $40.3 > 5.991$, H_0 is rejected.

3 a

Grade, x%	$x < 50$	$50 \le x < 60$	$60 \le x < 70$	$70 \le x < 80$	$80 \le x$
Expected frequency	6.83	68.92	148.50	68.92	6.83

b 4

c H_0: The grades are normally distributed with mean of 65% and standard deviation of 7.5%. H_1: The grades are not normally distributed with mean of 65% and standard deviation of 7.5%. $\chi^2 = 0.705$ and $p = 0.951$

d Since $0.951 > 0.1$ and $0.705 < 7.779$, H_0 is accepted

4 a

Height, h cm	$h < 235$	$235 \le h < 245$	$245 \le h < 255$	$255 \le h < 265$	$265 \le h$
Expected frequency	21.59	59.59	87.64	59.59	21.59

b 4

c H_0: The heights are normally distributed with mean of 250 cm and standard deviation of 11 cm. H_1: The heights are not normally distributed with mean of 250 cm and standard deviation of 11 cm. $\chi^2 = 8.02$ and $p = 0.0908$

d Since $0.0908 > 0.05$ and $8.02 < 9.488$, H_0 is accepted.

5 a

Lifespan, h hours	$h < 1000$	$1000 \le h < 1100$	$1100 \le h < 1200$	$1200 \le h < 1300$	$1300 \le h < 1400$	$1400 \le h$
Frequency	9.1	54.36	136.54	136.54	54.36	9.1

b 5

c H_0: The lifespan is normally distributed with mean of 1200 hours and standard deviation of 100 hours. H_1: The heights are not normally distributed with mean of 1200 hours and standard deviation of 100 hours. $\chi^2 = 78.7$ and $p = 1.5 \times 10^{-15}$

d Since $1.5 \times 10^{-15} < 0.05$ and $78.7 > 11.070$, H_0 is rejected.

Exercise 8F

1 a $P(S = 0) = 0.015625$, $P(S = 1) = 0.140625$, $P(S = 2) = 0.421875$, $P(S = 3) = 0.421875$

b

Number of seeds germinating	0	1	2	3
Expected frequency	0.78	7.03	21.09	21.09

c Expected value of no seeds germinating is less than 5. Hence, combine this with the data of 1 seed germinating:

Number of seeds germinating	0, 1	2	3
Observed frequency	15	15	20
Expected frequency	7.81	21.09	21.09

d 2

e H_0: The number of germinating seeds follows a binomial distribution. H_1: The number of germinating seeds does not follow a binomial distribution. $\chi^2 = 8.43$ and $p = 0.0147$

f Since $0.0147 < 0.05$ and $8.43 > 5.991$, H_0 is rejected.

2 a $P(S = 0) = 0.125$, $P(S = 1) = 0.375$, $P(S = 2) = 0.375$, $P(S = 3) = 0.125$

b None **c** 3

d H_0: The number of boys follows a binomial distribution. H_1: The number of boys does not follow a binomial distribution. $\chi^2 = 10.77$ and $p = 0.0130$

e Since $0.0130 > 0.01$ and $10.77 < 11.345$, H_0 is accepted.

3 a

Number of sixes	0	1	2
Expected frequency	173.61	69.44	6.94

b None **c** 2

d H_0: The number of 6s follows a binomial distribution. H_1: The number of sixes does not follow a binomial distribution. $\chi^2 = 28.1$ and $p = 7.7 \times 10^{-7}$

e Since $7.7 \times 10^{-7} < 0.05$ and $28.1 > 5.991$, H_0 is rejected.

4 a Since there are four answers, P(getting a question right) = 0.25

b, c

Number correct	0	1	2	3	4	5
Probability	0.2373	0.3955	0.2637	0.0879	0.0147	0.000 98
Expected frequency	118.65	197.75	131.84	43.95	7.324	0.4883

d Expected frequency of 5 correct answers is less than 5, so need to combine the data for 4 and 5 correct answers. The new table is:

Number correct	0	1	2	3	4,5
Observed frequency	38	66	177	132	87
Expected frequency	118.65	197.75	131.84	43.95	7.81

e 4

f H_0: The number of correct answers follows a binomial distribution. H_1: The number of correct answers does not follow a binomial distribution. $\chi^2 = 1137$ and $p = 0$

g H_0 is rejected as $0 < 0.05$ and $1137 > 9.488$

Exercise 8G

1 a $H_0: \bar{x}_1 = \bar{x}_2$ (there is no difference in the weights). $H_1: \bar{x}_1 < \bar{x}_2$ (there is a difference in the weights of the apples: apples from the shade weigh less).

b One-tailed

c $t = -0.687$ and $p = 0.251$

d Since $0.251 > 0.10$, H_0 is rejected.

2 a $H_0: \bar{x}_1 = \bar{x}_2$ (there is no difference between the weights of town and country babies). $H_1: \bar{x}_1 > \bar{x}_2$ (babies born in the country weigh more than babies born in the town).

b One-tailed

c $t = -0.191$ and $p = 0.575$

d Since $0.575 > 0.10$, H_0 is accepted.

3 a $H_0: \bar{x}_1 = \bar{x}_2$ (there is no difference between the lengths of the beans). $H_1: \bar{x}_1 \neq \bar{x}_2$ (there is a difference between the lengths of the beans).

b Two-tailed

c $t = -3.126$ and $p = 0.00584$

d Since $0.00584 < 0.05$, H_0 is rejected.

4 a $H_0: \bar{x}_1 = \bar{x}_2$ (there is no difference between the lifetimes of the bulbs).H_1: $\bar{x}_1 \neq \bar{x}_2$ (there is a difference between the lifetimes of the bulbs).

b Two-tailed

c $t = 0.3$ and $p = 0.769$

d Since $0.769 > 0.05$, H_0 is accepted.

5 a $H_0: \bar{x}_1 = \bar{x}_2$ (there is no difference between the weights of the girls and the boys). $H_1: \bar{x}_1 < \bar{x}_2$ (the boys weigh less than the girls).

b One-tailed test

c $t = -2.45$ and $p = 0.015$

d Since $0.015 < 0.05$, H_0 is rejected.

6 a $H_0: \bar{x}_1 = \bar{x}_2$ (there is no difference between the weight loss with and without the remedy). H_1: $\bar{x}_1 > \bar{x}_2$ (people lose more weight with the remedy than without it).

b One-tailed

c $t = 2.84$ and $p = 0.00539$

d Since $0.00539 < 0.01$, H_0 is rejected.

7 a $H_0: \bar{x}_1 = \bar{x}_2$ (there is no difference in the lengths of the sweetcorn cobs).

$H_1: \bar{x}_1 \neq \bar{x}_2$ (there is a difference in the lengths of the sweetcorn cobs).

b Two-tailed

c $t = 0.535$ and $p = 0.600$

d Since $0.600 > 0.10$, H_0 is accepted.

Chapter review

1 a −0.953

b Strong negative: the taller the person, the faster they are.

2 a Instead of quantitative data, ranks of tennis players are given.

b 0.850

c Strong positive; the higher the rank, the more aces they are likely to hit.

3 a $H_{0:}$ colours of the eggs laid are independent of the type of the hen. H_1: colours of the eggs laid are not independent of the type of hen.

b Probability that a hen chosen at random is Leghorn is $\frac{30}{9}$. Probability that an egg chosen at random is white is $\frac{42}{90}$.

If the two events are independent, the expected number of white eggs laid by Leghorn hens is

$90 \times \frac{30}{9} \times \frac{42}{90} = 14$

c 2

d $\chi^2 = 21.7$ and $p = 1.94 \times 10^{-5}$

e $21.7 > 5.991$, H_0 rejected

4 a $H_{0:}$ favourite colour of car is independent of gender; H_1: favourite colour of car is not independent of gender.

b Probability that a person chosen at random likes white cars is $\frac{22}{100}$. Probability that a person chosen at random is a male is $\frac{48}{100}$. If the two events are independent, the expected number of males who like white cars is $100 \times \frac{20}{100} \times \frac{48}{100} = 9.6$.

c 3

d $\chi^2 = 9.43$ and $p = 0.0241$

e $9.43 > 6.251$, H_0 rejected

5 a

Day	Sun	Mon	Tue	Wed	Thu	Fri	Sat
Expected frequency	20	20	20	20	20	20	20

b 6

c H_0: The data satisfies a uniform distribution. H_1: The data does not satisfy a uniform distribution. $\chi^2 = 10$ and $p = 0.1247$; since $10 < 12.592$, H_0 is accepted.

6 a

Number on die	1	2	3	4	5	6
Expected frequency	15	15	15	15	15	15

b 5

c H_0: The data satisfies a uniform distribution. H_1: The data does not satisfy a uniform distribution. $\chi^2 = 0.667$ and $p = 0.985$; since $0.667 < 15.086$, H_0 is accepted.

7 a Let $S \sim B(2, \frac{1}{2})$, then $P(S = 0) = 0.25$ and expected value is $0.25 \times 60 = 15$

b

Number of tails	0	1	2
Expected frequency	15	30	15

c All expected frequencies are higher than 5

d 2

e H_0: The number of tails follows a binomial distribution. H_1: The number of tails does not follow a binomial distribution. $\chi^2 = 1.2$ and $p = 0.549$

f Since $1.2 < 5.991$, H_0 is accepted.

8 a

Height, x cm	$x < 152$	$152 \le x < 156$	$156 \le x < 160$	$160 \le x < 164$	$164 \le x$
Expected frequency	33.40	120.87	191.46	120.87	33.40

b None **c** 4

d H_0: The heights are normally distributed with mean of 158 cm and standard deviation of 4 cm. H_1:The heights are not normally distributed with mean of 158 cm and standard deviation of 4 cm. $\chi^2 = 20.6$ and $p = 0.000\,380$

e Since $0.000\,380 < 0.10$ and $20.6 > 7.779$, H_0 is rejected.

9 a $H_0: \bar{x}_1 = \bar{x}_2$ $H_1: \bar{x}_1 \neq \bar{x}_2$.

b Two-tailed test

c $t = -0.421$ and $p = 0.678$

d $0.678 > 0.05$, H_0 accepted

Exam-style questions

10 a $H_0: \mu_1 = \mu_2$ (1 mark)

$H_1: \mu_1 < \mu_2$ (1 mark)

b One-tailed test (1 mark)

c t – value $= -0.706$ (2 marks)

p – value $= 0.244$ (1 mark)

d $0.244 > 0.05$ so accept H_0. (2 marks)

ie there is no significant difference in the results of the two schools.

11 a $H_0: \mu_1 = \mu_2$ (1 mark)

$H_1: \mu_1 > \mu_2$ (1 mark)

b t – value $= 1.735$ (2 marks)

p – value $= 0.0499$ (1 mark)

c $0.0499 < 0.05$ so reject H_0. (2 marks)

ie there is significant evidence to suggest that older students read fewer books.

12 a $H_0: \mu_1 = \mu_2$ (1 mark)

$H_1: \mu_1 \neq \mu_2$ (1 mark)

b t – value $= 1.942$ (2 marks)

p – value $= 0.0725$ (1 mark)

c $0.0725 < 0.1$ so reject H_0. (2 marks)

ie there is significant evidence to suggest that there is a difference in average battery length.

13 a

Age rank	5	4	8	1	6	7	2	3
Reaction rank	6	2.5	8	5	4	7	1	2.5

(3 marks)

b $r_S = 0.707$ (2 marks)

c r_S is positive and reasonably close to 1, (1 mark)

indicating a fairly strong positive correlation between a person's age and their reaction time. (1 mark)

d Include a greater number of participants. (1 mark)

Ensure the participants were spread equally throughout the age range. (1 mark)

14 a

Age rank	1	9.5	3	12	11	2	7	9.5	6	8	5	4
Hours rank	6.5	3	12	1	9	11	5	2	4	8	10	6.5

(3 marks)

b $r_S = -0.596$ (2 marks)

c Neeve is incorrect. (1 mark)

A value of $r_S = -0.596$ indicates a small but significant negative correlation between a person's age and the hours per week they watch TV. (1 mark)

However, you cannot say this is *causal*. (1 mark)

ie you cannot conclude that your age *affects* the amount of TV you watch.

15 A2 (1 mark)

B1 (1 mark)

C3 (1 mark)

D6 (1 mark)

E4 (1 mark)

F5 (1 mark)

16 a H_0: Type of burger favoured is independent of age. (1 mark)

H_1: Type of burger favoured is not independent of age. (1 mark)

b $(3 - 1) \times (4 - 1) = 6$ (1 mark)

c $\chi^2_{calc} = 12.314$ (2 marks)

p – value $= 0.0553$ (1 mark)

d $\chi^2_{calc} = 12.314 < 12.59$, therefore we accept H_0. (2 marks)

ie the type of burger favoured is independent of age.

17 a

	Smoker	Non-smoker	Totals
16-25 years	21.6	68.4	90
26-60 years	28.4	89.6	118
Totals	50	158	208

(3 marks)

b H_0: Smoking is independent of age (1 mark)

H_1: Smoking is not independent of age (1 mark)

$\chi^2_{calc} = 9.408$ (2 marks)

p – value $= 0.00216$ (1 mark)

$\chi^2_{calc} = 9.408 > 6.64$, therefore we reject H_0 and accept H_1 (2 marks)

ie smoking is not independent of age.

18 a H_0: Movie preference is independent of gender (1 mark)

H_1: Movie preference is not independent of gender (1 mark)

b $(5-1)\times(2-1)=4$ (1 mark)

c $\chi^2_{calc} = 11.111$ (2 marks)

p – value = 0.0253 (1 mark)

d $\chi^2_{calc} = 11.111 > 9.49$, therefore we reject H_0 and accept H_1 (2 marks)

ie movie preference is not independent of age.

19 a

Number of heads	0	1	2	3	4
Expected frequency	5	20	30	20	5

(2 marks)

b H_0: The data satisfies a binomial distribution (1 mark)

H_1: The data does not satisfy a binomial distribution (1 mark)

$\chi^2_{calc} = 7.583$ (2 marks)

p–value = 0.108 (1 mark)

$\chi^2_{calc} = 7.583 < 9.49$, therefore we accept H_0 (2 marks)

ie the observed data fits a binomial distribution.

20 a

Mark obtained	$0 \le x < 20$	$20 \le x < 40$	$40 \le x < 60$	$60 \le x < 80$	$80 \le x < 100$
Probability	0.004279	0.080233	0.365696	0.419444	0.121520
Expected frequency	2	22	97	111	32

(3 marks)

b Re-writing:

Mark obtained	$0 \le x < 40$	$40 \le x < 60$	$60 \le x < 80$	$80 \le x < 100$
Probability	0.0845	0.3657	0.4194	0.1215
Expected frequency	24	97	111	32

(2 marks)

Degrees of freedom = 3 (1 mark)

H_0: The data satisfies a normal distribution (1 mark)

H_1: The data does not satisfy a normal distribution (1 mark)

The critical value is $\chi^2_{(5\%)}(3) = 7.82$ (1 mark)

$\chi^2_{calc} = 10.47$ (2 marks)

p – value = 0.015 (1 mark)

$\chi^2_{calc} = 10.47 > 7.82$, therefore we reject H_0 and accept H_1 (2 marks)

ie the observed data does not fit a normal distribution with mean 62 and standard deviation 16.

Chapter 9

Skills check

1 a $2x^2 - 5x - 12$

b $14x^2 + 11x - 15$

2 a $(x+2)(2x+1)$

b $(x+3)(5x-2)$

3 a -3 **b** -11

Exercise 9A

1 a Vertex: $(-1.25, -1.125)$, x-intercepts: $(-0.5, 0)$, $(-2, 0)$, y-intercept: $(0, 2)$, domain = R, range = $\{y : y \ge -1.125\}$

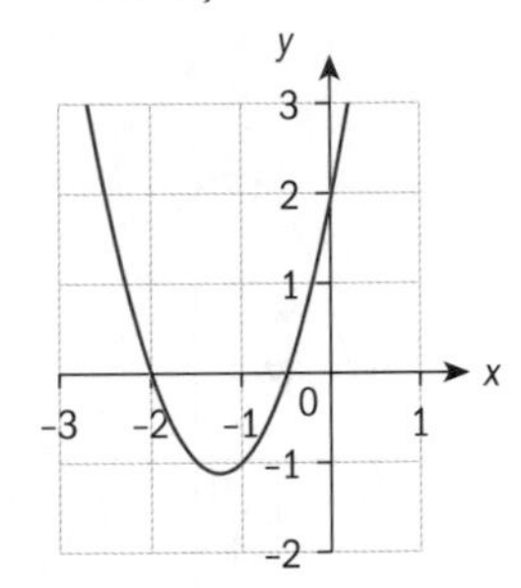

b Vertex = (3,16), x-intercepts = (7, 0), (−1, 0), y-intercept = (0, 7), domain = $\mathbb{R}$, range = $\{y : y \le 16\}$

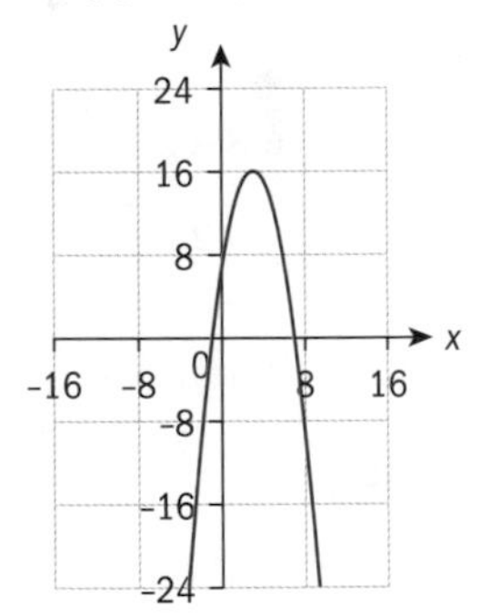

c Vertex = $\left(\frac{1}{6}, -\frac{49}{12}\right)$, x-intercepts = $(-1, 0)$ and $\left(\frac{4}{3}, 0\right)$, y-intercept = $(0, -4)$, domain = $\mathbb{R}$,

range = $\left\{y : y \ge -\frac{49}{12}\right\}$

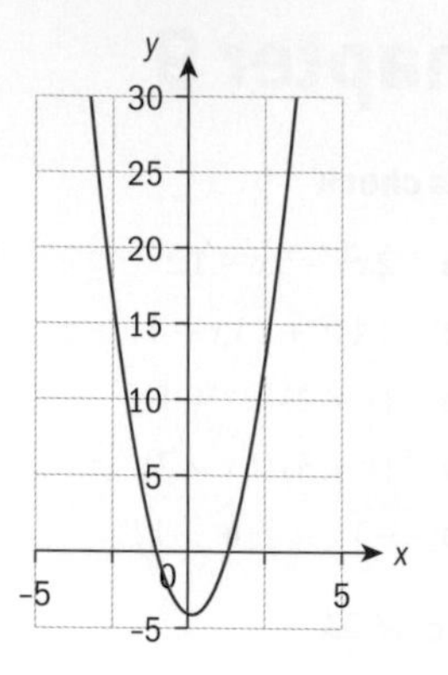

d Vertex = (−0.4, 12.8), x-intercepts = (1.2,0) and (−2,0), y-intercept = (0,12), domain = $\mathbb{R}$, range = $\{y{:}y \le 12.8\}$

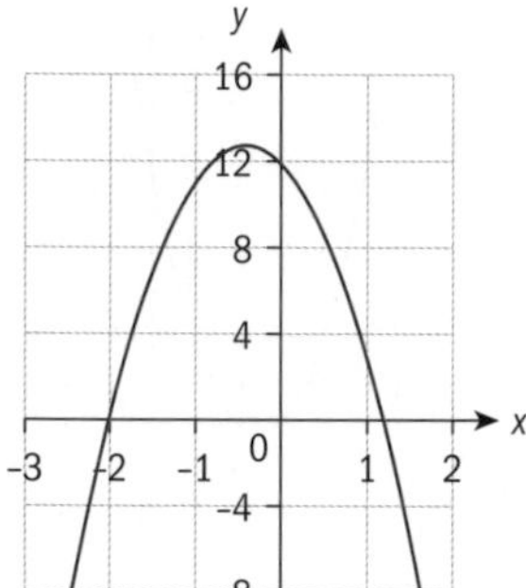

2 **a** (7.62, 0) and (−2.62, 0)

b (0, −8)

c

$$x = \frac{\left(2.5+\frac{5}{4}\sqrt{16.8}+2.5-\frac{5}{4}\sqrt{16.8}\right)}{2}$$

$= 2.5$

d (1.26, −9.88).

Exercise 9B

1 **a** $a = 1, b = -4, c = 2$

b $a = 2, b = -2, c = 3$

c $a = -2, b = 1, c = 1$

d $a = -1, b = 2, c = -3$

e $a = 5, b = 1, c = -10$

2 $a = 2, b = 2 - 18 = -16, c = -9$

Exercise 9C

1 **a** $h = 35 - x$

b $A = 35x - x^2$

c

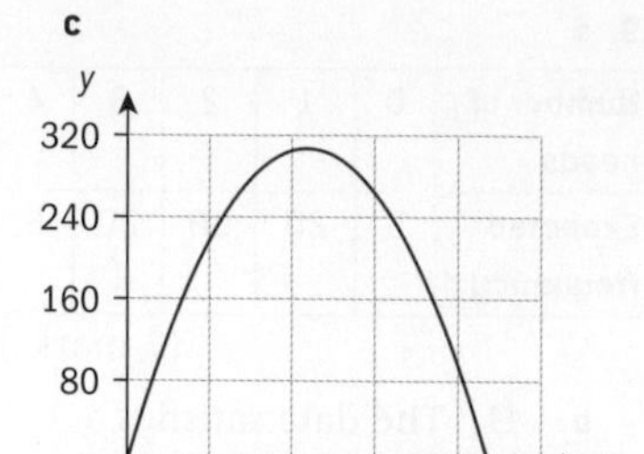

d (0, 0) and (35, 0)

e x must lie between 0 and 35

f $x = 17.5$ The largest area occurs when the width is 17.5 cm.

2 **a**

$a = 6, d = 10 - 6 = 4.$

$$S_n = \frac{n}{2}\left(2\times 6 + 4(n-1)\right)$$

$$= 2n(n-1) + 6n = 2n^2 + 4n$$

b $2n^2 + 4n - 880 = 0$, vertex = (−1, −882), x-intercepts = (−22,0) and (20,0), y-intercept = (0, −880)

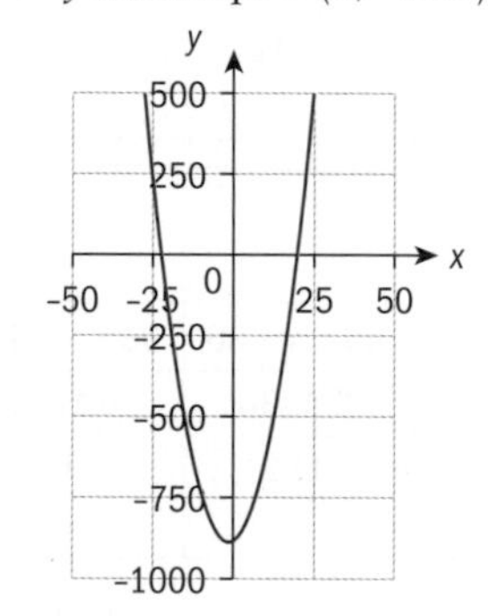

c (20,0)

d tells us the sum of the first 20 terms is 880

3 **a** $P(x) = (-0.12x^2 + 30x) - (0.1x^2 + 400) = -0.22x^2 + 30x - 400$

b Vertex = (68.2, 623), x-intercepts = (15.0,0) and (121, 0), y-intercept = (0, −400)

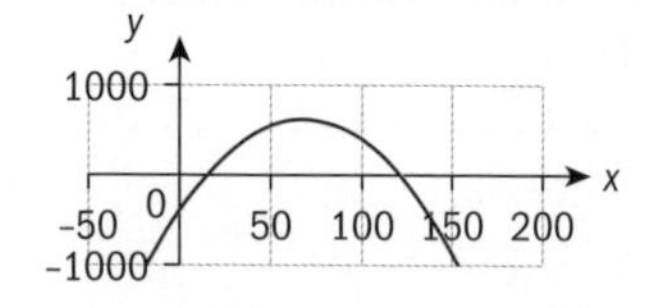

c The upper and lower bounds for the number of books that can be produced per week for the company to makes a profit

d The axis of symmetry is $x = 68.2$. This tells us that in order to maximize profit, 68 books should be produced per week.

4 **a** Vertex = (26, 5), x-intercepts = (6, 0) and (46, 0), y-intercept is (0, −3.45)

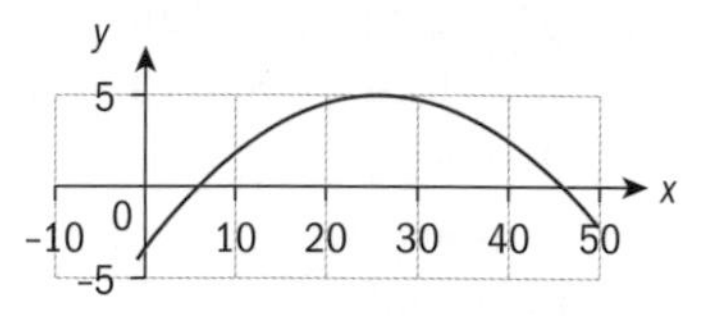

b The x-intercepts are (6, 0) and (46, 0), meaning ball is in the air between horizontal distances 6 m and 46 m.

c The line of symmetry is $x = \frac{6+46}{2} = 26$, meaning the ball is at maximum height when horizontal distance = 26 m.

5 **a** Vertex = (1.5,9), t-intercepts = (0,0) and (3,0), h-intercept = (0,0)

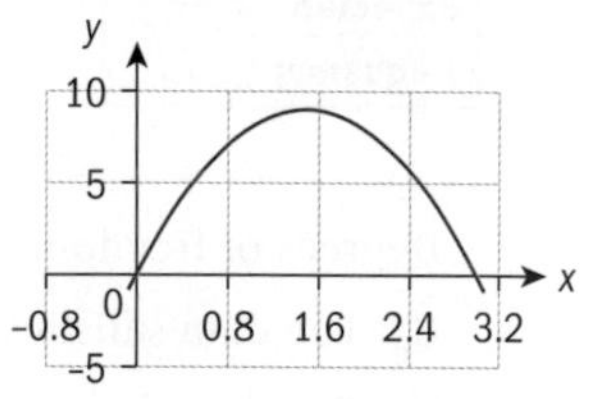

b t-intercepts = (0,0) and (3,0); ball in the air between 0 seconds and 3 seconds

c $x = \frac{3-0}{2} = 1.5$. The ball is at the maximum height after 1.5 seconds.

Exercise 9D

1 **a** **i** (2, −2) **ii** $x = 2$

b **i** (0.5,2.5) **ii** $x = 0.5$

c i (0.25, 1.125)
ii $x = 0.25$
d i (1, −2) ii $x = 1$
e i (−0.1, −10.05)
ii $x = -0.1$

2 a 5.05 m
b 31.7 m; distance from Zander that the ball hits the ground
c (0, 1); height at which the ball is hit is 1 m

3 a 4.3125 m
b $x = 7.5$
c $x = 16.8$; horizontal distance travelled by shot-put while travelling through the air
d (0, 1.5); height from which the shot put is thrown is 1.5 m.

4 a 6 m b $x = 10$
c $x = 0$ and $x = 20$; ball kicked at 0 m and lands at 20 m.

5 a 6.05 metres
b $x = 32.1$ and $x = 7.98$; parabola starts at 7.98 m and ends at 32.1 m.

6 a 8.01 m b 4 seconds

7 a $h = 50 - x$
b $A = x(50 - x)$
c

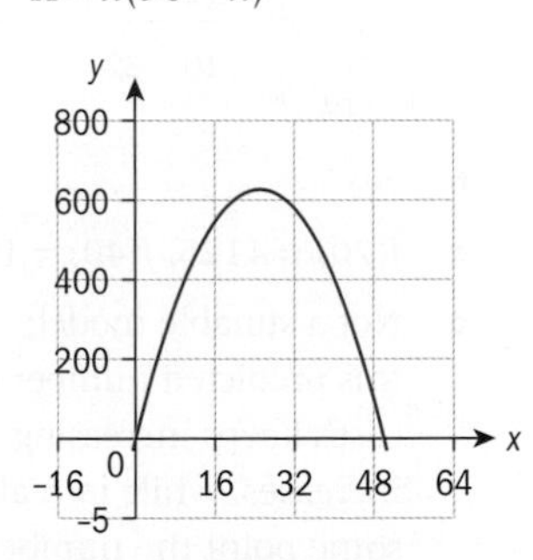

d (0, 0), (50, 0) e (0, 0)
f $x = 25$ g (25, 625)
h $A = 625\text{ cm}^2$ (25 cm × 25 cm)

8 a 96 m d 164 m
c 174 m

Exercise 9E

a i (0, 45) ii (5, 0), (9, 0)
iii (7, −4) iv $x = 7$
v

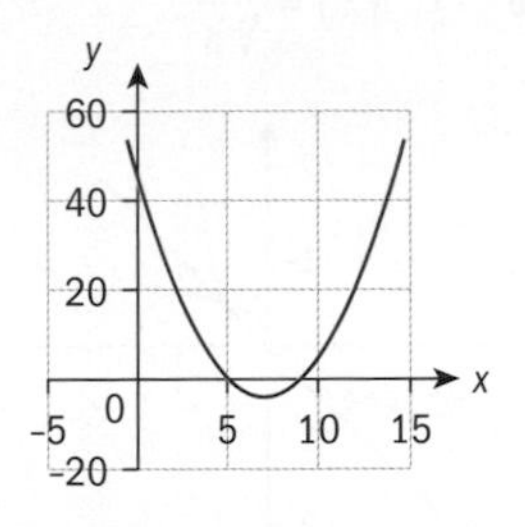

b i (0, −6)
ii (−3, 0), (1, 0)
iii (−1, −8) iv $x = -1$
v

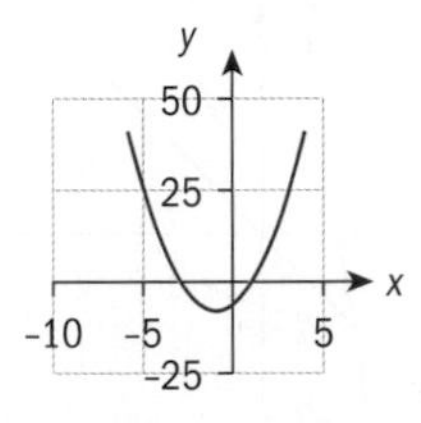

c i (0, −9)
ii (1, 0), (3, 0)
iii (2, 3) iv $x = 2$
v

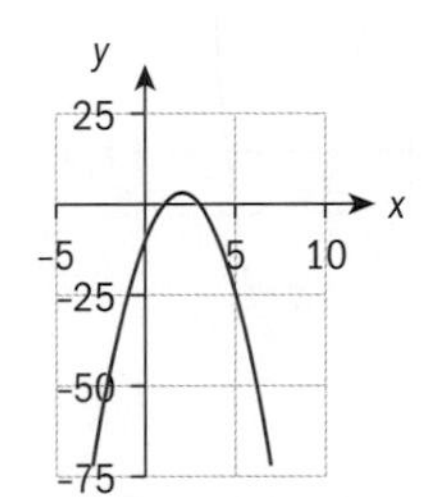

d i (0, 4)
ii (−1, 0), (2, 0)
iii (0.5, 4.5)
iv $x = 0.5$
v

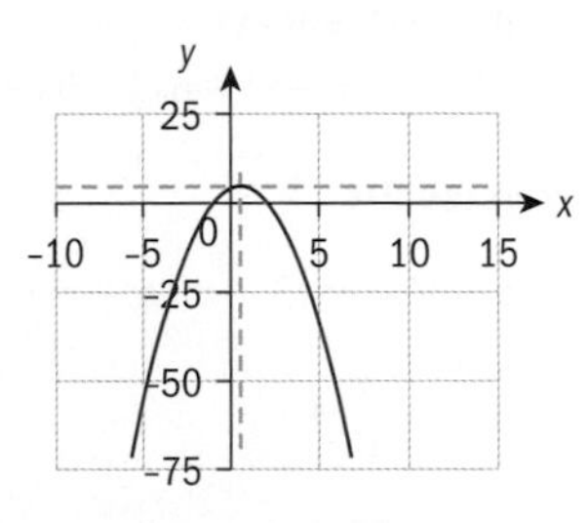

e i (0, −1)
ii (1, 0), (−0.5, 0)
iii (0.25, −1.125)
iv $x = 0.25$
v

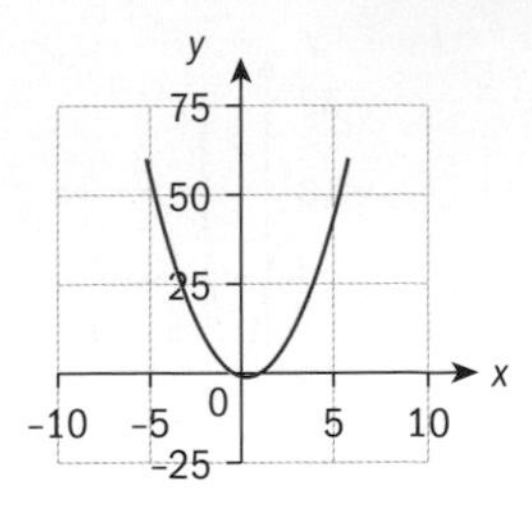

Exercise 9F

1 $y = x^2 + 3x - 4$
2 $y = -3x^2 + 3x + 6$
3 $y = 2x^2 + 7x - 4$
4 $y = 2x^2 - 12x + 22$
5 $y = -x^2 - 10x - 13$
6 $y = -2x^2 + 4x + 4$

Exercise 9G

1 x-intercept = (−3, 0), y-intercept = (0, 27)

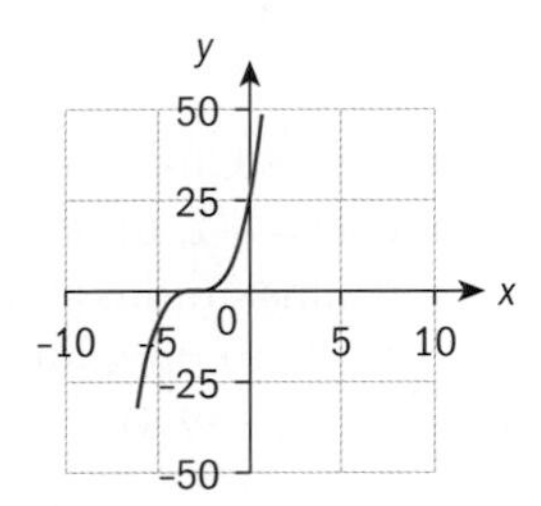

2 x-intercepts = (−1.15, 0), y-intercept = (0, 3), maximum = (−0.215, 3.11), minimum = (1.55, 0.369)

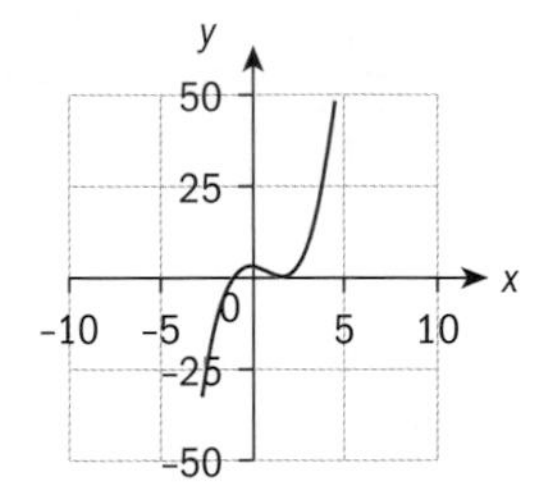

3 y-intercept = (0, 0), x-intercepts = (−2, 0), (0, 0) and (3, 0), maximum = (−1.12, 8.12), minimum = (1.79, −16.4)

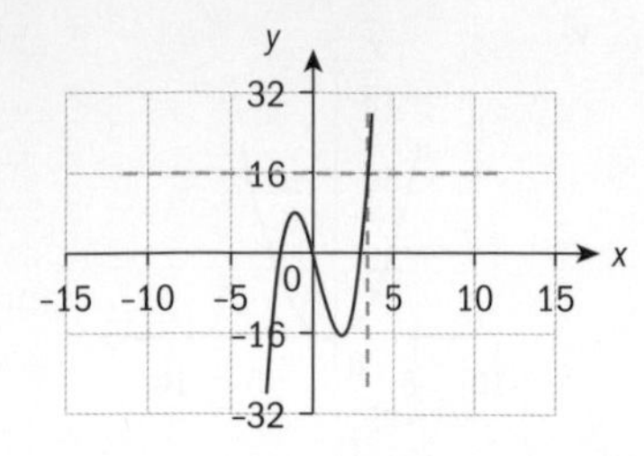

4 y-intercept = (0, 20), x-intercept = (−0.899, 0), no minimum or maximum points.

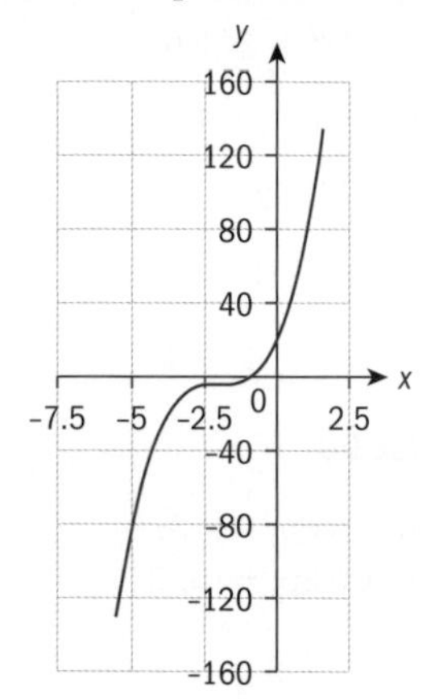

5 y-intercept = (0, 0), x-intercepts = (−1, 0), (0, 0) and (4, 0), maximum point = (−0.528, 3.39), minimum point = (2.53, −39.4)

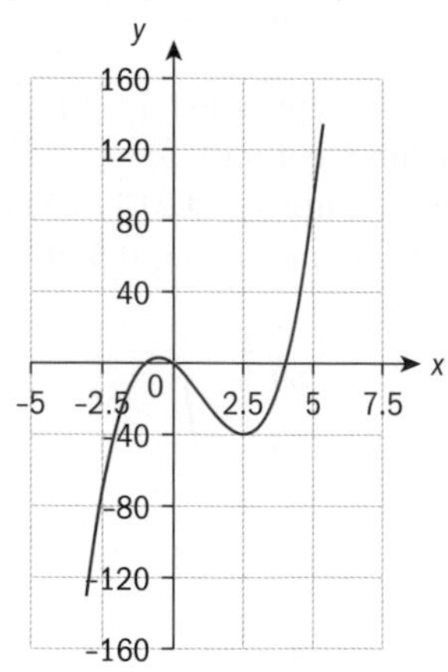

6 **a** $f^{-1}(x) = \sqrt[3]{x-3}$

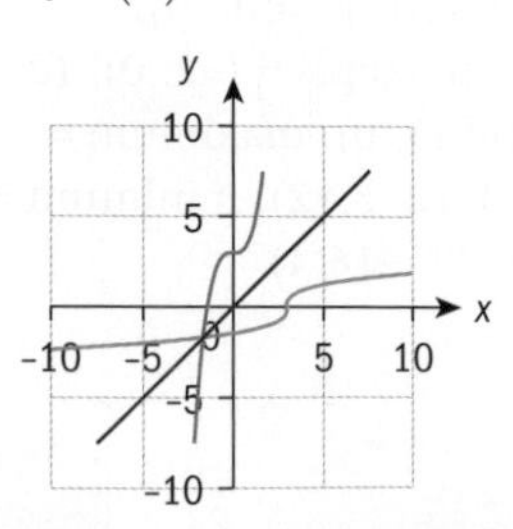

b $f^{-1}(x) = \sqrt[3]{\dfrac{x}{4}}$

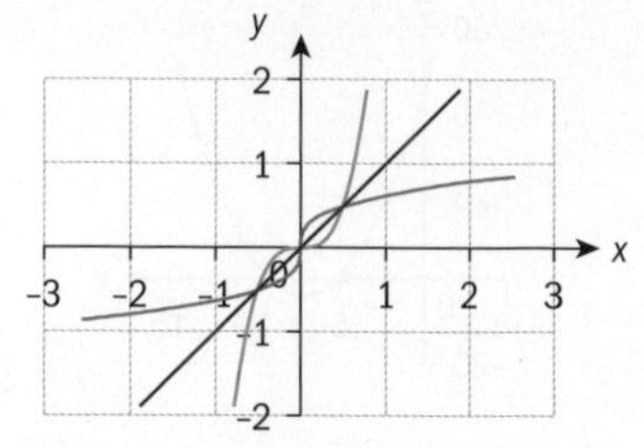

c $f^{-1}(x) = \sqrt[3]{\dfrac{x-1}{2}}$

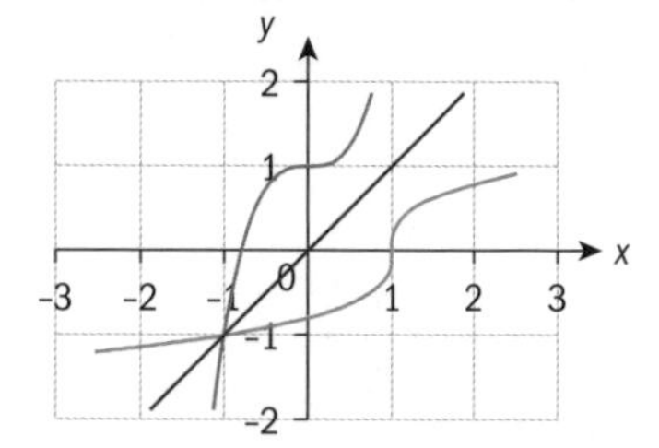

7

a x-intercepts = (5, 0), (2, 0), (−1, 0) y-intercept = (0, 10)

b (3.73, −10.4) and (0.268, 10.4)

c (2,0)

d $g(x) = f(-x)$
$= -x^3 - 6x^2 - 3x + 10$

Exercise 9H

1 **a** $V = 4x(3 - x)(4 - x)$

b

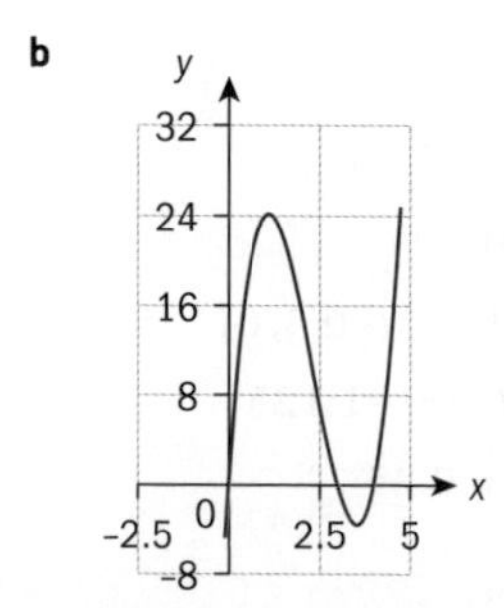

c (0, 0), (3, 0) and (4, 0)

d $0 < x < 3$

e Minimum = (3.54, −3.52), maximum = (1.13, 24.3)

f V cannot take the value of the local minimum as volume cannot be negative.

2 **a**

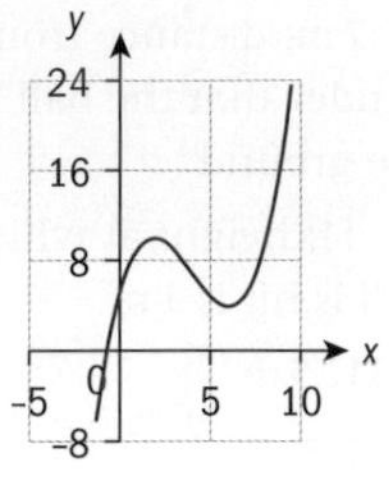

b Maximum = (1.92, 10.0), minimum = (6.08, 3.99)

c 6.01 m

3 **a**

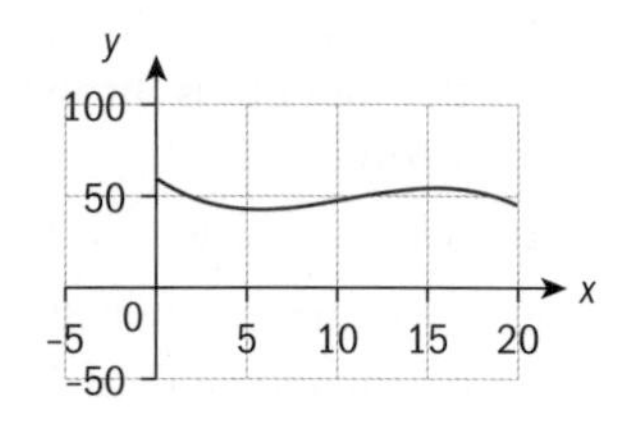

b 51 **c** 43

4 **a**

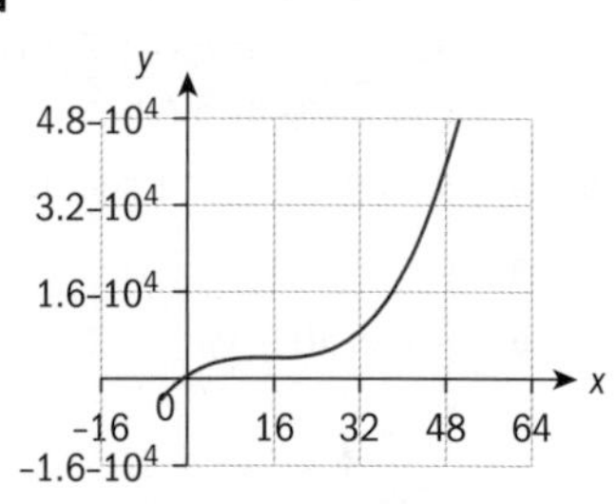

b No

c $f(20) = 4125$, $f(40) = 19\,625$

d Not a suitable model: this predicted number of cases keeps increasing as x increases, while in reality at some point the number of cases will start decreasing.

5 **a**

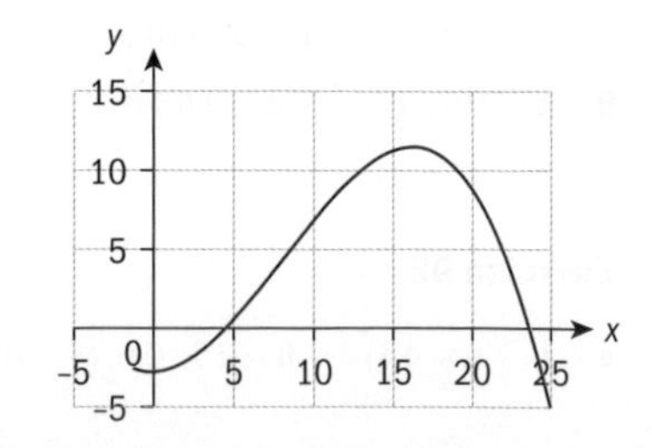

b Minimum = −2.76 °C, maximum 11.6 °C

c 6.89 °C

6 a 5.17

b Maximum = 13.2 minimum = 10.9

c Not a suitable model as after 33 minutes the number of bacteria continues to increase and never stops increasing.

7 This is a suitable model for the first 5 seconds, however after this the graph continues increasing indefinitely and so is no longer a suitable model after 5 seconds.

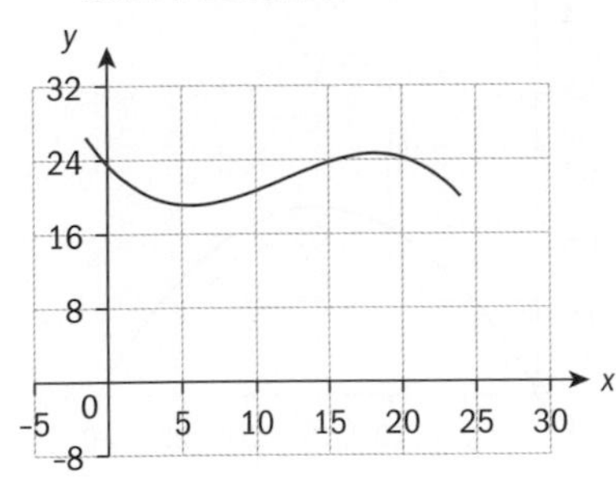

8 a Lowest = 19.2 °C, highest = 24.7 °C

b 20.7 °C c 11.7 hours

d Not useful as after 24 hours the graph just continues to decrease while it is very unlikely that the temperature will do this.

Exercise 9I

1 a (1.18,1.82)

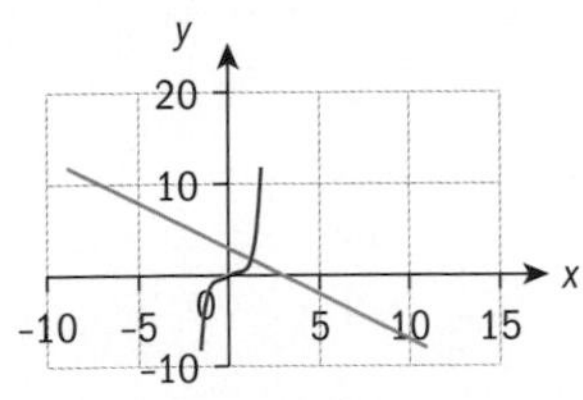

b (1, 6) and (−1, 6)

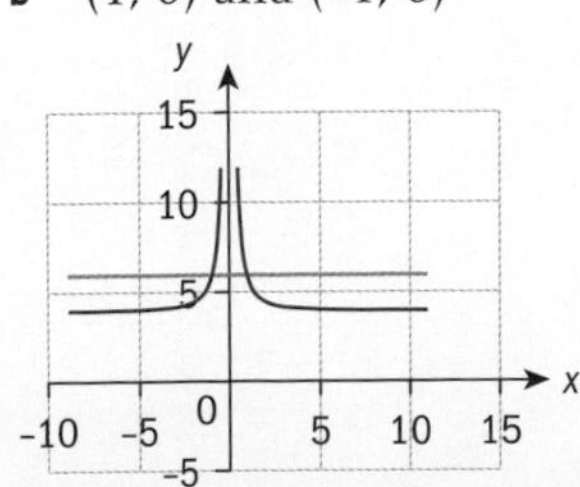

c (3, 10), (0, 1) and (−1, −2)

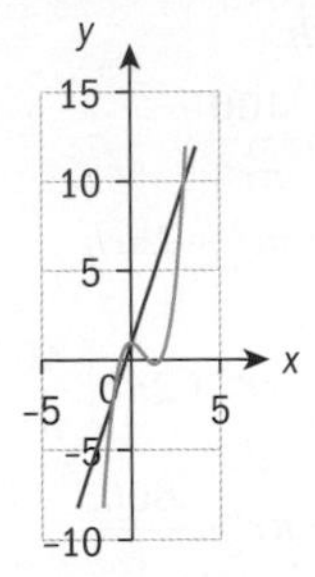

2 0.528 and 3.35 seconds

3 0.337 and 5.60 seconds

4 18.8 or 19 units and 38.9 or 39 units

Exercise 9J

1 $p = \frac{7}{4} \times n$, 70 AUD

2 a $d = 2.25t^2$ b 56.25 m

c 3.4 s

3 524 cm^3

Exercise 9K

1 a $m = \frac{3000}{w}$ b US$ 200

2 a $h = \frac{12}{p}$ b 1.2 hours

c 3 people

3 a $y = \frac{48}{x^2}$ b $y = \frac{4}{3}$

c $x = \pm 2$

4 $V = 40\,\text{m}^3$

5 8 pieces

6 a Decreased by a factor of 4

b Increases by a factor of 1.5625

7 a 6 b 64

Exercise 9L

1 −9

2 a $V = x(9 - 2x)(6 - 2x)$
$= x(54 - 30x + 4x^2)$
$= 4x^3 - 30x^2 + 54x$

b Maximum value of V is 28.5 cm^3

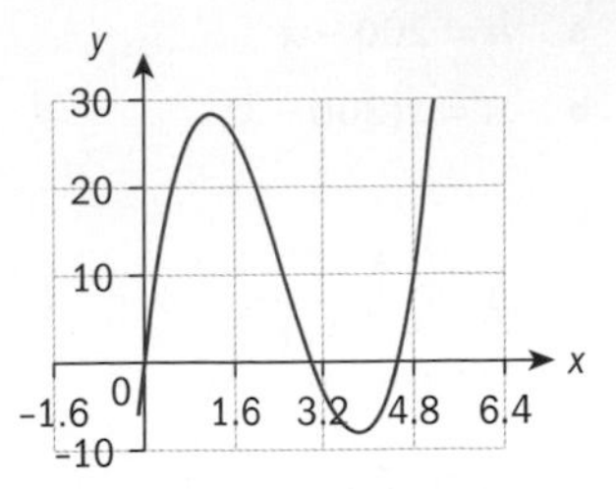

3 Maximum daily profit is $120 and 50 cakes are needed to be baked to make it.

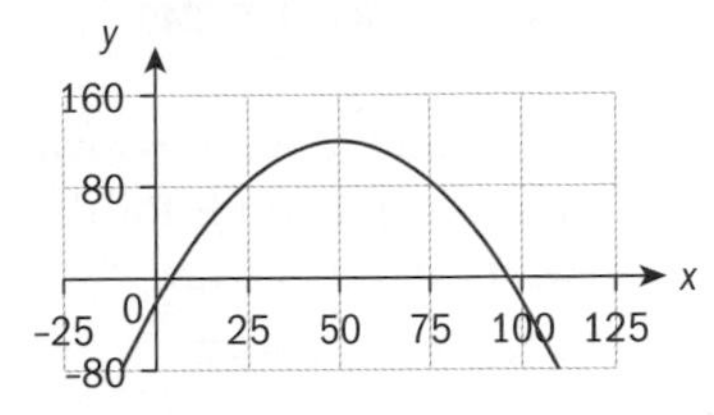

4 US$ 391.11

5 Maximum = 13.2 million, minimum = 10.9 million

6 The radius that gives the maximum volume is 16.3 cm, the height is 32.6 cm and this gives a maximum volume of 27 145 cm^3.

Chapter review

1 a Vertex = (0, 2), y-intercept = (0, 2), x-intercepts = $(\sqrt{2}, 0), (-\sqrt{2}, 0)$

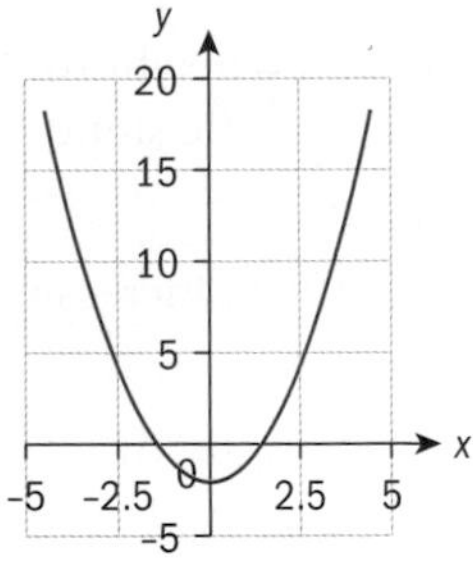

b y-intercept = (0, 4), x-intercepts = (1, 0) and (4, 0), vertex = (2.5, −2.25)

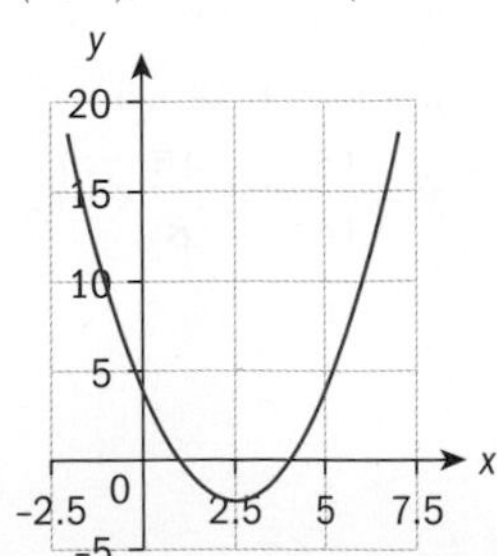

2 **a** $h = 200 - x$

b $A = x(200 - x)$

c

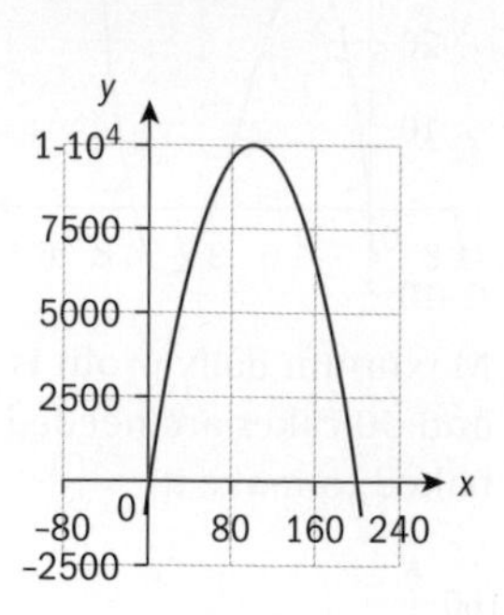

d (0, 0) and (200, 0) = lower and upper bounds for the width of the picture

3 **a** $a = 1$, $b = 6$, $c = -7$

b $(0, -7)$

c $(-7, 0)$ and $(1, 0)$

d $x = -3$

e $(-3, -16)$

4 **a**

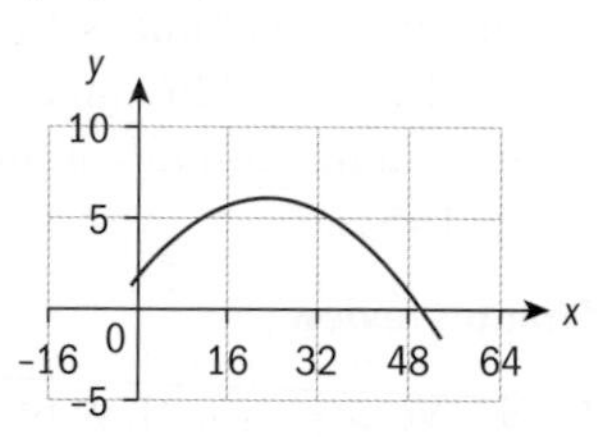

b 5.976 m **c** 6.04 m

d 50 m, horizontal distance javelin travels

5 **a** (0, 1.9) = height Anmol lets go of the stone

b 10.0 m

c $0.204\,\text{ms}^{-1}$, increasing

d 4 s

6 **a** **i** (2, 0), (4, 0)

ii $x = 3$ **iii** $(3, -3)$

b **i** $(-1, 0)$, $(5, 0)$

ii $x = 2$ **iii** $(2, -36)$

7 **a** $y = x^2 - x - 6$

b $y = -x^2 + 3x + 4$

c $y = \frac{15}{16}x^2 + \frac{45}{8}x - 5$

8 **a** The volume of a cylinder is $\pi r^2 h$.

b $h = \frac{400}{\pi r^2}$

c $A = \pi r^2 + 2\pi rh$

d $A = \pi r^2 + 2\pi r\left(\frac{400}{\pi r^2}\right)$

$= \pi r^2 + \frac{800}{r}$

e

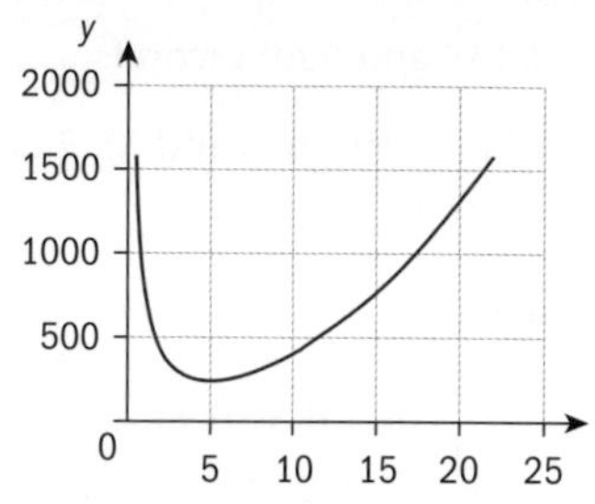

f Minimum area = $239\,\text{cm}^2$ when $r = 5.03$ cm

9 **a** y-intercept = $(0, -1)$, x-intercept = $\sqrt[3]{\frac{1}{2}}$

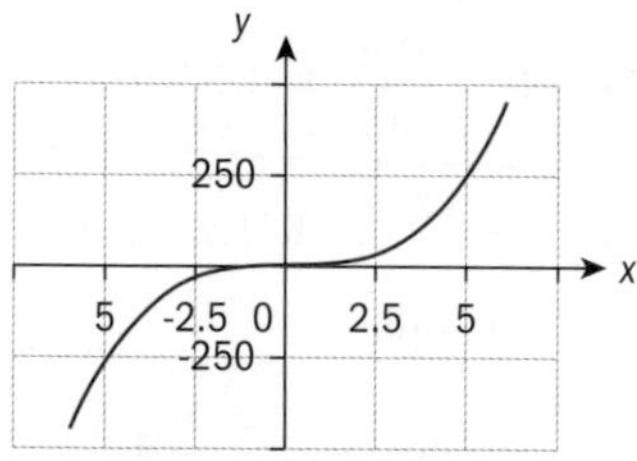

b y-intercept = (0, 3), x-intercepts are (1, 0), (−1, 0) and (3, 0)

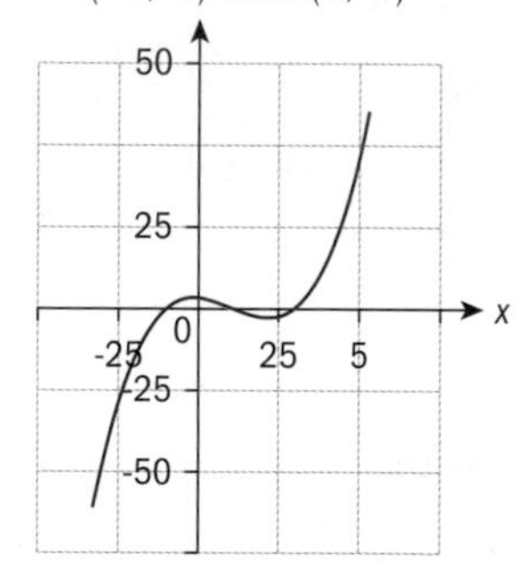

10 **a** $y = 2x^2 + 8x - 10$

b $y = -3x^2 + 6x + 24$

11 **a**

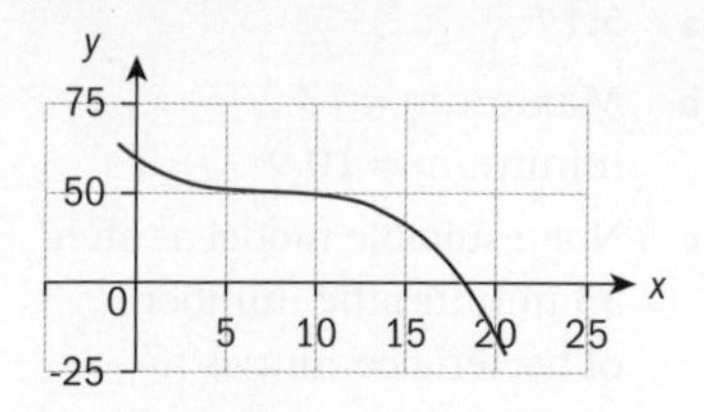

b 53 **c** 46

d 70 in 2000 **e** 0 in 2018

f 2001

12 **a,b**

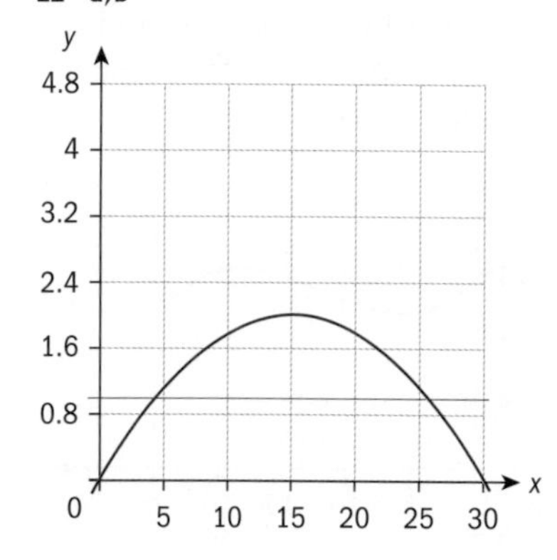

c (4.39, 1) and (25.65, 1)

d 21.3 m

13 **a** $d = 80t$ **b** 160 km

c 3.75 hours

14 **a** $p = \frac{12}{n}$ **b** 4

c 4

Exam-style questions

15 a Using GDC, (1 mark)

maximum height reached is 2.38 m (1 mark)

b $x = 11.98 \Rightarrow H(11.98) = 1.073...$ (1 mark)

$1.073... > 1.07$. So ball just goes passes over the net. (1 mark)

c Choose X as being the value when $y = 0$, (1 mark)

From GDC (or otherwise), $X = 14.9$ (1 mark)

This is sensible, since if $X > 14.9$, then $H < 0$, making no sense with regard to the model. (1 mark)

(The ball rebounds off the floor after this point, so the original curve would not model the motion of the ball any longer.)

16 $F \propto \frac{1}{r^2}$

$F = \frac{k}{r^2}$

So $k = Fr^2$ (1 mark)

$= 980 \times 6370^2$ (1 mark)

ie $F = \frac{980 \times 6370^2}{r^2}$

11 km above the earth's surface, $r = 6370 + 11 = 6381$ km (1 mark)

So $F = \frac{980 \times 6370^2}{6381^2} \approx 977\text{N}$ (1 mark)

17 a Substituting $t = 0$ into $P(t)$ gives $P(0) = £10\,850$ (2 marks)

b Maximum salary occurs when $t = 40$ and is £53 990 (2 marks)

c From GDC, $P(t) = 35 \Rightarrow t = 7, 24, 32$ years (3 marks)

d Salary decreases for $28.4 - 13.6 = 14.8$ years. (2 marks)

Proportion Mannie's salary was increasing was $\frac{40 - 14.8}{40} = 0.63$ (1 mark)

18 a $y = -0.340x^2 + 5.04x + 4.82$ (3 marks)

b $12.1 - 2.72 = 9.38$ months (2 marks)

c The maximum height of the quadratic curve is only 23.5°C, compared to 26°C in real life. (1 mark)

When $x = 1$, the quadratic curve gives a reading of 9.5°C, some way off the actual 13°C. (1 mark)

19 a

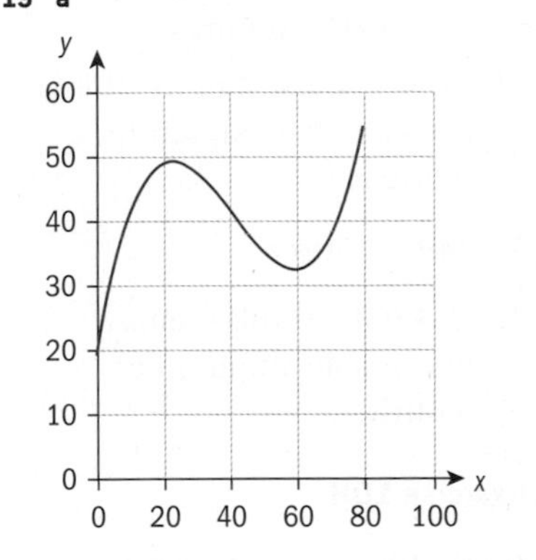

(2 marks)

b 50 ferrets (1 mark)

c $0 < t < 8.58$ and (2 marks)

$44.4 < t < 71.4$ (2 marks)

20 $l \propto \frac{1}{f} \Rightarrow l = \frac{k}{f}$

$k = lf = 400 \times 13 = 5200$ (2 marks)

When $l = 10$, (1 mark)

$f = \frac{k}{l} = \frac{5200}{10} = 520$ Hz (1 mark)

Chapter 10

Skills check

1 a 4^{11} **b** 5^6

2 a 0.72 **d** 42

3 a 30 **b** 39

Exercise 10A

1 a 2 **b** 49 152

2 a 1.2 **b** 3.72 m

c Unlikely to be accurate, as the plant cannot grow indefinitely or live forever

3 a 9 **b** $\frac{1}{3}$

c $\frac{1}{81}$

4 a $\frac{50}{10} = 5. \frac{10}{2} = 5$. The ratio between successive terms is the same, so it is a geometric sequence.

b 31 250

c $f_{365} = 2 \times 5^{364}$. This answer is unlikely to be correct as it is likely to be far larger than the population of the town.

5 $r = \frac{1}{4}$, $u_1 = -128$

Exercise 10B

1 a 230 000 **b** 349 801

c Probably, as population in the world is also increasing

2 US$2.44

3 33 079.14 Euros

4 a 10 222.08 Euros

b 8.59... years

Exercise 10C

1 a 3 **b** 59 048

2 a EGP 2760.20

b EGP 33 530.22

3 a 13 045 **b** 88 130

4 a $r=4,\ a_1=\frac{1}{2}$ b 8192

c 10922.5

5 a CNY 280800, CNY 292032

b CNY 1790903.38

6 a 76.50 Euros

b 390.30 Euros

7 1797.28

8 a CAD 1898.44

b CAD 26304.69

9 29524

10 a 580 b 8048

Exercise 10D

1 Misty

2 a Lao Kip 643291158.90

b 22.01 years

3 a 2.4% b 25 years

4 6 years

5 a 5307.29 Euros b 1.35%

Exercise 10E

1 a MYR 11032.28

b 15.1 years

2 a 4288.34 Euros

b $x = 3265$ Euros

c 21 years

3 a SGD 48736.12

b SGD 59265.14

4 Mrs Chang earns more.

5 a £596.33 b 18 years

6 6.95

7 a Kyle b 21 years

c 26 years

Exercise 10F

1 £77434.02. It is a good investment as she has a guaranteed interest rate for 30 years.

2 MXN 160806.67

3 $P = \frac{0.06 \times 25000}{1-(1+0.06)^{-10}} = 3396.70$ Euros

4 £112.68; any sensible reason for either choice.

5 £23752.11

6 US$273.24; saving for pension or any other sensible reason.

Exercise 10G

1 a 989.93 Euros

b 1590.98 Euros

c The difference in monthly payments is 601.05 Euros. In the 20-year case, the total amount paid is 240 × 989.93 = 237583.20 Euros so the total amount of interest payed is 87583.20 Euros. In the 10-year case the total amount payed is 120 × 1590.98 = 190917 and so the interest paid is 40917.60 Euros.

2 £1346.80 per month. This is a reasonable amount for a mortgage.

3 449.25 Euros

4 US$594.91; this seems a reasonable amount per month.

Exercise 10H

1 a (0, 2) b $y = 1$

c Increasing

2 a (0, −2) b $y = -3$

c Decreasing

3 a (0, 1) b $y = 0$

c Increasing

4 a (0, 3) b $y = 2$

c Increasing

5 a (0, −2) b $y = -5$

c Increasing

6 a (0, 7) b $y = 3$

c Decreasing

7 a (0, 4) b $y = -1$

c Increasing

8 a (0, 1) b $y = -1$

c Decreasing

Exercise 10I

1 a 2 m b 5.19 m

c 16.9 weeks

2 a 130 g b 67.9 g

c $y = 0$ d 4.27 years

3 a

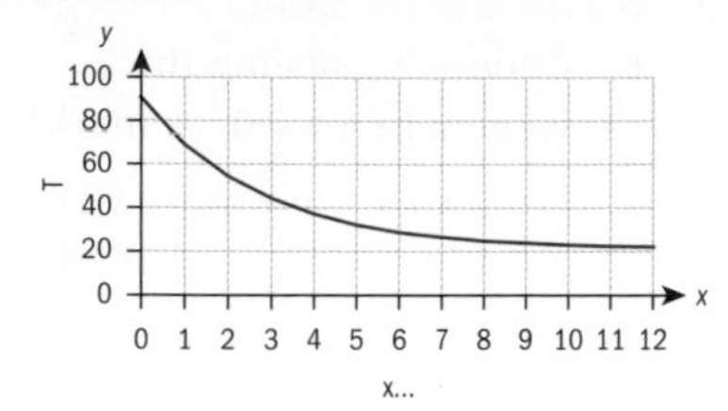

b 91 °C c 31.9 °C

d 2.37 min e $y = 21$

f 21 °C

4 a 5% b £26463.31

c 11.5 years d $y = 0$

5 a 10% b 66550000

c 2025

6 a

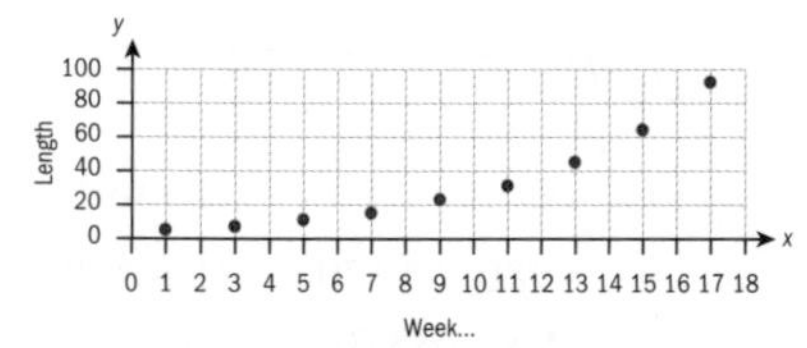

b Exponential

c $y = 4.23(1.20)^x$

d 26.2 cm. This is a reasonable estimate.

7 a

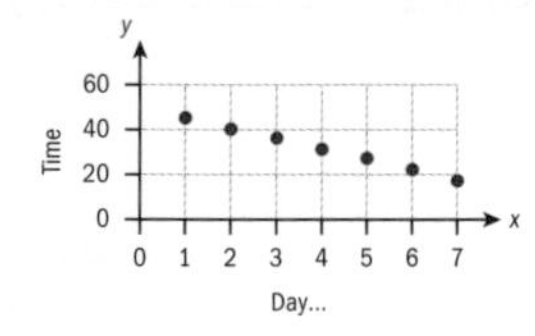

b Linear decay

c $y = -4.61x + 49.6$

d 3.5 min

Exercise 10J

1 a 1 m

b 2.30 weeks

2 a $f(2) = 100$

b 2.70 weeks

c $x = 3.30$ weeks

3 a 2.48 b 10

c 1000

4 a 1.60 b 66.7

c 1.46

5 a 5

b Around 5 minutes

Chapter review

1 a 2.5 b $\frac{78125}{64}$

c $\frac{130123}{64}$

2 $184.71

3 a 10% per year

b 300 000 Euros

c 643 076.64 Euros

4 a 11 b 177 146

5 a £9610.20

b £9719.76

6 a SGD 3453.80

b 22 years

7 2.35%

8 73.67 Euros

9 115.36 Euros

10 a 42 296 b 7 years

11 a 95 °C b 33.0 °C

c 7.46 min d 21 °C

12 a $y = 16$ b (0, 20)

13 a 1100 b 11 days

14 a 1.15 m b 29 weeks

Exam-style questions

15 $(6 + k) - k^2 = (12 - k) - (6 + k)$ (1 mark)

$k(k - 3) = 0$ (1 mark)

$k = 0$ or $k = 3$ (2 marks)

For $k = 3$ the sequence is constant (1 mark)

Hence, $k = 0$

$a_n = 6n - 6$ (1 mark)

16 a $100\,000 \times 0.05 \times 10$ = US$50 000 (2 marks)

b $100\,000 \times 1.05^{10} -$ 100 000 = US$62 889.46 (2 marks)

c $100\,000 \times \left(1 + \frac{5}{100 \times 12}\right)^{10 \times 12}$ $-100\,000$ = US$64 700.95 (2 marks)

17 a i Use the GDC finance function with (1 mark)

$n = 20$

$i\% = 12.8$

$PV = 20000$

$PMT =$

$FV = 0$

$PpY = 4$

$CpY = 4$

Answer PMT = SOL 1369.29 (2 d.p.) (1 mark)

ii $PMT \times 20$ = SOL 27385.89(2 d.p.) (2 marks)

iii SOL 7385.89 (2 d.p.) (1 mark)

b i $400 \times 60 + 0.1 \times 20000$ = SOL 26000 (2 marks)

ii 10% deposit = SOL 2000; borrowed SOL 18 000 (1 mark)

Use Financial APP with (1 mark)

$n = 60$

$i\% =$

$PV = 18000$

$PMT = -400$

$FV = 0$

$PpY = 12$

$CpY = 12$

R = 11.96 (2 d.p.) (1 mark)

c If Carlos has 10% of the money the option *B* is better as he can save SOL 1385.89. (2 marks)

(If he does not have the deposit amount then he must choose option A)

18 a $\mathbb{R}$ (1 mark)

b −2 (1 mark)

c 12 (1 mark)

d $y < 16$ (1 mark)

e $x > -2$ (1 mark)

f $y = 16$ (1 mark)

19 a

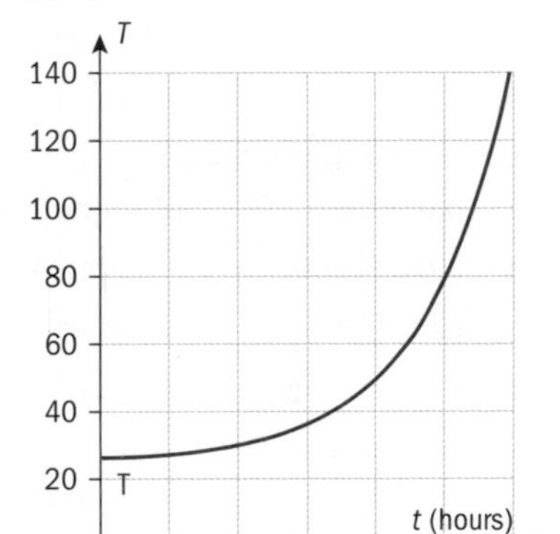

b $T(6) = 25 + e^{0.4 \times 6}$

36.0 °C (1 d.p.) (1 mark)

c Solve $25 + e^{0.4t} = 100$ (1 mark)

$t = 10.793...$ (1 mark)

10 hours 48 minutes (1 mark)

20 a

log*N*	2.06	2.25	2.68	2.95	3.20

(2 marks)

b

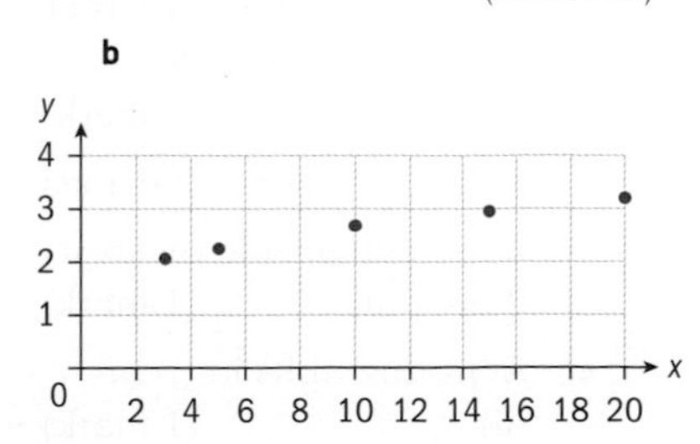

(Scales: 2 marks; Points plotted correctly: 1 mark; Labels: 1 mark)

c Linear model (1 mark)

d $\log N = a + bt \Rightarrow N(t) = 10^{a+bt}$ (2 marks)

$N(t) = \underbrace{10^a}_{A} \times \underbrace{(10^b)}_{B}^t$ (2 marks)

e Determine coefficients of regression line

$\log N = a + bt \Rightarrow$
$a = 1.9181..., b = 0.06716...$ (3 marks)

$A = 10^{1.9181...} = 82.8$ (1 mark)

$B = 10^{0.06716...} = 1.17$ (1 mark)

21 a $N(0) = 35$ (1 mark)

b $N(4) = 409.97$; 410 bacteria (2 marks)

c $N(t) > 1000$ (1 mark)

$t > 5.449...$ (1 mark)

d $0 \le t < 5.45$ (1 mark)

22 Option 1: $10\,000 \times 10 =$ US\$100 000 (1 mark)

Option 2: $(2500 \times 2 + 2000 \times 9) \times 5 =$ US\$115 000 (2 marks)

Option 3:
$100 \times \frac{2^{10}-1}{2-1} =$ US\$102 300 (2 marks)

Option 2 has the greatest total value (1 mark)

23 a i $T_5 = 73.205 \Rightarrow 73\,205$ taxis (2 marks)

ii $T_n = 100 \Rightarrow n = 8.27$ (1 mark)

This is in the 9th year after 2010, which is 2019. (1 mark)

b $P_{10} = 2.1873705...$ (2 marks)
2 187 000 people (nearest thousand) (1 mark)

c Adjusting units in (i) or (ii) (1 mark)

i $\frac{P_5 \times 10^6}{T_5 \times 10^3} = 28.4$ people per taxi; (2 marks)

ii $\frac{P_{10} \times 10^6}{T_{10} \times 10^3} = 18.6$ people per taxi. (1 mark)

d The model predicts a reduction in the number of people per taxi which may mean that the taxis are in use for less hours, or less taxis are used everyday. (1 mark)

24 a i $f(0) = \left(\frac{3}{2}\right)^{-1} + 2$

$= 2\frac{2}{3} = 2.67$ (1 mark)

ii $(2.67, 0)$ (1 mark)

b $x = \left(\frac{3}{2}\right)^{y-1} + 2$ (1 mark)

$\left(\frac{3}{2}\right)^{y-1} = x - 2$ (1 mark)

$y = \log_{\frac{3}{2}}(x-2) + 1$ (2 marks)

$g(x) = \log_{\frac{3}{2}}(x-2) + 1$

c i $y = 2$ (1 mark)

ii $x = 2$ (1 mark)

d $x > 2$ (1 mark)

e Use GDC solver or intersection of graphs (1 mark)

$x = 2.16$ (1 mark)

25 a i 2 min 16 sec (1 mark)

ii 2 minutes 6 seconds = 126 seconds (1 mark)

$126 \times 1.04^2 = 136.28...$ sec = 2 min 16 sec (3 marks)

b Katharina´s time: $(126 \times 2 + 9 \times 5) \times 5 = 1485$ sec = 24 min 45 sec (2 marks)

Carolina´s time:

$126 \times \frac{1.04^{10}-1}{1.04-1}$ (2 marks)
$= 1512.77$ sec
= 25 min 13 sec

c 28 seconds (1 mark)

26 a Each year, the new annual salary is obtained by multiplying the previous salary by the same constant. This defines a geometric sequence. (1 mark)

$R = 1.05$ (1 mark)

b $20 \times 1.05^2 =$ COP 22.05 million (2 marks)

c $20 \times \frac{1.05^{10}-1}{1.05-1}$ million
= COP 251.5578... (2 marks)

COP 251 558 000 (nearest thousand pesos) (1 mark)

d $\frac{2 \times 20 + 9x}{2} \times 10$
$= 251.5578...$ (2 marks)

$x = 1.145\,730\,01...$ million pesos (2 marks)

$x = 1\,145\,730$ pesos (nearest peso) (1 mark)

e Original offer annual salary: $u_n = 20 \times 1.05^{n-1}$ (1 mark)

Alternative scheme annual salary: $v_n = 20 + 1.14573(n-1)$ (1 mark)

Use GDC to tabulate values or solve $v_n > u_n$ (1 mark)

$v_n > u_n \Rightarrow n \le 7$, so his earnings are higher in the first 7 years. (1 mark)

Chapter 11

Skills check

1 PR = 12 m, $\hat{P} = 22.6°$, $\hat{Q} = 67.4°$

2 68.1 m

3 a −11.5

b 2.08

Exercise 11A

1 a 25.4°, 155°, 205°, 335°

b 73.4° and 286°

2 a i $\theta_1 = 45°$, 225°

ii $\theta_2 = 53.1°$, 127°

iii $\theta_3 = 95.7°$, 264°

b i (49.5, 49.5), (−49.5, −49.5)

ii (42.0, 56.0), (−42.0, 56.0)

iii (−6.95, 69.7), (−6.95, −69.7)

3 a 13°C

b Middle of February and November

4 a 12.8 m

b 7.7 hours, 10.3 hours, 19.7 hours and 22.3 hours

Exercise 11B

1 a Amplitude = 3, period = 90°, principal axis is $y = 1$, range = $\{y: -2 \le y \le 4\}$

b Amplitude = 0.5, period = 720°, principal axis is $y = -3$, range = $\{y: -3.5 \le y \le -2.5\}$

c Amplitude = 7.1, period = 120°, principal axis is $y = 1$, range = $\{y: -6.1 \le y \le 8.1\}$

d Amplitude = 5, period = 720°, principal axis is $y = 8.1$, range = $\{y: 3.1 \le y \le 13.1\}$

2 a $a = 2$ b $a = 1, b = 2$

c $a = 3, b = 0.5, d = -1$

3 a $b = 2$

b $a = -3, d = 1$

c $a = 2.5, b = 0.5, d = 0$

4 a $t = 3.1.\ s = -2.1.\ r = 45$

b i t ii s iii r

5 a $d = -0.8,\ a = 5.4,\ b = 90$

b a = highest amount above sea level, b = number of cycles in 6 hours. d = sea level.

6 a

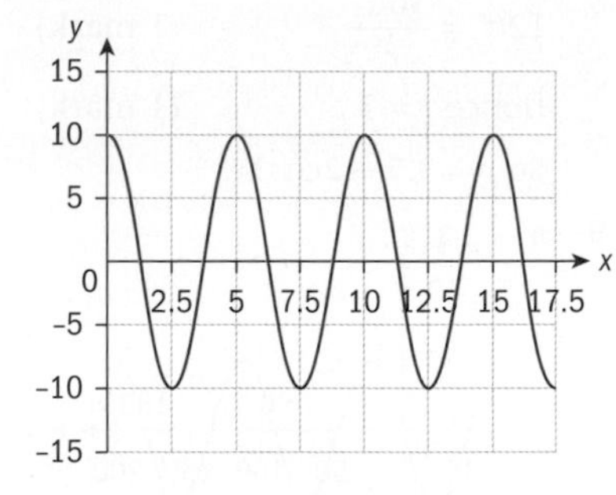

b $r = 10,\ s = 72$

7 0.25

Exercise 11C

1 a

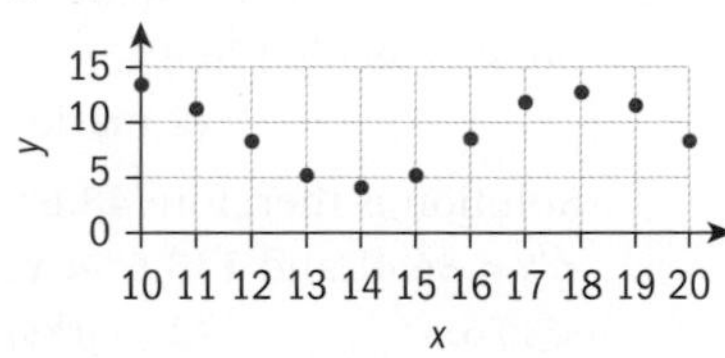

b $y = 4.65\sin(45x) + 8.75$

2 a

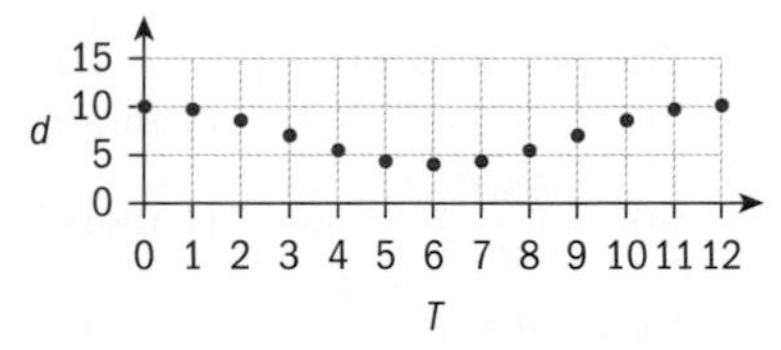

b $a = \dfrac{10.1 - 4}{2} = 3.05$,

$d = \dfrac{10.1 + 4}{2} = 7.05$,

$b = \dfrac{360}{12} = 30$,

so depth is modelled by $y = 3.05\cos(30x) + 7.05$

c Between 0820 and 1540 and between 2020 and 0340

3 a

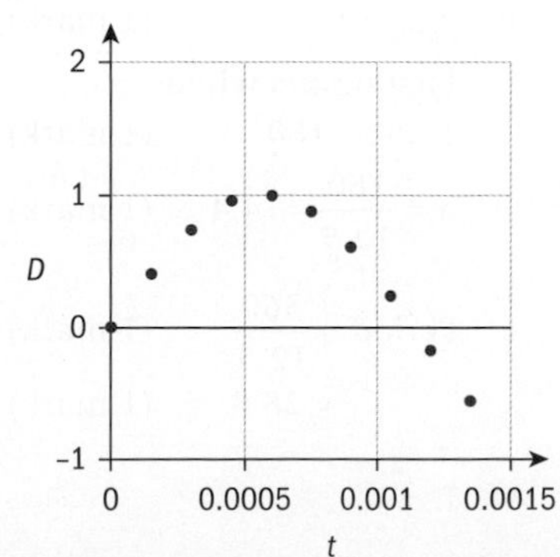

b $D(t) = 0.9961\sin(150\,000t)$

c 0.283 decibels

d −0.863 decibels

e Part **c** is more reliable as 0.000 11 falls within the given data range while 0.002 is outside it.

4 a $y = 2.395\cos(\alpha) + 7.705$

b $\text{AD} = \sqrt{65.05 - 36.96\cos(180 - \alpha - \beta)}$

c Model is a reasonable fit, but not accurate as cosine rule depends on two angles, not just one.

5 a So they can be graphed on the cartesian plane, which uses decimal notation.

Month(t)	Rise	Set
0	11.33	15.70
1	10.15	17.22
2	8.62	18.72
3	6.78	20.27
4	5.02	21.82
5	3.4	23.45
6	3.08	23.95
7	4.57	22.57
8	6.15	20.78
9	7.58	19.00
10	9.15	17.22
11	10.73	15.82

b $r(t) = 4.125\cos(30t) + 7.208$, $s(t) = -4.125\cos(30t) + 19.825$

c i Around 3 months or 90 days

ii About $\frac{1}{4}$ of the year

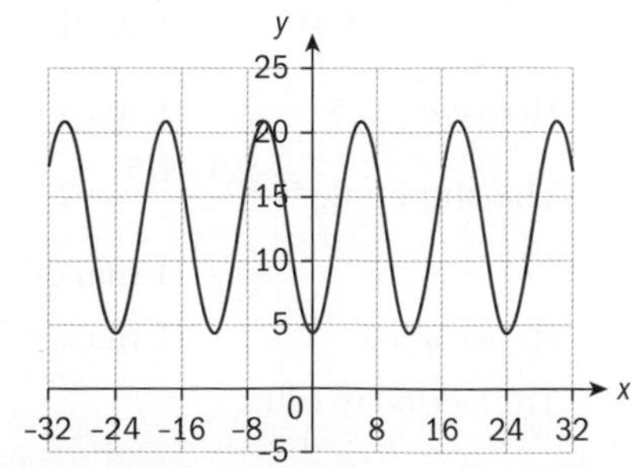

6 a $r(t) = 0.995\cos(30t) + 5.88$

b Between around 7 October and 24 March

7 a $y = -15\cos(6000t) + 15$

b $500\pi \text{ cm s}^{-1}$

Chapter review

1 $t = 499$ and 581

2 a $y = -\cos(3x) - 1$

b $y = 3\sin(0.5x) - 1$

c $y = -\sin(x) + 3$

3 $a(x)$ and $b(x)$ are not periodic.

$c(x) = 3(-2\cos(x)) = -6\cos(x)$. Period = 360°, amplitude = 6, principal axis is $y = 0$

$d(x) = -2\cos(3x)$. Period = 120°, amplitude = 2, principal axis is $y = 0$

4 a $y = -2\cos(2x) + 1$

b $\beta = -1$

5 a Maximum = 18.77 m, minimum = 17.83 m

b 19.63 s

6 a $y = 8\sin\left(\frac{30}{17}x\right) + 14$ or $y = -8\sin\left(\frac{30}{17}x\right) + 14$

b F = (204,14), A = (255, 22)

Exam-style questions

7 a Minimum is 1.8 (1 mark)

Maximum is 6.4 (1 mark)

b Using GDC (1 mark)

$x = 268°$ (1 mark)

$x = 452°$ (1 mark)

8 The principal axis is

$\frac{5.5+1.5}{2}(=3.5)$ (1 mark)

Hence $p = 3.5$ (1 mark)

The amplitude is $\frac{5.5-1.5}{2} = 2$ (1 mark)

Hence $q = 2$ (1 mark)

The period is 120°

$120° = \frac{360°}{r}$ (1 mark)

Hence $r = 3$ (1 mark)

So $y = 3.5 + 2\cos 3x$

9 a

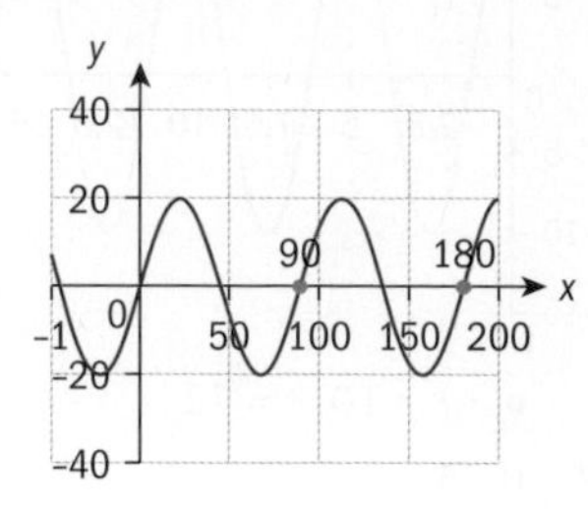

(3 mark)

b $y = 0$ (1 mark)

c $f(x) = -5$ when $x = 48.6°$, 86.4°, 138.6°, 176.4° (2 marks)

Solution is therefore $48.6° < x < 86.4°$ and $138.6° < x < 176.4°$ (2 marks)

10 The amplitude is 3 (1 mark)

Hence $p = 3$ (1 mark)

The period is $81° - (-9°) = 90°$ (1 mark)

$90° = \frac{360°}{q}$ (1 mark)

Hence $q = 4$ (1 mark)

So $y = 3\sin(4x + r)$

$y = 0$ when $x = 36°$ (equidistant from $-9°$ and 81°) (1 mark)

So $0 = 3\sin(144° + r)$ (1 mark)

So $\sin(144° + r) = 0$

and the first positive root is when $r = 36°$ (1 mark)

Therefore $y = 3\sin(4x + 36°)$

11 a 0.3 (1 mark)

b $y_{MIN} = 5.4$ (1 mark)

First occurs when

$12.5x = 180$ (1 mark)

$x = \frac{180}{12.5} = 14.4$ (1 mark)

c Period $= \frac{360}{12.5}$ (1 mark)

$= 28.8$ (1 mark)

12 a

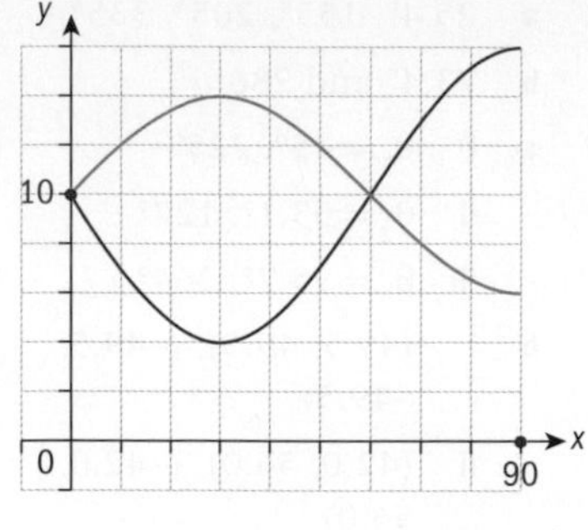

(2 marks)

b The principal axis is

$\frac{16+4}{2} = (10)$ (1 mark)

Hence $p = 10$ (1 mark)

The amplitude is

$\frac{16-4}{2}(=6)$ (1 mark)

Hence $q = 6$ (1 mark)

The period is $2 \times 60° = 120°$ (1 mark)

$120° = \frac{360°}{r}$ (1 mark)

Hence $r = 3$ (1 mark)

So $y = 10 - 6\sin 3x$

13 Amplitude = 2, so $b = 2$ (1 mark)

At (60°, 5), $5 = a + 2$ (because this is a max. point) (2 marks)

So $a = 3$ (1 mark)

Therefore $y = 3 + 2\sin cx$

By symmetry of the sin curve, there is a minimum point midway between the max points, ie at

$x = \frac{60+300}{2} = 180$

Since the amplitude = 2, the y-value at $x = 180$ is $y = 5 - 2 \times 2 = 1$

So the curve goes through the point (180°,1) (1 mark)

So $1 = 3 + 2\sin(180c)$ (1 mark)

$-1 = \sin(180c)$

$180c = 270$ (1 mark)

Therefore $c = \frac{3}{2}$ (1 mark)

Therefore $y = 3 + 2\sin\left(\frac{3x}{2}\right)$

14 a $D = \frac{22+12}{2}$ (1 mark)

$= 17$ (1 mark)

$A = \frac{22-12}{2}$ (1 mark)

$= 5$ (1 mark)

The period is $\frac{360}{B} = 24$

(1 mark)

Therefore $B = 15$ (1 mark)

So $T = 5\sin(15(t - C)) + 17$

At (3,12), $12 = 5\sin(15(t - C)) + 17$ (1 mark)

$-1 = \sin(15(3 - C))$

$15(3 - C) = -90$ (1 mark)

$C = 9$ (1 mark)

Therefore $T = 5\sin(15(t - 9)) + 17$

b Solving $T = 5\sin(15(t - 9)) + 17$ and $T = 20$ by GDC (1 mark)

Solutions are $T = 18.54$ and $T = 11.46$ (2 marks)

$18.54 - 11.46$

$= 7.08$ hours (7 hours 5 minutes) (1 mark)

Chapter 12

Skills check

1 a $3x^2 - 13x - 10$

b $6x^3 + 7x^2 - 5x$

2 $y = \frac{1}{2}(11 - x)$

3 Volume: $32\pi\,\text{cm}^3$, surface area: $16\pi + 8\sqrt{13}\pi\ \text{cm}^2$

4 a x^{-5} **b** x^{-1}

5 -10

Exercise 12A

a i $y' = 0$

ii $y' = 0$, stationary

iii No values, stationary

b i $y' = 4$

ii $y' = 4$, increasing

iii Any value of x

c i $\frac{df}{dx} = 6x$

ii $\frac{df}{dx} = 6$, increasing

iii $x > 0$

d i $\frac{df}{dx} = 10x - 3$

ii $\frac{df}{dx} = 7$, increasing

iii $x > \frac{3}{10}$

e i $\frac{df}{dx} = 12x^3 + 7$

ii $\frac{df}{dx} = 19$, increasing

iii $x > -0.836$ (3 s.f.)

f i $\frac{df}{dx} = 20x^3 - 6x + 2$

ii $\frac{df}{dx} = 16$, increasing

iii $x > -0.670$

g i $y'(x) = 4x + \frac{3}{x^2}$

ii $y' = 7$, increasing

iii $x > -\sqrt[3]{\frac{3}{4}}$

h i $y' = -\frac{18}{x^4} + 4$

ii $y' = -14$, decreasing

iii when $x > 1.46$ and when $x < -1.46$

i i $y' = 12x + 5$

ii $y' = 17$, increasing

iii $x > -\frac{5}{12}$

j i $f'(x) = 8x^3 - 16x - 10$

ii $f'(1) = -18$, decreasing

iii $x > 1.66$

k i $y' = -21x^{-4} + 32x^3 - 12x$

ii $y' = -1$, decreasing

iii $x > 1.01$

Exercise 12B

1 a $2\pi r$ **b** 4π

2 a $-0.112c + 5.6$

b When $c = 20$, $\frac{dP}{dc} = 3.36$, when $c = 60$, $\frac{dP}{dc} = -1.12$

c At the larger number of sales, selling more cupcakes will decrease profit, whilst it will increase profit at the lower value.

3 a $f'(t) = 160(t - 1)$

b Velocity of the bungee jumper

c $f'(0.5) = -80$, $f'(1.5) = 80$. Travelling at the same speed, but in opposite directions.

d $f(2) = 160 = f(0)$. The bungee jumper passes through the start point at the same speed that he left at. This is unrealistic, as some energy will be lost, for example, in overcoming air resistance.

4 $A: (-1, -2)$, $B: \left(\frac{1}{3}, \frac{22}{27}\right)$

5 7.5 m

Exercise 12C

1 $y = 12x - 22$

2 $y = 1$

3 a -4 **b** $\frac{1}{4}$

4 $y = -\frac{x}{2} + \frac{9}{2}$

5 Tangent: $y = 26x - 45$;

Normal: $y = \frac{92}{13} - \frac{x}{26}$

6 $y = -\frac{1}{4}x + \frac{9}{2}$ and $y = \frac{1}{4}x + \frac{9}{2}$, the fountain will be placed at $\left(0, \frac{9}{4}\right)$

7 a

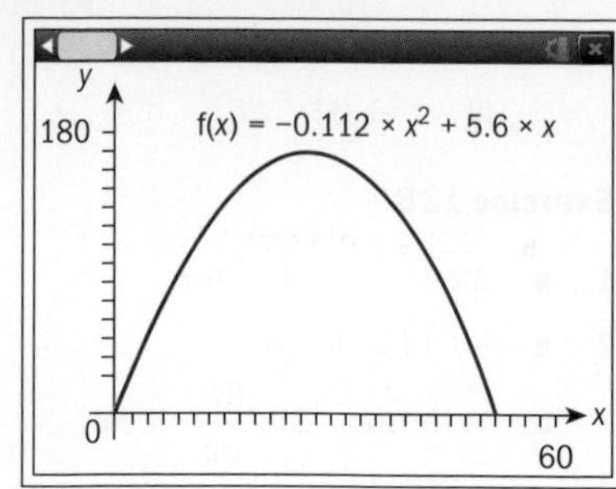

b $y = 65.5 - 0.446x$ and $y = 43.1 + 0.446x$

c (25.1, 54.3)

d Yes, position is within the park.

8 $a = 1, b = 9$

9 $k = 1, b = 5$

10 $a = 2, b = -5$

Exercise 12D

1 a $\frac{3}{4}$ **b** 2.10 (3 s.f.)

c $\frac{55}{36}$ **d** $\frac{-47}{196}$

e 2824 **f** 672

Exercise 12E

1 a $f'(t) = 7.25 - 3.75t$

b (1.93, 8.01)

c When $t < 1.93$, then $3.75t < 7.25$ so $f'(t) > 0$ and when $t > 1.93$, then $3.75t > 7.25$, so $f'(t) < 0$, hence $t = 1.93$ s is a maximum.

2 a

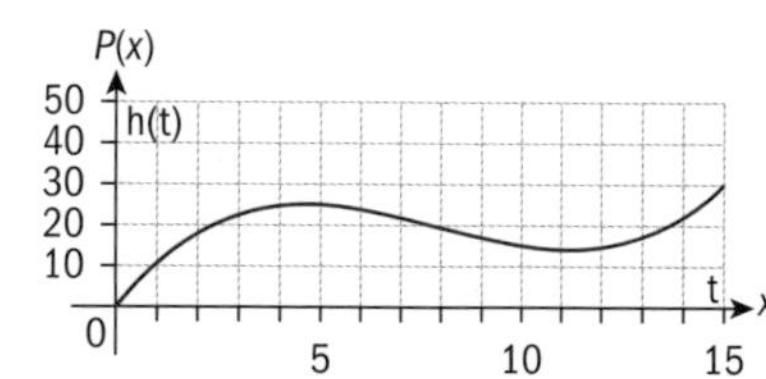

b 4.52 thousand dollars per million units sold

c $P'(3) = 3.26$, $P'(8) = -2.54$, $P'(13) = 3.66$. These represent the instantaneous rate of change of profit with respect to number of units sold.

d Negative when $4.66 < x < 11.2$ and positive when $x < 4.66$ and $x > 11.2$. This means that profit increases with more sales when $x < 4.66$ and $x > 11.2$ but profit will decrease with more sales when $4.66 < x < 11.2$.

e $x = 4.66, 11.2$

f At the points where $P'(x) = 0$ (ie $x = 4.66, 11.2$), then $P(x) = 25.1, 14.1$ (respectively). The sketch shows that the gradient function $P'(x)$ changes sign at these points, so they are (local) maxima and minima.

3 Maximum height = 4m (at $t = 6$s)

4 $n = 50$, maximum profit = \$120

5 a i 55 000 EUR, no parts

ii 51 667 EUR, 5000 parts

iii 46 667 EUR, 4000 parts

b They should adopt the first strategy

6

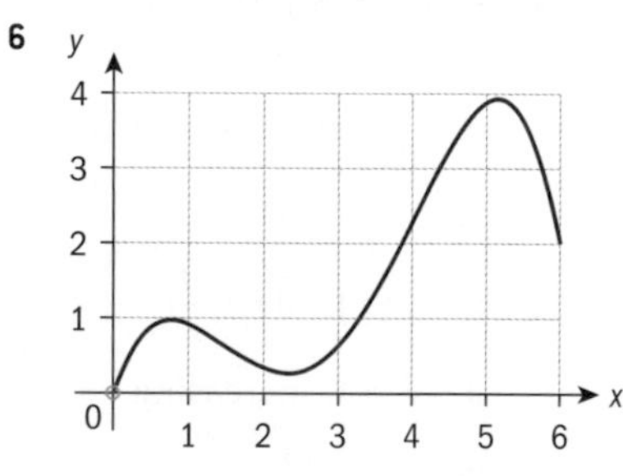

Maximum height = 392 m (3 s.f.)

Exercise 12F

1 a The volume of a cylinder is the product of its cross sectional area (in this case πr^2) and its height h. As the volume is 400 cm³, $400 = \pi r^2 h$.

b $A = 2\pi rh + \pi r^2$

c Using part **a**, express A_c as $A_c = \frac{2(\pi r^2 h)}{r} = 2 \times \frac{400}{r} = \frac{800}{r}$. Using this form, total surface area is $A = \pi r^2 + \frac{800}{r}$.

d

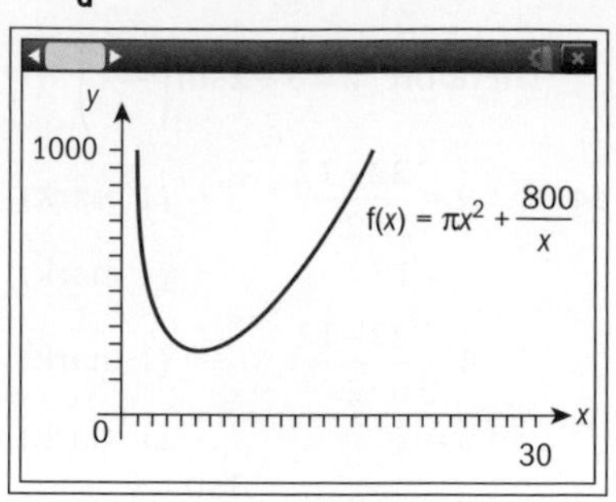

e $\min A = 239\text{ cm}^2$ at $r = 5.03$cm

f At 5 the gradient is negative and at 6 it is positive – so, it is a minimum.

2 $V_{max} = 27\,145\text{ cm}^3$ at $r = 16.3$ cm; at 16 the gradient is positive and at 17 it is negative, so it is a maximum.

3 a $l = 50 - x$

b $A = xl = x(50 - x)\text{ m}^2$

c $50 - 2x$

d $A_{max} = 625\text{ m}^2$, which occurs when garden is a 25 m × 25 m square

4 $V_{max} = 905\text{ cm}^3$, when $r = 12$ cm

5 a $V = x(20 - 2x)(24 - 2x)$

b $x = 3.62$ cm, $V = 774\text{ cm}^3$

6 a $V = \pi r^2 h$

b $A = 2\pi r^2 + \frac{600}{r}$

c $A = 248\text{ cm}^2$ when $r = 3.63$ cm and $h = 7.25$

7 Maximum profit is \$2700.29, when $n = 181$ goods sold per day

8 $f(29) = 391.10$ USD, when 29 units are sold

Chapter review

1 a $f'(x) = 0, m = 0$

b $y'(x) = 3, m = 3$

c $g'(x) = 4x - 4, m = 0$

d $y'(x) = 18x^2 - 6x + 1, m = 13$

e $f'(x) = -\frac{2}{x^2} + 3, m = 1$

f $f'(x) = -\frac{18}{x^4} + 4x, m = -14$

2 Normal: $y = 2 - x$, tangent: $y = x - 6$

3 $(3, -10)$

4 $(-1, -4)$

5 **a** $f'(t) = -1.667 + 0.0834t$

b $f'(12) = -0.666$ travelling downhill, $f'(32) = 1.00$ travelling uphill

c $t = 20.0$ s; minimum point because $f'(t) < 0$ when $t < 20$ and $f'(t) > 0$ when $t > 20$

6 **a** The volume of a cylinder of radius r and height h is $V = \pi r^2 h$. Given $V = 300\,\text{cm}^3$, $300 = \pi r^2 h$

b $S = 2\pi rh + 2\pi r^2$

c Using the expression from part **a**,

$$S = 2\pi r \times \frac{300}{\pi r^2} + 2\pi r^2$$

$$= 2\pi r^2 + \frac{600}{r}$$

d $4\pi r - \frac{600}{r^2}$

e The only stationary point of S is where

$4\pi r^3 = 600 \Rightarrow r = \sqrt[3]{\frac{150}{\pi}}$

$= 3.62\,\text{cm}$.

The corresponding surface area is $S = 248\,\text{cm}^2$ and

$h = \frac{300}{\pi r^2} = 7.26\,\text{cm}$ (all 3 s.f.)

Exam-style questions

7 **a** $\frac{dy}{dx} = -0.1x + 1.5$ (2 marks)

b Setting $\frac{dy}{dx} = 0$ (1 mark)

$-0.1x + 1.5 = 0$

$x = 15$ (1 mark)

$y(15) = -0.05 \times 15^2 + 1.5 \times 15 + 82$

$= 93.25\,\text{m}$ (1 mark)

c Evaluating $\frac{dy}{dx}$ at $x = 14.5$ and $x = 15.5$ (1 mark)

$\left.\frac{dy}{dx}\right|_{x=14.5} = 0.05$ and

$\left.\frac{dy}{dx}\right|_{x=15.5} = -0.05$ (1 mark)

Sign goes from positive to negative, therefore a maximum point (1 mark)

8 **a** $V = x(40 - 2x)(30 - 2x)$ (2 marks)

$= x(1200 - 60x - 80x + 4x^2)$

$= x(1200 - 140x + 4x^2)$ (1 mark)

$= 1200x - 140x^2 + 4x^3$

b $\frac{dV}{dx} = 1200 - 280x + 12x^2$ (2 marks)

c Setting $\frac{dV}{dx} = 0$ (1 mark)

$1200 - 280x + 12x^2 = 0$ (1 mark)

$12x^2 - 280x + 1200 = 0$

$3x^2 - 70x + 300 = 0$

$x^2 - \frac{70}{3}x + 100 = 0$

d Using GDC to solve $x^2 - \frac{70}{3}x + 100 = 0$ (1 mark)

$x = 5.66\,\text{cm}$ (discard $x = 17.7$ as this is impossible) (1 mark)

$V_{max} = 1200 \times 5.657 - 140 \times 5.657^2 + 4 \times 5.657^3$ (1 mark)

$= 3032\,\text{cm}^3$ (3030 to 3s.f.) (1 mark)

9 **a** Use of GDC (demonstrated by one correct value) (1 mark)

$a = -0.0534$ (1 mark)

$b = 1.09$ (1 mark)

$c = 7.48$ (1 mark)

$T = -0.0534h^2 + 1.09h + 7.48$

b $\frac{dT}{dh} = -0.107h + 1.09$ (2 marks)

c Setting $\frac{dT}{dh} = 0$ (1 mark)

$h = 10.2$ (1 mark)

d The maximum temperature usually occurs after midday, whereas this is only 10 hours after midnight. (1 mark)

10 $(-1.59, -13.2)$ (2 marks)

$(0.336, 3.81)$ (2 marks)

$(1.17, 1.97)$ (2 marks)

11 **a** $\frac{dy}{dx} = 2x^2 - 7x + 2$ (2 marks)

$\frac{dy}{dx} = -3$ (1 mark)

$2x^2 - 7x + 2 = -3$

$2x^2 - 7x + 5 = 0$

$(2x - 5)(x - 1) = 0$ (1 mark)

$x = \frac{5}{2}$ $y = -\frac{35}{24}$ (1 mark)

$x = 1$ $y = \frac{25}{6}$ (1 mark)

b $0.314 < x < 3.19$ (2 marks)

12 At $x = 1$, $y = \frac{3}{2}$ (1 mark)

$\frac{dy}{dx} = -2x^3$ (2 marks)

At $x = 1$ $\frac{dy}{dx} = -2$ (1 mark)

gradient of the normal is therefore $\frac{1}{2}$ (1 mark)

Equation of the normal is therefore $y - \frac{3}{2} = \frac{1}{2}(x - 1)$ (1 mark)

$y - \frac{3}{2} = \frac{1}{2}x - \frac{1}{2}$

$y = \frac{1}{2}x + 1$ (1 mark)

13 Substituting (2, −1) gives (1 mark)

$-1 = 4a + 2b + 3$ (1 mark)

$4a + 2b = -4$

$2a + b = -2$

$\frac{dy}{dx} = 2ax + b$ (1 mark)

$8 = 4a + b$ (1 mark)

Solving simultaneously gives (1 mark)

$a = 5$ (1 mark)

$b = -12$ (1 mark)

14 a $y = 20 - x$

$xy = x(20 - x) = 20x - x^2$ (1 mark)

Differentiate and set to zero: (1 mark)

$20 - 2x = 0$ (1 mark)

$x = 10$ (1 mark)

So $(xy)_{MAX} = 100$ (1 mark)

b $x^2 + y^2 = x^2 + (20 - x)^2$ (1 mark)

$= x^2 + x^2 - 40x + 400$

$= 2x^2 - 40x + 400$

Differentiate and set to zero: (1 mark)

$4x - 40 = 0$ (1 mark)

$x = 10$ (1 mark)

So $(x^2 + y^2)_{MAX} = 10^2 + 10^2 = 200$ (1 mark)

c $\left.\frac{dy}{dx}\right|_{x=9.5} = 4\times 9.5 - 40 = -2$ (1 mark)

$\left.\frac{dy}{dx}\right|_{x=10.5} = 4\times 10.5 - 40 = +2$ (1 mark)

The derivative goes from negative to positive, therefore this is a minimum (1 mark)

Chapter 13

Skills check

1 a 18 cm^2

b 14.1 cm^2

2 a $3x^2 + \frac{5}{2}$ 

b $8x - 1$

Exercise 13A

1 2250 m

2 a 24 m **b** 4.5 km

c 13.5 m

3 a 12 **b** 9

c 36 **d** 40

4 a and b

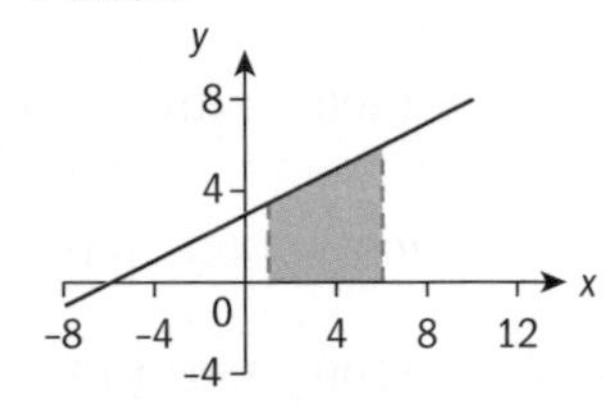

c 23.75

5 a and b

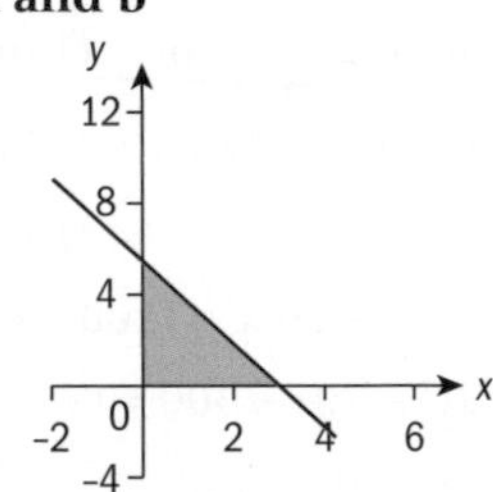

c 9

6 a and b

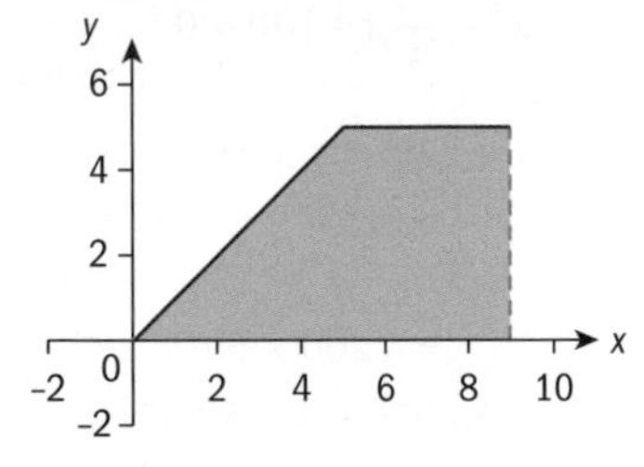

c 32.5

Exercise 13B

1 a i $\int_{-2}^{4} 5\,dx$

ii $\int_{0.5}^{3.5} (-2x + 8)\,dx$

iii $\int_{1}^{3} (-3x + 10)\,dx$

iv $\int_{0}^{4} 2x\,dx$

b i 30 **ii** 12

iii 8 **iv** 16

2 a

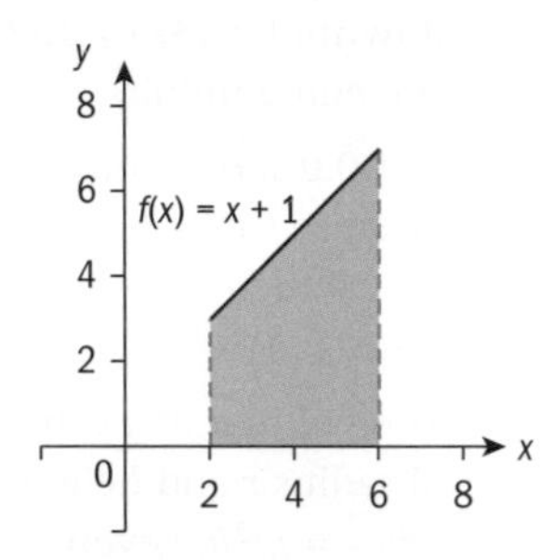

b

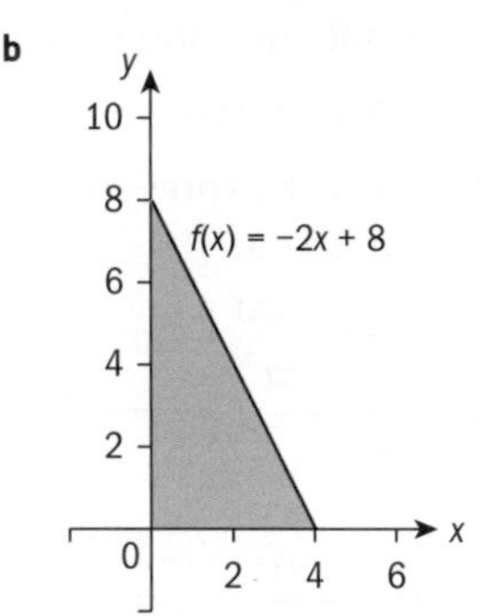

c

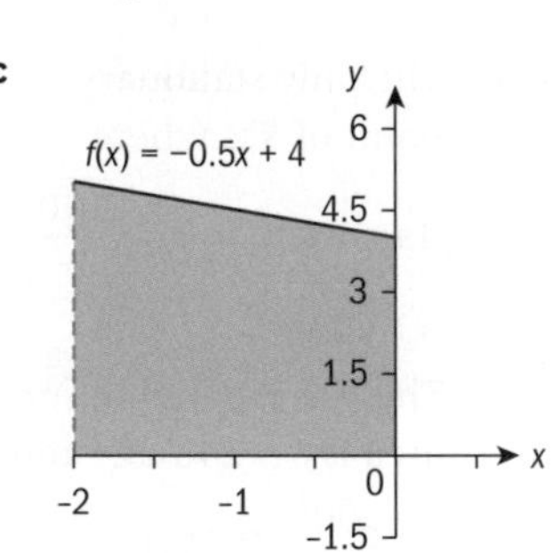

Exercise 13C

1 a i $\int_{-2}^{2} (x+1)^2 + 0.5\,dx$

ii $\frac{34}{3}$

b i $\int_{1}^{1.5} x^3\,dx$ **ii** $\frac{65}{64}$

c i $\int_{-1}^{3} \sqrt{x+1}\,dx$ **ii** 5.33

d i $\int_{-2}^{0} x^3 - 4x\,dx$ ii 4

e i $\int_{0}^{2} -x^2 + 2x\,dx$ ii $\frac{4}{3}$

f i $\int_{-1}^{3} -(x+1)(x-3)\,dx$

ii $\frac{32}{3}$

2 **a** i $\int_{2}^{4} x^2\,dx$ ii $\frac{56}{3}$

b i $\int_{-1}^{1} 2x\,dx$ ii 0

c i $\int_{-1}^{1} \frac{1}{1+x^2}\,dx$ ii 1.57

d i $\int_{0.5}^{3} \frac{1}{x}\,dx$ ii 1.79

e i $\int_{0}^{1} -(x-3)(x+2)\,dx$

ii $\frac{37}{6}$

f i $\int_{-2}^{0} -(x-3)(x+2)\,dx$ or

$\int_{0}^{3} -(x-3)(x+2)\,dx$

ii $\frac{22}{3}$ or 13.5

g i $\int_{-2}^{3} -(x-3)(x+2)\,dx$

ii $\frac{125}{6}$

h i $\int_{-2}^{4.5} -x^2 + 2x + 15\,dx$

ii 80.7

i i $\int_{-3}^{5} -x^2 + 2x + 15\,dx$

ii 85.4

j i $\int_{-1}^{\ln(3)} 3 - e^x\,dx$

ii $\ln 27 + \frac{1}{e}$

k i $\int_{-2-5^{\frac{1}{3}}}^{0} (x+2)^3 + 5\,dx$

ii $14 + \frac{15\sqrt[3]{5}}{4}$

3 **a** −2, 2 **b** (0, 4)

c 20

d i $\int_{0}^{2} -x^2 + 4\,dx$

ii $\frac{16}{3}$

e $\frac{76}{3}$

Exercise 13D

1 3

2 −0.8169

3 **a**

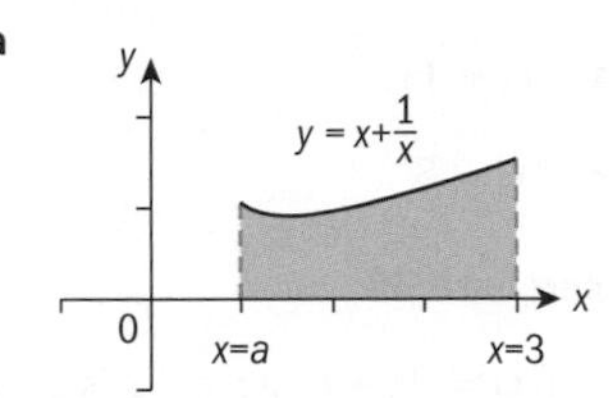

b 0.5693

4 **a** −1

b Area between curve $y = x^2$ and the x-axis between $x = -2$ and $x = -1$

5 **a** $b = 1$

b Area between curve $y = (1 + x^3)$ and the x-axis, between $x = -1$ and $x = 1$

6 **a** $t = 3$

b Area between curve $y = \sqrt{x+1}$ and the x-axis, between $x = -1$ and $x = 3$

Exercise 13E

1 55

2 18

3 28.75

4 9.85

5 11.6 (3 s.f.)

6 **a** 5.198 **b** 23.74

c 12.21 **d** 35

7 **a**

b i $\int_{3}^{6} -2(x-3)(x-6)\,dx$

ii 9

c 8.75

d 2.78% (3 s.f.)

8 **a**

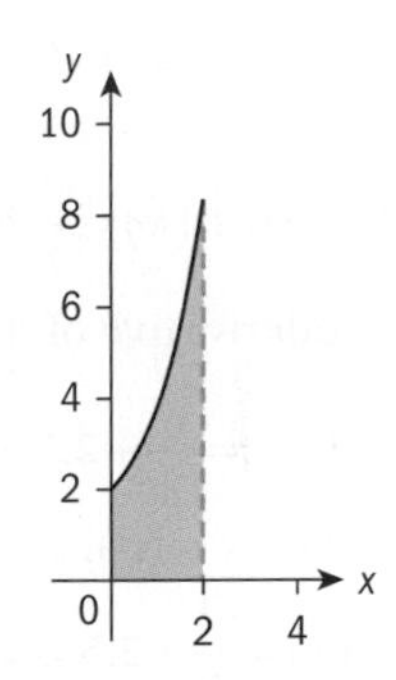

b i $\int_{0}^{2} 1 + e^x\,dx$

ii 8.389

c 8.474

d 1.01% (3 s.f.)

Exercise 13F

1 **a** i If $F(x) = x$, $F'(x) = 1 = f(x)$, so $F(x) = x$ is an antiderivative of $f(x) = 1$

ii If $F(x) = x + 3$, $F'(x) = 1 = f(x)$, so $F(x) = x + 3$ is an antiderivative of $f(x) = 1$

iii If $F(x) = x - 6$, $F'(x) = 1 = f(x)$, so $F(x) = x - 6$ is an antiderivative of $f(x) = 1$

b For example, $F_1(x) = x - 10$, $F_2(x) = x + 20$

c $F(x) = x + c$, where c is any number.

2 a i If $F(x) = x^2$, $F'(x) = 2x = f(x)$, so $F(x) = x^2$ is an antiderivative of $f(x) = 2x$

ii If $F(x) = x^2 + 1$, $F'(x) = 2x = f(x)$, so $F(x) = x^2 + 1$ is an antiderivative of $f(x) = 2x$

iii If $F(x) = x^2 - 4$, $F'(x) = 2x = f(x)$, so $F(x) = x^2 - 4$ is an antiderivative of $f(x) = 2x$

b For example, $F_1(x) = x^2 - 1$, $F_2(x) = x^2 + 2$

c $F(x) = x^2 + c$, where c is any number.

3 a i If $F(x)=\frac{x^2}{2}$, $F'(x) = x = f(x)$, so $F(x)=\frac{x^2}{2}$ is an antiderivative of $f(x) = x$

ii If $F(x)=\frac{x^2}{2}+2.5$, $F'(x) = x = f(x)$, so $F(x)=\frac{x^2}{2}+2.5$ is an antiderivative of $f(x) = x$

iii If $F(x)=\frac{x^2}{2}-12$, $F'(x) = x = f(x)$, so $F(x)=\frac{x^2}{2}-12$ is an antiderivative of $f(x) = x$

b $F(x)=\frac{x^2}{2}+c$, where c is any number.

4 a i If $F(x) = x^2 + x$, $F'(x) = 2x + 1 = f(x)$, so $F(x) = x^2 + x$ is an antiderivative of $f(x) = 2x + 1$

ii If $F(x) = x^2 + x - 3.2$, $F'(x) = 2x + 1 = f(x)$, so $F(x) = x^2 + x - 3.2$ is an antiderivative of $f(x) = 2x + 1$

iii If $F(x) = x^2 + x + 4$, $F'(x) = 2x + 1 = f(x)$, so $F(x) = x^2 + x + 4$ is an antiderivative of $f(x) = 2x + 1$

b $F(x) = x^2 + x + c$, where c is any number.

5 For example, $g(x) = 3x$

6 For example, $g(x) = -2x$

7 For example, $g(x)=\frac{x^2}{4}$

8 For example, $F(x)=\frac{x^3}{3}$ is an antiderivative of $f(x) = x^2$

9 a $A(x) = 1.25x^2$

b $A(x) = -0.25x^2 + 4x$

c $A(x) = \frac{x^3}{3}$

d $A(x)= \frac{x^3}{3}+x$

10 $4x$

11 $x(x + 1)$

12 $2x^2 - 8$

Exercise 13G

1 $f(x) = 2x + c$, where c is any real number

2 $f(x)=\frac{x^2}{2}+x+c$, where c is any real number

3 $f(x) = -x + c$, where c is any real number

4 $f(x) = x^2 + c$, where c is any real number

5 a $x + c, c\in \mathbb{R}$

b $6x + c, c\in \mathbb{R}$

c $\frac{x}{2}+c, c\in\mathbb{R}$

6 a $\frac{x^2}{2}+c, c\in\mathbb{R}$

b $x^2 + c, c\in \mathbb{R}$

c $\frac{5x^2}{2}+c, c\in\mathbb{R}$

d $\frac{x^2}{4}+c, c\in\mathbb{R}$

e $\frac{ax^2}{2}+c, c\in\mathbb{R}$

7 a $\frac{x^3}{3}+c, c\in\mathbb{R}$

b $x^3 + c, c\in \mathbb{R}$

c $\frac{4x^3}{3}+c, c\in\mathbb{R}$

d $\frac{x^3}{6}+c, c\in\mathbb{R}$

e $\frac{x^3}{9}+c, c\in\mathbb{R}$

f $\frac{x^3}{2}+c, c\in\mathbb{R}$

g $\frac{ax^3}{3}+c, c\in\mathbb{R}$

8 a $\frac{x^4}{4}+c, c\in\mathbb{R}$

b $x^4 + c, c\in \mathbb{R}$

c $\frac{x^4}{2}+c, c\in\mathbb{R}$

d $\frac{x^4}{5}+c, c\in\mathbb{R}$

e $\frac{x^4}{12}+c, c\in\mathbb{R}$

f $-\frac{x^4}{4}+c, c\in\mathbb{R}$

g $\frac{ax^4}{4}+c, c\in\mathbb{R}$

Exercise 13H

1 a $10x + c, c\in \mathbb{R}$

b $0.2x^3 + c, c\in \mathbb{R}$

c $\frac{x^6}{6}+c, c\in\mathbb{R}$

d $\int 7\,dx - \int 2x\,dx = 7x + c_1 - x^2 - c_2 = 7x - x^2 + c\ (c = c_1 - c_2 \in \mathbb{R})$

e $\int 1\,dx + \int 2x\,dx = x + c_1 + x^2 + c_2 = x + x^2 + c\ (c = c_1 + c_2 \in \mathbb{R})$

f $\int 5dx + \int x\,dx + \int -\frac{x^2}{3} dx = 5x + c_1 + \frac{x^2}{2} + c_2 - \frac{x^3}{6} + c_3 = 5x + \frac{x^2}{2} - \frac{x^3}{6} + c$

$(c = c_1 + c_2 \in \mathbb{R})$

g $\int(-x+\frac{3x^2}{4}+0.5)\,dx = \int -x\,dx + \int\frac{3x^2}{4}\,dx + \int 0.5\,dx = -\frac{x^2}{2}+\frac{x^3}{4}+0.5x+c$

h $\int\left(1-x+\frac{x^3}{2}\right)dx=\int 1\,dx - \int x\,dx+\int\frac{x^3}{2}dx=x+c_1 -\frac{x^2}{2}+c_2+\frac{x^4}{8}+c_3=x -\frac{x^2}{2}+\frac{x^4}{8}+c\ (c=c_1+c_2 +c_3\in\mathbb{R})$

i $\int(x^2-\frac{x}{2}+4)\,dx = \int x^2\,dx - \int\frac{x}{2}\,dx+\int 4\,dx = \frac{x^3}{3}+c_1 -\frac{x^2}{4}+c_2+4x+c_3 =\frac{x^3}{3}-\frac{x^2}{4}+4x+c$ $(c=c_1+c_2+c_3)$

j $-\frac{1}{2}x^{-1}+c$

k $-\frac{2}{3}x^{-3}+c$

l $2x^2-\frac{5}{2}x^{-2}+c$

2 **a** $2x-\frac{1}{3}$

b $\frac{x^3}{3}-\frac{x^2}{6}+4x+c$

3 $\frac{t^2}{2}-t^3+c$

4 $t^4-\frac{3t^2}{2}+t+c$

5 Any function F with gradient $3+x-\frac{x^2}{4}$ of the form $F(x)=\int\left(3+x-\frac{x^2}{4}\right)dx=3x +\frac{x^2}{2}-\frac{x^3}{12}+c(c\in\mathbb{R})$

6 Any function F with gradient $-x+0.5x^2$ of the form $F(x)=\int(-x+0.5x^2)\,dx=-\frac{x^2}{2}+\frac{x^3}{6}+c(c\in\mathbb{R})$

7 **a** x^2-x-2

b Any function $y=f(x)$ with $\frac{dy}{dx}=(x+1)(x-2)$ of the form $y=f(x)=\int(x^2-x+2)\,dx=\frac{x^3}{3}-\frac{x^2}{2}-2x+c$, where c is any real number

8 $\frac{x^4}{4}+\frac{2}{x}+x+c$

Exercise 13I

1 $f(x)=\frac{3x^2}{2}+\frac{4x^3}{3}-\frac{1}{6}$

2 $y=\frac{x^2}{2}+\frac{x^3}{15}+2x-\frac{259}{15}$

3 $f(x)=3x-\frac{x^2}{2}-3$

4 $f(x)=x^3-\frac{x^5}{15}+2$

5 $y(0.5)=-2.45$

6 **a** $f(x)=-\frac{x^2}{2}+3x+8$

b $f(2)=-\frac{2^2}{2}+3\times2+8=12$

7 $C(x)=\frac{x^2}{2}+\frac{2}{x}+x+145$

8 $C(x)=3x-2x^2+\frac{x^3}{3}+1000$

9 **a** $R(t)=\frac{t^3}{3}-80t+600$

b USD 1 113 600

Chapter review

1 **a** **i** $\int_{-1}^{4}0.5x+4\,dx$

ii 23.75

b **i** $\int_{0}^{3}(-3x+9)\,dx$

ii 13.5

2 **a** **i** $\int_{-2}^{3}(4-x)\,dx$ **ii** 17.5

b **i** $\int_{0}^{4}\sqrt{16-x^2}\,dx$

ii 12.6 (3 s.f.)

3 **a** **i** $\int_{-4}^{0}x^2\,dx$ **ii** $\frac{64}{3}$

b **i** $\int_{-1}^{4}((x-1)^2+3)\,dx$

ii $\frac{80}{3}$

c **i** $\int_{0}^{2}(-(x-4)(x+1))\,dx$

ii $\frac{34}{3}$

d **i** $\int_{-1}^{1.5}(8-x^3)\,dx$

ii 19.0

e **i** $\int_{-2}^{2}(x+3)(x-2)^2\,dx$

ii 42.7 (3 s.f.)

f **i** $\int_{0}^{3}(x+3)(x-2)^2\,dx$

ii 11.3 (3 s.f.)

g **i** $\int_{-1}^{2}\frac{3}{1+x^2}\,dx$

ii 5.68 (3 s.f.)

h **i** $\int_{-4}^{3}\left(1+\sqrt{3-x}\right)dx$

ii 19.3 (3 s.f.)

i **i** $\int_{-2}^{2}e^{-x}\,dx$

ii 7.25 (3 s.f.)

4 **a** **i**

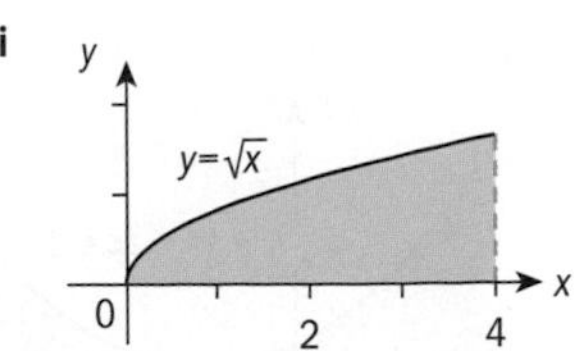

ii 5.33

b

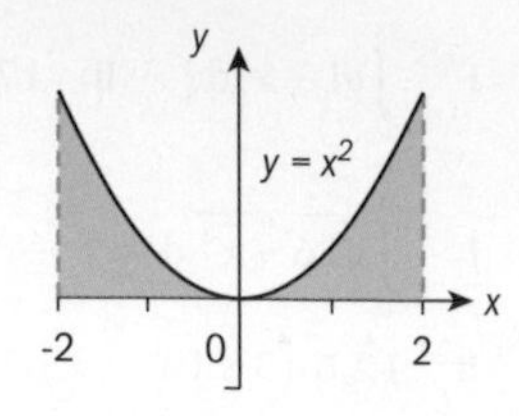

ii $\frac{16}{3}$

c i

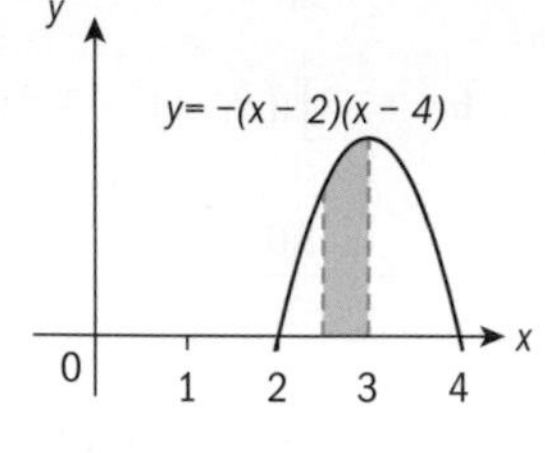

ii 0.458 (3 s.f.)

d i

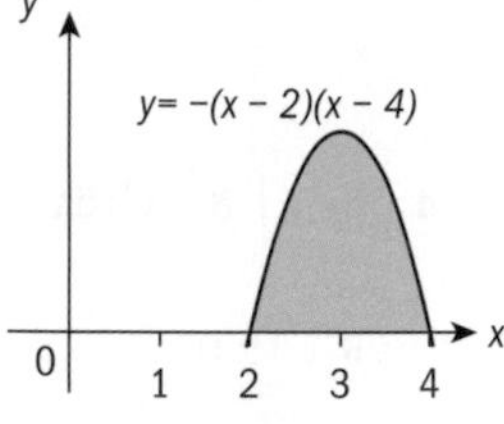

ii 1.33 (3 s.f.)

e i

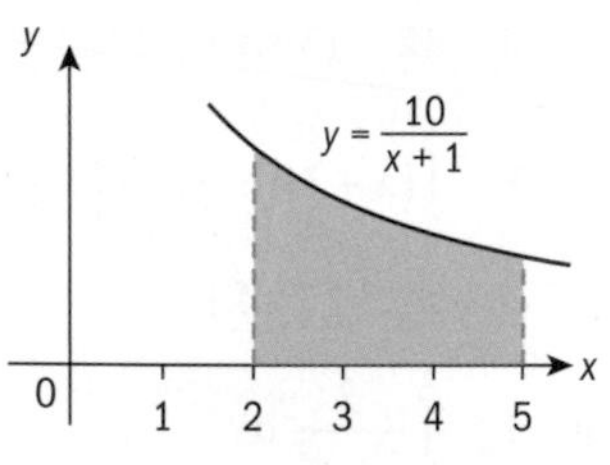

ii 6.93 (3 s.f.)

f i

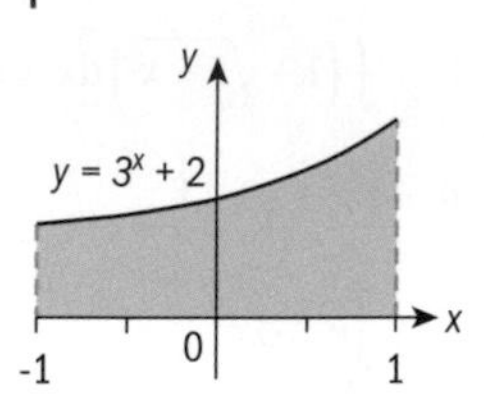

ii 6.43 (3 s.f.)

g i

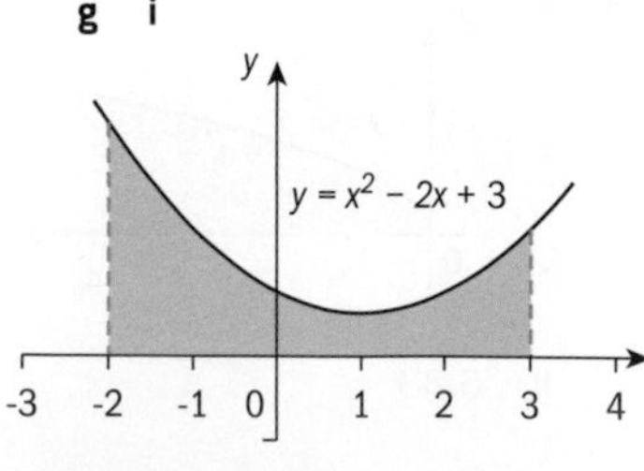

ii $\frac{65}{3}$

5 a $\int_{-1}^{4} -(x+1)(x-4)\,\mathrm{d}x$

b 20.8 (3 s.f.)

6 a $x = 0, 4$

b $\int_{0}^{4} x(x-4)^2\,\mathrm{d}x$

c 21.3 (3 s.f.)

7 a 0

b

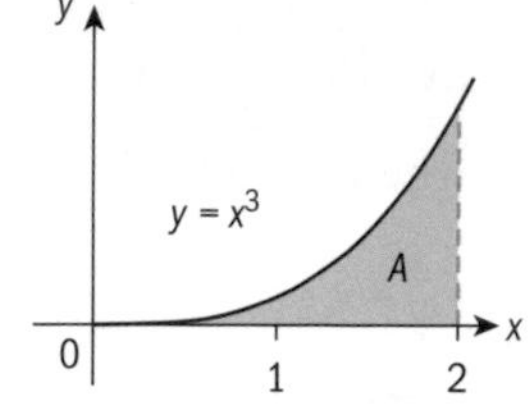

c $\int_{0}^{2} x^3\,\mathrm{d}x$

d 4

8 a

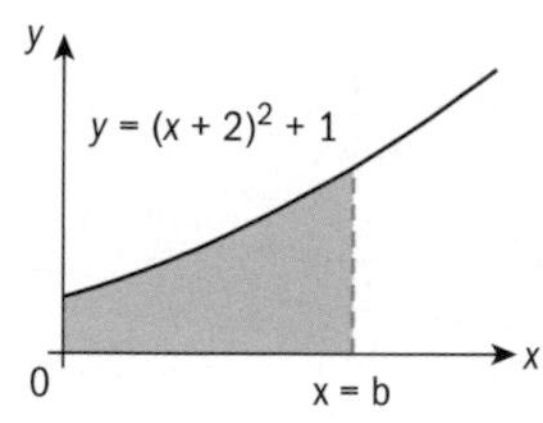

b 3

9 72

10 1.819

11 a

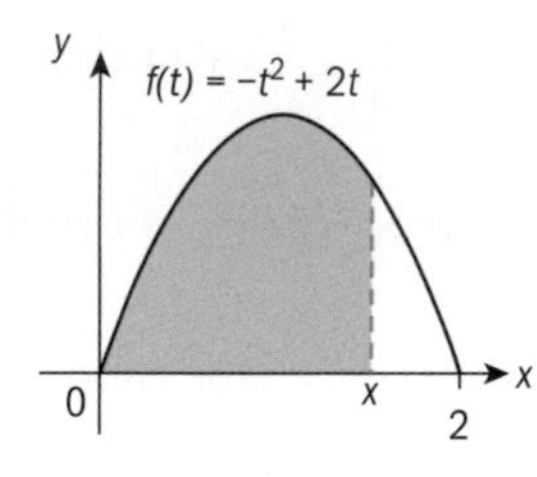

b $\int_{0}^{x} \left(-t^2 + 2t\right)\mathrm{d}t$

12 $2(x^2 - 9)$

13 $2x + \frac{x^2}{2} + c$, where $c \in \mathbb{R}$ is a constant of integration

14 $x + \frac{x^2}{2} - \frac{x^4}{16} + c$, where $c \in \mathbb{R}$ is a constant of integration

15 a 6.25

b $\int_{1}^{6} \left((x-4)^2 + 1\right)\mathrm{d}x$

c 22.9 (3 s.f.)

16 a Green line: $y = f_1(x) = 3x + 1$, purple line: $y = 3x$, pink line: $y = 3x - 2$

b $f_i(x)$ are anti-derivatives of $y = t(x)$ and hence $t(x) = f_i'(x) = 3$

c $3x + c$, where c is an arbitrary constant of integration

17 10.4 (3 s.f.)

18 a 0.7444

b i $\int_{0}^{1} \mathrm{e}^{-x^2}\,\mathrm{d}x$

ii 0.7468

c 0.329% (3 s.f.)

Exam-style questions

19 $5x - \frac{12x^3}{3} + \frac{4x^4}{4} (+c)$ (3 marks)

$= 5x - 4x^3 + x^4 + c$ (1 mark)

20 $\int \left(\frac{3}{2}x^2 + x + 3\right)\mathrm{d}x$ (3 marks)

$= \frac{x^3}{2} + \frac{x^2}{2} + 3x + c$

Substituting $x = -1$ and equating to $\frac{13}{2}$ (1 mark)

$-\frac{1}{2} + \frac{1}{2} - 3 + c = \frac{13}{2}$ (1 mark)

$c = \frac{19}{2}$ (1 mark)

$f(x) = \frac{x^3}{2} + \frac{x^2}{2} + 3x + \frac{19}{2}$

21 a $6x - x^2 = 10 - x$ (1 mark)

$x^2 - 7x + 10 = 0$

$(x - 2)(x - 5) = 0$ (1 mark)

Coordinates are (2,8) (1 mark)

and (5,5) (1 mark)

b

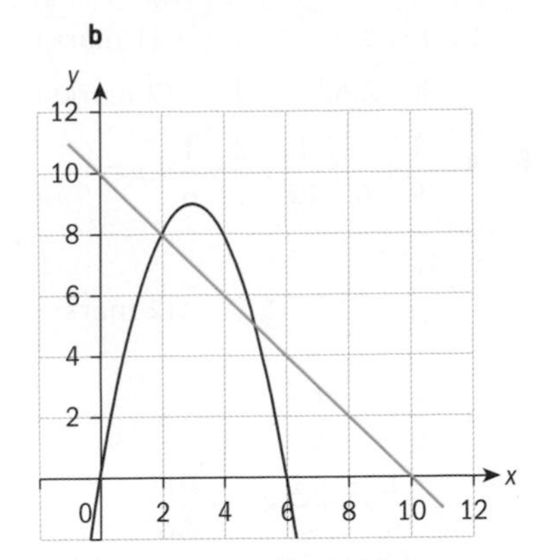

c Area under curve

$= \int_2^5 6x - x^2 \, dx$ (1 mark)

$= \left[3x^2 - \frac{x^3}{3}\right]_2^5$ (1 mark)

$= \frac{100}{3} - \frac{28}{3} = \frac{72}{3}$

(1 mark)

$= 24$ (1 mark)

Area under line

$= \frac{1}{2} \times (5+8) \times 3 = \frac{39}{2}$

(1 mark)

Required area $= 24 - \frac{39}{2}$

(1 mark)

$= \frac{9}{2}$ units^2 (1 mark)

[Note: Alternatively you can find

$\int_2^5 (6x - x^2) - (10 - x) dx$ to

give the same answer]

22 Displacement $= \int 0.6t^2 + 4t + 1 \, dt$

(1 mark)

$= 0.2t^3 + 2t^2 + t + k$ (1 mark)

Distance travelled in 3rd

second $= \int_2^3 (0.6t^2 + 4t + 1) \, dt$

(1 mark)

$= [0.2 \times 3^3 + 2 \times 3^2 + 3] - [0.2 \times 2^3 + 2 \times 2^2 + 2]$ (1 mark)

$= 14.8$ m (1 mark)

23 a

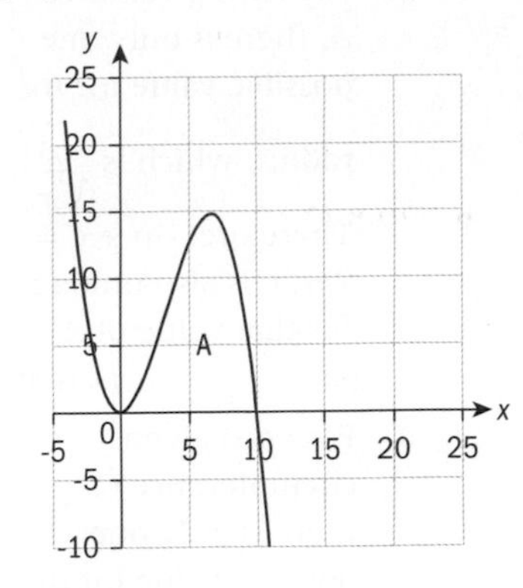

(3 marks)

b $\int_0^{10} -\frac{x^2}{10}(x - 10) \, dx$

(1 mark)

c $-\frac{x^2}{10}(x - 10) = x^2 - \frac{x^3}{10}$

$\int_0^{10} \left(x^2 - \frac{x^3}{10}\right) dx$ (1 mark)

$= \left[\frac{x^3}{3} - \frac{x^4}{40}\right]_0^{10}$ (2 marks)

$= \frac{1000}{3} - \frac{10\,000}{40}$ (1 mark)

$= \frac{1000}{3} - \frac{1000}{4}$

$= \frac{1000}{12}\left(= \frac{250}{3}\right)$ units^2

(1 mark)

24 a Use of GDC (1 mark)

$(-1, -1), (0, 0), (1, 1)$

(3 marks)

b

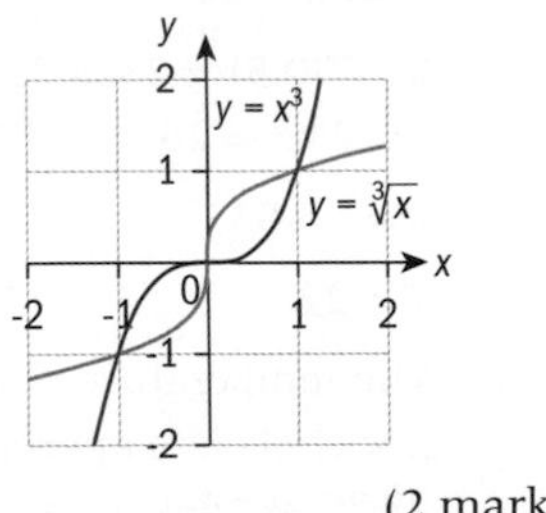

(2 marks)

c $\int_0^1 x^3 \, dx = \left[\frac{x^4}{4}\right]_0^1 = \frac{1}{4}$

(2 marks)

$\int_0^1 x^{\frac{1}{3}} \, dx = \left[\frac{3x^{\frac{4}{3}}}{4}\right]_0^1 = \frac{3}{4}$

(1 mark)

Positive area is therefore

$\frac{3}{4} - \frac{1}{4} = \frac{1}{2}$ (1 mark)

Therefore total area is

$2 \times \frac{1}{2} = 1$ unit^2 (2 marks)

25 a $A(1,0)$ (1 mark)

$B(6,0)$ (1 mark)

$C(3,4)$ (1 mark)

$D(6, 1)$ (1 mark)

b Area $= \frac{1}{2} \times 2 \times 4 + \int_3^6 \frac{36}{x^2} \, dx$

(3 marks)

$= 4 + 36 \int_3^6 x^{-2} \, dx$

$= 4 + 36\left[-\frac{1}{x}\right]_3^6$ (1 mark)

$= 4 + 36\left[-\frac{1}{6} - \left(-\frac{1}{3}\right)\right]$

(1 mark)

$= 4 + 36 \times \frac{1}{6} = 10$ units^2

(1 mark)

26 a

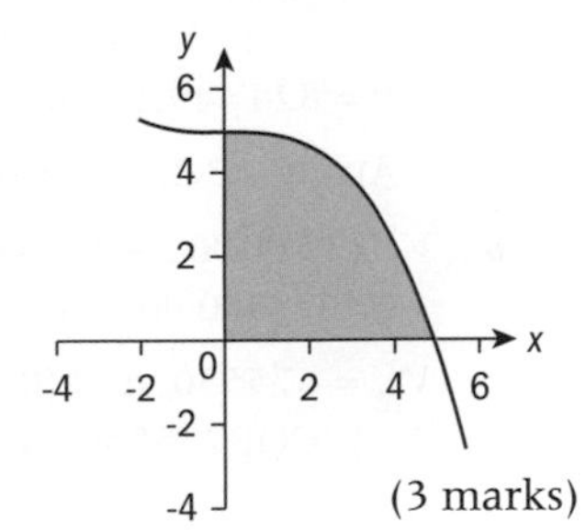

(3 marks)

b

x	0	1	2	3	4	5
y	5	4.96	4.68	3.92	2.44	0

(2 marks)

Area

$= \int_0^5 \left(5 - \frac{x^3}{25}\right) dx \approx \frac{1}{2}\left(\frac{5-0}{5}\right)$

$\left[5 + 0 + 2\left(\begin{array}{c}4.96 + 4.68 \\ +3.92 + 2.44\end{array}\right)\right]$

(2 marks)

$= 18.5$ units^2 (1 mark)

c The curve is concave down in the interval $0 \le x \le 5$, so each trapezium will be an underestimate. (1 mark)
Therefore the sum of the trapezia will also be an underestimate. (1 mark)

d $\int_0^5 \left(5 - \frac{x^3}{25}\right) dx = \left[5x - \frac{x^4}{100}\right]_0^5$ (2 marks)

$= 25 - \frac{625}{100} = 18.75 \text{ units}^2$ (2 marks)

e Percentage error
$= \frac{|18.5 - 18.75|}{18.75} \times 100 = 1.33\%$ (2 marks)

Paper 1

Exam-style questions

1 a $P\left(1 + \frac{6}{100 \times 12}\right)^{12 \times 10} = 1\,500\,000$ (1 mark)
$P = 824\,449.10$ (1 mark)
Ans: COP 824 449.10

b $V_0 = 1500000 - 824449.10 = 675550.90$ (1 mark)
$V_{10} = 675550.90 \times (0.9)^{10} = \text{COP}\,235550.03$ (2 marks)

2 a $d = 40$ (metres) (1 mark)

b EITHER
$200 + 40 \times 13 = 720\text{ m}$ (2 marks)
OR
$360 + 40 \times 9 = 720\text{ m}$ (2 marks)

c $S_{21} = \frac{2 \times 200 + 40 \times 20}{2} \times 21$ (2 marks)
$= 12\,600\text{ m}$ (1 mark)

3 a i For each given area A, there is only one possible value for the radius, which is $\sqrt{\frac{A}{\pi}}$.
Therefore, since $C = 2\pi r$, C is also unique for that value of A. (1 mark)

ii For each given circumference C, there is only one possible value for the radius, which is $\frac{C}{2\pi}$.
Therefore, $A = \pi r^2$, A is unique for that value of C, so C^{-1} exists. (1 mark)

b

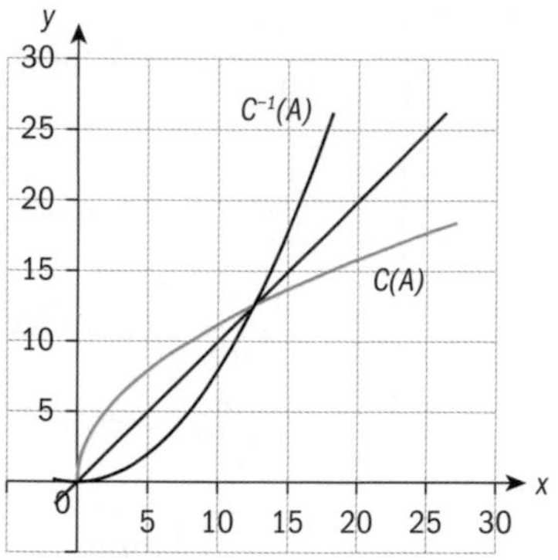

(2 marks)

c $A = C^{-1}(25) \Rightarrow C(A) = 25 \Rightarrow A = 49.7$ square units (2 marks)

4 a i $T(0) = 86 \Rightarrow 22 + a = 86 \Rightarrow a = 64$ (2 marks)

ii $T(0.5) = 28 \Rightarrow 22 + 64 \times 2^{0.5b} = 28$ (1 mark)
$b = -6.83$ (1 mark)

b $T = 22$ (1 mark)

c The temperature of the hot chocolate approaches 22 °C as t gets very large, indicating that the temperature of the room is 22°C. (1 mark)

5 a Discrete (1 mark)
Number of walks can be counted (1 mark)

b i 3 (1 mark)
ii 2.68 (3 s.f.) (2 marks)

6 a $\frac{1}{9} + \frac{1}{6} + \frac{1}{12} + \frac{2}{9} + \frac{1}{6} + p = 1$

$p = \frac{1}{4}$ (2 marks)

b $\frac{1}{9} \times 2 + \frac{1}{6} \times 3 + \frac{1}{12} \times 5 + \frac{2}{9} \times 1 + \frac{1}{6} \times 3 + \frac{1}{4} \times 4 = \frac{103}{36}$
(2.86, 3 sf) (2 marks)

7 a $\frac{8}{14}\left(= \frac{4}{7}\right)$ or 0.571 (3 s.f.) (1 mark)

b $\binom{5}{2}\left(\frac{4}{7}\right)^2\left(\frac{3}{7}\right)^3 = 0.257$ (3 s.f.) (3 marks)

8 a

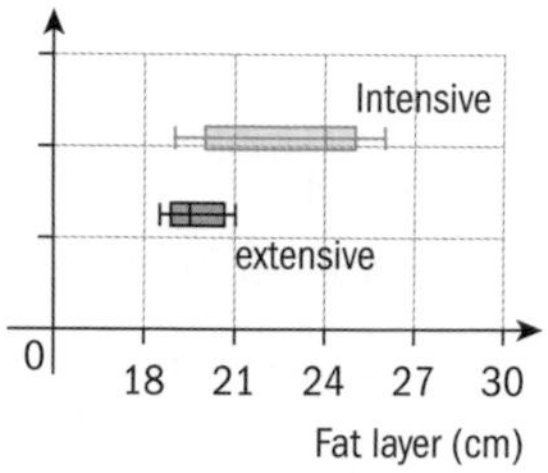

Scale and labelled axis (1 mark)
Correct diagrams (2 marks)

b The average layer of fat was thicker on the cows from the Intensive programme. (1 mark)
The interquartile range and range of fat layer sizes was also much greater for cows from the Intensive programme. (1 mark)

9 a $L = 10\log_{10}(10^{-6} \times 10^{12}) = 60\text{dB}$ (2 marks)
Yes as the value is between 0 and 140. (1 mark)

b $10\log_{10}(S \times 10^{12}) = 80$ (1 mark)
$S = 0.0001\text{ Wm}^{-2}$ (1 mark)

10 a H_0: Favourite colour is independent of gender. (1 mark)

b 0.0276 (1 mark)

c As 0.0276 < 0.05 the null hypothesis is rejected at 5% level of significance. (2 marks)

11 a Labelled axes (1 mark)
Each region (2 marks)

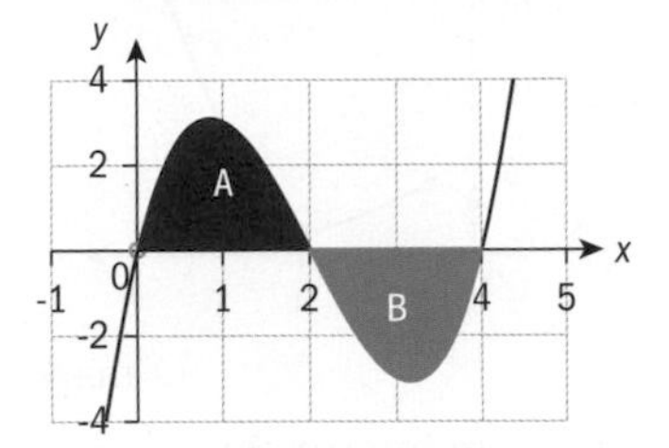

b Use of definite integration (1 mark)

i $\text{area}(A) = \int_0^2 f(x)\,dx = 4$ (1 mark)

ii $\text{area}(B) = \int_2^4 |f(x)|\,dx = 4$ (1 mark)

iii $\text{area}(A) + \text{area}(B) = 8$ (1 mark)

12 a i $H_0: \mu_1 = \mu_2$ (1 mark)

ii $H_1: \mu_1 \neq \mu_2$ (1 mark)

b $H_0: p = 0.209$ (3 sf) (2 marks)

c As 0.209 > 0.1 there is no evidence to reject the null hypothesis at the 10% level of significance. (2 marks)

13 a $AE = \sqrt{(2-4)^2 + (9-6)^2}$
$= 3.61$ (3 s.f.) (2 marks)

b Attempt to find perpendicular bisector of BE. (1 mark)

One technique is shown here:

$PB = PE \Rightarrow \sqrt{(x-2)^2 + (y-3)^2}$
$= \sqrt{(x-4)^2 + (y-6)^2}$ (1 mark)

Attempt to expand (1 mark)

$x^2 - 4x + 4 + y^2 - 6y + 9 =$
$x^2 - 8x + 16 + y^2 - 12y + 36$ (1 mark)

$4x + 6y - 39 = 0$ (1 mark)

c The cell corresponds to the region of the park that has E as the closest well. (1 mark)

14 a $AD^2 = 100^2 - 65^2 = 5775$ or $AD = 75.9934...$ (1 mark)

Area
$= \frac{\pi}{2}\left((50 + AD)^2 - AD^2\right)$
$= 15\,900\ \text{m}^2$ (3 marks)

b $C\hat{A}D$
$= \sin^{-1}\left(\frac{65}{100}\right) = 40.54...''$
$= 40.5°$ (2 marks)

c $B\hat{A}C = 180 - C\hat{A}D = 139.46$

Cosine rule

$\Rightarrow (BC)^2 = (AB)^2 + (AC)^2$
$-2AB \times AC \times \cos B\hat{A}C$ (1 mark)

$BC = \sqrt{50^2 + 100^2 - 2\times 50 \times 100\cos(139.46...)}$ (1 mark)

$BC = 142$ m (3 s.f.) (1 mark)

Paper 2

Exam-style questions

1 a

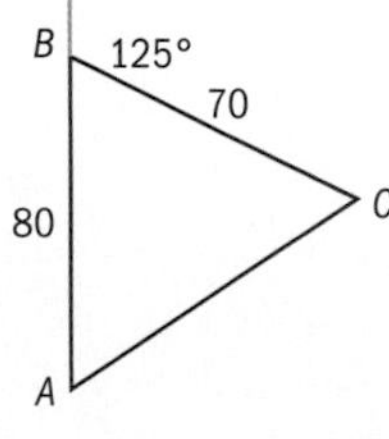

(2 marks)

b $A\hat{B}C = 180 - 125 = 55$ (1 mark)

$AC^2 = 80^2 + 70^2 - 2 \times 70 \times 80 \times \cos 55$ (1 mark)

$AC = 69.8279... = 69.8$ m (3 s.f.) (1 mark)

c $\text{Area} = \frac{1}{2} \times 80 \times 70 \times \sin 55$ (1 mark)

$= 2293.62...$
$= 2.29 \times 10^3$ m (3 s.f.) (1 mark)

d $\frac{69.8279}{\sin 55} = \frac{80}{\sin C} \Rightarrow C = 69.8$ (2 marks)

Bearing is $360 - 55 - 69.8$ (2 marks)

$= 235°$ (3 s.f.) (1 mark)

2 a

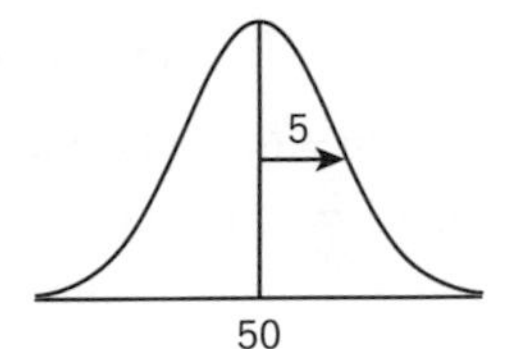

(3 marks)

b $P(45 < t < 55) = 0.683$ (3 s.f.) (2 marks)

c $P(t < 40) = 0.0228$ (3 s.f.) (2 marks)

d $P(t < M) = 0.75 \Rightarrow$
$M = 53.4$ (3 s.f.) mins (2 marks)

e $P(t < 40 | \text{gained medal})$
$= \frac{P(t < 40 \cap \text{gained medal})}{P(\text{gained medal})}$
$= \frac{0.02275...}{0.75} = 0.303$ (3 s.f.) (3 marks)

f $10\,000 \times P(t < 33) = 3.37...$
so 3 competitors (2 marks)

3 a 11 m (1 mark)

b $\frac{360}{30} = 12$ hours (2 marks)

c since $-1 \le \sin \le +1$

i 16 m at 03:00 and 15:00 (2 marks)

ii 6 m at 09:00 and 21:00 (2 marks)

d

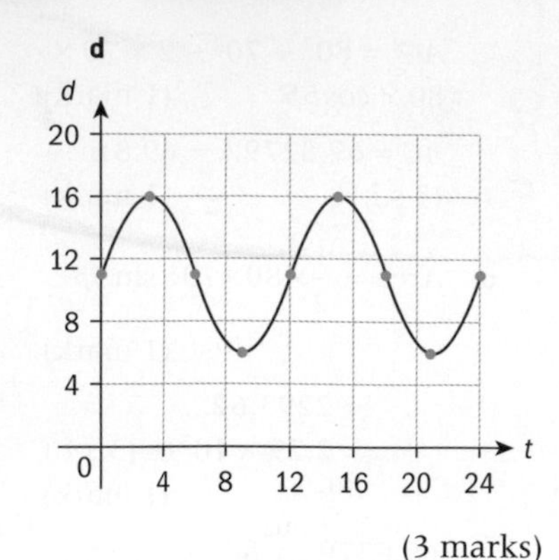

(3 marks)

e First two intersections between sine curve and $d = 14$ are 1.23... and 4.77...

So time period is $01{:}14 < t < 04{:}46$ a.m. (3 marks)

4 a i Binomial; $X \sim B\left(9, \frac{2}{3}\right)$ (3 marks)

ii $\mu = np = 6$ (1 mark)

iii $P(X = 7) = 0.234$ (3s.f.) (2 marks)

iv $P(X \geq 4) = 1 - P(X \leq 3) = 0.958$ (3s.f.) (2 marks)

d P(scoring at least once) > 0.99 ⇒ P(missing all shots) ≤ 0.01 (1 mark)

Let him take n shots, so require $\left(\frac{2}{3}\right)^n \leq 0.01$ (2 marks)

Solving by GDC ⇒ $n \geq 11.5$ so $n = 12$ is the minimum (2 marks)

5 a $Q_1 = 6, Q_3 = 9 \Rightarrow \text{IQR} = 3$ (2 marks)

$6 - 1.5 \times 3 = 1.5$ so value of h for student K is an outlier (2 marks)

b $r = 0.816$ (3s.f.) (2 marks)

c $s = 3.79h + 53.6$ (3s.f.) (2 marks)

d $3.79 \times 5.5 + 53.6 = 74$ to the nearest integer (2 marks)

6 a $h'(x) = x^2 - 5x + 4$ (2 marks)

b $x = 1$ or $x = 4$ (2 marks)

c max (1,6.83)
min (4,2.33) (4 marks)

d

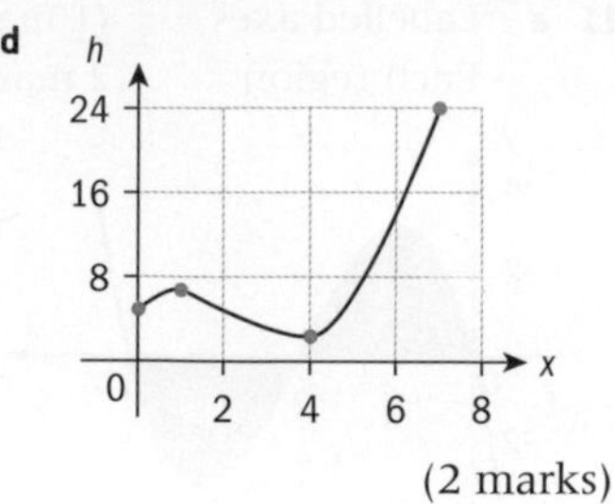

(2 marks)

e from intersection of $h = 10$ and graph, $5.90 < x \leq 7$ (3s.f.) (1 mark)

f from intersection of $y = x^2 - 5x + 4$ and $y = -1$, $1.38 < x < 3.62$ (3s.f.) (2 marks)

e

	A	B	C	D	E	F	G	H	I	J	K
Hour rank	1	2	3	4	5	6	7	8	9	10	11
Score rank	1	3	2	4	6	5	7	8	9	11	10

(2 marks)

f $r_s = 0.973$ (3s.f.) (2 marks)

g Spearman's is less sensitive to outliers like K, which distort the data, so Spearman's shows greater correlation. (1 mark)

Index